Intermediate Algebra

Concepts and Applications

Fifth Edition

Bittinger • Ellenbogen

Student's Solutions Manual

Judith A. Penna

Indiana University - Purdue University at Indianapolis

An imprint of Addison Wesley Longman, Inc.

Reading, Massachusetts • Menlo Park, California • New York • Harlow, England
Don Mills, Ontario • Sydney • Mexico City • Madrid • Amsterdam

Reproduced by Addison-Wesley Publishing Company Inc. from camera-ready copy supplied by the author.

ISBN 0-201-30502-X

1 2 3 4 5 6 7 8 9 10 CRS 00999897

Table of Contents

Chapter 1 1
Chapter 2 27
Chapter 3 65
Chapter 4 111
Chapter 5 141
Chapter 6 173
Chapter 7 213
Chapter 8 243
Chapter 9 297
Chapter 10 333
Chapter 11 363

Chapter 1

Algebra and Problem Solving

Exercise Set 1.1

1. Seven more than some number

Let n represent the number. Then we have

$n + 7$, or $7 + n$.

2. Let n represent the number; $n - 2$

3. Twelve times a number

Let t represent the number. Then we have

$12t$.

4. Let x represent the number; $2x$

5. Sixty-five percent of some number

Let x represent the number. Then we have

$0.65x$, or $\frac{65}{100}x$.

6. Let x represent the number; $0.39x$, or $\frac{39}{100}x$.

7. Nine less than twice a number

Let y represent the number. Then we have

$2y - 9$.

8. Let y represent the number; $\frac{1}{2}y + 4$, or $\frac{y}{2} + 4$

9. Eight more than ten percent of some number

Let s represent the number. Then we have

$0.1s + 8$

10. Let s represent the number; $0.06s - 5$, or $\frac{6}{100}s - 5$

11. One less than the difference of two numbers

Let m and n represent the numbers. Then we have

$m - n - 1$.

12. Let m and n represent the numbers;

$mn + 2$

13. Ninety miles per every four gallons of gas

We have

$90 \div 4$, or $\frac{90}{4}$.

14. $100 \div 60$, or $\frac{100}{60}$

15. Substitute and carry out the operations indicated.

$$\begin{aligned} 4x - y &= 4 \cdot 3 - 2 \\ &= 12 - 2 \\ &= 10 \end{aligned}$$

16. 19

17. Substitute and carry out the operations indicated.

$$\begin{aligned} 2c \div 3b &= 2 \cdot 6 \div 3 \cdot 4 \\ &= 12 \div 3 \cdot 4 \\ &= 4 \cdot 4 \\ &= 16 \end{aligned}$$

18. 9

19. Substitute and carry out the operations indicated.

$$\begin{aligned} 25 - r^2 + s &= 25 - 3^2 + 7 \\ &= 25 - 9 + 7 \\ &= 16 + 7 \\ &= 23 \end{aligned}$$

20. 11

21. Substitute and carry out the operations indicated.

$$\begin{aligned} 3n^2p + 2p^4 &= 3 \cdot 5^2 \cdot 3 + 2 \cdot 3^4 \\ &= 3 \cdot 25 \cdot 3 + 2 \cdot 81 \\ &= 75 \cdot 3 + 162 \\ &= 225 + 162 \\ &= 387 \end{aligned}$$

22. 280

23. Substitute and carry out the operations indicated.

$$\begin{aligned} 5x \div (2 + x - y) &= 5 \cdot 6 \div (2 + 6 - 2) \\ &= 5 \cdot 6 \div (8 - 2) \\ &= 5 \cdot 6 \div 6 \\ &= 30 \div 6 \\ &= 5 \end{aligned}$$

24. 3

25. Substitute and carry out the operations indicated.

$$\begin{aligned} 29 - (a - b)^2 &= 29 - (7 - 2)^2 \\ &= 29 - 5^2 \\ &= 29 - 25 \\ &= 4 \end{aligned}$$

26. 64

27. Substitute and carry out the operations indicated.

$$\begin{aligned} m + n(5 + n^2) &= 15 + 3(5 + 3^2) \\ &= 15 + 3(5 + 9) \\ &= 15 + 3 \cdot 14 \\ &= 15 + 42 \\ &= 57 \end{aligned}$$

28. 40

29. We substitute 5 for b and 7 for h and multiply:

$A = \frac{1}{2} \cdot b \cdot h = \frac{1}{2} \cdot 5 \cdot 7 = 17.5$ sq ft

30. 3.045 sq m

31. We substitute 4 for b and 3.2 for h and multiply:

$A = \frac{1}{2} \cdot b \cdot h = \frac{1}{2}(4)(3.2) = 6.4$ sq m

32. 9.2 sq ft

33. List the letters in the set: {a,e,i,o,u}, or {a,e,i,o,u,y}

34. {Sunday, Monday, Tuesday, Wednesday, Thursday, Friday, Saturday}

35. List the numbers in the set: $\{1, 3, 5, 7, \ldots\}$

36. $\{2, 4, 6, 8, \ldots\}$

37. List the numbers in the set: $\{7, 14, 21, 28, \ldots\}$

38. $\{10, 20, 30, 40, \ldots\}$

39. Specify the conditions under which a number is in the set: $\{x|x$ is an odd number between 10 and 30$\}$

40. $\{x|x$ is a multiple of 4 between 22 and 45$\}$

41. Specify the conditions under which a number is in the set: $\{x|x$ is a whole number less than 5$\}$

42. $\{x|x$ is an integer greater than -4 and less than 3$\}$

43. Specify the conditions under which a number is in the set: $\{n|n$ is a multiple of 5 between 7 and 79$\}$

44. $\{x|x$ is an even number between 9 and 99$\}$

45. Since 7.3 is not a natural number, the statement is false.

46. True

47. Since every member of the set of natural numbers is also a member of the set of whole numbers, the statement is true.

48. True

49. Since $\sqrt{8}$ is not a rational number, the statement is false.

50. False

51. Since every member of the set of irrational numbers is also a member of the set of real numbers, the statement is true.

52. True

53. Since 4.3 is not an integer, the statement is true.

54. True

55. Since every member of the set of rational numbers is also a member of the set of real numbers, the statement is true.

56. False

57. ◈

58. ◈

59. ◈

60. ◈

61. The product of the sum of two numbers and their difference

Let a and b represent the numbers. Then we have $(a + b)(a - b)$.

62. Let m and n represent the numbers;

$3(m + n)$

63. Half of the difference of two numbers

Let r and s represent the numbers. Then we have $\frac{1}{2}(r - s)$, or $\frac{r - s}{2}$.

64. Let x and y represent the numbers; $\frac{x - y}{x + y}$

65. The only whole number that is not also a natural number is 0. Using roster notation to name the set, we have $\{0\}$.

66. $\{-1, -2, -3, \ldots\}$

67. List the numbers in the set:

$\{5, 10, 15, 20, \ldots\}$

68. $\{3, 6, 9, 12, \ldots\}$

69. List the numbers in the set:

$\{\ldots, -4, -2, 0, 2, 4, \ldots\}$

70. $\{1, 3, 5, 7, \ldots\}$

71. Recall from geometry that when a right triangle has legs of length 2 and 3, the length of the hypotenuse is $\sqrt{2^2+3^2} = \sqrt{4+9} = \sqrt{13}$. We draw such a triangle:

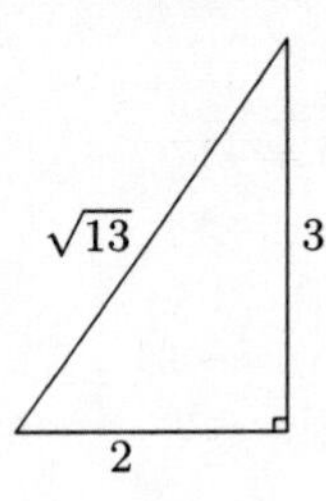

Exercise Set 1.2

1. $|-8| = 8$ $\quad -8$ is 8 units from 0.

2. 7

3. $|9| = 9$ $\quad$ 9 is 9 units from 0.

4. 12

5. $|-6.2| = 6.2$ $\quad -6.2$ is 6.2 units from 0.

6. 7.9

7. $|0| = 0$ $\quad$ 0 is 0 units from itself.

8. $3\frac{3}{4}$

9. $\left|1\frac{7}{8}\right| = 1\frac{7}{8}$ $\quad 1\frac{7}{8}$ is $1\frac{7}{8}$ units from 0.

10. 0.91

11. $|-4.21| = 4.21$ $\quad -4.21$ is 4.21 units from 0.

12. 5.309

13. $-8 \leq -2$

-8 is less than or equal to -2, a true statement since -8 is left of -2.

14. -1 is less than or equal to -5; false

15. $-7 > 1$

-7 is greater than 1, a false statement since -7 is left of 1.

16. 7 is greater than or equal to -2; true

17. $3 \geq -5$

3 is greater than or equal to -5, a true statement since -5 is left of 3.

18. 9 is less than or equal to 9; true

19. $-9 < -4$

-9 is less than -4, a true statement since -9 is left of -4.

20. 7 is greater than or equal to -8; true

21. $-4 \geq -4$

-4 is greater or equal to -4. Since $-4 = -4$ is true, $-4 \geq -4$ is true.

22. 2 is less than 2; false

23. $-5 < -5$

-5 is less than -5, a false statement since -5 does not lie to the left of itself.

24. -2 is greater than -12; true

25. $5 + 12$

Two positive numbers: Add the numbers, getting 17. The answer is positive, 17.

26. 16

27. $-4 + (-7)$

Two negative numbers: Add the absolute values, getting 11. The answer is negative, -11.

28. -11

29. $-5.9 + 2.7$

A negative and a positive number: The absolute values are 5.9 and 2.7. Subtract 2.7 from 5.9 to get 3.2. The negative number is farther from 0, so the answer is negative, -3.2.

30. 5.4

31. $\frac{2}{7} + \left(-\frac{3}{5}\right) = \frac{10}{35} + \left(-\frac{21}{35}\right)$

A positive and a negative number. The absolute values are $\frac{10}{35}$ and $\frac{21}{35}$. Subtract $\frac{10}{35}$ from $\frac{21}{35}$ to get $\frac{11}{35}$. The negative number is farther from 0, so the answer is negative, $-\frac{11}{35}$.

32. $-\frac{1}{40}$

33. $-4.9 + (-3.6)$

Two negative numbers: Add the absolute values, getting 8.5. The answer is negative, -8.5.

34. -9.6

35. $-\frac{1}{9} + \frac{2}{3} = -\frac{1}{9} + \frac{6}{9}$

A negative and a positive number. The absolute values are $\frac{1}{9}$ and $\frac{6}{9}$. Subtract $\frac{1}{9}$ from $\frac{6}{9}$ to get $\frac{5}{9}$. The positive number is farther from 0, so the answer is positive, $\frac{5}{9}$.

36. $\frac{3}{10}$

37. $0 + (-4.5)$

One number is zero: The sum is the other number, -4.5.

38. -3.19

39. $-7.24 + 7.24$

A negative and a positive number: The numbers have the same absolute value, 7.24, so the answer is 0.

40. 0

41. $15.9 + (-22.3)$

A positive and a negative number: The absolute values are 15.9 and 22.3. Subtract 15.9 from 22.3 to get 6.4. The negative number is farther from 0, so the answer is negative, -6.4.

42. -6.6

43. The opposite of 7.29 is -7.29, because $-7.29+7.29 = 0$.

44. -5.43

45. The opposite of $-4\frac{1}{3}$ is $4\frac{1}{3}$, because $-4\frac{1}{3} + 4\frac{1}{3} = 0$.

46. $-2\frac{3}{5}$

47. The opposite of 0 is 0, because $0 + 0 = 0$.

48. $2\frac{3}{4}$

49. If $x = 7$, then $-x = -7$. (The opposite of 7 is -7.)

50. -3

51. If $x = -2.7$, then $-x = -(-2.7) = 2.7$.
(The opposite of -2.7 is 2.7.)

52. 1.9

53. If $x = 1.79$, then $-x = -1.79$. (The opposite of 1.79 is -1.79.)

54. -3.14

55. If $x = 0$, then $-x = 0$. (The opposite of 0 is 0.)

56. 1

57. $9 - 7 = 9 + (-7)$ Change the sign and add.
$= 2$

58. 5

59. $4 - 9 = 4 + (-9)$ Change the sign and add.
$= -5$

60. -7

61. $-6-(-10) = -6+10$ Change the sign and add.
$= 4$

62. 6

63. $-4 - 13 = -4 + (-13) = -17$

64. -15

65. $2.7 - 5.8 = 2.7 + (-5.8) = -3.1$

66. -0.5

67. $-\frac{3}{5} - \frac{1}{2} = -\frac{3}{5} + \left(-\frac{1}{2}\right)$
$= -\frac{6}{10} + \left(-\frac{5}{10}\right)$ Finding a common denominator
$= -\frac{11}{10}$

68. $-\frac{13}{15}$

69. $-3.9 - (-6.8) = -3.9 + 6.8 = 2.9$

70. -1.1

71. $0 - (-7.9) = 0 + 7.9 = 7.9$

72. -5.3

73. $(-4)7$

Two numbers with unlike signs: Multiply their absolute values, getting 28. The answer is negative, -28.

74. -45

75. $(-3)(-8)$

Two numbers with the same sign: Multiply their absolute values, getting 24. The answer is positive, 24.

76. 56

77. $(4.2)(-5)$

Two numbers with unlike signs: Multiply their absolute values, getting 21. The answer is negative, -21.

78. -28

79. $\frac{3}{7}(-1)$

Two numbers with unlike signs: Multiply their absolute values, getting $\frac{3}{7}$. The answer is negative, $-\frac{3}{7}$.

80. $-\frac{2}{5}$

81. $15.2 \times 0 = 0$

82. 0

83. $(-3.2)\times(-1.7)$

Two numbers with the same sign: Multiply their absolute values, getting 5.44. The answer is positive, 5.44.

84. 8.17

85. $\frac{-10}{-2}$

Two numbers with the same sign: Divide their absolute values, getting 5. The answer is positive, 5.

86. 5

87. $\frac{-100}{20}$

Two numbers with unlike signs: Divide their absolute values, getting 5. The answer is negative, -5.

88. -10

89. $\frac{73}{-1}$

Two numbers with unlike signs: Divide their absolute values, getting 73. The answer is negative, -73.

90. -62

91. $\frac{0}{-7}=0$

92. 0

93. The reciprocal of 5 is $\frac{1}{5}$, because $5\cdot\frac{1}{5}=1$.

94. $\frac{1}{3}$

95. The reciprocal of -9 is $\frac{1}{-9}$, or $-\frac{1}{9}$, because $-9\left(-\frac{1}{9}\right)=1$.

96. $-\frac{1}{7}$

97. The reciprocal of $\frac{2}{3}$ is $\frac{3}{2}$, because $\frac{2}{3}\cdot\frac{3}{2}=1$.

98. $\frac{7}{4}$

99. The reciprocal of $-\frac{3}{11}$ is $-\frac{11}{3}$, because $-\frac{3}{11}\left(-\frac{11}{3}\right)=1$.

100. $-\frac{3}{7}$

101.
$$\frac{2}{3}\div\frac{4}{5}$$
$$=\frac{2}{3}\cdot\frac{5}{4}\quad\text{Multiplying by the reciprocal of } 4/5$$
$$=\frac{10}{12},\text{ or }\frac{5}{6}$$

102. $\frac{5}{21}$

103.
$$\left(-\frac{3}{5}\right)\div\frac{1}{2}$$
$$=-\frac{3}{5}\cdot\frac{2}{1}\quad\text{Multiplying by the reciprocal of } 1/2$$
$$=-\frac{6}{5}$$

104. $-\frac{12}{7}$

105.
$$\left(-\frac{2}{9}\right)\div(-8)$$
$$=-\frac{2}{9}\cdot\left(-\frac{1}{8}\right)\quad\text{Multiplying by the reciprocal of } -8$$
$$=\frac{2}{72},\text{ or }\frac{1}{36}$$

106. $\frac{1}{33}$

107.
$$\frac{12}{7}\div(-1)=\frac{12}{7}\cdot(-1)\quad\text{Multiplying by the reciprocal of } -1$$
$$=-\frac{12}{7}$$

108. $\frac{2}{7}$

109.
$$12-(9-3\cdot2^3)=12-(9-3\cdot8)\quad\text{Working within the parentheses first}$$
$$=12-(9-24)$$
$$=12-(-15)$$
$$=12+15$$
$$=27$$

110. -3

111. $\frac{5\cdot2-4^2}{27-2^4}=\frac{5\cdot2-16}{27-16}=\frac{10-16}{11}=\frac{-6}{11}$, or $-\frac{6}{11}$

112. $-\frac{4}{17}$

113. $\frac{3^4-(5-3)^4}{1-2^3}=\frac{3^4-2^4}{1-8}=\frac{81-16}{-7}=\frac{65}{-7}$, or $-\frac{65}{7}$

114. $\frac{55}{2}$

115. $5^3-[2(4^2-3^2-6)]^3=5^3-[2(16-9-6)]^3=5^3-[2\cdot1]^3=5^3-2^3=125-8=117$

116. 13

117. $|2^2-7|^3+1=|4-7|^3+1=|-3|^3+1=3^3+1=27+1=28$

118. 79

119. $30 - (-5)^2 + 15 \div (-3) \cdot 2$
$= 30 - 25 + 15 \div (-3) \cdot 2$ Evaluating the exponential expression
$= 30 - 25 - 5 \cdot 2$ Dividing
$= 30 - 25 - 10$ Multiplying
$= -5$ Subtracting

120. 0

121. $12 - \sqrt{7-3} + 4 \div 3 \cdot 2^3$
$= 12 - \sqrt{4} + 4 \div 3 \cdot 2^3$
$= 12 - 2 + 4 \div 3 \cdot 8$
$= 12 - 2 + \frac{4}{3} \cdot 8$
$= 12 - 2 + \frac{32}{3}$
$= 10 + \frac{32}{3}$
$= \frac{62}{3}$

122. $13\frac{1}{2}$, or $\frac{27}{2}$

123. Using the commutative law of addition, we have

$3x + 8y = 8y + 3x.$

Using the commutative law of multiplication, we have

$3x + 8y = x3 + 8y$
or $3x + 8y = 3x + y8$
or $3x + 8y = x3 + y8.$

124. $9 + ab$; $ba + 9$

125. Using the commutative law of multiplication, we have

$(7x)y = y(7x)$
or $(7x)y = (x7)y.$

126. $(ab)(-9)$; $-9(ba)$

127. $(3x)y$
$= 3(xy)$ Associative law of multiplication

128. $(-7a)b$

129. $x + (2y + 5)$
$= (x + 2y) + 5$ Associative law of addition

130. $3y + (4 + 10)$

131. $3(a + 1)$
$= 3 \cdot a + 3 \cdot 1$ Using the distributive law
$= 3a + 3$

132. $8x + 8$

133. $4(x - y)$
$= 4 \cdot x - 4 \cdot y$ Using the distributive law
$= 4x - 4y$

134. $9a - 9b$

135. $-5(2a + 3b)$
$= -5 \cdot 2a + (-5) \cdot 3b$
$= -10a - 15b$

136. $-6c - 10d$

137. $2a(b - c + d)$
$= 2a \cdot b - 2a \cdot c + 2a \cdot d$
$= 2ab - 2ac + 2ad$

138. $5xy - 5xz + 5xw$

139. $5x + 5y = 5 \cdot x + 5 \cdot y = 5(x + y)$

140. $7(a + b)$

141. $3p - 9 = 3 \cdot p - 3 \cdot 3 = 3(p - 3)$

142. $3(4x - 1)$

143. $7x - 21y = 7 \cdot x - 7 \cdot 3y = 7(x - 3y)$

144. $3(2y - 3)$

145. $2x - 2y + 2z = 2 \cdot x - 2 \cdot y + 2 \cdot z = 2(x - y + z)$

146. $3(x + y - z)$

147. Five less than seventy percent of a number

Let x represent the number. Then we have $0.7x - 5$, or $\frac{70}{100}x - 5$.

148. Let x represent the number; $\frac{1}{2}x + 2$

149. ◈

150. ◈

151. ◈

152. ◈

153. $(3 - 8)^2 + 9 = 34$

154. $2 \cdot (7 + 3^2 \cdot 5) = 104$

155. $5 \cdot 2^3 \div (3 - 4)^4 = 40$

156. $(2 - 7) \cdot 2^2 + 9 = -11$

157. Any value of a such that $a \leq -6.2$ satisfies the given conditions. The largest of these values is -6.2.

158.

$$\begin{aligned} & 5(a+bc) & \\ &= (a+bc)5 & \text{Commutative law of multiplication} \\ &= a5+(bc)5 & \text{Distributive law} \\ &= a5+(cb)5 & \text{Commutative law of multiplication} \\ &= a5+c(b5) & \text{Associative law of multiplication} \\ &= c(b5)+a5 & \text{Commutative law of addition} \end{aligned}$$

Exercise Set 1.3

1. $3x = 15$ and $2x = 10$

The equation $3x = 15$ is true only when $x = 5$. Similarly, $2x = 10$ is true only when $x = 5$. Since both equations have the same solution, they are equivalent.

2. Equivalent

3. $x + 5 = 11$ and $3x = 18$

Each equation has only one solution, the number 6. Thus the equations are equivalent.

4. Not equivalent

5. $13 - x = 4$ and $2x = 20$

When x is replaced by 9, the first equation is true, but the second equation is false. Thus the equations are not equivalent.

6. Equivalent

7. $5x = 2x$ and $\dfrac{4}{x} = 3$

When x is replaced by 0, the first equation is true, but the second equation is not defined. Thus the equations are not equivalent.

8. Not equivalent

9.

$$\begin{aligned} x - 5.2 &= 9.4 & \\ x - 5.2 + 5.2 &= 9.4 + 5.2 & \text{Addition principle; adding 5.2} \\ x + 0 &= 9.4 + 5.2 & \text{Law of opposites} \\ x &= 14.6 & \end{aligned}$$

Check:

$$\begin{array}{r|l} \multicolumn{2}{c}{x - 5.2 = 9.4} \\ \hline 14.6 - 5.2 & ?\ 9.4 \\ 9.4 & 9.4 \quad \text{TRUE} \end{array}$$

The solution is 14.6.

10. 6.9

11.

$$\begin{aligned} 9y &= 72 & \\ \frac{1}{9} \cdot 9y &= \frac{1}{9} \cdot 72 & \text{Multiplication principle; multiplying by } \frac{1}{9}\text{, the reciprocal of 9} \\ 1y &= 8 & \\ y &= 8 & \end{aligned}$$

Check:

$$\begin{array}{r|l} \multicolumn{2}{c}{9y = 72} \\ \hline 9 \cdot 8 & ?\ 72 \\ 72 & 72 \quad \text{TRUE} \end{array}$$

The solution is 8.

12. 9

13.

$$\begin{aligned} 4x - 12 &= 60 \\ 4x - 12 + 12 &= 60 + 12 \\ 4x &= 72 \\ \frac{1}{4} \cdot 4x &= \frac{1}{4} \cdot 72 \\ 1x &= \frac{72}{4} \\ x &= 18 \end{aligned}$$

Check:

$$\begin{array}{r|l} \multicolumn{2}{c}{4x - 12 = 60} \\ \hline 4 \cdot 18 - 12 & ?\ 60 \\ 72 - 12 & \\ 60 & 60 \quad \text{TRUE} \end{array}$$

The solution is 18.

14. 19

15.

$$\begin{aligned} 5y + 3 &= 28 \\ 5y + 3 + (-3) &= 28 + (-3) \\ 5y &= 25 \\ \frac{1}{5} \cdot 5y &= \frac{1}{5} \cdot 25 \\ 1y &= \frac{25}{5} \\ y &= 5 \end{aligned}$$

Check:

$$\begin{array}{r|l} \multicolumn{2}{c}{5y + 3 = 28} \\ \hline 5 \cdot 5 + 3 & ?\ 28 \\ 25 + 3 & \\ 28 & 28 \quad \text{TRUE} \end{array}$$

The solution is 5.

16. 9

17.

$$\begin{aligned} 2y - 11 &= 37 \\ 2y - 11 + 11 &= 37 + 11 \\ 2y &= 48 \\ \frac{1}{2} \cdot 2y &= \frac{1}{2} \cdot 48 \\ 1y &= \frac{48}{2} \\ y &= 24 \end{aligned}$$

Check:

$$\begin{array}{r|l} \multicolumn{2}{c}{2y - 11 = 37} \\ \hline 2 \cdot 24 - 11 & ?\ 37 \\ 48 - 11 & \\ 37 & 37 \quad \text{TRUE} \end{array}$$

The solution is 24.

18. 14

19. $4a + 5a = (4 + 5)a = 9a$

20. $12x$

21. $7rt - 9rt = (7 - 9)rt = -2rt$

22. $10ab$

23. $8x^2 + x^2 = (8 + 1)x^2 = 9x^2$

24. $8a^2$

25. $12a - a = (12 - 1)a = 11a$

26. $14x$

27. $t - 9t = (1 - 9)t = -8t$

28. $-5x$

29. $5x - 3x + 8x = (5 - 3 + 8)x = 10x$

30. $-6x$

31. $5x - 2x^2 + 3x$
$= 5x + 3x - 2x^2$ Commutative law of addition
$= (5 + 3)x - 2x^2$
$= 8x - 2x^2$

32. $13a - 5a^2$

33. $3a + 5a^2 - a + 4a^2$
$= 3a - a + 5a^2 + 4a^2$ Commutative law of addition
$= (3 - 1)a + (5 + 4)a^2$
$= 2a + 9a^2$

34. $14x + 2x^3 - 6x^2$

35. $4x - 7 + 18x + 25$
$= 4x + 18x - 7 + 25$
$= (4 + 18)x + (-7 + 25)$
$= 22x + 18$

36. $9p + 12$

37. $-7t^2 + 3t + 5t^3 - t^3 + 2t^2 - t$
$= (-7 + 2)t^2 + (3 - 1)t + (5 - 1)t^3$
$= -5t^2 + 2t + 4t^3$

38. $-12n + 6n^2 + 5n^3$

39. $a - (2a + 5)$
$= a - 2a - 5$
$= -a - 5$

40. $-4x - 9$

41. $4m - (3m - 1)$
$= 4m - 3m + 1$
$= m + 1$

42. $a + 3$

43. $3d - 7 - (5 - 2d)$
$= 3d - 7 - 5 + 2d$
$= 5d - 12$

44. $13x - 16$

45. $-2(x + 3) - 5(x - 4)$
$= -2x - 6 - 5x + 20$
$= -7x + 14$

46. $-15y - 45$

47. $5x - 7(2x - 3)$
$= 5x - 14x + 21$
$= -9x + 21$

48. $-12y + 24$

49. $9a - [7 - 5(7a - 3)]$
$= 9a - [7 - 35a + 15]$
$= 9a - [22 - 35a]$
$= 9a - 22 + 35a$
$= 44a - 22$

50. $47b - 51$

51. $5\{-2a + 3[4 - 2(3a + 5)]\}$
$= 5\{-2a + 3[4 - 6a - 10]\}$
$= 5\{-2a + 3[-6 - 6a]\}$
$= 5\{-2a - 18 - 18a\}$
$= 5\{-20a - 18\}$
$= -100a - 90$

52. $-721x - 728$

53. $2y + \{7[3(2y - 5) - (8y + 7)] + 9\}$
$= 2y + \{7[6y - 15 - 8y - 7] + 9\}$
$= 2y + \{7[-2y - 22] + 9\}$
$= 2y + \{-14y - 154 + 9\}$
$= 2y + \{-14y - 145\}$
$= 2y - 14y - 145$
$= -12y - 145$

54. $-11b + 217$

55. $5x + 2x = 56$
$7x = 56$
$\frac{1}{7} \cdot 7x = \frac{1}{7} \cdot 56$
$x = 8$

Check:

$$\begin{array}{c|c} 5x + 2x = 56 & \\ \hline 5 \cdot 8 + 2 \cdot 8 \ ? \ 56 & \\ 40 + 16 & \\ 56 & 56 \quad \text{TRUE} \end{array}$$

The solution is 8.

56. 12

57.
$$\begin{aligned} 9y - 7y &= 42 \\ 2y &= 42 \\ \frac{1}{2} \cdot 2y &= \frac{1}{2} \cdot 42 \\ y &= 21 \end{aligned}$$

Check:

$$\begin{array}{c|c} 9y - 7y = 42 & \\ \hline 9 \cdot 21 - 7 \cdot 21 \ ? \ 42 & \\ 189 - 147 & \\ 42 & 42 \quad \text{TRUE} \end{array}$$

The solution is 21.

58. 13

59.
$$\begin{aligned} -6y - 10y &= -32 \\ -16y &= -32 \\ -\frac{1}{16} \cdot (-16y) &= -\frac{1}{16} \cdot (-32) \\ y &= 2 \end{aligned}$$

Check:

$$\begin{array}{c|c} -6y - 10y = -32 & \\ \hline -6 \cdot 2 - 10 \cdot 2 \ ? \ -32 & \\ -12 - 20 & \\ -32 & -32 \quad \text{TRUE} \end{array}$$

The solution is 2.

60. −2

61.
$$\begin{aligned} 2(x + 6) &= 8x \\ 2x + 12 &= 8x \\ 2x + 12 - 2x &= 8x - 2x \\ 12 &= 6x \\ \frac{1}{6} \cdot 12 &= \frac{1}{6} \cdot 6x \\ 2 &= x \end{aligned}$$

Check:

$$\begin{array}{c|c} 2(x + 6) = 8x & \\ \hline 2(2 + 6) \ ? \ 8 \cdot 2 & \\ 2 \cdot 8 & 16 \\ 16 & 16 \quad \text{TRUE} \end{array}$$

The solution is 2.

62. 3

63.
$$\begin{aligned} 80 &= 10(3t + 2) \\ 80 &= 30t + 20 \\ 80 - 20 &= 30t + 20 - 20 \\ 60 &= 30t \\ \frac{1}{30} \cdot 60 &= \frac{1}{30} \cdot 30t \\ 2 &= t \end{aligned}$$

Check:

$$\begin{array}{c|c} 80 = 10(3t + 2) & \\ \hline 80 \ ? \ 10(3 \cdot 2 + 2) & \\ & 10(6 + 2) \\ & 10 \cdot 8 \\ 80 & 80 \quad \text{TRUE} \end{array}$$

The solution is 2.

64. 1

65.
$$\begin{aligned} 180(n - 2) &= 900 \\ 180n - 360 &= 900 \\ 180n - 360 + 360 &= 900 + 360 \\ 180n &= 1260 \\ \frac{1}{180} \cdot 180n &= \frac{1}{180} \cdot 1260 \\ n &= 7 \end{aligned}$$

Check:

$$\begin{array}{c|c} 180(n - 2) = 900 & \\ \hline 180(7 - 2) \ ? \ 900 & \\ 180 \cdot 5 & \\ 900 & 900 \quad \text{TRUE} \end{array}$$

The solution is 7.

66. 7

67.
$$\begin{aligned} 5y - (2y - 10) &= 25 \\ 5y - 2y + 10 &= 25 \\ 3y + 10 &= 25 \\ 3y + 10 - 10 &= 25 - 10 \\ 3y &= 15 \\ \frac{1}{3} \cdot 3y &= \frac{1}{3} \cdot 15 \\ y &= 5 \end{aligned}$$

Check:

$$\begin{array}{c|c} 5y - (2y - 10) = 25 & \\ \hline 5 \cdot 5 - (2 \cdot 5 - 10) \ ? \ 25 & \\ 25 - (10 - 10) & \\ 25 - 0 & \\ 25 & 25 \quad \text{TRUE} \end{array}$$

The solution is 5.

68. 7

69.
$$\begin{aligned} 7y-1&=23-5y\\ 7y-1+5y&=23-5y+5y\\ 12y-1&=23\\ 12y-1+1&=23+1\\ 12y&=24\\ \frac{1}{12}\cdot 12y&=\frac{1}{12}\cdot 24\\ y&=2 \end{aligned}$$

Check:

$$\begin{array}{l|l} \multicolumn{2}{c}{7y-1=23-5y} \\ \hline 7\cdot 2-1 \ ? & 23-5\cdot 2 \\ 14-1 & 23-10 \\ 13 & 13 \quad \text{TRUE} \end{array}$$

The solution is 2.

70. -6

71.
$$\begin{aligned} \frac{1}{5}+\frac{3}{10}x&=\frac{4}{5}\\ \frac{1}{5}+\frac{3}{10}x-\frac{1}{5}&=\frac{4}{5}-\frac{1}{5}\\ \frac{3}{10}x&=\frac{3}{5}\\ \frac{10}{3}\cdot\frac{3}{10}x&=\frac{10}{3}\cdot\frac{3}{5}\\ x&=2 \end{aligned}$$

Check:

$$\begin{array}{l|l} \multicolumn{2}{c}{\frac{1}{5}+\frac{3}{10}x=\frac{4}{5}} \\ \hline \frac{1}{5}+\frac{3}{10}\cdot 2 \ ? & \frac{4}{5} \\ \frac{1}{5}+\frac{3}{5} & \\ \frac{4}{5} & \frac{4}{5} \quad \text{TRUE} \end{array}$$

The solution is 2.

72. $\frac{37}{5}$

73.
$$\begin{aligned} 0.9y-0.7&=4.2\\ 0.9y-0.7+0.7&=4.2+0.7\\ 0.9y&=4.9\\ \frac{1}{0.9}(0.9y)&=\frac{1}{0.9}(4.9)\\ y&=\frac{4.9}{0.9}\\ y&=\frac{49}{9} \end{aligned}$$

Check:

$$\begin{array}{l|l} \multicolumn{2}{c}{0.9y-0.7=4.2} \\ \hline 0.9\left(\frac{49}{9}\right)-0.7 \ ? & 4.2 \\ 4.9-0.7 & \\ 4.2 & 4.2 \quad \text{TRUE} \end{array}$$

The solution is $\frac{49}{9}$.

74. 13

75.
$$\begin{aligned} 5r-2+3r&=2r+6-4r\\ 8r-2&=6-2r\\ 8r-2+2r&=6-2r+2r\\ 10r-2&=6\\ 10r-2+2&=6+2\\ 10r&=8\\ \frac{1}{10}\cdot 10r&=\frac{1}{10}\cdot 8\\ r&=\frac{8}{10}\\ r&=\frac{4}{5} \end{aligned}$$

Check:

$$\begin{array}{l|l} \multicolumn{2}{c}{5r-2+3r=2r+6-4r} \\ \hline 5\cdot\frac{4}{5}-2+3\cdot\frac{4}{5} \ ? & 2\cdot\frac{4}{5}+6-4\cdot\frac{4}{5} \\ \frac{20}{5}-\frac{10}{5}+\frac{12}{5} & \frac{8}{5}+\frac{30}{5}-\frac{16}{5} \\ \frac{22}{5} & \frac{22}{5} \quad \text{TRUE} \end{array}$$

The solution is $\frac{4}{5}$.

76. -8

77.
$$\begin{aligned} \frac{1}{8}(16y+8)-17&=-\frac{1}{4}(8y-16)\\ 2y+1-17&=-2y+4\\ 2y-16&=-2y+4\\ 2y-16+2y&=-2y+4+2y\\ 4y-16&=4\\ 4y-16+16&=4+16\\ 4y&=20\\ \frac{1}{4}\cdot 4y&=\frac{1}{4}\cdot 20\\ y&=5 \end{aligned}$$

Check:

$$\begin{array}{r|l} \frac{1}{8}(16y+8)-17 = & -\frac{1}{4}(8y-16) \\ \hline \frac{1}{8}(16\cdot 5+8)-17 \ ? & -\frac{1}{4}(8\cdot 5-16) \\ \frac{1}{8}(80+8)-17 & -\frac{1}{4}(40-16) \\ \frac{1}{8}\cdot 88-17 & -\frac{1}{4}\cdot 24 \\ 11-17 & -6 \\ -6 & -6 \qquad \text{TRUE} \end{array}$$

The solution is 5.

78. 6

79.
$$\begin{aligned} 5+2(x-3) &= 2[5-4(x+2)] \\ 5+2x-6 &= 2[5-4x-8] \\ 2x-1 &= 2[-4x-3] \\ 2x-1 &= -8x-6 \\ 2x-1+1 &= -8x-6+1 \\ 2x &= -8x-5 \\ 2x+8x &= -8x-5+8x \\ 10x &= -5 \\ \frac{1}{10}\cdot 10x &= \frac{1}{10}(-5) \\ x &= -\frac{1}{2} \end{aligned}$$

Check:

$$\begin{array}{r|l} 5+2(x-3) = & 2[5-4(x+2)] \\ \hline 5+2\left(-\frac{1}{2}-3\right) \ ? & 2\left[5-4\left(-\frac{1}{2}+2\right)\right] \\ 5+2\left(-\frac{7}{2}\right) & 2\left[5-4\left(\frac{3}{2}\right)\right] \\ 5-7 & 2[5-6] \\ -2 & 2[-1] \\ -2 & -2 \qquad \text{TRUE} \end{array}$$

The solution is $-\frac{1}{2}$.

80. $\frac{23}{8}$

81.
$$\begin{aligned} 4x-2x-2 &= 2x \\ 2x-2 &= 2x \\ -2x+2x-2 &= -2x+2x \\ -2 &= 0 \end{aligned}$$

Since the original equation is equivalent to the false equation $-2=0$, there is no solution. The solution set is $\emptyset$. The equation is a contradiction.

82. All real numbers; identity

83.
$$\begin{aligned} 2+9x &= 3(3x+1)-1 \\ 2+9x &= 9x+3-1 \\ 2+9x &= 9x+2 \\ 2+9x-9x &= 9x+2-9x \\ 2 &= 2 \end{aligned}$$

The original equation is equivalent to the equation $2=2$ which is true for all real numbers. Thus the solution set is the set of all real numbers. The equation is an identity.

84. $\emptyset$; contradiction

85.
$$\begin{aligned} -8x+5 &= 5-10x \\ -8x+5-5 &= 5-10x-5 \\ -8x &= -10x \\ -8x+10x &= -10x+10x \\ 2x &= 0 \\ \frac{1}{2}\cdot 2x &= \frac{1}{2}\cdot 0 \\ x &= 0 \end{aligned}$$

There is one solution, 0. The equation is conditional.

86. All real numbers; identity

87.
$$\begin{aligned} 2\{9-3[-2x-4]\} &= 12x+42 \\ 2\{9+6x+12\} &= 12x+42 \\ 2\{21+6x\} &= 12x+42 \\ 42+12x &= 12x+42 \\ 42+12x-12x &= 12x+42-12x \\ 42 &= 42 \end{aligned}$$

The original equation is equivalent to the equation $42=42$, which is true for all real numbers. Thus the solution set is the set of all real numbers. The equation is an identity.

88. 0; conditional

89. Roster notation: List the numbers in the set.

$\{1,2,3,4,5,6,7,8,9\}$

Set-builder notation: Specify the conditions under which a number is in the set.

$\{x|x$ is a positive integer less than $10\}$

90. $\{-8,-7,-6,-5,-4,-3,-2,-1\}$;

$\{x|x$ is a negative integer greater than $-9\}$

91.

92.

93.

94.

95.
$$\begin{aligned} 4.23x - 17.898 &= -1.65x - 42.454 \\ 5.88x - 17.898 &= -42.454 \\ 5.88x &= -24.556 \\ x &= -\frac{24.556}{5.88} \\ x &\approx -4.176190476 \end{aligned}$$

The check is left to the student. The solution is approximately -4.176190476.

96. 0.2140224

97.
$$\begin{aligned} x - \{3x - [2x - (5x - (7x - 1))]\} &= x + 7 \\ x - \{3x - [2x - (5x - 7x + 1)]\} &= x + 7 \\ x - \{3x - [2x - (-2x + 1)]\} &= x + 7 \\ x - \{3x - [2x + 2x - 1]\} &= x + 7 \\ x - \{3x - [4x - 1]\} &= x + 7 \\ x - \{3x - 4x + 1\} &= x + 7 \\ x - \{-x + 1\} &= x + 7 \\ x + x - 1 &= x + 7 \\ 2x - 1 &= x + 7 \\ x - 1 &= 7 \\ x &= 8 \end{aligned}$$

The check is left to the student. The solution is 8.

98. 4

99.
$$\begin{aligned} 17 - 3\{5 + 2[x - 2]\} + 4\{x - 3(x + 7)\} &= \\ 9\{x + 3[2 + 3(4 - x)]\} & \\ 17 - 3\{5 + 2x - 4\} + 4\{x - 3x - 21\} &= \\ 9\{x + 3[2 + 12 - 3x]\} & \\ 17 - 3\{1 + 2x\} + 4\{-2x - 21\} &= 9\{x + 3[14 - 3x]\} \\ 17 - 3 - 6x - 8x - 84 &= 9\{x + 42 - 9x\} \\ -14x - 70 &= 9\{-8 + 42\} \\ -14x - 70 &= -72x + 378 \\ 58x - 70 &= 378 \\ 58x &= 448 \\ x &= \frac{448}{58}, \text{ or } \frac{224}{29} \end{aligned}$$

The check is left to the student. The solution is $\frac{224}{29}$.

100. $\frac{19}{46}$

Exercise Set 1.4

1. ***Familiarize.*** There are two numbers involved, and we want to find both of them. We can let x represent the first number and note that the second number is 7 more than the first. Also, the sum of the numbers is 65.

Translate. The second number can be named $x+7$. We translate to an equation:

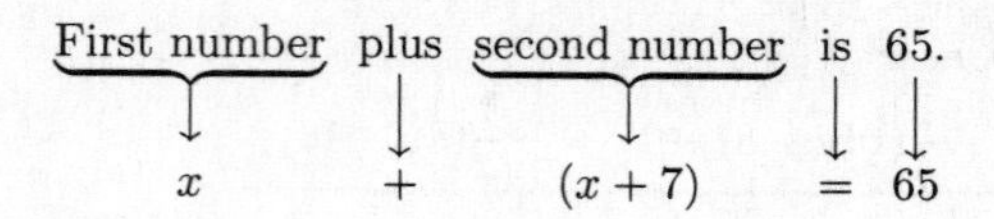

2. Let x and $x + 11$ represent the numbers;
$x + (x + 11) = 83$

3. ***Familiarize.*** Since the sidewalk's speed is 5 ft/sec and Alida's walking speed is 4 ft/sec, Alida will move at a speed of 5+4, or 9 ft/sec on the sidewalk. Let t = the time, in seconds, it takes her to walk the length of the moving sidewalk, 300 ft.

Translate. We will use the formula Distance = Speed × Time.

$$\underbrace{\text{Distance}} = \underbrace{\text{Speed}} \times \underbrace{\text{Time}}$$
$$300 = 9 \times t$$

4. Let t = the time, in hours, it will take Fran to swim 1.8 km upriver; $(5 - 2.3)t = 1.8$, or $2.7t = 1.8$

5. ***Familiarize.*** The plane's speed, traveling into the wind, is the difference between its speed in still air and the speed of the head wind: 390 − 65, or 325 km/h. Let t = the time, in hours, it will take the plane to travel 725 km into the wind.

Translate. We will use the formula Distance = Speed × Time.

$$\underbrace{\text{Distance}} = \underbrace{\text{Speed}} \times \underbrace{\text{Time}}$$
$$725 = 325 \times t$$

6. Let t = the boat's time, in hours; $(14 + 7)t = 56$, or $21t = 56$

7. ***Familiarize.*** There are three angle measures involved, and we want to find all three. We can let x represent the smallest angle measure and note that the second is one more than x and the third is one more than the second, or two more than x. We also note that the sum of the three angle measures must be 180°.

Translate. The three angle measures are x, $x + 1$, and $x + 2$. We translate to an equation:

$$\underbrace{\text{First}} \text{ plus } \underbrace{\text{second}} \text{ plus } \underbrace{\text{third}} \text{ is } 180°.$$
$$x + (x + 1) + (x + 2) = 180$$

8. Let w = the wholesale price; $1.5w + 0.25 = 1.99$

9. ***Familiarize.*** Let t represent the time required. Note that the plane must climb 29,000 − 8000, or 21,000 ft.

Translate.

$$\underbrace{\text{Speed}} \times \underbrace{\text{Time}} = \underbrace{\text{Distance}}$$
$$3500 \times t = 21{,}000$$

10. Let x represent the longer length; $x + \frac{2}{3}x = 10$

11. ***Familiarize.*** Let x represent the measure of the second angle. Then the first angle is three times x, and the third is 12° less than twice x. The sum of the three angle measures is 180°.

Translate. The first angle is $3x$, the second is x, and the third is $2x - 12$. Translate to an equation:

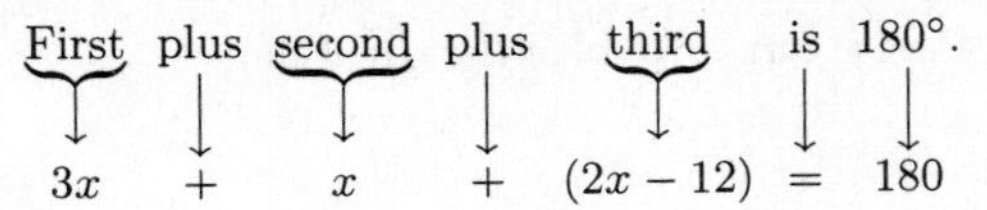

First plus second plus third is 180°.

$$3x + x + (2x - 12) = 180$$

12. Let x represent the measure of the second angle; $4x + x + (2x + 5) = 180$

13. ***Familiarize.*** Note that each odd integer is two more than the one preceding it. If we let n represent the first odd integer, then the second is 2 more than the first and the third is 2 more than the second, or 4 more than the first. We are told that the sum of the first, twice the second, and three times the third is 70.

Translate. The three odd integers are n, $n+2$, and $n+4$. Translate to an equation.

First plus two times second plus three times third is 70.

$$n + 2(n+2) + 3(n+4) = 70$$

14. Let x represent the first number; $2x + 3(x+2) = 76$

15. ***Familiarize.*** Recall that the perimeter of a square is 4 times the length of a side. Let s = the length of a side of the smaller square. Then $2s$ = the length of a side of the larger square. The sum of the two perimeters is 100 cm.

Translate.

Perimeter of smaller square plus perimeter of larger square is 100 cm.

$$4s + 4 \cdot 2s = 100$$

16. Let x represent the length of one piece;

$$\left(\frac{x}{4}\right)^2 = \left(\frac{100 - x}{4}\right)^2 + 144$$

17. ***Familiarize.*** If we let x represent the first number, then the second is six less than 3 times x and the third is two more than $\frac{2}{3}$ of the second. The sum of the three numbers is 172.

Translate.

First plus second plus third is 172.

$$x + 3x - 6 + \frac{2}{3}\left(3x - 6\right) + 2 = 172$$

18. Let x represent the price of the least expensive set; $(x + 20) + (x + 6 \cdot 20) = x + 12 \cdot 20$

19. ***Familiarize.*** After the next test there will be six test scores. The average of the six scores is their sum divided by 6. We let x represent the next test score.

Translate.

The average of the six scores is 88.

$$\frac{93 + 89 + 72 + 80 + 96 + x}{6} = 88$$

20. Let p = the population at the start of the three-year period; $1.12(1.12)(1.12)p = 50,577$

21. ***Familiarize.*** Let x = the unknown factor. Then the product of the two numbers, 125, is represented by $50x$.

Translate.

The product is 125.

$$50x = 125$$

Carry out. We solve the equation.

$$50x = 125$$
$$x = \frac{1}{50} \cdot 125$$
$$x = \frac{5}{2}, \text{ or } 2.5$$

Check. If the other number is $\frac{5}{2}$, the product is $50 \cdot \frac{5}{2}$, or 125. Our answer checks.

State. The other number is $\frac{5}{2}$, or 2.5.

22. 50.3

23. ***Familiarize.*** Let n = the number.

Translate. We reword the problem.

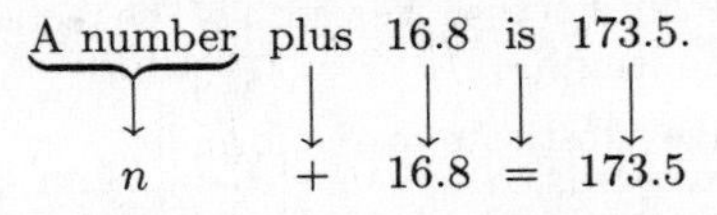

A number plus 16.8 is 173.5.

$$n + 16.8 = 173.5$$

Carry out. We solve the equation.

$$n + 16.8 = 173.5$$
$$n = 173.5 - 16.8$$
$$n = 156.7$$

Check. Since $156.7 + 16.8 = 173.5$, our answer checks.

State. The number is 156.7.

24. 320

25. ***Familiarize.*** Let y = the number.

Translate. We reword the problem.

$$\underbrace{\frac{1}{3}}_{\downarrow\ \frac{1}{3}} \text{ of } \underbrace{\text{a number}}_{\downarrow\ y} \text{ is } 456.$$

$$\frac{1}{3} \cdot y = 456$$

Carry out. We solve the equation.

$$\frac{1}{3}y = 456$$
$$y = 3 \cdot 456$$
$$y = 1368$$

Check. Since $\frac{1}{3} \cdot 1368 = 456$, our answer checks.

State. The number is 1368.

26. 49.2

27. ***Familiarize***. If we let x represent the smaller number, then the larger is $x+12$. The sum of the numbers is 114. We want to find the larger number.

Translate.

Smaller number plus larger number is 114.

$$x + (x + 12) = 114$$

Carry out.

$$x + x + 12 = 114$$
$$2x + 12 = 114$$
$$2x = 102$$
$$x = 51$$

If $x = 51$, then $x + 12 = 51 + 12$, or 63.

Check. The larger number, 63, is 12 more than the smaller number, 51. Also, $51 + 63 = 114$. The numbers check.

State. The larger number is 63.

28. 13.5

29. ***Familiarize***. Recall that the perimeter P of a rectangle with length l and width w is given by the formula $P = 2l+2w$. Let $x =$ the width of the rectangle. Then $2x =$ the length.

Translate. We substitute in the formula.

$$P = 2l + 2w$$
$$21 = 2 \cdot 2x + 2 \cdot x$$

Carry out. We solve the equation.

$$21 = 2 \cdot 2x + 2 \cdot x$$
$$21 = 4x + 2x$$
$$21 = 6x$$
$$\frac{1}{6} \cdot 21 = x$$
$$3.5 = x$$

If $x = 3.5$, then $2x = 2(3.5) = 7$.

Check. If the length of the rectangle is 7 m and the width is 3.5 m, then the perimeter is $2 \cdot 7$ m + $2(3.5$ m), or 14 m + 7 m, or 21 m. Our answer checks.

State. The length of the rectangle is 7 m, and the width is 3.5 m.

30. Length: 12 m, width: 4 m

31. The Familiarize and Translate steps were done in Exercise 4.

Carry out. We solve the equation.

$$2.7t = 1.8$$
$$t = \frac{1}{2.7}(1.8)$$
$$t = \frac{2}{3}$$

Check. At a speed of 2.7 km/h, in $\frac{2}{3}$ hr Fran swims $\frac{2}{3}(2.7)$, or 1.8 km. Our answer checks.

State. It will take Fran $\frac{2}{3}$ hr to swim 1.8 km upriver.

32. $33\frac{1}{3}$ sec

33. The Familiarize and Translate steps were done in Exercise 14.

Carry out.

$$2x + 3(x + 2) = 76$$
$$2x + 3x + 6 = 76$$
$$5x + 6 = 76$$
$$5x = 70$$
$$x = 14$$

If $x = 14$, the $x + 2 = 14 + 2$, or 16.

Check. 14 and 16 are consecutive even integers. Two times 14 plus three times 16 is $2 \cdot 14 + 3 \cdot 16$, or $28 + 48$, or 76. The numbers check.

State. The numbers are 14 and 16.

34. 9, 11, and 13

35. The Familiarize and Translate steps were done in Exercise 10.

Carry out.

$$x + \frac{2}{3} = 10$$
$$\frac{5}{3}x = 10$$
$$\frac{3}{5} \cdot \frac{5}{3}x = \frac{3}{5} \cdot 10$$
$$x = 6$$

If $x = 6$, $\frac{2}{3}x = \frac{2}{3} \cdot 6$, or 4.

Check. 4 is $\frac{2}{3}$ of 6. Also, $4 + 6 = 10$. The numbers check.

State. The wire should be cut into a 6 m piece and a 4 m piece.

36. 6 min

37. The Familiarize and Translate steps were done in Exercise 8.

Carry out. We solve the equation.

$$\begin{aligned} 1.5w + 0.25 &= 1.99 \\ 1.5w &= 1.74 \\ w &= \frac{1}{1.5}(1.74) \\ w &= 1.16 \end{aligned}$$

Check. If a wholesale price of $1.16 is raised 50%, we have $\$1.16 + 0.5(\$1.16)$, or $1.74. When 25 cents is added to this figure, we have $\$1.74 + \$0.25$, or $1.99. Our answer checks.

State. The tape's wholesale price is $1.16.

38. A $33\frac{1}{3}$-cm piece and a $66\frac{2}{3}$-cm piece

39.
$$\begin{aligned} 3[2x - (5 + 4x)] &= 5 - 7x \\ 3[2x - 5 - 4x] &= 5 - 7x \\ 3[-2x - 5] &= 5 - 7x \\ -6x - 15 &= 5 - 7x \\ x - 15 &= 5 \\ x &= 20 \end{aligned}$$

The solution is 20.

40. -19

41.
$$\begin{aligned} 4 + 2(a - 7) &= 2(a - 5) \\ 4 + 2a - 14 &= 2a - 10 \\ -10 + 2a &= 2a - 10 \\ -10 &= -10 \end{aligned}$$

All real numbers are solutions. The solution set is the set of all real numbers.

42. $\emptyset$

43. ◈

44. ◈

45. ◈

46. ◈

47. ***Familiarize.*** The average score on the first four tests is $\frac{83 + 91 + 78 + 81}{4}$, or 83.25. Let x = the number of points above this average that Tico scores on the next test. Then the score on the fifth test is $83.25 + x$.

Translate.

Average score on 5 tests | is 2 | more than | average score on 4 tests.

$$\frac{83+91+78+81+(83.25+x)}{5} = 2 + 83.25$$

Carry out. Carry out some algebraic manipulation.

$$\begin{aligned} \frac{83 + 91 + 78 + 81 + (83.25 + x)}{5} &= 2 + 83.25 \\ \frac{416.25 + x}{5} &= 85.25 \\ 416.25 + x &= 426.25 \\ x &= 10 \end{aligned}$$

Check. If Tico scores 10 points more than the average of the first four tests on the fifth test, his score will be $83.25 + 10$, or 93.25. Then the five-test average will be $\frac{83 + 91 + 78 + 81 + 93.25}{5}$, or 85.25. This is 2 points above the four-test average, so the answer checks.

State. Tico must score 10 points above the four-test average in order to raise the average 2 points.

48. 90 sq in

49. ***Familiarize.*** Let p = the price of the house in 1994. In 1995 real estate prices increased 6%, so the house was worth $p + 0.06p$, or $1.06p$. In 1996 prices increased 2%, so the house was then worth $1.06p + 0.02(1.06p)$, or $1.02(1.06p)$. In 1997 prices dropped 1%, so the value of the house became $1.02(1.06p) - 0.01(1.02)(1.06p)$, or $0.99(1.02)(1.06p)$.

Translate.

The price of the house in 1997 | was | $117,743.

$$0.99(1.02)(1.06p) = 117,743$$

Carry out. We carry out some algebraic manipulation.

$$\begin{aligned} 0.99(1.02)(1.06p) &= 117,743 \\ p &= \frac{117,743}{0.99(1.02)(1.06)} \\ p &\approx 110,000 \end{aligned}$$

Check. If the price of the house in 1994 was $110,000, then in 1995 it was worth $1.06(\$110,000)$, or $116,600. In 1996 it was worth $1.02(\$116,600)$, or $118,932, and in 1997 it was worth $0.99(\$118,932)$, or $117,743. Our answer checks.

State. The house was worth $110,000 in 1994.

50. $\frac{1}{1 - 0.01n}$, or $\frac{100}{100 - n}$

Exercise Set 1.5

1.
$$\begin{aligned} d &= rt \\ \frac{1}{r} \cdot d &= \frac{1}{r} \cdot rt \quad \text{Multiplying by } \frac{1}{r} \text{ on both sides} \\ \frac{d}{r} &= t \quad \text{Simplifying} \end{aligned}$$

2. $r = \frac{d}{t}$

3. $F = ma$

$\frac{1}{m} \cdot F = \frac{1}{m} \cdot ma$ Multiplying by $\frac{1}{m}$ on both sides

$\frac{F}{m} = a$ Simplifying

4. $w = \frac{A}{l}$

5. $W = EI$

$\frac{1}{I} \cdot W = \frac{1}{I} \cdot EI$ Multiplying by $\frac{1}{I}$ on both sides

$\frac{W}{I} = E$ Simplifying

6. $I = \frac{W}{E}$

7. $V = lwh$

$\frac{1}{lw} \cdot V = \frac{1}{lw} \cdot lwh$ Multiplying by $\frac{1}{lw}$ on both sides

$\frac{V}{lw} = h$ Simplifying

8. $r = \frac{I}{Pt}$

9. $L = \frac{k}{d^2}$

$d^2 \cdot L = d^2 \cdot \frac{k}{d^2}$ Multiplying by d^2 on both sides

$d^2 L = k$ Simplifying

10. $d^2 = \frac{k}{L}$

11. $G = w + 150n$

$G - w = 150n$ Subtracting w on both sides

$\frac{1}{150}(G - w) = \frac{1}{150} \cdot 150n$ Multiplying by $\frac{1}{150}$ on both sides

$\frac{G - w}{150} = n$ Simplifying

12. $t = \frac{P - b}{0.5}$

13. $2w + 2h + l = p$

$l = p - 2w - 2h$ Adding $-2w - 2h$ on both sides

14. $d^2 = \frac{km_1m_2}{g}$

15. $Ax + By = C$

$By = C - Ax$ Subtracting Ax on both sides

$\frac{1}{B} \cdot By = \frac{1}{B}(C - Ax)$ Multiplying by $\frac{1}{B}$ on both sides

$y = \frac{C - Ax}{B}$ Simplifying

16. $l = \frac{P - 2w}{2}$, or $\frac{P}{2} - w$

17. $C = \frac{5}{9}(F - 32)$

$\frac{9}{5} \cdot C = \frac{9}{5} \cdot \frac{5}{9}(F - 32)$ Multiplying by $\frac{9}{5}$ on both sides

$\frac{9}{5}C = F - 32$ Simplifying

$\frac{9}{5}C + 32 = F$ Adding 32 on both sides

18. $I = \frac{10}{3}T + 12,000$

19. $V = \frac{4}{3}\pi r^3$

$\frac{3}{4\pi} \cdot V = \frac{3}{4\pi} \cdot \frac{4}{3}\pi r^3$ Multiplying by $\frac{3}{4\pi}$ on both sides

$\frac{3V}{4\pi} = r^3$ Simplifying

20. $\pi = \frac{3V}{4r^3}$

21. $A = \frac{h}{2}(b_1 + b_2)$

$\frac{2}{h} \cdot A = \frac{2}{h} \cdot \frac{h}{2}(b_1 + b_2)$ Multiplying by $\frac{2}{h}$ on both sides

$\frac{2A}{h} = b_1 + b_2$ Simplifying

$\frac{2A}{h} - b_1 = b_2$ Subtracting b_1 on both sides

22. $h = \frac{2A}{b_1 + b_2}$

23. $F = \frac{mv^2}{r}$

$F \cdot \frac{r}{v^2} = \frac{mv^2}{r} \cdot \frac{r}{v^2}$ Multiplying by $\frac{r}{v^2}$ on both sides

$\frac{Fr}{v^2} = m$ Simplifying

24. $v^2 = \frac{rF}{m}$

25. $A = \frac{q_1 + q_2 + q_3}{n}$

$n \cdot A = n \cdot \frac{q_1 + q_2 + q_3}{n}$ Clearing the fraction

$nA = q_1 + q_2 + q_3$

$nA \cdot \frac{1}{A} = (q_1 + q_2 + q_3) \cdot \frac{1}{A}$ Multiplying by $\frac{1}{A}$ on both sides

$n = \frac{q_1 + q_2 + q_3}{A}$

26. $d = \dfrac{s+t}{r}$

27.
$$v = \frac{d_2 - d_1}{t}$$
$$t \cdot v = t \cdot \frac{d_2 - d_1}{t} \quad \text{Clearing the fraction}$$
$$tv = d_2 - d_1$$
$$tv \cdot \frac{1}{v} = (d_2 - d_1) \cdot \frac{1}{v} \quad \text{Multiplying by } \frac{1}{v} \text{ on both sides}$$
$$t = \frac{d_2 - d_1}{v}$$

28. $m = \dfrac{s_2 - s_1}{v}$

29.
$$v = \frac{d_2 - d_1}{t}$$
$$t \cdot v = t \cdot \frac{d_2 - d_1}{t} \quad \text{Clearing the fraction}$$
$$tv = d_2 - d_1$$
$$tv - d_2 = -d_1 \quad \text{Subtracting } d_2 \text{ on both sides}$$
$$-1 \cdot (tv - d_2) = -1 \cdot (-d_1) \quad \text{Multiplying by } -1 \text{ on both sides}$$
$$-tv + d_2 = d_1,$$
$$\text{or } d_2 - tv = d_1$$

30. $s_1 = s_2 - vm$

31.
$$r = m + mnp$$
$$r = m(1 + np) \quad \text{Factoring}$$
$$r \cdot \frac{1}{1+np} = m(1+np) \cdot \frac{1}{1+np}$$
$$\frac{r}{1+np} = m$$

32. $x = \dfrac{p}{1 - yz}$

33.
$$y = ab - ac^2$$
$$y = a(b - c^2) \quad \text{Factoring}$$
$$y \cdot \frac{1}{b-c^2} = a(b-c^2) \cdot \frac{1}{b-c^2}$$
$$\frac{y}{b-c^2} = a$$

34. $m = \dfrac{d}{n - p^3}$

35. ***Familiarize***. In an algebra book or a business math book we can find a formula for simple interest, $I = Prt$, when I is the interest, P is the principal, r is the interest rate, and t is the time, in years.

Translate. We want to find the amount that should be invested (the principal), so we solve the formula for P.

$$I = Prt$$
$$\frac{1}{rt} \cdot I = \frac{1}{rt} \cdot Prt$$
$$\frac{I}{rt} = P$$

Carry out. The model $P = \dfrac{I}{rt}$ can be used to find the principal that must be invested in order to earn a given amount of interest at a given rate for a given time. We substitute \$110 for I, 7% or 0.07 for r, and 1 for t.

$$P = \frac{I}{rt}$$
$$P = \frac{\$110}{0.07(1)}$$
$$P \approx \$1571.43$$

Check. We can repeat the calculation. The answer checks.

State. Yvonne should invest \$1571.43.

36. 6%

37. ***Familiarize***. In a geometry book or an algebra book we can find a formula for the area of a parallelogram, $A = bh$, where b is the base and h is the height.

Translate. We solve the formula for h.

$$A = bh$$
$$\frac{1}{b} \cdot A = \frac{1}{b} \cdot bh$$
$$\frac{A}{b} = h$$

Carry out. The model $h = \dfrac{A}{b}$ can be used to find the height of any parallelogram for which the area and base are known. We substitute 78 for A and 13 for b.

$$h = \frac{A}{b}$$
$$h = \frac{78}{13}$$
$$h = 6$$

Check. We repeat the calculation. The answer checks.

State. The height is 6 cm.

38. 12 cm

39. ***Familiarize***. We will use the model developed in Example 5, $m = \pi r^2 hD$ to find the weight of salt in a filled canister. We will add the weight of the empty canister, 28 grams, to this to find the total weight of the filled canister.

Translate. The total weight of the filled canister is given by the formula

$$m = \pi r^2 hD + 28.$$

Carry out. We substitute.

$$m = \pi r^2 hD + 28$$
$$m = \pi(4^2)(13.6)(2.16) + 28$$
$$m \approx 1504.6$$

Check. We repeat the calculation. The answer checks.

State. The filled canister will weigh about 1504.6 grams.

40. 235 grams

41. ***Familiarize.*** We will use Thurnau's model, $P = 9.337da - 299$.

Translate. Since we want to find the diameter of the fetus' head, we solve for d.

$$P = 9.337da - 299$$
$$P + 299 = 9.337da$$
$$\frac{P + 299}{9.337a} = d$$

Carry out. Substitute 1614 for P and 24.1 for a in the formula and calculate:

$$\frac{1614 + 299}{9.337(24.1)} = d$$
$$8.5 \approx d$$

Check. We repeat the calculation. The answer checks.

State. The diameter of the fetus' head at 29 weeks is about 8.5 cm.

42. 27.6 cm

43. ***Familiarize.*** The formula for the area of a trapezoid is $A = \frac{1}{2}h(b_1 + b_2)$, where A is the area, h is the height, and b_1 and b_2 are the bases.

Translate. The unknown dimension is the height, so we solve the formula for h. This was done in Exercise 22. We have

$$h = \frac{2A}{b_1 + b_2}.$$

Carry out. We substitute.

$$h = \frac{2A}{b_1 + b_2}$$
$$h = \frac{2 \cdot 90}{8 + 12}$$
$$h = \frac{180}{20}$$
$$h = 9$$

Check. We repeat the calculation. The answer checks.

State. The unknown dimension, the height of the trapezoid, is 9 ft.

44. 25 ft

45. ***Familiarize.*** The formula $A = P + Prt$ tells us how much a principal P, in dollars, will be worth when invested at simple interest at a rate r for t years.

Translate. We want to find the length of the investment period, so we solve the formula for t.

$$A = P + Prt$$
$$A - P = Prt$$
$$\frac{A - P}{Pr} = t$$

Carry out. Substitute 2608 for A, 1600 for P and 9% (or 0.09) for r.

$$\frac{A - P}{Pr} = t$$
$$\frac{2608 - 1600}{1600(0.09)} = t$$
$$\frac{1008}{144} = t$$
$$7 = t$$

Check. We repeat the calculation. The answer checks.

State. It will take 7 years for the investment to be worth $2608.

46. 6 years

47. ***Familiarize.*** We will use Goiten's model, $I = 1.08(T/N)$.

Translate. We solve the formula for T.

$$I = 1.08\left(\frac{T}{N}\right)$$
$$\frac{N}{1.08} \cdot I = \frac{N}{1.08} \cdot 1.08\left(\frac{T}{N}\right)$$
$$\frac{NI}{1.08} = T$$

Carry out. We substitute 25 for N and 20 for I.

$$T = \frac{NI}{1.08}$$
$$T = \frac{25 \cdot 20}{1.08}$$
$$T \approx 463$$

Check. We repeat the calculation. The answer checks.

State. The doctor should be prepared to spend about 463 min, or 7.7 hr, with patients.

48. 34

49. ***Familiarize.*** Recall that "percent" means "per hundred" or " × 0.01". Let n = the percent.

Translate.

What percent of 5800 is 4176?

$$n \times 0.01 \times 5800 = 4176$$

Carry out. We solve the equation.

$$n \times 0.01 \times 5800 = 4176$$
$$58n = 4176$$
$$n = \frac{4176}{58}$$
$$n = 72$$

Check. 72% of 5800 is 0.72(5800) = 4176. The answer checks.

State. 4176 is 72% of 5800.

50. $-8a + 13b$

51. $-72.5 - (-14.06) = -72.5 + 14.06 = -58.44$

52. -125

53. ◈

54. ◈

55. ◈

56. ◈

57. ***Familiarize.*** We will use the formulas for density and for the volume of a right circular cylinder. Note that the radius of a penny is $\frac{1.85 \text{ cm}}{2}$, or 0.925 cm.

Translate. Solving the formula $D = \frac{m}{V}$ for V, we get $V = \frac{m}{D}$. Also, the volume of a right circular cylinder with radius r and height h is given by $V = \pi r^2 h$, so we have $\pi r^2 h = \frac{m}{D}$. Solve for h by multiplying by $\frac{1}{\pi r^2}$ on both sides of the equation:

$$h = \frac{m}{\pi r^2 D}$$

Carry out. Substitute 8.93 for D, 177.6 for m, and 0.925 for r.

$$h = \frac{m}{\pi r^2 D}$$
$$h = \frac{177.6}{\pi(0.925)^2(8.93)}$$
$$h \approx 7.4$$

Check. We repeat the calculation. The answer checks.

State. A roll of pennies is about 7.4 cm tall.

58. About 610 cm

59.

$$s = v_i t + \frac{1}{2}at^2$$
$$s - v_i t = \frac{1}{2}at^2$$
$$2(s - v_i t) = at^2 \quad \text{Multiplying by 2 on both sides}$$
$$\frac{2(s - v_i t)}{t^2} = a,$$
$$\text{or } \frac{2s - 2v_i t}{t^2} = a$$

60. $l = \frac{A - w^2}{4w}$

61.

$$\frac{P_1V_1}{T_1} = \frac{P_2V_2}{T_2}$$
$$P_1V_1T_2 = P_2V_2T_1 \quad \text{Multiplying by } T_1T_2 \text{ on both sides}$$
$$T_2 = \frac{P_2V_2T_1}{P_1V_1} \quad \text{Multiplying by } \frac{1}{P_1V_1} \text{ on both sides}$$

62. $T_1 = \frac{P_1V_1T_2}{P_2V_2}$

63.

$$\frac{b}{a - b} = c$$
$$b = c(a - b) \quad \text{Multiplying by } a - b \text{ on both sides}$$
$$b = ac - bc$$
$$b + bc = ac \quad \text{Adding } bc \text{ on both sides}$$
$$b(1 + c) = ac \quad \text{Factoring}$$
$$b = \frac{ac}{1 + c} \quad \text{Multiplying by } \frac{1}{1 + c} \text{ on both sides}$$

64. $d = \frac{me^2}{f}$

Exercise Set 1.6

1. $7^5 \cdot 7^2 = 7^{5+2} = 7^7$

2. 2^{11}

3. $5^6 \cdot 5^3 = 5^{6+3} = 5^9$

4. 6^8

5. $a^3 \cdot a^0 = a^{3+0} = a^3$

6. x^5

7. $6x^5 \cdot 3x^2 = 6 \cdot 3 \cdot x^5 \cdot x^2 = 18x^{5+2} = 18x^7$

8. $8a^{10}$

9. $(-3m^4)(-7m^9) = (-3)(-7)m^4 \cdot m^9 = 21m^{4+9} = 21m^{13}$

10. $-14a^9$

11. $(x^3y^4)(x^7y^6z^0) = (x^3x^7)(y^4y^6)(z^0) = x^{3+7}y^{4+6} \cdot 1 = x^{10}y^{10}$

12. $m^{10}n^{12}$

13. $\frac{a^9}{a^3} = a^{9-3} = a^6$

14. x^9

15. $\frac{8x^7}{4x^4} = \frac{8}{4} \cdot x^{7-4} = 2x^3$

16. $4a^{16}$

17. $\frac{m^7n^9}{m^2n^5} = m^{7-2} \cdot n^{9-5} = m^5n^4$

18. m^8n^3

19. $\frac{35x^8y^5}{7x^2y} = \frac{35}{7} \cdot x^{8-2} \cdot y^{5-1} = 5x^6y^4$

20. $9x^6y^6$

21. $\frac{-49a^5b^{12}}{7a^2b^2} = \frac{-49}{7} \cdot a^{5-2} \cdot b^{12-2} = -7a^3b^{10}$

22. $-6ab^3$

23. $\frac{18x^6y^7z^9}{-6x^2yz^3} = \frac{18}{-6} \cdot x^{6-2} \cdot y^{7-1} \cdot z^{9-3} = -3x^4y^6z^6$

24. $-4x^4y^8z^{11}$

25. $(-3)^4 = -3(-3)(-3)(-3) = 81$

26. 64

27. $-3^4 = -3 \cdot 3 \cdot 3 \cdot 3 = -81$

28. -64

29. $(-5)^{-2} = \frac{1}{(-5)^2} = \frac{1}{-5(-5)} = \frac{1}{25}$

30. $\frac{1}{16}$

31. $-5^{-2} = -\frac{1}{5^2} = -\frac{1}{25}$

32. $-\frac{1}{16}$

33. $-2^{-5} = -\frac{1}{2^5} = -\frac{1}{32}$

34. $-\frac{1}{125}$

35. $-2^{-6} = -\frac{1}{2^6} = -\frac{1}{64}$

36. -1

37. $a^{-3} = \frac{1}{a^3}$

38. $\frac{1}{n^6}$

39. $(5x)^{-3} = \frac{1}{(5x)^3}$

40. $\frac{1}{(4xy)^5}$

41. $x^2y^{-3} = x^2 \cdot \frac{1}{y^3} = \frac{x^2}{y^3}$

42. $\frac{2a^2}{b^5}$

43. $x^2y^{-2} = x^2 \cdot \frac{1}{y^2} = \frac{x^2}{y^2}$

44. $\frac{a^2c^4}{b^3d^5}$

45. $\frac{y^{-5}}{x^2} = \frac{1}{x^2} \cdot y^{-5} = \frac{1}{x^2} \cdot \frac{1}{y^5} = \frac{1}{x^2y^5}$

46. $\frac{1}{3x^5z^4}$

47. $\frac{y^{-5}}{x^{-3}} = \frac{1}{x^{-3}} \cdot y^{-5} = x^3 \cdot \frac{1}{y^5} = \frac{x^3}{y^5}$

48. $\frac{y^7}{(7x)^4}$

49. $\frac{x^{-2}y^7}{z^{-4}} = x^{-2} \cdot y^7 \cdot \frac{1}{z^{-4}} = \frac{1}{x^2} \cdot y^7 \cdot z^4 = \frac{y^7z^4}{x^2}$

50. $\frac{x^2y^4}{z^3}$

51. $\frac{1}{3^4} = 3^{-4}$

52. 9^{-2}

53. $\frac{1}{(-16)^2} = (-16)^{-2}$

54. $(-8)^{-6}$

55. $x^5 = \frac{1}{x^{-5}}$

56. $\frac{1}{n^{-3}}$

57. $6x^2 = 6 \cdot \frac{1}{x^{-2}} = \frac{6}{x^{-2}}$

58. $\frac{-4}{y^{-5}}$

59. $\frac{1}{(5y)^3} = (5y)^{-3}$

60. $(5x)^{-5}$

61. $\frac{1}{3y^4} = \frac{1}{3} \cdot \frac{1}{y^4} = \frac{1}{3} \cdot y^{-4} = \frac{y^{-4}}{3}$

62. $\frac{b^{-3}}{4}$

63. $8^{-2} \cdot 8^{-4} = 8^{-2+(-4)} = 8^{-6}$, or $\frac{1}{8^6}$

64. 9^{-7}, or $\frac{1}{9^7}$

65. $b^2 \cdot b^{-5} = b^{2+(-5)} = b^{-3}$, or $\frac{1}{b^3}$

66. a

67. $a^{-3} \cdot a^4 \cdot a^2 = a^{-3+4+2} = a^3$

68. 1

69. $(14m^2n^3)(-2m^3n^2) = 14 \cdot (-2) \cdot m^2 \cdot m^3 \cdot n^3 \cdot n^2$
$= -28m^{2+3}n^{3+2}$
$= -28m^5n^5$

70. $-18x^7y$

71. $(-2x^{-3})(7x^{-8}) = -2 \cdot 7 \cdot x^{-3} \cdot x^{-8} = -14x^{-3+(-8)}$
$= -14x^{-11}$, or $\dfrac{-14}{x^{11}}$

72. $-24x^{-12}y$, or $-\dfrac{24y}{x^{12}}$

73. $(5a^{-2}b^{-3})(2a^{-4}b) = 5 \cdot 2 \cdot a^{-2} \cdot a^{-4} \cdot b^{-3} \cdot b$
$= 10a^{-2+(-4)}b^{-3+1}$
$= 10a^{-6}b^{-2}$, or $\dfrac{10}{a^6b^2}$

74. $6a^{-4}b^{-9}$, or $\dfrac{6}{a^4b^9}$

75. $\dfrac{10^{-3}}{10^6} = 10^{-3-6} = 10^{-9}$, or $\dfrac{1}{10^9}$

76. 12^{-12}, or $\dfrac{1}{12^{12}}$

77. $\dfrac{2^{-7}}{2^{-5}} = 2^{-7-(-5)} = 2^{-7+5} = 2^{-2}$, or $\dfrac{1}{2^2}$, or $\dfrac{1}{4}$

78. 9^2, or 81

79. $\dfrac{y^4}{y^{-5}} = y^{4-(-5)} = y^{4+5} = y^9$

80. a^5

81. $\dfrac{24a^5b^3}{-8a^4b} = \dfrac{24}{-8}a^{5-4}b^{3-1} = -3ab^2$

82. $3ab^{-3}$, or $\dfrac{3a}{b^3}$

83. $\dfrac{14a^4b^{-3}}{-8a^8b^{-5}} = \dfrac{14}{-8}a^{4-8}b^{-3-(-5)} = -\dfrac{7}{4}a^{-4}b^2$, or
$-\dfrac{7b^2}{4a^4}$

84. $-\dfrac{4}{3}x^9y^{-2}$, or $-\dfrac{4x^9}{3y^2}$

85. $\dfrac{-5x^{-2}y^4z^7}{30x^{-5}y^6z^{-3}} = \dfrac{-5}{30}x^{-2-(-5)}y^{4-6}z^{7-(-3)} =$
$-\dfrac{1}{6}x^3y^{-2}z^{10}$, or $-\dfrac{x^3z^{10}}{6y^2}$

86. $\dfrac{1}{3}a^{10}b^{-9}c^{-2}$, or $\dfrac{a^{10}}{3b^9c^2}$

87. $(x^4)^3 = x^{4\cdot 3} = x^{12}$

88. a^6

89. $(9^3)^{-4} = 9^{3(-4)} = 9^{-12}$, or $\dfrac{1}{9^{12}}$

90. 8^{-12}, or $\dfrac{1}{8^{12}}$

91. $(7^{-8})^{-5} = 7^{-8(-5)} = 7^{40}$

92. 6^{12}

93. $(6xy)^2 = 6^2x^2y^2 = 36x^2y^2$

94. $125a^3b^3$

95. $(a^3b)^4 = a^{3\cdot 4}b^4 = a^{12}b^4$

96. $x^{15}y^5$

97. $5(x^2y^2)^3 = 5(x^2)^3(y^2)^3 = 5x^6y^6$

98. $7a^6b^8$

99. $(7x^3y^{-4})^{-2} = 7^{-2}(x^3)^{-2}(y^{-4})^{-2} =$
$\dfrac{1}{7^2}x^{-6}y^8 = \dfrac{1}{49}x^{-6}y^8$, or $\dfrac{y^8}{49x^6}$

100. $\dfrac{x^{-4}y^{10}}{9}$, or $\dfrac{y^{10}}{9x^4}$

101. $(-8x^{-4}y^5z^2)^{-4} = (-8)^{-4}(x^{-4})^{-4}(y^5)^{-4}(z^2)^{-4} =$
$(-8)^{-4}x^{16}y^{-20}z^{-8} = \dfrac{x^{16}y^{-20}z^{-8}}{(-8)^4} = \dfrac{x^{16}y^{-20}z^{-8}}{4096}$,
or $\dfrac{x^{16}}{4096y^{20}z^8}$

102. $\dfrac{a^4b^{-6}c^{-2}}{36}$, or $\dfrac{a^4}{36b^6c^2}$

103. $\dfrac{(3x^3y^4)^3}{6xy^3} = \dfrac{3^3(x^3)^3(y^4)^3}{6xy^3} = \dfrac{27x^9y^{12}}{6xy^3} =$
$\dfrac{27}{6}x^{9-1}y^{12-3} = \dfrac{9}{2}x^8y^9$, or $\dfrac{9x^8y^9}{2}$

104. $\dfrac{5a^4b}{2}$, or $\dfrac{5a^4b}{2}$

105. $\left(\dfrac{-4x^4y^{-2}}{5x^{-1}y^4}\right)^{-4} = \left(\dfrac{-4}{5}x^{4-(-1)}y^{-2-4}\right)^{-4} =$
$\left(\dfrac{-4}{5}x^5y^{-6}\right)^{-4} = \dfrac{(-4)^{-4}(x^5)^{-4}(y^{-6})^{-4}}{5^{-4}} =$
$\dfrac{5^4x^{-20}y^{24}}{(-4)^4} = \dfrac{625x^{-20}y^{24}}{256}$, or $\dfrac{625y^{24}}{256x^{20}}$

106. $\dfrac{8x^9y^3}{27}$

107. $\left(\dfrac{3a^{-2}b^5}{9a^{-4}b^0}\right)^{-2} = \left(\dfrac{3}{9}a^{-2-(-4)}b^{5-0}\right)^{-2} =$

$\left(\dfrac{1}{3}a^2b^5\right)^{-2} = \left(\dfrac{1}{3}\right)^{-2}(a^2)^{-2}(b^5)^{-2} =$

$\dfrac{1}{\left(\dfrac{1}{3}\right)^2}a^{-4}b^{-10} = \dfrac{1}{\dfrac{1}{9}}a^{-4}b^{-10} = 9a^{-4}b^{-10}$, or $\dfrac{9}{a^4b^{10}}$

108. 1

(Any nonzero real number raised to the zero power is 1.)

109. $\left(\dfrac{4a^3b^{-9}}{2a^{-2}b^5}\right)^0 = 1$

(Any nonzero real number raised to the zero power is 1.)

110. $\dfrac{4x^{-4}y^{22}}{25}$, or $\dfrac{4y^{22}}{25x^4}$

111. $x + (2xy - z)^2 = 3 + (2 \cdot 3(-4) - (-20))^2 =$

$3 + (-24 + 20)^2 = 3 + (-4)^2 = 3 + 16 = 19$

112. -8

113. ***Familiarize***. Let x represent the first integer, $x+2$ the second, and $x+4$ the third.

Translate.

First plus second plus third is 183.

$x \quad + \quad (x+2) \quad + \quad (x+4) \quad = \quad 183$

Carry out.

$$x + x + 2 + x + 4 = 183$$
$$3x + 6 = 183$$
$$3x = 177$$
$$x = 59$$

If $x = 59$, then $x+2 = 59+2$, or 61, and $x+4 = 59+4$, or 63.

Check. 59, 61, and 63 are consecutive odd integers and their sum is $59 + 61 + 63$, or 183. The answer checks.

State. The integers are 59, 61, and 63.

114. 36, 38, and 40

115. ◈

116. ◈

117. ◈

118. ◈

119. $\dfrac{9a^{x-2}}{3a^{2x+2}} = \dfrac{9}{3} \cdot a^{x-2-(2x+2)} = 3a^{x-2-2x-2} = 3a^{-x-4}$

120. $-3x^{2a-1}$

121. $[7y(7-8)^{-2} - 8y(8-7)^{-2}]^{(-2)^2}$

$= [7y(7-8)^{-2} - 8y(8-7)^{-2}]^4$

$= [7y(-1)^{-2} - 8y(1)^{-2}]^4$

$= \left[\dfrac{7y}{(-1)^2} - \dfrac{8y}{1^2}\right]^4$

$= \left[\dfrac{7y}{1} - \dfrac{8y}{1}\right]^4$

$= [7y - 8y]^4$

$= [-y]^4$

$= y^4$

122. 8^{-2abc}

123. $(3^{a+2})^a = 3^{(a+2)(a)} = 3^{a^2+2a}$

124. 12^{6b-2ab}

125. $\dfrac{4x^{2a+3}y^{2b-1}}{2x^{a+1}y^{b+1}} = \dfrac{4}{2} \cdot x^{2a+3-(a+1)}y^{2b-1-(b+1)} =$

$2x^{2a+3-a-1}y^{2b-1-b-1} = 2x^{a+2}y^{b-2}$

126. $-5x^{2b}y^{-2a}$

127. $\dfrac{(2^{-2})^a \cdot (2^b)^{-a}}{(2^{-2})^{-b}(2^b)^{-2a}} = \dfrac{2^{-2a} \cdot 2^{-ab}}{2^{2b} \cdot 2^{-2ab}} = \dfrac{2^{-2a-ab}}{2^{2b-2ab}} =$

$2^{-2a-ab-(2b-2ab)} = 2^{-2a-ab-2b+2ab} = 2^{-2a-2b+ab}$

128. $-4x^{10}y^8$

129. $\dfrac{3^{q+3} - 3^2(3^q)}{3(3^{q+4})} = \dfrac{3^{q+3} - 3^{q+2}}{3^{q+5}} = \dfrac{3^{q+2}(3-1)}{3^{q+2}(3^3)} =$

$\dfrac{2}{3^3} = \dfrac{2}{27}$

130. $\dfrac{a^{-14ac}}{b^{27ac}}$

Exercise Set 1.7

1. $47{,}000{,}000{,}000$

$= \dfrac{47{,}000{,}000{,}000}{10^{10}} \cdot 10^{10}$ Multiplying by 1: $\dfrac{10^{10}}{10^{10}} = 1$

$= 4.7 \times 10^{10}$ This is scientific notation.

2. 2.6×10^{12}

3. $863{,}000{,}000{,}000{,}000{,}000$

$= \dfrac{863{,}000{,}000{,}000{,}000{,}000}{10^{17}} \cdot 10^{17}$

$= 8.63 \times 10^{17}$

4. 9.57×10^{17}

5. 0.000000016

$= \dfrac{0.000000016}{10^8} \cdot 10^8$ Multiplying by 1: $\dfrac{10^8}{10^8} = 1$

$= \dfrac{1.6}{10^8}$

$= 1.6 \times 10^{-8}$ Writing scientific notation

6. 2.63×10^{-7}

7. 0.00000000007

$= \dfrac{0.00000000007}{10^{11}} \cdot 10^{11}$

$= \dfrac{7}{10^{11}}$

$= 7 \times 10^{-11}$

8. 9×10^{-11}

9. $407,000,000,000$

$= \dfrac{407,000,000,000}{10^{11}} \cdot 10^{11}$

$= 4.07 \times 10^{11}$

10. 3.09×10^{12}

11. 0.000000603

$= \dfrac{0.000000603}{10^7} \cdot 10^7$

$= \dfrac{6.03}{10^7}$

$= 6.03 \times 10^{-7}$

12. 8.02×10^{-9}

13. $492,700,000,000$

$= \dfrac{492,700,000,000}{10^{11}} \cdot 10^{11}$

$= 4.927 \times 10^{11}$

14. 9.534×10^{11}

15. $4 \times 10^{-4} = 0.0004$ Moving the decimal point 4 places to the left

16. 0.00005

17. $6.73 \times 10^8 = 673,000,000$ Moving the decimal point 8 places to the right

18. $92,400,000$

19. $8.923 \times 10^{-10} = 0.0000000008923$ Moving the decimal point 10 places to the left

20. 0.07034

21. $9.03 \times 10^{10} = 90,300,000,000$ Moving the decimal point 10 places to the right

22. $1,010,000,000,000$

23. $4.037 \times 10^{-8} = 0.00000004037$ Moving the decimal point 8 places to the left

24. 0.000000003007

25. $8.007 \times 10^{12} = 8,007,000,000,000$ Moving the decimal point 12 places to the right

26. $90,010,000,000$

27. $(2.3 \times 10^6)(4.2 \times 10^{-11})$

$= (2.3 \times 4.2)(10^6 \times 10^{-11})$

$= 9.66 \times 10^{-5}$

$= 9.7 \times 10^{-5}$ Rounding to 2 significant digits

28. 3.4×10^{-4}

29. $(2.34 \times 10^{-8})(5.7 \times 10^{-4})$

$= (2.34 \times 5.7)(10^{-8} \times 10^{-4})$

$= 13.338 \times 10^{-12}$

$= (1.3338 \times 10^1) \times 10^{-12}$

$= 1.3338 \times (10^1 \times 10^{-12})$

$= 1.3338 \times 10^{-11}$

$= 1.3 \times 10^{-11}$ Rounding to 2 significant digits

30. 2.7×10^{-11}

31. $(3.2 \times 10^6)(2.6 \times 10^4) = (3.2 \times 2.6)(10^6 \times 10^4)$

$= 8.32 \times 10^{10}$

$= 8.3 \times 10^{10}$

(2 significant digits)

32. 3.14×10^{16}

33. $(3.01 \times 10^{-5})(6.5 \times 10^7)$

$= (3.01 \times 6.5)(10^{-5} \times 10^7)$

$= 19.565 \times 10^2$

$= (1.9565 \times 10^1) \times 10^2$

$= 1.9565 \times (10^1 \times 10^2)$

$= 1.9565 \times 10^3$

$= 2.0 \times 10^3$ Rounding to 2 significant digits

34. 3.1×10^{-4}

35. $(5.01 \times 10^{-7})(3.02 \times 10^{-6})$

$= (5.01 \times 3.02)(10^{-7} \times 10^{-6})$

$= 15.1302 \times 10^{-13}$

$= (1.51302 \times 10^1) \times 10^{-13}$

$= 1.51302 \times (10^1 \times 10^{-13})$

$= 1.51302 \times 10^{-12}$

$= 1.51 \times 10^{-12}$ Rounding to 3 significant digits

36. 6.34×10^{-15}

37. $\dfrac{5.1\times10^6}{3.4\times10^3}=\dfrac{5.1}{3.4}\times\dfrac{10^6}{10^3}$
$= 1.5\times10^3$

38. 2.5×10^3

39. $\dfrac{7.5\times10^{-9}}{2.5\times10^{-4}}=\dfrac{7.5}{2.5}\times\dfrac{10^{-9}}{10^{-4}}$
$= 3.0\times10^{-5}$ (2 significant digits)

40. 5.0×10^{-4}

41. $\dfrac{3.2\times10^{-7}}{8.0\times10^{8}}=\dfrac{3.2}{8.0}\times\dfrac{10^{-7}}{10^{8}}$
$= 0.40\times10^{-15}$ (2 significant digits)
$= (4.0\times10^{-1})\times10^{-15}$
$= 4.0\times10^{-16}$

42. 3.0×10^{11}

43. $\dfrac{9.36\times10^{-11}}{3.12\times10^{11}}=\dfrac{9.36}{3.12}\times\dfrac{10^{-11}}{10^{11}}$
$= 3.00\times10^{-22}$ (3 significant digits)

44. 2.00×10^{10}

45. $\dfrac{6.12\times10^{19}}{3.06\times10^{-7}}=\dfrac{6.12}{3.06}\times\dfrac{10^{19}}{10^{-7}}$
$= 2.00\times10^{26}$ (3 significant digits)

46. 2.0×10^{-16}

47. $\dfrac{780,000,000\times0.00071}{0.000005}$
$= \dfrac{(7.8\times10^8)\times(7.1\times10^{-4})}{5\times10^{-6}}$
$= \dfrac{7.8\times7.1}{5}\times\dfrac{10^8\times10^{-4}}{10^{-6}}$
$= 11.076\times10^{10}$
$= 1.1076\times10^1\times10^{10}$
$= 1.1076\times(10^1\times10^{10})$
$= 1\times10^{11}$ (1 significant digit)

48. 3.2×10

49. $5.9\times10^{23}+2.4\times10^{23}$
$= (5.9+2.4)\times10^{23}$
$= 8.3\times10^{23}$

50. 7.2×10^{-34}

51. ***Familiarize.*** The distance Venus travels in one orbit can be approximated by the circumference of a circle whose radius is the average distance from the sun to Venus, 1.08×10^8 km. Recall that the formula for the circumference of a circle is $C=2\pi r$, where r is the radius.

Translate. Substitute 1.08×10^8 for r in the formula
$C=2\pi r$
$C=2\pi\times1.08\times10^8$

Carry out. Do the calculation. Use a calculator with a π key.
$C=2\pi\times1.08\times10^8$
$\approx6.79\times10^8$ Rounding to 2 significant digits

Check. Repeat the calculation.

State. Venus travels about 6.79×10^8 km in one orbit.

52. 4.5×10^{-3} kg, or 4.5 g

53. ***Familiarize.*** First we will find the number n of \$5 bills in \$4,540,000 worth of \$5 bills. Then we will find the weight w of a \$5 bill. Recall that 1 ton = 2000 lb.

Translate. To find the number of \$5 bills in \$4,540,000 worth of \$5 bills we divide:
$n=\dfrac{4,540,000}{5}$.
Then we divide again to find the weight w of a \$5 bill:
$w=\dfrac{2000}{n}$.

Carry out. We begin by finding n.
$n=\dfrac{4,540,000}{5}=\dfrac{4.54\times10^6}{5}=0.908\times10^6=$
$(9.08\times10^{-1})\times10^6=9.08\times10^5$
Now we find w.
$w=\dfrac{2000}{n}=\dfrac{2000}{9.08\times10^5}=\dfrac{2\times10^3}{9.08\times10^5}\approx$
$0.220\times10^{-2}=(2.20\times10^{-1})\times10^{-2}=$
2.20×10^{-3} (Our answer has 3 significant digits.)

Check. We recheck the translation and calculations.

State. A \$5 bill weighs about 2.20×10^{-3} lb.

54. 5.8×10^8 miles

55. ***Familiarize.*** From Example 7 we know that 1 light year = 5.88×10^{12} mi. Let y = the number of light years from the earth to Sirius.

Translate. The distance from the earth to Sirius is y light years or $(5.88\times10^{12})y$ mi. It is also given by 4.704×10^{13} mi. We write an equation:
$(5.88\times10^{12})y=4.704\times10^{13}$

Carry out. We solve the equation.

$$(5.88 \times 10^{12})y = 4.704 \times 10^{13}$$

$$y = \frac{4.704 \times 10^{13}}{5.88 \times 10^{12}}$$

$$y = \frac{4.704}{5.88} \times \frac{10^{13}}{10^{12}}$$

$y = 0.800 \times 10$ The answer must have 3 significant digits.

$$y = 8.00$$

Check. Since light travels 5.88×10^{12} mi in one year, in 8.00 yr it will travel $8.00 \times 5.88 \times 10^{12} = 4.704 \times 10^{13}$ mi, the distance from the earth to Sirius. The answer checks.

State. It is 8.00 light years from the earth to Sirius.

56. 1.00×10^5 light years

57. ***Familiarize.*** We are told that 1 Angstrom = 10^{-10} m, 1 parsec $\approx$ 3.26 light years, and 1 light year = 9.46×10^{15} m. Let a represent the number of Angstroms in one parsec.

Translate. The length of one parsec is $a \times 10^{-10}$ m. It can also be expressed as 3.26 light years, or $3.26 \times 9.46 \times 10^{15}$ m. Since these quantities represent the same number, we can write the equation.

$$a \times 10^{-10} = 3.26 \times 9.46 \times 10^{15}.$$

Carry out. Solve the equation:

$$a \times 10^{-10} = 3.26 \times 9.46 \times 10^{15}$$

$$a \times 10^{-10} \times \frac{1}{10^{-10}} = 3.26 \times 9.46 \times 10^{15} \times \frac{1}{10^{-10}}$$

$$a = \frac{3.26 \times 9.46 \times 10^{15}}{10^{-10}}$$

$$= (3.26 \times 9.46) \times \frac{10^{15}}{10^{-10}}$$

$$= 30.8396 \times 10^{25}$$

$$= (3.08396 \times 10) \times 10^{25}$$

$$= 3.08396 \times (10 \times 10^{25})$$

$= 3.08 \times 10^{26}$ Rounding to 3 significant digits

Check. We recheck the translation and calculation.

State. There are about 3.08×10^{26} Angstroms in one parsec.

58. 3.08×10^{13} km

59. ***Familiarize.*** We have a very long cylinder. Its length is the average distance from the earth to the sun, 1.5×10^{11} m, and the diameter of its base is 3 Å. We will use the formula for the volume of a cylinder, $V = \pi r^2 h$. (See Example 8.)

Translate. We will express all distances in Angstroms.

Height (length) : $1.5 \times 10^{11}\text{ m} = \frac{1.5 \times 10^{11}}{10^{-10}}$ Å, or 1.5×10^{21} Å

Diameter: 3 Å

The radius is half the diameter:

Radius: $\frac{1}{2} \times 3$ Å $= 1.5$ Å

Now substitute into the formula (using 3.14 for π):

$$V = \pi r^2 h$$

$$V = 3.14 \times 1.5^2 \times 1.5 \times 10^{21}$$

Carry out. Do the calculations.

$$V = 3.14 \times 1.5^2 \times 1.5 \times 10^{21}$$

$$= 10.5975 \times 10^{21}$$

$$= 1.05975 \times 10^{22}$$

$= 1 \times 10^{22}$ Rounding to 1 significant digit

Check. We recheck the translation and the calculations.

State. The volume of the sunbeam is about 1×10^{22} cu Å.

60. 3×10^{22} cu Å

61. ***Familiarize.*** First we will find d, the number of drops in a pound. Then we will find b, the number of bacteria in a drop of U.S. mud.

Translate. To find d we convert 1 pound to drops:

$$d = 1\text{ lb} \cdot \frac{16\text{oz}}{1\text{ lb}} \cdot \frac{6\text{ tsp}}{1\text{ oz}} \cdot \frac{60\text{ drops}}{1\text{ tsp}}.$$

Then we divide to find b:

$$b = \frac{4.55 \times 10^{11}}{d}.$$

Carry out. We do the calculations.

$$d = 1\text{ lb} \cdot \frac{16\text{oz}}{1\text{ lb}} \cdot \frac{6\text{ tsp}}{1\text{ oz}} \cdot \frac{60\text{ drops}}{1\text{ tsp}}$$

$$= 5760\text{ drops}$$

Now we find b.

$$b = \frac{4.55 \times 10^{11}}{5760} = \frac{4.55 \times 10^{11}}{5.760 \times 10^3} \approx 0.790 \times 10^8 \approx$$

$$(7.90 \times 10^{-1}) \times 10^8 = 7.90 \times 10^7$$

(Our answer must have 3 significant digits.)

Check. If there are about 7.90×10^7 bacteria in a drop of U.S. mud, then in a pound there are about

$$\frac{7.90 \times 10^7}{1\text{ drop}} \cdot \frac{60\text{ drops}}{1\text{ tsp}} \cdot \frac{6\text{ tsp}}{1\text{ oz}} \cdot \frac{16\text{ oz}}{1\text{ lb}} =$$

$\frac{45,504 \times 10^7}{1\text{ lb}} \approx 4.55 \times 10^{11}$ bacteria per pound. The answer checks.

State. About 7.90×10^7 bacteria live in a drop of U.S. mud.

62. 1.5×10^{12} mi

63. ***Familiarize.*** First we will find the distance C around Jupiter at the equator, in km. Then we will use the formula Speed $\times$ Time = Distance to find the speed s at which Jupiter's equator is spinning.

Translate. We will use the formula for the circumference of a circle to find the distance around Jupiter at the equator:

$$C = \pi d = \pi(1.43 \times 10^5).$$

Then we find the speed s at which Jupiter's equator is spinning:

$$\underbrace{\text{Speed}}_{\downarrow\, s} \times \underbrace{\text{Time}}_{\downarrow\, 10} = \underbrace{\text{Distance}}_{\downarrow\, C}$$

$$s \times 10 = C$$

Carry out. First we find C.

$C = \pi(1.43 \times 10^5) \approx 4.49 \times 10^5$

Then we find s.

$$s \times 10 = C$$
$$s \times 10 = 4.49 \times 10^5$$
$$s = \frac{4.49 \times 10^5}{10}$$
$$s = 4.49 \times 10^4$$

Check. At 4.49×10^4 km/h, in 10 hr, Jupiter's equator travels $4.49 \times 10^4 \times 10$, or 4.49×10^5 km. A circle with circumference 4.49×10^5 km has a diameter of $\dfrac{4.49 \times 10^5}{\pi} \approx 1.43 \times 10^5$ km. The answer checks.

State. Jupiter's equator spins at a speed of about 4.49×10^4 km/h.

64. \$67,000

65. $-\frac{5}{6} - \left(-\frac{3}{4}\right) = -\frac{5}{6} + \frac{3}{4} = -\frac{10}{12} + \frac{9}{12} = -\frac{1}{12}$

66. 30.96

67.
$$\begin{aligned} -2(x-3) - 3(4-x) &= -2x + 6 - 12 + 3x \\ &= (-2x + 3x) + (6 - 12) \\ &= x - 6 \end{aligned}$$

68. $\frac{7}{4}$

69. ◈

70. ◈

71. ◈

72. ◈

73. The larger number is the one in which the power of ten has the larger exponent. Since -90 is larger than -91, $8 \cdot 10^{-90}$ is larger than $9 \cdot 10^{-91}$.

$$\begin{aligned} 8 \cdot 10^{-90} - 9 \cdot 10^{-91} &= 10^{-90}(8 - 9 \cdot 10^{-1}) \\ &= 10^{-90}(8 - 0.9) \\ &= 7.1 \times 10^{-90} \end{aligned}$$

74. 1.25×10^{22}

75.
$$\begin{aligned} (4096)^{0.05}(4096)^{0.2} &= 4096^{0.25} \\ &= (2^{12})^{0.25} \\ &= 2^3 \\ &= 8 \end{aligned}$$

76. 1

77. ***Familiarize.*** Observe that there are 2^{n-1} grains of sand on the nth square of the chessboard. Let g represent this quantity. Recall that a chessboard has 64 squares. Note also that $2^{10} \approx 10^3$.

Translate. We write the equation

$$g = 2^{n-1}.$$

To find the number of grains of sand on the last (or 64th) square, substitute 64 for n: $g = 2^{64-1}$

Carry out. Do the calculations, expressing the result in scientific notation.

$$\begin{aligned} g = 2^{64-1} = 2^{63} &= 2^3(2^{10})^6 \\ &\approx 2^3(10^3)^6 \approx 8 \times 10^{18} \end{aligned}$$

Check. Recheck the translation and the calculations.

State. Approximately 8×10^{18} grains of sand are required for the last square.

Chapter 2

Graphs, Functions, and Linear Equations

Exercise Set 2.1

1.

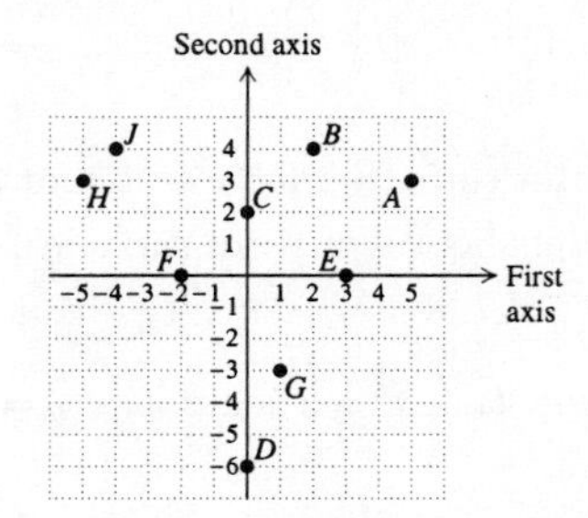

$A(5,3)$ is 5 units right and 3 units up.

$B(2,4)$ is 2 units right and 4 units up.

$C(0,2)$ is 0 units left or right and 2 units up.

$D(0,-6)$ is 0 units left or right and 6 units down.

$E(3,0)$ is 3 units right and 0 units up or down.

$F(-2,0)$ is 2 units left and 0 units up or down.

$G(1,-3)$ is 1 unit right and 3 units down.

$H(-5,3)$ is 5 units left and 3 units up.

$J(-4,4)$ is 4 units left and 4 units up.

2.

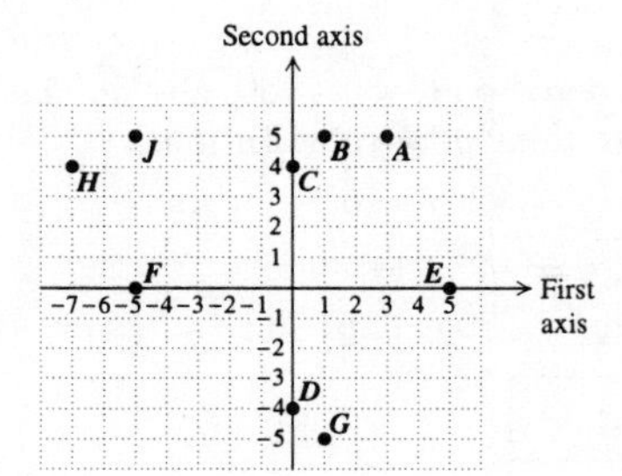

3.

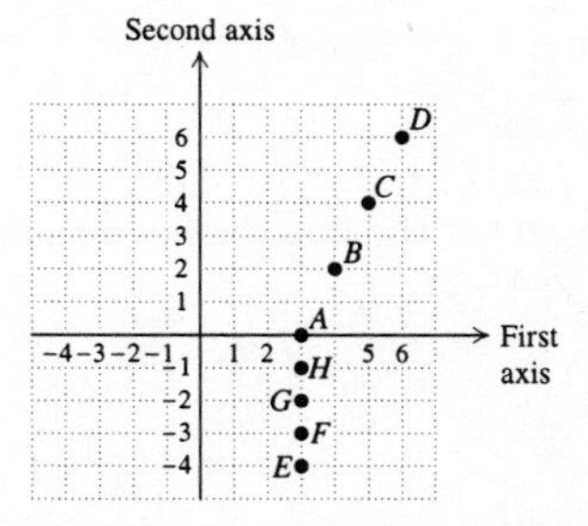

$A(3,0)$ is 3 units right and 0 units up or down.

$B(4,2)$ is 4 units right and 2 units up.

$C(5,4)$ is 5 units right and 4 units up.

$D(6,6)$ is 6 units right and 6 units up.

$E(3,-4)$ is 3 units right and 4 units down.

$F(3,-3)$ is 3 units right and 3 units down.

$G(3,-2)$ is 3 units right and 2 units down.

$H(3,-1)$ is 3 units right and 1 unit down.

4.

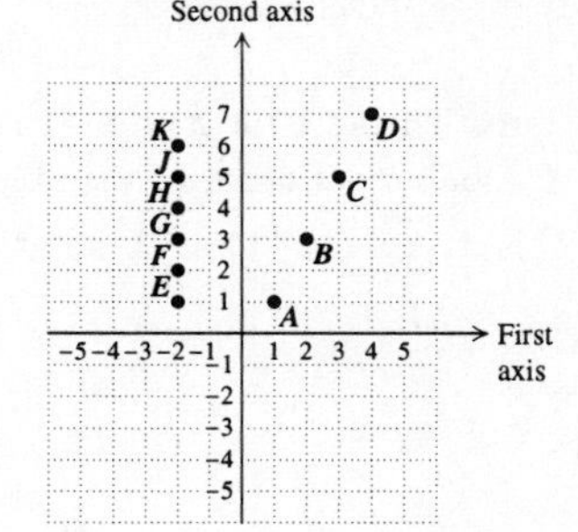

5.

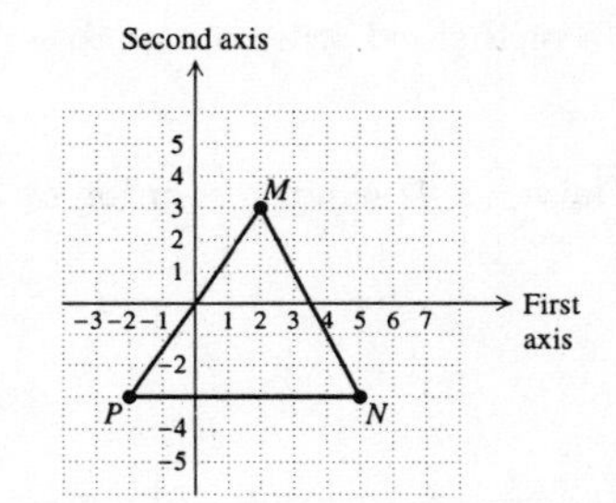

A triangle is formed. The area of a triangle is found by using the formula $A = \frac{1}{2}bh$. In this triangle the base and height are 7 units and 6 units, respectively.

$$A = \frac{1}{2}bh = \frac{1}{2} \cdot 7 \cdot 6 = \frac{42}{2} = 21 \text{ square units}$$

6.

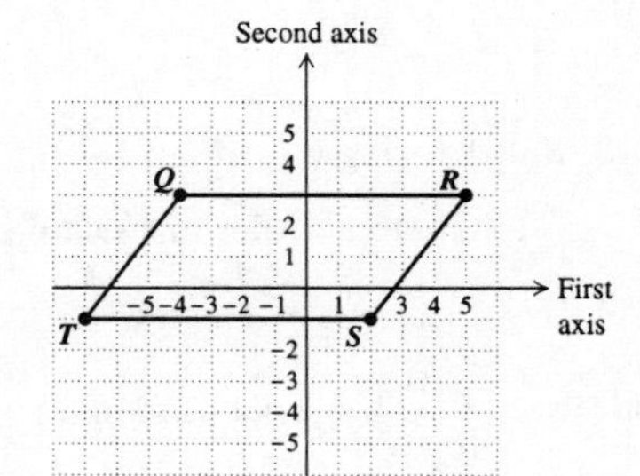

A parallelogram is formed.

$A = 36$ square units

7. Both coordinates are negative, so the point $(-3,-5)$ is in quadrant III.

8. I

9. The first coordinate is negative and the second positive, so the point $(-6,1)$ is in quadrant II.

10. IV

11. Both coordinates are positive, so the point $\left(3, \frac{1}{2}\right)$ is in quadrant I.

12. III

13. The first coordinate is positive and the second negative, so the point $(7, -0.2)$ is in quadrant IV.

14. II

15. $y = 2x - 3$

$-1 \; ? \; 2 \cdot 1 - 3$ Substituting 1 for x and -1 for y
$\quad\quad -1$ (alphabetical order of variables)

Since $-1 = -1$ is true, $(1, -1)$ is a solution of $y = 2x - 3$.

16. Yes

17. $3s + t = 4$

$3 \cdot 3 + 4 \; ? \; 4$ Substituting 3 for s and 4 for t
$9 + 4$ (alphabetical order of variables)
13

Since $13 = 4$ is false, $(3, 4)$ is not a solution of $3s + t = 4$.

18. No

19. $4x - y = 7$

$4 \cdot 3 - 5 \; ? \; 7$ Substituting 3 for x and 5 for y
$12 - 5$ (alphabetical order of variables)
7

Since $7 = 7$ is true, $(3, 5)$ is a solution of $4x - y = 7$.

20. Yes

21. $2a + 5b = 3$

$2 \cdot 0 + 5 \cdot \frac{3}{5} \; ? \; 3$ Substituting 0 for a and $\frac{3}{5}$ for b
$0 + 3$ (alphabetical order of variables)
3

Since $3 = 3$ is true, $\left(0, \frac{3}{5}\right)$ is a solution of $2a + 5b = 3$.

22. Yes

23. $4r + 3s = 5$

$4 \cdot 2 + 3 \cdot (-1) \; ? \; 5$ Substituting 2 for r and -1 for s
$8 - 3$ (alphabetical order of variables)
5

Since $5 = 5$ is true, $(2, -1)$ is a solution of $4r + 3s = 5$.

24. Yes

25. $3x - 2y = -4$

$3 \cdot 3 - 2 \cdot 2 \; ? \; -4$ Substituting 3 for x and 2 for y
$9 - 4$ (alphabetical order of variables)
5

Since $5 = -4$ is false, $(3, 2)$ is not a solution of $3x - 2y = -4$.

26. No

27. $y = 3x^2$

$3 \; ? \; 3(-1)^2$ Substituting -1 for x and 3 for y
$3 \cdot 1$ (alphabetical order of variables)
3

Since $3 = 3$ is true, $(-1, 3)$ is a solution of $y = 3x^2$.

28. No

29. $5s^2 - t = 7$

$5(2)^2 - 3 \; ? \; 7$ Substituting 2 for s and 3 for t
$5 \cdot 4 - 3$ (alphabetical order of variables)
$20 - 3$
17

Since $17 = 7$ is false, $(2, 3)$ is not a solution of $5s^2 - t = 7$.

30. Yes

31. $y = -2x$

To find an ordered pair, we choose any number for x and then determine y by substitution.

When $x = 0$, $y = -2 \cdot 0 = 0$.

When $x = 3$, $y = -2 \cdot 3 = -6$.

When $x = -2$, $y = -2 \cdot (-2) = 4$.

x	y	(x, y)
0	0	$(0, 0)$
3	-6	$(3, -6)$
-2	4	$(-2, 4)$

Plot these points, draw the line they determine, and label the graph $y = -2x$.

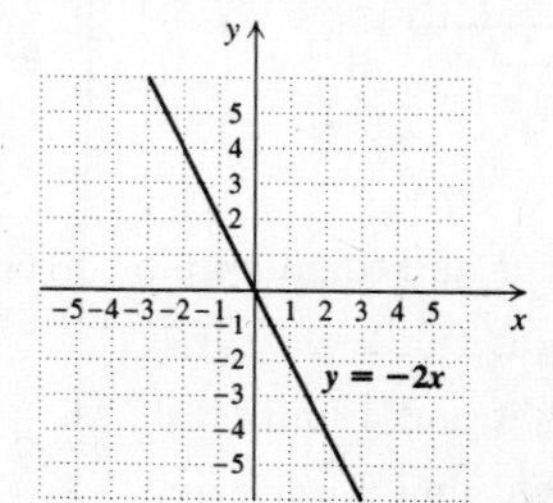

32.

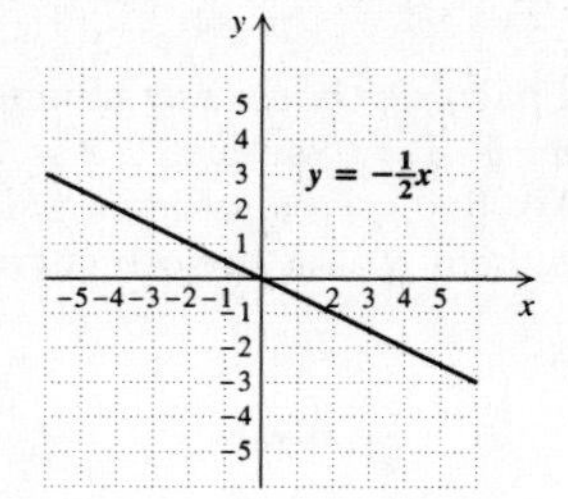

33. $y = x + 3$

To find an ordered pair, we choose any number for x and then determine y. For example, if we choose 1 for x, then $y = 1 + 3$, or 4. We find several ordered pairs, plot them, and draw the line.

x	y	(x, y)
1	4	$(1, 4)$
2	5	$(2, 5)$
-1	2	$(-1, 2)$
-3	0	$(-3, 0)$

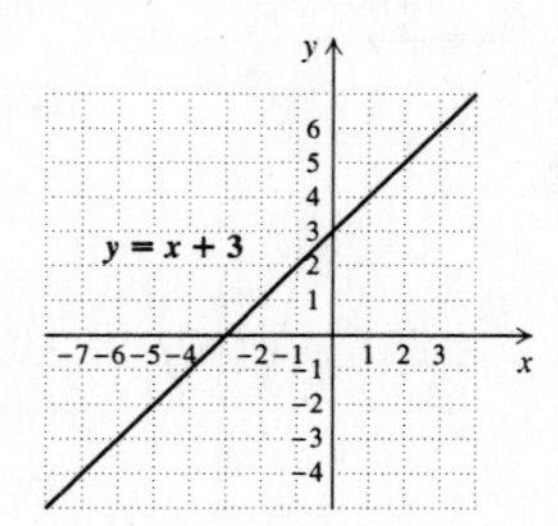

34.

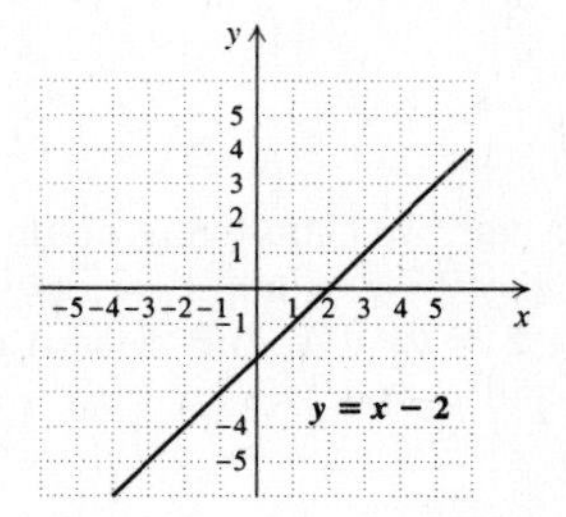

35. $y = 3x - 2$

To find an ordered pair, we choose any number for x and then determine y. For example, if $x = 2$, then $y = 3 \cdot 2 - 2 = 6 - 2 = 4$. We find several ordered pairs, plot them, and draw the line.

x	y	(x, y)
2	4	$(2, 4)$
0	-2	$(0, -2)$
-1	-5	$(-1, -5)$
1	1	$(1, 1)$

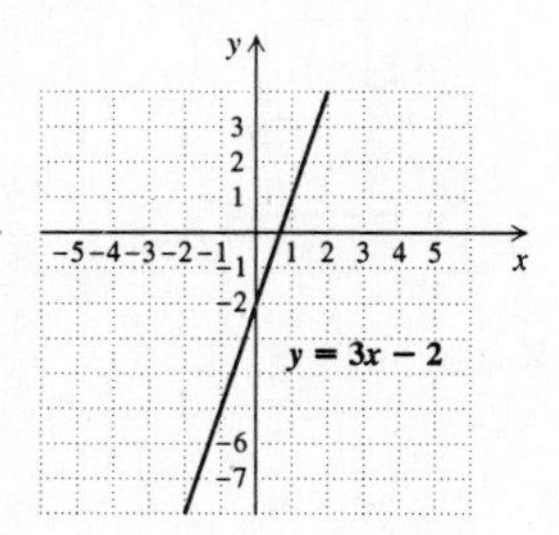

36.

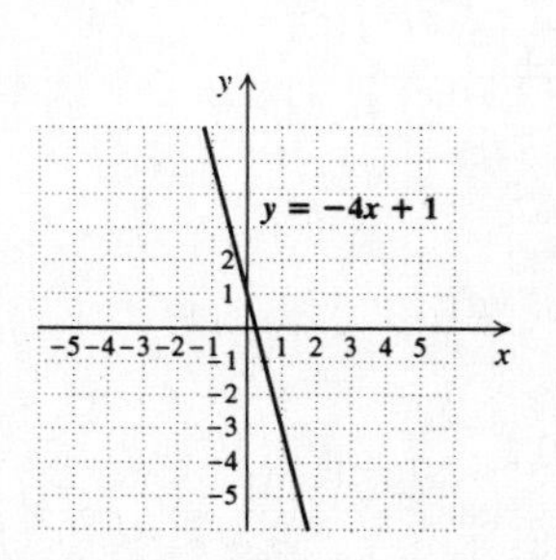

37. $y = -2x + 3$

To find an ordered pair, we choose any number for x and then determine y. For example, if $x = 1$, then $y = -2 \cdot 1 + 3 = -2 + 3 = 1$. We find several ordered pairs, plot them, and draw the line.

x	y
1	1
3	-3
-1	5
0	3

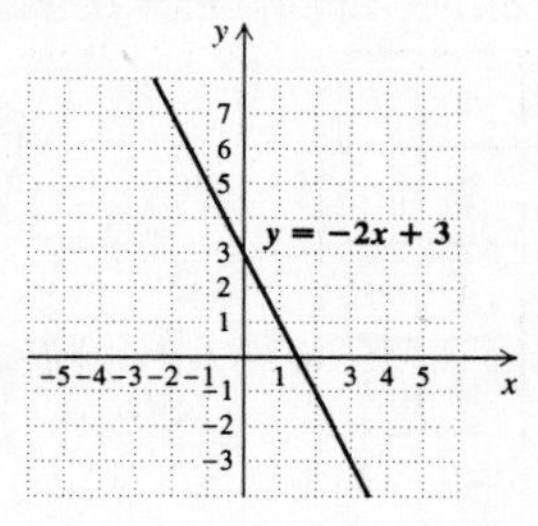

38.

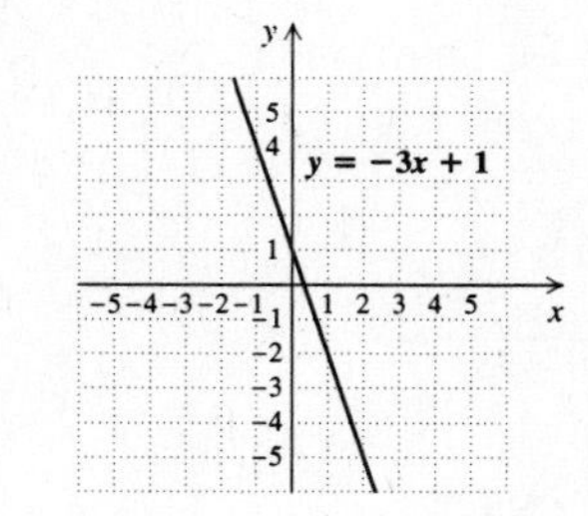

39. $y = \dfrac{2}{3}x + 1$

To find an ordered pair, we choose any number for x and then determine y. For example, if $x = 3$, then $y = \dfrac{2}{3} \cdot 3 + 1 = 2 + 1 = 3$. We find several ordered pairs, plot them, and draw the line.

x	y
3	3
0	1
-3	-1

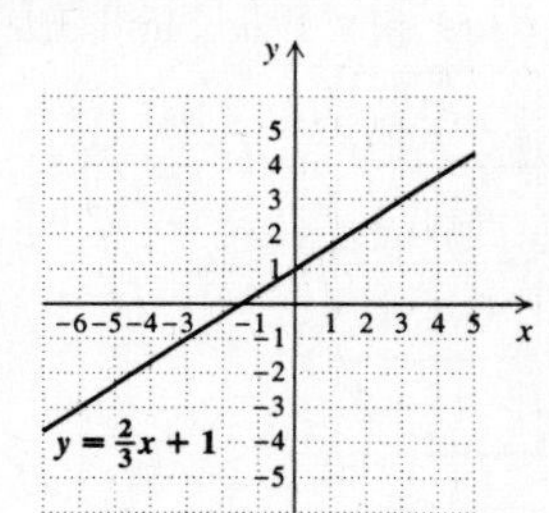

40.

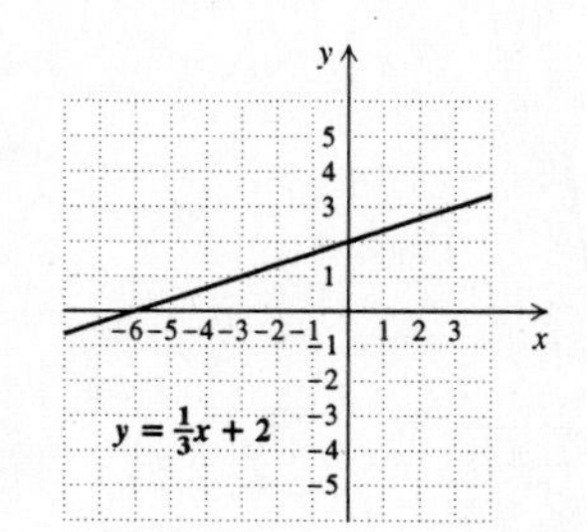

41. $y = -\frac{3}{2}x + 1$

To find an ordered pair, we choose any number for x and then determine y. For example, if $x = 2$, then $y = -\frac{3}{2} \cdot 2 + 1 = -3 + 1 = -2$. We find several ordered pairs, plot them, and draw the line.

x	y
2	-2
4	-5
0	1
-2	4

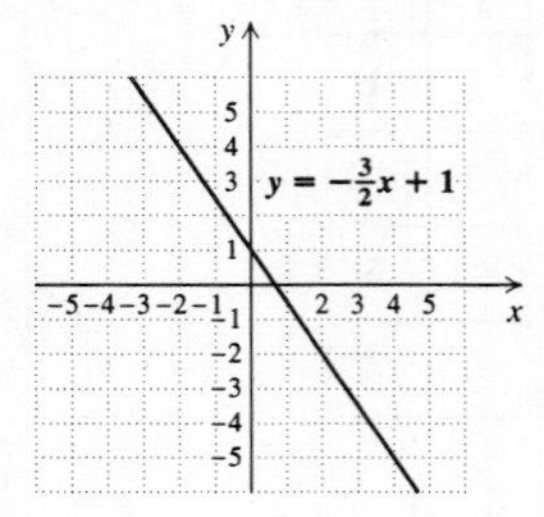

42.

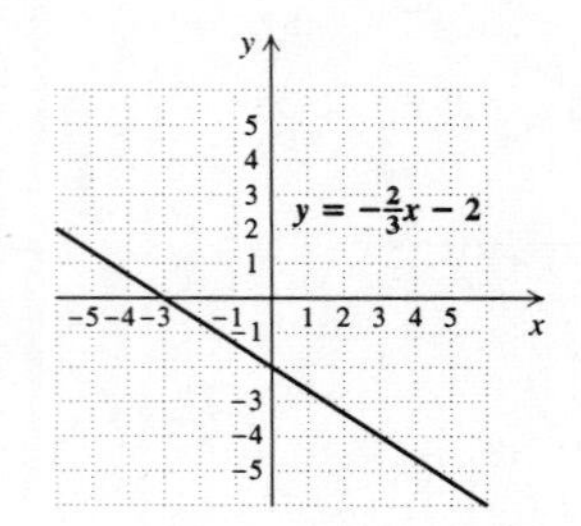

43. $y = \frac{3}{4}x + 1$

To find an ordered pair, we choose any number for x and then determine y. For example, if $x = 4$, then $y = \frac{3}{4} \cdot 4 + 1 = 3 + 1 = 4$. We find several ordered pairs, plot them, and draw the line.

x	y
4	4
0	1
-4	-2

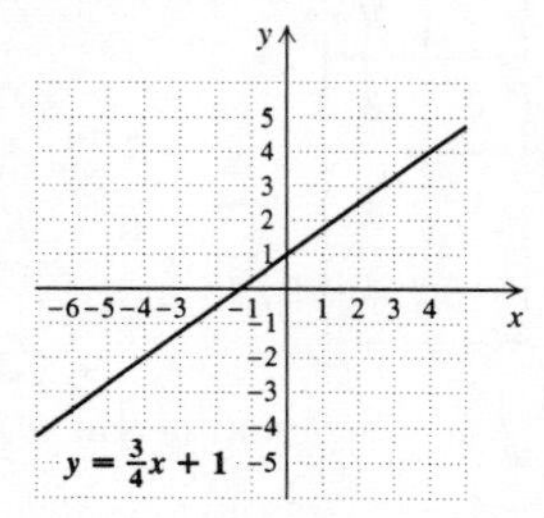

44.

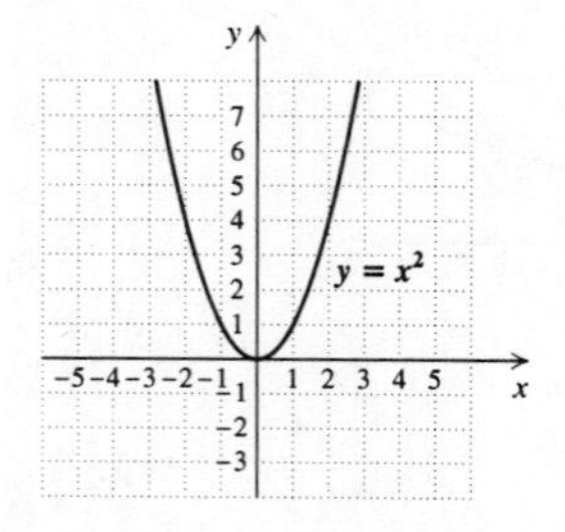

45. $y = -x^2$

To find an ordered pair, we choose any number for x and then determine y. For example, if $x = 2$, then $y = -(2)^2 = -4$. We find several ordered pairs, plot them, and connect them with a smooth curve.

x	y
2	-4
1	-1
0	0
-1	-1
-2	-4

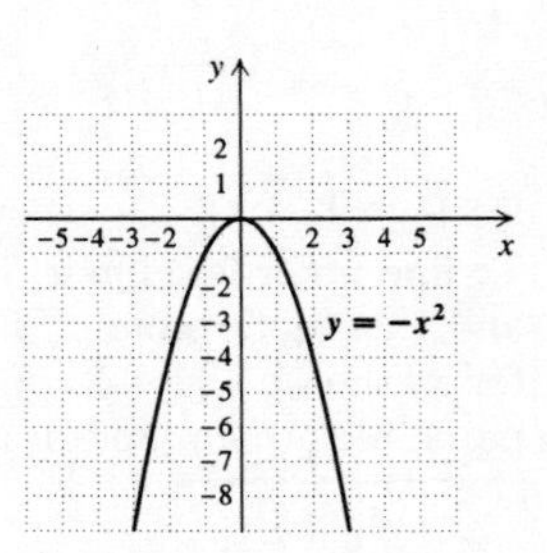

46.

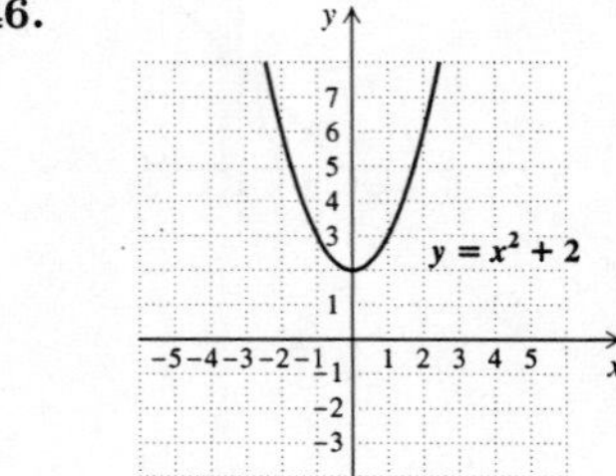

47. $y = x^2 - 2$

To find an ordered pair, we choose any number for x and then determine y. For example, if $x = 2$, then $y = 2^2 - 2 = 4 - 2 = 2$. We find several ordered pairs, plot them, and connect them with a smooth curve.

x	y
2	2
1	-1
0	-2
-1	-1
-2	2

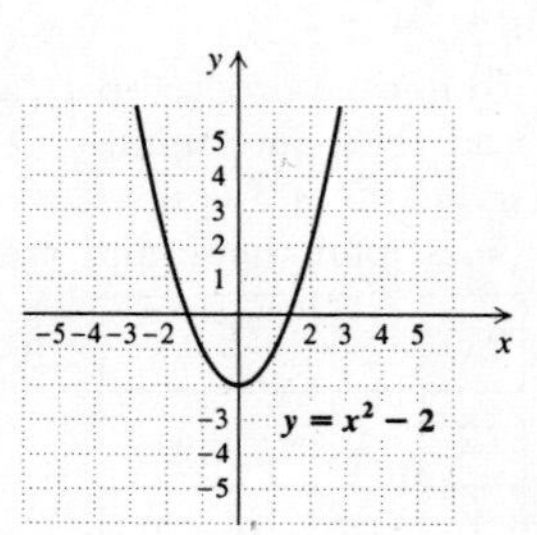

48.

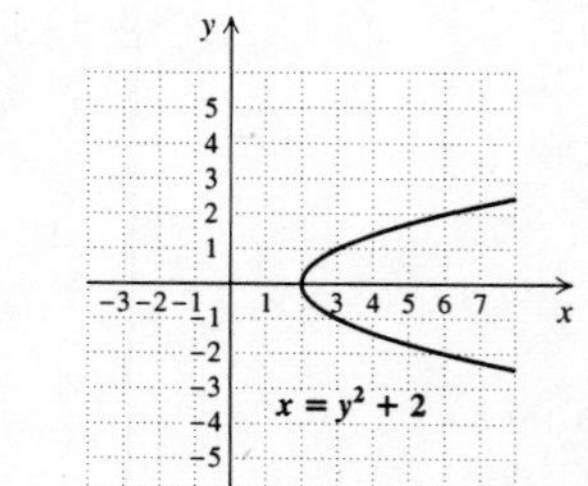

49. $y = |x| + 2$

We select x-values and find the corresponding y-values. The table lists some ordered pairs. We plot these points.

x	y
3	5
1	3
0	2
-1	3
-3	5

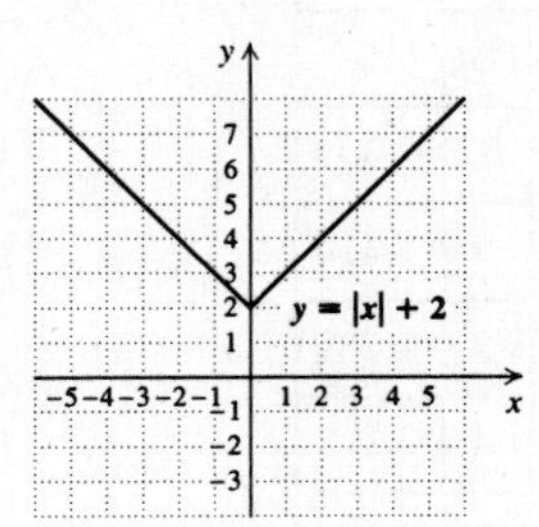

Note that the graph is V-shaped, centered at (0,2).

50.

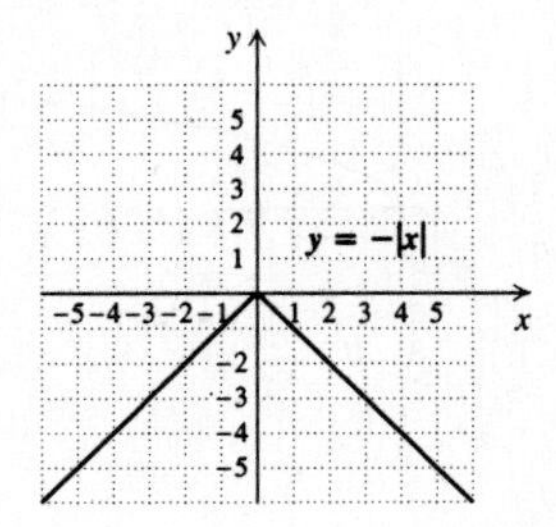

51. $y = 3 - x^2$

To find an ordered pair, we choose any number for x and then determine y. For example, if $x = 2$, then $y = 3 - 2^2 = 3 - 4 = -1$. We find several ordered pairs, plot them, and connect them with a smooth curve.

x	y
2	-1
1	2
0	3
-1	2
-2	-1

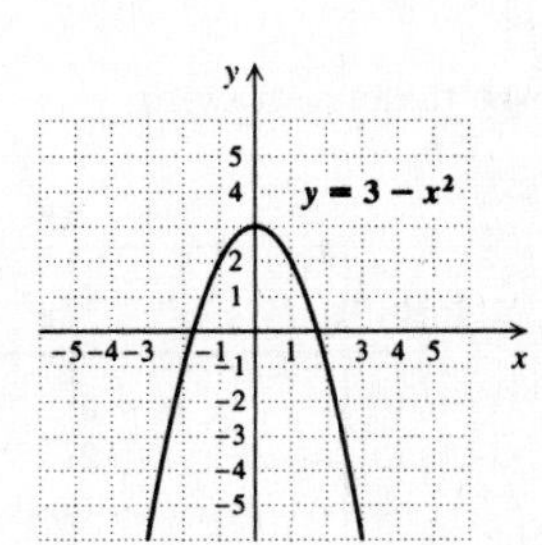

52.

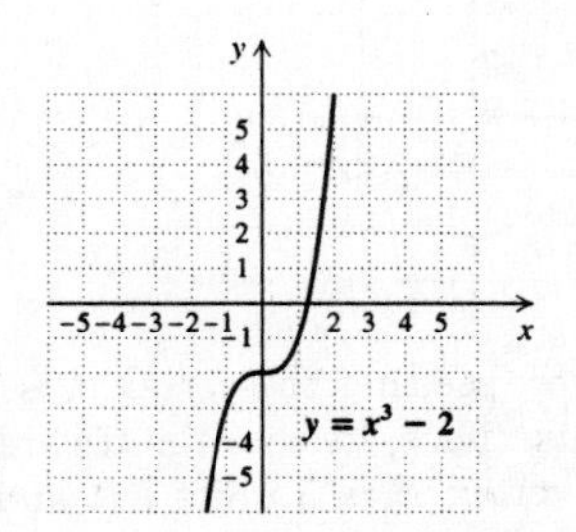

53. $y = -\frac{1}{x}$

We select x-values and find the corresponding y-values. The table lists some ordered pairs. We plot these points.

x	y
4	$-\frac{1}{4}$
2	$-\frac{1}{2}$
1	-1
$\frac{1}{2}$	-2
$-\frac{1}{2}$	2
-1	1
-2	$\frac{1}{2}$
-4	$\frac{1}{4}$

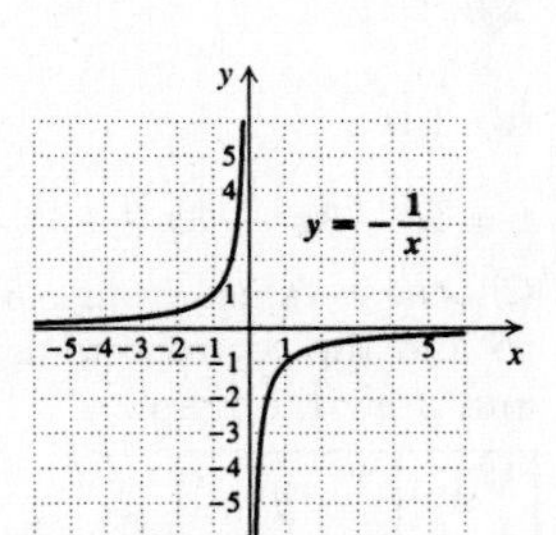

Note that we cannot use 0 as a first-coordinate, since $-1/0$ is undefined. Thus the graph has two branches, one on each side of the y-axis.

54.

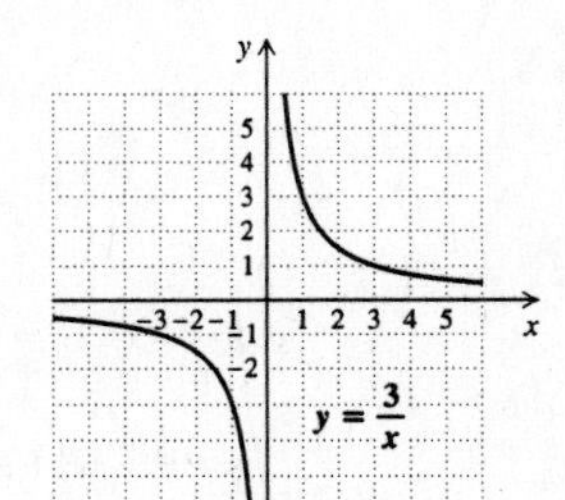

55. ***Familiarize***. The formula for the area of a triangle with base b and height h is $A = \frac{1}{2}bh$.

Translate. Substitute 200 for A and 16 for b in the formula.

$$A = \frac{1}{2}bh$$

$$200 = \frac{1}{2} \cdot 16 \cdot h$$

Carry out. We solve the equation.

$$200 = \frac{1}{2} \cdot 16 \cdot h$$

$200 = 8h$ Multiplying

$25 = h$ Dividing by 8 on both sides

Check. The area of a triangle with base 16 ft and height 25 ft is $\frac{1}{2} \cdot 16 \cdot 25$, or 200 ft^2. The answer checks.

State. The seed can fill a triangle that is 25 ft tall.

56. 11%

57. $-3.9-(-2.5)=-3.9+2.5=-1.4$

58. 5

59.

60.

61.

62.

63.

64. (a), (d)

65. $y=x^3+3x^2+3x+1$

Choose x-values from -3 to 1 and use a calculator to find the corresponding y-values. Plot the points and draw the graph.

x	y
-3	-8
-2.5	-3.375
-2	-1
-1.5	-0.125
-1	0
-0.5	0.125
0	1
0.5	3.375
0.75	5.359375
1	8

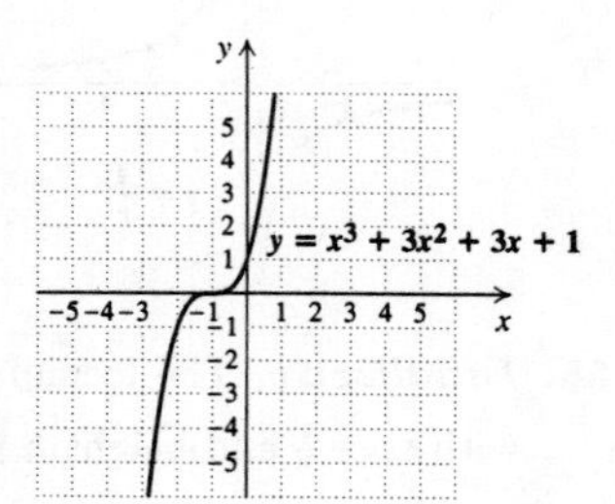

66.

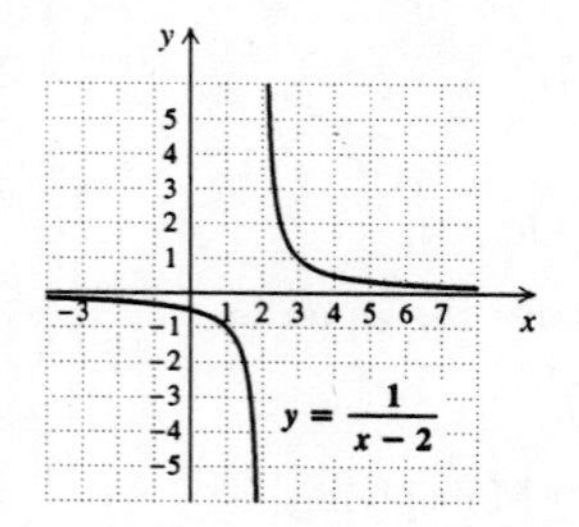

67. $y=\sqrt{x}$

Choose x-values from 0 to 10 and use a calculator to find the corresponding y-values. Plot the points and draw the graph.

x	y
0	0
1	1
2	1.414
3	1.732
4	2
5	2.236
6	2.449
7	2.646
8	2.828
9	3
10	3.162

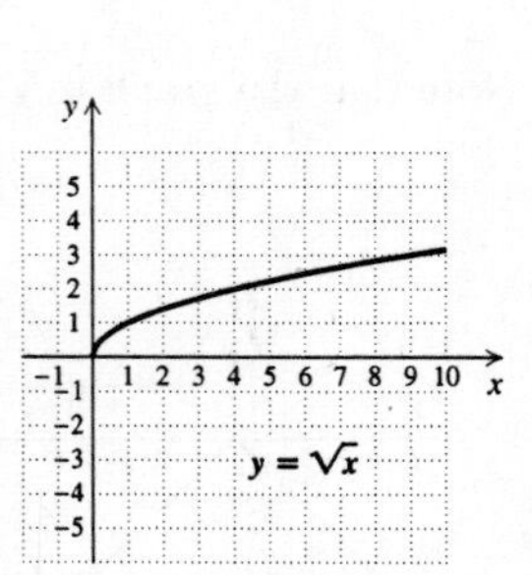

68. 

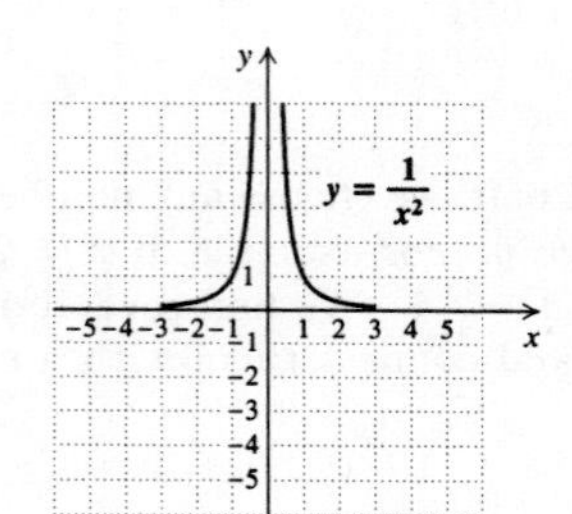

69. We make a drawing.

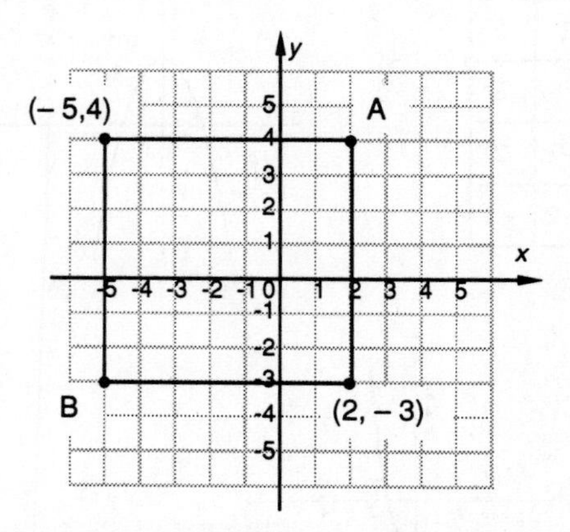

From the drawing we see that the vertex A is 2 units right and up 4 units. Thus, its coordinates are $(2,4)$. We also see that vertex B is 5 units left and down 3 units, so its coordinates are $(-5,-3)$. We also see that the distance between any pair of adjacent vertices is 7 units. The area of a square whose side has length 7 units is $7\cdot7$, or 49 square units.

70. $(-1, -2)$

71. a) Graph IV seems most appropriate for this situation. It reflects driving speeds on local streets for the first 10 and last 5 minutes and freeway cruising speeds from 10 through 30 minutes.

b) Graph III seems most appropriate for this situation. It reflects driving speeds on local streets for the first 10 minutes, an express train speed for the next 20 minutes, and walking speeds for the final 5 minutes.

c) Graph I seems most appropriate for this situation. It reflects walking speeds for the first 10 and last 5 minutes and express bus speeds from 10 through 30 minutes.

d) Graph II seems most appropriate for this situation. It reflects that the speed was 0 mph for the first 10 minutes, the time spent waiting at the bus stop. Then it shows driving speeds that fall to 0 mph several times during the next 20 minutes, indicating that the school bus stops for other students during this period of time. Finally, it shows a walking speed for the last 5 minutes.

72. a) III; b) II; c) I; d) IV

73. a), b), c) See the answer section in the text.

74. a)

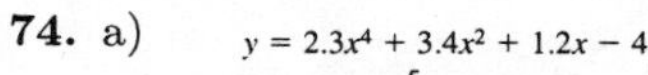

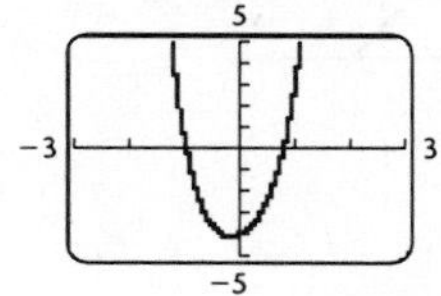

X	Y1
0	−4
.1	−3.846
.2	−3.62
.3	−3.315
.4	−2.917
.5	−2.406
.6	−1.758

X = 0

b) $y = 12.3x - 3.5$

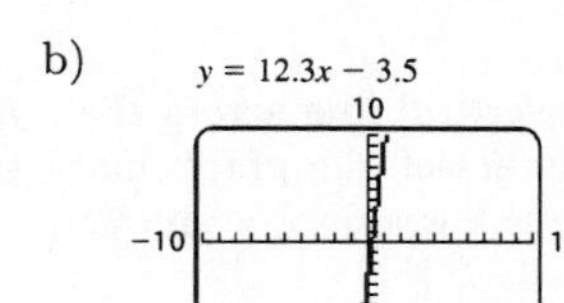

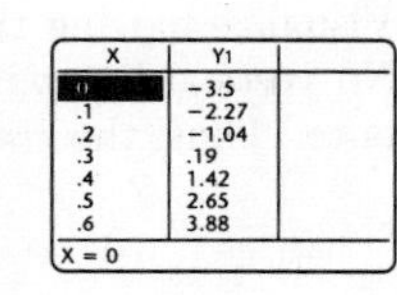

X	Y1
0	−3.5
.1	−2.27
.2	−1.04
.3	.19
.4	1.42
.5	2.65
.6	3.88

X = 0

c) $y = 3(x + 2.3)^2 + 2.3$

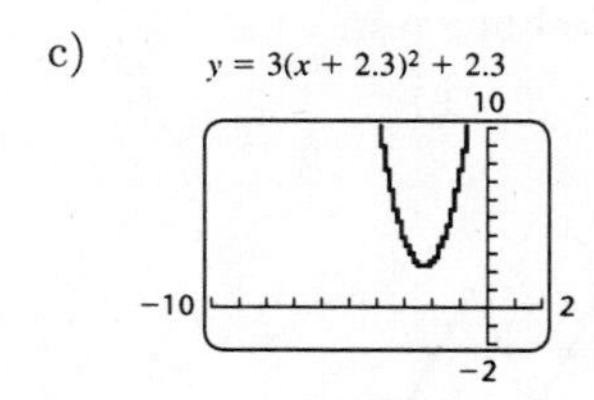

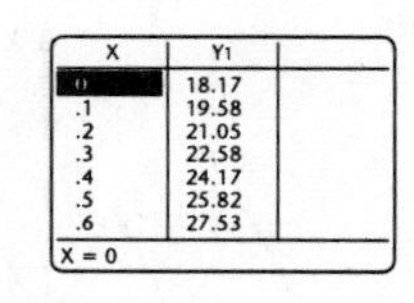

X	Y1
0	18.17
.1	19.58
.2	21.05
.3	22.58
.4	24.17
.5	25.82
.6	27.53

X = 0

Exercise Set 2.2

1. The correspondence is not a function, because a member of the domain (3) corresponds to more than one member of the range.

2. Yes

3. The correspondence is a function, because each member of the domain corresponds to just one member of the range.

4. Yes

5. The correspondence is a function, because each member of the domain corresponds to just one member of the range.

6. No

7. This correspondence is a function, because each person in the family has only one eye color.

8. A relation but not a function

9. The correspondence is not a function, since it is reasonable to assume that at least one avenue is intersected by more than one road.

The correspondence is a relation, since it is reasonable to assume that each avenue is intersected by at least one road.

10. Function

11. This correspondence is a function, because each number in the domain, when squared and then increased by 4, corresponds to only one number in the range.

12. Function

13. a) Locate 1 on the horizontal axis and then find the point on the graph for which 1 is the first coordinate. From that point, look to the vertical axis to find the corresponding y-coordinate, 3. Thus, $f(1) = 3$.

b) The domain is the set of all x-values in the graph. It is $\{-4, -3, -2, -1, 0, 1, 2\}$.

c) To determine which member(s) of the domain are paired with 2, locate 2 on the vertical axis. From there look left and right to the graph to find any points for which 2 is the second coordinate. Two such points exist, $(-2, 2)$ and $(0, 2)$. Thus, the x-values for which $f(x) = 2$ are -2 and 0.

d) The range is the set of all y-values in the graph. It is $\{1, 2, 3, 4\}$.

14. (a) $f(1) = 1$; (b) $\{-3, -1, 1, 3, 5\}$; (c) 3; (d) $\{-1, 0, 1, 2, 3\}$

15. a) Locate 1 on the horizontal axis and then find the point on the graph for which 1 is the first coordinate. From that point, look to the vertical axis to find the corresponding y-coordinate, about 2.5 or $\frac{5}{2}$. Thus, $f(1) \approx \frac{5}{2}$.

b) The set of all x-values in the graph extends from -3 to 5, so the domain is $\{x | -3 \le x \le 5\}$.

c) To determine which member(s) of the domain are paired with 2, locate 2 on the vertical axis. From there look left and right to the graph to find any points for which 2 is the second coordinate. One such point exists. Its first coordinate appears to be about $2\frac{1}{3}$ or $\frac{7}{3}$. Thus, the x-value for which $f(x) = 2$ is about $\frac{7}{3}$.

d) The set of all y-values in the graph extends from 1 to 4, so the range is $\{y|1 \leq y \leq 4\}$.

16. (a) About $\frac{13}{5}$; (b) $\{x| - 2 \leq x \leq 3\}$; (c) about $\frac{7}{4}$; (d) $\{y|1 \leq y \leq 5\}$

17. a) Locate 1 on the horizontal axis and the find the point on the graph for which 1 is the first coordinate. From that point, look to the vertical axis to find the corresponding y-coordinate. It appears to be about $2\frac{1}{4}$, or $\frac{9}{4}$. Thus, $f(1) \approx \frac{9}{4}$.

b) The set of all x-values in the graph extends from -4 to 3, so the domain is $\{x| - 4 \leq x \leq 3\}$.

c) To determine which member(s) of the domain are paired with 2, locate 2 on the vertical axis. From there look left and right to the graph to find any points for which 2 is the second coordinate. One such point exists. Its first coordinate is about 0, so the x-value for which $f(x) = 2$ is about 0.

d) The set of all y-values in the graph extends from -5 to 4, so the range is $\{y| - 5 \leq y \leq 4\}$.

18. (a) About $-\frac{7}{3}$; (b) $\{x| - 4 \leq x \leq 2\}$; (c) about -2; (d) $\{y| - 3 \leq y \leq 3\}$

19. a) Locate 1 on the horizontal axis and then find the point on the graph for which 1 is the first coordinate. From that point, look to the vertical axis to find the corresponding y-coordinate, about 1.5 or $\frac{3}{2}$. Thus, $f(1) \approx \frac{3}{2}$.

b) The set of all x-values in the graph extends from -5 to 2, so the domain is $\{x| - 5 \leq x \leq 2\}$.

c) To determine which member(s) of the domain are paired with 2, locate 2 on the vertical axis. From there look left and right to the graph to find any points for which 2 is the second coordinate. One such point exists. Its first coordinate appears to be about $1\frac{1}{6}$, or $\frac{7}{6}$. Thus, the x-value for which $f(x) = 2$ is about $\frac{7}{6}$.

d) The set of all y-values in the graph extends from -3 to 5, so the range is $\{y| - 3 \leq y \leq 5\}$.

20. (a) About $\frac{9}{2}$; (b) $\{x| - 3 \leq x \leq 4\}$; (c) about $-\frac{2}{3}$ and 3; (d) $\{y| - 4 \leq y \leq 5\}$

21. a) Locate 1 on the horizontal axis and then find the point on the graph for which 1 is the first coordinate. From that point, look to the vertical axis to find the corresponding y-coordinate, 2. Thus, $f(1) = 2$.

b) The set of all x-values in the graph extends from -4 to 4, so the domain is $\{x| - 4 \leq x \leq 4\}$.

c) To determine which member(s) of the domain are paired with 2, locate 2 on the vertical axis. From there look left and right to the graph to find any points for which 2 is the second coordinate. All points in the set $\{x|0 < x \leq 2\}$ satisfy this condition. These are the x-values for which $f(x) = 2$.

d) The domain is the set of all y-values in the graph. It is $\{1, 2, 3, 4\}$.

22. (a) 1; (b) $\{x| - 4 < x \leq 5\}$; (c) $\{x|2 < x \leq 5\}$; (d) $\{-1, 1, 2\}$

23. We can use the vertical line test:

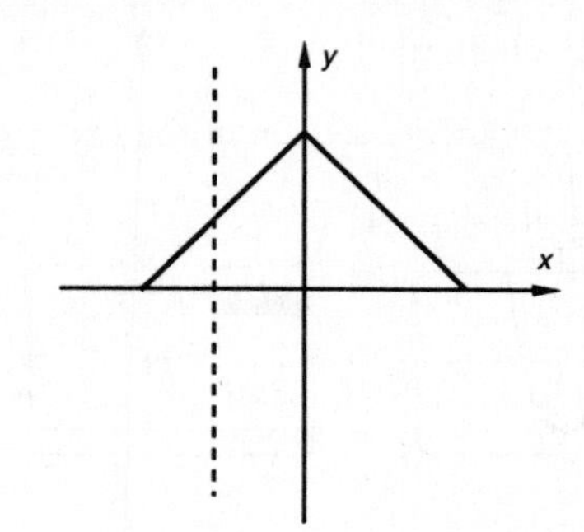

Visualize moving this vertical line across the graph. No vertical line will intersect the graph more than once. Thus, the graph is a graph of a function.

24. No

25. We can use the vertical line test:

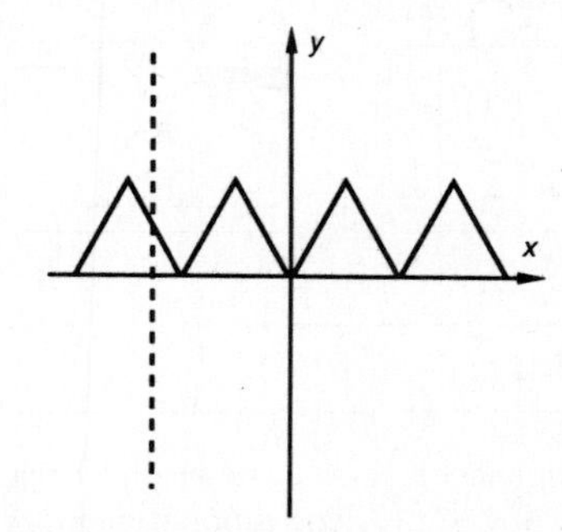

Visualize moving this vertical line across the graph. No vertical line will intersect the graph more than once. Thus, the graph is a graph of a function.

26. No

27. We can use the vertical line test.

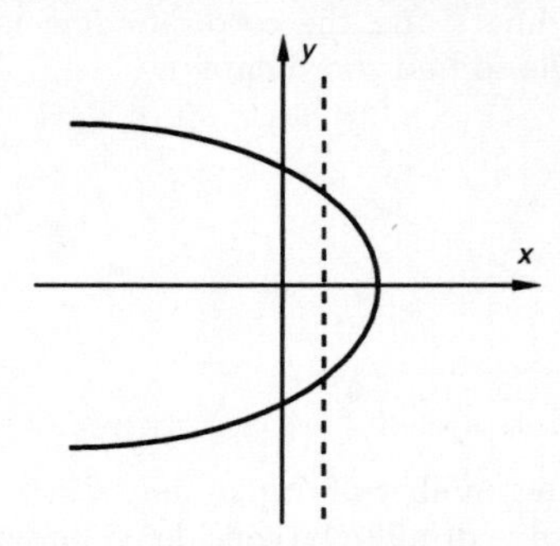

It is possible for a vertical line to intersect the graph more than once. Thus this is not the graph of a function.

28. Yes

29. We can use the vertical line test.

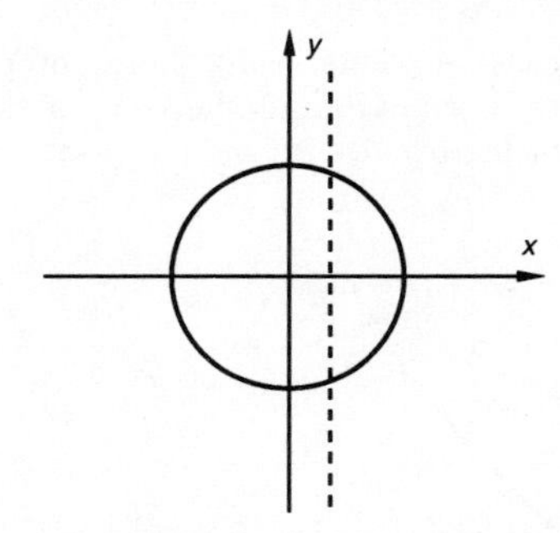

It is possible for a vertical line to intersect the graph more than once. Thus this is not a graph of a function.

30. Yes

31. $g(x) = x + 1$

a) $g(0) = 0 + 1 = 1$

b) $g(-4) = -4 + 1 = -3$

c) $g(-7) = -7 + 1 = -6$

d) $g(8) = 8 + 1 = 9$

e) $g(a + 2) = a + 2 + 1 = a + 3$

32. (a) 0; (b) 4; (c) -7; (d) -8; (e) $a - 5$

33. $f(n) = 5n^2 + 4n$

a) $f(0) = 5 \cdot 0^2 + 4 \cdot 0 = 0 + 0 = 0$

b) $f(-1) = 5(-1)^2 + 4(-1) = 5 - 4 = 1$

c) $f(3) = 5 \cdot 3^2 + 4 \cdot 3 = 45 + 12 = 57$

d) $f(t) = 5t^2 + 4t$

e) $f(2a) = 5(2a)^2 + 4 \cdot 2a = 5 \cdot 4a^2 + 8a = 20a^2 + 8a$

34. (a) 0; (b) 5; (c) 21; (d) $3t^2 - 2t$; (e) $12a^2 - 4a$

35. $f(x) = \dfrac{x-3}{2x-5}$

a) $f(0) = \dfrac{0-3}{2 \cdot 0 - 5} = \dfrac{-3}{0-5} = \dfrac{-3}{-5} = \dfrac{3}{5}$

b) $f(4) = \dfrac{4-3}{2 \cdot 4 - 5} = \dfrac{1}{8-5} = \dfrac{1}{3}$

c) $f(-1) = \dfrac{-1-3}{2(-1)-5} = \dfrac{-4}{-2-5} = \dfrac{-4}{-7} = \dfrac{4}{7}$

d) $f(3) = \dfrac{3-3}{2 \cdot 3 - 5} = \dfrac{0}{6-5} = \dfrac{0}{1} = 0$

e) $f(x+2) = \dfrac{x+2-3}{2(x+2)-5} = \dfrac{x-1}{2x+4-5} = \dfrac{x-1}{2x-1}$

36. (a) $\dfrac{26}{25}$; (b) $\dfrac{2}{9}$; (c) $-\dfrac{5}{12}$; (d) $-\dfrac{7}{3}$; (e) $\dfrac{3x+5}{2x+11}$

37. a) $f(x) = \dfrac{2}{x-3}$

Since $\dfrac{2}{x-3}$ cannot be computed when the denominator is 0, we find the x-value that causes $x - 3$ to be 0:

$x - 3 = 0$

$x = 3$ Adding 3 on both sides

Thus, 3 is not in the domain of f, while all other real numbers are. The domain of f is $\{x|x \text{ is a real number and } x \neq 3\}$.

b) $f(x) = \dfrac{7}{5-x}$

Since $\dfrac{7}{5-x}$ cannot be computed when the denominator is 0, we find the x-value that causes $5 - x$ to be 0:

$5 - x = 0$

$5 = x$ Adding x on both sides

Thus, 5 is not in the domain of f, while all other real numbers are. The domain of f is $\{x|x \text{ is a real number and } x \neq 5\}$.

c) $f(x) = 2x + 1$

Since we can compute $2x+1$ for any real number x, the domain is the set of all real numbers.

d) $f(x) = x^2 + 3$

Since we can compute x^2+3 for any real number x, the domain is the set of all real numbers.

e) $f(x) = \dfrac{3}{2x-5}$

Since $\dfrac{3}{2x-5}$ cannot be computed when the denominator is 0, we find the x-value that causes $2x - 5$ to be 0:

$2x - 5 = 0$

$2x = 5$

$x = \dfrac{5}{2}$

Thus, $\dfrac{5}{2}$ is not in the domain of f, while all other real numbers are. The domain of f is $\left\{x|x \text{ is a real number and } x \neq \dfrac{5}{2}\right\}$.

f) $f(x) = |3x - 4|$

Since we can compute $|3x - 4|$ for any real number x, the domain is the set of all real numbers.

38. (a) $\{x | x \text{ is a real number and } x \neq 1\}$;

(b) the set of all real numbers;

(c) $\{x | x \text{ is a real number and } x \neq -3\}$;

(d) $\left\{x | x \text{ is a real number and } x \neq -\frac{4}{3}\right\}$;

(e) the set of all real numbers;

(f) the set of all real numbers

39. $A(s) = s^2 \frac{\sqrt{3}}{4}$

$A(4) = 4^2 \frac{\sqrt{3}}{4} = 4\sqrt{3} \approx 6.93$

The area is $4\sqrt{3}$ cm$^2 \approx 6.93$ cm^2.

40. $9\sqrt{3}$ in^2

41. $V(r) = 4\pi r^2$

$V(3) = 4\pi(3)^2 = 36\pi$

The area is 36π in$^2 \approx 113.10$ in^2.

42. 314.16 cm^2

43. $F(C) = \frac{9}{5}C + 32$

$F(-10) = \frac{9}{5}(-10) + 32 = -18 + 32 = 14$

The equivalent temperature is 14° F.

44. 41° F

45. $H(x) = 2.75x + 71.48$

$H(32) = 2.75(32) + 71.48 = 159.48$

The predicted height is 159.48 cm.

46. 167.73 cm

47. Locate the point that is directly above 225. Then estimate its second coordinate by moving horizontally from the point to the vertical axis. The rate is about 75 per 10,000 men.

48. 125

49. Locate the point that is directly above 1992. Then estimate its second coordinate by moving horizontally from the point to the vertical axis, keeping in mind that units on the vertical axis are given in thousands. The number of minivans sold in 1992 was approximately 1000 thousand, or 1,000,000.

50. 850 thousand

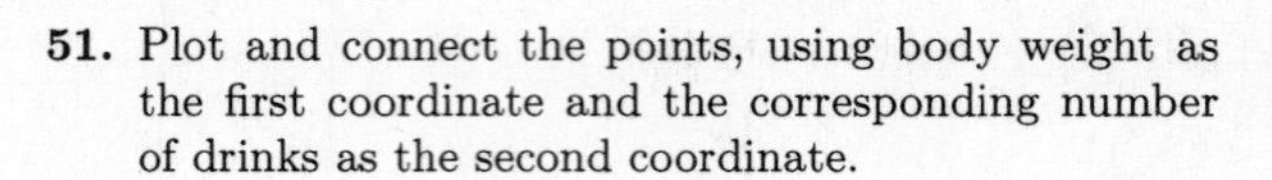

51. Plot and connect the points, using body weight as the first coordinate and the corresponding number of drinks as the second coordinate.

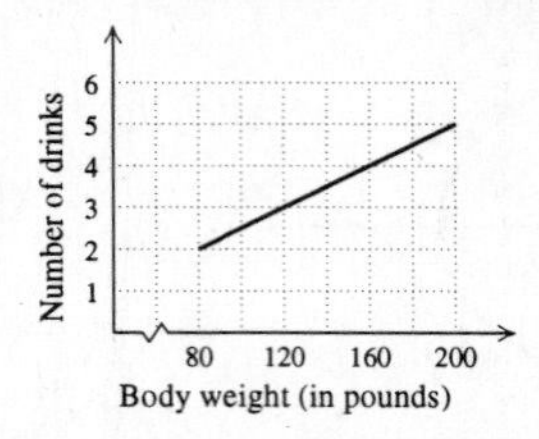

To estimate the number of drinks that a 140-lb person would have to drink to be considered intoxicated, first locate the point that is directly above 140. Then estimate its second coordinate by moving horizontally from the point to the vertical axis. Read the approximate function value there. The estimated number of drinks is 3.5.

52. 3

53. Plot and connect the points, using the counter reading as the first coordinate and the time of tape as the second coordinate.

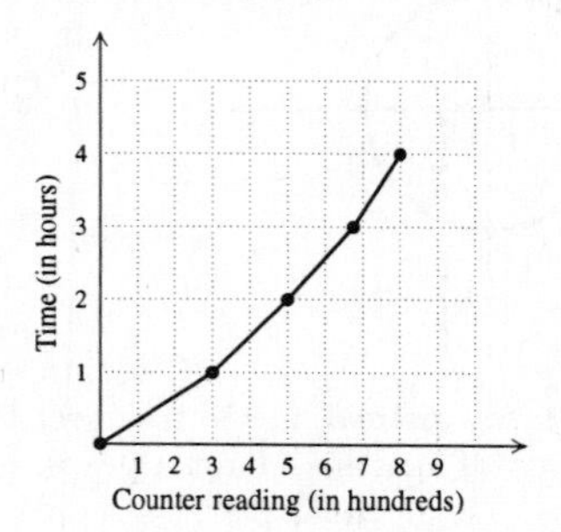

To estimate the time elapsed when the counter has reached 600, first locate the point that is directly above 600. Then estimate its second coordinate by moving horizontally from the point to the vertical axis. Read the approximate function value there. The time elapsed is about 2.5 hr.

54. About 0.5 hr

55. Plot and connect the points, using the year as the first coordinate and the population as the second.

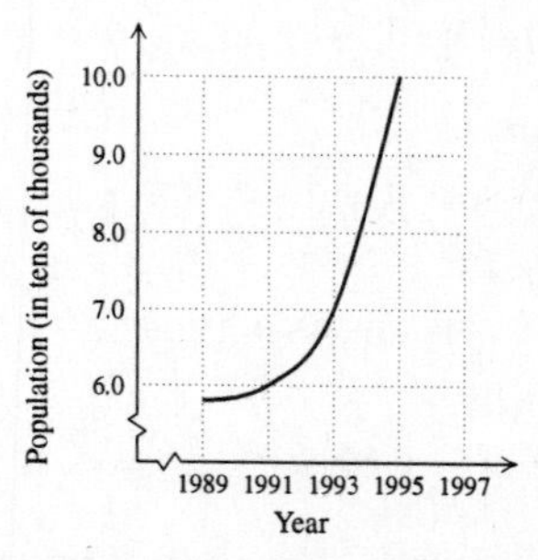

To estimate what the population was in 1992, first locate the point that is directly above 1992. Then estimate its second coordinate by moving horizontally from the point to the vertical axis. Read the approximate function value there. The population was about 64,000.

56. About 150,000

57. Plot and connect the points, using the year as the first coordinate and the sales total as the second coordinate.

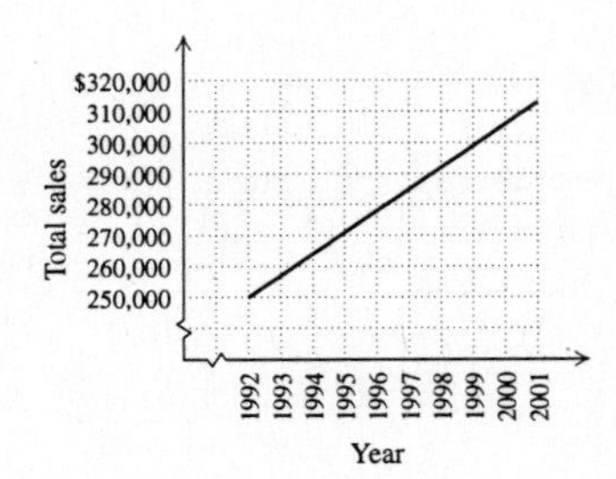

To predict the total sales for 2001, first locate the point directly above 2001. Then estimate its second coordinate by moving horizontally to the vertical axis. Read the approximate function value there. The predicted 2001 sales total is about $313,000.

58. About $271,000

59. ***Familiarize***. If x represents the first integer, then $x+2$ represents the second integer, and $x+4$ represents the third.

Translate. We write an equation.

First integer	plus	two times the second	plus	three times the third	is	124.
↓	↓	↓	↓	↓	↓	↓
x	$+$	$2(x+2)$	$+$	$3(x+4)$	$=$	124

Carry out. Solve the equation.

$$\begin{aligned} x+2(x+2)+3(x+4) &= 124 \\ x+2x+4+3x+12 &= 124 \\ 6x+16 &= 124 \\ 6x &= 108 \\ x &= 18 \end{aligned}$$

If $x=18$, then $x+2=18+2$, or 20, and $x+4=18+4$, or 22.

Check. 18, 20, and 22 are consecutive even integers. Also, $18+2\cdot 20+3\cdot 22=18+40+66=124$. The numbers check.

State. The integers are 18, 20, and 22.

60. 20.175% increase

61.

$$\begin{aligned} S &= 2lh+2lw+2wh \\ S-2wh &= 2lh+2lw \\ S-2wh &= l(2h+2w) \\ \frac{S-2wh}{2h+2w} &= l \end{aligned}$$

62. $w=\dfrac{S-2lh}{2l+2h}$

63. $3(x^2-5x)+2(x-7)=3x^2-15x+2x-14=3x^2-13x-14$

64. $-12a^6b^{13}$

65.

66.

67.

68.

69. To find $f(g(-4))$, we first find $g(-4)$:

$g(-4)=2(-4)+5=-8+5=-3.$

Then $f(g(-4))=f(-3)=3(-3)^2-1=3\cdot 9-1=27-1=26.$

To find $g(f(-4))$, we first find $f(-4)$:

$f(-4)=3(-4)^2-1=3\cdot 16-1=48-1=47.$

Then $g(f(-4))=g(47)=2\cdot 47+5=94+5=99.$

70. 26; 9

71. Locate the highest point on the graph. Then move horizontally to the vertical axis and read the corresponding pressure. It is about 22 mm.

72. About 2 minutes, 50 seconds.

73.

74. 1 every 3 minutes

75.

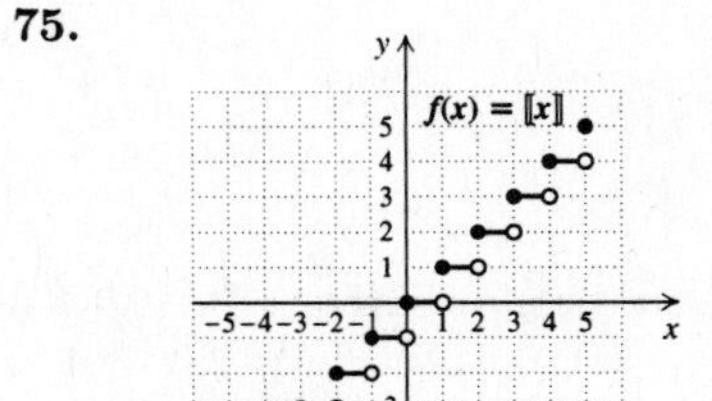

76. $g(x)=\dfrac{15}{4}x-\dfrac{13}{4}$

77. Graph the energy expenditures for walking and for bicycling on the same axes. Using the information given we plot and connect the points $\left(2\frac{1}{2},210\right)$ and $\left(3\frac{3}{4},300\right)$ for walking. We use the points $\left(5\frac{1}{2},210\right)$ and $(13,660)$ for bicycling.

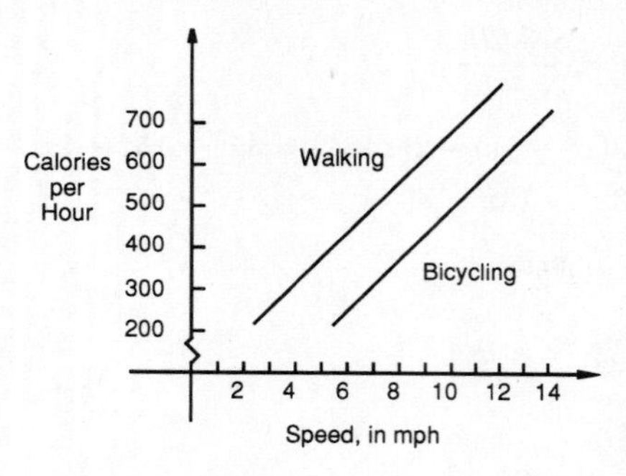

From the graph we see that walking $4\frac{1}{2}$ mph burns about 350 calories per hour and bicycling 14 mph burns about 725 calories per hour. Walking for two hours at $4\frac{1}{2}$ mph, then, would burn about $2 \cdot 350$, or 700 calories. Thus, bicycling 14 mph for one hour burns more calories than walking $4\frac{1}{2}$ mph for two hours.

Exercise Set 2.3

1. $y = 4x + 5$
$\uparrow \quad \uparrow$
$y = mx + b$

The slope is 4, and the y-intercept is $(0, 5)$.

2. Slope is 5; y-intercept is $(0, 3)$.

3. $f(x) = -2x - 6$
$\uparrow \quad \uparrow$
$f(x) = mx + b$

The slope is -2, and the y-intercept is $(0, -6)$.

4. Slope is -5; y-intercept is $(0, 7)$.

5. $y = -\frac{3}{8}x - 0.2$
$\uparrow \quad \uparrow$
$y = mx + b$

The slope is $-\frac{3}{8}$, and the y-intercept is $(0, -0.2)$.

6. Slope is $\frac{15}{7}$; y-intercept is $(0, 2.2)$.

7. $g(x) = 0.5x - 9$
$\uparrow \quad \uparrow$
$g(x) = mx + b$

The slope is 0.5, and the y-intercept is $(0, -9)$.

8. Slope is -3.1; y-intercept is $(0, 5)$.

9. $y = 43x + 197$
$\uparrow \quad \uparrow$
$y = mx + b$

The slope is 43, and the y-intercept is $(0, 197)$.

10. Slope is -52; y-intercept is $(0, 700)$.

11. Use the slope-intercept equation, $f(x) = mx + b$, with $m = \frac{2}{3}$ and $b = -7$.

$$f(x) = mx + b$$
$$f(x) = \frac{2}{3}x + (-7)$$
$$f(x) = \frac{2}{3}x - 7$$

12. $f(x) = -\frac{3}{4}x + 5$

13. Use the slope-intercept equation, $f(x) = mx + b$, with $m = -4$ and $b = 2$.

$$f(x) = mx + b$$
$$f(x) = -4x + 2$$

14. $f(x) = 2x - 1$

15. Use the slope-intercept equation, $f(x) = mx + b$, with $m = -\frac{7}{9}$ and $b = 3$.

$$f(x) = mx + b$$
$$f(x) = -\frac{7}{9}x + 3$$

16. $f(x) = -\frac{4}{11}x + 9$

17. Use the slope-intercept equation, $f(x) = mx + b$, with $m = 5$ and $b = \frac{1}{2}$.

$$f(x) = mx + b$$
$$f(x) = 5x + \frac{1}{2}$$

18. $f(x) = 6x + \frac{2}{3}$

19. Slope $= \dfrac{\text{change in } y}{\text{change in } x} = \dfrac{5-9}{4-6} = \dfrac{-4}{-2} = 2$

20. $\frac{4}{3}$

21. Slope $= \dfrac{\text{change in } y}{\text{change in } x} = \dfrac{-4-8}{9-3} = \dfrac{-12}{6} = -2$

22. $\frac{3}{26}$

23. Slope $= \dfrac{\text{change in } y}{\text{change in } x} = \dfrac{8.7 - 12.4}{-5.2 - (-16.3)} = \dfrac{-3.7}{11.1} = -\dfrac{37}{111} = -\dfrac{1}{3}$

24. $\frac{98}{269}$

25. We can use the coordinates of any two points on the line. We'll use $(0, 0)$ and $(2, 1)$.

$$\text{Slope} = \frac{\text{change in } y}{\text{change in } x} = \frac{1-0}{2-0} = \frac{1}{2}$$

The weight is increasing at the rate of 1 pound per 2 bags of feed, or $\frac{1}{2}$ pound per bag of feed.

26. The distance is increasing at the rate of 3 miles per hour.

27. We can use the coordinates of any two points on the line. Let's use $(0,5)$ and $(4,6)$.

$$\text{Slope} = \frac{\text{change in } y}{\text{change in } x} = \frac{6-5}{4-0} = \frac{1}{4}$$

The distance is increasing at the rate of 1 km per 4 minutes, or $\frac{1}{4}$ km per minute.

28. The number of pages read is increasing at the rate of 75 pages per day.

29. We can use the coordinates of any two points on the line. We'll use $(0,100)$ and $(9,40)$.

$$\text{Slope} = \frac{\text{change in } y}{\text{change in } x} = \frac{40-100}{9-0} = \frac{-60}{9} = -\frac{20}{3},$$

or $-6\frac{2}{3}$

The distance is decreasing at the rate of 20 m per 3 seconds, or $6\frac{2}{3}$ m per second.

30. The value is decreasing at the rate of \$700 per year.

31. The skier's speed is given by $\dfrac{\text{change in distance}}{\text{change in time}}$.

Note that the skier reaches the 12-km mark 45 min after the 3-km mark was reached or after $15+45$, or 60 min. We will express time in hours: 15 min = 0.25 hr and 60 min = 1 hr. Then

$$\frac{\text{Change in distance}}{\text{change in time}} = \frac{12-3}{1-0.25} = \frac{9}{0.75} = 12.$$

The speed is 12 km/h.

32. 10 km/h

33. The rate of production is given by

$$\frac{\text{change in amount refined}}{\text{change in time}} = \frac{8.1-4.5}{6-0} = \frac{3.6}{6} = 0.6$$

The rate of production is 0.6 ton/hr.

34. $\frac{5}{96}$ of the house per hour

35. The average rate of descent is given by

$\dfrac{\text{change in altitude}}{\text{change in time}}$. We will express time in minutes:

$$1\frac{1}{2}\text{ hr} = \frac{3}{2}\text{ hr} \cdot \frac{60\text{ min}}{1\text{ hr}} = 90\text{ min}$$

2 hr, 10 min = 2 hr + 10 min =

$$2\text{ hr} \cdot \frac{60\text{ min}}{1\text{ hr}} + 10\text{ min} = 120\text{ min} + 10\text{ min} = 130\text{ min}$$

Then

$$\frac{\text{change in altitude}}{\text{change in time}} = \frac{0-12,000}{130-90} = \frac{-12,000}{40} =$$

-300.

The average rate of descent is 300 ft/min.

36. $\frac{71}{70}$ million/yr

37. $y = \frac{5}{2}x + 1$

Slope is $\frac{5}{2}$; y-intercept is $(0,1)$.

From the y-intercept, we go *up* 5 units and to the *right* 2 units. This gives us the point $(2,6)$. We can now draw the graph.

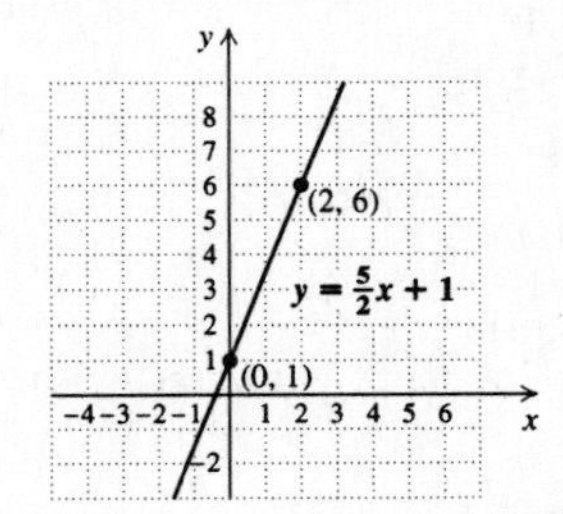

As a check, we can rename the slope and find another point.

$$\frac{5}{2} = \frac{5}{2} \cdot \frac{-1}{-1} = \frac{-5}{-2}$$

From the y-intercept, we go *down* 5 units and to the *left* 2 units. This gives us the point $(-2,-4)$. Since $(-2,-4)$ is on the line, we have a check.

38. Slope is $\frac{2}{5}$; y-intercept is $(0,4)$.

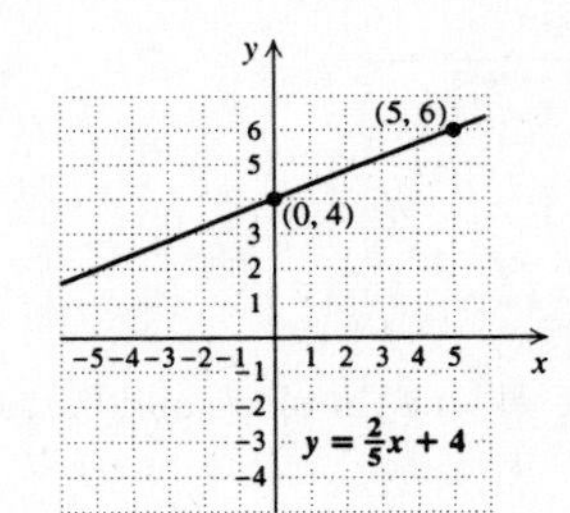

39. $f(x) = -\frac{5}{2}x + 4$

Slope is $-\frac{5}{2}$, or $\frac{-5}{2}$; y-intercept is $(0,4)$.

From the y-intercept, we go *down* 5 units and to the *right* 2 units. This gives us the point $(2,-1)$. We can now draw the graph.

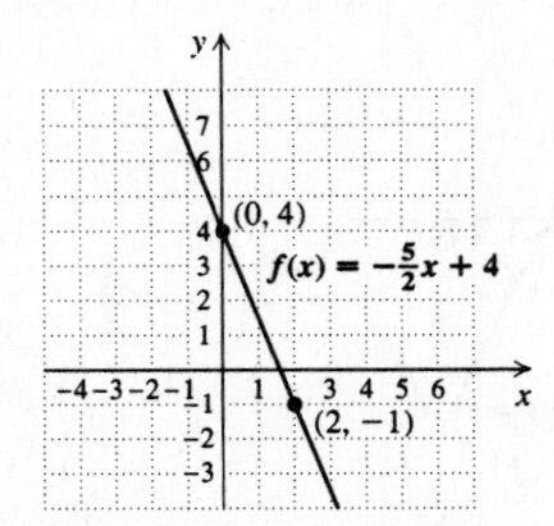

As a check, we can rename the slope and find another point.

$$\frac{-5}{2} = \frac{-5}{2} \cdot \frac{2}{2} = \frac{-10}{4}$$

From the y-intercept, we go *down* 10 units and to the *right* 4 units. This gives us the point $(4, -6)$. Since $(4, -6)$ is on the line, we have a check.

40. Slope is $-\frac{2}{5}$; y-intercept is $(0, 3)$.

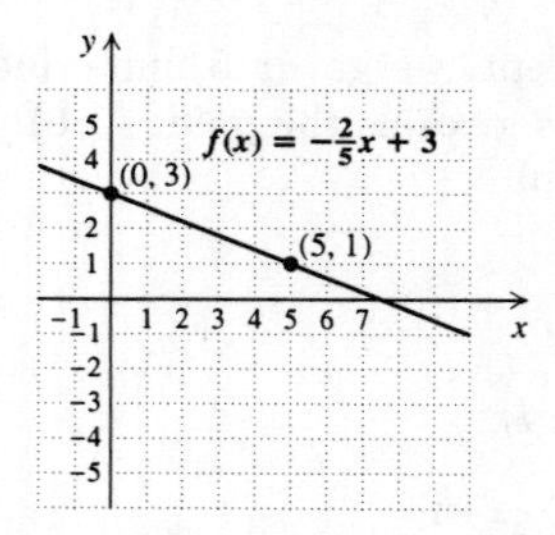

41. Convert to a slope-intercept equation.

$$2x - y = 5$$
$$-y = -2x + 5$$
$$y = 2x - 5$$

Slope is 2, or $\frac{2}{1}$; y-intercept is $(0, -5)$.

From the y-intercept, we go *up* 2 units and to the *right* 1 unit. This gives us the point $(1, -3)$. We can now draw the graph.

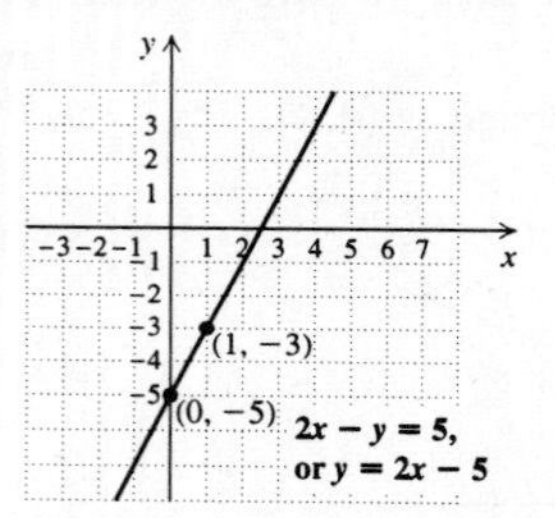

As a check, we can rename the slope and find another point.

$$2 = \frac{2}{1} \cdot \frac{3}{3} = \frac{6}{3}$$

From the y-intercept, we go *up* 6 units and to the *right* 3 units. This gives us the point $(3, 1)$. Since $(3, 1)$ is on the line, we have a check.

42. Slope is -2; y-intercept is $(0, 4)$.

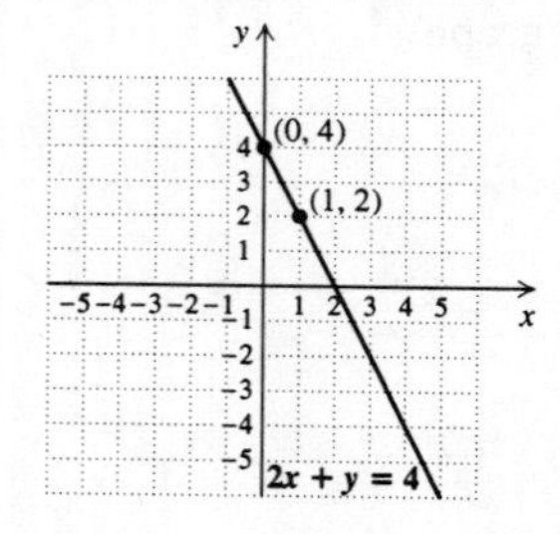

43. $f(x) = \frac{1}{3}x + 6$

Slope is $\frac{1}{3}$; y-intercept is $(0, 6)$.

From the y-intercept, we go *up* 1 unit and to the *right* 3 units. This gives us the point $(3, 7)$. We can now draw the graph.

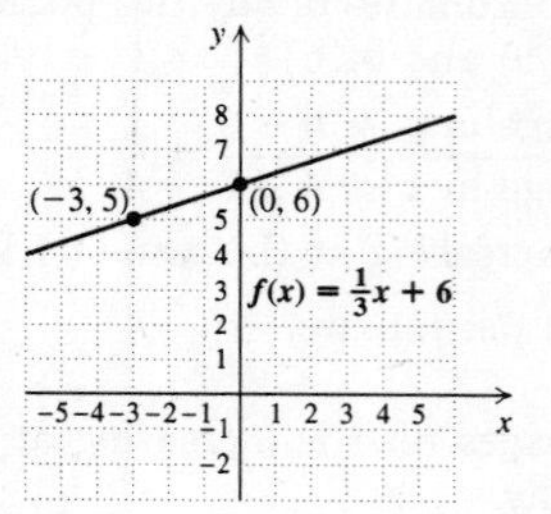

As a check, we can rename the slope and find another point.

$$\frac{1}{3} = \frac{1}{3} \cdot \frac{-1}{-1} = \frac{-1}{-3}$$

From the y-intercept, we go *down* 1 unit and to the *left* 3 units. This gives us the point $(-3, 5)$. Since $(-3, 5)$ is on the line, we have a check.

44. Slope is -3; y-intercept is $(0, 6)$.

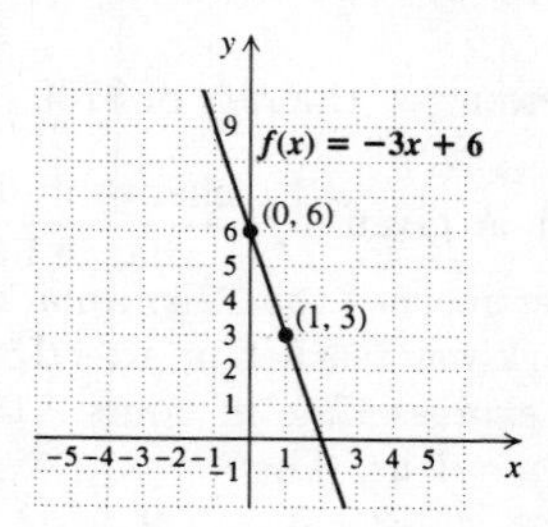

45. Convert to a slope intercept equation:

$$7y + 2x = 7$$
$$7y = -2x + 7$$
$$y = \frac{1}{7}(-2x + 7)$$
$$y = -\frac{2}{7}x + 1$$

Slope is $-\frac{2}{7}$ or $\frac{-2}{7}$; y-intercept is $(0, 1)$.

From the y-intercept we go *down* 2 units and to the *right* 7 units. This gives us the point $(7, -1)$. We can now draw the graph.

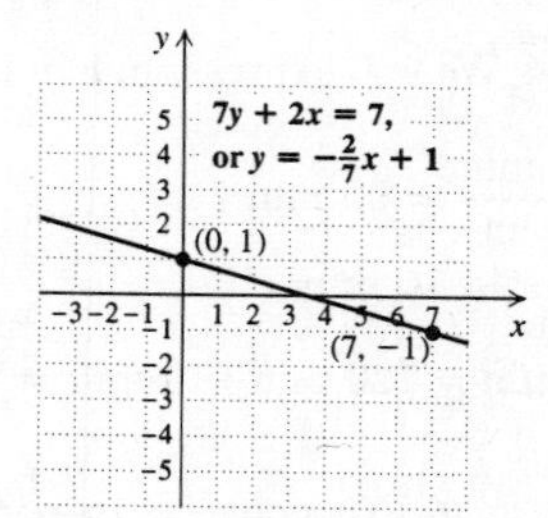

As a check we can rename the slope and find another point.

$$\frac{-2}{7} = \frac{-2}{7} \cdot \frac{-1}{-1} = \frac{2}{-7}$$

From the y-intercept, we go *up* 2 units and to the *left* 7 units. This gives us the point $(-7, 3)$. Since $(-7, 3)$ is on the line, we have a check.

46. Slope is $\frac{1}{4}$; y-intercept is $(0, -5)$.

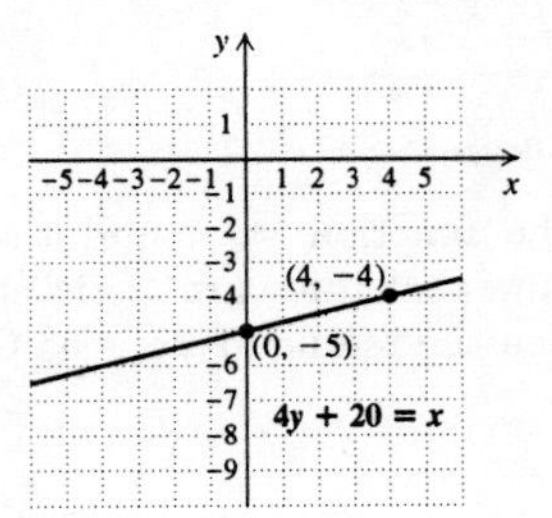

47. $f(x) = -0.25x + 2$

Slope is -0.25, or $\frac{-1}{4}$; y-intercept is $(0, 2)$.

From the y-intercept, we go *down* 1 unit and to the *right* 4 units. This gives us the point $(4, 1)$. We can now draw the graph.

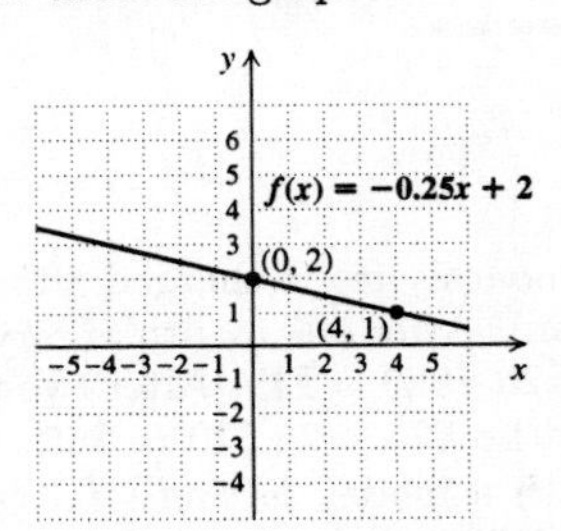

As a check, we can rename the slope and find another point.

$$\frac{-1}{4} = \frac{-1}{4} \cdot \frac{-1}{-1} = \frac{1}{-4}$$

From the y-intercept, we go *up* 1 unit and to the *left* 4 units. This gives us the point $(-4, 3)$. Since $(-4, 3)$ is on the line, we have a check.

48. Slope is 1.5, or $\frac{3}{2}$; y-intercept is $(0, -3)$.

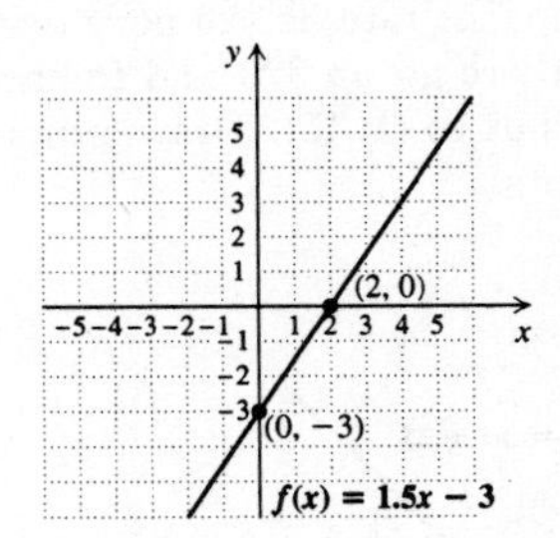

49. Convert to a slope-intercept equation.

$$4x - 5y = 10$$
$$-5y = 4x + 10$$
$$y = \frac{4}{5}x - 2$$

Slope is $\frac{4}{5}$; y-intercept is $(0, -2)$.

From the y-intercept, we go *up* 4 units and to the *right* 5 units. This gives us the point $(5, 2)$. We can now draw the graph.

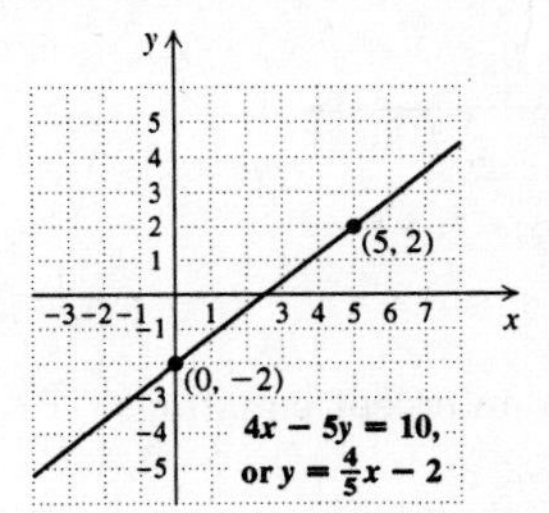

As a check, we choose some other value for x, say -5, and determine y:

$$y = \frac{4}{5}(-5) - 2 = -4 - 2 = -6$$

We plot the point $(-5, -6)$ and see that it *is* on the line.

50. Slope is $-\frac{5}{4}$; y-intercept is $(0, 1)$.

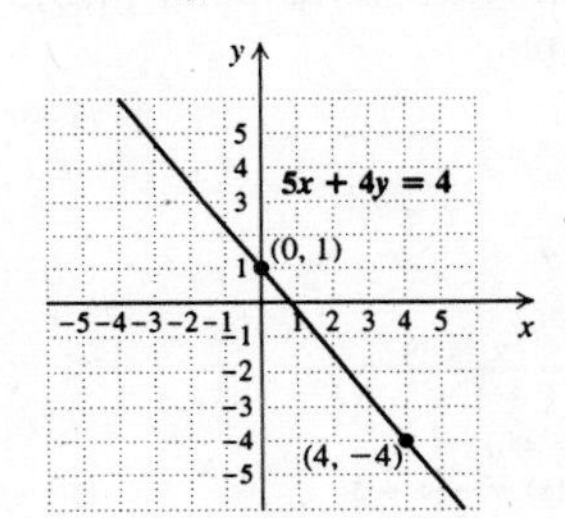

51. $f(x) = \frac{5}{4}x - 2$

Slope is $\frac{5}{4}$; y-intercept is $(0, -2)$.

From the y-intercept, we go *up* 5 units and to the *right* 4 units. This gives us the point $(4, 3)$. We can now draw the graph.

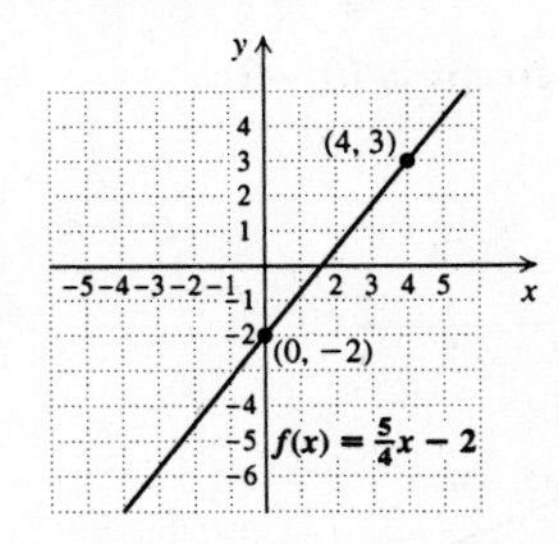

As a check, we choose some other value for x, say -2, and determine $f(x)$:

$$f(x) = \frac{5}{4}(-2) - 2 = -\frac{5}{2} - 2 = -\frac{9}{2}$$

We plot the point $\left(-2, -\frac{9}{2}\right)$ and see that it *is* on the line.

52. Slope is $\frac{4}{3}$; y-intercept is $(0, 2)$.

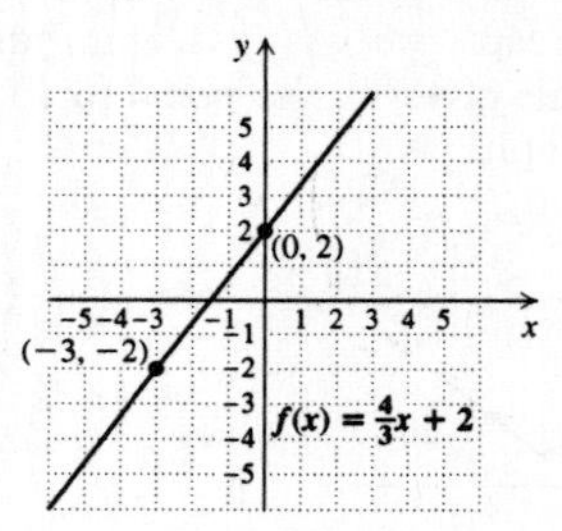

53. Convert to a slope-intercept equation:

$$12 - 4f(x) = 3x$$
$$-4f(x) = 3x - 12$$
$$f(x) = -\frac{1}{4}(3x - 12)$$
$$f(x) = -\frac{3}{4}x + 3$$

Slope is $-\frac{3}{4}$, or $\frac{-3}{4}$; y-intercept is $(0, 3)$.

From the y-intercept, we go *down* 3 units and to the *right* 4 units. This gives us the point $(4, 0)$. We can now draw the graph.

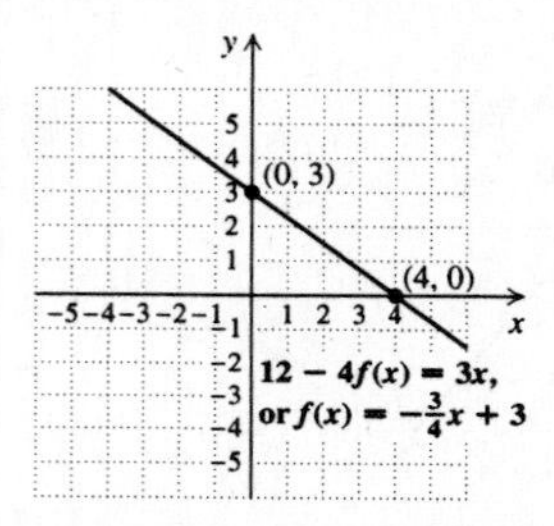

As a check, we choose some other value for x, say -4, and determine $f(x)$:

$$f(-4) = -\frac{3}{4}(-4) + 3 = 3 + 3 = 6$$

We plot the point $(-4, 6)$ and see that it *is* on the line.

54. Slope is $-\frac{2}{5}$; y-intercept is $(0, -3)$.

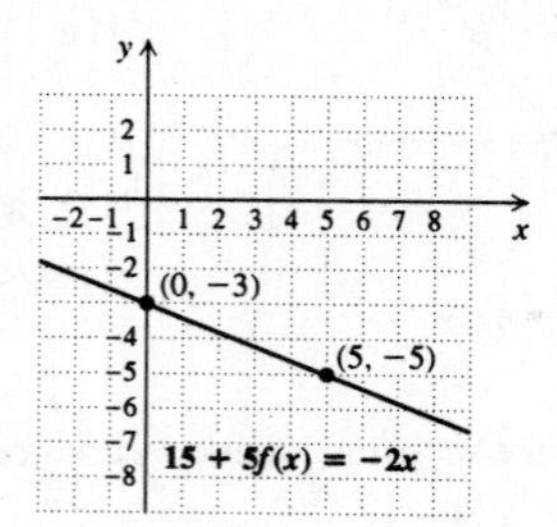

55. a) $N(t) = 7.2t - 32$

Slope is 7.2, or $\frac{72}{10}$, or $\frac{36}{5}$; vertical intercept is $(0, -32)$.

From the vertical intercept go *up* 36 units and *right* 5 units. This gives us the point $(5, 4)$. We can now draw the graph.

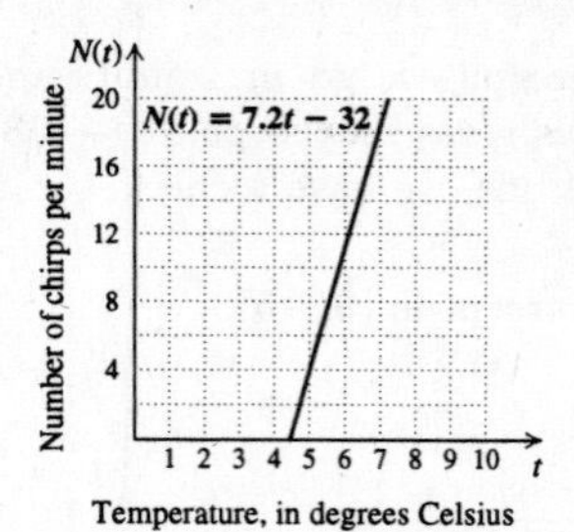

b) In graphing the function we found the point $(5, 4)$, so we know that there are 4 cricket chirps per minute when the temperature is 5° C.

56. a)

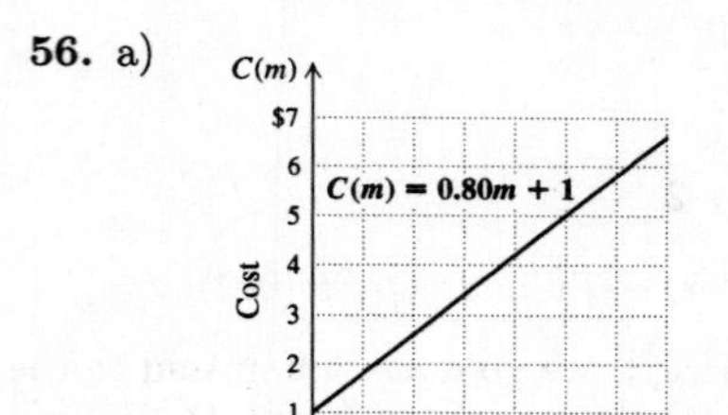

b) About \$4.25

c) About 5 min

57. ***Familiarize***. A monthly fee is charged after the installation fee is paid. After one month of service, the total cost will be \$25 + \$20 = \$45. After two months, the total cost will be $\$25 + 2 \cdot \$20 = \$65$. We can generalize this with a model, letting $C(t)$ represent the total cost, in dollars, for t months of service.

Translate. The total cost consists of the \$25 installation fee plus an additional \$20 for each month of service. Thus,

$$C(t) = 25 + 20t,$$

where $t \geq 0$ (since there cannot be a negative number of months).

Carry out. First write the model in slope-intercept form: $C(t) = 20t + 25$. The vertical intercept is $(0, 25)$ and the slope, or rate, is \$20 per month. Plot $(0, 25)$ and from there go *up* \$20 and to the *right* 1 month. This takes us to $(1, 45)$. Draw a line passing through both points.

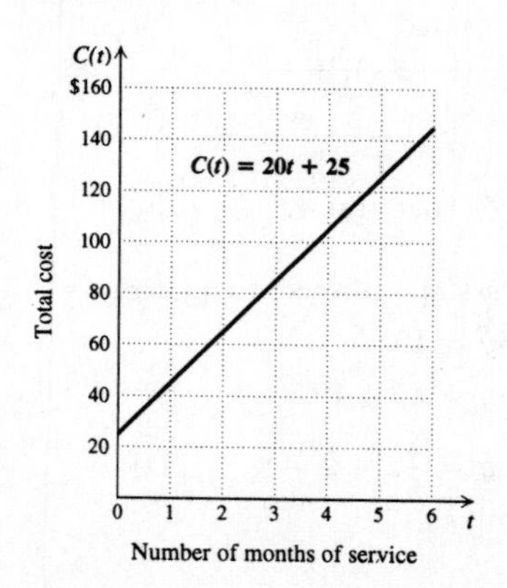

To find the total cost for 6 months, find $C(6)$:

$$C(t) = 20t + 25$$
$$C(6) = 20 \cdot 6 + 25 \quad \text{Substituting 6 for } t$$
$$= 120 + 25$$
$$= 145$$

Check. Note that $(2, 65)$ and $(6, 145)$ are on the graph, as expected from the Familiarize and Carry out steps.

State. The model $C(t) = 20t + 25$ can be used to determine the total cost, in dollars, for t months of basic cable TV service. The total cost for 6 months of service is \$145.

58. $C(t) = 30t + 35$

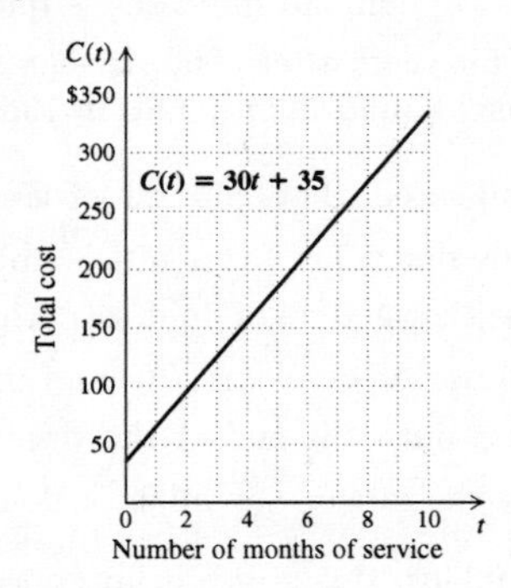

\$275

59. ***Familiarize***. A monthly fee is charged after the purchase of the phone. After one month of service, the total cost will be $\$60 + \$40 = \$100$. After two months, the total cost will be $\$60 + 2 \cdot \$40 = \$140$. We can generalize this with a model, letting $C(t)$ represent the total cost, in dollars, for t months of service.

Translate. The total cost consists of the \$60 purchase price of the phone plus an additional \$40 for each month of service. Thus,

$$C(t) = 60 + 40t.$$

Carry out. First write the model in slope-intercept form: $C(t) = 40t + 60$. The vertical intercept is $(0, 60)$ and the slope, or rate, is \$40 per month. Plot $(0, 60)$ and from there go *up* \$40 and to the *right* 1 month. This takes us to $(1, 100)$. Draw a line passing through both points.

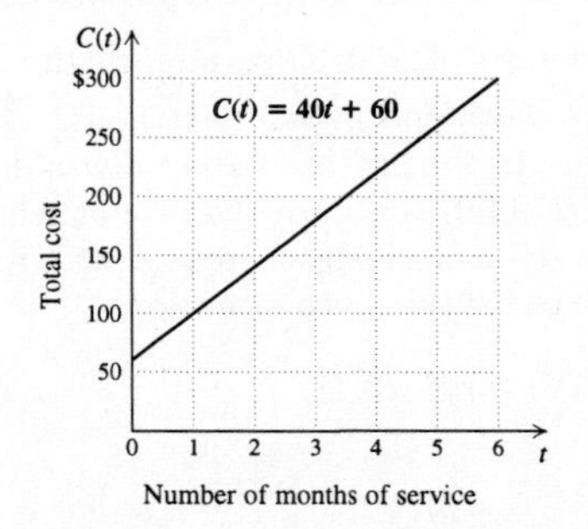

To find the total cost for $5\frac{1}{2}$ months of service, find $C(5.5)$:

$$C(t) = 40t + 60$$
$$C(t) = 40(5.5) + 60$$
$$= 220 + 60$$
$$= 280$$

The domain of C is $\{t|t \geq 0\}$ since there cannot be a negative number of months.

Check. Note that $(2, 140)$ and $(5.5, 280)$ are on the graph, as expected from the Familiarize and Carry out steps.

State. The model $C(t) = 40t + 60$ can be used to determine the total cost, in dollars, for t months of cellular phone service under the economy plan. The total cost for $5\frac{1}{2}$ months of service is \$280. The domain of C is $\{t|t \geq 0\}$.

60. $C(t) = 25t + 50$

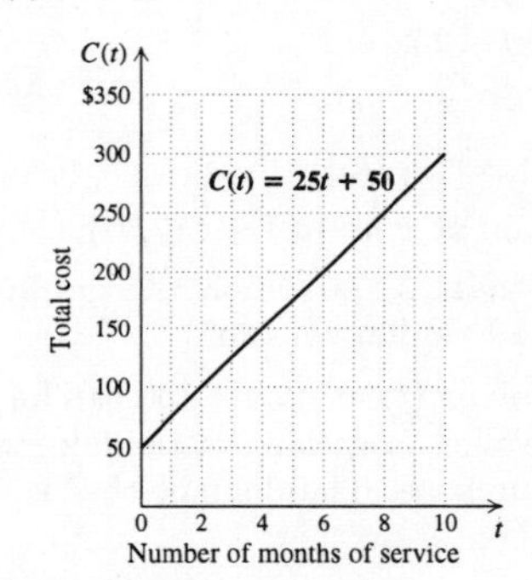

\$162.50; $\{t|t \geq 0\}$

61. ***Familiarize***. The fax machine's value decreases from its purchase price of \$750 by \$25 each month. After one month, its value will be $\$750 - \$25 = \$725$. After two months, its value will be $\$750 - 2 \cdot \$25 = \$700$. This can be generalized with a model, letting $F(t)$ represent the value of the fax machine, in dollars, t months after its purchase.

Translate. The value of the fax machine consists of the \$750 purchase price less \$25 for each month after its purchase. Thus,

$$F(t) = 750 - 25t.$$

Carry out. First write the model in slope-intercept form: $F(t) = -25t + 750$. The vertical intercept is $(0, 750)$ and the slope, or rate, is $-\$25$ per month. Plot $(0, 750)$ and from there go *down* \$25 and to the *right* 1 month. This takes us to $(1, 725)$. Draw a line passing through both points.

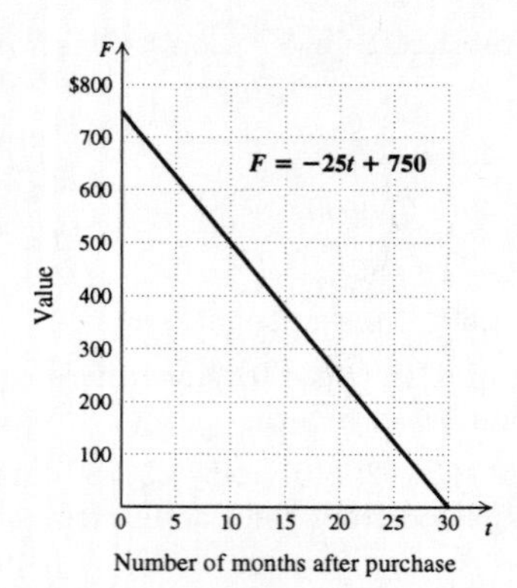

We know that the domain of F cannot contain negative values of t, since there cannot be a negative number of months. In addition, the domain cannot contain values of t that produce negative values of F, since the value of the fax machine cannot be less than \$0. We find t when $F(t) = 0$:

$$F(t) = 750 - 25t$$
$$0 = 750 - 25t$$
$$25t = 750$$
$$t = 30$$

Thus, the domain of F is $\{t|0 \le t \le 30\}$.

Check. Note that $(2, 700)$ is on the graph, as expected from the Familiarize step.

State. The model $F(t) = -25t + 750$ can be used to determine the value, in dollars, of the fax machine t months after purchase. The domain of F is $\{t|0 \le t \le 30\}$.

62. $C(t) = -50t + 3800$

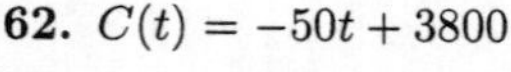
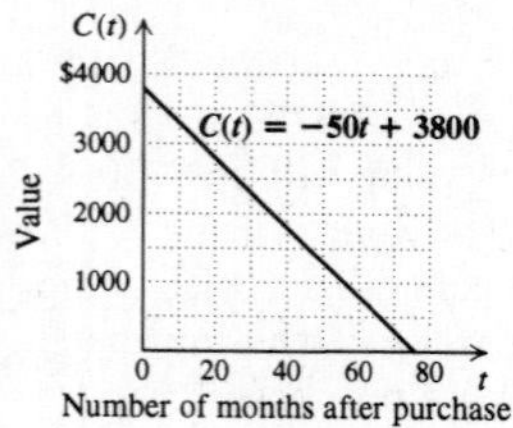

$\{t|0 \le t \le 76\}$

63. $C(d) = 0.75d + 2$ is of the form $y = mx + b$ with $m = 0.75$ and $b = 2$.

0.75 signifies that the cost per mile of a taxi ride is \$0.75.

2 signifies that the minimum cost of a taxi ride is \$2.

64. 0.1522 signifies that the price increases \$0.1522 per year, for years since 1990; 4.29 signifies that the average cost of a movie ticket in 1990 was \$4.29.

65. $f(t) = 2.6t + 17.8$ is of the form $y = mx + b$ with $m = 2.6$ and $b = 17.8$.

2.6 signifies that sales increase \$2.6 billion per year, for years after 1975.

17.8 signifies that sales in 1975 were \$17.8 billion.

66. 0.3 signifies that the cost per mile of renting the truck is \$0.30; 20 signifies that the minimum cost is \$20.

67. $A(t) = \frac{3}{20}t + 72$ is of the form $y = mx + b$ with $m = \frac{3}{20}$ and $b = 72$.

$\frac{3}{20}$ signifies that the life expectancy of American women increases $\frac{3}{20}$ yr per year for years after 1950.

72 signifies that the life expectancy of American women in 1950 was 72 years.

68. $\frac{1}{5}$ signifies that the demand increases $\frac{1}{5}$ quadrillion joules per year for years after 1960; 20 signifies that the demand was 20 quadrillion joules in 1960.

69. a) Graph II indicated that 200 ml of fluid was dripped in the first 3 hr, a rate of $\frac{200}{3}$ ml/hr. It also indicates that 400 ml of fluid was dripped in the next 3 hr, a rate of $\frac{400}{3}$ ml/hr, and that this rate continues until the end of the time period shown. Since the rate of $\frac{400}{3}$ ml/hr is double the rate of $\frac{200}{3}$ ml/hr, this graph is appropriate for the given situation.

b) Graph IV indicates that 300 ml of fluid was dripped in the first 2 hr, a rate of 300/2, or 150 ml/hr. In the next 2 hr, 200 ml was dripped. This is a rate of 200/2, or 100 ml/hr. Then 100 ml was dripped in the next 3 hr, a rate of 100/3, or $33\frac{1}{3}$ ml/hr. Finally, in the remaining 2 hr, 0 ml of fluid was dripped, a rate of 0/2, or 0 ml/hr. Since the rate at which the fluid was given decreased as time progressed and eventually became 0, this graph is appropriate for the given situation.

c) Graph I is the only graph that shows a constant rate for 5 hours, in this case from 3 PM to 8 PM. Thus, it is appropriate for the given situation.

d) Graph III indicates that 100 ml of fluid was dripped in the first 4 hr, a rate of 100/4, or 25 ml/hr. In the next 3 hr, 200 ml was dripped. This is a rate of 200/3, or $66\frac{2}{3}$ ml/hr. Then 100 ml was dripped in the next hour, a rate of 100 ml/hr. In the last hour 200 ml was dripped, a rate of 200 ml/hr. Since the rate at which the fluid was given gradually increased, this graph is appropriate for the given situation.

70. (a) III; (b) IV; (c) I; (d) II

71. $9\{2x-3[5x+2(-3x+y^0-2)]\}$

$= 9\{2x-3[5x+2(-3x+1-2)]\} \quad (y^0=1)$

$= 9\{2x-3[5x+2(-3x-1)]\}$

$= 9\{2x-3[5x-6x-2]\}$

$= 9\{2x-3[-x-2]\}$

$= 9\{2x+3x+6\}$

$= 9\{5x+6\}$

$= 45x+54$

72. $-\dfrac{27}{14}$

73. $(5x^3y^4)^2 = 5^2(x^3)^2(y^4)^2 = 25x^{3\cdot 2}y^{4\cdot 2} = 25x^6y^8$

74. 5.3×10^{10}

75.

76.

77.

78.

79. We first solve for y.

$$rx + py = s$$
$$py = -rx + s$$
$$y = -\frac{r}{p}x + \frac{s}{p}$$

The slope is $-\dfrac{r}{p}$, and the y-intercept is $\left(0, \dfrac{s}{p}\right)$.

80. Slope: $-\dfrac{r}{r+p}$; y-intercept: $\left(0, \dfrac{s}{r+p}\right)$

81. See the answer section in the text.

82. False.

83. Let $c=2$ and $d=3$. Then $f(cd)=f(2\cdot 3)=f(6)=m\cdot 6+b=6m+b$, but $f(c)f(d)=f(2)f(3)=(m\cdot 2+b)(m\cdot 3+b)=6m^2+5mb+b^2$. Thus, the given statement is false.

84. False.

85. Let $c=5$ and $d=2$. Then $f(c-d)=f(5-2)=f(3)=m\cdot 3+b=3m+b$, but $f(c)-f(d)=f(5)-f(2)=(m\cdot 5+b)-(m\cdot 2+b)=5m+b-2m-b=3m$. Thus, the given statement is false.

86. $-\dfrac{31}{4}$

87. $C(n) = 5n + 17.5$, for $1 \le n \le 10$,

$C(n) = 6n + 17.5$, for $11 \le n \le 20$,

$C(n) = 7n + 17.5$, for $21 \le n \le 30$,

$C(n) = 8n + 17.5$ for $n \ge 31$

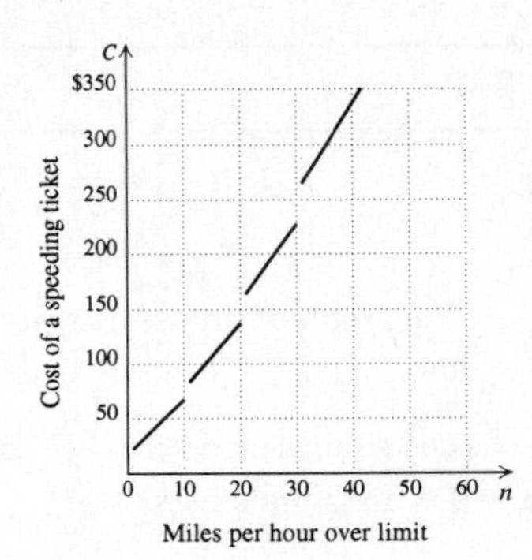

88. (a) $-\dfrac{5c}{4b}$; (b) undefined; (c) $\dfrac{a+d}{f}$

89. a) Graph III indicates that the first 2 mi and the last 3 mi were traveled in approximately the same length of time and at a fairly rapid rate. The mile following the first two miles was traveled at a much slower rate. This could indicate that the first two miles were driven, the next mile was swum and the last three miles were driven, so this graph is most appropriate for the given situation.

b) The slope in Graph IV decreases at 2 mi and again at 3 mi. This could indicate that the first two miles were traveled by bicycle, the next mile was run, and the last 3 miles were walked, so this graph is most appropriate for the given situation.

c) The slope in Graph I decreases at 2 mi and then increases at 3 mi. This could indicate that the first two miles were traveled by bicycle, the next mile was hiked, and the last three miles were traveled by bus, so this graph is most appropriate for the given situation.

d) The slope in Graph II increases at 2 mi and again at 3 mi. This could indicate that the first two miles were hiked, the next mile was run, and the last three miles were traveled by bus, so this graph is most appropriate for the given situation.

90.

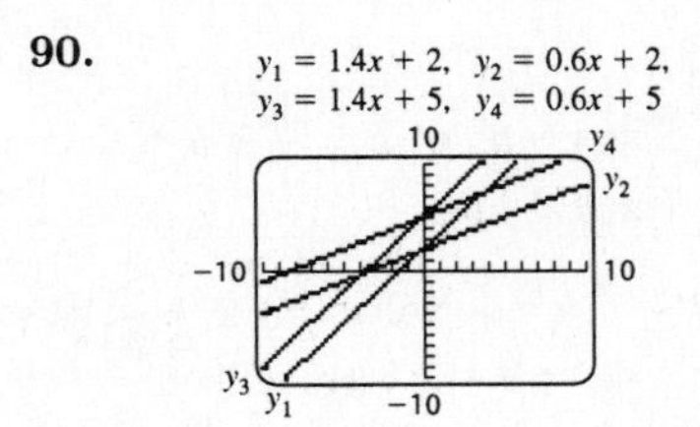

91.

92.

Exercise Set 2.4

1. $5x - 6 = 15$

$5x = 21$

$x = \frac{21}{5}$ The graph of $x = \frac{21}{5}$ is a vertical line.

Since $5x - 6 = 15$ is equivalent to $x = \frac{21}{5}$, the slope of the line $5x - 6 = 15$ is undefined.

2. $\frac{3}{5}$

3. $3x = 12 + y$

$3x - 12 = y$

$y = 3x - 12$ $(y = mx + b)$

The slope is 3.

4. Undefined

5. $5y = 6$

$y = \frac{6}{5}$ The graph of $y = \frac{6}{5}$ is a horizontal line.

Since $5y = 6$ is equivalent to $y = \frac{6}{5}$, the slope of the line $5y = 6$ is 0.

6. 0

7. $5x - 7y = 30$

$-7y = -5x + 30$

$y = \frac{5}{7}x - \frac{30}{7}$ $(y = mx + b)$

The slope is $\frac{5}{7}$.

8. $\frac{2}{3}$

9. $12 - 4x = 9 + x$

$3 = 5x$

$\frac{3}{5} = x$ The graph of $x = \frac{3}{5}$ is a vertical line.

Since $12 - 4x = 9 + x$ is equivalent to $x = \frac{3}{5}$, the slope of the line $12 - 4x = 9 + x$ is undefined.

10. Undefined

11. $2y - 4 = 35 + x$

$2y = x + 39$

$y = \frac{1}{2}x + \frac{39}{2}$ $(y = mx + b)$

The slope is $\frac{1}{2}$.

12. -2

13. $3y + x = 3y + 2$

$x = 2$ The graph of $x = 2$ is a vertical line.

Since $3y + x = 3y + 2$ is equivalent to $x = 2$, the slope of the line $3y + x = 3y + 2$ is undefined.

14. Undefined

15. $4y + 8x = 6$

$4y = -8x + 6$

$y = -2x + \frac{3}{2}$ $(y = mx + b)$

The slope is -2.

16. $-\frac{6}{5}$

17. $y - 6 = 14$

$y = 20$ The graph of $y = 20$ is a horizontal line.

Since $y - 6 = 14$ is equivalent to $y = 20$, the slope of the line $y - 6 = 14$ is 0.

18. 0

19. $3y - 2x = 5 + 9y - 2x$

$3y = 5 + 9y$

$-6y = 5$

$y = -\frac{5}{6}$ The graph of $y = -\frac{5}{6}$ is a horizontal line.

Since $3y - 2x = 5 + 9y - 2x$ is equivalent to $y = -\frac{5}{6}$, the slope of the line $3y - 2x = 5 + 9y - 2x$ is 0.

20. 0

21. $7x - 3y = -2x + 1$

$-3y = -9x + 1$

$y = 3x - \frac{1}{3}$ $(y = mx + b)$

The slope is 3.

22. $\frac{3}{2}$

23. Graph $y = 4$.

This is a horizontal line that crosses the y-axis at $(0, 4)$. If we find some ordered pairs, note that, for any x-value chosen, y must be 4.

x	y
-2	4
0	4
3	4

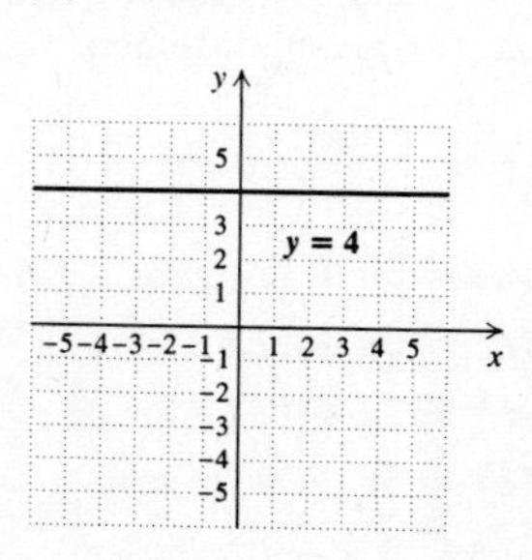

24.

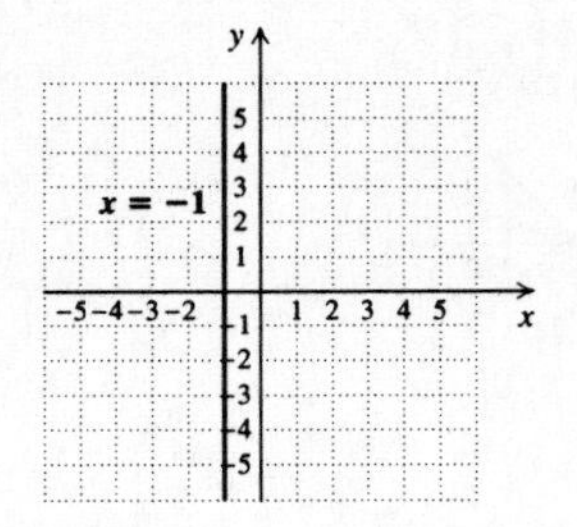

25. Graph $x = 2$.

This is a vertical line that crosses the x-axis at $(2, 0)$. If we find some ordered pairs, note that, for any y-value chosen, x must be -2.

x	y
-2	-1
-2	0
-2	2

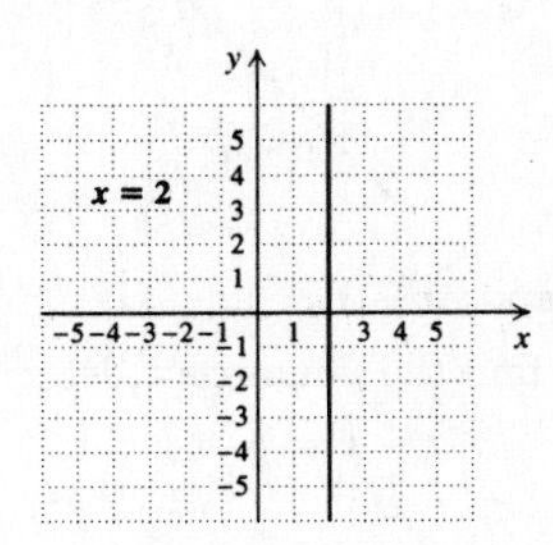

26.

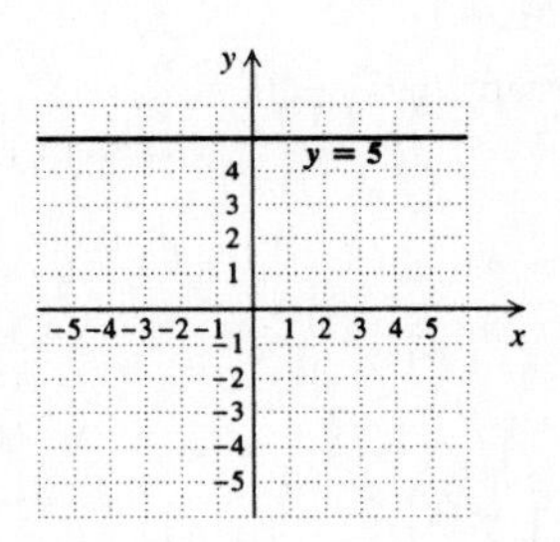

27. Graph $4 \cdot f(x) = 20$.

First solve for $f(x)$.

$$4 \cdot f(x) = 20$$
$$f(x) = 5$$

This is a horizontal line that crosses the vertical axis at $(0, 5)$.

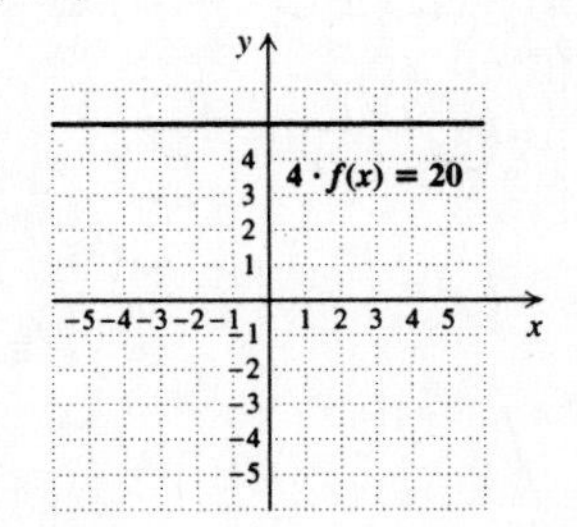

28.

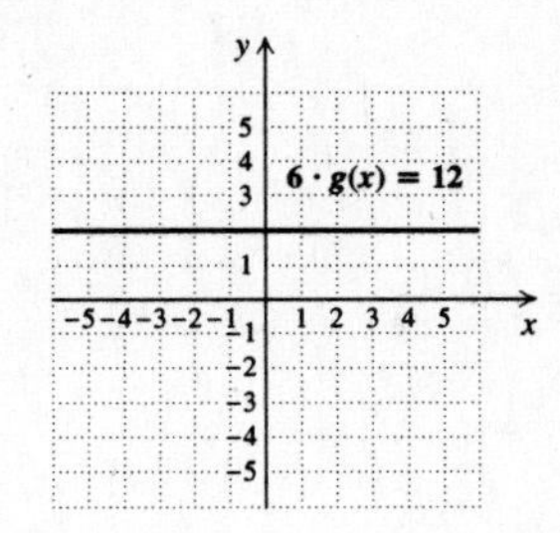

29. Graph $3x = -15$.

Since y does not appear, we solve for x.

$$3x = -15$$
$$x = -5$$

This is a vertical line that crosses the x-axis at $(-5, 0)$.

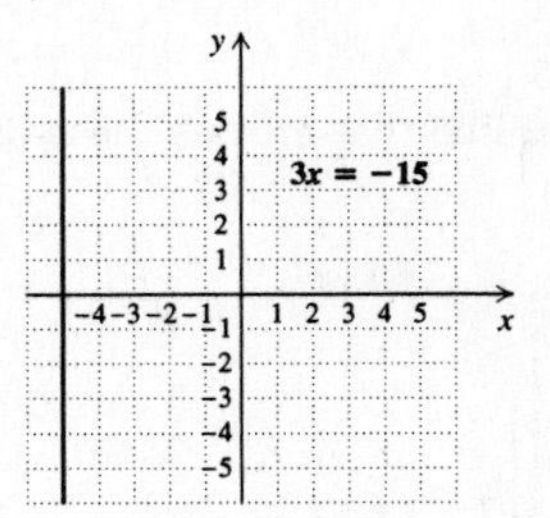

30.

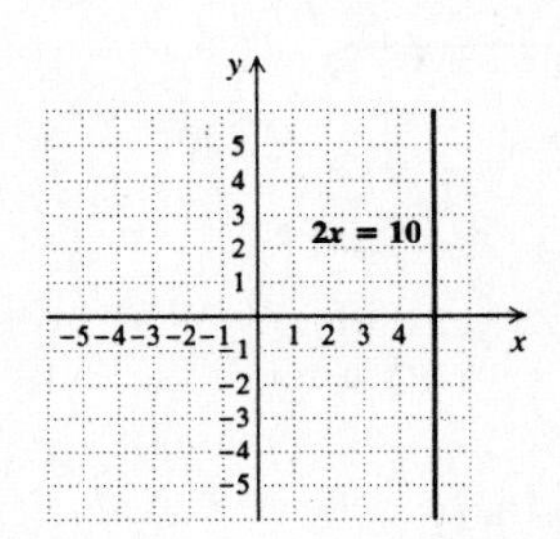

31. Graph $4 \cdot g(x) + 3x = 12 + 3x$.

First solve for $g(x)$.

$$4 \cdot g(x) + 3x = 12 + 3x$$
$$4 \cdot g(x) = 12 \quad \text{Subtracting } 3x \text{ on both sides}$$
$$g(x) = 3$$

This is a horizontal line that crosses the vertical axis at $(0, 3)$.

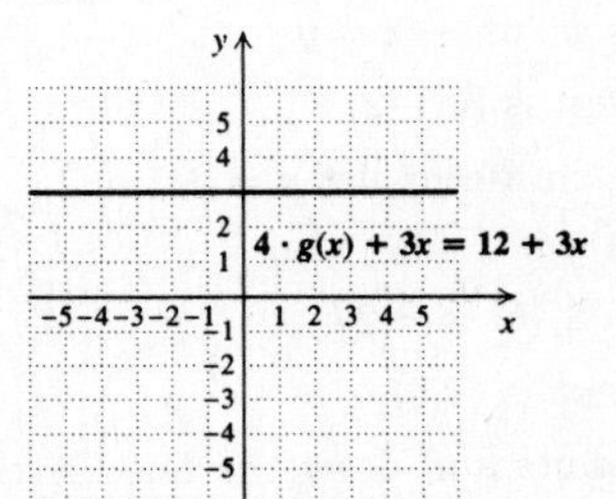

32.

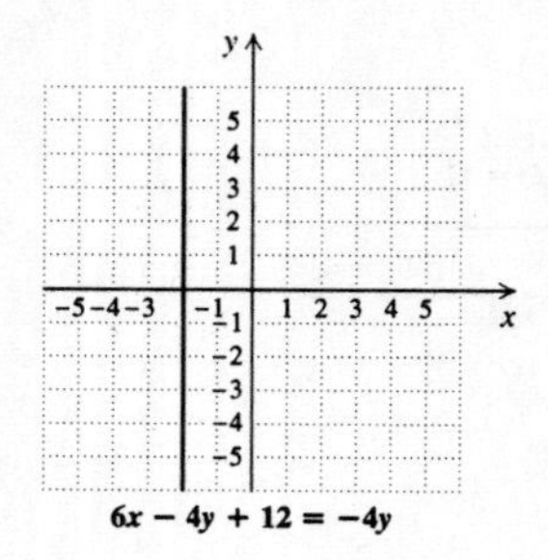

33. Graph $7 - 3x = 4 + 2x$.

Since y does not appear, we solve for x.

$$7 - 3x = 4 + 2x$$
$$7 - 5x = 4$$
$$-5x = -3$$
$$x = \frac{3}{5}$$

This is a vertical line that crosses the x-axis at $\left(\frac{3}{5}, 0\right)$.

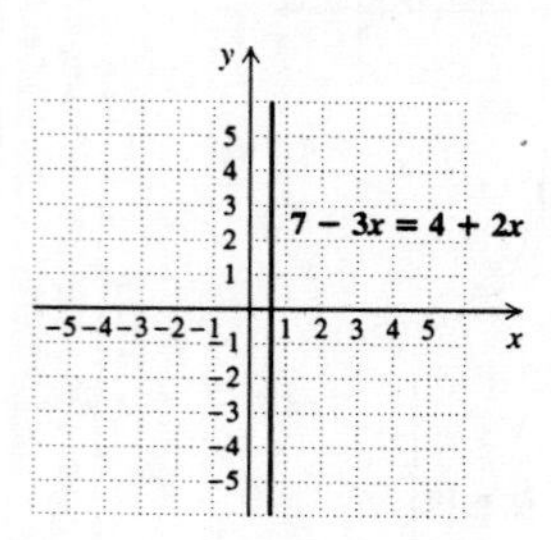

34.

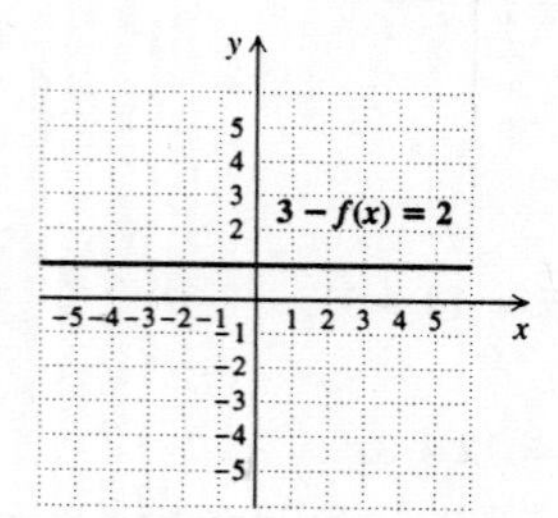

35. Graph $x - 2 = y$.

To find the y-intercept, let $x = 0$.

$$x - 2 = y$$
$$0 - 2 = y, \text{ or } -2 = y$$

The y-intercept is $(0, -2)$.

To find the x-intercept, let $y = 0$.

$$x - 2 = y$$
$$x - 2 = 0, \text{ or } x = 2$$

The x-intercept is (2,0).

Plot these points and draw the line. A third point could be used as a check.

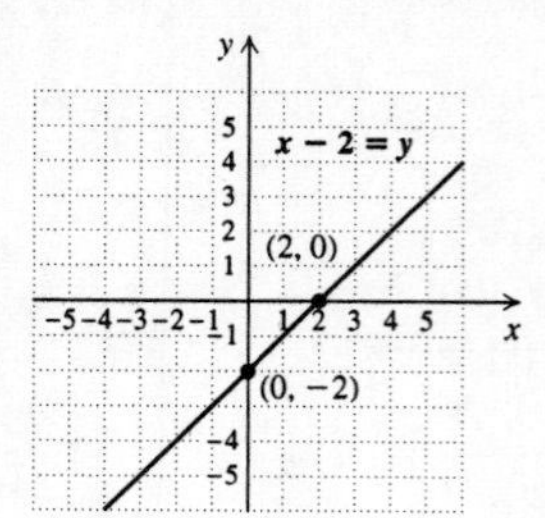

36.

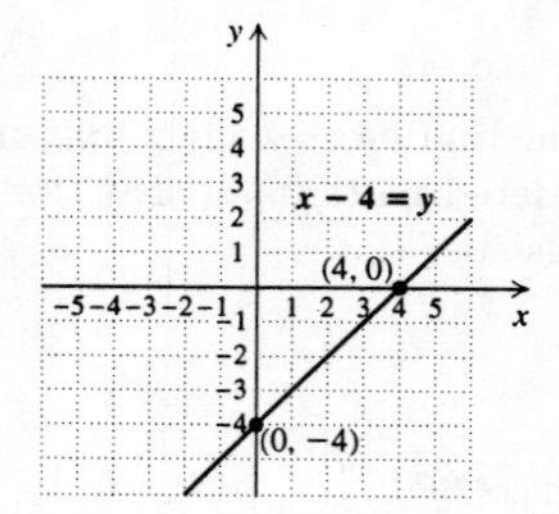

37. Graph $3x - 1 = y$.

To find the y-intercept, let $x = 0$.

$$3x - 1 = y$$
$$3 \cdot 0 - 1 = y, \text{ or } -1 = y$$

The y-intercept is $(0, -1)$.

To find the x-intercept, let $y = 0$.

$$3x - 1 = y$$
$$3x - 1 = 0$$
$$3x = 1$$
$$x = \frac{1}{3}$$

The x-intercept is $\left(\frac{1}{3}, 0\right)$.

Plot these points and draw the line. A third point could be used as a check.

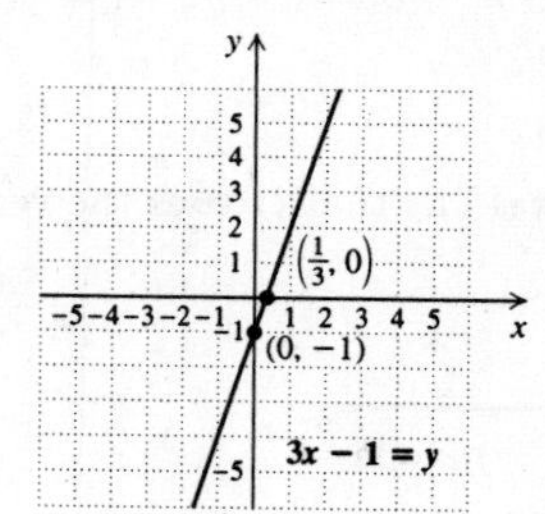

38.

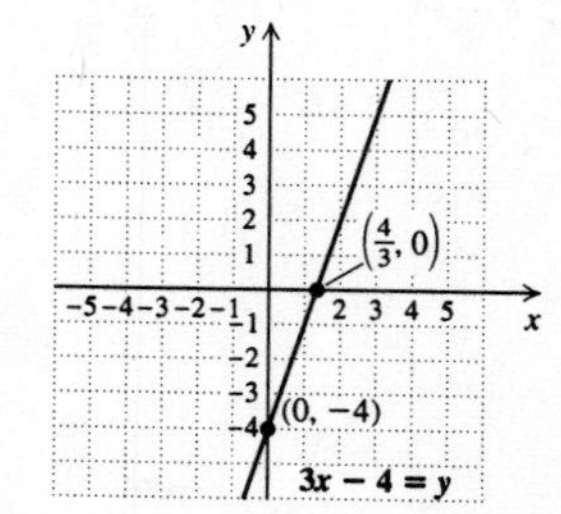

39. Graph $5x - 4y = 20$.

To find the y-intercept, let $x = 0$.

$$\begin{aligned} 5x - 4y &= 20 \\ 5 \cdot 0 - 4y &= 20 \\ -4y &= 20 \\ y &= -5 \end{aligned}$$

The y-intercept is $(0, -5)$.

To find the x-intercept, let $y = 0$.

$$\begin{aligned} 5x - 4y &= 20 \\ 5x - 4 \cdot 0 &= 20 \\ 5x &= 20 \\ x &= 4 \end{aligned}$$

The x-intercept is $(4, 0)$.

Plot these points and draw the line. A third point could be used as a check.

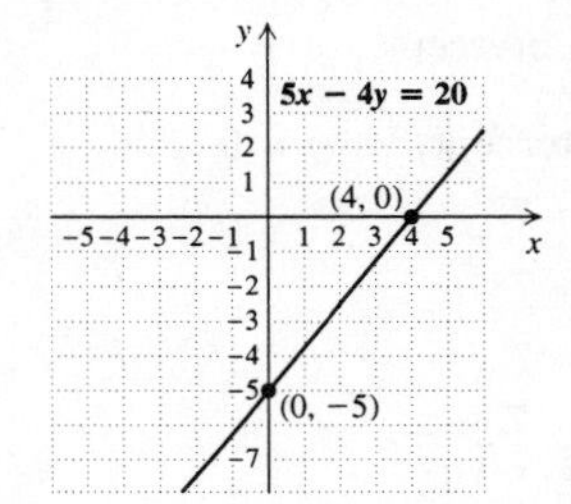

40.

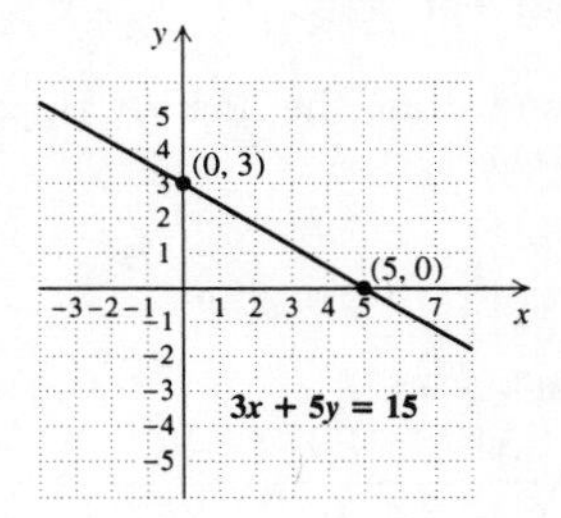

41. Graph $f(x) = -2 - 2x$, or $f(x) = -2x - 2$.

Because the function is in slope-intercept form, we know that the y-intercept is $(0, -2)$. To find the x-intercept, replace $f(x)$ with 0 and solve for x.

$$\begin{aligned} 0 &= -2 - 2x \\ 2x &= -2 \\ x &= -1 \end{aligned}$$

The x-intercept is $(-1, 0)$.

Plot these points and draw the line. As a check, note that the line's slope is -2, as expected.

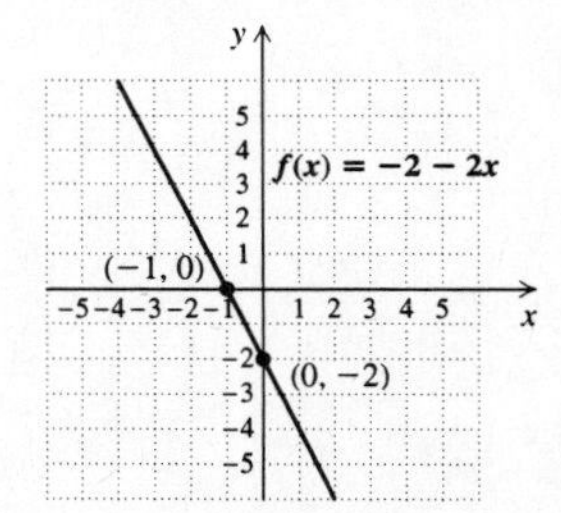

42.

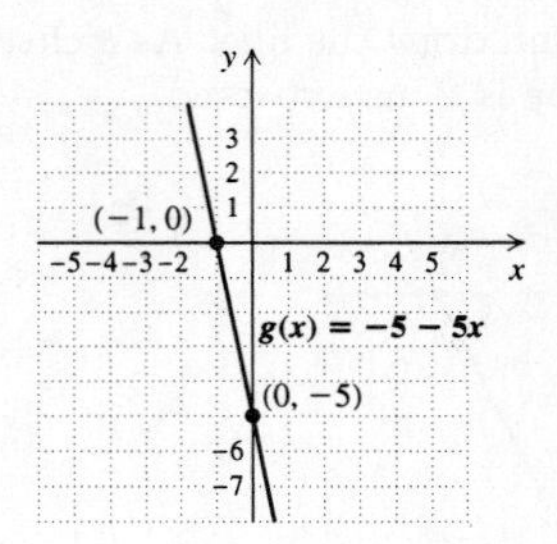

43. Graph $5y = -15 + 3x$.

To find the y-intercept, let $x = 0$.

$$\begin{aligned} 5y &= -15 + 3x \\ 5y &= -15 + 3 \cdot 0 \\ 5y &= -15 \\ y &= -3 \end{aligned}$$

$(0, -3)$ is the y-intercept.

To find the x-intercept, let $y = 0$.

$$\begin{aligned} 5y &= -15 + 3x \\ 5 \cdot 0 &= = -15 + 3x \\ 15 &= 3x \\ 5 &= x \end{aligned}$$

$(5, 0)$ is the x-intercept.

Plot these points and draw the line. A third point could be used as a check.

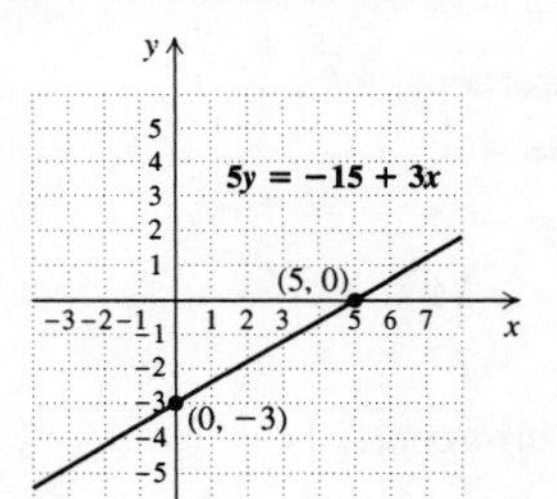

44.

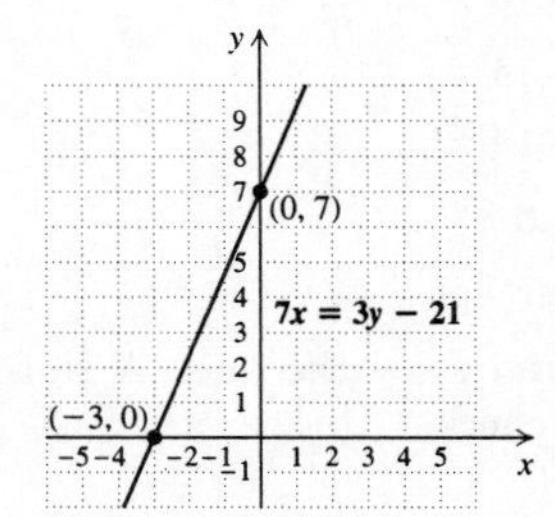

45. Graph $g(x) = 2x - 9$.

Because the function is in slope-intercept form, we know that the y-intercept is $(0, -9)$. To find the x-intercept, replace $f(x)$ with 0 and solve for x.

$$\begin{aligned} 0 &= 2x - 9 \\ 9 &= 2x \\ \frac{9}{2} &= x \end{aligned}$$

The x-intercept is $\left(\frac{9}{2}, 0\right)$.

Plot these points and draw the line. As a check, note that the line's slope is 2, as expected.

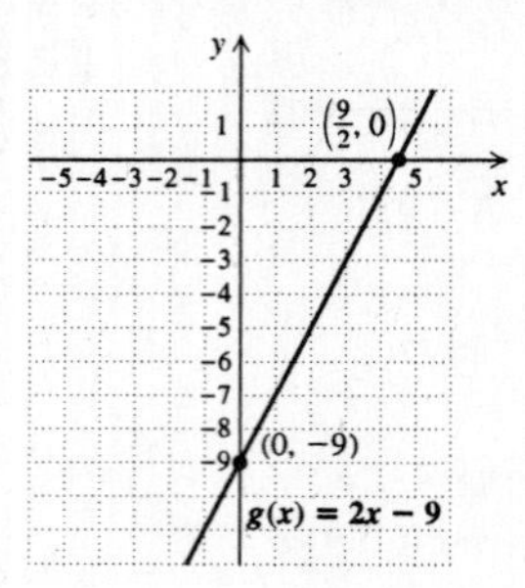

46.

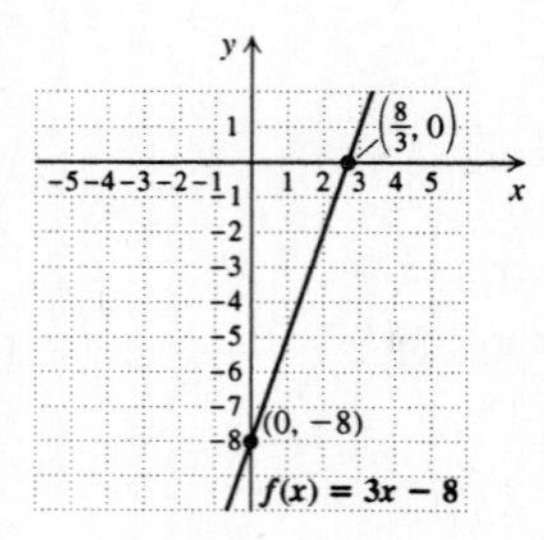

47. $1.4y - 3.5x = -9.8$

$14y - 35x = -98$ Multiplying by 10

$2y - 5x = -14$ Multiplying by $\frac{1}{7}$

Graph $2y - 5x = -14$.

To find the y-intercept, let $x = 0$.

$$2y - 5x = -14$$
$$2y - 5 \cdot 0 = -14$$
$$2y = -14$$
$$y = -7$$

$(0, -7)$ is the y-intercept.

To find the x-intercept, let $y = 0$.

$$2y - 5x = -14$$
$$2 \cdot 0 - 5x = -14$$
$$-5x = -14$$
$$x = 2.8$$

$(2.8, 0)$ is the x-intercept.

Plot these points and draw the line. A third point could be used as a check.

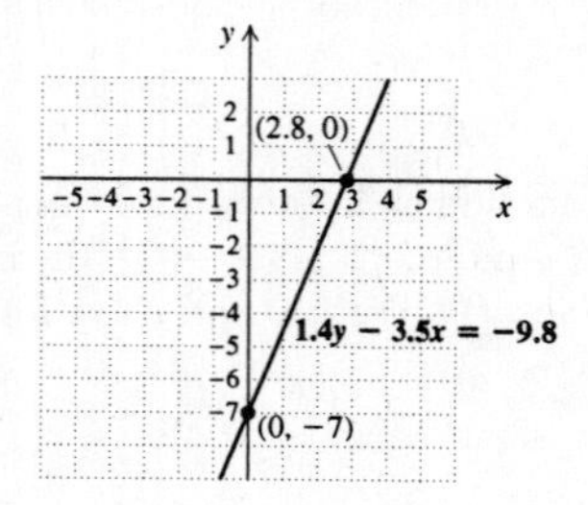

48.

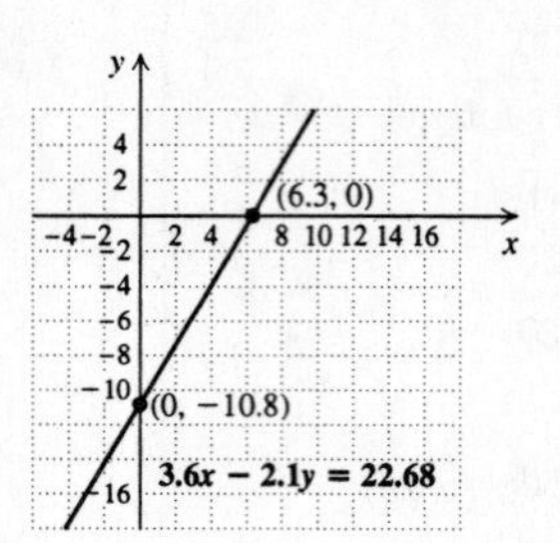

49. Graph $5x + 2y = 7$

To find the y-intercept, let $x = 0$.

$$5x + 2y = 7$$
$$5 \cdot 0 + 2y = 7$$
$$2y = 7$$
$$y = \frac{7}{2}$$

$\left(0, \frac{7}{2}\right)$ is the y-intercept.

To find the x-intercept, let $y = 0$.

$$5x + 2y = 7$$
$$5x + 2 \cdot 0 = 7$$
$$5x = 7$$
$$x = \frac{7}{5}$$

$\left(\frac{7}{5}, 0\right)$ is the x-intercept.

Plot these points and draw the line. A third point could be used as a check.

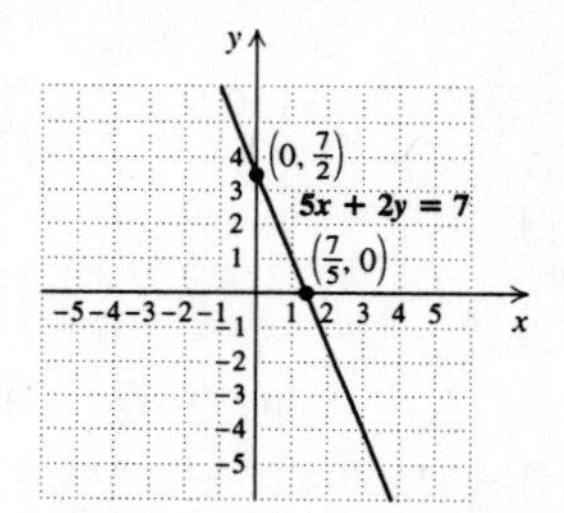

50.

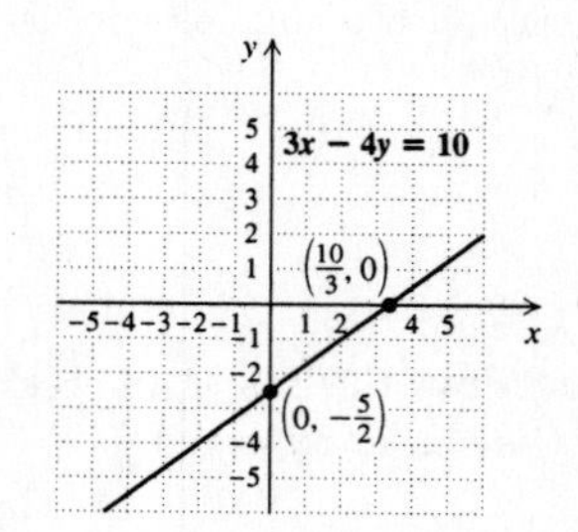

51. $x - 3 = 4$

Graph $f(x) = x - 3$ and $g(x) = 4$ on the same grid.

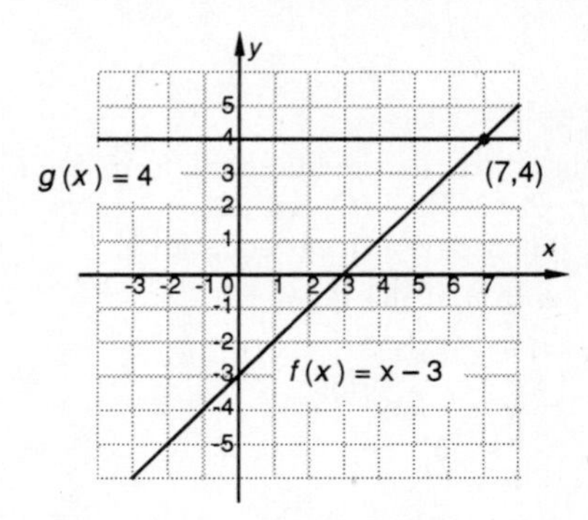

The lines appear to intersect at $(7, 4)$, so the solution is apparently 7.

To check we solve the equation algebraically.

$$x - 3 = 4$$
$$x = 7$$

The solution is 7.

52. 2

53. $2x + 1 = 7$

Graph $f(x) = 2x + 1$ and $g(x) = 7$ on the same grid.

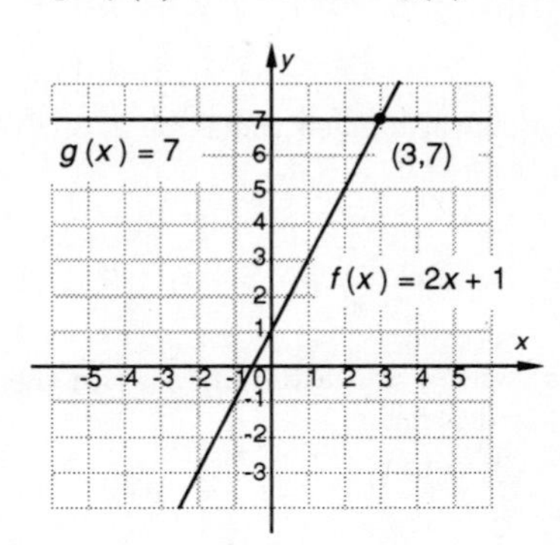

The lines appear to intersect at $(3, 7)$, so the solution is apparently 3.

To check we solve the equation algebraically.

$$2x + 1 = 7$$
$$2x = 6$$
$$x = 3$$

The solution is 3.

54. 2

55. $\frac{1}{3}x - 2 = 1$

Graph $f(x) = \frac{1}{3}x - 2$ and $g(x) = 1$ on the same grid.

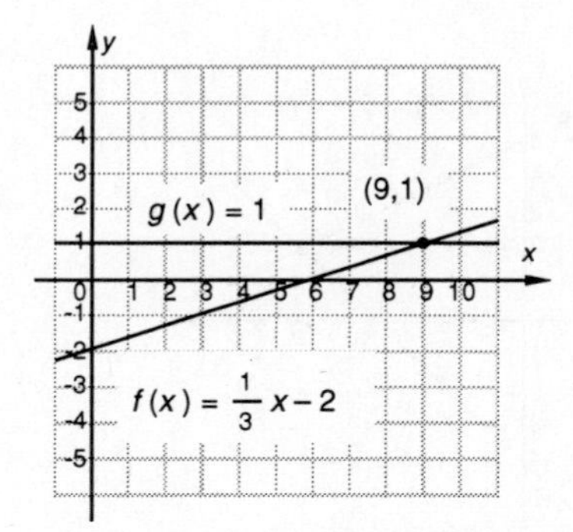

The lines appear to intersect at $(9, 1)$, so the solution is apparently 9.

To check we solve the equation algebraically.

$$\frac{1}{3}x - 2 = 1$$
$$\frac{1}{3}x = 3$$
$$x = 9$$

The solution is 9.

56. -8

57. $x + 3 = 5 - x$

Graph $f(x) = x + 3$ and $g(x) = 5 - x$ on the same grid.

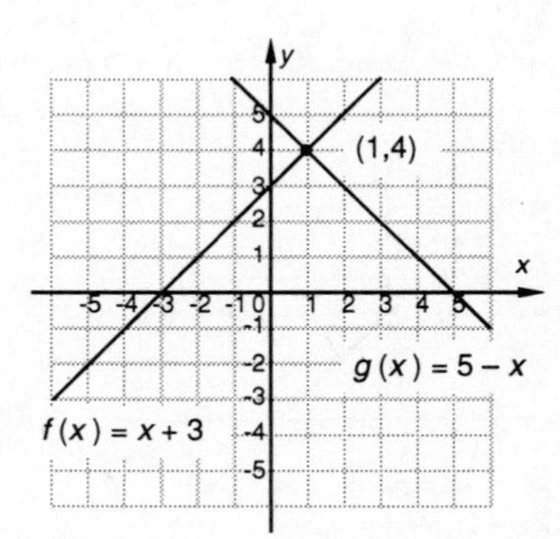

The lines appear to intersect at $(1, 4)$, so the solution is apparently 1.

To check we solve the equation algebraically.

$$x + 3 = 5 - x$$
$$2x + 3 = 5$$
$$2x = 2$$
$$x = 1$$

The solution is 1.

58. -2

59. $5 - \frac{1}{2}x = x - 4$

Graph $f(x) = 5 - \frac{1}{2}x$ and $g(x) = x - 4$ on the same grid.

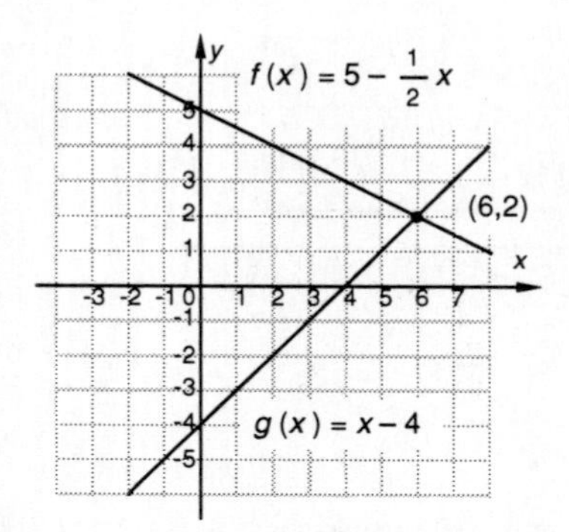

The lines appear to intersect at $(6, 2)$, so the solution is apparently 6.

To check we solve the equation algebraically.

$$5 - \frac{1}{2}x = x - 4$$
$$5 = \frac{3}{2}x - 4$$
$$9 = \frac{3}{2}x$$
$$6 = x$$

The solution is 6.

60. 4

61. $2x - 1 = -x + 3$

Graph $f(x) = 2x - 1$ and $g(x) = -x + 3$ on the same grid.

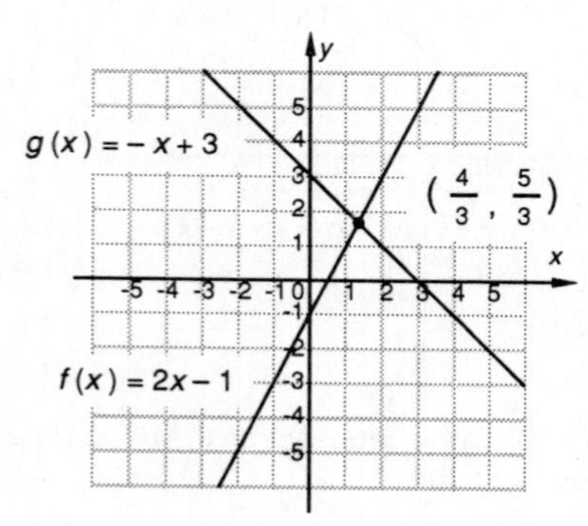

The lines appear to intersect at $\left(\frac{4}{3}, \frac{5}{3}\right)$, so the solution is apparently $\frac{4}{3}$.

To check we solve the equation algebraically.

$$2x - 1 = -x + 3$$
$$3x - 1 = 3$$
$$3x = 4$$
$$x = \frac{4}{3}$$

The solution is $\frac{4}{3}$.

62. $\frac{4}{3}$

63. $5x - 3y = 15$

This equation is in the standard form for a linear equation, $Ax + By = C$, with $A = 5$, $B = -3$, and $C = 15$. Thus, it is a linear equation.

Solve for y to find the slope.

$$5x - 3y = 15$$
$$-3y = -5x + 15$$
$$y = \frac{5}{3}x - 5$$

The slope is $\frac{5}{3}$.

64. Linear; $-\frac{3}{5}$

65. $16 + 4y = 0$

$$4y = -16$$

This equation is in the standard form for a linear equation, $Ax + By = C$, with $A = 0$, $B = 4$, and $C = -16$. Thus, it is a linear equation.

Solve for y to find the slope.

$$4y = -16$$
$$y = -4$$

This is a horizontal line, so the slope is 0. (We can think of this as $y = 0 \cdot x - 4$.)

66. Linear

67. $3g(x) = 6x^2$

Replace $g(x)$ with y and attempt to write the equation in standard form.

$$3y = 6x^2$$
$$-6x^2 + 3y = 0$$

The equation is not linear, because it has an x^2-term.

68. Linear; $-\frac{1}{2}$

69.

$$3y = 7xy - 5$$
$$-7xy + 3y = -5$$

The equation is not linear, because it has an xy-term.

70. Not linear

71. $6y - \frac{4}{x} = 0$

$6xy - 4 = 0$ Multiplying by x on both sides

The equation is not linear, because it has an xy-term.

72. Not linear

73. $\dfrac{f(x)}{x} = x^2$

Replace $f(x)$ with y and attempt to write the equation in standard form.

$$\frac{y}{x} = x^2$$
$$y = x^3$$
$$-x^3 + y = 0$$

The equation is not linear, because it has an x^3-term.

74. Linear; 2

75. $$f = \frac{F(c - v_0)}{c - v_s}$$

$f(c - v_s) = F(c - v_0)$ Multiplying by $c - v_s$ on both sides

$\dfrac{f(c - v_s)}{c - v_0} = F$ Dividing by $c - v_0$ on both sides

76. 4 mi

77. $9x - 15y = 3 \cdot 3x - 3 \cdot 5y = 3(3x - 5y)$

78. $3a(4 + 7b)$

79. ◈

80. ◈

81. ◈

82. ◈

83. The line contains the points $(5, 0)$ and $(0, -4)$. We use the points to find the slope.

Slope $= \dfrac{-4 - 0}{0 - 5} = \dfrac{-4}{-5} = \dfrac{4}{5}$

Then the slope-intercept equation is $y = \dfrac{4}{5}x - 4$. We rewrite this equation in standard form.

$y = \dfrac{4}{5}x - 4$

$5y = 4x - 20$ Multiplying by 5 on both sides

$-4x + 5y = -20$ Standard form

This equation can also be written as $4x - 5y = 20$.

84. $\left(-\dfrac{b}{m}, 0\right)$

85. $rx + 3y = p - s$

The equation is in standard form with $A = r$, $B = 3$, and $C = p - s$. It is linear.

86. Linear

87. Try to put the equation in standard form.

$$r^2x = py + 5$$
$$r^2x - py = 5$$

The equation is in standard form with $A = r^2$, $B = -p$, and $C = 5$. It is linear.

88. Linear

89. Let equation A have intercepts $(a, 0)$ and $(0, b)$. Then equation B has intercepts $(2a, 0)$ and $(0, b)$.

Slope of $A = \dfrac{b - 0}{0 - a} = -\dfrac{b}{a}$

Slope of $B = \dfrac{b - 0}{0 - 2a} = -\dfrac{b}{2a} = \dfrac{1}{2}\left(-\dfrac{b}{a}\right)$

The slope of equation B is $\dfrac{1}{2}$ the slope of equation A.

90. $a = 5$, $b = 1$

91. First write the equation in standard form.

$ax + 3y = 5x - by + 8$

$ax - 5x + 3y + by = 8$ Adding $-5x + by$ on both sides

$(a - 5)x + (3 + b)y = 8$ Factoring

If the graph is a vertical line, then the coefficient of y is 0.

$$3 + b = 0$$
$$b = -3$$

Then we have $(a - 5)x = 8$.

If the line passes through $(4, 0)$, we have:

$(a - 5)4 = 8$ Substituting 4 for x

$$a - 5 = 2$$
$$a = 7$$

92. $0.\overline{571428}$

93. Graph $y_1 = 4x - 1$ and $y_2 = 3 - 2x$ in the same window and use the Zoom and Trace features or the Intersect feature to find the first coordinate of the point of intersection, 0.66666667.

We check by solving the equation algebraically.

$$4x - 1 = 3 - 2x$$
$$6x - 1 = 3$$
$$6x = 4$$
$$x = \frac{2}{3} = 0.\overline{6} \approx 0.66666667$$

94. 3.5

95. Graph $y_1 = 8 - 7x$ and $y_2 = -2x - 5$ in the same window and use the Zoom and Trace features or the Intersect feature to find the first coordinate of the point of intersection, 2.6.

We check by solving the equation algebraically.

$$8 - 7x = -2x - 5$$
$$8 - 5x = -5$$
$$-5x = -13$$
$$x = 2.6$$

Exercise Set 2.5

1. $y - y_1 = m(x - x_1)$ Point-slope equation
   $y - 3 = 4(x - 2)$ Substituting 4 for m, 2 for x_1, and 3 for y_1

We graph the equation by plotting $(2, 3)$, counting off a slope of $\frac{4}{1}$, and drawing the line.

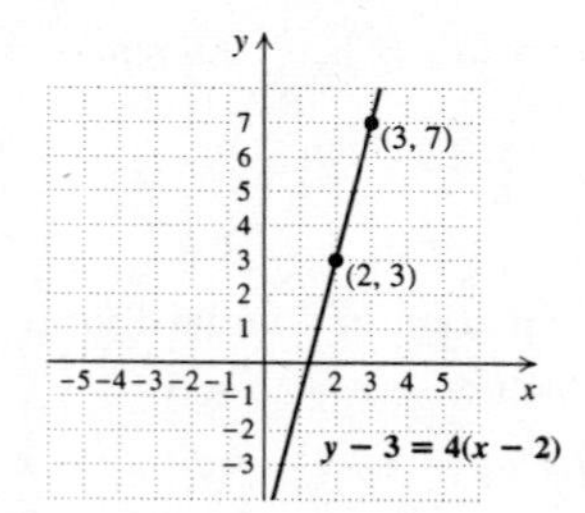

2. $y - 4 = 5(x - 7)$

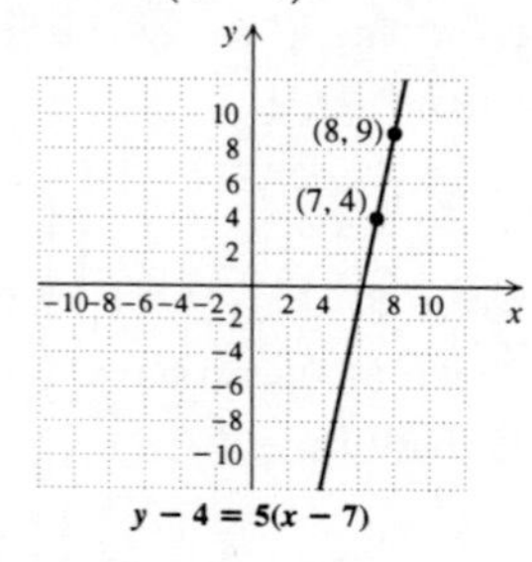

3. $y - y_1 = m(x - x_1)$ Point-slope equation
   $y - 7 = -2(x - 4)$ Substituting -2 for m, 4 for x_1, and 7 for y_1

We graph the equation by plotting $(4, 7)$, counting off a slope of $\frac{-2}{1}$, and drawing the line.

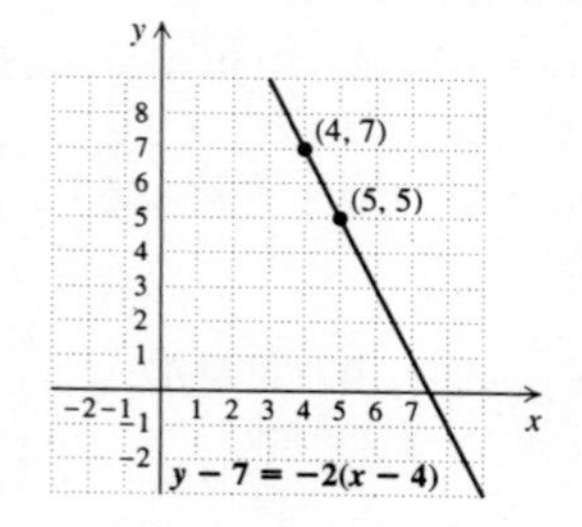

4. $y - 3 = -3(x - 7)$

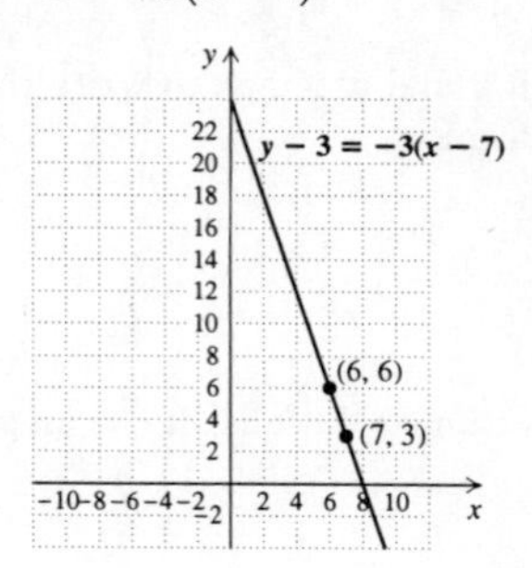

5. $y - y_1 = m(x - x_1)$ Point-slope equation
   $y - (-4) = 3[x - (-2)]$ Substituting 3 for m, -2 for x_1, and -4 for y_1

We graph the equation by plotting $(-2, -4)$, counting off a slope of $\frac{3}{1}$, and drawing the line.

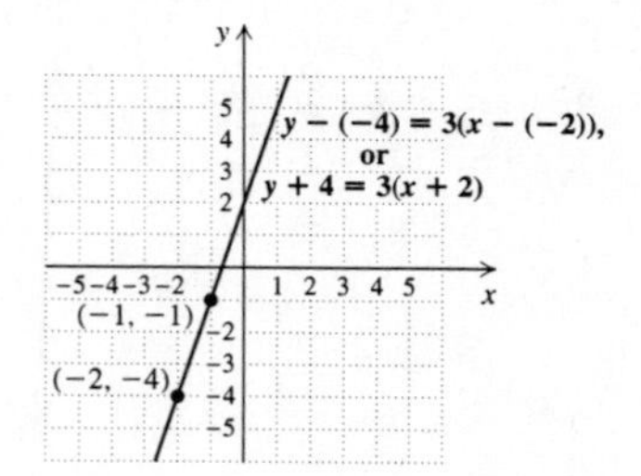

6. $y - (-7) = 1 \cdot [x - (-5)]$, or $y - (-7) = x - (-5)$

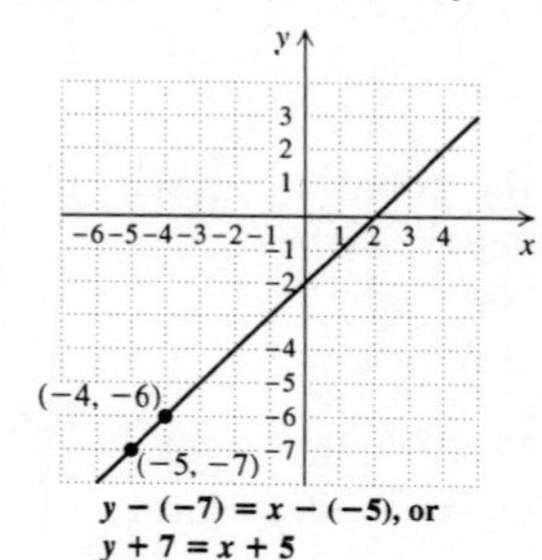

7. $y - y_1 = m(x - x_1)$ Point-slope equation
   $y - 0 = -2(x - 8)$ Substituting -2 for m, 8 for x_1, and 0 for y_1

We graph the equation by plotting $(8, 0)$, counting off a slope of $\frac{2}{-1}$, and drawing the line.

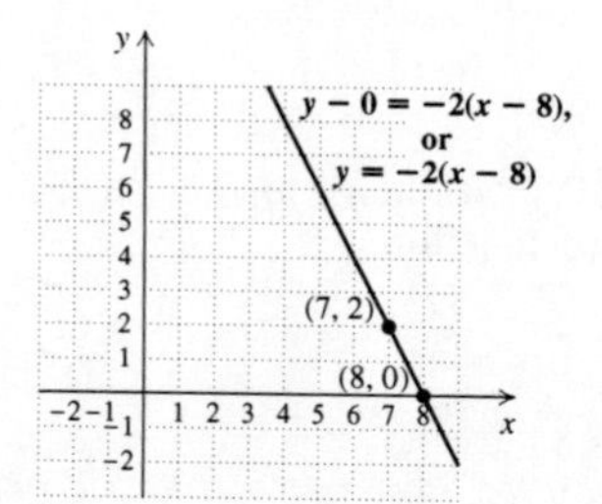

8. $y - 0 = -3[x - (-2)]$

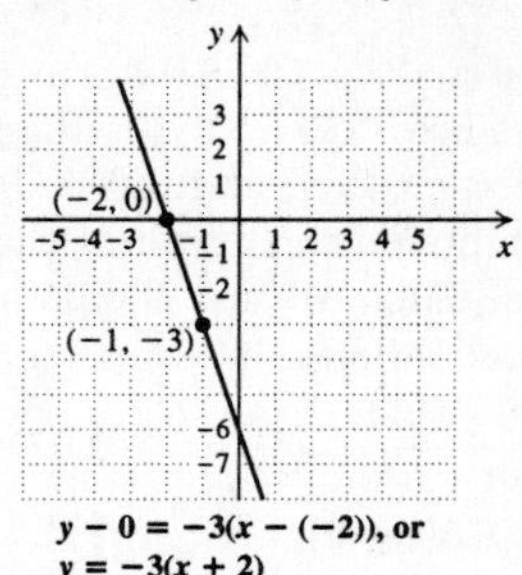

9. $y - y_1 = m(x - x_1)$ Point-slope equation

$y - 8 = \frac{2}{5}[x - (-3)]$ Substituting $\frac{2}{5}$ for m, -3 for x_1, and 8 for y_1

We graph the equation by plotting $(-3, 8)$, counting off a slope of $\frac{2}{5}$, and drawing the line.

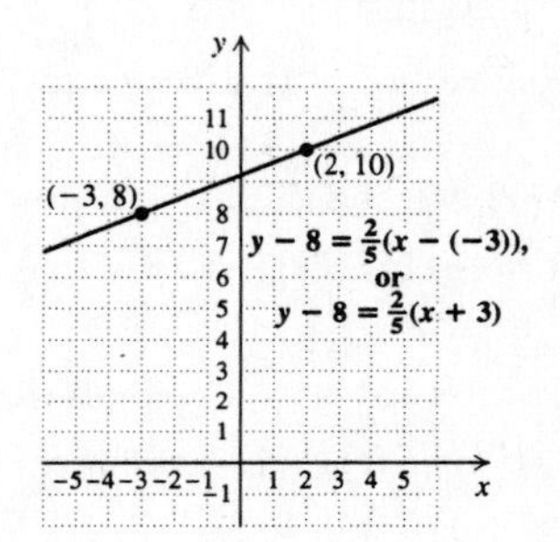

10. $y - (-5) = \frac{3}{4}(x - 1)$ or $y + 5 = \frac{3}{4}(x - 1)$

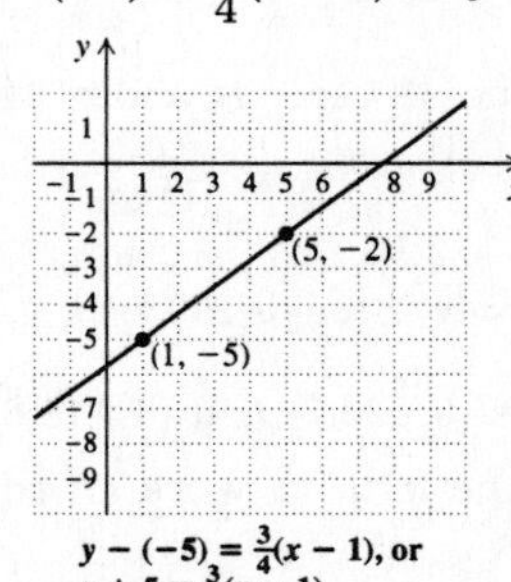

11. $y - 3 = \frac{2}{7}(x - 1)$

$y - y_1 = m(x - x_1)$

$m = \frac{2}{7}$, $x_1 = 1$, and $y_1 = 3$, so the slope m is $\frac{2}{7}$ and a point (x_1, y_1) on the graph is $(1, 3)$.

12. 9; $(2, 4)$

13. $y + 2 = -5(x - 7)$

$y - (-2) = -5(x - 7)$

$y - y_1 = m(x - x_1)$

$m = -5$, $x_1 = 7$, and $y_1 = -2$, so the slope m is -5 and a point (x_1, y_1) on the graph is $(7, -2)$.

14. $-\frac{2}{9}$; $(-5, 1)$

15. $y - 1 = -\frac{5}{3}(x + 2)$

$y - 1 = -\frac{5}{3}[x - (-2)]$

$y - y_1 = m(x - x_1)$

$m = -\frac{5}{3}$, $x_1 = -2$, and $y_1 = 1$, so the slope m is $-\frac{5}{3}$ and a point (x_1, y_1) on the graph is $(-2, 1)$.

16. -4; $(9, -7)$

17. $y - y_1 = m(x - x_1)$ Point-slope equation

$y - (-3) = 5(x - 2)$ Substituting 5 for m, 2 for x_1, and -3 for y_1

$y + 3 = 5x - 10$ Simplifying

$y = 5x - 13$ Subtracting 3 on both sides

$f(x) = 5x - 13$ Using function notation

18. $f(x) = -4x + 1$

19. $y - y_1 = m(x - x_1)$ Point-slope equation

$y - (-7) = -\frac{2}{3}(x - 4)$ Substituting $-\frac{2}{3}$ for m, 4 for x_1, and -7 for y_1

$y + 7 = -\frac{2}{3}x + \frac{8}{3}$ Simplifying

$y = -\frac{2}{3}x - \frac{13}{3}$ Subtracting 7 on both sides

$f(x) = -\frac{2}{3}x - \frac{13}{3}$ Using function notation

20. $f(x) = -\frac{1}{5}x + \frac{3}{5}$

21. $y - y_1 = m(x - x_1)$ Point-slope equation

$y - (-4) = -0.6[x - (-3)]$ Substituting -0.6 for m, -3 for x_1, and -4 for y_1

$y + 4 = -0.6(x + 3)$

$y + 4 = -0.6x - 1.8$

$y = -0.6x - 5.8$

$f(x) = -0.6x - 5.8$ Using function notation

22. $f(x) = 2.3x - 14.2$

23. First find the slope of the line:

$$m = \frac{6 - 4}{5 - 1} = \frac{2}{4} = \frac{1}{2}$$

Use the point-slope equation with $m = \frac{1}{2}$ and $(1, 4) = (x_1, y_1)$. (We could let $(5, 6) = (x_1, y_1)$ instead and obtain an equivalent equation.)

$$y-4=\frac{1}{2}(x-1)$$
$$y-4=\frac{1}{2}x-\frac{1}{2}$$
$$y=\frac{1}{2}x+\frac{7}{2}$$
$$f(x)=\frac{1}{2}x+\frac{7}{2} \quad \text{Using function notation}$$

24. $f(x)=-\frac{5}{2}x+11$

25. First find the slope of the line:

$$m=\frac{3-(-3)}{6.5-2.5}=\frac{3+3}{4}=\frac{6}{4}=1.5$$

Use the point-slope equation with $m = 1.5$ and $(6.5,3)=(x_1,y_1)$.

$$y-3=1.5(x-6.5)$$
$$y-3=1.5x-9.75$$
$$y=1.5x-6.75$$
$$f(x)=1.5x-6.75 \quad \text{Using function notation}$$

26. $f(x)=0.6x-2.5$

27. First find the slope of the line:

$$m=\frac{-2-1}{0-6}=\frac{-3}{-6}=\frac{1}{2}$$

Use the point-slope equation with $m=\frac{1}{2}$ and $(6,1)=(x_1,y_1)$.

$$y-1=\frac{1}{2}(x-6)$$
$$y-1=\frac{1}{2}x-3$$
$$y=\frac{1}{2}x-2$$
$$f(x)=\frac{1}{2}x-2 \quad \text{Using function notation}$$

28. $f(x)=\frac{5}{2}x+\frac{15}{2}$

29. First find the slope of the line:

$$m=\frac{-6-(-3)}{-4-(-2)}=\frac{-6+3}{-4+2}=\frac{-3}{-2}=\frac{3}{2}$$

Use the point-slope equation with $m=\frac{3}{2}$ and $(-2,-3)=(x_1,y_1)$.

$$y-(-3)=\frac{3}{2}[x-(-2)]$$
$$y+3=\frac{3}{2}(x+2)$$
$$y+3=\frac{3}{2}x+3$$
$$y=\frac{3}{2}x$$
$$f(x)=\frac{3}{2}x \quad \text{Using function notation}$$

30. $f(x)=3x+5$

31. a) We form pairs of the type (t,R) where t is the number of years since 1930 and R is the record. We have two pairs, $(0,46.8)$ and $(40,43.8)$. These are two points on the graph of the linear function we are seeking. We use the point-slope form to write an equation relating R and t:

$$m=\frac{43.8-46.8}{40-0}=\frac{-3}{40}=-0.075$$
$$R-46.8=-0.075(t-0)$$
$$R-46.8=-0.075t$$
$$R=-0.075t+46.8$$
$$R(t)=-0.075t+46.8 \quad \text{Using function notation}$$

b) 1999 is 69 years since 1930, so to predict the record in 1999, we find $R(69)$:

$$R(69)=-0.075(69)+46.8$$
$$=41.625$$

The predicted record is 41.625 seconds in 1999.

2002 is 72 years since 1930, so to predict the record in 2002, we find $R(72)$:

$$R(72)=-0.075(72)+46.8$$
$$=41.4$$

The predicted record is 41.4 seconds in 2002.

c) Substitute 40 for $R(t)$ and solve for t:

$$40=-0.075t+46.8$$
$$-6.8=-0.075t$$
$$91\approx t$$

The record will be 40 seconds about 91 years after 1930, or in 2021.

32. (a) $R(t)=-0.0075t+3.85$; (b) 3.34 min, 3.31 min; (c) one-third of the way through 2003

33. a) We form the pairs $(0,132.7)$ and $(6,178.6)$.

Use the point-slope form to write an equation relating A and t:

$$m=\frac{178.6-132.7}{6-0}=\frac{45.9}{6}=7.65$$
$$A-132.7=7.65(t-0)$$
$$A-132.7=7.65t$$
$$A=7.65t+132.7$$
$$A(t)=7.65t+132.7 \quad \text{Using function notation}$$

b) 2002 is 16 years since 1986, so we find $A(16)$:

$$A(16)=7.65(16)+132.7$$
$$=255.1$$

We predict that the amount of PAC contributions in 2002 will be \$255.1 million.

34. (a) $A(p)=-2.5p+26.5$; (b) 11.5 million lb

35. a) We form the pairs $(0, 32.9)$ and $(3, 45.0)$. Use the point-slope form to write an equation relating N and t:

$$m = \frac{45.0 - 32.9}{3 - 0} = \frac{12.1}{3} = \frac{121}{30}$$

$$N - 45 = \frac{121}{30}(t - 3)$$

$$N - 45 = \frac{121}{30}t - \frac{121}{10}$$

$$N = \frac{121}{30}t + \frac{329}{10}$$

$$N(t) = \frac{121}{30}t + \frac{329}{10} \quad \text{Using function notation}$$

b) 2001 is 11 years since 1990, so we find $N(11)$:

$$N(11) = \frac{121}{30} \cdot 11 + \frac{329}{10}$$

$$\approx 77.3$$

We predict that Americans will recycle about 77.3 million tons of garbage in 2001.

36. (a) $A(p) = 2p - 11$; (b) 1 million lb

37. a) We form the pairs $(0, 76.4)$ and $(4, 74.9)$.

Use the point-slope form to write an equation relating A and t:

$$m = \frac{74.9 - 76.4}{4 - 0} = \frac{-1.5}{4} = -0.375$$

$$A - 76.4 = -0.375(t - 0)$$

$$A - 76.4 = -0.375t$$

$$A = -0.375t + 76.4$$

$$A(t) = -0.375t + 76.4 \quad \text{Using function notation}$$

b) 2002 is 12 years after 1990, so we find $A(12)$:

$$A(12) = -0.375(12) + 76.4$$

$$= 71.9$$

We predict that there will be 71.9 million acres of land in the national park system in 2002.

38. (a) $P(d) = 0.03d + 1$; (b) 21.7 atm

39. a) We form the pairs $(0, 78.8)$ and $(5, 79.7)$.

Use the point-slope form to write an equation relating E and t:

$$m = \frac{79.7 - 78.8}{5 - 0} = \frac{0.9}{5} = 0.18$$

$$E - 78.8 = 0.18(t - 0)$$

$$E - 78.8 = 0.18t$$

$$E = 0.18t + 78.8$$

$$E(t) = 0.18t + 78.8 \quad \text{Using function notation}$$

b) 2004 is 14 years after 1990, so we find $E(14)$:

$$E(14) = 0.18(14) + 78.8$$

$$= 81.32$$

We predict that the life expectancy of females in the United States in 2004 will be 81.32 years.

40. (a) $E(t) = 0.2t + 71.8$; (b) 74.6 years

41. We first solve for y and determine the slope of each line.

$$x + 6 = y$$

$$y = x + 6 \quad \text{Reversing the order}$$

The slope of $y = x + 6$ is 1.

$$y - x = -2$$

$$y = x - 2$$

The slope of $y = x - 2$ is 1.

The slopes are the same; the lines are parallel.

42. Yes

43. We first solve for y and determine the slope of each line.

$$y + 3 = 5x$$

$$y = 5x - 3$$

The slope of $y = 5x - 3$ is 5.

$$3x - y = -2$$

$$3x + 2 = y$$

$$y = 3x + 2 \quad \text{Reversing the order}$$

The slope of $y = 3x + 2$ is 3.

The slopes are not the same; the lines are not parallel.

44. No

45. We determine the slope of each line.

The slope of $f(x) = 3x + 9$ is 3.

$$2y = 6x - 2$$

$$y = 3x - 1$$

The slope of $y = 3x - 1$ is 3.

The slopes are the same; the lines are parallel.

46. Yes

47. First solve the equation for y and determine the slope of the given line.

$$x + 2y = 6 \quad \text{Given line}$$

$$2y = -x + 6$$

$$y = -\frac{1}{2}x + 3$$

The slope of the given line is $-\frac{1}{2}$.

The slope of every line parallel to the given line must also be $-\frac{1}{2}$. We find the equation of the line with slope $-\frac{1}{2}$ and containing the point $(3, 7)$.

$$y - y_1 = m(x - x_1) \quad \text{Point-slope equation}$$

$$y - 7 = -\frac{1}{2}(x - 3) \quad \text{Substituting}$$

$$y - 7 = -\frac{1}{2}x + \frac{3}{2}$$

$$y = -\frac{1}{2}x + \frac{17}{2}$$

48. $y = 3x + 3$

49. First solve the equation for y and determine the slope of the given line.

$5x - 7y = 8$ Given line

$5x - 8 = 7y$

$\frac{5}{7}x - \frac{8}{7} = y$

$y = \frac{5}{7}x - \frac{8}{7}$

The slope of the given line is $\frac{5}{7}$.

The slope of every line parallel to the given line must also be $\frac{5}{7}$. We find the equation of the line with slope $\frac{5}{7}$ and containing the point $(2, -1)$.

$y - y_1 = m(x - x_1)$ Point-slope equation

$y - (-1) = \frac{5}{7}(x - 2)$ Substituting

$y + 1 = \frac{5}{7}x - \frac{10}{7}$

$y = \frac{5}{7}x - \frac{17}{7}$

50. $y = -2x - 13$

51. First solve the equation for y and determine the slope of the given line.

$3x - 9y = 2$ Given line

$3x - 2 = 9y$

$\frac{1}{3}x - \frac{2}{9} = y$

The slope of the given line is $\frac{1}{3}$.

The slope of every line parallel to the given line must also be $\frac{1}{3}$. We find the equation of the line with slope $\frac{1}{3}$ and containing the point $(-6, 2)$.

$y - y_1 = m(x - x_1)$ Point-slope equation

$y - 2 = \frac{1}{3}[x - (-6)]$ Substituting

$y - 2 = \frac{1}{3}(x + 6)$

$y - 2 = \frac{1}{3}x + 2$

$y = \frac{1}{3}x + 4$

52. $y = -\frac{5}{2}x - \frac{35}{2}$

53. First solve the equation for y and determine the slope of the given line.

$3x + 2y = -7$ Given line

$2y = -3x - 7$

$y = -\frac{3}{2}x - \frac{7}{2}$

The slope of the given line is $-\frac{3}{2}$.

The slope of every line parallel to the given line must also be $-\frac{3}{2}$. We find the equation of the line with slope $-\frac{3}{2}$ and containing the point $(-3, -2)$.

$y - y_1 = m(x - x_1)$ Point-slope equation

$y - (-2) = -\frac{3}{2}[x - (-3)]$ Substituting

$y + 2 = -\frac{3}{2}(x + 3)$

$y + 2 = -\frac{3}{2}x - \frac{9}{2}$

$y = -\frac{3}{2}x - \frac{13}{2}$

54. $y = \frac{6}{5}x + \frac{39}{5}$

55. We determine the slope of each line.

The slope of $f(x) = 4x - 5$ is 4.

$4y = 8 - x$

$4y = -x + 8$

$y = -\frac{1}{4}x + 2$

The slope of $4y = 8 - x$ is $-\frac{1}{4}$.

The product of their slopes is $4\left(-\frac{1}{4}\right)$, or -1; the lines are perpendicular.

56. No

57. We determine the slope of each line.

$x + 2y = 5$

$2y = -x + 5$

$y = -\frac{1}{2}x + \frac{5}{2}$

The slope of $x + 2y = 5$ is $-\frac{1}{2}$.

$2x + 4y = 8$

$4y = -2x + 8$

$y = -\frac{1}{2}x + 2.$

The slope of $2x + 4y = 8$ is $-\frac{1}{2}$.

The product of their slopes is $\left(-\frac{1}{2}\right)\left(-\frac{1}{2}\right)$, or $\frac{1}{4}$; the lines are not perpendicular. For the lines to be perpendicular, the product must be -1.

58. Yes

59. First solve the equation for y and determine the slope of the given line.

$2x + y = -3$ Given line

$y = -2x - 3$

The slope of the given line is -2.

The slope of a perpendicular line is given by the opposite of the reciprocal of -2, $\frac{1}{2}$.

We find the equation of the line with slope $\frac{1}{2}$ containing the point $(2, 5)$.

$$y - y_1 = m(x - x_1) \quad \text{Point-slope equation}$$
$$y - 5 = \frac{1}{2}(x - 2) \quad \text{Substituting}$$
$$y - 5 = \frac{1}{2}x - 1$$
$$y = \frac{1}{2}x + 4$$

60. $y = -3x + 12$

61. First solve the equation for y and determine the slope of the given line.

$$3x + 4y = 5 \quad \text{Given line}$$
$$4y = -3x + 5$$
$$y = -\frac{3}{4}x + \frac{5}{4}$$

The slope of the given line is $-\frac{3}{4}$.

The slope of a perpendicular line is given by the opposite of the reciprocal of $-\frac{3}{4}$, $\frac{4}{3}$.

We find the equation of the line with slope $\frac{4}{3}$ and containing the point $(3, -2)$.

$$y - y_1 = m(x - x_1) \quad \text{Point-slope equation}$$
$$y - (-2) = \frac{4}{3}(x - 3) \quad \text{Substituting}$$
$$y + 2 = \frac{4}{3}x - 4$$
$$y = \frac{4}{3}x - 6$$

62. $y = -\frac{2}{5}x - \frac{31}{5}$

63. First solve the equation for y and determine the slope of the given line.

$$2x + 5y = 7 \quad \text{Given line}$$
$$5y = -2x + 7$$
$$y = -\frac{2}{5}x + \frac{7}{5}$$

The slope of the given line is $-\frac{2}{5}$.

The slope of a perpendicular line is given by the opposite of the reciprocal of $-\frac{2}{5}$, $\frac{5}{2}$.

We find the equation of the line with slope $\frac{5}{2}$ and containing the point $(0, 9)$.

$$y - y_1 = m(x - x_1) \quad \text{Point-slope equation}$$
$$y - 9 = \frac{5}{2}(x - 0) \quad \text{Substituting}$$
$$y - 9 = \frac{5}{2}x$$
$$y = \frac{5}{2}x + 9$$

64. $y = -2x - 10$

65. First solve the equation for y and find the slope of the given line.

$$3x - 5y = 6$$
$$-5y = -3x + 6$$
$$y = \frac{3}{5}x - \frac{6}{5}$$

The slope of the given line is $\frac{3}{5}$. The slope of a perpendicular line is given by the opposite of the reciprocal of $\frac{3}{5}$, $-\frac{5}{3}$.

We find the equation of the line with slope $-\frac{5}{3}$ and containing the point $(-4, -7)$.

$$y - y_1 = m(x - x_1) \quad \text{Point-slope equation}$$
$$y - (-7) = -\frac{5}{3}[x - (-4)]$$
$$y + 7 = -\frac{5}{3}(x + 4)$$
$$y + 7 = -\frac{5}{3}x - \frac{20}{3}$$
$$y = -\frac{5}{3}x - \frac{41}{3}$$

66. $y = -\frac{2}{7}x + \frac{27}{7}$

67. ***Familiarize***. We find the sales tax by multiplying the listed price by the tax rate. We let x represent the price and $5\%x$ represent the sales tax.

Translate.

$$\underbrace{\text{Listed price}}_{\downarrow\; x} + \underbrace{\text{Sales tax}}_{\downarrow\; 5\%x} = \underbrace{\text{Total price}}_{\downarrow\; \$36.75}$$

$$x + 5\%x = \$36.75$$

Carry out. We solve the equation.

$$x + 0.05x = 36.75$$
$$1.05x = 36.75$$
$$x = \frac{36.75}{1.05}$$
$$x = 35$$

Check. If the listed price is \$35, then the sales tax is $5\% \cdot \$35$, or \$1.75. The total price is $35 + 1.75$, or \$36.75. The value checks.

State. The price of the radio before the tax is \$35.

68. 69 points

69. ***Familiarize.*** Let n represent the number.

Translate.

$$\begin{array}{ccccc} 15\% & \text{of} & \underbrace{\text{what number}} & = & 12.4? \\ \downarrow & \downarrow & \downarrow & \downarrow & \downarrow \\ 15\% & \times & n & = & 12.4 \end{array}$$

Carry out. We solve the equation.

$15\%n = 12.4$

$0.15n = 12.4$

$15n = 1240$ Multiplying by 100 to clear decimals

$n = \frac{1240}{15}$

$n = \frac{248}{3}$, or $82\frac{2}{3}$

Check. 15% of $82\frac{2}{3}$ is $0.15\left(\frac{248}{3}\right) = 12.4$. The result checks.

State. 15% of $82\frac{2}{3}$ is 12.4.

70. $-\frac{13}{12}$

71. ◈

72. ◈

73. ◈

74. ◈

75. ***Familiarize.*** The value C of the computer, in dollars, after t months can be modeled by a line that contains the points (0,2500) and (5, 2150).

Translate. We find an equation relating C and t.

$m = \frac{2150 - 2500}{5 - 0} = \frac{-350}{5} = -70$

$C - 2500 = -70(t - 0)$

$C - 2500 = -70t$

$C = -70t + 2500$

Carry out. Using function notation we have $C(t) = -70t+2500$. To find the value after 8 months we find $C(8)$:

$C(8) = -70 \cdot 8 + 2500$

$= 1940$

Check. We can repeat our calculations. We could also graph the function and determine that (8, 1940) is on the graph.

State. The computer's value after 8 months will be $1940.

76. $340

77. ***Familiarize.*** The total operating costs C, in dollars, for t months can be modeled by a linear equation containing the points (4, 7500) and (7, 9250).

Translate. We find an equation relating C and t.

$m = \frac{9250 - 7500}{7 - 4} = \frac{1750}{3}$

$C - 7500 = \frac{1750}{3}(t - 4)$

$C - 7500 = \frac{1750}{3}t - \frac{7000}{3}$

$C = \frac{1750}{3}t + \frac{15,500}{3}$

Carry out. Using function notation we have $C(t) = \frac{1750}{3}t + \frac{15,500}{3}$. To predict the total costs after 10 months, we find $C(10)$:

$C(10) = \frac{1750}{3} \cdot 10 + \frac{15,500}{3}$

$= 11,000$

Check. We can repeat the calculations. We could also graph the function and determine that (10, 11, 000) is on the graph.

State. We predict that the total operating costs after 10 months will be $11,000.

78. $1350

79. ***Familiarize.*** We form the pairs (32, 0) and (212, 100) and plot these data points, choosing suitable scales on the two axes. We will let F represent the Fahrenheit temperature and C represent the Celsius temperature.

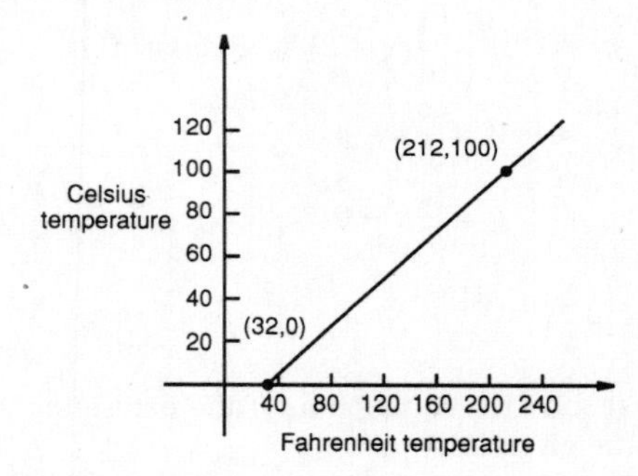

Translate. We seek an equation relating C and F. We find the slope and use the point-slope form:

$m = \frac{100 - 0}{212 - 32} = \frac{100}{180} = \frac{5}{9}$

$C - 0 = \frac{5}{9}(F - 32)$

$C = \frac{5}{9}(F - 32)$

$C = \frac{5}{9}F - \frac{160}{9}$

Carry out. Using function notation, we have

$C(F) = \frac{5}{9}F - \frac{160}{9}$.

To find the Celsius temperature that corresponds to 70° F, we find

$$C(70) = \frac{5}{9}(70) - \frac{160}{9}$$
$$\approx 21.1$$

Check. We observe that the result seems reasonable since $(70, 21.1)$ appears to lie on the graph. We can also repeat the calculations.

State. A temperature of about 21.1° C corresponds to 70° F.

80. (a) $f(x) = \frac{1}{3}x + \frac{10}{3}$; (b) $\frac{13}{3}$; (c) 290

81. a) We have two pairs, $(3, -5)$ and $(7, -1)$. Use the point-slope form:

$$m = \frac{-1-(-5)}{7-3} = \frac{-1+5}{4} = \frac{4}{4} = 1$$
$$y - (-5) = 1(x-3)$$
$$y + 5 = x - 3$$
$$y = x - 8$$
$$g(x) = x - 8 \quad \text{Using function notation}$$

b) $g(-2) = -2 - 8 = -10$

c) $g(a) = a - 8$

If $g(a) = 75$, we have

$$a - 8 = 75$$
$$a = 83.$$

82. $-\frac{40}{9}$

83. Find the slope of $7y - kx = 9$:

$$7y - kx = 9$$
$$7y = kx + 9$$
$$y = \frac{k}{7}x + \frac{9}{7}$$

The slope is $\frac{k}{7}$.

Find the slope of the line containing $(2, -1)$ and $(-4, 5)$.

$$m = \frac{5-(-1)}{-4-2} = \frac{6}{-6} = -1$$

If the lines are perpendicular, the product of their slopes must be -1:

$$\frac{k}{7}(-1) = -1$$
$$-\frac{k}{7} = -1$$
$$k = 7 \quad \text{Multiplying by } -7 \text{ on both sides}$$

84. 

85. 

Exercise Set 2.6

1. Since $f(2) = -3 \cdot 2 + 1 = -5$, and $g(2) = 2^2 + 2 = 6$, we have $f(2) + g(2) = -5 + 6 = 1$.

2. 7

3. Since $f(5) = -3 \cdot 5 + 1 = -14$ and $g(5) = 5^2 + 2 = 27$, we have $f(5) - g(5) = -14 - 27 = -41$.

4. -29

5. Since $f(-1) = -3(-1) + 1 = 4$ and $g(-1) = (-1)^2 + 2 = 3$, we have $f(-1) \cdot g(-1) = 4 \cdot 3 = 12$.

6. 42

7. Since $f(-4) = -3(-4) + 1 = 13$ and $g(-4) = (-4)^2 + 2 = 18$, we have $f(-4)/g(-4) = 13/18$.

8. $-\frac{8}{11}$

9. Since $g(1) = 1^2 + 2 = 3$ and $f(1) = -3 \cdot 1 + 1 = -2$, we have $g(1) - f(1) = 3 - (-2) = 3 + 2 = 5$.

10. $-\frac{6}{5}$

11. Since $g(0) = 0^2 + 2 = 2$ and $f(0) = -3 \cdot 0 + 1 = 1$, we have $g(0)/f(0) = 2/1 = 2$.

12. 55

13.
$$(F+G)(x) = F(x) + G(x)$$
$$= x^2 - 3 + 4 - x$$
$$= x^2 - x + 1$$

14. $a^2 - a + 1$

15. Using our work in Exercise 13, we have

$$(F+G)(-4) = (-4)^2 - (-4) + 1$$
$$= 16 + 4 + 1$$
$$= 21.$$

16. 31

17.
$$(F-G)(x) = F(x) - G(x)$$
$$= x^2 - 3 - (4 - x)$$
$$= x^2 - 3 - 4 + x$$
$$= x^2 + x - 7$$

Then we have

$$(F-G)(3) = 3^2 + 3 - 7$$
$$= 9 + 3 - 7$$
$$= 5.$$

18. -1

19. $(F \cdot G)(x) = F(x) \cdot G(x)$

$$= (x^2 - 3)(4 - x)$$
$$= 4x^2 - x^3 - 12 + 3x$$

Then we have

$$(F \cdot G)(-3) = 4(-3)^2 - (-3)^3 - 12 + 3(-3)$$
$$= 4 \cdot 9 - (-27) - 12 - 9$$
$$= 36 + 27 - 12 - 9$$
$$= 42.$$

20. 104

21. $(F/G)(x) = F(x)/G(x)$

$$= \frac{x^2 - 3}{4 - x}$$

Then we have

$$(F/G)(0) = \frac{0^2 - 3}{4 - 0} = -\frac{3}{4}.$$

22. $-\frac{2}{3}$

23. Using our work in Exercise 21, we have

$$(F/G)(-2) = \frac{(-2)^2 - 3}{4 - (-2)} = \frac{4 - 3}{4 + 2} = \frac{1}{6}.$$

24. $-\frac{2}{5}$

25. $F(7)$ is the value of F seven years after January, 1986, or in January, 1993. Locate 1993 on the horizontal axis, go up to the graph of F, then go across to the vertical axis and read the corresponding value. It is 1550. This indicates that as of January, 1993, there were 1550 new cases of AIDS per month among people born in 1960 or later.

26. 3450; in January, 1991 there were 3450 new cases of AIDS per month among people born before 1960.

27. $t = 6$ pertains to function values six years after January, 1986, or in January, 1992. From the graph we see that $T(6) \approx 4800$, $G(6) \approx 3550$, and $F(6) \approx 1250$. Then we have

$$(G + F)(6) = G(6) + F(6)$$
$$\approx 3550 + 1250$$
$$\approx 4800$$
$$\approx T(6).$$

28. $(T - F)(4) = T(4) - F(4) \approx 4100 - 800 \approx 3300 \approx G(4)$

29. $(n + l)(92) = n(92) + l(92)$

From the middle line of the graph, we can see that $n(92) + l(92) \approx 44$ million.

This represents the total number of passengers serviced by Newark and LaGuardia airports in 1992.

30. 49; a total of about 49 million passengers used Kennedy and LaGuardia airports in 1994.

31. $(k - l)(92) = k(92) - l(92)$

$$\approx 28 - 20$$
$$\approx 8 \text{ million}$$

This represents how many more passengers used Kennedy airport than LaGuardia airport in 1992.

32. 9; about 9 million more passengers used Kennedy airport than Newark airport in 1989.

33. $(n + l + k)(93) = n(93) + l(93) + k(93)$

From the top line of the graph, we can see that $n(93) + l(93) + k(93) \approx 72$ million.

This represents the number of passengers serviced by Newark, LaGuardia, and Kennedy airports in 1993.

34. 65; a total of about 65 million passengers used Newark, LaGuardia, and Kennedy airports in 1983.

35. The domain of f and of g is all real numbers. Thus, Domain of $f + g$ = Domain of $f - g$ = Domain of $f \cdot g = \{x|x \text{ is a real number}\}$.

36. $\{x|x \text{ is a real number}\}$

37. Because division by 0 is undefined, we have

Domain of $f = \{x|x \text{ is a real number and } x \neq 2\}$,

and

Domain of $g = \{x|x \text{ is a real number}\}$.

Thus, Domain of $f + g$ = Domain of $f - g$ = Domain of $f \cdot g = \{x|x \text{ is a real number and } x \neq 2\}$.

38. $\{x|x \text{ is a real number and } x \neq 4\}$

39. Because division by 0 is undefined, we have

Domain of $f = \{x|x \text{ is a real number and } x \neq 0\}$,

and

Domain of $g = \{x|x \text{ is a real number}\}$.

Thus, Domain of $f + g$ = Domain of $f - g$ = Domain of $f \cdot g = \{x|x \text{ is a real number and } x \neq 0\}$.

40. $\{x|x \text{ is a real number and } x \neq 0\}$

41. Because division by 0 is undefined, we have

Domain of $f = \{x|x \text{ is a real number and } x \neq 1\}$,

and

Domain of $g = \{x|x \text{ is a real number}\}$.

Thus, Domain of $f + g$ = Domain of $f - g$ = Domain of $f \cdot g = \{x|x \text{ is a real number and } x \neq 1\}$.

42. $\{x|x \text{ is a real number and } x \neq 5\}$

43. Because division by 0 is undefined, we have

Domain of $f = \{x|x \text{ is a real number and } x \neq 2\}$,

and

Domain of $g = \{x|x \text{ is a real number and } x \neq 4\}$.

Thus, Domain of $f + g$ = Domain of $f - g$ = Domain of $f \cdot g = \{x|x \text{ is a real number and } x \neq 2 \text{ and } x \neq 4\}$.

44. $\{x|x \text{ is a real number and } x \neq 3 \text{ and } x \neq 2\}$

45. Domain of f = Domain of g =

$\{x|x \text{ is a real number}\}$.

Since $g(x) = 0$ when $x - 3 = 0$, we have $g(x) = 0$ when $x = 3$. We conclude that Domain of $f/g = \{x|x \text{ is a real number and } x \neq 3\}$.

46. $\{x|x \text{ is a real number and } x \neq 5\}$

47. Domain of f = Domain of g =

$\{x|x \text{ is a real number}\}$.

Since $g(x) = 0$ when $2x - 8 = 0$, we have $g(x) = 0$ when $x = 4$. We conclude that Domain of $f/g = \{x|x \text{ is a real number and } x \neq 4\}$.

48. $\{x|x \text{ is a real number and } x \neq 3\}$

49. Domain of $f = \{x|x \text{ is a real number and } x \neq 4\}$.

Domain of $g = \{x|x \text{ is a real number}\}$.

Since $g(x) = 0$ when $5 - x = 0$, we have $g(x) = 0$ when $x = 5$. We conclude that Domain of $f/g = \{x|x \text{ is a real number and } x \neq 4 \text{ and } x \neq 5\}$.

50. $\{x|x \text{ is a real number and } x \neq 2 \text{ and } x \neq 7\}$

51. Domain of $f = \{x|x \text{ is a real number and } x \neq -1\}$.

Domain of $g = \{x|x \text{ is a real number}\}$.

Since $g(x) = 0$ when $2x + 5 = 0$, we have $g(x) = 0$ when $x = -\frac{5}{2}$. We conclude that Domain of $f/g =$

$\left\{x \middle| x \text{ is a real number and } x \neq -1 \text{ and } x \neq -\frac{5}{2}\right\}$.

52. $\left\{x \middle| x \text{ is a real number and } x \neq 2 \text{ and } x \neq -\frac{7}{3}\right\}$

53. $(F + G)(5) = F(5) + G(5) = 1 + 3 = 4$

$(F + G)(7) = F(7) + G(7) = -1 + 4 = 3$

54. 0; 2

55. From the graph we see that Domain of

$F = \{x|0 \leq x \leq 9\}$ and Domain of

$G = \{x|3 \leq x \leq 10\}$. Then Domain of

$F + G = \{x|3 \leq x \leq 9\}$. Since $G(x)$ is never 0, Domain of $F/G = \{x|3 \leq x \leq 9\}$.

56. $\{x|3 \leq x \leq 9\}$; $\{x|3 \leq x \leq 9\}$;
$\{x|3 \leq x \leq 9 \text{ and } x \neq 6 \text{ and } x \neq 8\}$

57. We use $(F + G)(x) = F(x) + G(x)$.

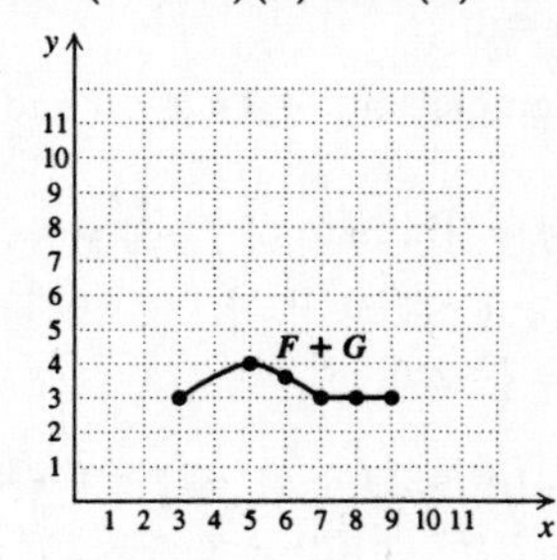

58.

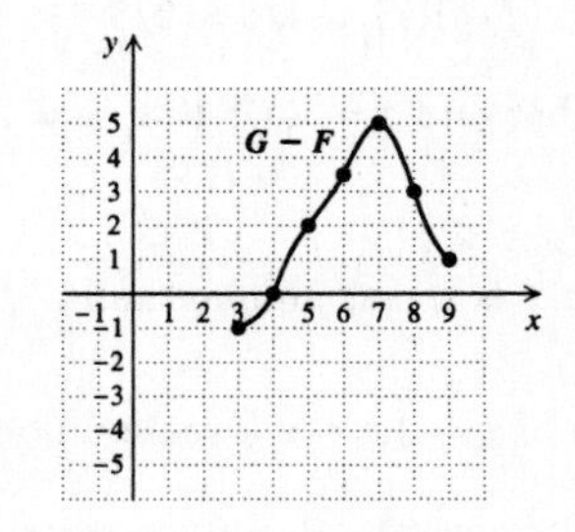

59.
$$5x - 7 = 0$$
$$5x = 7 \quad \text{Adding 7 on both sides}$$
$$x = \frac{7}{5} \quad \text{Multiplying by } \frac{1}{5} \text{ on both sides}$$

The solution is $\frac{7}{5}$.

60. -37

61. ***Familiarize***. The average score on n tests is the sum of the n test scores divided by n. If the average on 4 tests is 78.5, then the sum of the 4 scores is 4(78.5), or 314. Let s represent the score required on the fifth test to raise the average to 80.

Translate.

Sum of 5 scores divided by 5 is 80.

$$(314 + s) \div 5 = 80$$

Carry out. Solve the equation.

$$\frac{314 + s}{5} = 80$$
$$314 + s = 400$$
$$s = 86$$

Check. If the sum of the 5 scores is 314+86, or 400, then the average is 400/5, or 80. The value checks.

State. A score of 86 is needed on the fifth test to raise the average to 80.

62. 6.31×10^{-6}

63. ◈

64. ◈

65. The problem states that Domain of $m =$ $\{x| -1 < x < 5\}$. Since $n(x) = 0$ when $2x - 3 = 0$, we have $n(x) = 0$ when $x = \frac{3}{2}$. We conclude that Domain of $m/n =$
$\left\{x|x \text{ is a real number and } -1<x<5 \text{ and } x \neq \frac{3}{2}\right\}$.

66. Domain of $f + g =$ Domain of $f - g =$ Domain of $f \cdot g = \{-2, -1, 0, 1\}$;
domain of $f/g = \{-2, 0, 1\}$.

67. $f(-2) = 1$, $g(-2) = 4$, $(f + g)(-2) = 1 + 4 = 5$
$f(0) = 3$, $g(0) = 5$, $(f \cdot g)(0) = 3 \cdot 5 = 15$
$f(1) = 4$, $g(1) = 6$, $(f/g)(1) = 4/6 = 2/3$

68. $\{x|x$ is a real number and $x \neq 4$, $x \neq 3$, $x \neq 2$, and $x \neq -2\}$

69. Domain of $f = \left\{x \middle| x \text{ is a real number and } x \neq -\frac{5}{2}\right\}$; domain of $g = \{x|x$ is a real number and $x \neq -3\}$; $g(x) = 0$ when $x^4 - 1 = 0$, or when $x = 1$ or $x = -1$.

Then domain of $f/g = \left\{x \middle| x \text{ is a real number and } x \neq -\frac{5}{2}, x \neq -3, x \neq 1, \text{ and } x \neq -1\right\}$.

70. Answers may vary. $f(x) = \frac{1}{x+2}$, $g(x) = \frac{1}{x-5}$

71. Answers may vary.

72.

73.

74.

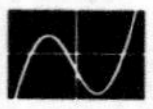

75. Because $y_2 = 0$ when $x = 3$, the domain of $y_3 =$ $\{x|x$ is a real number and $x \neq 3\}$. Since the graph produced using Connected mode contains the line $x = 3$, it does not represent y_3 accurately. The domain of the graph produced using Dot mode does not include 3, so it represents y_3 more accurately.

76.

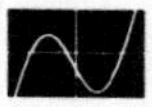

Chapter 3

Systems of Equations and Problem Solving

Exercise Set 3.1

1. ***Familiarize.*** Let x = the larger number and y = the smaller number.

   Translate.

   The difference between two numbers is 11.

   Rewording:

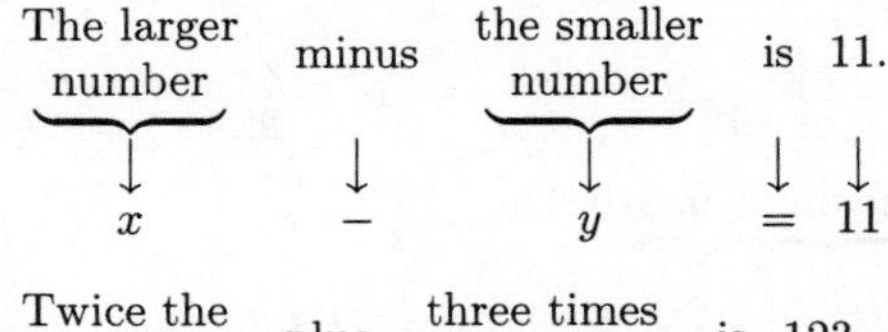

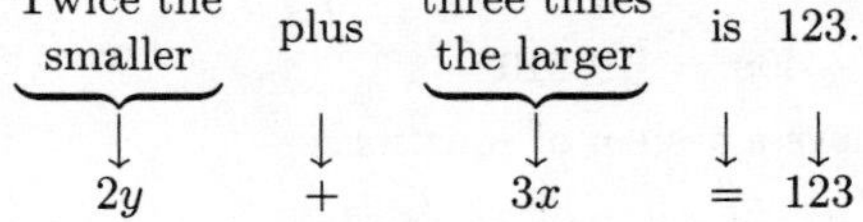

   We have a system of equations:

   $x - y = 11,$

   $3x + 2y = 123$

2. Let x = the first number and y = the second number.

   $x + y = -42,$

   $x - y = 52$

3. ***Familiarize.*** Let x = the number of less expensive brushes sold and y = the number of more expensive brushes sold.

   Translate. We organize the information in a table.

Kind of brush	Less expen-sive	More expen-sive	Total
Number sold	x	y	45
Price	\$8.50	\$9.75	
Amount taken in	$8.50x$	$9.75y$	398.75

   The "Number sold" row of the table gives us one equation:

   $x + y = 45$

   The "Amount taken in" row gives us a second equation:

   $8.50x + 9.75y = 398.75$

   We have a system of equations:

   $x + y = 45,$

   $8.50x + 9.75y = 398.75$

   We can multiply the second equation on both sides by 100 to clear the decimals:

   $x + y = 45,$

   $850x + 975y = 39,875$

4. Let x = the number of solid color neckwarmers sold and y = the number of print ones sold.

   $x + y = 40,$

   $9.9x + 12.75y = 421.65$

5. ***Familiarize.*** Let x = the measure of one angle and y = the measure of the other angle.

   Translate.

   Two angles are supplementary.

   Rewording: $\underbrace{\text{The sum of the measures}}_{x+y}$ $\underset{=}{\text{is}}$ $\underset{180}{180°.}$

   One angle is 3° less than twice the other.

   Rewording: $\underbrace{\text{One angle}}_{x}$ $\underset{=}{\text{is}}$ $\underbrace{\text{twice the other angle}}_{2y}$ $\underset{-}{\text{minus}}$ $\underset{3}{3°.}$

   We have a system of equations:

   $x + y = 180,$

   $x = 2y - 3$

6. Let x = the measure of the first angle and y = the measure of the second angle.

   $x + y = 90,$

   $x + \frac{1}{2}y = 64$

7. ***Familiarize.*** Let g = the number of field goals and t = the number of free throws Amma made.

   Translate. We organize the information in a table.

Kind of shot	Field goal	Free throw	Total
Number scored	g	t	18
Points per score	2	1	
Points scored	$2g$	t	30

   From the "Number scored" row of the table we get one equation:

   $g + t = 18$

   The "Points scored" row gives us another equation:

   $2g + t = 30$

We have a system of equations:

$$g + t = 18,$$
$$2g + t = 30$$

8. Let x = the number of children's plates and y = the number of adult's plates served.

$$x + y = 250,$$
$$3.5x + 7y = 1347.5$$

9. ***Familiarize.*** Let h = the number of vials of Humulin Insulin sold and n = the number of vials of Novolin Insulin sold.

Translate. We organize the information in a table.

Brand	Humulin	Novolin	Total
Number sold	h	n	65
Price	\$15.75	\$12.95	
Amount taken in	$15.75h$	$12.95n$	959.35

The "Number sold" row of the table gives us one equation:

$$h + n = 65$$

The "Amount taken in" row gives us a second equation:

$$15.75h + 12.95n = 959.35$$

We have a system of equations:

$$h + n = 65$$
$$15.75h + 12.95n = 959.35$$

We can multiply the second equation on both sides by 100 to clear the decimals:

$$h + n = 65$$
$$1575h + 1295n = 95,935$$

10. Let l = the length, in feet, and w = the width, in feet.

$$2l + 2w = 288,$$
$$l = w + 44$$

11. ***Familiarize.*** The tennis court is a rectangle with perimeter 228 ft. Let l = the length, in feet, and w = width, in feet. Recall that for a rectangle with length l and width w, the perimeter P is given by $P = 2l + 2w$.

Translate. The formula for perimeter gives us one equation:

$$2l + 2w = 228$$

The statement relating width and length gives us another equation:

The width is 42 ft less than the length.

$$w \quad = l - 42$$

We have a system of equations:

$$2l + 2w = 228,$$
$$w = l - 42$$

12. Let x = the number of 2-pointers scored and y = the number of 3-pointers scored.

$$x + y = 40,$$
$$2x + 3y = 89$$

13. ***Familiarize.*** Let l = the number of units of lumber produced and p = the number of units of plywood produced. Then the total profit from the lumber is $25l$ and the total profit from the plywood is $40p$.

Translate. The mill turns out twice as many units of plywood as lumber.

Rewording:

The number of units of plywood | is | twice the number of units of lumber.

$$p \quad = \quad 2l$$

The profit | is | \$10,920.

$$25l + 40p = 10,920$$

We have a system of equations:

$$p = 2l$$
$$25l + 40p = 10,920$$

14. Let x = the number of general interest films rented and y = the number of children's films rented.

$$x + y = 77,$$
$$3x + 1.5y = 213$$

15. ***Familiarize.*** Let w = the number of wins and t = the number of ties. Then the total number of points received from wins was $2w$ and the total number of points received from ties was t.

Translate.

The total number of points received | was | 60.

$$2w + t \quad = \quad 60$$

The number of wins | was | 9 | more than | the number of ties.

$$w \quad = \quad 9 \quad + \quad t$$

We have a system of equations:

$$2w + t = 60,$$
$$w = 9 + t$$

16. Let x = the number of 30-sec commercials and y = the number of 60-sec commercials.

$$x + y = 12,$$
$$30x + 60y = 600$$

17. ***Familiarize***. Let x = the number of ounces of lemon juice and y = the number of ounces of linseed oil to be used.

Translate.

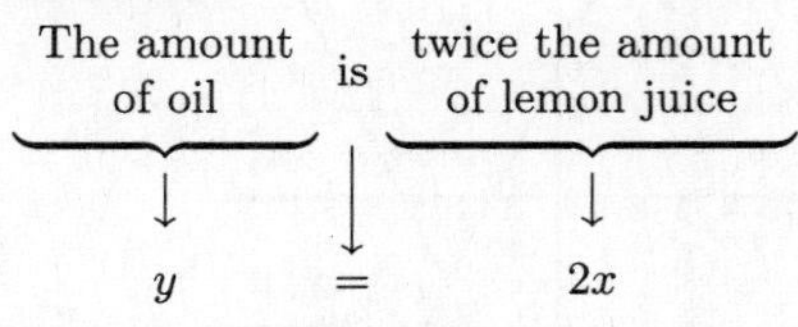

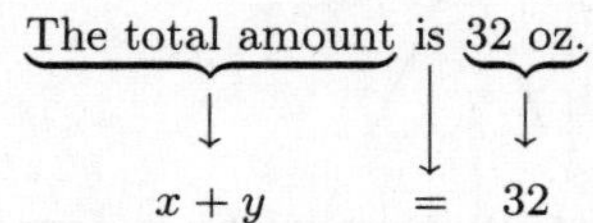

We have a system of equations:

$y = 2x,$

$x + y = 32$

18. Let c = the number of coach-class seats and f = the number of first-class seats.

$c + f = 152,$

$c = 5 + 6f$

19. We use alphabetical order for the variables. We replace x by 1 and y by 2.

$$\begin{array}{c|l} 4x - y = 2 & \\ \hline 4 \cdot 1 - 2 \ ? & 2 \\ 4 - 2 & \\ 2 & 2 \ \text{TRUE} \end{array} \qquad \begin{array}{c|l} 10x - 3y = 4 & \\ \hline 10 \cdot 1 - 3 \cdot 2 \ ? & 4 \\ 10 - 6 & \\ 4 & 4 \ \text{TRUE} \end{array}$$

The pair $(1, 2)$ makes both equations true, so it is a solution of the system.

20. Yes

21. We use alphabetical order for the variables. We replace x by 2 and y by 5.

$$\begin{array}{c|l} y = 3x - 1 & \\ \hline 5 \ ? & 3 \cdot 2 - 1 \\ & 6 - 1 \\ 5 & 5 \quad \text{TRUE} \end{array} \qquad \begin{array}{c|l} 2x + y = 4 & \\ \hline 2 \cdot 2 + 5 \ ? & 4 \\ 4 + 5 & \\ 9 & 4 \ \text{FALSE} \end{array}$$

The pair $(2, 5)$ is not a solution of $2x + y = 4$. Therefore, it is not a solution of the system of equations.

22. No

23. We replace x by 1 and y by 5.

$$\begin{array}{c|l} x + y = 6 & \\ \hline 1 + 5 \ ? & 6 \\ 6 & 6 \ \text{TRUE} \end{array} \qquad \begin{array}{c|l} y = 2x + 3 & \\ \hline 5 \ ? & 2 \cdot 1 + 3 \\ & 2 + 3 \\ 5 & 5 \quad \text{TRUE} \end{array}$$

The pair $(1, 5)$ makes both equations true, so it is a solution of the system.

24. Yes

25. We replace x by 3 and y by 1.

$$\begin{array}{c|l} 3x + 4y = 13 & \\ \hline 3 \cdot 3 + 4 \cdot 1 \ ? & 13 \\ 9 + 4 & \\ 13 & 13 \quad \text{TRUE} \end{array}$$

$$\begin{array}{c|l} 5x - 4y = 11 & \\ \hline 5 \cdot 3 - 4 \cdot 1 \ ? & 11 \\ 15 - 4 & \\ 11 & 11 \quad \text{TRUE} \end{array}$$

The pair $(3, 1)$ makes both equations true, so it is a solution of the system.

26. No

27. Graph both equations.

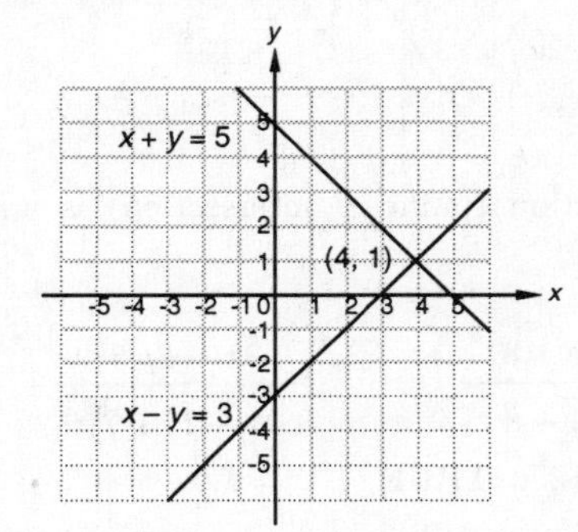

The solution (point of intersection) is apparently $(4, 1)$.

Check:

$$\begin{array}{c|l} x - y = 3 & \\ \hline 4 - 1 \ ? & 3 \\ 3 & 3 \quad \text{TRUE} \end{array} \qquad \begin{array}{c|l} x + y = 5 & \\ \hline 4 + 1 \ ? & 5 \\ 5 & 5 \quad \text{TRUE} \end{array}$$

The solution is $(4, 1)$.

28. (3,1)

29. Graph the equations.

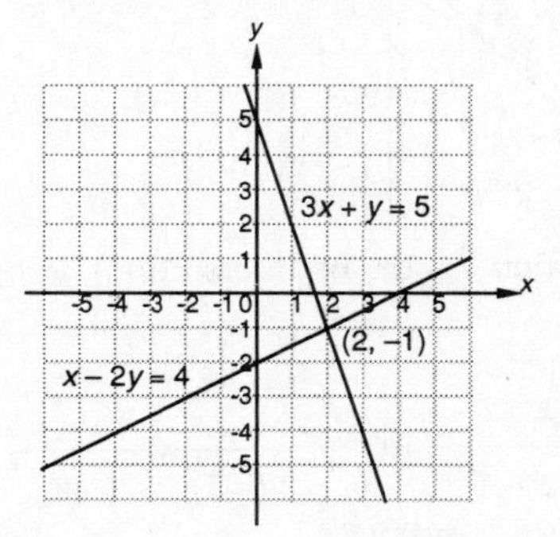

The solution (point of intersection) is apparently $(2, -1)$.

Check:

$3x + y = 5$		$x - 2y = 4$	
$3\cdot 2+(-1)$	? 5	$2-2(-1)$	? 4
$6 - 1$		$2 + 2$	
5	5 TRUE	4	4 TRUE

The solution is $(2, -1)$.

30. (3,2)

31. Graph both equations.

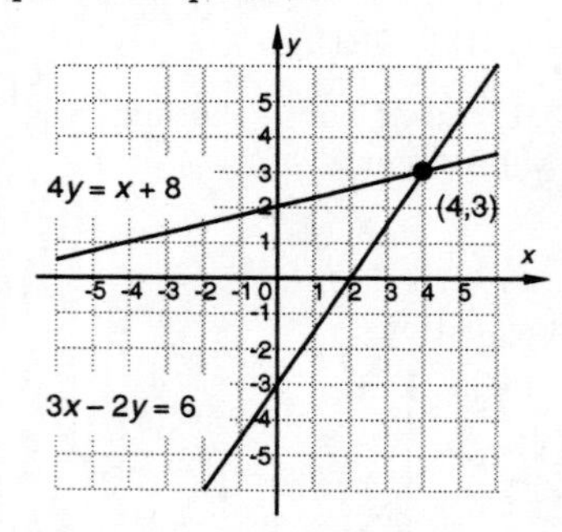

The solution (point of intersection) is apparently $(4, 3)$.

Check:

$4y = x + 8$		$3x - 2y = 6$	
$4 \cdot 3$	? $4 + 8$	$3 \cdot 4 - 2 \cdot 3$	? 6
12	12 TRUE	$12 - 6$	
		6	6 TRUE

The solution is $(4, 3)$.

32. $(1, -5)$

33. Graph both equations.

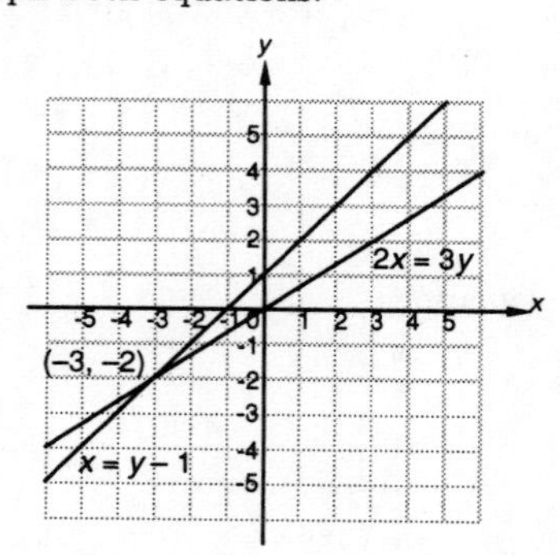

The solution (point of intersection) is apparently $(-3, -2)$.

Check:

$x = y - 1$		$2x = 3y$	
-3	? $-2 - 1$	$2(-3)$	? $3(-2)$
-3	-3 TRUE	-6	-6 TRUE

The solution is $(-3, -2)$.

34. (2,1)

35. Graph both equations.

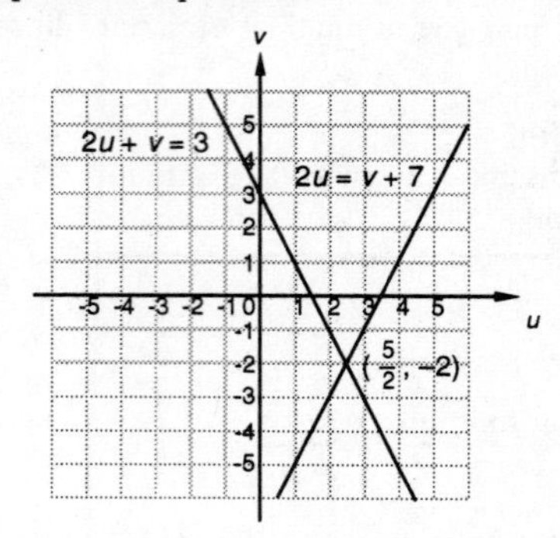

The solution (point of intersection) is apparently $\left(\frac{5}{2}, -2\right)$.

Check:

$2u + v = 3$		$2u = v + 7$	
$2 \cdot \frac{5}{2}$	? 3	$2 \cdot \frac{5}{2}$	? $-2 + 7$
$5 - 2$		5	5 TRUE
3	3 TRUE		

The solution is $\left(\frac{5}{2}, -2\right)$.

36. $(4, -5)$

37. Graph both equations.

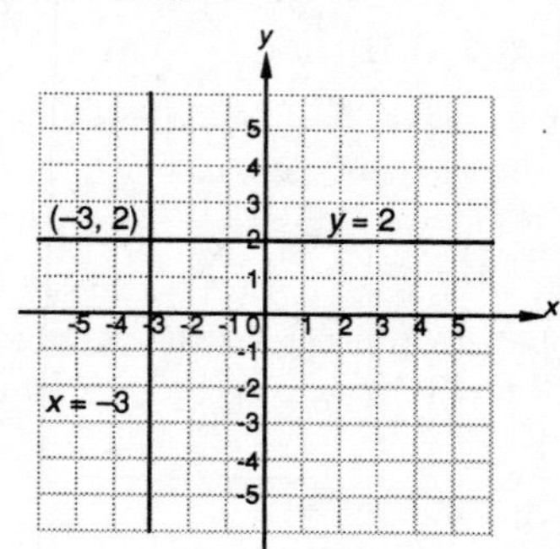

The ordered pair $(-3, 2)$ checks in both equations. It is the solution.

38. $\left(\frac{1}{2}, 3\right)$

39. Graph both equations.

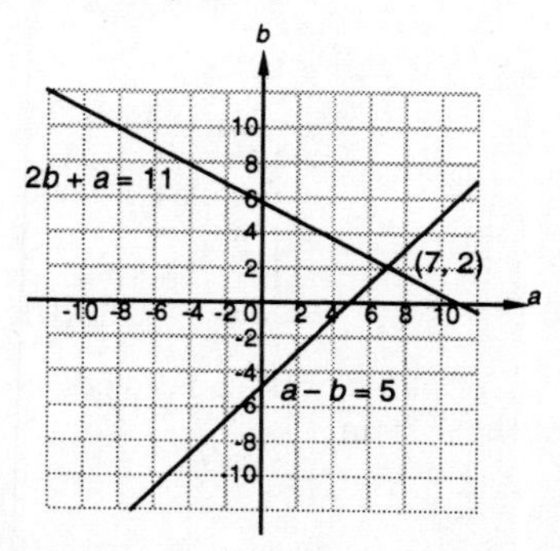

The solution (point of intersection) is apparently $(7,2)$.

Check:

$2b+a=11$		$a-b=5$	
$2\cdot 2+7$? 11		$7-2$? 5	
$4+7$		5	5 TRUE
11	11 TRUE		

The solution is (7,2).

40. $(3,-2)$

41. Graph both equations.

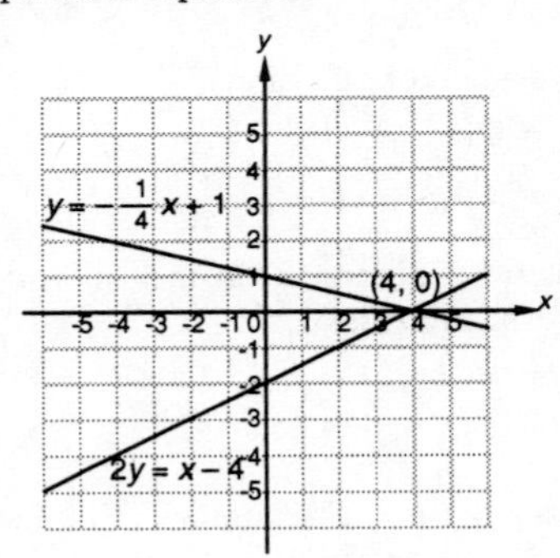

The solution (point of intersection) is apparently $(4,0)$.

Check:

$y=-\frac{1}{4}x+1$		$2y=x-4$	
0 ? $-\frac{1}{4}\cdot 4+1$		$2\cdot 0$? $4-4$	
$-1+1$		0	0 TRUE
0	0 TRUE		

The solution is (4,0).

42. No solution

43. Graph both equations.

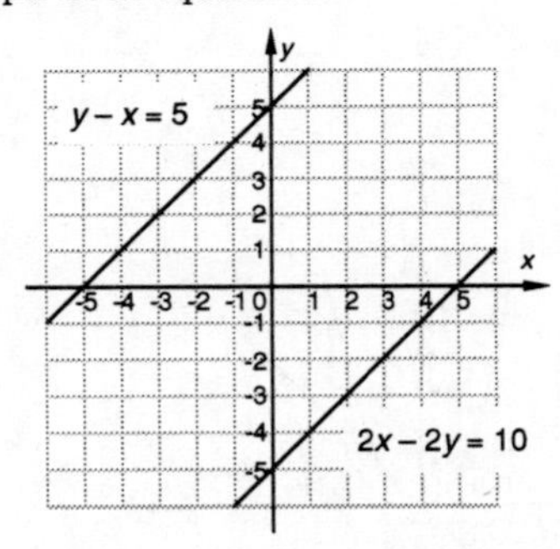

The lines are parallel. The system has no solution.

44. $(3,-4)$

45. Graph both equations.

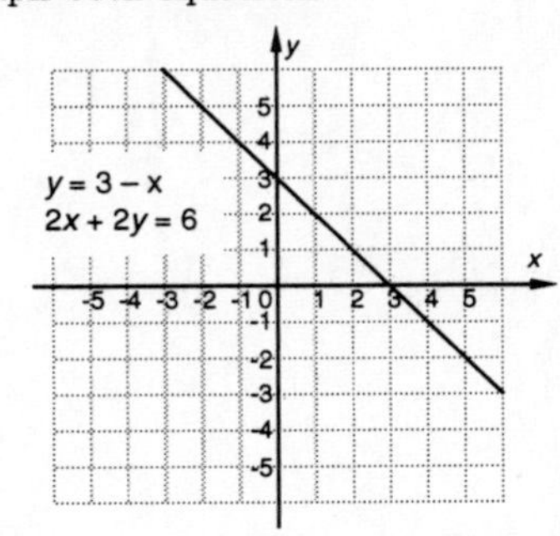

The graphs are the same. Any solution of one equation is a solution of the other. Each equation has infinitely many solutions. The solution set is the set of all pairs (x,y) for which $y=3-x$, or $\{(x,y)|y=3-x\}$. (In place of $y=3-x$ we could have used $2x+2y=6$ since the two equations are equivalent.)

46. $\{(x,y)|2x-3y=6\}$

47. A system of equations is consistent if it has at least one solution. Of the systems under consideration, only the one in Exercise 43 has no solution. Therefore, all except the system in Exercise 43 are consistent.

48. All except 42

49. A system of two equations in two variables is dependent if it has infinitely many solutions. Only the system in Exercise 45 is dependent.

50. 46

51. $3x+4=x-2$

$2x+4=-2$ Adding $-x$ on both sides

$2x=-6$ Adding -4 on both sides

$x=-3$ Multiplying by $\frac{1}{2}$ on both sides

The solution is -3.

52. -35

53. $4x-5x=8x-9+11x$

$-x=19x-9$ Collecting like terms

$-20x=-9$ Adding $-19x$ on both sides

$x=\frac{9}{20}$ Multiplying by $-\frac{1}{20}$ on both sides

The solution is $\frac{9}{20}$.

54. $b=a-4Q$

55. $3x-21=3\cdot x-3\cdot 7=3(x-7)$

56. $a(2+b-c)$

57. 

58.

59.

60. 1983, 1990, and 1991

61. 1989

62. 1987

63. a) There are many correct answers. One can be found by expressing the sum and difference of the two numbers:

$$x + y = 6,$$
$$x - y = 4$$

b) There are many correct answers. For example, write an equation in two variables. Then write a second equation by multiplying the left side of the first equation by one nonzero constant and multiplying the right side by another nonzero constant.

$$x + y = 1,$$
$$2x + 2y = 3$$

c) There are many correct answers. One can be found by writing an equation in two variables and then writing a nonzero constant multiple of that equation:

$$x + y = 1,$$
$$2x + 2y = 2$$

64. a) Answers may vary; $(4, -5)$; (b) Infinitely many

65. Substitute 4 for x and -5 for y in the first equation:

$$A(4) - 6(-5) = 13$$
$$4A + 30 = 13$$
$$4A = -17$$
$$A = -\frac{17}{4}$$

Substitute 4 for x and -5 for y in the second equation:

$$4 - B(-5) = -8$$
$$4 + 5B = -8$$
$$5B = -12$$
$$B = -\frac{12}{5}$$

We have $A = -\frac{17}{4}$, $B = -\frac{12}{5}$.

66. Let x = Burl's age now and y = his son's age now.

$$x = 2y,$$
$$x - 10 = 3(y - 10)$$

67. ***Familiarize.*** Let x = the number of years Lou has taught and y = the number of years Juanita has taught. Two years ago, Lou and Juanita had taught $x - 2$ and $y - 2$ years, respectively.

Translate.

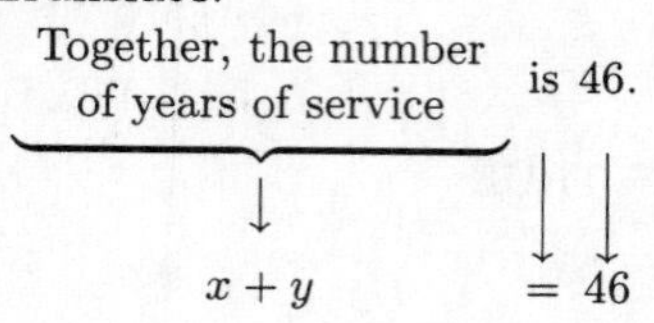

Two years ago
Lou had taught 2.5 times as many years as Juanita.

$$x - 2 = 2.5(y - 2)$$

We have a system of equations:

$$x + y = 46,$$
$$x - 2 = 2.5(y - 2)$$

68. Let l = the original length, in inches, and w = the original width, in inches.

$$2l + 2w = 156,$$
$$l = 4(w - 6)$$

69. ***Familiarize.*** Let b = the number of ounces of baking soda and v = the number of ounces of vinegar to be used. The amount of baking soda in the mixture will be four times the amount of vinegar.

Translate.

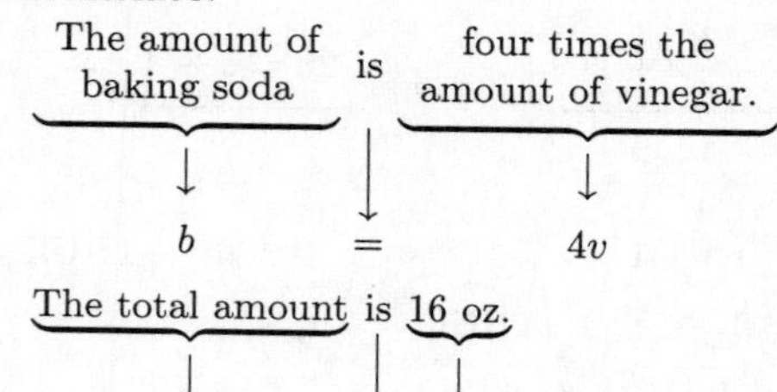

We have a system of equations.

$$b = 4v,$$
$$b + v = 16$$

70. $(-5, 5)$ $(3, 3)$

71. Graph both equations.

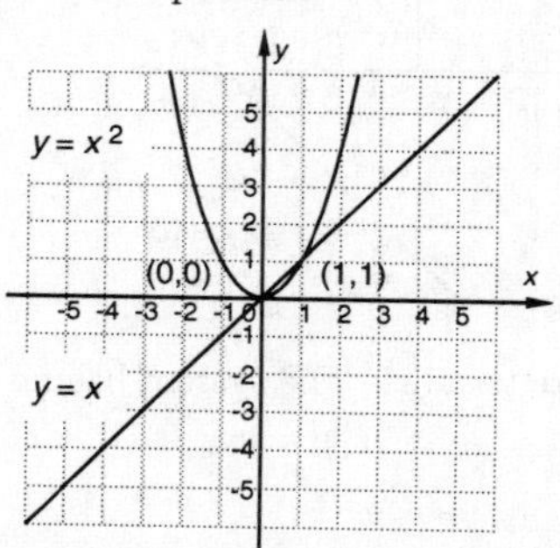

The solutions are apparently $(0, 0)$ and $(1, 1)$. Both pairs check.

72. $(-0.39, -1.10)$

73. $(0.07, -7.95)$

74. $(-0.13, 0.67)$

75. $(0.02, 1.25)$

Exercise Set 3.2

1. $3x + 5y = 3,\quad (1)$
$x = 8 - 4y \quad (2)$

We substitute $8 - 4y$ for x in the first equation and solve for y.

$$\begin{aligned} 3x + 5y &= 3 && (1)\\ 3(8 - 4y) + 5y &= 3 && \text{Substituting}\\ 24 - 12y + 5y &= 3\\ 24 - 7y &= 3\\ -7y &= -21\\ y &= 3 \end{aligned}$$

Next we substitute 3 for y in either equation of the original system and solve for x.

$$\begin{aligned} x &= 8 - 4y && (2)\\ x &= 8 - 4\cdot 3 && \text{Substituting}\\ x &= 8 - 12\\ x &= -4 \end{aligned}$$

We check the ordered pair $(-4, 3)$.

$$\begin{array}{r|l} \multicolumn{2}{c}{3x + 5y = 3}\\ \hline 3(-4) + 5\cdot 3 \ ? & 3\\ -12 + 15 & \\ 3 & 3 \quad \text{TRUE} \end{array}$$

$$\begin{array}{r|l} \multicolumn{2}{c}{x = 8 - 4y}\\ \hline -4 \ ? & 8 - 4\cdot 3\\ & 8 - 12\\ -4 & -4 \quad \text{TRUE} \end{array}$$

Since $(-4, 3)$ checks, it is the solution.

2. $(2, -3)$

3. $3x - 6 = y,\quad (1)$
$9x - 2y = 3 \quad (2)$

We substitute $3x - 6$ for y in the second equation and solve for x.

$$\begin{aligned} 9x - 2y &= 3 && (2)\\ 9x - 2(3x - 6) &= 3 && \text{Substituting}\\ 9x - 6x + 12 &= 3\\ 3x + 12 &= 3\\ 3x &= -9\\ x &= -3 \end{aligned}$$

Next we substitute -3 for x in either equation of the original system and solve for y.

$$\begin{aligned} 3x - 6 &= y && (1)\\ 3(-3) - 6 &= y && \text{Substituting}\\ -9 - 6 &= y\\ -15 &= y \end{aligned}$$

We check the ordered pair $(-3, -15)$.

$$\begin{array}{r|l} \multicolumn{2}{c}{3x - 6 = y}\\ \hline 3(-3) - 6 \ ? & -15\\ -9 - 6 & \\ -15 & -15 \quad \text{TRUE} \end{array}$$

$$\begin{array}{r|l} \multicolumn{2}{c}{9x - 2y = 3}\\ \hline 9(-3) - 2(-15) \ ? & 3\\ -27 + 30 & \\ 3 & 3 \quad \text{TRUE} \end{array}$$

Since $(-3, -15)$ checks, it is the solution.

4. $\left(\frac{21}{5}, \frac{12}{5}\right)$

5. $4x + y = 1,\quad (1)$
$x - 2y = 16 \quad (2)$

We solve the second equation for x.

$$\begin{aligned} x - 2y &= 16 && (2)\\ x &= 2y + 16 && (3) \end{aligned}$$

We substitute $2y + 16$ for x in the first equation and solve for y.

$$\begin{aligned} 4x + y &= 1 && (1)\\ 4(2y + 16) + y &= 1 && \text{Substituting}\\ 8y + 64 + y &= 1\\ 9y + 64 &= 1\\ 9y &= -63\\ y &= -7 \end{aligned}$$

Now we substitute -7 for y in Equation (1), (2), or (3). It is easiest to use Equation (3) since it is already solved for y.

$$x = 2(-7) + 16 = 2$$

We check the ordered pair $(2, -7)$.

$$\begin{array}{r|l} \multicolumn{2}{c}{4x + y = 1}\\ \hline 4\cdot 2 + (-7) \ ? & 1\\ 8 - 7 & \\ 1 & 1 \quad \text{TRUE} \end{array}$$

$$\begin{array}{r|l} \multicolumn{2}{c}{x - 2y = 16}\\ \hline 2 - 2(-7) \ ? & 16\\ 2 + 14 & \\ 16 & 16 \quad \text{TRUE} \end{array}$$

Since $(2, -7)$ checks, it is the solution.

6. $(2, -2)$

7. $-3b + a = 7$, (1)

$5a + 6b = 14$ (2)

We solve the first equation for a.

$$\begin{aligned} -3b + a &= 7 && (1) \\ a &= 3b + 7 && (3) \end{aligned}$$

Substitute $3b + 7$ for a in the second equation and solve for b.

$$\begin{aligned} 5a + 6b &= 14 && (2) \\ 5(3b + 7) + 6b &= 14 && \text{Substituting} \\ 15b + 35 + 6b &= 14 \\ 21b + 35 &= 14 \\ 21b &= -21 \\ b &= -1 \end{aligned}$$

Now substitute -1 for b in Equation (3).

$$a = 3(-1) + 7 = 4$$

We check the ordered pair $(4, -1)$.

$$\begin{array}{r|l} \multicolumn{2}{c}{-3b + a = 7} \\ \hline -3(-1) + 4 & ?\ 7 \\ 3 + 4 & \\ 7 & 7 \quad \text{TRUE} \end{array}$$

$$\begin{array}{r|l} \multicolumn{2}{c}{5a + 6b = 14} \\ \hline 5 \cdot 4 + 6(-1) & ?\ 14 \\ 20 - 6 & \\ 14 & 14 \quad \text{TRUE} \end{array}$$

Since $(4, -1)$ checks, it is the solution.

8. $(-2, 1)$

9. $5p + 7q = 1$, (1)

$4p - 2q = 16$ (2)

We solve the second equation for q.

$$\begin{aligned} 4p - 2q &= 16 && (2) \\ -2q &= -4p + 16 \\ q &= 2p - 8 && (3) \end{aligned}$$

Substitute $2p - 8$ for q in the first equation and solve for p.

$$\begin{aligned} 5p + 7q &= 1 && (1) \\ 5p + 7(2p - 8) &= 1 && \text{Substituting} \\ 5p + 14p - 56 &= 1 \\ 19p - 56 &= 1 \\ 19p &= 57 \\ p &= 3 \end{aligned}$$

Now we substitute 3 for p in Equation (3).

$$q = 2 \cdot 3 - 8 = -2$$

We check the ordered pair $(3, -2)$.

$$\begin{array}{r|l} \multicolumn{2}{c}{5p + 7q = 1} \\ \hline 5 \cdot 3 + 7(-2) & ?\ 1 \\ 15 - 14 & \\ 1 & 1 \quad \text{TRUE} \end{array}$$

$$\begin{array}{r|l} \multicolumn{2}{c}{4p - 2q = 16} \\ \hline 4 \cdot 3 - 2(-2) & ?\ 16 \\ 12 + 4 & \\ 16 & 16 \quad \text{TRUE} \end{array}$$

Since $(3, -2)$ checks, it is the solution.

10. $\left(\frac{1}{2}, \frac{1}{2}\right)$

11. $5x + 3y = 4$, (1)

$x - 4y = 3$ (2)

We solve the second equation for x.

$$\begin{aligned} x - 4y &= 3 && (2) \\ x &= 4y + 3 && (3) \end{aligned}$$

Substitute $4y + 3$ for x in the first equation and solve for y.

$$\begin{aligned} 5x + 3y &= 4 && (1) \\ 5(4y + 3) + 3y &= 4 && \text{Substituting} \\ 20y + 15 + 3y &= 4 \\ 23y + 15 &= 4 \\ 23y &= -11 \\ y &= -\frac{11}{23} \end{aligned}$$

Now we substitute $-\frac{11}{23}$ for y in Equation (3).

$$x = 4\left(-\frac{11}{23}\right) + 3 = -\frac{44}{23} + 3 = \frac{25}{23}$$

The ordered pair $\left(\frac{25}{23}, -\frac{11}{23}\right)$ checks in both equations. It is the solution.

12. $\left(\frac{19}{8}, \frac{1}{8}\right)$

13. $y - 2x = 1$, (1)

$2x - 3 = y$ (2)

We substitute $2x - 3$ for y in the first equation and solve for x.

$$\begin{aligned} y - 2x &= 1 && (1) \\ 2x - 3 - 2x &= 1 && \text{Substituting} \\ -3 &= 1 && \text{Collecting like terms} \end{aligned}$$

We have a false equation. Therefore, there is no solution.

14. No solution

15. $\begin{array}{r} x + 3y = \ \ 7 \quad (1) \\ -x + 4y = \ \ 7 \quad (2) \\ \hline 0 + 7y = 14 \end{array}$ Adding

$7y = 14$

$y = 2$

Substitute 2 for y in one of the original equations and solve for x.

$x + 3y = 7 \quad (1)$

$x + 3 \cdot 2 = 7$ Substituting

$x + 6 = 7$

$x = 1$

Check:

$x + 3y = 7$			$-x + 4y = 7$		
$1 + 3 \cdot 2$	?	7	$-1 + 4 \cdot 2$	?	7
$1 + 6$			$-1 + 8$		
7		7 TRUE	7		7 TRUE

Since $(1, 2)$ checks, it is the solution.

16. $(2, 7)$

17. $\begin{array}{r} 2x + \ \ y = 6 \quad (1) \\ x - \ \ y = 3 \quad (2) \\ \hline 3x + \ \ 0 = 9 \end{array}$ Adding

$3x = 9$

$x = 3$

Substitute 3 for x in one of the original equations and solve for y.

$2x + y = 6 \quad (1)$

$2 \cdot 3 + y = 6$ Substituting

$6 + y = 6$

$y = 0$

We obtain $(3, 0)$. This checks, so it is the solution.

18. $(10, 2)$

19. $\begin{array}{r} 6x - 3y = \ \ 18 \quad (1) \\ 6x + 3y = -12 \quad (2) \\ \hline 12x + \ \ 0 = \ \ \ 6 \end{array}$ Adding

$x = \frac{1}{2}$

Substitute $\frac{1}{2}$ for x in Equation (2) and solve for y.

$6x + 3y = -12$

$6\left(\frac{1}{2}\right) + 3y = -12$ Substituting

$3 + 3y = -12$

$3y = -15$

$y = -5$

We obtain $\left(\frac{1}{2}, -5\right)$. This checks, so it is the solution.

20. $(-1, 2)$

21. $3x + 2y = 3, \quad (1)$

$9x - 8y = -2 \quad (2)$

We multiply Equation (1) by 4 to make two terms become opposites.

$\begin{array}{r} 12x + 8y = 12 \\ 9x - 8y = -2 \\ \hline 21x + \ \ 0 = 10 \end{array}$ Multiplying (1) by 4; Adding

$x = \frac{10}{21}$

Substitute $\frac{10}{21}$ for x in Equation (1) and solve for y.

$3x + 2y = 3$

$3\left(\frac{10}{21}\right) + 2y = 3$ Substituting

$\frac{10}{7} + 2y = 3$

$2y = \frac{11}{7}$

$y = \frac{11}{14}$

We obtain $\left(\frac{10}{21}, \frac{11}{14}\right)$. This checks, so it is the solution.

22. $\left(\frac{128}{31}, -\frac{17}{31}\right)$

23. $5x - 7y = -16, \quad (1)$

$2x + 8y = 26 \quad (2)$

We multiply twice to make two terms become opposites.

From (1): $40x - 56y = -128$ Multiplying by 8

From (2): $14x + 56y = \ \ 182$ Multiplying by 7

$54x + \ \ 0 = \ \ 54$ Adding

$x = 1$

Substitute 1 for x in Equation (2) and solve for y.

$2x + 8y = 26$

$2 \cdot 1 + 8y = 26$ Substituting

$2 + 8y = 26$

$8y = 24$

$y = 3$

We obtain $(1, 3)$. This checks, so it is the solution.

24. $(6, 2)$

25. $0.7x - 0.3y = 0.5,$

$-0.4x + 0.7y = 1.3$

We first multiply each equation by 10 to clear decimals.

$7x - 3y = 5, \quad (1)$

$-4x + 7y = 13 \quad (2)$

We multiply so that the y-terms can be eliminated.

$$\begin{array}{lrl} \text{From (1):} & 49x - 21y = 35 & \text{Multiplying by 7} \\ \text{From (2):} & -12x + 21y = 39 & \text{Multiplying by 3} \\ \hline & 37x + \quad 0 = 74 & \text{Adding} \\ & x = 2 & \end{array}$$

Substitute 2 for x in one of the equations in which the decimals were cleared and solve for y.

$$\begin{aligned} -4x + 7y &= 13 \quad (2) \\ -4 \cdot 2 + 7y &= 13 \quad \text{Substituting} \\ -8 + 7y &= 13 \\ 7y &= 21 \\ y &= 3 \end{aligned}$$

We obtain $(2, 3)$. This checks, so it is the solution.

26. $\left(\frac{140}{13}, -\frac{50}{13}\right)$

27. $6x + 7y = 9, \quad (1)$

$8x + 9y = 11 \quad (2)$

We multiply so that the x-terms can be eliminated.

$$\begin{array}{lrl} \text{From (1):} & 24x + 28y = 36 & \text{Multiplying by 4} \\ \text{From (2):} & -24x - 27y = -33 & \text{Multiplying by } -3 \\ \hline & 0 + \quad y = 3 & \text{Adding} \\ & y = 3 & \end{array}$$

Substitute 3 for y in Equation (1) and solve for x.

$$\begin{aligned} 6x + 7y &= 9 \\ 6x + 7 \cdot 3 &= 9 \quad \text{Substituting} \\ 6x + 21 &= 9 \\ 6x &= -12 \\ x &= -2 \end{aligned}$$

We obtain $(-2, 3)$. This checks, so it is the solution.

28. $(3, -1)$

29. $\frac{1}{3}x + \frac{1}{5}y = 7, \quad (1)$

$\frac{1}{6}x - \frac{2}{5}y = -4 \quad (2)$

First we multiply each equation by the LCM of the denominators to clear the fractions.

$5x + 3y = 105, \quad (3) \quad \text{Multiplying (1) by 15}$

$5x - 12y = -120 \quad (4) \quad \text{Multiplying (2) by 30}$

Then we multiply Equation (4) by -1 so that the x-terms can be eliminated.

$$\begin{array}{rl} 5x + 3y = 105 & (3) \\ -5x + 12y = 120 & \text{Multiplying (4) by } -1 \\ \hline 0 + 15y = 225 & \text{Adding} \\ y = 15 & \end{array}$$

Substitute 15 for y in one of the equations in which fractions were cleared and solve for y.

$$\begin{aligned} 5x + 3y &= 105 \quad (3) \\ 5x + 3 \cdot 15 &= 105 \quad \text{Substituting} \\ 5x + 45 &= 105 \\ 5x &= 60 \\ x &= 12 \end{aligned}$$

We obtain $(12, 15)$. This checks, so it is the solution.

30. $\left(\frac{110}{19}, -\frac{12}{19}\right)$

31. $6x + 10y = 14, \quad (1)$

$3x + 5y = 7 \quad (2)$

We multiply Equation (2) by -2.

$$\begin{array}{r} 6x + 10y = 14, \\ -6x - 10y = -14 \\ \hline 0 = 0 \end{array}$$

We have a true equation. If a pair solves Equation (1), then it will also solve Equation (2). The system is dependent and the solution set is infinite. It can be expressed as $\{(x, y) | 3x + 5y = 7\}$.

32. $\{(x, y) | -4x + 2y = 5\}$

33. $a - 2b = 16, \quad (1)$

$b + 3 = 3a \quad (2)$

We will use the substitution method. First solve Equation (1) for a.

$$\begin{aligned} a - 2b &= 16 \\ a &= 2b + 16 \quad (3) \end{aligned}$$

Now substitute $2b + 16$ for a in Equation (2) and solve for b.

$$\begin{aligned} b + 3 &= 3a \quad (2) \\ b + 3 &= 3(2b + 16) \quad \text{Substituting} \\ b + 3 &= 6b + 48 \\ -45 &= 5b \\ -9 &= b \end{aligned}$$

Substitute -9 for b in Equation (3).

$a = 2(-9) + 16 = -2$

We obtain $(-2, -9)$. This checks, so it is the solution.

34. $\left(\frac{1}{2}, -\frac{1}{2}\right)$

35. $10x + y = 306, \quad (1)$

$10y + x = 90 \quad (2)$

We will use the substitution method. First solve Equation (1) for y.

$$\begin{aligned} 10x + y &= 306 \\ y &= -10x + 306 \quad (3) \end{aligned}$$

Now substitute $-10x+306$ for y in Equation (2) and solve for y.

$$\begin{aligned} 10y + x &= 90 \quad (2) \\ 10(-10x+306) + x &= 90 \quad \text{Substituting} \\ -100x + 3060 + x &= 90 \\ -99x + 3060 &= 90 \\ -99x &= -2970 \\ x &= 30 \end{aligned}$$

Substitute 30 for x in Equation (3).

$y = -10 \cdot 30 + 306 = 6$

We obtain $(30, 6)$. This checks, so it is the solution.

36. $\left(-\frac{4}{3}, -\frac{19}{3}\right)$

37. $3y = x - 2$, (1)

$x = 2 + 3y$ (2)

We will use the substitution method. Substitute $2 + 3y$ for x in the first equation and solve for y.

$$\begin{aligned} 3y &= x - 2 \quad (1) \\ 3y &= 2 + 3y - 2 \quad \text{Substituting} \\ 3y &= 3y \quad \text{Collecting like terms} \end{aligned}$$

We get a true equation. The system is dependent and the solution set is infinite. It can be expressed as $\{(x, y)|x = 2 + 3y\}$.

38. No solution

39. $2x - 7y = 9$, (1)

$7y - 2x = -5$ (2)

$$\begin{aligned} 2x - 7y &= 9 \quad (1) \\ -2x + 7y &= -5 \quad \text{Rewriting (2)} \\ \hline 0 &= 4 \quad \text{Adding} \end{aligned}$$

We get a false equation, so the system has no solution.

40. No solution

41. $0.05x + 0.25y = 22$, (1)

$0.15x + 0.05y = 24$ (2)

We first multiply each equation by 100 to clear decimals.

$5x + 25y = 2200$

$15x + 5y = 2400$

We multiply by -5 on both sides of the second equation and add.

$$\begin{aligned} 5x + 25y &= 2200 \\ -75x - 25y &= -12,000 \quad \text{Multiplying (2) by } -5 \\ \hline -70x &= -9800 \quad \text{Adding} \\ x &= \frac{-9800}{-70} \\ x &= 140 \end{aligned}$$

Substitute 140 for x in one of the equations in which the decimals were cleared and solve for y.

$$\begin{aligned} 5x + 25y &= 2200 \quad (1) \\ 5 \cdot 140 + 25y &= 2200 \quad \text{Substituting} \\ 700 + 25y &= 2200 \\ 25y &= 1500 \\ y &= 60 \end{aligned}$$

We obtain $(140, 60)$. This checks, so it is the solution.

42. $(10, 5)$

43. ***Familiarize***. Recall the formula for simple interest, $I = Prt$, where I = interest, P = principal, r = interest rate, and t = time (in years).

Translate. Substitute \$17.60 for I, \$320 for P, and $\frac{1}{2}$ for t.

$$\begin{aligned} I &= Prt \\ 17.60 &= 320r\left(\frac{1}{2}\right) \end{aligned}$$

Carry out. Solve the equation.

$$\begin{aligned} 17.60 &= 320r\left(\frac{1}{2}\right) \\ 17.60 &= 160r \\ \frac{17.60}{160} &= r \\ 0.11 &= r \end{aligned}$$

Check. $320(0.11)\left(\frac{1}{2}\right) = 17.60$. The answer checks.

State. The simple interest rate is 0.11, or 11%.

44. $-15y - 39$

45. $3x - 14 = x + 2(x - 7)$

$$\begin{aligned} 3x - 14 &= x + 2x - 14 \\ 3x - 14 &= 3x - 14 \\ -14 &= -14 \quad \text{Adding } -3x \text{ on both sides} \end{aligned}$$

We get a true equation, so any real number is a solution. The solution set is the set of all real numbers.

46. $\emptyset$

47. ◈

48. ◈

49. ◈

50. ◈

51. First write $f(x) = mx + b$ as $y = mx + b$. Then substitute 1 for x and 2 for y to get one equation and also substitute -3 for x and 4 for y to get a second equation:

$2 = m \cdot 1 + b$

$4 = m(-3) + b$

Solve the resulting system of equations.

$2 = m + b$

$4 = -3m + b$

Multiply the second equation by -1 and add.

$$\begin{aligned} 2 &= m + b \\ -4 &= 3m - b \\ \hline -2 &= 4m \\ -\frac{1}{2} &= m \end{aligned}$$

Substitute $-\frac{1}{2}$ for m in the first equation and solve for b.

$2 = -\frac{1}{2} + b$

$\frac{5}{2} = b$

Thus, $m = -\frac{1}{2}$ and $b = \frac{5}{2}$.

52. $p = 2,\ q = -\frac{1}{3}$

53. Substitute -4 for x and -3 for y in both equations and solve for a and b.

$-4a - 3b = -26,$ (1)

$-4b + 3a = 7$ (2)

$$\begin{aligned} -12a - 9b &= -78 \quad \text{Multiplying (1) by 3} \\ 12a - 16b &= 28 \quad \text{Multiplying (2) by 4} \\ \hline -25b &= -50 \\ b &= 2 \end{aligned}$$

Substitute 2 for b in Equation (2).

$-4 \cdot 2 + 3a = 7$

$3a = 15$

$a = 5$

Thus, $a = 5$ and $b = 2$.

54. $\left(\frac{a+2b}{7}, \frac{a-5b}{7}\right)$

55. $\frac{x+y}{2} - \frac{x-y}{5} = 1,$

$\frac{x-y}{2} + \frac{x+y}{6} = -2$

After clearing fractions we have:

$3x + 7y = 10,$ (1)

$4x - 2y = -12$ (2)

$$\begin{aligned} 6x + 14y &= 20 \quad \text{Multiplying (1) by 2} \\ 28x - 14y &= -84 \quad \text{Multiplying (2) by 7} \\ \hline 34x &= -64 \\ x &= -\frac{32}{17} \end{aligned}$$

Substitute $-\frac{32}{17}$ for x in Equation (1).

$3\left(-\frac{32}{17}\right) + 7y = 10$

$7y = \frac{266}{17}$

$y = \frac{38}{17}$

The solution is $\left(-\frac{32}{17}, \frac{38}{17}\right)$.

56. $(23.118879, -12.039964)$

57. $\frac{2}{x} + \frac{1}{y} = 0,$ or $2 \cdot \frac{1}{x} + \frac{1}{y} = 0,$

$\frac{5}{x} + \frac{2}{y} = -5$ $\quad 5 \cdot \frac{1}{x} + 2 \cdot \frac{1}{y} = -5$

Substitute u for $\frac{1}{x}$ and v for $\frac{1}{y}$.

$2u + v = 0,$ (1)

$5u + 2v = -5$ (2)

$$\begin{aligned} -4u - 2v &= 0 \quad \text{Multiplying (1) by } -2 \\ 5u + 2v &= -5 \quad (2) \\ \hline u &= -5 \end{aligned}$$

Substitute -5 for u in Equation (1).

$2(-5) + v = 0$

$-10 + v = 0$

$v = 10$

If $u = -5$, then $\frac{1}{x} = -5$. Thus $x = -\frac{1}{5}$.

If $v = 10$, then $\frac{1}{y} = 10$. Thus $y = \frac{1}{10}$.

The solution is $\left(-\frac{1}{5}, \frac{1}{10}\right)$.

58. $\left(-\frac{1}{4}, -\frac{1}{2}\right)$

59.

Exercise Set 3.3

1. The Familiarize and Translate steps were done in Exercise 1 of Exercise Set 3.1

Carry out. We solve the system of equations

$x - y = 11,$ (1)

$3x + 2y = 123$ (2)

where x = the larger number and y = the smaller number. We use elimination.

$$\begin{aligned} 2x - 2y &= 22 \quad \text{Multiplying (1) by 2} \\ 3x + 2y &= 123 \\ \hline 5x &= 145 \\ x &= 29 \end{aligned}$$

Substitute 29 for x in (1) and solve for y.

$$29 - y = 11$$
$$-y = -18$$
$$y = 18$$

Check. The difference between the numbers is $29 - 18$, or 11. Also $2 \cdot 18 + 3 \cdot 29 = 36 + 87 = 123$. The numbers check.

State. The larger number is 29, and the smaller is 18.

2. 5, -47

3. The Familiarize and Translate steps were done in Exercise 3 of Exercise Set 3.1

Carry out. We solve the system of equations

$$x + y = 45, \quad (1)$$
$$850x + 975y = 39,875 \quad (2)$$

where x = the number of less expensive brushes sold and y = the number of more expensive brushes sold. We use elimination. Begin by multiplying Equation (1) by -850.

$$-850x - 850y = -38,250 \quad \text{Multiplying (1)}$$
$$850x + 975y = 39,875$$
$$125y = 1625$$
$$y = 13$$

Substitute 13 for y in (1) and solve for x.

$$x + 13 = 45$$
$$x = 32$$

Check. The number of brushes sold is $32 + 13$, or 45. The amount taken in was $\$8.50(32) + \$9.75(13) = \$272 + \$126.75 = \$398.75$. The answer checks.

State. 32 of the less expensive brushes were sold, and 13 of the more expensive brushes were sold.

4. 31 solid, 9 print

5. The Familiarize and Translate steps were done in Exercise 5 of Exercise Set 3.1

Carry out. We solve the system of equations

$$x + y = 180, \quad (1)$$
$$x = 2y - 3 \quad (2)$$

where x = the measure of one angle and y = the measure of the other angle. We use substitution.

Substitute $2y - 3$ for x in (1) and solve for y.

$$2y - 3 + y = 180$$
$$3y - 3 = 180$$
$$3y = 183$$
$$y = 61$$

Now substitute 61 for y in (2).

$$x = 2 \cdot 61 - 3 = 122 - 3 = 119$$

Check. The sum of the angle measures is $119° + 61°$, or $180°$, so the angles are supplementary. Also $2 \cdot 61° - 3° = 122° - 3° = 119°$. The answer checks.

State. The measures of the angles are 119° and 61°.

6. 38°, 52°

7. The Familiarize and Translate steps were done in Exercise 7 of Exercise Set 3.1

Carry out. We solve the system of equations

$$g + t = 18, \quad (1)$$
$$2g + t = 30 \quad (2)$$

where g = the number of field goals and t = the number of free throws Amma made. We use elimination.

$$-g - t = -18 \quad \text{Multiplying (1) by } -1$$
$$2g + t = 30$$
$$g = 12$$

Substitute 12 for g in (1) and solve for t.

$$12 + t = 18$$
$$t = 6$$

Check. The total number of scores was $12+6$, or 18. The total number of points was $2 \cdot 12 + 6 = 24 + 6 = 30$. The answer checks.

State. Amma made 12 field goals and 6 free throws.

8. 115 children's plates, 135 adult's plates

9. The Familiarize and Translate steps were done in Exercise 9 of Exercise Set 3.1

Carry out. We solve the system of equations

$$h + n = 65, \quad (1)$$
$$1575h + 1295n = 95,935 \quad (2)$$

where h = the number of vials of Humulin Insulin sold and n = the number of vials of Novolin Insulin sold. We use elimination.

$$-1295h - 1295n = -84,175 \quad \text{Multiplying (1) by } -1295$$
$$1575h + 1295n = 95,935$$
$$280h = 11,760$$
$$h = 42$$

Substitute 42 for h in (1) and solve for n.

$$42 + n = 65$$
$$n = 23$$

Check. A total of $42 + 23$, or 65 vials, was sold. The amount collected was $\$15.75(42) + \$12.95(23) = \$661.50 + \$297.85 = \$959.35$. The answer checks.

State. 42 vials of Humulin Insulin and 23 vials of Novolin Insulin were sold.

10. Width: 50 ft, length: 94 ft

11. The Familiarize and Translate steps were done in Exercise 11 of Exercise Set 3.1

Carry out. We solve the system of equations

$$2l + 2w = 228, \quad (1)$$
$$w = l - 42 \quad (2)$$

where l = the length, in feet, and w = the width, in feet, of the tennis court. We use substitution.

Substitute $l - 42$ for w in (1) and solve for l.

$$\begin{aligned} 2l + 2(l-42)w &= 228 \\ 2l + 2l - 84 &= 228 \\ 4l - 84 &= 228 \\ 4l &= 312 \\ l &= 78 \end{aligned}$$

Now substitute 78 for l in (2).

$$w = 78 - 42 = 36$$

Check. The perimeter is $2 \cdot 78 \text{ ft} + 2 \cdot 36 \text{ ft} = 156 \text{ ft} + 72 \text{ ft} = 228 \text{ ft}$. The width, 36 ft, is 42 ft less than the length, 78 ft. The answer checks.

State. The length of the tennis court is 78 ft, and the width is 36 ft.

12. 31 two-point, 9 three-point

13. The Familiarize and Translate steps were done in Exercise 13 of Exercise Set 3.1.

Carry out. We solve the system of equations

$$\begin{aligned} p &= 2l, & (1) \\ 25l + 40p &= 10,920 & (2) \end{aligned}$$

where l = the number of units of lumber produced and p = the number of units of plywood produced. We use substitution.

Substitute $2l$ for p in (2) and solve for l.

$$\begin{aligned} 25l + 40 \cdot 2l &= 10,920 \\ 25l + 80l &= 10,920 \\ 105l &= 10,920 \\ l &= 104 \end{aligned}$$

Substitute 104 for l in (1).

$$p = 2 \cdot 104 = 208$$

Check. The number of units of plywood, 208, is twice the number of units of lumber, 104. The profit is $\$25 \cdot 104 + \$40 \cdot 208 = \$2600 + \$8320 = \$10,920$. The answer checks.

State. 104 units of lumber and 208 units of plywood must be produced.

14. 65 general interest, 12 children's

15. The Familiarize and Translate steps were done in Exercise 15 of Exercise Set 3.1.

Carry out. We solve the system of equations

$$\begin{aligned} 2w + t &= 60, & (1) \\ w &= 9 + t & (2) \end{aligned}$$

where w = the number of wins and t = the number of ties. We use substitution.

Substitute $9 + t$ for w in (1) and solve for t.

$$\begin{aligned} 2(9+t) + t &= 60 \\ 18 + 2t + t &= 60 \\ 18 + 3t &= 60 \\ 3t &= 42 \\ t &= 14 \end{aligned}$$

Now substitute 14 for t in (2).

$$w = 9 + 14 = 23$$

Check. The total number of points is $2 \cdot 23 + 14 = 46 + 14 = 60$. The number of wins, 23, is nine more than the number of ties, 14. The answer checks.

State. The Wildcats had 23 wins and 14 ties.

16. 4 30-sec, 8 60-sec

17. The Familiarize and Translate steps were done in Exercise 17 of Exercise Set 3.1.

Carry out. We solve the system of equations

$$\begin{aligned} y &= 2x, & (1) \\ x + y &= 32 & (2) \end{aligned}$$

where x = the number of ounces of lemon juice and y = the number of ounces of linseed oil to be used. We use substitution.

Substitute $2x$ for y in (2) and solve for x.

$$\begin{aligned} x + 2x &= 32 \\ 3x &= 32 \\ x &= \frac{32}{3}, \text{ or } 10\frac{2}{3} \end{aligned}$$

Now substitute $\frac{32}{3}$ for x in (1).

$$y = 2 \cdot \frac{32}{3} = \frac{64}{3}, \text{ or } 21\frac{1}{3}$$

Check. The amount of oil, $\frac{64}{3}$ oz, is twice the amount of lemon juice, $\frac{32}{3}$ oz. The mixture contains $\frac{32}{3} \text{ oz} + \frac{64}{3} \text{ oz} = \frac{96}{3} \text{ oz} = 32$ oz. The answer checks.

State. $10\frac{2}{3}$ oz of lemon juice and $21\frac{1}{3}$ oz of linseed oil are needed.

18. 131 coach-class, 21 first-class

19. ***Familiarize.*** Let x = the number of scientific calculators and y = the number of graphing calculators ordered.

Translate. We organize the information in a table.

	Scientific calculators	Graphing calculators	Total order
Number ordered	x	y	45
Price	\$9	\$58	
Total cost	$9x$	$58y$	1728

We get one equation from the "Numbered ordered" row of the table:

$x + y = 45$

The "Total cost" row yields a second equation:

$9x + 58y = 1728$

We have translated to a system of equations:

$$x + y = 45, \quad (1)$$
$$9x + 58y = 1728 \quad (2)$$

Carry out. We solve the system of equations using the elimination method.

$$\begin{aligned} -9x - 9y &= -405 \quad \text{Multiplying (1) by } -9 \\ 9x + 58y &= 1728 \\ \hline 49y &= 1323 \\ y &= 27 \end{aligned}$$

Now substitute 27 for y in (1) and solve for x.

$$x + 27 = 45$$
$$x = 18$$

Check. The total number of calculators ordered was 18+27, or 45. The total cost of the order was $\$9 \cdot 18 + \$58 \cdot 27 = \$162 + \$1566 = \$1728$. The answer checks.

State. 18 scientific calculators and 27 graphing calculators were ordered.

20. 17 buckets, 11 dinners

21. ***Familiarize.*** Let k = the number of pounds of Kenyan French Roast coffee and s = the number of pounds of Sumatran coffee to be used in the mixture. The value of the mixture will be \$8.40(20), or \$168.

Translate. We organize the information in a table.

	Kenyan	Sumatran	Mixture
Number of pounds	k	s	20
Price per pound	\$9	\$8	\$8.40
Value of coffee	$9k$	$8s$	168

The "Number of pounds" row of the table gives us one equation:

$k + s = 20$

The "Value of coffee" row yields a second equation:

$9k + 8s = 168$

We have translated to a system of equations:

$$k + s = 20, \quad (1)$$
$$9k + 8s = 168 \quad (2)$$

Carry out. We use the elimination method to solve the system of equations.

$$\begin{aligned} -8k - 8s &= -160 \quad \text{Multiplying (1) by } -8 \\ 9k + 8s &= 168 \\ \hline k \quad &= 8 \end{aligned}$$

Substitute 8 for k in (1) and solve for s.

$$8 + s = 20$$
$$s = 12$$

Check. The total mixture contains 8 lb + 12 lb, or 20 lb. Its value is $\$9 \cdot 8 + \$8 \cdot 12 = \$72 + \$96 = \$168$. The answer checks.

State. 8 lb of Kenyan French Roast coffee and 12 lb of Sumatran coffee should be used.

22. 20 lb of cashews, 30 lb of Brazil nuts

23. ***Familiarize.*** Let x = the number of pounds of Deep Thought Granola and y = the number of pounds of Oat Dream Granola to be used in the mixture. The amount of nuts and dried fruit in the mixture is 19%(20 lb), or 0.19(20 lb) = 3.8 lb.

Translate. We organize the information in a table.

	Deep Thought	Oat Dream	Mixture
Number of pounds	x	y	20
Percent of nuts and dried fruit	25%	10%	19%
Amount of nuts and dried fruit	$0.25x$	$0.1y$	3.8 lb

We get one equation from the "Number of pounds" row of the table:

$x + y = 20$

The last row of the table yields a second equation:

$0.25x + 0.1y = 3.8$

After clearing decimals, we have the problem translated to a system of equations:

$$x + y = 20, \quad (1)$$
$$25x + 10y = 380 \quad (2)$$

Carry out. We use the elimination method to solve the system of equations.

$$\begin{aligned} -10x - 10y &= -200 \quad \text{Multiplying (1) by } -10 \\ 25x + 10y &= 380 \\ \hline 15x \quad &= 180 \\ x &= 12 \end{aligned}$$

Substitute 12 for x in (1) and solve for y.

$$12 + y = 20$$
$$y = 8$$

Check. The amount of the mixture is 12 lb + 8 lb, or 20 lb. The amount of nuts and dried fruit in the mixture is 0.25(12 lb) + 0.1(8 lb) = 3 lb + 0.8 lb = 3.8 lb. The answer checks.

State. 12 lb of Deep Thought Granola and 8 lb of Oat Dream Granola should be mixed.

24. 5 lb of each

25. ***Familiarize***. Let x = the number of liters of 25% solution and y = the number of liters of 50% solution to be used. The mixture contains 40%(10 L), or 0.4(10 L) = 4 L of acid.

Translate. We organize the information in a table.

	25% solution	50% solution	Mixture
Number of liters	x	y	10
Percent of acid	25%	50%	40%
Amount of acid	$0.25x$	$0.5y$	4 L

We get one equation from the "Number of liters" row of the table.

$$x + y = 10$$

The last row of the table yields a second equation.

$$0.25x + 0.5y = 4$$

After clearing decimals, we have the problem translated to a system of equations:

$$x + y = 10, \quad (1)$$
$$25x + 50y = 400 \quad (2)$$

Carry out. We use the elimination method to solve the system of equations.

$$-25x - 25y = -250 \quad \text{Multiplying (1) by } -25$$
$$25x + 50y = 400$$
$$25y = 150$$
$$y = 6$$

Substitute 6 for y in (1) and solve for x.

$$x + 6 = 10$$
$$x = 4$$

Check. The total amount of the mixture is 4 lb + 6 lb, or 10 lb. The amount of acid in the mixture is 0.25(4 L) + 0.5(6 L) = 1 L + 3 L = 4 L. The answer checks.

State. 4 L of the 25% solution and 6 L of the 50% solution should be mixed.

26. 150 lb of soybean meal, 220 lb of corn meal

27. ***Familiarize***. Let x = the amount of the 6% loan and y = the amount of the 9% loan. Recall that the formula for simple interest is

$$\text{Interest} = \text{Principal} \cdot \text{Rate} \cdot \text{Time}.$$

Translate. We organize the information in a table.

	6% loan	9% loan	Total
Principal	x	y	\$12,000
Interest Rate	6%	9%	
Time	1 yr	1 yr	
Interest	$0.06x$	$0.09y$	\$855

The "Principal" row of the table gives us one equation:

$$x + y = 12,000$$

The last row of the table yields another equation:

$$0.06x + 0.09y = 855$$

After clearing decimals, we have the problem translated to a system of equations:

$$x + y = 12,000 \quad (1)$$
$$6x + 9y = 85,500 \quad (2)$$

Carry out. We use the elimination method to solve the system of equations.

$$-6x - 6y = -72,000 \quad \text{Multiplying (1) by } -6$$
$$6x + 9y = 85,500$$
$$3y = 13,500$$
$$y = 4500$$

Substitute 4500 for y in (1) and solve for x.

$$x + 4500 = 12,000$$
$$x = 7500$$

Check. The loans total \$7500 + \$4500, or \$12,000. The total interest is 0.06(\$7500) + 0.09(\$4500) = \$450 + \$405 = \$855. The answer checks.

State. The 6% loan was for \$7500, and the 9% loan was for \$4500.

28. \$6800 at 9%, \$8200 at 10%

29. ***Familiarize***. From the bar graph we see that whole milk is 4% milk fat, milk for cream cheese is 8% milk fat, and cream is 30% milk fat. Let x = the number of pounds of whole milk and y = the number of pounds of cream to be used. The mixture contains 8%(200 lb), or 0.08(200 lb) = 16 lb of milk fat.

Translate. We organize the information in a table.

	Whole milk	Cream	Mixture
Number of pounds	x	y	200
Percent of milk fat	4%	30%	8%
Amount of milk fat	$0.04x$	$0.3y$	16 lb

We get one equation from the " Number of pounds" row of the table:

$$x + y = 200$$

The last row of the table yields a second equation:

$$0.04x + 0.3y = 16$$

After clearing decimals, we have the problem translated to a system of equations:

$$x + y = 200, \quad (1)$$
$$4x + 30y = 1600 \quad (2)$$

Carry out. We use the elimination method to solve the system of equations.

$$-4x - 4y = -800 \quad \text{Multiplying (1) by } -4$$
$$4x + 30y = 1600$$
$$26y = 800$$
$$y = \frac{400}{13}, \text{ or } 30\frac{10}{13}$$

Substitute $\frac{400}{13}$ for y in (1) and solve for x.

$$x + \frac{400}{13} = 200$$
$$x = \frac{2200}{13}, \text{ or } 169\frac{3}{13}$$

Check. The total amount of the mixture is $\frac{2200}{13}$ lb $+ \frac{400}{13}$ lb $= \frac{2600}{13}$ lb $= 200$ lb. The amount of milk fat in the mixture is $0.04\left(\frac{2200}{13}\text{ lb}\right) + 0.3\left(\frac{400}{13}\text{ lb}\right) = \frac{88}{13}$ lb $+ \frac{120}{13}$ lb $= \frac{208}{13}$ lb $= 16$ lb.

The answer checks.

State. $169\frac{3}{13}$ lb of whole milk and $30\frac{10}{13}$ lb of cream should be mixed.

30. 12.5 L of Arctic Antifreeze, 7.5 L of Frost-No-More

31. ***Familiarize.*** Let l = the length, in feet, and w = the width, in feet. Recall that the formula for the perimeter P of a rectangle with length l and width w is $P = 2l + 2w$.

Translate.

The perimeter is 860 ft.

$$2l + 2w = 860$$

The length is 100 ft. more than the width.

$$l = 100 + w$$

We have translated to a system of equations:

$$2l + 2w = 860, \quad (1)$$
$$l = 100 + w \quad (2)$$

Carry out. We use the substitution method to solve the system of equations.

Substitute $100 + w$ for l in (1) and solve for w.

$$2(100 + w) + 2w = 860$$
$$200 + 2w + 2w = 860$$
$$200 + 4w = 860$$
$$4w = 660$$
$$w = 165$$

Now substitute 165 for w in (2).

$$l = 100 + 165 = 265$$

Check. The perimeter is $2 \cdot 265$ ft $+ 2 \cdot 165$ ft $= 530$ ft $+ 330$ ft $= 860$ ft. The length, 265 ft, is 100 ft more than the width, 165 ft. The answer checks.

State. The length is 265 ft, and the width is 165 ft.

32. Length: 76 m, width: 19 m

33. ***Familiarize.*** Let x = the number of \$5 bills and y = the number of \$1 bills. The total value of the \$5 bills is $5x$, and the total value of the \$1 bills is $1 \cdot y$, or y.

Translate.

The total number of bills is 22.

$$x + y = 22$$

The total value of the bills is \$50.

$$5x + y = 50$$

We have a system of equations:

$$x + y = 22, \quad (1)$$
$$5x + y = 50 \quad (2)$$

Carry out. We use the elimination method.

$$-x - y = -22 \quad \text{Multiplying (1) by } -1$$
$$5x + y = 50$$
$$4x = 28$$
$$x = 7$$

$$7 + y = 22 \quad \text{Substituting 7 for } x \text{ in (1)}$$
$$y = 15$$

Check. Total number of bills: $7 + 15 = 22$

Total value of bills: $\$5 \cdot 7 + \$1 \cdot 15 = \$35 + \$15 = \$50$.

The numbers check.

State. There are 7 \$5 bills and 15 \$1 bills.

34. 17 quarters, 13 fifty-cent pieces

35. ***Familiarize.*** We first make a drawing.

Slow train
d kilometers 75 km/h $(t + 2)$ hr

Fast train
d kilometers 125 km/h t hr

From the drawing we see that the distances are the same. Now complete the chart.

$$d = r \cdot t$$

	Distance	Rate	Time	
Slow train	d	75	$t+2$	$\rightarrow d = 75(t+2)$
Fast train	d	125	t	$\rightarrow d = 125t$

Translate. Using $d = rt$ in each row of the table, we get a system of equations:

$$d = 75(t + 2),$$
$$d = 125t$$

Carry out. We solve the system of equations.

$$125t = 75(t+2) \quad \text{Using substitution}$$
$$125t = 75t + 150$$
$$50t = 150$$
$$t = 3$$

Then $d = 125t = 125 \cdot 3 = 375$

Check. At 125 km/h, in 3 hr the fast train will travel $125 \cdot 3 = 375$ km. At 75 km/h, in 3 + 2, or 5 hr the slow train will travel $75 \cdot 5 = 375$ km. The numbers check.

State. The trains will meet 375 km from the station.

36. 3 hr

37. ***Familiarize.*** We first make a drawing. Let $d =$ the distance and $r =$ the speed of the canoe in still water. Then when the canoe travels downstream its speed is $r+6$, and its speed upstream is $r-6$. From the drawing we see that the distances are the same.

Downstream, 6 mph current

d mi, $r+6$, 4 hr

Upstream, 6 mph current

d mi, $r-6$, 10 hr

Organize the information in a table.

	Distance	Rate	Time
With current	d	$r+6$	4
Against current	d	$r-6$	10

Translate. Using $d = rt$ in each row of the table, we get a system of equations:

$$d = 4(r+6), \qquad d = 4r + 24,$$
or
$$d = 10(r-6) \qquad d = 10r - 60$$

Carry out. Solve the system of equations.

$$4r + 24 = 10r - 60 \quad \text{Using substitution}$$
$$24 = 6r - 60$$
$$84 = 6r$$
$$14 = r$$

Check. When $r = 14$, then $r + 6 = 14 + 6 = 20$, and the distance traveled in 4 hr is $4 \cdot 20 = 80$ km. Also, $r - 6 = 14 - 6 = 8$, and the distance traveled in 10 hr is $8 \cdot 10 = 80$ km. The answer checks.

State. The speed of the canoe in still water is 14 km/h.

38. 24 mph

39. ***Familiarize.*** We make a drawing. Note that the plane's speed traveling toward London is 360 + 50, or 410 mph, and the speed traveling toward New York City is 360 − 50, or 310 mph. Also, when the plane is d mi from New York City, it is $3458 - d$ mi from London.

New York City 310 mph t hours — t hours London 410 mph

3458 mi

d — 3458 mi $-d$

Organize the information in a table.

	Distance	Rate	Time
Toward NYC	d	310	t
Toward London	$3458-d$	410	t

Translate. Using $d = rt$ in each row of the table, we get a system of equations:

$$d = 310t, \quad (1)$$
$$3458 - d = 410t \quad (2)$$

Carry out. We solve the system of equations.

$$3458 - 310t = 410t \quad \text{Using substitution}$$
$$3458 = 720t$$
$$4.8028 \approx t$$

Substitute 4.8028 for t in (1).

$$d \approx 310(4.8028) \approx 1489$$

Check. If the plane is 1489 mi from New York City, it can return to New York City, flying at 310 mph, in $1489/310 \approx 4.8$ hr. If the plane is 3458 − 1489, or 1969 mi from London, it can fly to London, traveling at 410 mph, in $1969/410 \approx 4.8$ hr. Since the times are the same, the answer checks.

State. The point of no return is about 1489 mi from New York City.

40. 1524 mi

41.
$$-3(x-7) - 2[x - (4+3x)]$$
$$= -3(x-7) - 2[x - 4 - 3x] \quad \text{Removing the inner-most parentheses}$$
$$= -3(x-7) - 2(-2x-4) \quad \text{Simplifying}$$
$$= -3x + 21 + 4x + 8 \quad \text{Removing parentheses}$$
$$= x + 29 \quad \text{Simplifying}$$

42. $\frac{27}{4}$

43. We use the point-slope equation.

$$y - y_1 = m(x - x_1)$$
$$y - (-5) = -\frac{3}{4}(x-2) \quad \text{Substituting}$$
$$y + 5 = -\frac{3}{4}x + \frac{3}{2}$$
$$y = -\frac{3}{4}x - \frac{7}{2}$$

44.

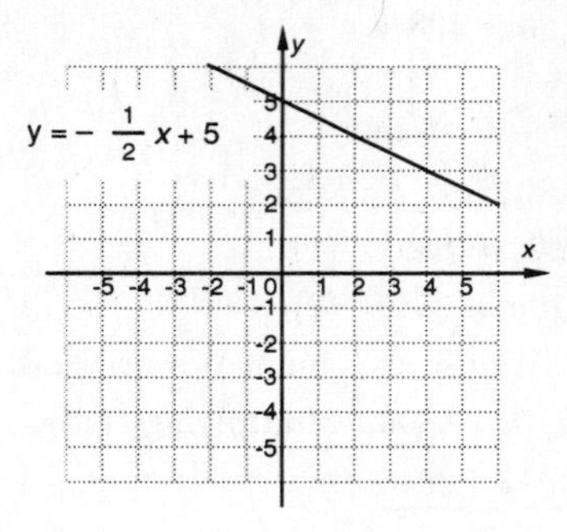

45. First find the slope of $2x + 3y = 7$..

$$2x + 3y = 7$$
$$3y = -2x + 7$$
$$y = -\frac{2}{3}x + \frac{7}{3} \quad \text{Slope-intercept form}$$

The slope of a linear function whose graph is perpendicular to the graph of $y = -\frac{2}{3}x + \frac{7}{3}$ is $\frac{3}{2}$ $\left(\text{because } -\frac{2}{3} \cdot \frac{3}{2} = -1\right)$. We find a linear function with slope $\frac{3}{2}$ and y-intercept $(0, -5)$:

$f(x) = \frac{3}{2}x - 5.$

46. 10

47. ◈

48. ◈

49. ◈

50. ◈

51. The Familiarize and Translate steps were done in Exercise 66 of Exercise Set 3.1.

Carry out. We solve the system of equations

$$x = 2y, \quad (1)$$
$$x + 20 = 3y \quad (2)$$

where x = Burl's age now and y = his son's age now.

$$2y + 20 = 3y \quad \text{Substituting } 2y \text{ for } x \text{ in (2)}$$
$$20 = y$$

$$x = 2 \cdot 20 \quad \text{Substituting 20 for } y \text{ in (1)}$$
$$x = 40$$

Check. Burl's age now, 40, is twice his son's age now, 20. Ten years ago Burl was 30 and his son was 10, and $30 = 3 \cdot 10$. The numbers check.

State. Now Burl is 40 and his son is 20.

52. Lou: 32 years, Juanita: 14 years

53. The Familiarize and Translate steps were done in Exercise 68 of Exercise Set 3.1.

Carry out. We solve the system of equations

$$2l + 2w = 156, \quad (1)$$
$$l = 4(w - 6) \quad (2)$$

where l = length, in inches, and w = width, in inches.

$$2 \cdot 4(w - 6) + 2w = 156 \quad \text{Substituting } 4(w-6) \text{ for } l \text{ in (1)}$$
$$8w - 48 + 2w = 156$$
$$10w - 48 = 156$$
$$10w = 204$$
$$w = \frac{204}{10}, \text{ or } \frac{102}{5}$$

$$l = 4\left(\frac{102}{5} - 6\right) \quad \text{Substituting } \frac{102}{5} \text{ for } w \text{ in (2)}$$
$$l = 4\left(\frac{102}{5} - \frac{30}{5}\right)$$
$$l = 4\left(\frac{72}{5}\right)$$
$$l = \frac{288}{5}$$

Check. The perimeter of a rectangle with width $\frac{102}{5}$ in. and length $\frac{288}{5}$ in. is

$2\left(\frac{288}{5}\right) + 2\left(\frac{102}{5}\right) = \frac{576}{5} + \frac{204}{5} = \frac{780}{5} = 156$ in.

If 6 in. is cut off the width, the new width is $\frac{102}{5} - 6 = \frac{102}{5} - \frac{30}{5} = \frac{72}{5}$. The length, $\frac{288}{5}$, is $4\left(\frac{72}{5}\right)$. The numbers check.

State. The original piece of posterboard had width $\frac{102}{5}$ in. and length $\frac{288}{5}$ in.

54. $\frac{64}{5}$ oz of baking soda, $\frac{16}{5}$ oz of vinegar

55. ***Familiarize.*** Let x = the amount of the original solution that remains after some of the original solution is drained and replaced with pure antifreeze. Let y = the amount of the original solution that is drained and replaced with pure antifreeze.

Translate. We organize the information in a table. Keep in mind that the table contains information regarding the solution *after* some of the original solution is drained and replaced with pure antifreeze.

	Original Solution	Pure Anti-freeze	New Mixture
Amount of solution	x	y	16 L
Percent of antifreeze	30%	100%	50%
Amount of antifreeze in solution	$0.3x$	$1 \cdot y$, or y	$0.5(16)$, or 8

The "Amount of solution" row gives us one equation: $x + y = 16$

The last row gives us a second equation: $0.3x + y = 8$

After clearing the decimal we have the following system of equations:

$$x + y = 16, \quad (1)$$
$$3x + 10y = 80 \quad (2)$$

Carry out. We use the elimination method.

$$\begin{aligned} -3x - 3y &= -48 \quad \text{Multiplying (1) by } -3 \\ 3x + 10y &= 80 \\ \hline 7y &= 32 \\ y &= \frac{32}{7}, \text{ or } 4\frac{4}{7} \end{aligned}$$

Although the problem only asks for the amount of pure antifreeze added, we will also find x in order to check.

$$x + 4\frac{4}{7} = 16 \quad \text{Substituting } 4\frac{4}{7} \text{ for } y \text{ in (1)}$$
$$x = 11\frac{3}{7}$$

Check. Total amount of new mixture: $11\frac{3}{7} + 4\frac{4}{7} = 16$ L

Amount of antifreeze in new mixture:
$0.3\left(11\frac{3}{7}\right) + 4\frac{4}{7} = \frac{3}{10} \cdot \frac{80}{7} + \frac{32}{7} = \frac{56}{7} = 8$ L

The numbers check.

State. Michelle should drain $4\frac{4}{7}$ L of the original solution and replace it with pure antifreeze.

56. 4 km

57. ***Familiarize.*** Let x = the number of members who ordered one book and y = the number of members who ordered two books. Note that the y members ordered a total of $2y$ books.

Translate.

The number of books sold was 880.

$$x + 2y = 880$$

Total sales were \$9840.

$$12x + 20y = 9840$$

We have a system of equations.

$$x + 2y = 880, \quad (1)$$
$$12x + 20y = 9840 \quad (2)$$

Carry out. We use the elimination method.

$$\begin{aligned} -10x - 20y &= -8800 \quad \text{Multiplying (1) by } -10 \\ 12x + 20y &= 9840 \\ \hline 2x &= 1040 \\ x &= 520 \end{aligned}$$

Substitute 520 for x in (1) and solve for y.

$$520 + 2y = 880$$
$$2y = 360$$
$$y = 180$$

Check. Total number of books sold: $520 + 2 \cdot 180 = 520 + 360 = 880$

Total sales: $\$12 \cdot 520 + \$20 \cdot 180 = \$6240 + \$3600 = \$9840$

The answer checks.

State. 180 members ordered two books.

58. 82

59. ***Familiarize.*** We first make a drawing. Let r_1 = the speed of the first train and r_2 = the speed of the second train. If the first train leaves at 9 A.M. and the second at 10 A.M., we have:

Train 1 r_1, 3 hr; Train 2 r_2, 2 hr
Union Station — Central Station
216 km

If the second train leaves at 9 A.M. and the first at 10:30 A.M. we have:

Train 1 r_1, $\frac{3}{2}$ hr; Train 2 r_2, 3 hr
Union Station — Central Station
216 km

The total distance traveled in each case is 216 km and is equal to the sum of the distances traveled by each train.

Translate. We will use the formula $d = rt$. For each situation we have:

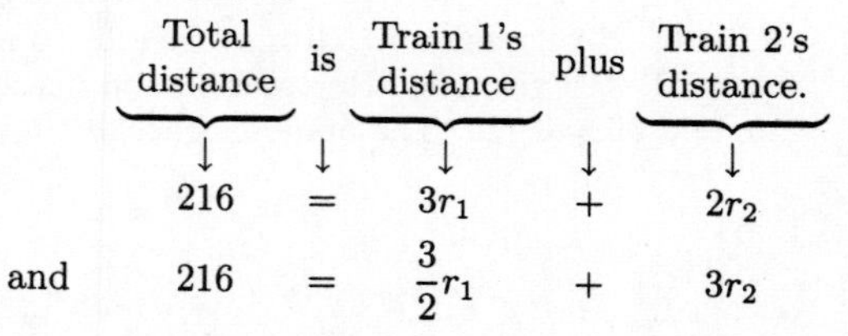

Clearing the fraction, we have this system:

$$216 = 3r_1 + 2r_2, \quad (1)$$
$$432 = 3r_1 + 6r_2 \quad (2)$$

Carry out. Solve the system of equations.

$$\begin{aligned} -216 &= -3r_1 - 2r_2 \quad \text{Multiplying (1) by } -1 \\ 432 &= 3r_1 + 6r_2 \\ \hline 216 &= \phantom{-3r_1 + {}} 4r_2 \\ 54 &= \phantom{-3r_1 + {}} r_2 \end{aligned}$$

$$216 = 3r_1 + 2(54) \quad \text{Substituting 54 for } r_2 \text{ in (1)}$$
$$216 = 3r_1 + 108$$
$$108 = 3r_1$$
$$36 = r_1$$

Check. If Train 1 travels for 3 hr at 36 km/h and Train 2 travels for 2 hr at 54 km/h, the total distance traveled is $3 \cdot 36 + 2 \cdot 54 = 108 + 108 = 216$ km. If Train 1 travels for $\frac{3}{2}$ hr at 36 km/h and Train 2 travels for 3 hr at 54 km/h, then the total distance traveled is $\frac{3}{2} \cdot 36 + 3 \cdot 54 = 54 + 162 = 216$ km. The numbers check.

State. The speed of the first train is 36 km/h, and the speed of the second train is 54 km/h.

60. City: 261 miles, hihgway: 204 miles

61. ***Familiarize.*** Let $x =$ the number of gallons of pure brown and $y =$ the number of gallons of neutral stain that should be added to the original 0.5 gal. Note that a total of 1 gal of stain needs to be added to bring the amount of stain up to 1.5 gal. The original 0.5 gal of stain contains 20%(0.5 gal), or 0.2(0.5 gal) = 0.1 gal of brown stain. The final solution contains 60%(1.5 gal), or 0.6(1.5 gal) = 0.9 gal of brown stain. This is composed of the original 0.1 gal and the x gal that are added.

Translate.

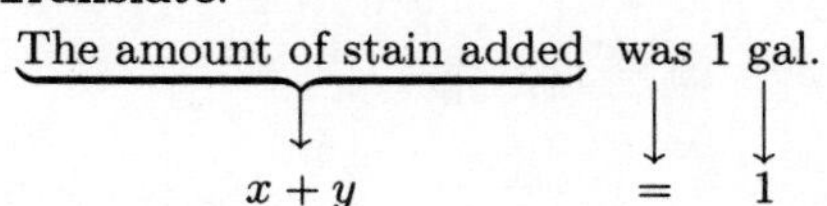

The amount of brown stain in the final solution is 0.9 gal.

$$0.1 + x \quad = \quad 0.9$$

We have a system of equations.

$$x + y = 1, \quad (1)$$
$$0.1 + x = 0.9 \quad (2)$$

Carry out. First we solve (2) for x.

$$0.1 + x = 0.9$$
$$x = 0.8$$

Then substitute 0.8 for x in (1) and solve for y.

$$0.8 + y = 1$$
$$y = 0.2$$

Check. Total amount of stain: $0.5 + 0.8 + 0.2 = 1.5$ gal

Total amount of brown stain: $0.1 + 0.8 = 0.9$ gal

Total amount of neutral stain: $0.8(0.5) + 0.2 = 0.4 + 0.2 = 0.6 \text{ gal} = 0.4(1.5 \text{ gal})$

The answer checks.

State. 0.8 gal of pure brown and 0.2 gal of neutral stain should be added.

62. 3 girls, 4 boys

63. The 1.5 gal mixture contains $0.1 + x$ gal of pure brown stain. (See Exercise 61.). Thus, the function $P(x) = \dfrac{0.1 + x}{1.5}$ gives the percentage of brown in the mixture as a decimal quantity. Using Trace and Zoom or the Table feature, we confirm that when $x = 0.8$, then $P(x) = 0.6$ or 60%.

Exercise Set 3.4

1. Substitute $(2, -1, -2)$ into the three equations, using alphabetical order.

$$\begin{array}{r|l} \multicolumn{2}{c}{x + y - 2z = 5} \\ \hline 2 + (-1) - 2(-2) & ?\ 5 \\ 2 - 1 + 4 & \\ 5 & 5 \quad \text{TRUE} \end{array}$$

$$\begin{array}{r|l} \multicolumn{2}{c}{2x - y - z = 7} \\ \hline 2 \cdot 2 - (-1) - (-2) & ?\ 7 \\ 4 + 1 + 2 & \\ 7 & 7 \quad \text{TRUE} \end{array}$$

$$\begin{array}{r|l} \multicolumn{2}{c}{-x - 2y + 3z = 6} \\ \hline -2 - 2(-1) + 3(-2) & ?\ 6 \\ -2 + 2 - 6 & \\ -6 & 6 \quad \text{FALSE} \end{array}$$

The triple $(2, -1, -2)$ does not make the third equation true, so it is not a solution of the system.

2. Yes

3.
$$2x - y + z = 10, \quad (1)$$
$$4x + 2y - 3z = 10, \quad (2)$$
$$x - 3y + 2z = 8 \quad (3)$$

1., 2. The equations are already in standard form with no fractions or decimals.

3. Use Equations (1) and (2) to eliminate y:

$$\begin{aligned} 4x - 2y + 2z &= 20 \quad \text{Multiplying (1) by 2} \\ 4x + 2y - 3z &= 10 \quad (2) \\ \hline 8x - z &= 30 \quad (4) \end{aligned}$$

4. Use a different pair of equations and eliminate y:

$$\begin{array}{rl} -6x + 3y - 3z = -30 & \text{Multiplying (1) by } -3 \\ x - 3y + 2z = 8 & (3) \\ \hline -5x \quad - z = -22 & (5) \end{array}$$

5. Now solve the system of Equations (4) and (5).

$$8x - z = 30 \quad (4)$$
$$-5x - z = -22 \quad (5)$$

$$\begin{array}{rl} 8x - z = 30 & (4) \\ 5x + z = 22 & \text{Multiplying (5) by } -1 \\ \hline 13x \quad = 52 & \\ x = 4 & \end{array}$$

$$\begin{aligned} 8 \cdot 4 - z &= 30 \quad \text{Substituting in (4)} \\ 32 - z &= 30 \\ -z &= -2 \\ z &= 2 \end{aligned}$$

6. Substitute in one of the original equations to find y.

$$\begin{aligned} 2 \cdot 4 - y + 2 &= 10 \quad \text{Substituting in (1)} \\ 10 - y &= 10 \\ -y &= 0 \\ y &= 0 \end{aligned}$$

We obtain $(4, 0, 2)$. This checks, so it is the solution.

4. $(1, 2, 3)$

5.
$$\begin{aligned} x - y + z &= 6, \quad (1) \\ 2x + 3y + 2z &= 2, \quad (2) \\ 3x + 5y + 4z &= 4 \quad (3) \end{aligned}$$

1., 2. The equations are already in standard form with no fractions or decimals.

3., 4. We eliminate y from two different pairs of equations.

$$\begin{array}{rl} 3x - 3y + 3z = 18 & \text{Multiplying (1) by 3} \\ 2x + 3y + 2z = 2 & (2) \\ \hline 5x \quad + 5z = 20 & (4) \end{array}$$

$$\begin{array}{rl} 5x - 5y + 5z = 30 & \text{Multiplying (1) by 5} \\ 3x + 5y + 4z = 4 & (3) \\ \hline 8x \quad + 9z = 34 & (5) \end{array}$$

5. Now solve the system of Equations (4) and (5).

$$5x + 5z = 20 \quad (4)$$
$$8x + 9z = 34 \quad (5)$$

$$\begin{array}{rl} 45x + 45z = 180 & \text{Multiplying (4) by 9} \\ -40x - 45z = -170 & \text{Multiplying (5) by } -5 \\ \hline 5x \quad = 10 & \\ x = 2 & \end{array}$$

$$\begin{aligned} 5 \cdot 2 + 5z &= 20 \quad \text{Substituting in (4)} \\ 10 + 5z &= 20 \\ 5z &= 10 \\ z &= 2 \end{aligned}$$

6. Substitute in one of the original equations to find y.

$$\begin{aligned} 2 - y + 2 &= 6 \quad \text{Substituting in (1)} \\ 4 - y &= 6 \\ -y &= 2 \\ y &= -2 \end{aligned}$$

We obtain $(2, -2, 2)$. This checks, so it is the solution.

6. $(-1, 5, -2)$

7.
$$\begin{aligned} 6x - 4y + 5z &= 31, \quad (1) \\ 5x + 2y + 2z &= 13, \quad (2) \\ x + y + z &= 2 \quad (3) \end{aligned}$$

1., 2. The equations are already in standard form with no fractions or decimals.

3., 4. We eliminate y from two different pairs of equations.

$$\begin{array}{rl} 6x - 4y + 5z = 31 & (1) \\ 4x + 4y + 4z = 8 & \text{Multiplying (3) by 4} \\ \hline 10x \quad + 9z = 39 & (4) \end{array}$$

$$\begin{array}{rl} 5x + 2y + 2z = 13 & (2) \\ -2x - 2y - 2z = -4 & \text{Multiplying (3) by } -2 \\ \hline 3x \quad = 9 & \\ x = 3 & \end{array}$$

5. When we used Equations (2) and (3) to eliminate y, we also eliminated z and found that $x = 3$. Substitute 3 for x in Equation (4) to find z.

$$\begin{aligned} 10 \cdot 3 + 9z &= 39 \quad \text{Substituting in (4)} \\ 30 + 9z &= 39 \\ 9z &= 9 \\ z &= 1 \end{aligned}$$

6. Substitute in one of the original equations to find y.

$$\begin{aligned} 3 + y + 1 &= 2 \quad \text{Substituting in (3)} \\ y + 4 &= 2 \\ y &= -2 \end{aligned}$$

We obtain $(3, -2, 1)$. This checks, so it is the solution.

8. $(3, 1, 2)$

9.
$$\begin{aligned} x + y + z &= 0, \quad (1) \\ 2x + 3y + 2z &= -3, \quad (2) \\ -x + 2y - 3z &= -1 \quad (3) \end{aligned}$$

1., 2. The equations are already in standard form with no fractions or decimals.

3., 4. We eliminate x from two different pairs of equations.

$$\begin{array}{rll} -2x - 2y - 2z = & 0 & \text{Multiplying (1) by } -2 \\ 2x + 3y + 2z = & -3 & (2) \\ \hline y \quad\quad = & -3 & \end{array}$$

We eliminated not only x but also z and found that $y = -3$.

5., 6. Substitute -3 for y in two of the original equations to produce a system of two equations in two variables. Then solve this system.

$$x - 3 + z = 0 \quad \text{Substituting in (1)}$$
$$-x + 2(-3) - 3z = -1 \quad \text{Substituting in (3)}$$

Simplifying we have

$$\begin{array}{r} x + \ \ z = 3 \\ -x - 3z = 5 \\ \hline -2z = 8 \\ z = -4 \end{array}$$

$$x - 3 - 4 = 0 \quad \text{Substituting in (1)}$$
$$x - 7 = 0$$
$$x = 7$$

We obtain $(7, -3, -4)$. This checks, so it is the solution.

10. $(-3, -4, 2)$

11. $2x + y - 3z = -4$, (1)
$4x - 2y + z = 9$, (2)
$3x + 5y - 2z = 5$ (3)

1., 2. The equations are already in standard form with no fractions or decimals.

3., 4. We eliminate z from two different pairs of equations.

$$\begin{array}{rll} 2x + y - 3z = & -4 & (1) \\ 12x - 6y + 3z = & 27 & \text{Multiplying (2) by 3} \\ \hline 14x - 5y \quad\quad = & 23 & (4) \end{array}$$

$$\begin{array}{rll} 8x - 4y + 2z = & 18 & \text{Multiplying (2) by 2} \\ 3x + 5y - 2z = & 5 & (3) \\ \hline 11x + y \quad\quad = & 23 & (5) \end{array}$$

5. Now solve the system of Equations (4) and (5).

$$14x - 5y = 23 \quad (4)$$
$$11x + y = 23 \quad (5)$$

$$\begin{array}{rll} 14x - 5y = & 23 & (4) \\ 55x + 5y = & 115 & \text{Multiplying (5) by 5} \\ \hline 69x \quad = & 138 & \\ x = & 2 & \end{array}$$

$$11 \cdot 2 + y = 23 \quad \text{Substituting in (5)}$$
$$22 + y = 23$$
$$y = 1$$

6. Substitute in one of the original equations to find z.

$$4 \cdot 2 - 2 \cdot 1 + z = 9 \quad \text{Substituting in (2)}$$
$$6 + z = 9$$
$$z = 3$$

We obtain $(2, 1, 3)$. This checks, so it is the solution.

12. $(2, 4, 1)$

13. $2x + y + 2z = 11$, (1)
$3x + 2y + 2z = 8$, (2)
$x + 4y + 3z = 0$ (3)

1., 2. The equations are already in standard form with no fractions or decimals.

3., 4. We eliminate x from two different pairs of equations.

$$\begin{array}{rll} 2x + y + 2z = & 11 & (1) \\ -2x - 8y - 6z = & 0 & \text{Multiplying (3) by } -2 \\ \hline -7y - 4z = & 11 & (4) \end{array}$$

$$\begin{array}{rll} 3x + 2y + 2z = & 8 & (2) \\ -3x - 12y - 9z = & 0 & \text{Multiplying (3) by } -3 \\ \hline -10y - 7z = & 8 & (5) \end{array}$$

5. Now solve the system of Equations (4) and (5).

$$-7y - 4z = 11 \quad (4)$$
$$-10y - 7z = 8 \quad (5)$$

$$\begin{array}{rll} -49y - 28z = & 77 & \text{Multiplying (4) by 7} \\ 40y + 28z = & -32 & \text{Multiplying (5) by } -4 \\ \hline -9y \quad\quad = & 45 & \\ y = & -5 & \end{array}$$

$$-7(-5) - 4z = 11 \quad \text{Substituting in (4)}$$
$$35 - 4z = 11$$
$$-4z = -24$$
$$z = 6$$

6. Substitute in one of the original equations to find x.

$$x + 4(-5) + 3 \cdot 6 = 0 \quad \text{Substituting in (3)}$$
$$x - 2 = 0$$
$$x = 2$$

We obtain $(2, -5, 6)$. This checks, so it is the solution.

14. $(-3, 0, 4)$

15. $-2x + 8y + 2z = 4$, (1)
$x + 6y + 3z = 4$, (2)
$3x - 2y + z = 0$ (3)

1., 2. The equations are already in standard form with no fractions or decimals.

3., 4. We eliminate z from two different pairs of equations.

$$\begin{array}{rll} -2x + 8y + 2z = 4 & (1) \\ -6x + 4y - 2z = 0 & \text{Multiplying (3) by } -2 \\ \hline -8x + 12y \quad = 4 & (4) \end{array}$$

$$\begin{array}{rll} x + 6y + 3z = 4 & (2) \\ -9x + 6y - 3z = 0 & \text{Multiplying (3) by } -3 \\ \hline -8x + 12y \quad = 4 & (5) \end{array}$$

5. Now solve the system of Equations (4) and (5).

$$\begin{array}{rl} -8x + 12y = 4 & (4) \\ -8x + 12y = 4 & (5) \end{array}$$

$$\begin{array}{rll} -8x + 12y = 4 & (4) \\ 8x - 12y = -4 & \text{Multiplying (5) by } -1 \\ \hline 0 = 0 & (6) \end{array}$$

Equation (6) indicates that Equations (1), (2), and (3) are dependent. (Note that if Equation (1) is subtracted from Equation (2), the result is Equation (3).) We could also have conlcuded that the equations are dependent by observing that Equations (4) and (5) are identical.

16. The equations are dependent.

17.
$$\begin{array}{rl} a + 2b + c = 1, & (1) \\ 7a + 3b - c = -2, & (2) \\ a + 5b + 3c = 2 & (3) \end{array}$$

1., 2. The equations are already in standard form with no fractions or decimals.

3., 4. We eliminate c from two different pairs of equations.

$$\begin{array}{rl} a + 2b + c = 1 & (1) \\ 7a + 3b - c = -2 & (2) \\ \hline 8a + 5b \quad = -1 & (4) \end{array}$$

$$\begin{array}{rl} 21a + 9b - 3c = -6 & \text{Multiplying (2) by 3} \\ a + 5b + 3c = 2 & \\ \hline 22a + 14b \quad = -4 & (5) \end{array}$$

5. Now solve the system of Equations (4) and (5).

$$\begin{array}{rl} 8a + 5b = -1 & (4) \\ 22a + 14b = -4 & (5) \end{array}$$

$$\begin{array}{rl} 112a + 70b = -14 & \text{Multiplying (4) by 14} \\ -110a - 70b = 20 & \text{Multiplying (5) by } -5 \\ \hline 2a \quad = 6 & \\ a = 3 & \end{array}$$

$$\begin{array}{rl} 8 \cdot 3 + 5b = -1 & \text{Substituting in (4)} \\ 24 + 5b = -1 & \\ 5b = -25 & \\ b = -5 & \end{array}$$

6. Substitute in one of the original equations to find c.

$$\begin{array}{rl} 3 + 2(-5) + c = 1 & \text{Substituting in (1)} \\ -7 + c = 1 & \\ c = 8 & \end{array}$$

We obtain $(3, -5, 8)$. This checks, so it is the solution.

18. $\left(\frac{1}{2}, 4, -6\right)$

19.
$$\begin{array}{rl} 5x + 3y + \frac{1}{2}z = & \frac{7}{2}, \\ 0.5x - 0.9y - 0.2z = & 0.3, \\ 3x - 2.4y + 0.4z = & -1 \end{array}$$

1. All equations are already in standard form.

2. Multiply the first equation by 2 to clear the fractions. Also, multiply the second and third equations by 10 to clear the decimals.

$$\begin{array}{rl} 10x + 6y + z = 7, & (1) \\ 5x - 9y - 2z = 3, & (2) \\ 30x - 24y + 4z = -10 & (3) \end{array}$$

3., 4. We eliminate z from two different pairs of equations.

$$\begin{array}{rl} 20x + 12y + 2z = 14 & \text{Multiplying (1) by 2} \\ 5x - 9y - 2z = 3 & (2) \\ \hline 25x + 3y \quad = 17 & (4) \end{array}$$

$$\begin{array}{rl} 10x - 18y - 4z = 6 & \text{Multiplying(2) by 2} \\ 30x - 24y + 4z = -10 & (3) \\ \hline 40x - 42y \quad = -4 & (5) \end{array}$$

5. Now solve the system of Equations (4) and (5).

$$\begin{array}{rl} 25x + 3y = 17 & (4) \\ 40x - 42y = -4 & (5) \end{array}$$

$$\begin{array}{rl} 350x + 42y = 238 & \text{Multiplying (4) by 14} \\ 40x - 42y = -4 & (5) \\ \hline 390x \quad = 234 & \\ x = \frac{3}{5} & \end{array}$$

$$\begin{array}{rl} 25\left(\frac{3}{5}\right) + 3y = 17 & \text{Substituting in (4)} \\ 15 + 3y = 17 & \\ 3y = 2 & \\ y = \frac{2}{3} & \end{array}$$

6. Substitute in one of the original equations to find z.

$$10\left(\frac{3}{5}\right)+6\left(\frac{2}{3}\right)+z=7 \quad \text{Substituting in (1)}$$
$$6+4+z=7$$
$$10+z=7$$
$$z=-3$$

We obtain $\left(\frac{3}{5},\frac{2}{3},-3\right)$. This checks, so it is the solution.

20. $\left(\frac{1}{2},\frac{1}{3},\frac{1}{6}\right)$

21.
$$\begin{aligned} 3p \quad\quad + 2r &= 11, \quad (1)\\ q - 7r &= 4, \quad (2)\\ p - 6q \quad\quad &= 1 \quad (3) \end{aligned}$$

1., 2. The equations are already in standard form with no fractions or decimals.

3., 4. Note that there is no q in Equation (1). We will use Equations (2) and (3) to obtain another equation with no q-term.

$$\begin{aligned} 6q - 42r &= 24 \quad \text{Multiplying (2) by 6}\\ p - 6q \quad\quad &= 1 \quad (3)\\ \hline p \quad\quad - 42r &= 25 \quad (4) \end{aligned}$$

5. Solve the system of Equations (1) and (4).

$$\begin{aligned} 3p + 2r &= 11 \quad (1)\\ p - 42r &= 25 \quad (4) \end{aligned}$$

$$\begin{aligned} 3p + 2r &= 11 \quad (1)\\ -3p + 126r &= -75 \quad \text{Multiplying (4) by } -3\\ \hline 128r &= -64\\ r &= -\frac{1}{2} \end{aligned}$$

$$3p+2\left(-\frac{1}{2}\right)=11 \quad \text{Substituting in (1)}$$
$$3p-1=11$$
$$3p=12$$
$$p=4$$

6. Substitute in Equation (2) or (3) to find q.

$$q-7\left(-\frac{1}{2}\right)=4 \quad \text{Substituting in (2)}$$
$$q+\frac{7}{2}=4$$
$$q=\frac{1}{2}$$

We obtain $\left(4,\frac{1}{2},-\frac{1}{2}\right)$. This checks, so it is the solution.

22. $\left(\frac{1}{2},\frac{2}{3},-\frac{5}{6}\right)$

23.
$$\begin{aligned} x + y + z &= 105, \quad (1)\\ 10y - z &= 11, \quad (2)\\ 2x - 3y \quad\quad &= 7 \quad (3) \end{aligned}$$

1., 2. The equations are already in standard form with no fractions or decimals.

3., 4. Note that there is no z in Equation (3). We will use Equations (1) and (2) to obtain another equation with no z-term.

$$\begin{aligned} x + y + z &= 105 \quad (1)\\ 10y - z &= 11 \quad (2)\\ \hline x + 11y \quad\quad &= 116 \quad (4) \end{aligned}$$

5. Now solve the system of Equations (3) and (4).

$$\begin{aligned} 2x - 3y &= 7 \quad (3)\\ x + 11y &= 116 \quad (4) \end{aligned}$$

$$\begin{aligned} 2x - 3y &= 7 \quad (3)\\ -2x - 22y &= -232 \quad \text{Multiplying (4) by } -2\\ \hline -25y &= -225\\ y &= 9 \end{aligned}$$

$$x+11\cdot 9=116 \quad \text{Substituting in (4)}$$
$$x+99=116$$
$$x=17$$

6. Substitute in Equation (1) or (2) to find z.

$$17+9+z=105 \quad \text{Substituting in (1)}$$
$$26+z=105$$
$$z=79$$

We obtain $(17,9,79)$. This checks, so it is the solution.

24. $(15,33,9)$

25.
$$\begin{aligned} 2a - 3b \quad\quad &= 2, \quad (1)\\ 7a \quad\quad + 4c &= \frac{3}{4}, \quad (2)\\ -3b + 2c &= 1 \quad (3) \end{aligned}$$

1. The equations are already in standard form.

2. Multiply Equation (2) by 4 to clear the fraction. The resulting system is

$$\begin{aligned} 2a - 3b \quad\quad &= 2, \quad (1)\\ 28a \quad\quad + 16c &= 3, \quad (4)\\ -3b + 2c &= 1 \quad (3) \end{aligned}$$

3. Note that there is no b in Equation (2). We will use Equations (1) and (3) to obtain another equation with no b-term.

$$\begin{aligned} 2a - 3b \quad\quad &= 2 \quad (1)\\ 3b - 2c &= -1 \quad \text{Multiplying (3) by } -1\\ \hline 2a \quad\quad - 2c &= 1 \quad (5) \end{aligned}$$

5. Now solve the system of Equations (4) and (5).

$28a + 16c = 3$ (4)
$2a - 2c = 1$ (5)

$28a + 16c = 3$ (4)
$16a - 16c = 8$ Multiplying (5) by 8
$44a = 11$

$a = \frac{1}{4}$

$2 \cdot \frac{1}{4} - 2c = 1$ Substituting $\frac{1}{4}$ for a in (5)

$\frac{1}{2} - 2c = 1$

$-2c = \frac{1}{2}$

$c = -\frac{1}{4}$

6. Substitute in Equation (1) or (2) to find b.

$2\left(\frac{1}{4}\right) - 3b = 2$ Substituting $\frac{1}{4}$ for a in (1)

$\frac{1}{2} - 3b = 2$

$-3b = \frac{3}{2}$

$b = -\frac{1}{2}$

We obtain $\left(\frac{1}{4}, -\frac{1}{2}, -\frac{1}{4}\right)$. This checks, so it is the solution.

26. $(3, 4, -1)$

27. $x + y + z = 180,$ (1)
$y = 2 + 3x,$ (2)
$z = 80 + x$ (3)

1. Only Equation (1) is in standard form. Rewrite the system with all equations in standard form.

$x + y + z = 180,$ (1)
$-3x + y = 2,$ (4)
$-x + z = 80$ (5)

2. There are no fractions or decimals.

3., 4. Note that there is no z in Equation (4). We will use Equations (1) and (5) to obtain another equation with no z-term.

$x + y + z = 180$ (1)
$x - z = -80$ Multiplying (5) by -1
$2x + y = 100$ (6)

5. Now solve the system of Equations (4) and (6).

$-3x + y = 2$ (4)
$2x + y = 100$ (5)

$-3x + y = 2$ (4)
$-2x - y = -100$ Multiplying (6) by -1
$-5x = -98$

$x = \frac{98}{5}$

$-3 \cdot \frac{98}{5} + y = 2$ Substituting $\frac{98}{5}$ for x in (4)

$-\frac{294}{5} + y = 2$

$y = \frac{304}{5}$

6. Substitute in Equation (1) or (5) to find z.

$-\frac{98}{5} + z = 80$ Substituting $\frac{98}{5}$ for x in (5)

$z = \frac{498}{5}$

We obtain $\left(\frac{98}{5}, \frac{304}{5}, \frac{498}{5}\right)$. This checks, so it is the solution.

28. $(2, 5, -3)$

29. $x + y = 0,$ (1)
$x + z = 1,$ (2)
$2x + y + z = 2$ (3)

1., 2. The equations are already in standard form with no fractions or decimals.

3., 4. Note that there is no z in Equation (1). We will use Equations (2) and (3) to obtain another equation with no z-term.

$-x - z = -1$ Multiplying (2) by -1
$2x + y + z = 2$ (3)
$x + y = 1$ (4)

5. Now solve the system of Equations (1) and (4).

$x + y = 0$ (1)
$x + y = 1$ (4)

$x + y = 0$ (1)
$-x - y = -1$ Multiplying (4) by -1
$0 = -1$ Adding

We get a false equation. There is no solution.

30. No solution

31. $y + z = 1,$ (1)
$x + y + z = 1,$ (2)
$x + 2y + 2z = 2$ (3)

1., 2. The equations are already in standard form with no fractions or decimals.

3., 4. Note that there is no x in Equation (1). We will use Equations (2) and (3) to obtain another equation with no x-term.

$$\begin{array}{rl} -x - y - z = -1 & \text{Multiplying (2) by } -1 \\ x + 2y + 2z = 2 & (3) \\ \hline y + z = 1 & (4) \end{array}$$

Equations (1) and (4) are identical. This means that Equations (1), (2), and (3) are dependant. (We have seen that if Equation (2) is multiplied by -1 and added to Equation (3), the result is Equation (1).)

32. The equations are dependent.

33.
$$f(x) = 2x + 7$$
$$f(a+1) = 2(a+1) + 7 = 2a + 2 + 7 = 2a + 9$$

34. $\{x|x \text{ is a real number and } x \neq 7\}$

35.
$$K = \frac{1}{2}t(a - b)$$
$$\frac{2K}{t} = a - b \quad \text{Multiplying by } \frac{2}{t}$$
$$\frac{2K}{t} - a = -b$$
$$-\frac{2K}{t} + a = b \quad \text{Multiplying by } -1$$

This result can also be expressed as $b = \dfrac{at - 2K}{t}$.

36. $a = \dfrac{2K}{t} + b$, or $\dfrac{2K + bt}{t}$

37. ◈

38. ◈

39. ◈

40. ◈

41.
$$\frac{x+2}{3} - \frac{y+4}{2} + \frac{z+1}{6} = 0,$$
$$\frac{x-4}{3} + \frac{y+1}{4} - \frac{z-2}{2} = -1,$$
$$\frac{x+1}{2} + \frac{y}{2} + \frac{z-1}{4} = \frac{3}{4}$$

1., 2. We clear fractions and write each equation in standard form.

To clear fractions, we multiply both sides of each equation by the LCM of its denominators. The LCM's are 6, 12, and 4, respectively.

$$6\left(\frac{x+2}{3} - \frac{y+4}{2} + \frac{z+1}{6}\right) = 6 \cdot 0$$
$$2(x+2) - 3(y+4) + (z+1) = 0$$
$$2x + 4 - 3y - 12 + z + 1 = 0$$
$$2x - 3y + z = 7$$

$$12\left(\frac{x-4}{3} + \frac{y+1}{4} - \frac{z-2}{2}\right) = 12 \cdot (-1)$$
$$4(x-4) + 3(y+1) - 6(z-2) = -12$$
$$4x - 16 + 3y + 3 - 6z + 12 = -12$$
$$4x + 3y - 6z = -11$$

$$4\left(\frac{x+1}{2} + \frac{y}{2} + \frac{z-1}{4}\right) = 4 \cdot \frac{3}{4}$$
$$2(x+1) + 2(y) + (z-1) = 3$$
$$2x + 2 + 2y + z - 1 = 3$$
$$2x + 2y + z = 2$$

The resulting system is

$$\begin{array}{ll} 2x - 3y + z = 7, & (1) \\ 4x + 3y - 6z = -11, & (2) \\ 2x + 2y + z = 2 & (3) \end{array}$$

3., 4. We eliminate z from two different pairs of equations.

$$\begin{array}{rl} 12x - 18y + 6z = 42 & \text{Multiplying (1) by 6} \\ 4x + 3y - 6z = -11 & (2) \\ \hline 16x - 15y \quad = 31 & \text{(4) Adding} \end{array}$$

$$\begin{array}{rl} 2x - 3y + z = 7 & (1) \\ -2x - 2y - z = -2 & \text{Multiplying (3) by } -1 \\ \hline -5y \quad = 5 & \text{(5) Adding} \end{array}$$

5. Solve (5) for y: $-5y = 5$

$$y = -1$$

Substitute -1 for y in (4):

$$16x - 15(-1) = 31$$
$$16x + 15 = 31$$
$$16x = 16$$
$$x = 1$$

6. Substitute 1 for x and -1 for y in (1):

$$2 \cdot 1 - 3(-1) + z = 7$$
$$5 + z = 7$$
$$z = 2$$

We obtain $(1, -1, 2)$. This checks, so it is the solution.

42. $(1, -2, 4, -1)$

43.
$$\begin{array}{ll} w + x - y + z = 0, & (1) \\ w - 2x - 2y - z = -5, & (2) \\ w - 3x - y + z = 4, & (3) \\ 2w - x - y + 3z = 7 & (4) \end{array}$$

The equations are already in standard form with no fractions or decimals.

Start by eliminating z from three different pairs of equations.

$$\begin{array}{rl} w + x - y + z = 0 & (1) \\ w - 2x - 2y - z = -5 & (2) \\ \hline 2w - x - 3y \quad = -5 & \text{(5) Adding} \end{array}$$

$$\begin{array}{rll} w - 2x - 2y - z = -5 & (2) & \\ w - 3x - y + z = 4 & (3) & \\ \hline 2w - 5x - 3y \quad = -1 & (6) & \text{Adding} \end{array}$$

$$\begin{array}{rl} 3w - 6x - 6y - 3z = -15 & \text{Multiplying (2) by 3} \\ 2w - x - y + 3z = 7 & (4) \\ \hline 5w - 7x - 7y \quad = -8 & (7) \quad \text{Adding} \end{array}$$

Now solve the system of equations (5), (6), and (7).

$2w - x - 3y = -5,$ (5)

$2w - 5x - 3y = -1,$ (6)

$5w - 7x - 7y = -8.$ (7)

$$\begin{array}{rl} 2w - x - 3y = -5 & (5) \\ -2w + 5x + 3y = 1 & \text{Multiplying (6) by } -1 \\ \hline 4x \quad = -4 & \\ x \quad = -1 & \end{array}$$

Substituting -1 for x in (5) and (7) and simplifying, we have

$2w - 3y = -6,$ (8)

$5w - 7y = -15.$ (9)

Now solve the system of Equations (8) and (9).

$$\begin{array}{rl} 10w - 15y = -30 & \text{Multiplying (8) by 5} \\ -10w + 14y = 30 & \text{Multiplying (9) by } -2 \\ \hline -y = 0 & \\ y = 0 & \end{array}$$

Substitute 0 for y in Equation (8) or (9) and solve for w.

$$\begin{array}{rl} 2w - 3 \cdot 0 = -6 & \text{Substituting in (8)} \\ 2w = -6 & \\ w = -3 & \end{array}$$

Substitute in one of the original equations to find z.

$$\begin{array}{rl} -3 - 1 - 0 + z = 0 & \text{Substituting in (1)} \\ -4 + z = 0 & \\ z = 4 & \end{array}$$

We obtain $(-3, -1, 0, 4)$. This checks, so it is the solution.

44. $\left(-1, \frac{1}{5}, -\frac{1}{2}\right)$

45. $\frac{2}{x} + \frac{2}{y} - \frac{3}{z} = 3,$

$\frac{1}{x} - \frac{2}{y} - \frac{3}{z} = 9,$

$\frac{7}{x} - \frac{2}{y} + \frac{9}{z} = -39$

Let u represent $\frac{1}{x}$, v represent $\frac{1}{y}$, and w represent $\frac{1}{z}$. Substituting, we have

$2u + 2v - 3w = 3,$ (1)

$u - 2v - 3w = 9,$ (2)

$7u - 2v + 9w = -39$ (3)

1., 2. The equations in u, v, and w are in standard form with no fractions or decimals.

3., 4. We eliminate v from two different pairs of equations.

$$\begin{array}{rll} 2u + 2v - 3w = 3 & (1) & \\ u - 2v - 3w = 9 & (2) & \\ \hline 3u \quad - 6w = 12 & (4) & \text{Adding} \end{array}$$

$$\begin{array}{rll} 2u + 2v - 3w = 3 & (1) & \\ 7u - 2v + 9w = -39 & (3) & \\ \hline 9u \quad + 6w = -36 & (5) & \text{Adding} \end{array}$$

5. Now solve the system of Equations (4) and (5).

$$\begin{array}{rl} 3u - 6w = 12, & (4) \\ 9u + 6w = -36 & (5) \\ \hline 12u \quad = -24 & \\ u = -2 & \end{array}$$

$$\begin{array}{rl} 3(-2) - 6w = 12 & \text{Substituting in (4)} \\ -6 - 6w = 12 & \\ -6w = 18 & \\ w = -3 & \end{array}$$

6. Substitute in Equation (1), (2), or (3) to find v.

$$\begin{array}{rl} 2(-2) + 2v - 3(-3) = 3 & \text{Substituting in (1)} \\ 2v + 5 = 3 & \\ 2v = -2 & \\ v = -1 & \end{array}$$

Solve for x, y, and z. We substitute -2 for u, -1 for v, and -3 for w.

$$\begin{array}{lll} u = \frac{1}{x} & v = \frac{1}{y} & w = \frac{1}{z} \\ -2 = \frac{1}{x} & -1 = \frac{1}{y} & -3 = \frac{1}{z} \\ x = \frac{1}{2} & y = -1 & z = -\frac{1}{3} \end{array}$$

We obtain $\left(-\frac{1}{2}, -1, -\frac{1}{3}\right)$. This checks, so it is the solution.

46. 12

47. $5x - 6y + kz = -5,$ (1)

$x + 3y - 2z = 2,$ (2)

$2x - y + 4z = -1$ (3)

Eliminate y from two different pairs of equations.

$$\begin{array}{rl} 5x - 6y + \quad kz = -5 & (1) \\ 2x + 6y - \quad 4z = 4 & \text{Multiplying (2) by 2} \\ \hline 7x \quad + (k-4)z = -1 & (4) \end{array}$$

$$\begin{array}{rl} x + 3y - 2z = 2 & (2) \\ 6x - 3y + 12z = -3 & \text{Multiplying (3) by 3} \\ \hline 7x \quad + 10z = -1 & (5) \end{array}$$

Solve the system of Equations (4) and (5).

$$\begin{array}{rl} 7x + (k-4)z = -1 & (4) \\ 7x + \quad 10z = -1 & (5) \end{array}$$

$$\begin{array}{rl} -7x - (k-4)z = 1 & \text{Multiplying (4) by } -1 \\ 7x + \quad 10z = -1 & (5) \\ \hline (-k+14)z = 0 & (6) \end{array}$$

The system is dependent for the value of k that makes Equation (6) true. This occurs when $-k+14$ is 0. We solve for k:

$$-k + 14 = 0$$
$$14 = k$$

48. $3x + 4y + 2z = 12$

49. $z = b - mx - ny$

Three solutions are $(1,1,2)$, $(3,2,-6)$, and $\left(\frac{3}{2},1,1\right)$. We substitute for x, y, and z and then solve for b, m, and n.

$$2 = b - m - n,$$
$$-6 = b - 3m - 2n,$$
$$1 = b - \frac{3}{2}m - n$$

1., 2. Write the equations in standard form. Also, clear the fraction in the last equation.

$$\begin{array}{rl} b - m - n = 2, & (1) \\ b - 3m - 2n = -6, & (2) \\ 2b - 3m - 2n = 2 & (3) \end{array}$$

3., 4. Eliminate b from two different pairs of equations.

$$\begin{array}{rl} b - m - n = 2 & (1) \\ -b + 3m + 2n = 6 & \text{Multiplying (2) by } -1 \\ \hline 2m + n = 8 & \text{(4)} \quad \text{Adding} \end{array}$$

$$\begin{array}{rl} -2b + 2m + 2n = -4 & \text{Multiplying (1) by } -2 \\ 2b - 3m - 2n = 2 & (3) \\ \hline -m \quad = -2 & \text{(5)} \quad \text{Adding} \end{array}$$

5. We solve Equation (5) for m:

$$-m = -2$$
$$m = 2$$

Substitute in Equation (4) and solve for n.

$$2 \cdot 2 + n = 8$$
$$4 + n = 8$$
$$n = 4$$

6. Substitute in one of the original equations to find b.

$$b - 2 - 4 = 2 \quad \text{Substituting 2 for } m \text{ and 4 for } n \text{ in (1)}$$
$$b - 6 = 2$$
$$b = 8$$

The solution is $(8,2,4)$, so the equation is $z = 8 - 2x - 4y$.

Exercise Set 3.5

1. ***Familiarize.*** Let x = the first number, y = the second number, and z = the third number.

Translate.

The sum of three numbers is 57.

$$x + y + z = 57$$

The second is 3 more than the first.

$$y = 3 + x$$

The third is 6 more than the first.

$$z = 6 + x$$

We now have a system of equations.

$$\begin{array}{lcl} x + y + z = 57, & \text{or} & x + y + z = 57, \\ y = 3 + x & & -x + y \quad = 3, \\ z = 6 + x & & -x \quad + z = 6 \end{array}$$

Carry out. Solving the system we get $(16,19,22)$.

Check. The sum of the three numbers is $16+19+22$, or 57. The second number, 19, is three more than the first number, 16. The third number, 22, is 6 more than the first number, 16. The numbers check.

State. The numbers are 16, 19, and 22.

2. 4, 2, -1

3. ***Familiarize.*** Let x = the first number, y = the second number, and z = the third number.

Translate.

The sum of three numbers is 26.

$$x + y + z = 26$$

Twice the first minus the second is the third less 2

$$2x - y = z - 2$$

The third is the second minus 3 times the first.

$$z = y - 3x$$

We now have a system of equations.

$$\begin{array}{lcl} x + y + z = 26, & \text{or} & x + y + z = 26, \\ 2x - y = z - 2, & & 2x - y - z = -2, \\ z = y - 3x & & 3x - y + z = 0 \end{array}$$

Carry out. Solving the system we get $(8, 21, -3)$.

Check. The sum of the numbers is $8 + 21 - 3$, or 26. Twice the first minus the second is $2 \cdot 8 - 21$, or -5, which is 2 less than the third. The second minus three times the first is $21 - 3 \cdot 8$, or -3, which is the third. The numbers check.

State. The numbers are 8, 21, and -3.

4. 17, 9, 79

5. ***Familiarize***. We first make a drawing.

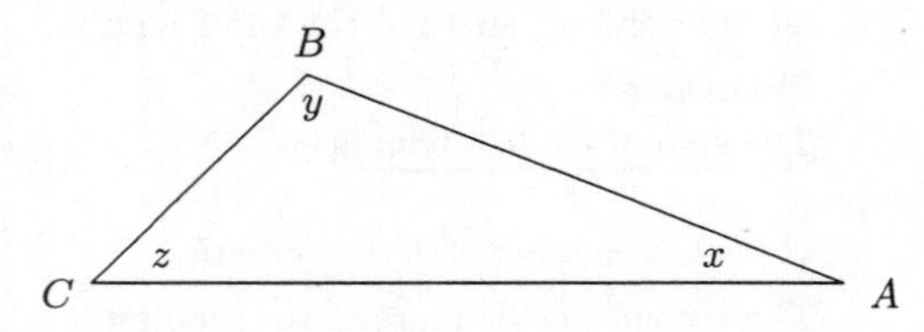

We let x, y, and z represent the measures of angles A, B, and C, respectively. The measures of the angles of a triangle add up to 180°.

Translate.

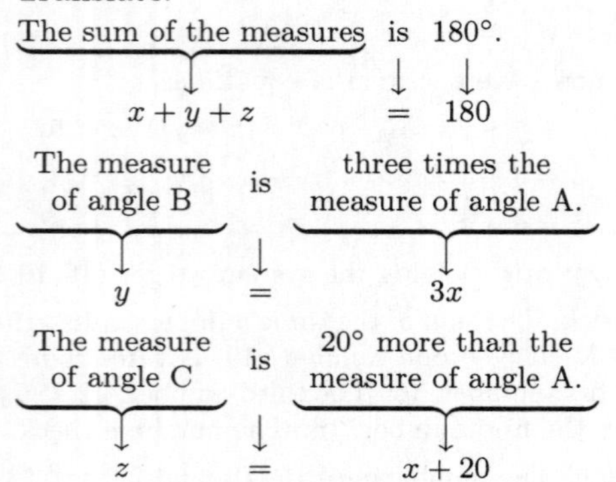

We now have a system of equations.

$$x + y + z = 180,$$
$$y = 3x,$$
$$z = x + 20$$

Carry out. Solving the system we get $(32, 96, 52)$.

Check. The sum of the measures is $32° + 96° + 52°$, or 180°. Three times the measure of angle A is $3 \cdot 32°$, or 96°, the measure of angle B. 20° more than the measure of angle A is $32° + 20°$, or 52°, the measure of angle C. The numbers check.

State. The measures of angles A, B, and C are 32°, 96°, and 52°, respectively.

6. A: 25°, B: 50°, C: 105°

7. ***Familiarize***. Let x = the cost of automatic transmission, y = the cost of power door locks, and z = the cost of air conditioning. The prices of the options are added to the basic price of \$12,685.

Translate.

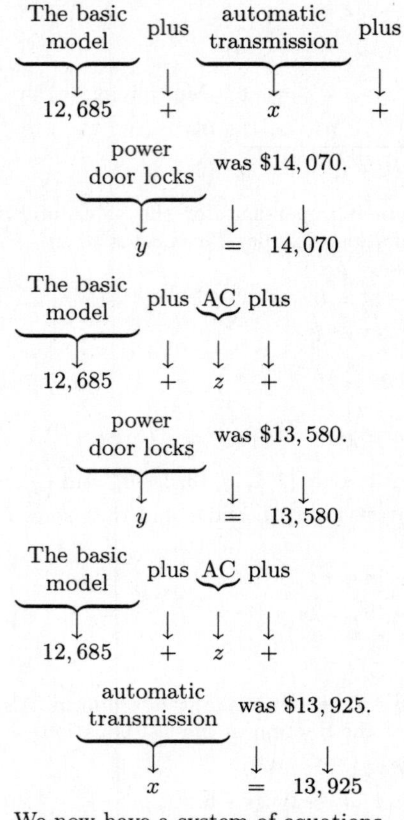

We now have a system of equations.

$$12,685 + x + y = 14,070,$$
$$12,685 + z + y = 13,580,$$
$$12,685 + z + x = 13,925$$

Carry out. Solving the system we get $(865, 520, 375)$.

Check. The basic model with automatic transmission and power door locks costs \$12,685 + \$865 + \$520, or \$14,070. The basic model with AC and power door locks costs \$12,685 + \$375 + \$520, or \$13,580. The basic model with AC and automatic transmission costs \$12,685+\$375+\$865, or \$13,925. The numbers check.

State. Automatic transmission costs \$865, power door locks cost \$520, and AC costs \$375.

8. Sven: 220, Tillie: 250, Isaiah: 270

9. ***Familiarize***. It helps to organize the information in a table. We let x, y, and z represent the weekly productions of the individual machines.

Machines Working	A	B	C
Weekly Production	x	y	z

Machines Working	A & B	B & C	A, B, & C
Weekly Production	3400	4200	5700

Translate. From the table, we obtain three equations.

$x+y+z=5700$ (All three machines working)

$x+y \quad =3400$ (A and B working)

$y+z=4200$ (B and C working)

Carry out. Solving the system we get $(1500, 1900, 2300)$.

Check. The sum of the weekly productions of machines A, B & C is $1500+1900+2300$, or 5700. The sum of the weekly productions of machines A and B is $1500+1900$, or 3400. The sum of the weekly productions of machines B and C is $1900+2300$, or 4200. The numbers check.

State. In a week Machine A can polish 1500 lenses, Machine B can polish 1900 lenses, and Machine C can polish 2300 lenses.

10. Elrod: 20, Dot: 24, Wendy: 30

11. ***Familiarize.*** Let x = the number of 10-oz cups, y = the number of 14-oz cups, and z = the number of 20-oz cups that Kyle filled. Note that five 96-oz pots contain $5\cdot 96$ oz, or 480 oz of coffee. Also, x 10-oz cups contain a total of $10x$ oz of coffee and bring in $\$0.95x$, y 14-oz cups contain $14y$ oz and bring in $\$1.15y$, and z 20-oz cups contain $20z$ oz and bring in $\$1.50z$.

Translate.

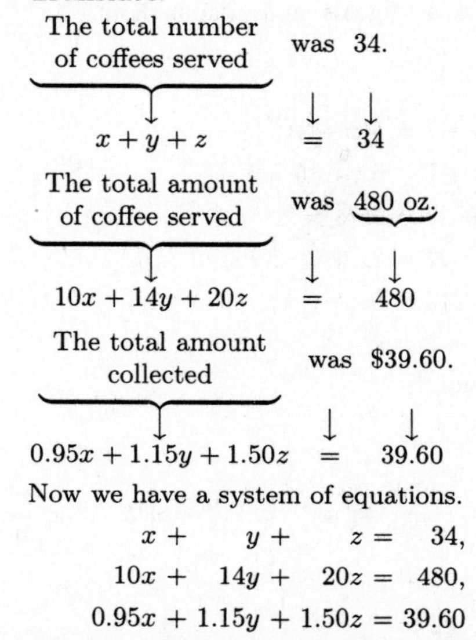

Now we have a system of equations.

$$x+y+z=34,$$
$$10x+14y+20z=480,$$
$$0.95x+1.15y+1.50z=39.60$$

Carry out. Solving the system we get $(8, 20, 6)$.

Check. The total number of coffees served was $8+20+6$, or 34, The total amount of coffee served was $10\cdot 8+14\cdot 20+20\cdot 6=80+280+120=480$ oz. The total amount collected was $\$0.95(8)+\$1.15(20)+\$1.50(6)=\$7.60+\$23.00+\$9.00=\$39.60$. The numbers check.

State. Kyle filled 8 10-oz cups, 20 14-oz cups, and 6 20-oz cups.

12. 15 small, 30 medium, 10 large

13. ***Familiarize.*** Let x = the amount invested in the first fund, y = the amount invested in the second fund, and z = the amount invested in the third fund. Then the earnings from the investments were $0.1x$, $0.06y$, and $0.15z$.

Translate.

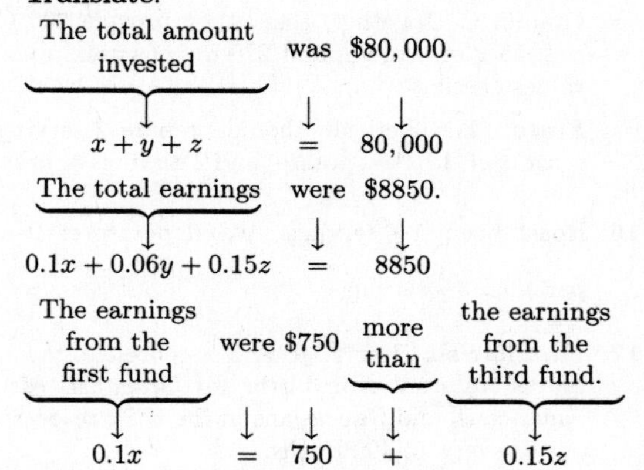

Now we have a system of equations.

$$x+y+z=80{,}000$$
$$0.1x+0.06y+0.15z=8850,$$
$$0.1x=750+0.15z$$

Carry out. Solving the system we get $(45{,}000,\ 10{,}000,\ 25{,}000)$.

Check. The total investment was $\$45{,}000+\$10{,}000+\$25{,}000$, or \$80,000. The total earnings were $0.1(\$45{,}000)+0.06(10{,}000)+0.15(25{,}000)=\$4500+\$600+\$3750=\$8850$. The earnings from the first fund, \$4500, were \$750 more than the earnings from the second fund, \$3750.

State. \$45,000 was invested in the first fund, \$10,000 in the second fund, and \$25,000 in the third fund.

14. Newspaper: \$41.1 billion, television: \$36 billion, radio: \$7.7 billion

15. ***Familiarize.*** Let r = the number of servings of roast beef, p = the number of baked potatoes, and b = the number of servings of broccoli. Then r servings of roast beef contain $300r$ Calories, $20r$ g of protein, and no vitamin C. In p baked potatoes there are $100p$ Calories, $5p$ g of protein, and $20p$ mg of vitamin C. And b servings of broccoli contain $50b$ Calories, $5b$ g of protein, and $100b$ mg of vitamin C. The patient requires 800 Calories, 55 g of protein, and 220 mg of vitamin C.

Translate. Write equations for the total number of calories, the total amount of protein, and the total amount of vitamin C.

$$\begin{aligned} 300r + 100p + \ 50b &= 800 \quad \text{(Calories)} \\ 20r + \quad 5p + \quad 5b &= \ 55 \quad \text{(protein)} \\ 20p + 100b &= 220 \quad \text{(vitamin C)} \end{aligned}$$

We now have a system of equations.

Carry out. Solving the system we get $(2, 1, 2)$.

Check. Two servings of roast beef provide 600 Calories, 40 g of protein, and no vitamin C. One baked potato provides 100 Calories, 5 g of protein, and 20 mg of vitamin C. And 2 servings of broccoli provide 100 Calories, 10 g of protein, and 200 mg of vitamin C. Together, then, they provide 800 Calories, 55 g of protein, and 220 mg of vitamin C. The values check.

State. The dietician should prepare 2 servings of roast beef, 1 baked potato, and 2 servings of broccoli.

16. Roast beef: $1\frac{1}{8}$ servings, baked potatoes: $2\frac{3}{4}$, asparagus: $3\frac{3}{4}$ servings

17. ***Familiarize.*** Let x, y, and z represent the number of fraternal twin births for Orientals, African-Americans, and Caucasians in the U.S., respectively, out of every 15,400 births.

Translate. Out of every 15,400 births, we have the following statistics:

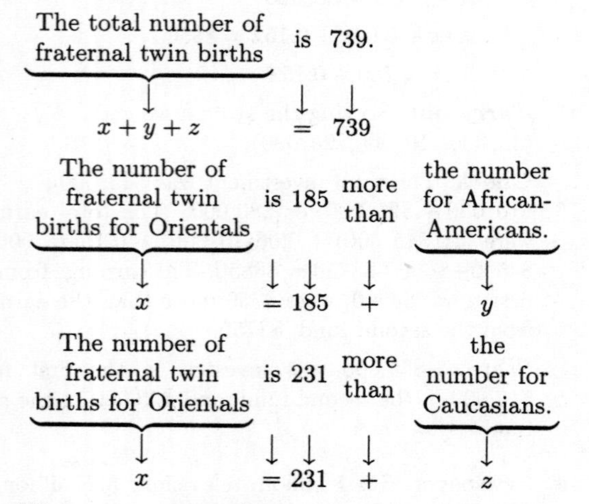

We have a system of equations.

$$x + y + z = 739,$$
$$x = 185 + y,$$
$$x = 231 + y$$

Carry out. Solving the system we get $(385, 200, 154)$.

Check. The total of the numbers is 739. Also 385 is 185 more than 200, and it is 231 more than 154.

State. Out of every 15,400 births, there are 385 births of fraternal twins for Orientals, 200 for African-Americans, and 154 for Caucasians.

18. Man: 3.6, woman: 18.1, one-year old child: 50

19. ***Familiarize.*** Let x, y, and z represent the number of 2-point field goals, 3-point field goals, and 1-point foul shots made, respectively. The total number of points scored from each of these types of goals is $2x$, $3y$, and z.

Translate.

The total number of points | was | 92.

$$2x + 3y + z = 92$$

The total number of baskets | was | 50.

$$x + y + z = 50$$

The number of 2-pointers | was 19 | more than | the number of foul shots.

$$x = 19 + z$$

Now we have a system of equations.

$$2x + 3y + z = 92,$$
$$x + y + z = 50,$$
$$x = 19 + z$$

Carry out. Solving the system we get $(32, 5, 13)$.

Check. The total number of points was $2 \cdot 32 + 3 \cdot 5 + 13 = 64 + 15 + 13 = 92$. The number of baskets was $32 + 5 + 13$, or 50. The number of 2-pointers, 32, was 19 more than the number of foul shots, 13. The numbers check.

State. The Knicks made 32 two-point field goals, 5 three-point field goals, and 13 foul shots.

20. 1869

21.
$$\begin{aligned} 3(5 - x) + 7 &= 5(x + 3) - 9 \\ 15 - 3x + 7 &= 5x + 15 - 9 \\ -3x + 22 &= 5x + 6 \\ 22 &= 8x + 6 \\ 16 &= 8x \\ 2 &= x \end{aligned}$$

The solution is 2.

22. 8

23. $\dfrac{(a^2b^3)^5}{a^7b^{16}} = \dfrac{a^{10}b^{15}}{a^7b^{16}} = a^{10-7}b^{15-16} = a^3b^{-1}$, or $\dfrac{a^3}{b}$

24. $y = -\frac{3}{5}x - 7$

25. $g(x) = \dfrac{x-5}{x+7}$

We cannot compute $g(x)$ when the denominator is 0. We solve an equation to determine when this occurs.

$$x + 7 = 0$$
$$x = -7$$

Thus, the domain of g is $\{x|x \text{ is a real number and } x \neq -7\}$.

26. 76

27. ◈

28. ◈

29. ***Familiarize.*** Let x = the one's digit, y = the ten's digit, and z = the hundred's digit. Then the number is represented by $100z + 10y + x$. When the digits are reversed, the resulting number is represented by $100x + 10y + z$.

Translate.

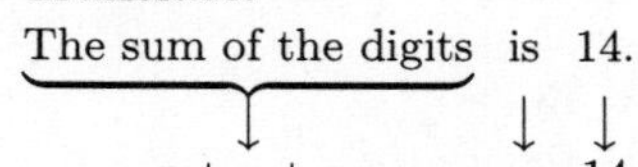

$x + y + z \quad = \quad 14$

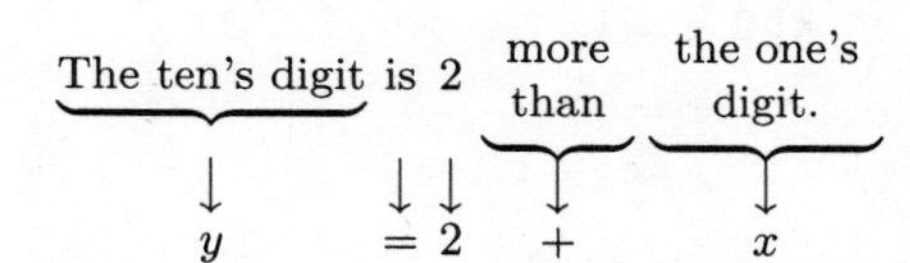

$y \quad = 2 \quad + \quad x$

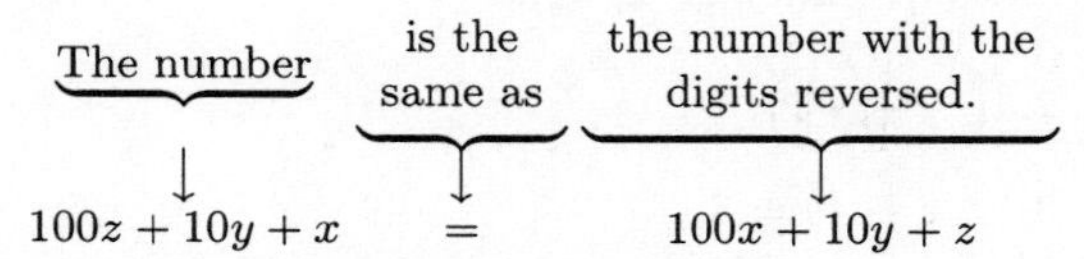

$100z + 10y + x \quad = \quad 100x + 10y + z$

Now we have a system of equations.

$$x + y + z = 14,$$
$$y = 2 + x,$$
$$100z + 10y + x = 100x + 10y + z$$

Carry out. Solving the system we get $(4, 6, 4)$.

Check. If the number is 464, then the sum of the digits is $4+6+4$, or 14. The ten's digit, 6, is 2 more than the one's digit, 4. If the digits are reversed the number is unchanged The result checks.

State. The number is 464.

30. 20

31. ***Familiarize.*** Let x = the number of adults, y = the number of students, and z = the number of children in attendance.

Translate. The given information gives rise to two equations.

The total number in attendance was 100.

$$x + y + z = 100$$

The total amount taken in was \$100.

$$10x + 3y + 0.5z = 100$$

Now we have a system of equations.

$$x + y + z = 100,$$
$$10x + 3y + 0.5z = 100$$

Multiply the second equation by 2 to clear the decimal:

$$x + y + z = 100, \quad (1)$$
$$20x + 6y + z = 200. \quad (2)$$

Carry out. We use the elimination method.

$$-x - y - z = -100 \quad \text{Multiplying (1) by } -1$$
$$20x + 6y + z = 200 \quad (2)$$
$$\overline{19x + 5y \quad = 100} \quad (3)$$

In (3), note that 5 is a factor of both $5y$ and 100. Therefore, 5 must also be a factor of $19x$, and hence of x, since 5 is not a factor of 19. Then for some positive integer n, $x = 5n$. (We require $n > 0$, since the number of adults clearly cannot be negative and must also be nonzero since the exercise states that the audience consists of *adults*, students, and children.) We have

$$19 \cdot 5n + 5y = 100, \text{ or}$$
$$19n + y = 20. \quad \text{Dividing by 5 on both sides}$$

Since n and y must both be positive, $n = 1$. Otherwise, $19n+y$ would be greater than 20. Then $x = 5 \cdot 1$, or 5.

$$19 \cdot 5 + 5y = 100 \quad \text{Substituting in (3)}$$
$$95 + 5y = 100$$
$$5y = 5$$
$$y = 1$$

$$5 + 1 + z = 100 \quad \text{Substituting in (1)}$$
$$6 + z = 100$$
$$z = 94$$

Check. The number of people in attendance was $5 + 1 + 94$, or 100. The amount of money taken in was $\$10 \cdot 5 + \$3 \cdot 1 + \$0.50(94) = \$50 + \$3 + \$47 = \$100$. The numbers check.

State. There were 5 adults, 1 student, and 94 children.

32. 35

33. ***Familiarize***. We first make a drawing with additional labels.

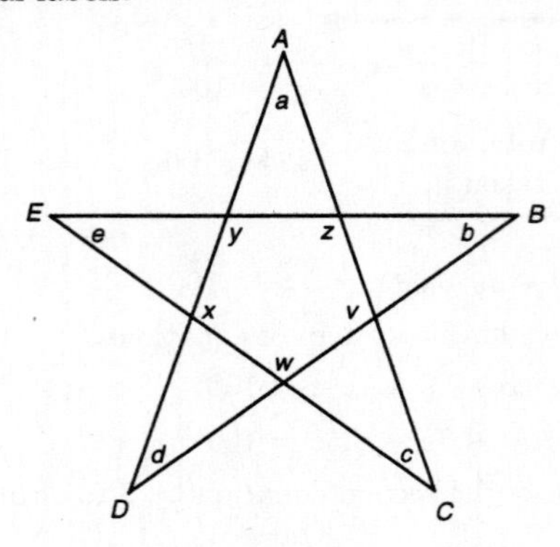

We let a, b, c, d, and e represent the angle measures at the tips of the star. We also label the interior angles of the pentagon v, w, x, y, and z. We recall the following geometric fact:

The sum of the measures of the interior angles of a polygon of n sides is given by $(n-2)180°$.

Using this fact we know:

1. The sum of the angle measures of a triangle is $(3-2)180°$, or $180°$.
2. The sum of the angle measures of a pentagon is $(5-2)180°$, or $3(180°)$.

Translate. Using fact (1) listed above we obtain a system of 5 equations.

$$a+v+d=180$$
$$b+w+e=180$$
$$c+x+a=180$$
$$d+y+b=180$$
$$e+z+c=180$$

Carry out. Adding we obtain

$$2a+2b+2c+2d+2e+v+w+x+y+z=5(180)$$

$$2(a+b+c+d+e)+(v+w+x+y+z)=5(180)$$

Using fact (2) listed above we substitute $3(180)$ for $(v+w+x+y+z)$ and solve for $(a+b+c+d+e)$.

$$\begin{aligned} 2(a+b+c+d+e)+3(180) &= 5(180) \\ 2(a+b+c+d+e) &= 2(180) \\ a+b+c+d+e &= 180 \end{aligned}$$

Check. We should repeat the above calculations.

State. The sum of the angle measures at the tips of the star is $180°$.

Exercise Set 3.6

1. $5x - 3y = 13,$
$4x + y = 7$

Write a matrix using only the constants.

$$\left[\begin{array}{cc|c} 5 & -3 & 13 \\ 4 & 1 & 7 \end{array}\right]$$

Multiply row 2 by 5 to make the first number in row 2 a multiple of 5.

$$\left[\begin{array}{cc|c} 5 & -3 & 13 \\ 20 & 5 & 35 \end{array}\right] \quad \text{New Row 2} = 5(\text{Row 2})$$

Multiply row 1 by -4 and add it to row 2.

$$\left[\begin{array}{cc|c} 5 & -3 & 13 \\ 0 & 17 & -17 \end{array}\right] \quad \text{New Row 2} = -4(\text{Row 1}) + \text{Row 2}$$

Reinserting the variables, we have

$$5x - 3y = 13, \quad (1)$$
$$17y = -17. \quad (2)$$

Solve Equation (2) for y.

$$17y = -17$$
$$y = -1$$

Back-substitute -1 for y in Equation (1) and solve for x.

$$\begin{aligned} 5x - 3y &= 13 \\ 5x - 3(-1) &= 13 \\ 5x + 3 &= 13 \\ 5x &= 10 \\ x &= 2 \end{aligned}$$

The solution is $(2, -1)$.

2. $\left(-\frac{1}{3}, -4\right)$

3. $x + 4y = 8,$
$3x + 5y = 3$

We first write a matrix using only the constants.

$$\left[\begin{array}{cc|c} 1 & 4 & 8 \\ 3 & 5 & 3 \end{array}\right]$$

Multiply the first row by -3 and add it to the second row.

$$\left[\begin{array}{cc|c} 1 & 4 & 8 \\ 0 & -7 & -21 \end{array}\right] \quad \text{New Row 2} = -3(\text{Row 1}) + \text{Row 2}$$

Reinserting the variables, we have

$$x + 4y = 8, \quad (1)$$
$$-7y = -21. \quad (2)$$

Solve Equation (2) for y.

$$-7y = -21$$
$$y = 3$$

Back-substitute 3 for y in Equation (1) and solve for x.

$$\begin{aligned} x + 4 \cdot 3 &= 8 \\ x + 12 &= 8 \\ x &= -4 \end{aligned}$$

The solution is $(-4, 3)$.

4. $(-3, 2)$

5. $6x - 2y = 4,$
$7x + y = 13$

Write a matrix using only the constants.

$$\left[\begin{array}{cc|c} 6 & -2 & 4 \\ 7 & 1 & 13 \end{array}\right]$$

Multiply the second row by 6 to make the first number in row 2 a multiple of 6.

$$\left[\begin{array}{cc|c} 6 & -2 & 4 \\ 42 & 6 & 78 \end{array}\right] \quad \text{New Row 2} = 6(\text{Row 2})$$

Now multiply the first row by -7 and add it to the second row.

$$\left[\begin{array}{cc|c} 6 & -2 & 4 \\ 0 & 20 & 50 \end{array}\right] \quad \text{New Row 2} = -7(\text{Row 1}) + \text{Row 2}$$

Reinserting the variables, we have

$$6x - 2y = 4, \quad (1)$$
$$20y = 50. \quad (2)$$

Solve Equation (2) for y.

$$20y = 50$$
$$y = \frac{5}{2}$$

Back-substitute $\frac{5}{2}$ for y in Equation (1) and solve for x.

$$6x - 2y = 4$$
$$6x - 2\left(\frac{5}{2}\right) = 4$$
$$6x - 5 = 4$$
$$6x = 9$$
$$x = \frac{3}{2}$$

The solution is $\left(\frac{3}{2}, \frac{5}{2}\right)$.

6. $\left(-1, \frac{5}{2}\right)$

7. $4x - y - 3z = 1,$
$8x + y - z = 5,$
$2x + y + 2z = 5$

Write a matrix using only the constants.

$$\left[\begin{array}{ccc|c} 4 & -1 & -3 & 1 \\ 8 & 1 & -1 & 5 \\ 2 & 1 & 2 & 5 \end{array}\right]$$

First interchange rows 1 and 3 so that each number below the first number in the first row is a multiple of that number.

$$\left[\begin{array}{ccc|c} 2 & 1 & 2 & 5 \\ 8 & 1 & -1 & 5 \\ 4 & -1 & -3 & 1 \end{array}\right]$$

Multiply row 1 by -4 and add it to row 2.

Multiply row 1 by -2 and add it to row 3.

$$\left[\begin{array}{ccc|c} 2 & 1 & 2 & 5 \\ 0 & -3 & -9 & -15 \\ 0 & -3 & -7 & -9 \end{array}\right]$$

Multiply row 2 by -1 and add it to row 3.

$$\left[\begin{array}{ccc|c} 2 & 1 & 2 & 5 \\ 0 & -3 & -9 & -15 \\ 0 & 0 & 2 & 6 \end{array}\right]$$

Reinserting the variables, we have

$$2x + y + 2z = 5, \quad (1)$$
$$-3y - 9z = -15, \quad (2)$$
$$2z = 6. \quad (3)$$

Solve (3) for z.

$$2z = 6$$
$$z = 3$$

Back-substitute 3 for z in (2) and solve for y.

$$-3y - 9z = -15$$
$$-3y - 9(3) = -15$$
$$-3y - 27 = -15$$
$$-3y = 12$$
$$y = -4$$

Back-substitute 3 for z and -4 for y in (1) and solve for x.

$$2x + y + 2z = 5$$
$$2x + (-4) + 2(3) = 5$$
$$2x - 4 + 6 = 5$$
$$2x = 3$$
$$x = \frac{3}{2}$$

The solution is $\left(\frac{3}{2}, -4, 3\right)$.

8. $\left(2, \frac{1}{2}, -2\right)$

9. $p - 2q - 3r = 3,$
$2p - q - 2r = 4,$
$4p + 5q + 6r = 4$

We first write a matrix using only the constants.

$$\left[\begin{array}{ccc|c} 1 & -2 & -3 & 3 \\ 2 & -1 & -2 & 4 \\ 4 & 5 & 6 & 4 \end{array}\right]$$

$$\left[\begin{array}{ccc|c} 1 & -2 & -3 & 3 \\ 0 & 3 & 4 & -2 \\ 0 & 13 & 18 & -8 \end{array}\right] \quad \begin{array}{l} \text{New Row 2} = -2(\text{Row 1}) + \text{Row 2} \\ \text{New Row 3} = -4(\text{Row 1}) + \text{Row 3} \end{array}$$

$$\left[\begin{array}{ccc|c} 1 & -2 & -3 & 3 \\ 0 & 3 & 4 & -2 \\ 0 & 39 & 54 & -24 \end{array}\right] \quad \text{New Row 3} = 3(\text{Row 3})$$

$$\left[\begin{array}{ccc|c} 1 & -2 & -3 & 3 \\ 0 & 3 & 4 & -2 \\ 0 & 0 & 2 & 2 \end{array}\right]$$ New Row 3 = −13(Row 2)+ Row 3

Reinstating the variables, we have

$$\begin{aligned} p - 2q - 3r &= 3, \quad (1) \\ 3q + 4r &= -2, \quad (2) \\ 2r &= 2 \quad (3) \end{aligned}$$

Solve (3) for r.

$$2r = 2$$
$$r = 1$$

Back-substitute 1 for r in (2) and solve for q.

$$\begin{aligned} 3q + 4 \cdot 1 &= -2 \\ 3q + 4 &= -2 \\ 3q &= -6 \\ q &= -2 \end{aligned}$$

Back-substitute −2 for q and 1 for r in (1) and solve for p.

$$\begin{aligned} p - 2(-2) - 3 \cdot 1 &= 3 \\ p + 4 - 3 &= 3 \\ p + 1 &= 3 \\ p &= 2 \end{aligned}$$

The solution is $(2, -2, 1)$.

10. $(-1, 2, -2)$

11.
$$\begin{aligned} 3p \qquad + 2r &= 11, \\ q - 7r &= 4, \\ p - 6q \qquad &= 1 \end{aligned}$$

We first write a matrix using only the constants.

$$\left[\begin{array}{ccc|c} 3 & 0 & 2 & 11 \\ 0 & 1 & -7 & 4 \\ 1 & -6 & 0 & 1 \end{array}\right]$$

$$\left[\begin{array}{ccc|c} 1 & -6 & 0 & 1 \\ 0 & 1 & -7 & 4 \\ 3 & 0 & 2 & 11 \end{array}\right]$$ Interchange Row 1 and Row 3

$$\left[\begin{array}{ccc|c} 1 & -6 & 0 & 1 \\ 0 & 1 & -7 & 4 \\ 0 & 18 & 2 & 8 \end{array}\right]$$ New Row 3 = −3(Row 1) + Row 3

$$\left[\begin{array}{ccc|c} 1 & -6 & 0 & 1 \\ 0 & 1 & -7 & 4 \\ 0 & 0 & 128 & -64 \end{array}\right]$$ New Row 3 = −18(Row 2) + Row 3

Reinserting the variables, we have

$$\begin{aligned} p - 6q \qquad &= 1, \quad (1) \\ q - \quad 7r &= 4, \quad (2) \\ 128r &= -64. \quad (3) \end{aligned}$$

Solve (3) for r.

$$128r = -64$$
$$r = -\frac{1}{2}$$

Back-substitute $-\frac{1}{2}$ for r in (2) and solve for q.

$$\begin{aligned} q - 7r &= 4 \\ q - 7\left(-\frac{1}{2}\right) &= 4 \\ q + \frac{7}{2} &= 4 \\ q &= \frac{1}{2} \end{aligned}$$

Back-substitute $\frac{1}{2}$ for q in (1) and solve for p.

$$\begin{aligned} p - 6 \cdot \frac{1}{2} &= 1 \\ p - 3 &= 1 \\ p &= 4 \end{aligned}$$

The solution is $\left(4, \frac{1}{2}, -\frac{1}{2}\right)$.

12. $\left(\frac{1}{2}, \frac{2}{3}, -\frac{5}{6}\right)$

13. We will rewrite the equations with the variables in alphabetical order:

$$\begin{aligned} -2w + 2x + 2y - 2z &= -10, \\ w + x + y + z &= -5, \\ 3w + x - y + 4z &= -2, \\ w + 3x - 2y + 2z &= -6 \end{aligned}$$

Write a matrix using only the constants.

$$\left[\begin{array}{cccc|c} -2 & 2 & 2 & -2 & -10 \\ 1 & 1 & 1 & 1 & -5 \\ 3 & 1 & -1 & 4 & -2 \\ 1 & 3 & -2 & 2 & -6 \end{array}\right]$$

$$\left[\begin{array}{cccc|c} -1 & 1 & 1 & -1 & -5 \\ 1 & 1 & 1 & 1 & -5 \\ 3 & 1 & -1 & 4 & -2 \\ 1 & 3 & -2 & 2 & -6 \end{array}\right]$$ New Row 1 = $\frac{1}{2}$(Row 1)

$$\left[\begin{array}{cccc|c} -1 & 1 & 1 & -1 & -5 \\ 0 & 2 & 2 & 0 & -10 \\ 0 & 4 & 2 & 1 & -17 \\ 0 & 4 & -1 & 1 & -11 \end{array}\right]$$ New Row 2 = Row 1 + Row 2; New Row 3 = 3(Row 1) +Row 3; New Row 4 = Row 1 + Row 4

$$\left[\begin{array}{cccc|c} -1 & 1 & 1 & -1 & -5 \\ 0 & 2 & 2 & 0 & -10 \\ 0 & 0 & -2 & 1 & 3 \\ 0 & 0 & -5 & 1 & 9 \end{array}\right]$$ New Row 3 = −2(Row 2) + Row 3; New Row 4 = −2(Row 2) + Row 4

$$\left[\begin{array}{cccc|c} -1 & 1 & 1 & -1 & -5 \\ 0 & 2 & 2 & 0 & -10 \\ 0 & 0 & -2 & 1 & 3 \\ 0 & 0 & -10 & 2 & 18 \end{array}\right]$$ New Row 4 = 2(Row 4)

$$\left[\begin{array}{cccc|c} -1 & 1 & 1 & -1 & -5 \\ 0 & 2 & 2 & 0 & -10 \\ 0 & 0 & -2 & 1 & 3 \\ 0 & 0 & 0 & -3 & 3 \end{array}\right]$$ New Row 4 = −5(Row 3) + Row 4

Reinserting the variables, we have

$$\begin{aligned} -w + x + y - z &= -5, && (1) \\ 2x + 2y &= -10, && (2) \\ -2y + z &= 3, && (3) \\ -3z &= 3. && (4) \end{aligned}$$

Solve (4) for z.

$$-3z = 3$$
$$z = -1$$

Back-substitute -1 for z in (3) and solve for y.

$$-2y + (-1) = 3$$
$$-2y = 4$$
$$y = -2$$

Back-substitute -2 for y in (2) and solve for x.

$$2x + 2(-2) = -10$$
$$2x - 4 = -10$$
$$2x = -6$$
$$x = -3$$

Back-substitute -3 for x, -2 for y, and -1 for z in (1) and solve for w.

$$-w + (-3) + (-2) - (-1) = -5$$
$$-w - 3 - 2 + 1 = -5$$
$$-w - 4 = -5$$
$$-w = -1$$
$$w = 1$$

The solution is $(1, -3, -2, -1)$.

14. (7,4,5,6)

15. ***Familiarize***. Let d = the number of dimes and n = the number of nickels. The value of d dimes is $\$0.10d$, and the value of n nickels is $\$0.05n$.

Translate.

Total number of coins is 34.

$$d + n = 34$$

Total value of coins is \$1.90.

$$0.10d + 0.05n = 1.90$$

After clearing decimals, we have this system.

$$d + n = 34,$$
$$10d + 5n = 190$$

Carry out. Solve using matrices.

$$\left[\begin{array}{cc|c} 1 & 1 & 34 \\ 10 & 5 & 190 \end{array}\right]$$

$$\left[\begin{array}{cc|c} 1 & 1 & 34 \\ 0 & -5 & -150 \end{array}\right]$$ New Row 2 = −10(Row 1) + Row 2

Reinserting the variables, we have

$$d + n = 34, \quad (1)$$
$$-5n = -150 \quad (2)$$

Solve (2) for n.

$$-5n = -150$$
$$n = 30$$

$$d + 30 = 34 \quad \text{Back-substituting}$$
$$d = 4$$

Check. The sum of the two numbers is 34. The total value is $\$0.10(4) + \$0.50(30) = \$0.40 + \$1.50 = \$1.90$. The numbers check.

State. There are 4 dimes and 30 nickels.

16. 21 dimes, 22 quarters

17. ***Familiarize***. We let x represent the number of pounds of the \$4.05 kind and y represent the number of pounds of the \$2.70 kind of granola. We organize the information in a table.

Granola	Number of pounds	Price per pound	Value
\$4.05 kind	x	\$4.05	$\$4.05x$
\$2.70 kind	y	\$2.70	$\$2.70y$
Mixture	15	\$3.15	$\$3.15 \times 15$ or \$47.25

Translate.

Total number of pounds is 15.

$$x + y = 15$$

Total value of mixture is \$47.25.

$$4.05x + 2.70y = 47.25$$

After clearing decimals, we have this system:

$$x + y = 15,$$
$$405x + 270y = 4725$$

Carry out. Solve using matrices.

$$\left[\begin{array}{cc|c} 1 & 1 & 15 \\ 405 & 270 & 4725 \end{array}\right]$$

$$\left[\begin{array}{cc|c} 1 & 1 & 15 \\ 0 & -135 & -1350 \end{array}\right] \quad \text{New Row 2} = -405(\text{Row 1}) + \text{Row 2}$$

Reinserting the variables, we have

$$x + y = 15, \quad (1)$$
$$-135y = -1350 \quad (2)$$

Solve (2) for y.

$$-135y = -1350$$
$$y = 10$$

Back-substitute 10 for y in (1) and solve for x.

$$x + 10 = 15$$
$$x = 5$$

Check. The sum of the numbers is 15. The total value is \$4.05(5) + \$2.70(10), or \$20.25 + \$27.00, or \$47.25. The numbers check.

State. 5 pounds of the \$4.05 per lb granola and 10 pounds of the \$2.70 per lb granola should be used.

18. 14 pounds of nuts, 6 pounds of oats

19. ***Familiarize***. We let x, y, and z represent the amounts invested at 7%, 8%, and 9%, respectively. Recall the formula for simple interest:

Interest = Principal × Rate × Time

Translate. We organize the imformation in a table.

	First Invest-ment	Second Invest-ment	Third Invest-ment	Total
P	x	y	z	\$2500
R	7%	8%	9%	
T	1 yr	1 yr	1 yr	
I	$0.07x$	$0.08y$	$0.09z$	\$212

The first row gives us one equation:

$$x + y + z = 2500$$

The last row gives a second equation:

$$0.07x + 0.08y + 0.09z = 212$$

Amount invested at 9% → z; is → $=$; \$1100 → \$1100; more than → $+$; amount invested at 8%. → y

$$z = \$1100 + y$$

After clearing decimals, we have this system:

$$x + y + z = 2500,$$
$$7x + 8y + 9z = 21,200,$$
$$-y + z = 1100$$

Carry out. Solve using matrices.

$$\left[\begin{array}{ccc|c} 1 & 1 & 1 & 2500 \\ 7 & 8 & 9 & 21,200 \\ 0 & -1 & 1 & 1100 \end{array}\right]$$

$$\left[\begin{array}{ccc|c} 1 & 1 & 1 & 2500 \\ 0 & 1 & 2 & 3700 \\ 0 & -1 & 1 & 1100 \end{array}\right] \quad \text{New Row 2} = -7(\text{Row 1}) + \text{Row 2}$$

$$\left[\begin{array}{ccc|c} 1 & 1 & 1 & 2500 \\ 0 & 1 & 2 & 3700 \\ 0 & 0 & 3 & 4800 \end{array}\right] \quad \text{New Row 3} = \text{Row 2} + \text{Row 3}$$

Reinserting the variables, we have

$$x + y + z = 2500, \quad (1)$$
$$y + 2z = 3700, \quad (2)$$
$$3z = 4800 \quad (3)$$

Solve (3) for z.

$$3z = 4800$$
$$z = 1600$$

Back-substitute 1600 for z in (2) and solve for y.

$$y + 2 \cdot 1600 = 3700$$
$$y + 3200 = 3700$$
$$y = 500$$

Back-substitute 500 for y and 1600 for z in (1) and solve for x.

$$x + 500 + 1600 = 2500$$
$$x + 2100 = 2500$$
$$x = 400$$

Check. The total investment is \$400 + \$500 + \$1600, or \$2500. The total interest is 0.07(\$400) + 0.08(\$500) + 0.09(\$1600) = \$28 + \$40 + \$144 = \$212. The amount invested at 9%, \$1600, is \$1100 more than the amount invested at 8%, \$500. The numbers check.

State. \$400 is invested at 7%, \$500 is invested at 8%, and \$1600 is invested at 9%.

20. \$500 at 8%, \$400 at 9%, \$2300 at 10%

21.

$$0.1x - 12 = 3.6x - 2.34 - 4.9x$$
$$10x - 1200 = 360x - 234 - 490x \quad \text{Multiplying by 100 to clear decimals}$$
$$10x - 1200 = -130x - 234$$
$$140x - 1200 = -234$$
$$140x = 966$$
$$x = \frac{966}{140}, \text{ or } \frac{69}{10}, \text{ or } 6.9$$

The solution is $\frac{69}{10}$, or 6.9.

22. -20

23.
$$\begin{aligned}4(9-x)-6(8-3x) &= 5(3x+4)\\ 36-4x-48+18x &= 15x+20\\ -12+14x &= 15x+20\\ -12 &= x+20\\ -32 &= x\end{aligned}$$

The solution is -32.

24. $b=\dfrac{c}{5+a}$

25. ◈

26. ◈

27. ***Familiarize.*** Let w, x, y, and z represent the thousand's, hundred's, ten's, and one's digits, respectively.

Translate.

The sum of the digits is 10.

$$w+x+y+z = 10$$

Twice the sum of the thousand's and ten's digits is the sum of the hundred's and one's digits less one.

$$2(w+y) = x+z-1$$

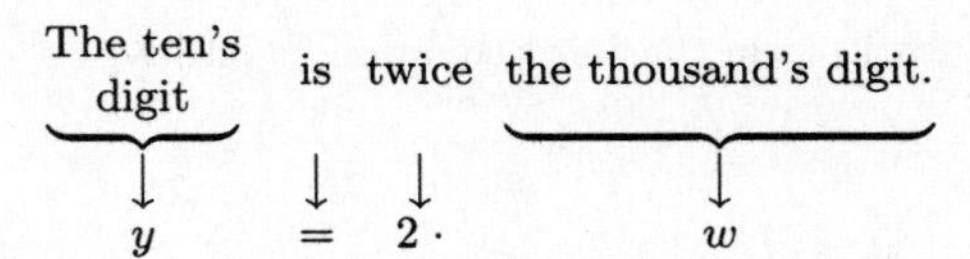

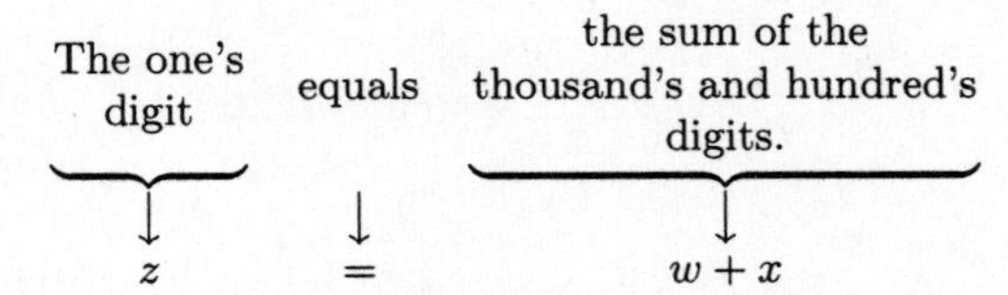

We have a system of equations which can be written as

$$\begin{aligned}w+x+y+z &= 10,\\ 2w-x+2y-z &= -1,\\ -2w+y &= 0,\\ w+x-z &= 0.\end{aligned}$$

Carry out. We can use matrices to solve the system. We get $(1,3,2,4)$.

Check. The sum of the digits is 10. Twice the sum of 1 and 2 is 6. This is one less than the sum of 3 and 4. The ten's digit, 2, is twice the thousand's digit, 1. The one's digit, 4, equals $1+3$. The numbers check.

State. The number is 1324.

28. $x=\dfrac{ce-bf}{ae-bd}$, $y=\dfrac{af-cd}{ae-bd}$

Exercise Set 3.7

1. $\begin{vmatrix}5 & 1\\ 2 & 4\end{vmatrix}=5\cdot4-2\cdot1=20-2=18$

2. -13

3. $\begin{vmatrix}6 & -9\\ 2 & 3\end{vmatrix}=6\cdot3-2(-9)=18+18=36$

4. 29

5. $\begin{vmatrix}1 & 4 & 0\\ 0 & -1 & 2\\ 3 & -2 & 1\end{vmatrix}$

$$\begin{aligned}&=1\begin{vmatrix}-1 & 2\\ -2 & 1\end{vmatrix}-0\begin{vmatrix}4 & 0\\ -2 & 1\end{vmatrix}+3\begin{vmatrix}4 & 0\\ -1 & 2\end{vmatrix}\\ &=1[-1\cdot1-(-2)\cdot2]-0+3[4\cdot2-(-1)\cdot0]\\ &=1\cdot3-0+3\cdot8\\ &=3-0+24\\ &=27\end{aligned}$$

6. 1

7. $\begin{vmatrix}-1 & -2 & -3\\ 3 & 4 & 2\\ 0 & 1 & 2\end{vmatrix}$

$$\begin{aligned}&=-1\begin{vmatrix}4 & 2\\ 1 & 2\end{vmatrix}-3\begin{vmatrix}-2 & -3\\ 1 & 2\end{vmatrix}+0\begin{vmatrix}-2 & -3\\ 4 & 2\end{vmatrix}\\ &=-1[4\cdot2-1\cdot2]-3[-2\cdot2-1(-3)]+0\\ &=-1\cdot6-3\cdot(-1)+0\\ &=-6+3+0\\ &=-3\end{aligned}$$

8. 3

9. $\begin{vmatrix}-4 & -2 & 3\\ -3 & 1 & 2\\ 3 & 4 & -2\end{vmatrix}$

$$\begin{aligned}&=-4\begin{vmatrix}1 & 2\\ 4 & -2\end{vmatrix}-(-3)\begin{vmatrix}-2 & 3\\ 4 & -2\end{vmatrix}+3\begin{vmatrix}-2 & 3\\ 1 & 2\end{vmatrix}\\ &=-4[1(-2)-4\cdot2]+3[-2(-2)-4\cdot3]+\\ &\qquad 3(-2\cdot2-1\cdot3)\\ &=-4(-10)+3(-8)+3(-7)\\ &=40-24-21=-5\end{aligned}$$

10. -6

11. $5x + 8y = 1,$

$3x + 7y = 5$

We compute D, D_x, and D_y.

$$D = \begin{vmatrix} 5 & 8 \\ 3 & 7 \end{vmatrix} = 35 - 24 = 11$$

$$D_x = \begin{vmatrix} 1 & 8 \\ 5 & 7 \end{vmatrix} = 7 - 40 = -33$$

$$D_y = \begin{vmatrix} 5 & 1 \\ 3 & 5 \end{vmatrix} = 25 - 3 = 22$$

Then,

$$x = \frac{D_x}{D} = \frac{-33}{11} = -3$$

and

$$y = \frac{D_y}{D} = \frac{22}{11} = 2.$$

The solution is $(-3, 2)$.

12. $(2, 0)$

13. $5x - 4y = -3,$

$7x + 2y = 6$

We compute D, D_x, and D_y.

$$D = \begin{vmatrix} 5 & -4 \\ 7 & 2 \end{vmatrix} = 10 - (-28) = 38$$

$$D_x = \begin{vmatrix} -3 & -4 \\ 6 & 2 \end{vmatrix} = -6 - (-24) = 18$$

$$D_y = \begin{vmatrix} 5 & -3 \\ 7 & 6 \end{vmatrix} = 30 - (-21) = 51$$

Then,

$$x = \frac{D_x}{D} = \frac{18}{38} = \frac{9}{19}$$

and

$$y = \frac{D_y}{D} = \frac{51}{38}.$$

The solution is $\left(\frac{9}{19}, \frac{51}{38}\right)$.

14. $\left(-\frac{25}{2}, -\frac{11}{2}\right)$

15. $3x - y + 2z = 1,$

$x - y + 2z = 3,$

$-2x + 3y + z = 1$

We compute D, D_x, and D_y.

$$D = \begin{vmatrix} 3 & -1 & 2 \\ 1 & -1 & 2 \\ -2 & 3 & 1 \end{vmatrix}$$

$$= 3\begin{vmatrix} -1 & 2 \\ 3 & 1 \end{vmatrix} - 1\begin{vmatrix} -1 & 2 \\ 3 & 1 \end{vmatrix} - 2\begin{vmatrix} -1 & 2 \\ -1 & 2 \end{vmatrix}$$

$$= 3(-7) - 1(-7) - 2(0)$$

$$= -21 + 7 - 0$$

$$= -14$$

$$D_x = \begin{vmatrix} 1 & -1 & 2 \\ 3 & -1 & 2 \\ 1 & 3 & 1 \end{vmatrix}$$

$$= 1\begin{vmatrix} -1 & 2 \\ 3 & 1 \end{vmatrix} - 3\begin{vmatrix} -1 & 2 \\ 3 & 1 \end{vmatrix} + 1\begin{vmatrix} -1 & 2 \\ -1 & 2 \end{vmatrix}$$

$$= 1(-7) - 3(-7) + 1(0)$$

$$= -7 + 21 + 0$$

$$= 14$$

$$D_y = \begin{vmatrix} 3 & 1 & 2 \\ 1 & 3 & 2 \\ -2 & 1 & 1 \end{vmatrix}$$

$$= 3\begin{vmatrix} 3 & 2 \\ 1 & 1 \end{vmatrix} - 1\begin{vmatrix} 1 & 2 \\ 1 & 1 \end{vmatrix} - 2\begin{vmatrix} 1 & 2 \\ 3 & 2 \end{vmatrix}$$

$$= 3 \cdot 1 - 1(-1) - 2(-4)$$

$$= 3 + 1 + 8$$

$$= 12$$

Then,

$$x = \frac{D_x}{D} = \frac{14}{-14} = -1$$

and

$$y = \frac{D_y}{D} = \frac{12}{-14} = -\frac{6}{7}.$$

Substitute in the third equation to find z.

$$-2(-1) + 3\left(-\frac{6}{7}\right) + z = 1$$

$$2 - \frac{18}{7} + z = 1$$

$$-\frac{4}{7} + z = 1$$

$$z = \frac{11}{7}$$

The solution is $\left(-1, -\frac{6}{7}, \frac{11}{7}\right)$.

16. $\left(\frac{3}{2}, \frac{13}{14}, \frac{33}{14}\right)$

17. $2x - 3y + 5z = 27,$

$x + 2y - z = -4,$

$5x - y + 4z = 27$

We compute D, D_x, and D_y.

$$D = \begin{vmatrix} 2 & -3 & 5 \\ 1 & 2 & -1 \\ 5 & -1 & 4 \end{vmatrix}$$

$$= 2\begin{vmatrix} 2 & -1 \\ -1 & 4 \end{vmatrix} - 1\begin{vmatrix} -3 & 5 \\ -1 & 4 \end{vmatrix} + 5\begin{vmatrix} -3 & 5 \\ 2 & -1 \end{vmatrix}$$

$$= 2(7) - 1(-7) + 5(-7)$$
$$= 14 + 7 - 35$$
$$= -14$$

$$D_x = \begin{vmatrix} 27 & -3 & 5 \\ -4 & 2 & -1 \\ 27 & -1 & 4 \end{vmatrix}$$

$$= 27\begin{vmatrix} 2 & -1 \\ -1 & 4 \end{vmatrix} - (-4)\begin{vmatrix} -3 & 5 \\ -1 & 4 \end{vmatrix} + 27\begin{vmatrix} -3 & 5 \\ 2 & -1 \end{vmatrix}$$
$$= 27(7) + 4(-7) + 27(-7)$$
$$= 189 - 28 - 189$$
$$= -28$$

$$D_y = \begin{vmatrix} 2 & 27 & 5 \\ 1 & -4 & -1 \\ 5 & 27 & 4 \end{vmatrix}$$

$$= 2\begin{vmatrix} -4 & -1 \\ 27 & 4 \end{vmatrix} - 1\begin{vmatrix} 27 & 5 \\ 27 & 4 \end{vmatrix} + 5\begin{vmatrix} 27 & 5 \\ -4 & -1 \end{vmatrix}$$
$$= 2(11) - 1(-27) + 5(-7)$$
$$= 22 + 27 - 35$$
$$= 14$$

Then,

$$x = \frac{D_x}{D} = \frac{-28}{-14} = 2,$$

and

$$y = \frac{D_y}{D} = \frac{14}{-14} = -1.$$

We substitute in the second equation to find z.

$$2 + 2(-1) - z = -4$$
$$2 - 2 - z = -4$$
$$-z = -4$$
$$z = 4$$

The solution is $(2, -1, 4)$.

18. $(-3, 2, 1)$

19. $r - 2s + 3t = 6,$

$2r - s - t = -3,$

$r + s + t = 6$

We compute D, D_r, and D_s.

$$D = \begin{vmatrix} 1 & -2 & 3 \\ 2 & -1 & -1 \\ 1 & 1 & 1 \end{vmatrix}$$

$$= 1\begin{vmatrix} -1 & -1 \\ 1 & 1 \end{vmatrix} - 2\begin{vmatrix} -2 & 3 \\ 1 & 1 \end{vmatrix} + 1\begin{vmatrix} -2 & 3 \\ -1 & -1 \end{vmatrix}$$

$$= 1(0) - 2(-5) + 1(5)$$
$$= 0 + 10 + 5$$
$$= 15$$

$$D_r = \begin{vmatrix} 6 & -2 & 3 \\ -3 & -1 & -1 \\ 6 & 1 & 1 \end{vmatrix}$$

$$= 6\begin{vmatrix} -1 & -1 \\ 1 & 1 \end{vmatrix} - (-3)\begin{vmatrix} -2 & 3 \\ 1 & 1 \end{vmatrix} + 6\begin{vmatrix} -2 & 3 \\ -1 & -1 \end{vmatrix}$$
$$= 6(0) + 3(-5) + 6(5)$$
$$= 0 - 15 + 30$$
$$= 15$$

$$D_s = \begin{vmatrix} 1 & 6 & 3 \\ 2 & -3 & -1 \\ 1 & 6 & 1 \end{vmatrix}$$

$$= 1\begin{vmatrix} -3 & -1 \\ 6 & 1 \end{vmatrix} - 2\begin{vmatrix} 6 & 3 \\ 6 & 1 \end{vmatrix} + 1\begin{vmatrix} 6 & 3 \\ -3 & -1 \end{vmatrix}$$
$$= 1(3) - 2(-12) + 1(3)$$
$$= 3 + 24 + 3$$
$$= 30$$

Then,

$$r = \frac{D_r}{D} = \frac{15}{15} = 1,$$

and

$$s = \frac{D_s}{D} = \frac{30}{15} = 2.$$

Substitute in the third equation to find t.

$$1 + 2 + t = 6$$
$$3 + t = 6$$
$$t = 3$$

The solution is $(1, 2, 3)$.

20. $(3, 4, -1)$

21. $0.5x - 2.34 + 2.4x = 7.8x - 9$

$$2.9x - 2.34 = 7.8x - 9$$
$$6.66 = 4.9x$$
$$\frac{6.66}{4.9} = x$$
$$\frac{666}{490} = x$$
$$\frac{333}{245} = x$$

22. -12

23. ***Familiarize***. We first make a drawing.

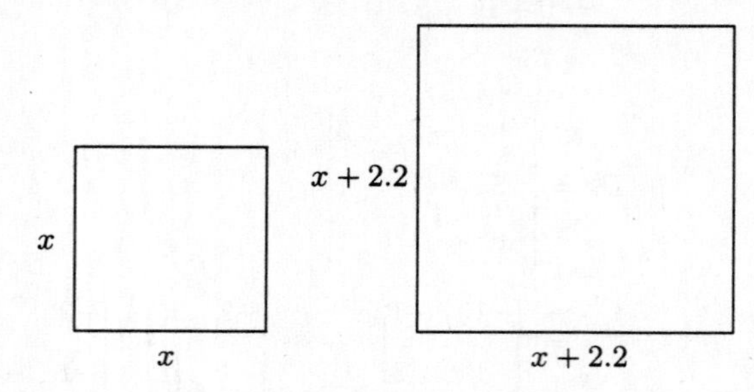

Let x represent the length of a side of the smaller square and $x + 2.2$ the length of a side of the larger square. The perimeter of the smaller square is $4x$. The perimeter of the larger square is $4(x + 2.2)$.

Translate.

The sum of the perimeters is 32.8 ft.

$$4x + 4(x + 2.2) = 32.8$$

Carry out. We solve the equation.

$$4x + 4x + 8.8 = 32.8$$
$$8x = 24$$
$$x = 3$$

Check. If $x = 3$ ft, then $x + 2.2 = 5.2$ ft. The perimeters are $4 \cdot 3$, or 12 ft, and $4(5.2)$, or 20.8 ft. The sum of the two perimeters is $12+20.8$, or 32.8 ft. The values check.

State. The wire should be cut into two pieces, one measuring 12 ft and the other 20.8 ft.

24. 12

25. $$\begin{vmatrix} 2 & x & -1 \\ -1 & 3 & 2 \\ -2 & 1 & 1 \end{vmatrix} = -12$$

$$2\begin{vmatrix} 3 & 2 \\ 1 & 1 \end{vmatrix} - (-1)\begin{vmatrix} x & -1 \\ 1 & 1 \end{vmatrix} + (-2)\begin{vmatrix} x & -1 \\ 3 & 2 \end{vmatrix} = -12$$

$$2(1) + 1(x + 1) - 2(2x + 3) = -12$$
$$2 + x + 1 - 4x - 6 = -12$$
$$-3x - 3 = -12$$
$$-3x = -9$$
$$x = 3$$

26. 10

27. $$\begin{vmatrix} x & y & 1 \\ x_1 & y_1 & 1 \\ x_2 & y_2 & 1 \end{vmatrix} = 0$$

is equivalent to

$$x\begin{vmatrix} y_1 & 1 \\ y_2 & 1 \end{vmatrix} - x_1\begin{vmatrix} y & 1 \\ y_2 & 1 \end{vmatrix} + x_2\begin{vmatrix} y & 1 \\ y_1 & 1 \end{vmatrix} = 0$$

or

$$x(y_1 - y_2) - x_1(y - y_2) + x_2(y - y_1) = 0$$

or

$$xy_1 - xy_2 - x_1y + x_1y_2 + x_2y - x_2y_1 = 0. \quad (1)$$

Since the slope of the line through (x_1, y_1) and (x_2, y_2) is $\dfrac{y_2 - y_1}{x_2 - x_1}$, an equation of the line through (x_1, y_1) and (x_2, y_2) is

$$y - y_1 = \frac{y_2 - y_1}{x_2 - x_1}(x - x_1)$$

which is equivalent to

$$(x_2 - x_1)(y - y_1) = (y_2 - y_1)(x - x_1)$$

or

$$x_2y - x_2y_1 - x_1y + x_1y_1 = y_2x - y_2x_1 - y_1x + y_1x_1$$

or

$$x_2y - x_2y_1 - x_1y - xy_2 + x_1y_2 + xy_1 = 0. \quad (2)$$

Equations (1) and (2) are equivalent.

28. If $a_1x + b_1y = c_1$ and $a_2x + b_2y = c_2$ are dependent, then one equation is a multiple of the other. That is, $a_1 = ka_2$ and $b_1 = kb_2$ for some constant k. Then

$$\begin{vmatrix} a_1 & b_1 \\ a_2 & b_2 \end{vmatrix} = \begin{vmatrix} ka_2 & kb_2 \\ a_2 & b_2 \end{vmatrix}$$
$$= ka_2(b_2) - a_2(kb_2)$$
$$= 0.$$

Exercise Set 3.8

1. $C(x) = 25x + 270,000 \qquad R(x) = 70x$

a) $P(x) = R(x) - C(x)$
$$= 70x - (25x + 270,000)$$
$$= 70x - 25x - 270,000$$
$$= 45x - 270,000$$

b) To find the break-even point we solve the system

$$R(x) = 70x,$$
$$C(x) = 25x + 270,000.$$

Since both $R(x)$ and $C(x)$ are in dollars and they are equal at the break-even point, we can rewrite the system:

$$d = 70x, \quad (1)$$
$$d = 25x + 270,000 \quad (2)$$

We solve using substitution.

$70x = 25x + 270,000$ Substituting $65x$ for d in (2)

$45x = 270,000$

$x = 6000$

Thus, 6000 units must be produced and sold in order to break even.

2. (a) $P(x) = 20x - 300,000$; (b) 15,000 units

3. $C(x) = 10x + 120,000$ $\quad R(x) = 60x$

a) $P(x) = R(x) - C(x)$

$= 60x - (10x + 120,000)$

$= 60x - 10x - 120,000$

$= 50x - 120,000$

b) Solve the system

$R(x) = 60x,$

$C(x) = 10x + 120,000.$

Since both $R(x)$ and $C(x)$ are in dollars and they are equal at the break-even point, we can rewrite the system:

$d = 60x,$ $\quad$ (1)

$d = 10x + 120,000$ $\quad$ (2)

We solve using substitution.

$60x = 10x + 120,000$ $\quad$ Substituting $60x$ for d in (2)

$50x = 120,000$

$x = 2400$

Thus, 2400 units must be produced and sold in order to break even.

4. (a) $P(x) = 55x - 49,500$: (b) 900 units

5. $C(x) = 20x + 10,000$ $\quad R(x) = 100x$

a) $P(x) = R(x) - C(x)$

$= 100x - (20x + 10,000)$

$= 100x - 20x - 10,000$

$= 80x - 10,000$

b) Solve the system

$R(x) = 100x,$

$C(x) = 20x + 10,000.$

Since both $R(x)$ and $C(x)$ are in dollars and they are equal at the break-even point, we can rewrite the system:

$d = 100x,$ $\quad$ (1)

$d = 20x + 10,000$ $\quad$ (2)

We solve using substitution.

$100x = 20x + 10,000$ $\quad$ Substituting $100x$ for d in (2)

$80x = 10,000$

$x = 125$

Thus, 125 units must be produced and sold in order to break even.

6. (a) $P(x) = 45x - 22,500$; (b) 500 units

7. $C(x) = 22x + 16,000$ $\quad R(x) = 40x$

a) $P(x) = R(x) - C(x)$

$= 40x - (22x + 16,000)$

$= 40x - 22x - 16,000$

$= 18x - 16,000$

b) Solve the system

$R(x) = 40x,$

$C(x) = 22x + 16,000.$

Since both $R(x)$ and $C(x)$ are in dollars and they are equal at the break-even point, we can rewrite the system:

$d = 40x,$ $\quad$ (1)

$d = 22x + 16,000$ $\quad$ (2)

We solve using substitution.

$40x = 22x + 16,000$ $\quad$ Substituting $40x$ for d in (2)

$18x = 16,000$

$x \approx 889$ units

Thus, 889 units must be produced and sold in order to break even.

8. (a) $P(x) = 40x - 75,000$; (b) 1875 units

9. $C(x) = 50x + 195,000$ $\quad R(x) = 125x$

a) $P(x) = R(x) - C(x)$

$= 125x - (50x + 195,000)$

$= 125x - 50x - 195,000$

$= 75x - 195,000$

b) Solve the system

$R(x) = 125x,$

$C(x) = 50x + 195,000.$

Since $R(x) = C(x)$ at the break-even point, we can rewrite the system:

$R(x) = 125x,$ $\quad$ (1)

$R(x) = 50x + 195,000$ $\quad$ (2)

We solve using substitution.

$125x = 50x + 195,000$ $\quad$ Substituting $125x$ for $R(x)$ in (2)

$75x = 195,000$

$x = 2600$

To break even 2600 units must be produced and sold.

10. (a) $P(x) = 94x - 928,000$; (b) 9873 units

11. $D(p) = 1000 - 10p,$

$S(p) = 230 + p$

Since both demand and supply are quantities, the system can be rewritten:

$q = 1000 - 10p,$ $\quad$ (1)

$q = 230 + p$ $\quad$ (2)

Substitute $1000 - 10p$ for q in (2) and solve.

$1000 - 10p = 230 + p$

$770 = 11p$

$70 = p$

The equilibrium price is \$70 per unit. To find the equilibrium quantity we substitute \$70 into either $D(p)$ or $S(p)$.

$D(70) = 1000 - 10 \cdot 70 = 1000 - 700 = 300$

The equilibrium quantity is 300 units.

The equilibrium point is $(\$70, 300)$.

12. $(\$10, 1400)$

13. $D(p) = 760 - 13p,$

$S(p) = 430 + 2p$

Rewrite the system:

$q = 760 - 13p,\quad (1)$

$q = 430 + 2p\quad (2)$

Substitute $760 - 13p$ for q in (2) and solve.

$760 - 13p = 430 + 2p$

$330 = 15p$

$22 = p$

The equilibrium price is \$22 per unit.

To find the equilibrium quantity we substitute \$22 into either $D(p)$ or $S(p)$.

$S(22) = 430 + 2(22) = 430 + 44 = 474$

The equilibrium quantity is 474 units.

The equilibrium point is $(\$22, 474)$.

14. $(\$10, 370)$

15. $D(p) = 7500 - 25p,$

$S(p) = 6000 + 5p$

Rewrite the system:

$q = 7500 - 25p,\quad (1)$

$q = 6000 + 5p\quad (2)$

Substitute $7500 - 25p$ for q in (2) and solve.

$7500 - 25p = 6000 + 5p$

$1500 = 30p$

$50 = p$

The equilibrium price is \$50 per unit.

To find the equilibrium quantity we substitute \$50 into either $D(p)$ or $S(p)$.

$D(50) = 7500 - 25(50) = 7500 - 1250 = 6250$

The equilibrium quantity is 6250 units.

The equilibrium point is $(\$50, 6250)$.

16. $(\$40, 7600)$

17. $D(p) = 1600 - 53p,$

$S(p) = 320 + 75p$

Rewrite the system:

$q = 1600 - 53p,\quad (1)$

$q = 320 + 75p\quad (2)$

Substitute $1600 - 53p$ for q in (2) and solve.

$1600 - 53p = 320 + 75p$

$1280 = 128p$

$10 = p$

The equilibrium price is \$10 per unit.

To find the equilibrium quantity we substitute \$10 into either $D(p)$ or $S(p)$.

$S(10) = 320 + 75(10) = 320 + 750 = 1070$

The equilibrium quantity is 1070 units.

The equilibrium point is $(\$10, 1070)$.

18. $(\$36, 4060)$

19. a) $C(x) = \text{Fixed costs} + \text{Variable costs}$

$C(x) = 22,500 + 40x,$

where x is the number of lamps produced.

b) Each lamp sells for \$85. The total revenue is 85 times the number of lamps sold. We assume that all lamps produced are sold.

$R(x) = 85x$

c) $P(x) = R(x) - C(x)$

$P(x) = 85x - (22,500 + 40x)$

$= 85x - 22,500 - 40x$

$= 45x - 22,500$

d) $P(3000) = 45(3000) - 22,500$

$= 135,000 - 22,500$

$= 112,500$

The company will realize a profit of \$112,500 when 3000 lamps are produced and sold.

$P(400) = 45(400) - 22,500$

$= 18,000 - 22,500$

$= -4500$

The company will realize a \$4500 loss when 400 lamps are produced and sold.

e) Solve the system

$R(x) = 85x,$

$C(x) = 22,500 + 40x.$

Since both $R(x)$ and $C(x)$ are in dollars and they are equal at the break-even point, we can rewrite the system:

$d = 85x,\quad (1)$

$d = 22,500 + 40x\quad (2)$

We solve using substitution.

$85x = 22,500 + 40x$ Substituting $85x$ for d in (2)

$45x = 22,500$

$x = 500$

The firm will break even if it produces and sells 500 lamps and takes in a total of $R(500) = 85 \cdot 500 = \$42,500$ in revenue. Thus, the break-even point is (500 lamps, \$42,500).

20. (a) $C(x) = 125,100 + 750x$; (b) $R(x) = 1050x$; (c) $P(x) = 300x - 125,100$; (d) \$5100 loss, \$84,900 profit; (e) (417 computers, \$437,850).

21. a) $C(x) = \text{Fixed costs} + \text{Variable costs}$

$C(x) = 16,404 + 6x,$

where x is the number of caps produced, in dozens.

b) Each dozen caps sell for \$18. The total revenue is 18 times the number of caps sold, in dozens. We assume that all caps produced are sold.

$R(x) = 18x$

c) $P(x) = R(x) - C(x)$

$$\begin{aligned} P(x) &= 18x - (16,404 + 6x) \\ &= 18x - 16,404 - 6x \\ &= 12x - 16,404 \end{aligned}$$

d)
$$\begin{aligned} P(3000) &= 12(3000) - 16,404 \\ &= 36,000 - 16,404 \\ &= 19,596 \end{aligned}$$

The company will realize a profit of \$19,596 when 3000 dozen caps are produced and sold.

$$\begin{aligned} P(1000) &= 12(1000) - 16,404 \\ &= 12,000 - 16,404 \\ &= -4404 \end{aligned}$$

The company will realize a \$4404 loss when 1000 dozen caps are produced and sold.

e) Solve the system

$R(x) = 18x,$

$C(x) = 16,404 + 6x.$

Since both $R(x)$ and $C(x)$ are in dollars and they are equal at the break-even point, we can rewrite the system:

$d = 18x,$ (1)

$d = 16,404 + 6x$ (2)

We solve using substitution.

$18x = 16,404 + 6x$ Substituting $18x$ for d in (2)

$12x = 16,404$

$x = 1367$

The firm will break even if it produces and sells 1367 dozen caps and takes in a total of $R(1367) = 18 \cdot 1367 = \$24,606$ in revenue. Thus, the break-even point is (1367 dozen caps, \$24,606).

22. (a) $C(x) = 10,000 + 20x$; (b) $R(x) = 100x$; (c) $P(x) = 80x - 10,000$; (d) \$150,000 profit, \$6000 loss; (e) (125 sport coats, \$12.500)

23. $y - 3 = \dfrac{2}{5}(x - 1)$

The equation of the line is in point-slope form. We see that the line has slope $\dfrac{2}{5}$ and contains the point $(1, 3)$. Plot $(1, 3)$. Then go up two units and right 5 units to find another point on the line, $(6, 5)$. A third point can be found as a check.

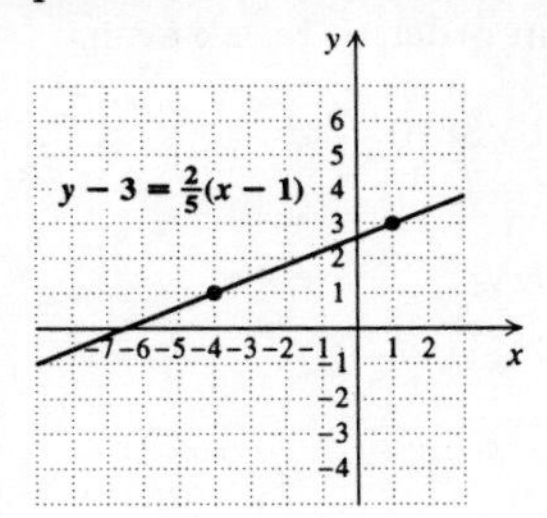

24. 5, 6

25,
$$\begin{aligned} 9x &= 5x - \{3(2x - 7) - 4\} \\ 9x &= 5x - \{6x - 21 - 4\} \\ 9x &= 5x - \{6x - 25\} \\ 9x &= 5x - 6x + 25 \\ 9x &= -x + 25 \\ 10x &= 25 \\ x &= \frac{25}{10} \\ x &= \frac{5}{2}, \text{ or } 2.5 \end{aligned}$$

26. $t = \dfrac{rw - v}{-s}$, or $\dfrac{v - rw}{s}$

27. ◈

28. ◈

29. Using the given information we know that $C(x) = 15,400 + 100x$, where x is the number of pairs of speakers produced, and $R(x) = 250x$. Then

$$\begin{aligned} P(x) &= R(x) - C(x) \\ &= 250x - (15,400 + 100x) \\ &= 250x - 15,400 - 100x \\ &= 150x - 15,400 \end{aligned}$$

The fixed costs of two new facilities are $2 \cdot \$15,400$, or \$30,800. We find the value of x for which the profit $P(x)$ is \$30,800:

$$\begin{aligned} 150x - 15,400 &= 30,800 \\ 150x &= 46,200 \\ x &= 308 \end{aligned}$$

Thus, 308 pairs of speakers must be produced and sold in order to have enough profit to cover the fixed costs of two new facilities.

30. (\$5, 300)

31. a) Use a grapher to find the first coordinate of the point of intersection of $y_1 = -14.97x + 987.35$ and $y_2 = 98.55x - 5.13$, to the nearest hundredth. It is 8.74, so the price per unit that should be charged is \$8.74.

b) Use a grapher to find the first coordinate of the point of intersection of $y_1 = 87,985 + 5.15x$ and $y_2 = 8.74x$. It is about 24, 508.4, so 24,509 units must be sold in order to break even.

32. (a) 4526 units; (b) $870

Chapter 4

Inequalities and Linear Programming

Exercise Set 4.1

1. $x - 2 \geq 6$

-4: We substitute and get $-4 - 2 \geq 6$, or $-6 \geq 6$, a false sentence. Therefore, -4 is not a solution.

0: We substitute and get $0 - 2 \geq 6$, or $-2 \geq 6$, a false sentence. Therefore, 0 is not a solution.

4: We substitute and get $4 - 2 \geq 6$, or $2 \geq 6$, a false sentence. Therefore, 4 is not a solution.

8: We substitute and get $8 - 2 \geq 6$, or $6 \geq 6$, a true sentence. Therefore, 8 is a solution.

2. Yes; yes; no; no

3. $t - 8 > 2t - 3$

0: We substitute and get $0 - 8 > 2 \cdot 0 - 3$, or $-8 > -3$, a false sentence. Therefore, 0 is not a solution.

-8: We substitute and get $-8 - 8 > 2(-8) - 3$, or $-16 > -19$, a true sentence. Therefore, -8 is a solution.

-9: We substitute and get $-9 - 8 > 2(-9) - 3$, or $-17 > -21$, a true sentence. Therefore, -9 is a solution.

-3: We substitute and get $-3 - 8 > 2(-3) - 3$, or $-11 > -9$, a false sentence. Therefore, -3 is not a solution.

4. No; yes; yes; no

5. $y < 5$

Graph: The solutions consist of all real numbers less than 5, so we shade all numbers to the left of 5 and use an open circle at 5 to indicate that it is not a solution.

Set builder notation: $\{y|y < 5\}$

Interval notation: $(-\infty, 5)$

6.

-2 -1 0 1 2 3 4 5 6 7 8

$\{x|x > 4\}, (4, \infty)$

7. $x \geq -4$

Graph: We shade all numbers to the right of -4 and use a solid endpoint at -4 to indicate that it is also a solution.

-5 -4 -3 -2 -1 0 1 2 3 4 5

Set builder notation: $\{x|x \geq -4\}$

Interval notation: $[-4, \infty)$

8.

-1 0 1 2 3 4 5 6 7 8 9

$\{t|t \leq 6\}, (-\infty, 6]$

9. $t > -2$

Graph: We shade all numbers to the right of -2 and use an open circle at -2 to indicate that it is not a solution.

-5 -4 -3 -2 -1 0 1 2 3 4 5

Set builder notation: $\{t|t > -2\}$

Interval notation: $(-2, \infty)$

10.

-7 -6 -5 -4 -3 -2 -1 0 1 2 3

$\{y|y < -3\}, (-\infty, -3)$

11. $x \leq -5$

Graph: We shade all numbers to the left of -5 and use a solid endpoint at -5 to indicate that it is also a solution.

-8 -7 -6 -5 -4 -3 -2 -1 0 1 2

Set builder notation: $\{x|x \leq -5\}$

Interval notation: $(-\infty, -5]$

12.

-9 -8 -7 -6 -5 -4 -3 -2 -1 0 1

$\{x|x \geq -6\}, [-6, \infty)$

13.

$$\begin{aligned} x + 8 &> 3 \\ x + 8 + (-8) &> 3 + (-8) \quad \text{Adding } -8 \\ x &> -5 \end{aligned}$$

The solution set is $\{x|x > -5\}$, or $(-5, \infty)$.

-9 -8 -7 -6 -5 -4 -3 -2 -1 0 1

14. $\{x|x > -3\}$, or $(-3, \infty)$

-5 -4 -3 -2 -1 0 1 2 3 4 5

15.

$$\begin{aligned} a + 7 &\leq -13 \\ a + 7 + (-7) &\leq -13 + (-7) \quad \text{Adding } -7 \\ a &\leq -20 \end{aligned}$$

The solution set is $\{a|a \leq -20\}$, or $(-\infty, -20]$.

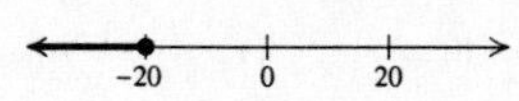

16. $\{a|a \leq -21\}$, or $(-\infty, -21]$

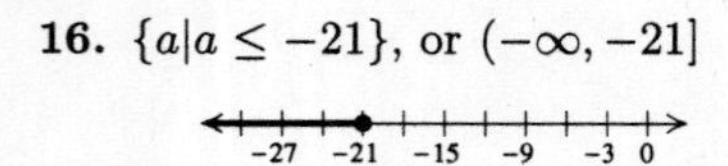

17.
$$x - 9 \leq 10$$
$$x - 9 + 9 \leq 10 + 9 \quad \text{Adding 9}$$
$$x \leq 19$$

The solution set is $\{x|x \leq 19\}$, or $(-\infty, 19]$.

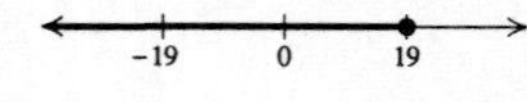

18. $\{t|t \geq -5\}$, or $[-5, \infty)$

19.
$$y - 9 > -18$$
$$y - 9 + 9 > -18 + 9 \quad \text{Adding 9}$$
$$y > -9$$

The solution set is $\{y|y > -9\}$, or $(-9, \infty)$.

20. $\{y|y > -6\}$, or $(-6, \infty)$

21.
$$y - 18 \leq -14$$
$$y - 18 + 18 \leq -4 + 18 \quad \text{Adding 18}$$
$$y \leq 14$$

The solution set is $\{y|y \leq 14\}$, or $(-\infty, 14]$.

22. $\{x|x \leq 9\}$, or $(-\infty, 9]$

23.
$$9t < -81$$
$$\frac{1}{9} \cdot 9t < \frac{1}{9}(-81) \quad \text{Multiplying by } \frac{1}{9}$$
$$t < -9$$

The solution set is $\{t|t < -9\}$, or $(-\infty, -9)$.

24. $\{x|x \geq 3\}$, or $[3, \infty)$

25.
$$0.5x < 25$$
$$\frac{1}{0.5}(0.5x) < \frac{1}{0.5}(25) \quad \text{Multiplying by } \frac{1}{0.5}$$
$$x < \frac{25}{0.5}$$
$$x < 50$$

The solution set is $\{x|x < 50\}$, or $(-\infty, 50)$.

26. $\{x|x < -60\}$, or $(-\infty, 60)$

27.
$$-8y \leq 3.2$$
$$-\frac{1}{8}(-8y) \geq -\frac{1}{8}(3.2) \quad \text{Multiplying by } -\frac{1}{8} \text{ and reversing the inequality symbol}$$
$$y \geq -0.4$$

The solution set is $\{y|y \geq -0.4\}$, or $[-0.4, \infty)$.

28. $\{x|x \leq 0.9\}$, or $(-\infty, 0.9]$

29.
$$-\frac{5}{6}y \leq -\frac{3}{4}$$
$$-\frac{6}{5}\left(-\frac{5}{6}y\right) \geq -\frac{6}{5}\left(-\frac{3}{4}\right) \quad \text{Multiplying by } -\frac{6}{5} \text{ and reversing the inequality symbol}$$
$$y \geq \frac{9}{10}$$

The solution set is $\left\{y \middle| y \geq \frac{9}{10}\right\}$, or $\left[\frac{9}{10}, \infty\right)$.

30. $\left\{x \middle| x \leq \frac{5}{6}\right\}$, or $\left(-\infty, \frac{5}{6}\right]$

31.
$$5y + 13 > 28$$
$$5y + 13 + (-13) > 28 + (-13) \quad \text{Adding } -13$$
$$5y > 15$$
$$\frac{1}{5} \cdot 5y > \frac{1}{5} \cdot 15 \quad \text{Multiplying by } \frac{1}{5}$$
$$y > 3$$

The solution set is $\{y|y > 3\}$, or $(3, \infty)$.

32. $\{x|x < 6\}$, or $(-\infty, 6)$

33.
$$-9x + 3x \geq -24$$
$$-6x \geq -24 \quad \text{Combining like terms}$$
$$-\frac{1}{6}(-6x) \leq -\frac{1}{6}(-24) \quad \text{Multiplying by } -\frac{1}{6} \text{ and reversing the inequality symbol}$$
$$x \leq 4$$

The solution set is $\{x|x \leq 4\}$, or $(-\infty, 4]$.

34. $\{y|y \le -3\}$, or $(-\infty, -3]$

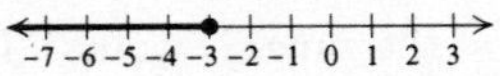

35. $f(x) = 8x - 9$, $g(x) = 3x - 11$

$$\begin{aligned} f(x) &< g(x) \\ 8x - 9 &< 3x - 11 \\ 5x - 9 &< -11 && \text{Adding } -3x \\ 5x &< -2 && \text{Adding } 9 \\ x &< -\frac{2}{5} && \text{Multiplying by } \frac{1}{5} \end{aligned}$$

The solution set is $\left\{x \middle| x < -\frac{2}{5}\right\}$, or $\left(-\infty, -\frac{2}{5}\right)$.

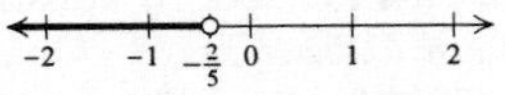

36. $\left\{x \middle| x > \frac{2}{3}\right\}$, or $\left(\frac{2}{3}, \infty\right)$

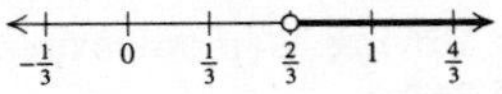

37. $f(x) = 0.4x + 5$, $g(x) = 1.2x - 4$

$$\begin{aligned} g(x) &\ge f(x) \\ 1.2x - 4 &\ge 0.4x + 5 \\ 0.8x - 4 &\ge 5 && \text{Adding } -0.4x \\ 0.8x &\ge 9 && \text{Adding } 4 \\ x &\ge 11.25 && \text{Multiplying by } \frac{1}{0.8} \end{aligned}$$

The solution set is $\{x|x \ge 11.25\}$, or $[11.25, \infty)$.

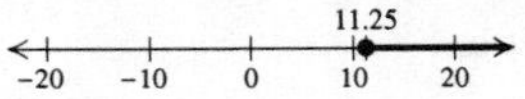

38. $\left\{x \middle| x \ge \frac{1}{2}\right\}$, or $\left[\frac{1}{2}, \infty\right)$

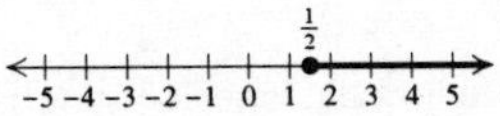

39. $4(3y - 2) \ge 9(2y + 5)$

$$\begin{aligned} 12y - 8 &\ge 18y + 45 \\ -6y - 8 &\ge 45 \\ -6y &\ge 53 \\ y &\le -\frac{53}{6} \end{aligned}$$

The solution set is $\left\{y \middle| y \le -\frac{53}{6}\right\}$, or $\left(-\infty, -\frac{53}{6}\right]$.

40. $\{m|m \le 3.3\}$, or $(-\infty, 3.3]$

41. $3(2 - 5x) + 2x < 2(4 + 2x)$

$$\begin{aligned} 6 - 15x + 2x &< 8 + 4x \\ 6 - 13x &< 8 + 4x \\ 6 - 17x &< 8 \\ -17x &< 2 \\ x &> -\frac{2}{17} \end{aligned}$$

The solution set is $\left\{x \middle| x > -\frac{2}{17}\right\}$, or $\left(-\frac{2}{17}, \infty\right)$.

42. $\left\{y \middle| y < \frac{13}{185}\right\}$, or $\left(-\infty, \frac{13}{185}\right)$

43. $5[3m - (m + 4)] > -2(m - 4)$

$$\begin{aligned} 5(3m - m - 4) &> -2(m - 4) \\ 5(2m - 4) &> -2(m - 4) \\ 10m - 20 &> -2m + 8 \\ 12m - 20 &> 8 \\ 12m &> 28 \\ m &> \frac{28}{12} \\ m &> \frac{7}{3} \end{aligned}$$

The solution set is $\left\{m \middle| m > \frac{7}{3}\right\}$, or $\left(\frac{7}{3}, \infty\right)$.

44. $\left\{x \middle| x \le -\frac{23}{2}\right\}$, or $\left(-\infty, -\frac{23}{2}\right]$

45. $19 - (2x + 3) \le 2(x + 3) + x$

$$\begin{aligned} 19 - 2x - 3 &\le 2x + 6 + x \\ 16 - 2x &\le 3x + 6 \\ 16 - 5x &\le 6 \\ -5x &\le -10 \\ x &\ge 2 \end{aligned}$$

The solution set is $\{x|x \ge 2\}$, or $[2, \infty)$.

46. $\{c|c \le 1\}$, or $(-\infty, 1]$

47. $\frac{1}{4}(8y + 4) - 17 < -\frac{1}{2}(4y - 8)$

$$\begin{aligned} 2y + 1 - 17 &< -2y + 4 \\ 2y - 16 &< -2y + 4 \\ 4y - 16 &< 4 \\ 4y &< 20 \\ y &< 5 \end{aligned}$$

The solution set is $\{y|y < 5\}$, or $(-\infty, 5)$.

48. $\{x|x > 6\}$, or $(6, \infty)$

49. $$\begin{aligned}2[4-2(3-x)]-1 &\ge 4[2(4x-3)+7]-25\\ 2[4-6+2x]-1 &\ge 4[8x-6+7]-25\\ 2[-2+2x]-1 &\ge 4[8x+1]-25\\ -4+4x-1 &\ge 32x+4-25\\ 4x-5 &\ge 32x-21\\ -28x-5 &\ge -21\\ -28x &\ge -16\\ x &\le \frac{-16}{-28}, \text{ or } \frac{4}{7}\end{aligned}$$

The solution set is $\left\{x\middle|x \le \frac{4}{7}\right\}$, or $\left(-\infty, \frac{4}{7}\right]$.

50. $\left\{t\middle|t \ge -\frac{27}{19}\right\}$, or $\left[-\frac{27}{19}, \infty\right)$

51. ***Familiarize.*** Let m = the mileage. Then the rental cost for the unlimited mileage plan is \$55 and for the other plan is \$29 + \$0.40m.

Translate. We write an inequality stating that the rental cost under the unlimited mileage plan is less than the cost under the other plan.

$55 < 29 + 0.4m$

Carry out.

$$\begin{aligned}55 &< 29 + 0.4m\\ 26 &< 0.4m\\ 65 &< m\end{aligned}$$

Check. We can do a partial check by substituting a value for m greater than 65. When $m = 66$, the cost of the second plan is \$29 + \$0.40(66) = \$55.40. This is more than \$55, the cost of the unlimited mileage plan. We cannot check all possible values for m, so we stop here.

State. The unlimited mileage plan would save money for mileages greater than 65 miles.

52. Mileages less than or equal to 150 miles

53. ***Familiarize.*** Let v = the blue book value of the car. Since the car was not replaced, we know that \$9200 does not exceed 80% of the blue book value.

Translate. We write an inequality stating that \$9200 does not exceed 80% of the blue book value.

$9200 \le 0.8v$

Carry out.

$$\begin{aligned}9200 &\le 0.8v\\ 11,500 &\le v \quad \text{Multiplying by } \frac{1}{0.8}\end{aligned}$$

Check. We can do a partial check by substituting a value for v greater than 11,500. When $v = 11,525$, then 80% of v is $0.8(11,525)$, or \$9220. This is greater than \$9200; that is, \$9200 does not exceed this amount. We cannot check all possible values for v, so we stop here.

State. The blue book value of the car is \$11,500 or more.

54. Calls longer than 1.5 minutes

55. ***Familiarize.*** Let t = the time, in hours. Then the cost of a move with Musclebound Movers is \$85+\$40t and with Champion Moving is \$60t.

Translate. We write an inequality stating that Champion Moving is more expensive than Musclebound Movers.

$60t > 85 + 40t$

Carry out.

$$\begin{aligned}60t &> 85 + 40t\\ 20t &> 85\\ t &> 4.25\end{aligned}$$

Check. We can do a partial check by substituting a value for t greater than 4.25. When $t = 4.5$, Champion Moving costs \$60(4.5), or \$270, and Musclebound Movers cost \$85+\$40(4.5), or \$265, so Champion Moving is more expensive. We cannot check all possible values for t, so we stop here.

State. Champion Moving is more expensive for times greater than 4.25 hr.

56. Gross sales greater than \$7000

57. ***Familiarize.*** Let c = the number of checks per month. Then the Anywhere plan will cost \$0.20c per month and the Acu-checking plan will cost \$2 + \$0.12c per month.

Translate. We write an inequality stating that the Acu-checking plan costs less than the Anywhere plan.

$2 + 0.12c < 0.20c$

Carry out.

$$\begin{aligned}2 + 0.12c &< 0.20c\\ 2 &< 0.08c\\ 25 &< c\end{aligned}$$

Check. We can do a partial check by substituting a value for c less than 25 and a value for c greater than 25. When $c = 24$, the Acu-checking plan costs \$2+\$0.12(24), or \$4.88, and the Anywhere plan costs \$0.20(24), or \$4.80, so the Anywhere plan is less expensive. When $c = 26$, the Acu-checking plan costs \$2+\$0.12(26), or \$5.12, and the Anywhere plan costs \$0.20(26), or \$5.20, so Acu-checking is less expensive. We cannot check all possible values for c, so we stop here.

State. The Acu-checking plan costs less for more than 25 checks per month.

58. Values of n greater than $85\frac{5}{7}$

59. ***Familiarize.*** Let m = the amount of the medical bills. Then under plan A Giselle would pay \$50 + 0.2(m − \$50). Under plan B she would pay \$250 + 0.1(m − \$250).

Translate. We write an inequality stating than the cost of plan B is less than the cost of plan A.

$$250 + 0.1(m - 250) < 50 + 0.2(m - 50)$$

Carry out.

$$250 + 0.1(m - 250) < 50 + 0.2(m - 50)$$
$$250 + 0.1m - 25 < 50 + 0.2m - 10$$
$$225 + 0.1m < 40 + 0.2m$$
$$185 + 0.1m < 0.2m$$
$$185 < 0.1m$$
$$1850 < m$$

Check. We can do a partial check by substituting a value for m less than \$1850 and a value for m greater than \$1850. When $m = \$1840$, plan A costs \$50 + 0.2(\$1840 − \$50), or \$408, and plan B costs \$250+0.1(\$1840−\$250), or \$409. When $m = \$1860$, plan A costs \$50+0.2(\$1860−\$50), or \$412, nd plan B costs \$250 + 0.1(\$1860 − \$250), or \$411, so plan B will save Giselle money. We cannot check all possible values for m, so we stop here.

State. Plan B will save Giselle money for medical bills greater than \$1850.

60. Parties of more than 80

61. ***Familiarize.*** Organize the information in a table. Let $x =$ the amount invested at 6%. Then $20,000 - x =$ the amount invested at 8%.

Amount invested		Rate of interest	Time
x		6%	1 yr
$20,000 - x$		8%	1 yr
Total	\$20,000		

Amount invested		Interest ($I = Prt$)
x		$0.06x$
$20,000 - x$		$0.08(20,000 - x)$
Total	\$20,000	\$1500 or more

Translate. Use the information in the table to write an inequality:

$$0.06x + 0.08(20,000 - x) \geq 1500$$

Carry out.

$$0.06x + 0.08(20,000 - x) \geq 1500$$
$$0.06x + 1600 - 0.08x \geq 1500$$
$$1600 - 0.02x \geq 1500$$
$$-0.02 \geq -100$$
$$x \leq 5000$$

Check. For $x = \$5000$, 6%(\$5000) = 0.06(\$5000), or \$300, and 8%(20,000 − \$5000) = 0.08(\$15,000), or \$1200. The total interest earned is \$300 + \$1200, or \$1500. We also calculate for some amount less than \$5000 and for some amount greater than \$5000. For $x = \$4000$, 6%(\$4000) = 0.06(\$4000), or \$240, and 8%(\$20,000 − \$4000) = 0.08(\$16,000), or \$1280. The total interest earned is \$240 + \$1280, or \$1520. For $x = \$6000$, 6%(\$6000) = 0.06(\$6000), or \$360, and 8%(\$20,000−\$6000) = 0.08(\$14,000), or \$1120. The total interest earned is \$360 + \$1120, or \$1480. For these values the inequality, $x \leq 5000$, gives correct results.

State. To make at least \$1500 interest per year, \$5000 is the most that can be invested at 6%.

62. (a) Fahrenheit temperatures less than 1945.4°;

(b) Fahrenheit temperatures less than 1761.44°

63. a) ***Familiarize.*** Find the values of x for which $R(x) < C(x)$.

Translate.

$$26x < 90,000 + 15x$$

Carry out.

$$11x < 90,000$$
$$x < 8181\frac{9}{11}$$

Check. $R\left(8181\frac{9}{11}\right) = \$212,727.27 = C\left(8181\frac{9}{11}\right)$.

Calculate $R(x)$ and $C(x)$ for some x greater than $8181\frac{9}{11}$ and for some x less than $8181\frac{9}{11}$.

Suppose $x = 8200$:

$$R(x) = 26(8200) = 213,200 \quad \text{and}$$
$$C(x) = 90,000 + 15(8200) = 213,000.$$

In this case $R(x) > C(x)$.

Suppose $x = 8000$:

$$R(x) = 26(8000) = 208,000 \quad \text{and}$$
$$C(x) = 90,000 + 15(8000) = 210,000.$$

In this case $R(x) < C(x)$.

Then for $x < 8181\frac{9}{11}$, $R(x) < C(x)$.

State. We will state the result in terms of integers, since the company cannot sell a fraction of a radio. For 8181 or fewer radios the company loses money.

b) Our check in part a) shows that for $x > 8181\frac{9}{11}$, $R(x) > C(x)$ and the company makes a profit. Again, we will state the result in terms of an integer. For more than 8182 radios the company makes money.

64. (a) $\{p|p < 10\}$; (b) $\{p|p > 10\}$

65. $f(x) = 2x - 1$

x	$f(x)$, or y
-2	-5
0	-1
2	3

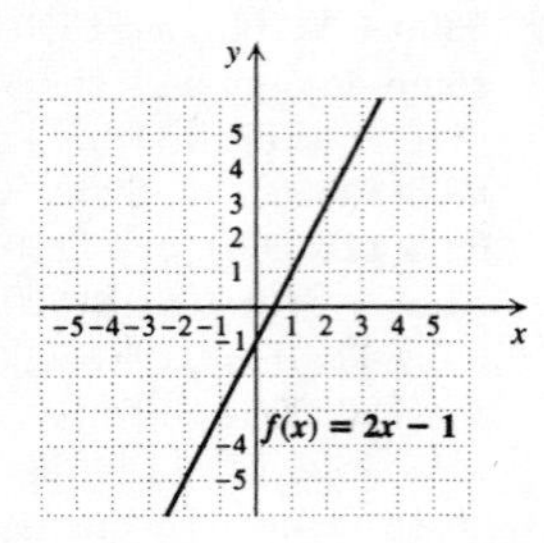

66.

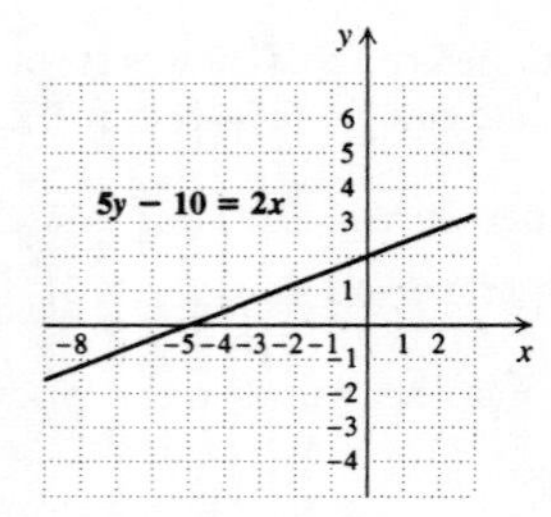

67.

$$\begin{aligned} -3x + 5 &= 11 \\ -3x &= 6 \quad \text{Subtracting 5 on both sides} \\ x &= -2 \quad \text{Dividing by } -3 \text{ on both sides} \end{aligned}$$

The solution is -2.

68. $-\dfrac{8}{5}$

69. $|-16| = 16$ $\quad -16$ is 6 units from 0

70. -4

71.

72.

73.

74. $\left\{x \middle| x \leq \dfrac{2}{a-1}\right\}$

75.

$$\begin{aligned} 6by - 4y &\leq 7by + 10 \\ -by - 4y &\leq 10 \\ y(-b - 4) &\leq 10 \\ y &\geq \frac{10}{-b-4}, \text{ or } -\frac{10}{b+4} \end{aligned}$$

We reversed the inequality symbol when we divided because when $b > 0$, then $-b - 4 < 0$.

The solution set is $\left\{y \middle| y \geq -\dfrac{10}{b+4}\right\}$.

76. $\left\{y \middle| y \geq \dfrac{2a + 5b}{b(a-2)}\right\}$

77.

$$\begin{aligned} c(6x - 4) &< d(3 + 2x) \\ 6cx - 4c &< 3d + 2dx \\ 6cx - 2dx - 4c &< 3d \\ 6cx - 2dx &< 4c + 3d \\ x(6c - 2d) &< 4c + 3d \\ x &< \frac{4c + 3d}{6c - 2d} \end{aligned}$$

The inequality symbol remained unchanged when we divided because when $3c > d$, then $6c - 2d > 0$.

The solution set is $\left\{x \middle| x < \dfrac{4c + 3d}{6c - 2d}\right\}$.

78. $\left\{x \middle| x > \dfrac{4m - 2c}{d - (5c + 2m)}\right\}$

79.

$$\begin{aligned} a(3 - 4x) + cx &< d(5x + 2) \\ 3a - 4ax + cx &< 5dx + 2d \\ 3a - 4ax + cx - 5dx &< 2d \\ -4ax + cx - 5dx &< -3a + 2d \\ x[c - (4a + 5d)] &< -3a + 2d \\ x &< \frac{-3a+2d}{c-(4a+5d)} \end{aligned}$$

The inequality symbol remained unchanged when we divided because when $c > 4a + 5d$, then $c - (4a + 5d) > 0$.

The solution set is $\left\{x \middle| x < \dfrac{-3a + 2d}{c - (4a + 5d)}\right\}$.

80. False. If $a = 2$, $b = 3$, $c = 4$, $d = 5$, then $2 - 4 = 3 - 5$.

81. False, because $-3 < -2$, but $9 > 4$.

82.

83.

84. The set of all real numbers

85.

$$\begin{aligned} x + 8 &< 3 + x \\ 8 &< 3 \quad \text{Subtracting } x \end{aligned}$$

We get a false inequality, so the solution set is $\emptyset$.

86. $\{x | x \text{ is a real number and } x \neq 0\}$

87. a) The graph of y_1 lies above the graph of y_2 for x-values to the left of the point of intersection, or in the interval $(-\infty, 4)$.

b) The graph of y_2 lies on or below the graph of y_3 for x-values at and to the right of the point of intersection, or in the interval $[2, \infty)$.

c) The graph of y_3 lies on or above the graph of y_1 at and to the right of the point of intersection, or in the interval $[3.2, \infty)$.

88. 

Exercise Set 4.2

1. $\{9, 10, 11\} \cap \{9, 11, 13\}$

   The numbers 9 and 11 are common to both sets, so the intersection is $\{9, 11\}$.

2. $\{2, 4, 8, 9, 10\}$

3. $\{1, 5, 10, 15\} \cup \{5, 15, 20\}$

   The numbers in either or both sets are 1, 5, 10, 15, and 20, so the union is $\{1, 5, 10, 15, 20\}$.

4. $\{5\}$

5. {a, b, c, d} ∩ {b, f, g}

   The only letter common to both sets is b, so the intersection is {b}.

6. {a, b, c}

7. {r, s, t} ∪ {r, u, t, s, v}

   The letters in either or both sets are r, s, t, u, and v, so the union is {r, s, t, u, v}.

8. {m, o, p}

9. $\{2, 5, 7, 9\} \cap \{5, 7\}$

   The numbers 5 and 7 are common to both sets, so the intersection is $\{5, 7\}$.

10. $\{1, 4, 5, 6, 8, 9\}$

11. $\{3, 5, 7\} \cup \emptyset$

    The numbers in either or both sets are 3, 5, and 7, so the union is $\{3, 5, 7\}$.

12. $\emptyset$

13. $2 < x < 7$

    This inequality is an abbreviation for the conjunction $2 < x$ *and* $x < 7$. The graph is the intersection of two separate solution sets: $\{x|2 < x\} \cap \{x|x < 7\} = \{x|2 < x < 7\}$.

    Interval notation: $(2, 7)$

14. 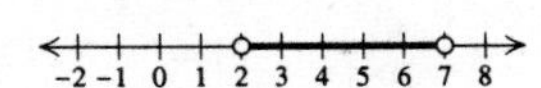

    $[0, 4]$

15. $-6 \le y \le -2$

    This inequality is an abbreviation for the conjunction $-6 \le y$ *and* $y \le -2$.

    Interval notation: $[-6, -2]$

16. $[-9, -5)$

17. $x < -2$ *or* $x > 1$

    The graph of this disjunction is the union of the graphs of the individual solution sets $\{x|x < -2\}$ and $\{x|x > 1\}$.

    Interval notation: $(-\infty, -2) \cup (1, \infty)$

18. $(-\infty, -2) \cup (3, \infty)$

19. $x \le -1$ *or* $x > 4$

    Interval notation: $(-\infty, -1] \cup (4, \infty)$

20. $(-\infty, -5] \cup (2, \infty)$

21. $-3 \le -x < 5$

    $3 \ge x > -5$ Multiplying by -1 and reversing the inequality symbols

    $-5 < x \le 3$ Rewriting

    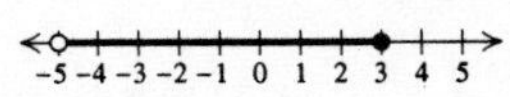

    Interval notation: $(-5, 3]$

22. $(-7, -2)$

23. $x > -2$ *and* $x < 4$

    This conjunction can be abbreviated as $-2 < x < 4$.

    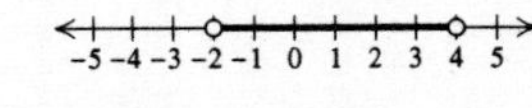

    Interval notation: $(-2, 4)$

24. $(-3, 1]$

25. $5 > a$ *or* $a > 7$

    Interval notation: $(-\infty, 5) \cup (7, \infty)$

26. $(-\infty, -3) \cup [2, \infty)$

27. $x \geq 5$ *or* $-x \geq 4$

Multiplying the second inequality by -1 and reversing the inequality symbols, we get $x \geq 5$ *or* $x \leq -4$.

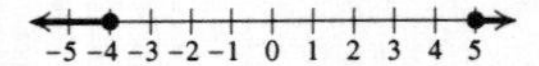

Interval notation: $(-\infty - 4] \cup [5, \infty)$

28.

$(-\infty, -6) \cup (-3, \infty)$

29. $5 > x$ *and* $x \geq -6$

This conjunction can be abbreviated as $-6 \leq x < 5$.

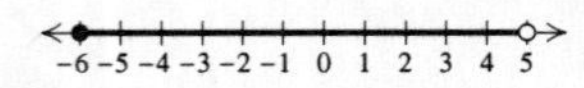

Interval notation: $[-6, 5)$

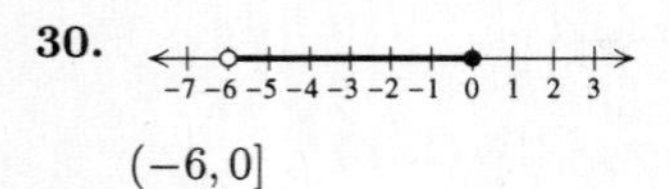

$(-6, 0]$

31. $x < 7$ *and* $x \geq 3$

This conjunction can be abbreviated as $3 \leq x < 7$.

Interval notation: $[3, 7)$

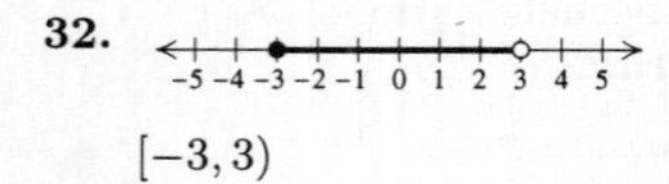

$[-3, 3)$

33. $t < 2$ *or* $t < 5$

The graph of this disjunction is the union of the graphs of the individual solution sets: $\{t|t < 2\} \cup \{t|t < 5\}$. This is the set $\{t|t < 5\}$.

Interval notation: $(-\infty, 5)$

34.

$(-1, \infty)$

35. $x > -1$ *or* $x \leq 3$

The graph of this disjunction is the union of the graphs of the individual solution sets:

$\{x|x > -1\} \cup \{x|x \leq 3\} =$ the set of all real numbers.

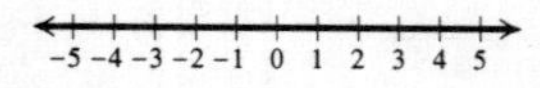

Interval notation: $(-\infty, \infty)$

36.

$(-\infty, \infty)$

37. $x \geq 5$ *and* $x > 7$

The graph of this conjunction is the intersection of two separate solution sets: $\{x|x \geq 5\} \cap \{x|x > 7\} = \{x|x > 7\}$.

Interval notation: $(7, \infty)$

38.

$(-\infty, -4]$

39.
$$-3 < t + 2 < 7$$
$$-3 - 2 < t < 7 - 2$$
$$-5 < t < 5$$

The solution set is $\{t| -5 < t < 5\}$, or $(-5, 5)$.

40. $\{t| -3 < t \leq 4\}$, or $(-3, 4]$

41.
$$2 < x + 3 \quad \text{and} \quad x + 1 \leq 5$$
$$-1 < x \quad \text{and} \quad x \leq 4$$

We can abbreviate the answer as $-1 < x \leq 4$. The solution set is $\{x| -1 < x \leq 4\}$, or $(-1, 4]$.

42. $\{x| -3 < x < 7\}$, or $(-3, 7)$

43.
$$-5 \leq 2a - 1 \quad \text{and} \quad 3a + 1 < 7$$
$$-4 \leq 2a \quad \text{and} \quad 3a < 6$$
$$-2 \leq a \quad \text{and} \quad a < 2$$

We can abbreviate the answer as $-2 \leq a < 2$. The solution set is $\{a| -2 \leq a < 2\}$, or $[-2, 2)$.

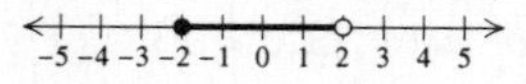

44. $\{n| -2 \leq n \leq 4\}$, or $[-2, 4]$

45.
$$x + 7 \leq -2 \quad \text{or} \quad x + 7 \geq 5$$
$$x \leq -9 \quad \text{or} \quad x \geq -2$$

The solution set is $\{x|x \leq -9$ *or* $x \geq -2\}$, or $(-\infty, -9] \cup [-2, \infty)$.

46. $\{x|x < -8$ *or* $x \geq -1\}$, or $(-\infty, -8) \cup [-1, \infty)$

47. $2 \le 3x - 1 \le 8$

$3 \le 3x \le 9$

$1 \le x \le 3$

The solution set is $\{x|1 \le x \le 3\}$, or $[1, 3]$.

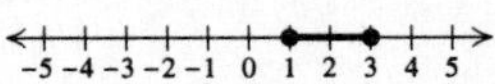

48. $\{x|1 \le x \le 5\}$, or $[1, 5]$

49. $-18 \le -2x - 7 < 0$

$-11 \le -2x < 7$

$\frac{11}{2} \ge x > -\frac{7}{2}$, or

$-\frac{7}{2} < x \le \frac{11}{2}$

The solution set is $\left\{x \middle| -\frac{7}{2} < x \le \frac{11}{2}\right\}$, or $\left(-\frac{7}{2}, \frac{11}{2}\right]$.

50. $\left\{t \middle| -4 < t \le -\frac{10}{3}\right\}$, or $\left(-4, -\frac{10}{3}\right]$

51. $3x - 1 \le 2$ *or* $3x - 1 \ge 8$

$3x \le 3$ *or* $3x \ge 9$

$x \le 1$ *or* $x \ge 3$

The solution set is $\{x|x \le 1 \text{ or } x \ge 3\}$, or $(-\infty, 1] \cup [3, \infty)$.

52. $\{x|x \le 1 \text{ or } x \ge 5\}$, or $(-\infty, 1] \cup [5, \infty)$

53. $2x - 7 < -1$ *or* $2x - 7 > 1$

$2x < 6$ *or* $2x > 8$

$x < 3$ *or* $x > 4$

The solution set is $\{x|x < 3 \text{ or } x > 4\}$, or $(-\infty, 3) \cup (4, \infty)$.

54. $\left\{x \middle| x < -4 \text{ or } x > \frac{2}{3}\right\}$, or $(-\infty, -4) \cup \left(\frac{2}{3}, \infty\right)$

55. $6 > 2a - 1$ *or* $-4 \le -3a + 2$

$7 > 2a$ *or* $-6 \le -3a$

$\frac{7}{2} > a$ *or* $2 \ge a$

The solution set is $\left\{a \middle| \frac{7}{2} > a\right\} \cup \{a|2 \ge a\} =$ $\left\{a \middle| \frac{7}{2} > a\right\}$, or $\left\{a \middle| a < \frac{7}{2}\right\}$, or $\left(-\infty, \frac{7}{2}\right)$.

56. The set of all real numbers, or $(-\infty, \infty)$

57. $a + 4 < -1$ *and* $3a - 5 < 7$

$a < -5$ *and* $3a < 12$

$a < -5$ *and* $a < 4$

The solution set is $\{a|a < -5\} \cap \{a|a < 4\} =$ $\{a|a < -5\}$, or $(-\infty, -5)$.

58. $\{a|a > 4\}$, or $(4, \infty)$

59. $3x + 2 < 2$ *or* $4 - 2x < 14$

$3x < 0$ *or* $-2x < 10$

$x < 0$ *or* $x > -5$

The solution set is $\{x|x < 0\} \cup \{x|x > -5\}$ = the set of all real numbers, or $(-\infty, \infty)$.

60. $\{x|x \le -2 \text{ or } x > 3\}$, or $(-\infty, -2] \cup (3, \infty)$

61. $2t - 7 \le 5$ *or* $5 - 2t > 3$

$2t \le 12$ *or* $-2t > -2$

$t \le 6$ *or* $t < 1$

The solution set is $\{t|t \le 6\} \cup \{t|t < 1\} = \{t|t \le 6\}$, or $(-\infty, 6]$.

62. $\{a|a \ge -1\}$, or $[-1, \infty)$

63. $f(x) = \frac{7}{x - 5}$

$f(x)$ cannot be computed when the denominator is 0. Since $x - 5 = 0$ is equivalent to $x = 5$, we have Domain of $f = \{x|x \text{ is a real number and } x \ne 5\} =$ $(-\infty, 5) \cup (5, \infty)$.

64. $(-\infty,-3)\cup(-3,\infty)$

65. $f(x)=\sqrt{x+4}$

The expression $\sqrt{x+4}$ is not a real number when $x+4$ is negative. Thus, the domain of f is the set of all x-values for which $x+4\geq 0$. Since $x+4\geq 0$ is equivalent to $x\geq -4$, we have Domain of $f=[-4,\infty)$.

66. $[-7,\infty)$

67. $f(x)=\dfrac{x+3}{2x-5}$

$f(x)$ cannot be computed when the denominator is 0. Since $2x-5=0$ is equivalent to $x=\dfrac{5}{2}$, we have Domain of $f=\left\{x\middle|x \text{ is a real number and } x\neq\dfrac{5}{2}\right\}$, or $\left(-\infty,\dfrac{5}{2}\right)\cup\left(\dfrac{5}{2},\infty\right)$.

68. $\left(-\infty,-\dfrac{4}{3}\right)\cup\left(-\dfrac{4}{3},\infty\right)$

69. $f(x)=\sqrt{12-3x}$

The expression $\sqrt{12-3x}$ is not a real number when $12-3x$ is negative. Thus, the domain of f is the set of all x-values for which $12-3x\geq 0$. Since $12-3x\geq 0$ is equivalent to $x\leq 4$, we have Domain of $f=(-\infty,4]$.

70. $(-\infty,2]$

71.
$$\begin{aligned} 2x-3y&=7, &&(1)\\ 3x+2y&=-10 &&(2)\end{aligned}$$

We will use the elimination method.

$$\begin{aligned} 4x-6y&=14 &&\text{Multiplying (1) by 2}\\ 9x+6y&=-30 &&\text{Multiplying (2) by 3}\\ \hline 13x&=-16 &&\text{Adding}\\ x&=-\frac{16}{13} \end{aligned}$$

Substitute $-\dfrac{16}{13}$ for x in (1) and solve for y.

$$\begin{aligned} 2\left(-\frac{16}{13}\right)-3y&=7\\ -\frac{32}{13}-3y&=7\\ \frac{32}{13}-\frac{32}{13}-3y&=\frac{32}{13}+\frac{91}{13}\\ -3y&=\frac{123}{13}\\ y&=-\frac{41}{13} \end{aligned}$$

These numbers check, so the solution is $\left(-\dfrac{16}{13},-\dfrac{41}{13}\right)$.

72. $-\dfrac{8}{3}$

73.
$$\begin{aligned} 5(2x+3)&=3(x-4)\\ 10x+15&=3x-12\\ 7x+15&=-12\\ 7x&=-27\\ x&=-\frac{27}{7} \end{aligned}$$

The solution is $-\dfrac{27}{7}$.

74.

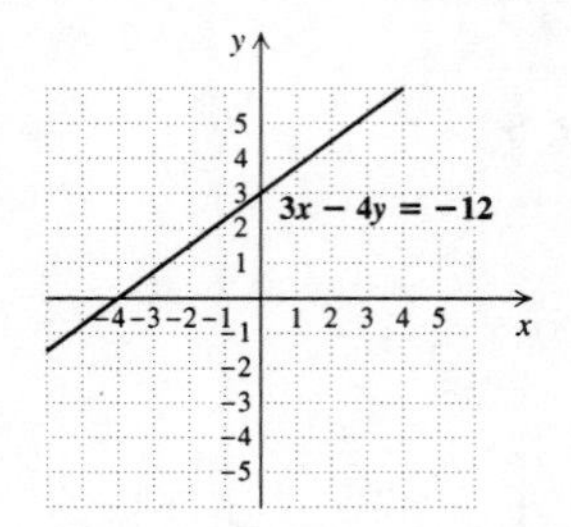

75. Graph: $f(x)=5$

The graph of any constant function $f(x)=c$ is a horizontal line that crosses the vertical axis at $(0,c)$. Thus, the graph of $f(x)=5$ is a horizontal line that crosses the vertical axis at $(0,5)$.

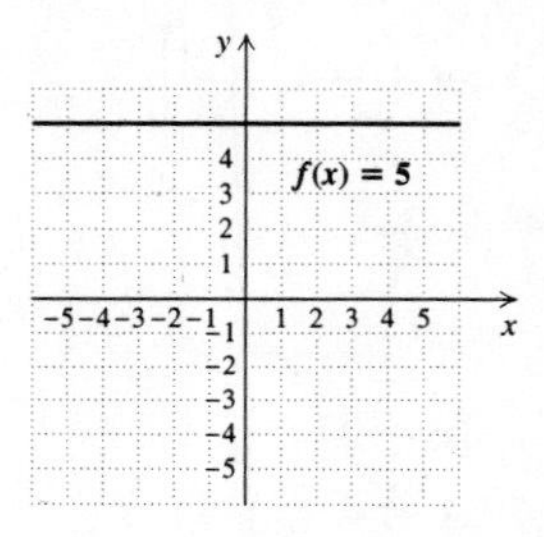

76.

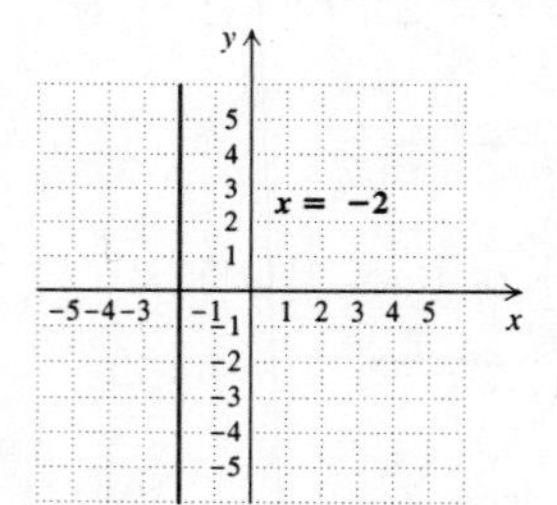

77. ◈

78. ◈

79. ◈

80. ◈

81. From the graph we observe that the values of x for which $2x-5>-7$ *and* $2x-5<7$ are $\{x|-1<x<6\}$, or $(-1,6)$.

82. $\{x|x<-3 \text{ or } x>6\}$, or $(-\infty,-3)\cup(6,\infty)$

83. ***Familiarize.*** Let c = the number of crossings per month. Then at the \$3 per crossing rate, the total cost of c crossings is \$3c. A six-month pass costs \$15/6, or \$2.50 per month. The additional 50¢ per crossing toll brings the total cost of c crossings to \$2.50 + \$0.50c. A one-year pass costs \$60/12, or \$5 per month regardless of the number of crossings.

Translate. We write an inequality that states that the cost of c crossings per month using the six-month pass is less than the cost using the \$3 per crossing toll and is less than the cost using the one-year pass.

$$2.50 + 0.50c < 3c \textit{ and } 2.50 + 0.50c < 5$$

Carry out. We solve the inequality.

$$2.50 + 0.50c < 3c \quad \textit{and} \quad 2.50 + 0.50c < 5$$
$$2.50 < 2.5c \quad \textit{and} \quad 0.50c < 2.50$$
$$1 < c \quad \textit{and} \quad c < 5$$

This result can be written as $1 < c < 5$.

Check. When we substitute values of c less than 1, between 1 and 5, and greater than 5, we find that the result checks. Since we cannot check every possible value of c, we stop here.

State. For more than 1 crossing but less than 5 crossings per month the six-month pass is the most economical choice.

84. Sizes between 6 and 13

85. Solve $1 \le P(d) = 7$, or $1 \le 1 + \dfrac{d}{33} \le 7$.

$$1 \le 1 + \frac{d}{33} \le 7$$
$$0 \le \frac{d}{33} \le 6$$
$$0 \le d \le 198$$

Thus, 0 ft $\le d \le$ 198 ft.

86. (a) $1945.4° \le F < 4820°$; (b) $1761.44° \le F < 3956°$

87. Solve $5.0 \le w(t) \le 5.25$, or $5.0 \le 0.05t + 4.3 \le 5.25$

$$5.0 \le 0.05t + 4.3 \le 5.25$$
$$0.7 \le 0.05t \le 0.95$$
$$14 \le t \le 19$$

Thus, from 14 to 19 years after 1991, or from 2005 to 2010, waste production will range from 5.0 to 5.25 pounds per person per day.

88. $1965 \le y \le 1981$

89. $4a - 2 \le a + 1 \le 3a + 4$

$$4a - 2 \le a + 1 \quad \textit{and} \quad a + 1 \le 3a + 4$$
$$3a \le 3 \quad \textit{and} \quad -3 \le 2a$$
$$a \le 1 \quad \textit{and} \quad -\frac{3}{2} \le a$$

The solution set is $\left\{a\middle| -\frac{3}{2} \le a \le 1\right\}$, or $\left[-\frac{3}{2}, 1\right]$.

-2 $-\frac{3}{2}$ -1 $-\frac{1}{2}$ 0 $\frac{1}{2}$ 1 $\frac{3}{2}$ 2

90. $\left\{m\middle|m < \frac{6}{5}\right\}$, or $\left(-\infty, \frac{6}{5}\right)$

$\frac{6}{5}$ -2 -1 0 1 2

91. $x - 10 < 5x + 6 \le x + 10$

$$-10 < 4x + 6 \le 10$$
$$-16 < 4x \le 4$$
$$-4 < x \le 1$$

The solution set is $\{x| -4 < x \le 1\}$, or $(-4, 1]$.

-6 -5 -4 -3 -2 -1 0 1 2 3 4

92. $\left\{x\middle| -\frac{1}{8} < x < \frac{1}{2}\right\}$, or $\left(-\frac{1}{8}, \frac{1}{2}\right)$

$-\frac{1}{8}$ 0 $\frac{1}{2}$ 1

93. If $-b < -a$, then $-1(-b) > -1(-a)$, or $b > a$, or $a < b$. The statement is true.

94. False

95. Let $a = 5$, $c = 12$, and $b = 2$. Then $a < c$ and $b < c$, but $a \not< b$. The given statement is false.

96. True

97. $f(x) = \dfrac{\sqrt{5 + 2x}}{x - 1}$

The expression $\sqrt{5 + 2x}$ is not a real number when $5 + 2x$ is negative. Then for $5 + 2x \ge 0$, or for $x \ge -\frac{5}{2}$, the numerator of $f(x)$ is a real number. In addition, $f(x)$ cannot be computed when the denominator is 0. Since $x - 1 = 0$ is equivalent to $x = 1$, we have Domain of $f = \left\{x\middle|x \ge -\frac{5}{2} \textit{ and } x \ne 1\right\}$, or $\left[-\frac{5}{2}, 1\right) \cup (1, \infty)$.

98. $\left\{x\middle|x \le \frac{3}{4} \textit{ and } x \ne -7\right\}$, or $(-\infty, -7) \cup \left(-7, \frac{3}{4}\right]$

99. Observe that the graph of y_2 lies on or above the graph of y_1 and below the graph of y_3 for x in the interval $[-3, 4)$.

100.

101.

102.

Exercise Set 4.3

1. $|x| = 7$

$x = -7$ *or* $x = 7$ Using the absolute-value principle

The solution set is $\{-7, 7\}$.

2. $\{-9, 9\}$

3. $|x| = -5$

The absolute value of a number is always nonnegative. Therefore, the solution set is $\emptyset$.

4. $\emptyset$

5. $|y| = 8.6$

$y = -8.6$ *or* $y = 8.6$ Using the absolute-value principle

The solution set is $\{-8.6, 8.6\}$.

6. $\{0\}$

7. $|m| = 0$

$m = 0$

$\{0\}$

The only number whose absolute value is 0 is 0. The solution set is $\{0\}$.

8. $\{-5.5, 5.5\}$

9. $|5x + 2| = 3$

$5x + 2 = -3$ *or* $5x + 2 = 3$ Absolute-value principle

$5x = -5$ *or* $5x = 1$

$x = -1$ *or* $x = \frac{1}{5}$

The solution set is $\left\{-1, \frac{1}{5}\right\}$.

10. $\left\{-\frac{1}{2}, \frac{7}{2}\right\}$

11. $|7x - 2| = -9$

Absolute value is always nonnegative, so the equation has no solution. The solution set is $\emptyset$.

12. $\emptyset$

13. $|2y| - 5 = 13$

$|2y| = 18$ Adding 5

$2y = -18$ *or* $2y = 18$

$y = -9$ *or* $y = 9$

The solution set is $\{-9, 9\}$.

14. $\{-8, 8\}$

15. $7|z| + 2 = 16$ Adding -2

$7|z| = 14$ Multiplying by $\frac{1}{7}$

$|z| = 2$

$z = -2$ *or* $z = 2$

The solution set is $\{-2, 2\}$.

16. $\left\{-\frac{11}{5}, \frac{11}{5}\right\}$

17. $\left|\frac{4-5x}{6}\right| = 7$

$\frac{4-5x}{6} = -7$ *or* $\frac{4-5x}{6} = 7$

$4 - 5x = -42$ *or* $4 - 5x = 42$

$-5x = -46$ *or* $-5x = 38$

$x = \frac{46}{5}$ *or* $x = -\frac{38}{5}$

The solution set is $\left\{-\frac{38}{5}, \frac{46}{5}\right\}$.

18. $\{-7, 8\}$

19. $|t - 7| + 3 = 4$ Adding -3

$|t - 7| = 1$

$t - 7 = -1$ *or* $t - 7 = 1$

$t = 6$ *or* $t = 8$

The solution set is $\{6, 8\}$.

20. $\{-12, 2\}$

21. $3|2x - 5| - 7 = -1$

$3|2x - 5| = 6$

$|2x - 5| = 2$

$2x - 5 = -2$ *or* $2x - 5 = 2$

$2x = 3$ *or* $2x = 7$

$x = \frac{3}{2}$ *or* $x = \frac{7}{2}$

The solution set is $\left\{\frac{3}{2}, \frac{7}{2}\right\}$.

22. $\left\{-\frac{1}{3}, 3\right\}$

23. $|3x - 4| = 8$

$3x - 4 = -8$ *or* $3x - 4 = 8$

$3x = -4$ *or* $3x = 12$

$x = -\frac{4}{3}$ *or* $x = 4$

The solution set is $\left\{-\frac{4}{3}, 4\right\}$.

24. $\left\{-\frac{3}{2}, \frac{17}{2}\right\}$

25. $|x| - 2 = 6.3$

$|x| = 8.3$

$x = -8.3$ *or* $x = 8.3$

The solution set is $\{-8.3, 8.3\}$.

26. $\{-11, 11\}$

27. $\left|\frac{3x-2}{5}\right| = 2$

$\frac{3x-2}{5} = -2$ *or* $\frac{3x-2}{5} = 2$

$3x - 2 = -10$ *or* $3x - 2 = 10$

$3x = -8$ *or* $3x = 12$

$x = -\frac{8}{3}$ *or* $x = 4$

The solution set is $\left\{-\frac{8}{3}, 4\right\}$.

28. $\{-1, 2\}$

29. The distance between 25 and 14 is $|25 - 14| = |11| = 11$, or $|14 - 25| = |-11| = 11$.

30. 15

31. The distance between -9 and 24 is $|24 - (-9)| = |24 + 9| = |33| = 33$, or $|-9 - 24| = |-33| = 33$.

32. 19

33. The distance between -8 and -42 is $|-8-(-42)| = |-8+42| = |34| = 34$, or $|-42-(-8)| = |-42+8| = |-34| = 34$.

34. 27

35. $|x + 4| = |2x - 7|$

$x + 4 = 2x - 7$ *or* $x + 4 = -(2x - 7)$

$4 = x - 7$ *or* $x + 4 = -2x + 7$

$11 = x$ *or* $3x + 4 = 7$

$3x = 3$

$x = 1$

The solution set is $\{1, 11\}$.

36. $\left\{-\frac{11}{2}, \frac{1}{4}\right\}$

37. $|x - 9| = |x + 6|$

$x - 9 = x + 6$ *or* $x - 9 = -(x + 6)$

$-9 = 6$ *or* $x - 9 = -x - 6$

False – yields no solution; $2x - 9 = -6$

$2x = 3$

$x = \frac{3}{2}$

The solution set is $\left\{\frac{3}{2}\right\}$.

38. $\left\{-\frac{1}{2}\right\}$

39. $|5t + 7| = |4t + 3|$

$5t + 7 = 4t + 3$ *or* $5t + 7 = -(4t + 3)$

$t + 7 = 3$ *or* $5t + 7 = -4t - 3$

$t = -4$ *or* $9t + 7 = -3$

$9t = -10$

$t = -\frac{10}{9}$

The solution set is $\left\{-4, -\frac{10}{9}\right\}$.

40. $\left\{-\frac{3}{5}, 5\right\}$

41. $|n - 3| = |3 - n|$

$n - 3 = 3 - n$ *or* $n - 3 = -(3 - n)$

$2n - 3 = 3$ *or* $n - 3 = -3 + n$

$2n = 6$ *or* $-3 = -3$

$n = 3$ True for all real values of n

The solution set is the set of all real numbers.

42. The set of all real numbers

43. $|7 - a| = |a + 5|$

$7 - a = a + 5$ *or* $7 - a = -(a + 5)$

$7 = 2a + 5$ *or* $7 - a = -a - 5$

$2 = 2a$ *or* $7 = -5$

$1 = a$ False

The solution set is $\{1\}$.

44. $\left\{-\frac{1}{2}\right\}$

45. $\left|\frac{1}{2}x - 5\right| = \left|\frac{1}{4}x + 3\right|$

$\frac{1}{2}x - 5 = \frac{1}{4}x + 3$ *or* $\frac{1}{2}x - 5 = -\left(\frac{1}{4}x + 3\right)$

$\frac{1}{4}x - 5 = 3$ *or* $\frac{1}{2}x - 5 = -\frac{1}{4}x - 3$

$\frac{1}{4}x = 8$ *or* $\frac{3}{4}x - 5 = -3$

$x = 32$ *or* $\frac{3}{4}x = 2$

$x = \frac{8}{3}$

The solution set is $\left\{32, \frac{8}{3}\right\}$.

46. $\left\{-\frac{48}{37}, -\frac{144}{5}\right\}$

47. $|a| \le 6$

$-6 \le a \le 6$ Part (b)

The solution set is $\{a| -6 \le a \le 6\}$, or $[-6, 6]$.

-7 -6 -5 -4 -3 -2 -1 0 1 2 3 4 5 6 7

48. $\{x| -2 < x < 2\}$, or $(-2, 2)$

49. $|x| > 7$

$x < -7 \text{ or } 7 < x$ Part (c)

The solution set is $\{x|x < -7 \text{ or } x > 7\}$, or $(-\infty, -7) \cup (7, \infty)$.

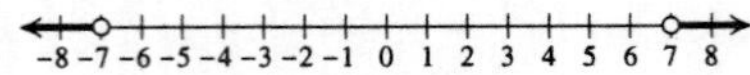

50. $\{a|a \leq -3 \text{ or } a \geq 3\}$, or $(-\infty, -3] \cup [3, \infty)$

51. $|t| > 0$

$t < 0 \text{ or } 0 < t$ Part (c)

The solution set is $\{t|t < 0 \text{ or } t > 0\}$, or $\{t|t \neq 0\}$, or $(-\infty, 0) \cup (0, \infty)$.

52. $\{t|t \leq -1.7 \text{ or } t \geq 1.7\}$, or $(-\infty, -1.7] \cup [1.7, \infty)$

53. $|x - 3| < 5$

$-5 < x - 3 < 5$ Part (b)

$-2 < x < 8$

The solution set is $\{x| -2 < x < 8\}$, or $(-2, 8)$.

54. $\{x| -2 < x < 4\}$, or $(-2, 4)$

55. $|x + 2| \leq 5$

$-5 \leq x + 2 \leq 5$ Part (b)

$-7 \leq x \leq 3$ Adding -2

The solution set is $\{x| -7 \leq x \leq 3\}$, or $[-7, 3]$.

56. $\{x| -5 \leq x \leq -3\}$, or $[-5, -3]$

57. $|x - 3| + 2 > 7$

$|x - 3| > 5$ Adding -2

$x - 3 < -5 \text{ or } 5 < x - 3$ Part (c)

$x < -2 \text{ or } 8 < x$

The solution set is $\{x|x < -2 \text{ or } x > 8\}$, or $(-\infty, -2) \cup (8, \infty)$.

58. The set of all real numbers, or $(-\infty, \infty)$

59. $|2y - 7| > -1$

Since absolute value is never negative, any value of $2y - 7$, and hence any value of y, will satisfy the inequality. The solution set is the set of all real numbers, or $(-\infty, \infty)$.

60. $\left\{y \middle| y < -\frac{4}{3} \text{ or } y > 4\right\}$, or $\left(-\infty, -\frac{4}{3}\right) \cup (4, \infty)$

61. $|3a - 4| + 2 \geq 7$

$|3a - 4| \geq 5$ Adding -2

$3a - 4 \leq -5 \text{ or } 5 \leq 3a - 4$ Part (c)

$3a \leq -1 \text{ or } 9 \leq 3a$

$a \leq -\frac{1}{3} \text{ or } 3 \leq a$

The solution set is $\left\{a \middle| a \leq -\frac{1}{3} \text{ or } a \geq 3\right\}$, or $\left(-\infty, -\frac{1}{3}\right] \cup [3, \infty)$.

62. $\{a|a \leq -1 \text{ or } a \geq 6\}$, or $(-\infty, -1] \cup [6, \infty)$

63. $|y - 3| < 12$

$-12 < y - 3 < 12$ Part (b)

$-9 < y < 15$ Adding 3

The solution set is $\{y| -9 < y < 15\}$, or $(-9, 15)$.

64. $\{p| -1 < p < 5\}$ or $(-1, 5)$

65. $9 - |x + 4| \leq 5$

$-|x + 4| \leq -4$

$|x + 4| \geq 4$ Multiplying by -1

$x + 4 \leq -4 \text{ or } 4 \leq x + 4$ Part (c)

$x \leq -8 \text{ or } 0 \leq x$

The solution set is $\{x|x \leq -8 \text{ or } x \geq 0\}$, or $(-\infty, -8] \cup [0, \infty)$.

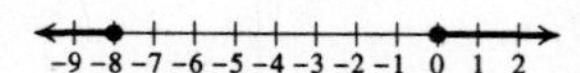

66. $\{x|x \le 2 \text{ or } x \ge 8\}$, or $(-\infty, 2] \cup [8, \infty)$

67. $|4 - 3y| > 8$

$4 - 3y < -8$ or $8 < 4 - 3y$ Part (c)

$-3y < -12$ or $4 < -3y$ Adding -4

$y > 4$ or $-\frac{4}{3} > y$ Multiplying by $-\frac{1}{3}$

The solution set is $\left\{y \middle| y < -\frac{4}{3} \text{ or } y > 4\right\}$, or $\left(-\infty, -\frac{4}{3}\right) \cup (4, \infty)$.

68. $\emptyset$

69. $|3 - 4x| < -5$

Absolute value is always nonnegative, so the inequality has no solution. The solution set is $\emptyset$.

70. $\left\{a \middle| -\frac{7}{2} \le a \le 6\right\}$, or $\left[-\frac{7}{2}, 6\right]$

71. $\left|\frac{2 - 5x}{4}\right| \ge \frac{2}{3}$

$\frac{2 - 5x}{4} \le -\frac{2}{3}$ or $\frac{2}{3} \le \frac{2 - 5x}{4}$ Part (c)

$2 - 5x \le -\frac{8}{3}$ or $\frac{8}{3} \le 2 - 5x$ Multiplying by 4

$-5x \le -\frac{14}{3}$ or $\frac{2}{3} \le -5x$ Adding -2

$x \ge \frac{14}{15}$ or $-\frac{2}{15} \ge x$ Multiplying by $-\frac{1}{5}$

The solution set is $\left\{x \middle| x \le -\frac{2}{15} \text{ or } x \ge \frac{14}{15}\right\}$, or $\left(-\infty, -\frac{2}{15}\right] \cup \left[\frac{14}{15}, \infty\right)$.

72. $\left\{x \middle| x < -\frac{43}{24} \text{ or } x > \frac{9}{8}\right\}$, or $\left(-\infty, -\frac{43}{24}\right) \cup \left(\frac{9}{8}, \infty\right)$

73. $|m + 5| + 9 \le 16$

$|m + 5| \le 7$ Adding -9

$-7 \le m + 5 \le 7$

$-12 \le m \le 2$

The solution set is $\{m| -12 \le m \le 2\}$, or $[-12, 2]$.

74. $\{t|t \le 6 \text{ or } t \ge 8\}$, or $(-\infty, 6] \cup [8, \infty)$

75. $25 - 2|a + 3| > 19$

$-2|a + 3| > -6$

$|a + 3| < 3$ Multiplying by $-\frac{1}{2}$

$-3 < a + 3 < 3$ Part (b)

$-6 < a < 0$

The solution set is $\{a| -6 < a < 0\}$, or $(-6, 0)$.

76. $\left\{a \middle| -\frac{13}{2} < a < \frac{5}{2}\right\}$, or $\left(-\frac{13}{2}, \frac{5}{2}\right)$

77. $|2x - 3| \le 4$

$-4 \le 2x - 3 \le 4$ Part (b)

$-1 \le 2x \le 7$ Adding 3

$-\frac{1}{2} \le x \le \frac{7}{2}$ Multiplying by $\frac{1}{2}$

The solution set is $\left\{x \middle| -\frac{1}{2} \le x \le \frac{7}{2}\right\}$, or $\left[-\frac{1}{2}, \frac{7}{2}\right]$.

78. $\left\{x \middle| -1 \le x \le \frac{1}{5}\right\}$, or $\left[-1, \frac{1}{5}\right]$

79. $2 + |3x - 4| \ge 13$

$|3x - 4| \ge 11$

$3x - 4 \le -11$ or $11 \le 3x - 4$ Part (c)

$3x \le -7$ or $15 \le 3x$

$x \le -\frac{7}{3}$ or $5 \le x$

The solution set is $\left\{x \middle| x \le -\frac{7}{3} \text{ or } x \ge 5\right\}$, or $\left(-\infty, -\frac{7}{3}\right] \cup [5, \infty)$.

80. $\left\{x \middle| x \le -\frac{23}{9} \text{ or } x \ge 3\right\}$, or $\left(-\infty, -\frac{23}{9}\right] \cup [3, \infty)$

81. $7 + |2x - 1| < 16$

$|2x - 1| < 9$

$-9 < 2x - 1 < 9$ Part (b)

$-8 < 2x < 10$

$-4 < x < 5$

The solution set is $\{x | -4 < x < 5\}$, or $(-4, 5)$.

82. $\left\{x \middle| -\dfrac{16}{3} < x < 4\right\}$, or $\left(-\dfrac{16}{3}, 4\right)$

83. ***Familiarize.*** Let x = the number of children's dinners and y = the number of adult's dinners served. The $\$3x$ is collected for children's dinners, and $\$8y$ is collected for adult's dinners.

Translate. The information that 250 dinners were served gives us one equation.

$x + y = 250$

The information that a total of $1410 was collected gives us a second equation.

$3x + 8y = 1410$

We have a system of equations.

$$x + y = 250, \quad (1)$$
$$3x + 8y = 1410 \quad (2)$$

Carry out. We use the elimination method. Begin by multiplying Equation (1) by -3 and adding.

$$\begin{aligned} -3x - 3y &= -750 \\ 3x + 8y &= 1410 \\ \hline 5y &= 660 \\ y &= 132 \end{aligned}$$

Substitute 132 for y in Equation (1) and solve for x.

$$x + 132 = 250$$
$$x = 118$$

Check. A total of 118 + 132, or 250 dinners were served. The amount collected was $\$3 \cdot 118 + \$8 \cdot 132 = \$354 + \$1056 = \$1410$. The numbers check.

State. 118 children's dinners and 132 adult's dinners were served.

84. 26,640 m^2

85. $$\begin{aligned} 2x^2 + 3y &= 2 \cdot 5^2 + 3 \cdot 6 \\ &= 2 \cdot 25 + 3 \cdot 6 \\ &= 50 + 18 \\ &= 68 \end{aligned}$$

86. 29

87.

88.

89.

90.

91. From the definition of absolute value, $|2x-5| = 2x-5$ only when $2x - 5 \geq 0$. Solve $2x - 5 \geq 0$.

$$2x - 5 \geq 0$$
$$2x \geq 5$$
$$x \geq \frac{5}{2}$$

The solution set is $\left\{x \middle| x \geq \dfrac{5}{2}\right\}$, or $\left[\dfrac{5}{2}, \infty\right)$.

92. $\{x | x \geq -3\}$, or $[-3, \infty)$

93. $|7x - 2| = x + 4$

From the definition of absolute value, we know $x + 4 \geq 0$, or $x \geq -4$. So we have $x \geq -4$ *and*

$$7x - 2 = x + 4 \quad or \quad 7x - 2 = -(x + 4)$$
$$6x = 6 \quad or \quad 7x - 2 = -x - 4$$
$$x = 1 \quad or \quad 8x = -2$$
$$x = -\frac{1}{4}$$

The solution set is $\left\{x \middle| x \geq -4 \text{ and } x = 1 \text{ or } x = -\dfrac{1}{4}\right\}$, or $\left\{1, -\dfrac{1}{4}\right\}$.

94. $\{x | x \text{ is a real number}\}$, or $(-\infty, \infty)$

95. $2 \leq |x - 1| \leq 5$

$2 \leq |x - 1|$ *and* $|x - 1| \leq 5$.

For $2 \leq |x - 1|$:

$x - 1 \leq -2$ *or* $2 \leq x - 1$

$x \leq -1$ *or* $3 \leq x$

The solution set of $2 \leq |x-1|$ is $\{x | x \leq -1 \text{ or } x \geq 3\}$.

For $|x - 1| \leq 5$:

$-5 \leq x - 1 \leq 5$

$-4 \leq x \leq 6$

The solution set of $|x - 1| \leq 5$ is $\{x | -4 \leq x \leq 6\}$.

The solution set of $2 \leq |x - 1| \leq 5$ is

$\{x | x \leq -1 \text{ or } x \geq 3\} \cap \{x | -4 \leq x \leq 6\}$

$= \{x | -4 \leq x \leq -1 \text{ or } 3 \leq x \leq 6\}$, *or*

$[-4, -1] \cup [3, 6]$.

96. $|x| < 3$

97. Using part (b), we find that $-5 \leq y \leq 5$ is equivalent to $|y| \leq 5$.

98. $|x| \geq 6$

99. $x < -4$ *or* $4 < x$

$|x| > 4$ Using part (c)

100. $|x+3| > 5$

101. $-5 < x < 1$

$-3 < x+2 < 3$ Adding 2

$|x+2| < 3$ Using part (b)

102. $|x-7| < 2$, or $|7-x| < 2$

103. The distance from x to 5 is $|x-5|$ or $|5-x|$, so we have $|x-5| > 1$ or $|5-x| > 1$.

104. $\left\{d \middle| 5\frac{1}{2} \text{ ft} \le d \le 6\frac{1}{2} \text{ ft}\right\}$, or $\left[5\frac{1}{2} \text{ ft}, 6\frac{1}{2} \text{ ft}\right]$

105. Graph $g(x) = 4$ or the same axes as $f(x) = |2x-6|$.

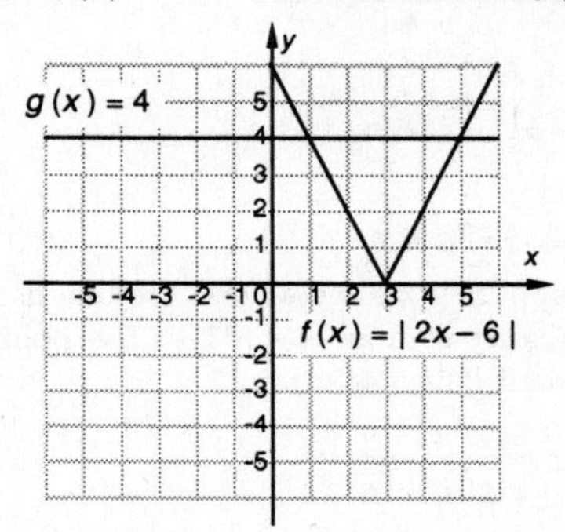

The solution set consists of the x-values for which $(x, f(x))$ is on or below the horizontal line $g(x) = 4$. These x-values comprise the interval $[1, 5]$.

106.

107.

108.

109.

110. 

Exercise Set 4.4

1. We replace x by -4 and y by 2.

$$\begin{array}{c|c} \multicolumn{2}{c}{2x+3y < -1} \\ \hline 2(-4)+3\cdot 2 & ? \; -1 \\ -8+6 & \\ -2 & -1 \quad \text{TRUE} \end{array}$$

Since $-2 < -1$ is true, $(-4, 2)$ is a solution.

2. No

3. We replace x by 8 and y by 14.

$$\begin{array}{c|c} \multicolumn{2}{c}{2y-3x \ge 9} \\ \hline 2\cdot 14-3\cdot 8 & ? \; 9 \\ 28-24 & \\ 4 & 9 \quad \text{FALSE} \end{array}$$

Since $4 > 9$ is false, $(8, 14)$ is not a solution.

4. Yes

5. Graph: $y < \frac{1}{2}x$

We first graph the line $y = \frac{1}{2}x$. We draw the line dashed since the inequality symbol is <. To determine which half-plane to shade, test a point not on the line. We try $(0, 1)$:

$$\begin{array}{c|c} \multicolumn{2}{c}{y < \frac{1}{2}x} \\ \hline 1 & ? \; \frac{1}{2}\cdot 0 \\ 1 & 0 \quad \text{FALSE} \end{array}$$

Since $1 < 0$ is false, (0.1) is not a solution, nor are any points in the half-plane containing $(0, 1)$. The points in the other half-plane are solutions, so we shade that half-plane and obtain the graph.

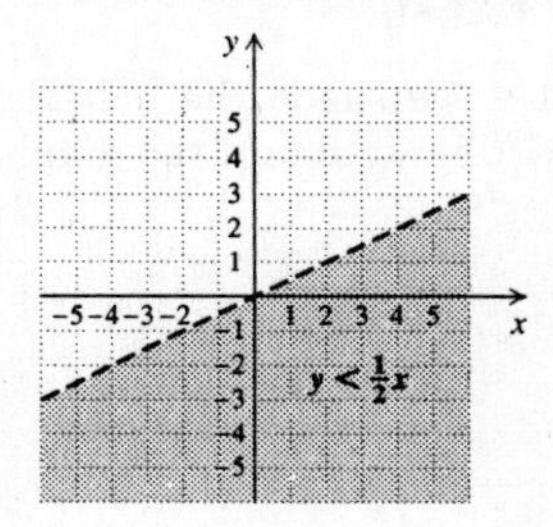

6.

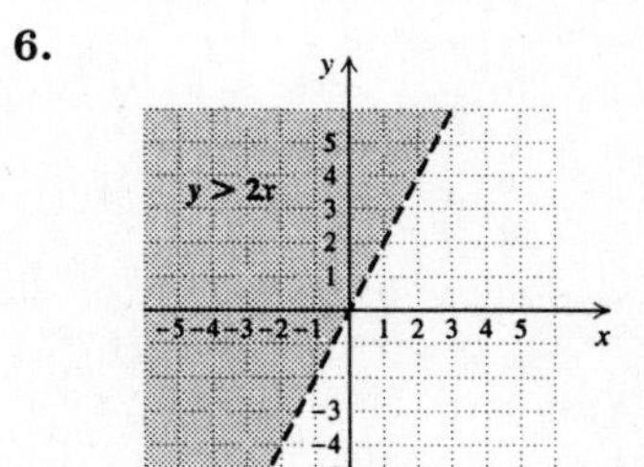

7. Graph: $y \le x - 4$

First graph the line $y = x - 4$. Draw it solid since the inequality symbol is $\le$. Test the point $(0, 0)$ to determine if it is a solution.

$$\begin{array}{c|c} \multicolumn{2}{c}{y \le x-4} \\ \hline 0 & ? \; 0-4 \\ 0 & -4 \quad \text{FALSE} \end{array}$$

Since $0 \le -4$ is false, we shade the half-plane that does not contain $(0, 0)$ and obtain the graph.

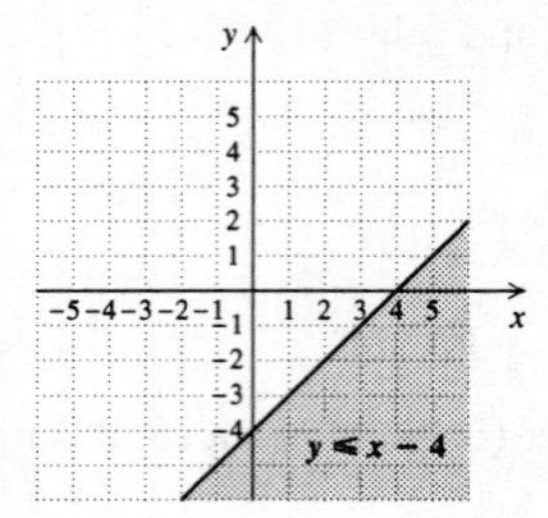

8.

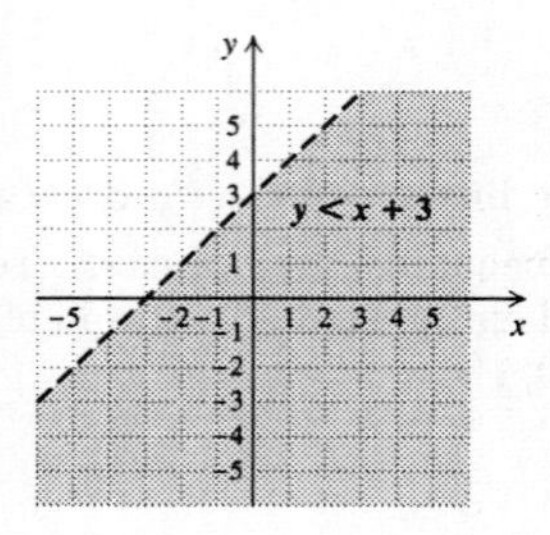

9. Graph: $y \geq x + 4$

First graph the line $y = x + 4$. Draw it solid since the inequality symbol is $\geq$. Test the point $(0, 0)$ to determine if it is a solution.

$$\begin{array}{c|c} \multicolumn{2}{c}{y \geq x + 4} \\ \hline 0 \ ? & 0 + 4 \\ 0 & 4 \end{array} \quad \text{FALSE}$$

Since $0 \geq 4$ is false, we shade the half-plane that does not contain $(0, 0)$ and obtain the graph.

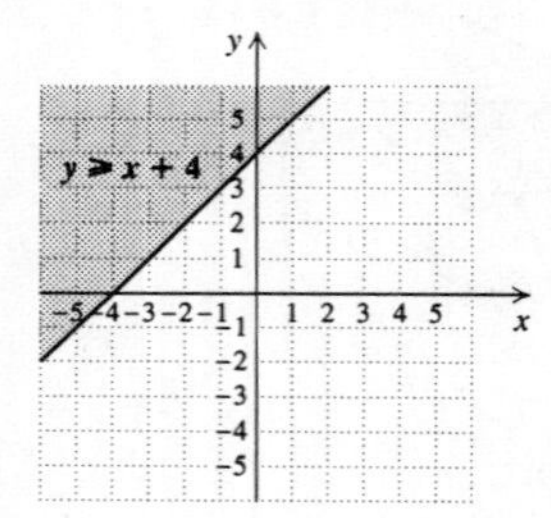

10.

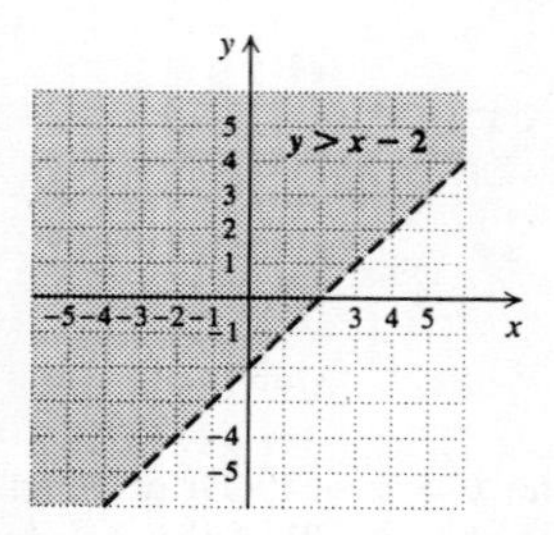

11. Graph: $x - y \geq 5$

First graph the line $x - y = 5$. Draw a solid line since the inequality symbol is $\geq$. Test the point $(0, 0)$ to determine if it is a solution.

$$\begin{array}{c|c} \multicolumn{2}{c}{x - y \geq 5} \\ \hline 0 - 0 \ ? & 5 \\ 0 & 5 \end{array} \quad \text{FALSE}$$

Since $0 \geq 5$ is false, we shade the half-plane that does not contain $(0, 0)$ and obtain the graph.

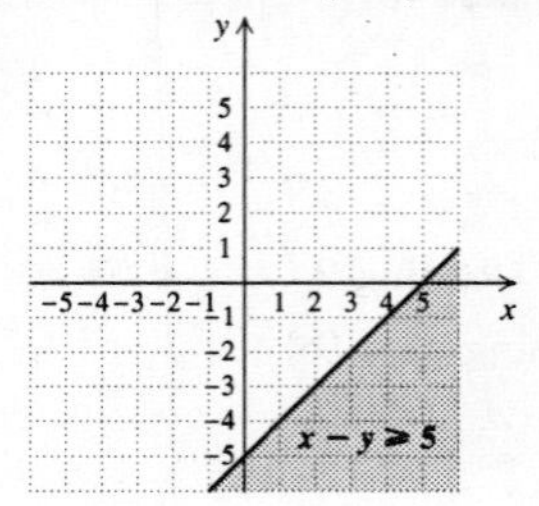

12.

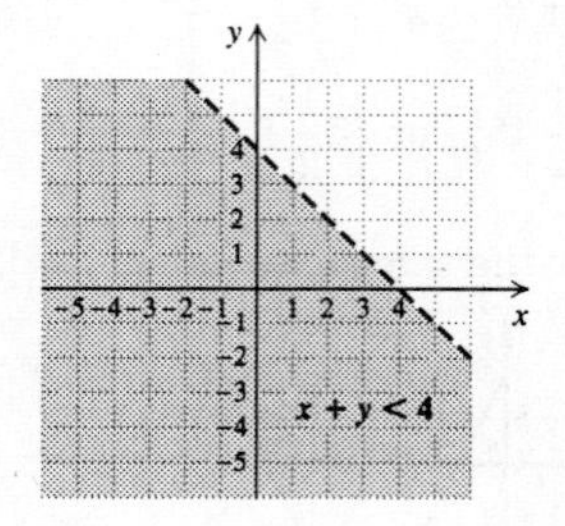

13. Graph: $2x + 3y < 6$

First graph $2x + 3y = 6$. Draw the line dashed since the inequality symbol is $<$. Test the point $(0, 0)$ to determine if it is a solution.

$$\begin{array}{c|c} \multicolumn{2}{c}{2x + 3y < 6} \\ \hline 2 \cdot 0 + 3 \cdot 0 \ ? & 6 \\ 0 & 6 \end{array} \quad \text{TRUE}$$

Since $0 < 6$ is true, we shade the half-plane containing $(0, 0)$ and obtain the graph.

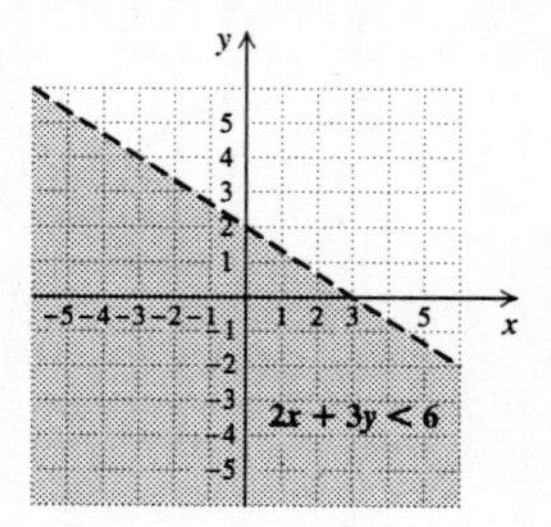

14.

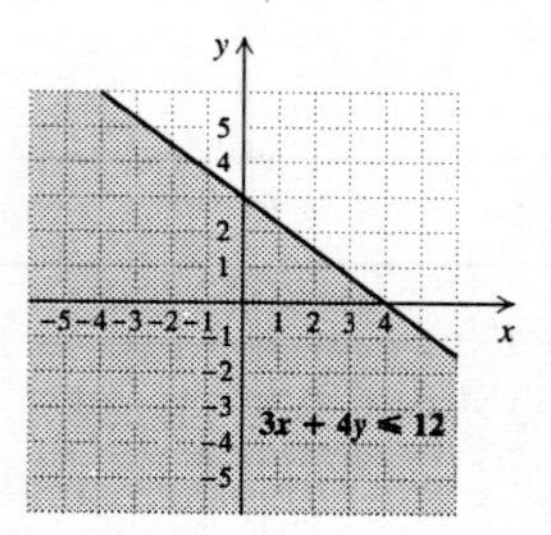

15. Graph: $2x - y \leq 4$

We first graph $2x - y = 4$. Draw the line solid since the inequality symbol is $\leq$. Test the point $(0, 0)$ to determine if it is a solution.

$$\begin{array}{c|c} 2x - y \le 4 & \\ \hline 2 \cdot 0 - 0 \ ? & 4 \\ 0 & 4 \quad \text{TRUE} \end{array}$$

Since $0 \le 4$ is true, we shade the half-plane containing $(0, 0)$ and obtain the graph.

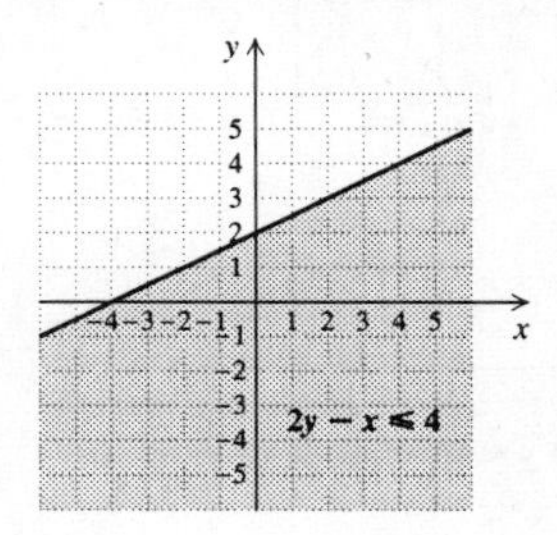

16.

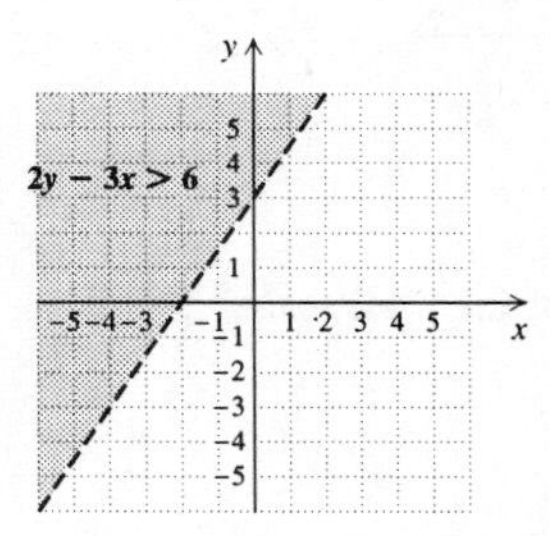

17. Graph: $2x - 2y \ge 8 + 2y$

$$2x - 4y \ge 8$$

First graph $2x - 4y = 8$. Draw the line solid since the inequality symbol is $\ge$. Test the point $(0, 0)$ to determine if it is a solution.

$$\begin{array}{c|c} 2x - 4y \ge 8 & \\ \hline 2 \cdot 0 - 4 \cdot 0 \ ? & 8 \\ 0 & 8 \quad \text{FALSE} \end{array}$$

Since $0 \ge 8$ is false, we shade the half-plane that does not contain $(0, 0)$ and obtain the graph.

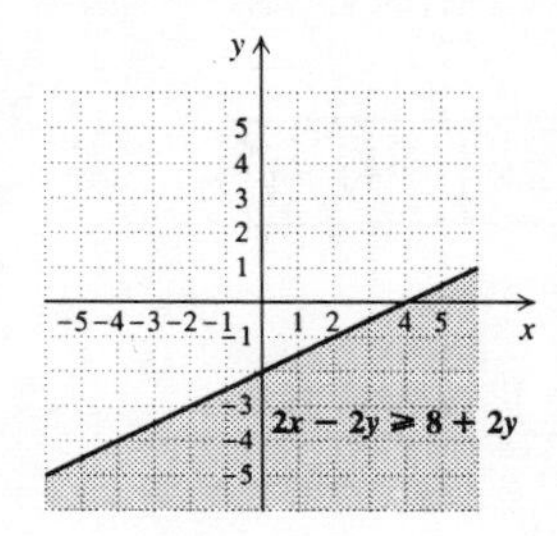

18.

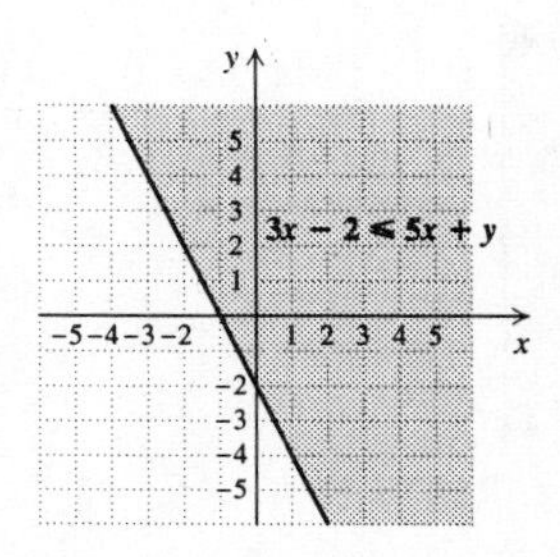

19. Graph: $y \ge 3$

We first graph $y = 3$. Draw the line solid since the inequality symbol is $\ge$. Test the point $(0, 0)$ to determine if it is a solution.

$$\begin{array}{c} y \ge 3 \\ \hline 0 \ ? \ 3 \quad \text{FALSE} \end{array}$$

Since $0 \ge 3$ is false, we shade the half-plane that does not contain $(0, 0)$ and obtain the graph.

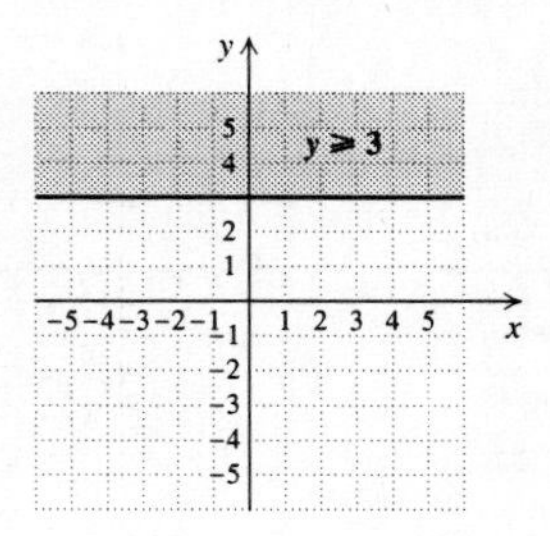

20.

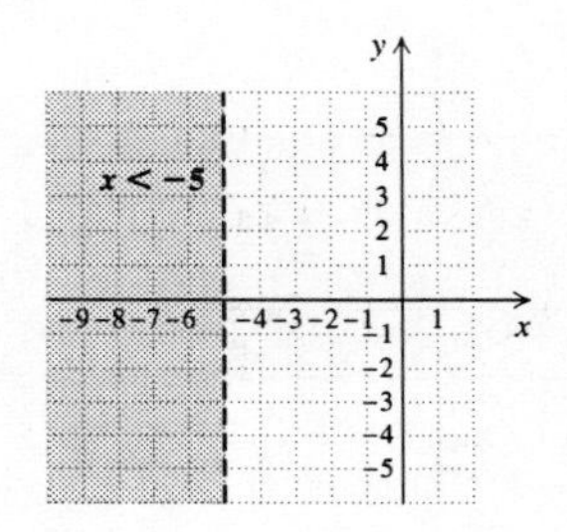

21. Graph: $x \le 6$

We first graph $x = 6$. We draw the line solid since the inequality symbol is $\le$. Test the point $(0, 0)$ to determine if it is a solution.

$$\begin{array}{c} x \le 6 \\ \hline 0 \ ? \ 6 \quad \text{TRUE} \end{array}$$

Since $0 \le 6$ is true, we shade the half-plane containing $(0, 0)$ and obtain the graph.

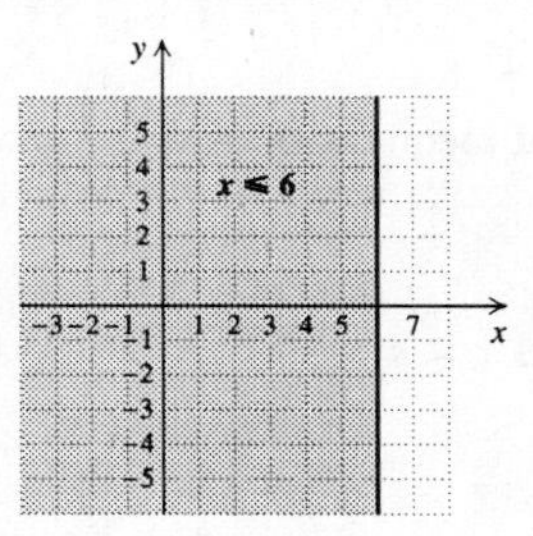

22.

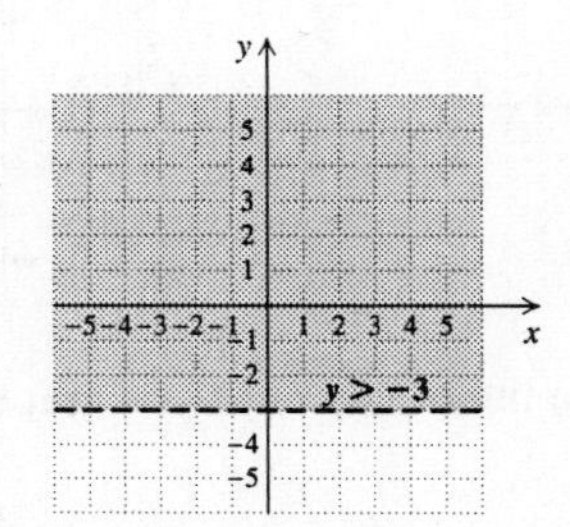

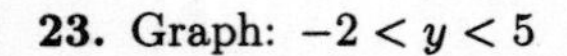

23. Graph: $-2 < y < 5$

This is a system of inequalities:

$$-2 < y,$$
$$y < 5$$

We graph the equation $-2 = y$ and see that the graph of $-2 < y$ is the half-plane above the line $-2 = y$. We also graph $y = 5$ and see that the graph of $y < 5$ is the half-plane below the line $y = 5$.

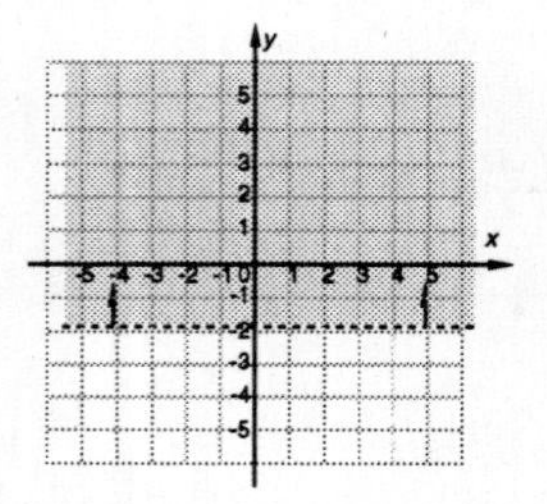

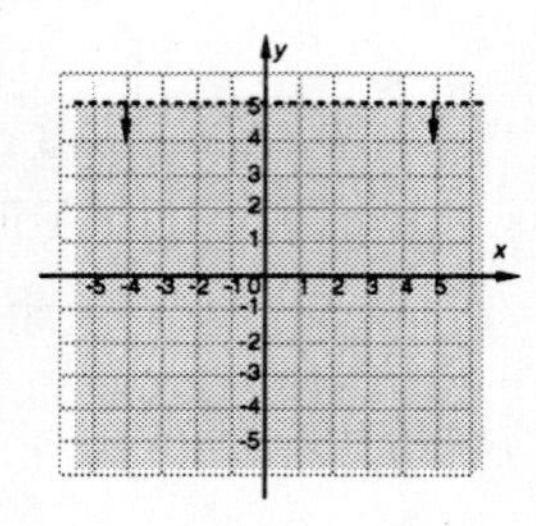

Finally, we shade the intersection of these graphs.

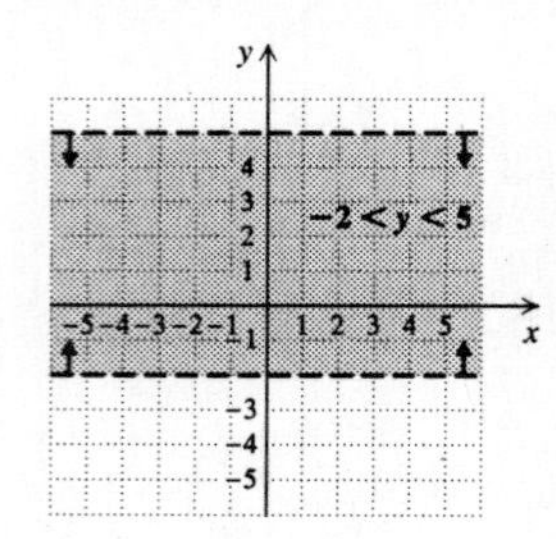

24.

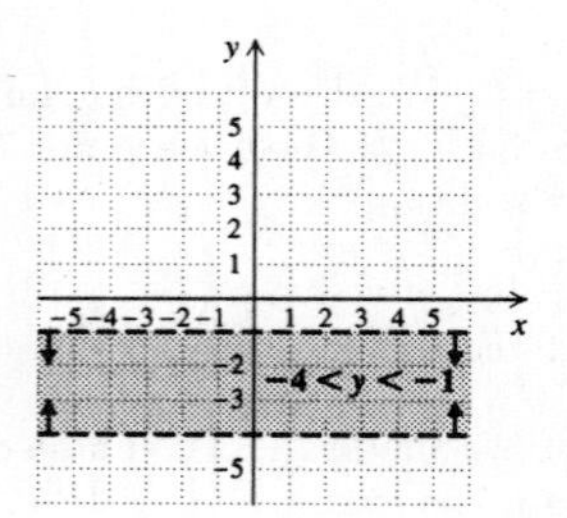

25. Graph: $-4 \le x \le 4$

This is a system of inequalities:

$$-4 \le x,$$
$$x \le 4$$

Graph $-4 \le x$ and $x \le 4$.

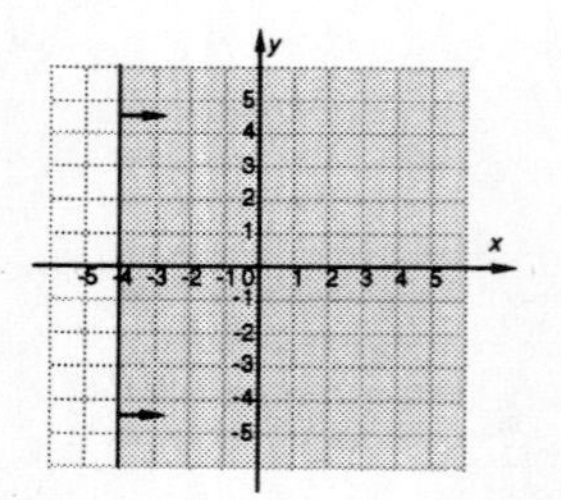

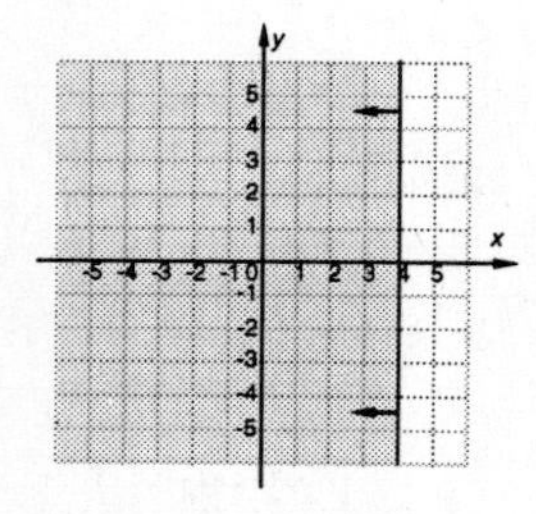

Then we shade the intersection of these graphs.

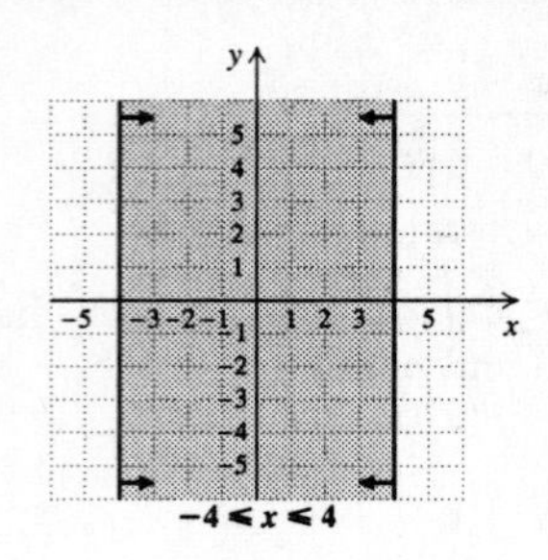

26.

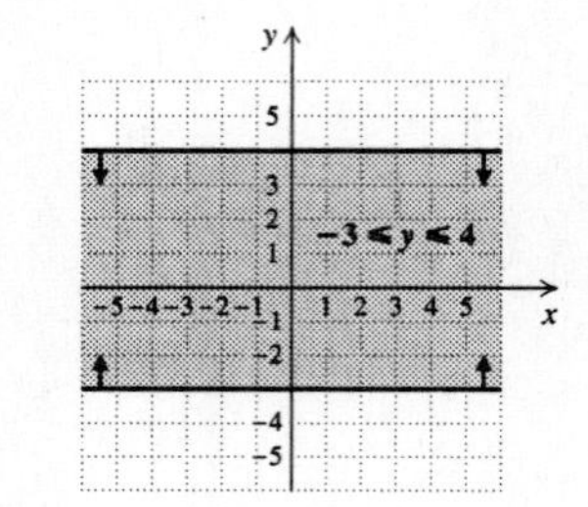

27. Graph: $0 \le y \le 3$

This is a system of inequalities:

$$0 \le y,$$
$$y \le 3$$

Graph $0 \le y$ and $y \le 3$.

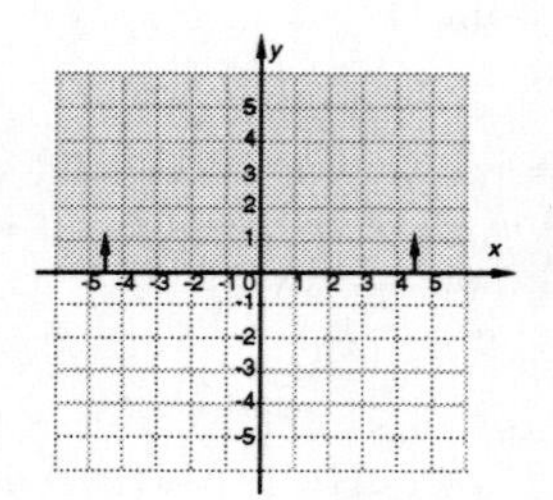

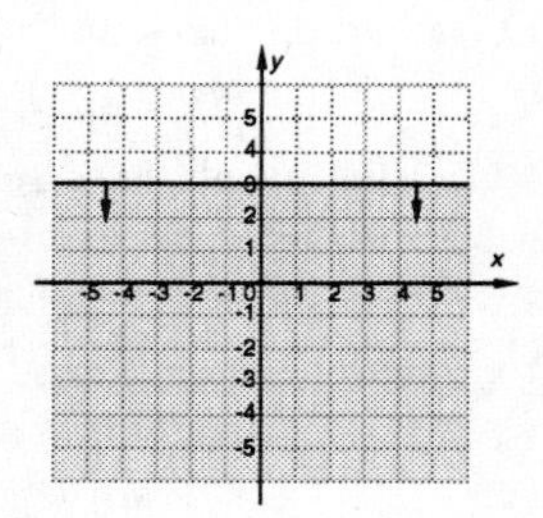

Then we shade the intersection of these graphs.

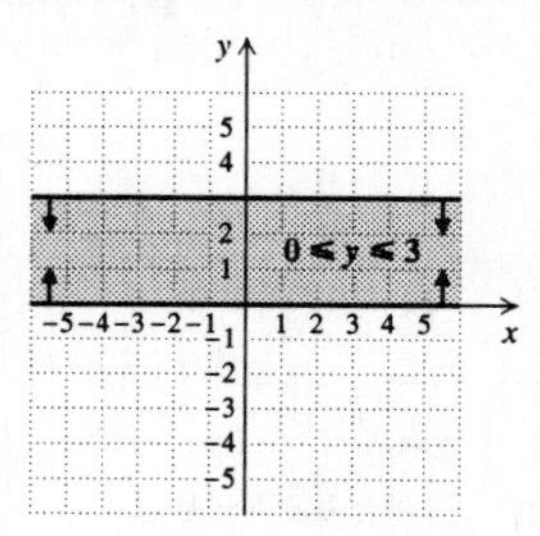

28.

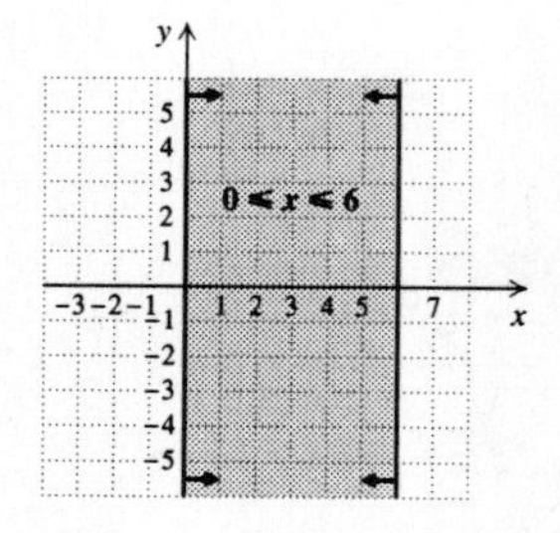

29. Graph: $y < x$,

$y > -x + 5$

We graph the lines $y = x$ and $y = -x + 5$, using dashed lines. We indicate the region for each inequality by the arrows at the ends of the lines. Note where the regions overlap and shade the region of solutions.

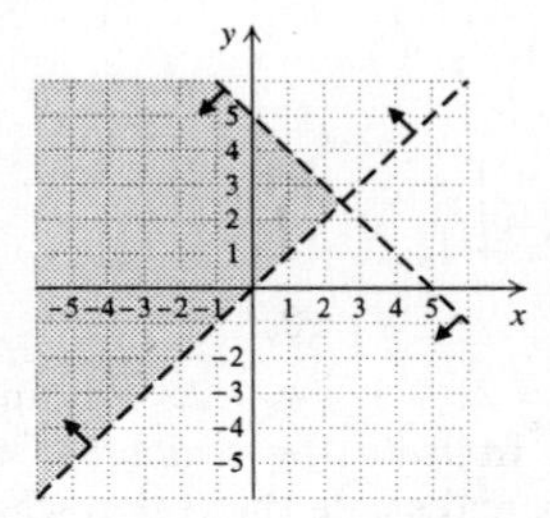

30.

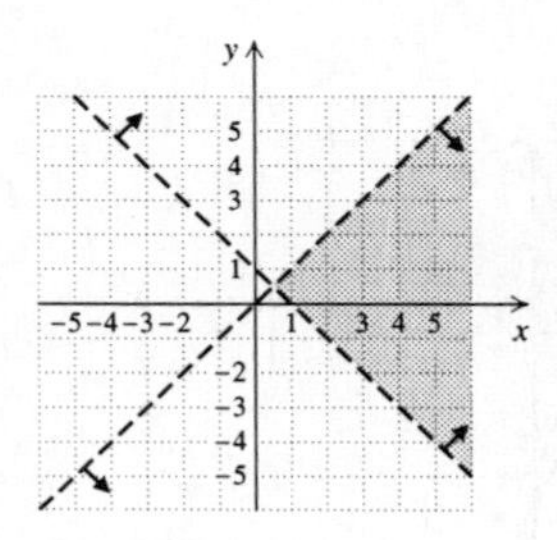

31. Graph: $y \geq x$,

$y \leq -x + 2$

Graph $y = x$ and $y = -x + 2$, using solid lines. Indicate the region for each inequality by arrows, and shade the region where they overlap.

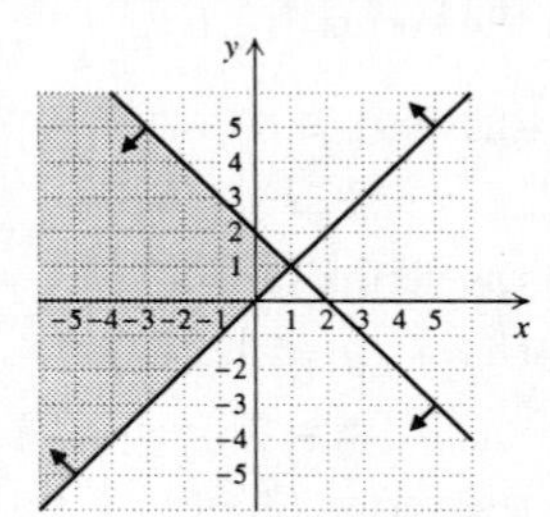

32.

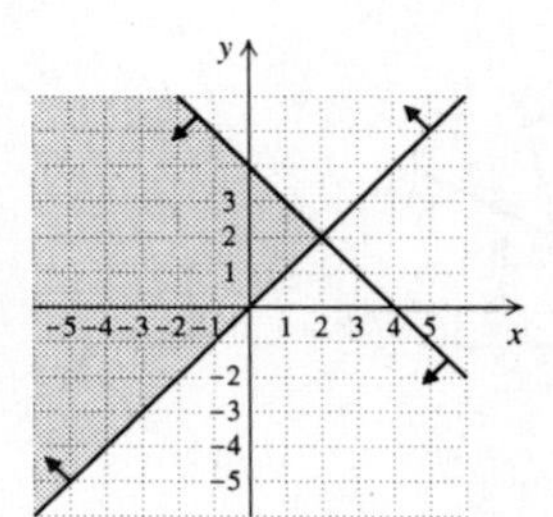

33. Graph: $y \leq -2$,

$x \geq -1$

Graph $y = -2$ and $x = -1$ using solid lines. Indicate the region for each inequality by arrows, and shade the region where they overlap.

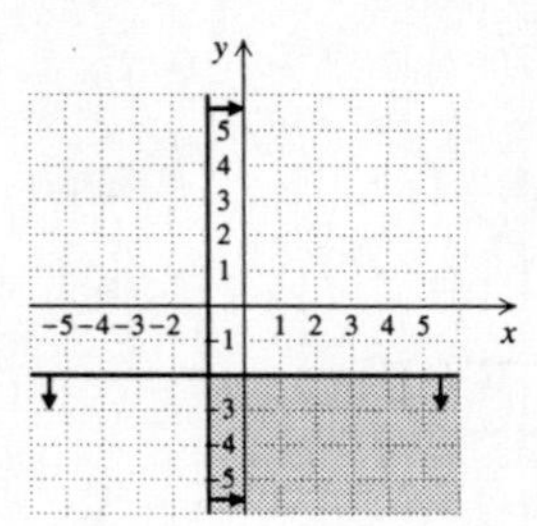

34.

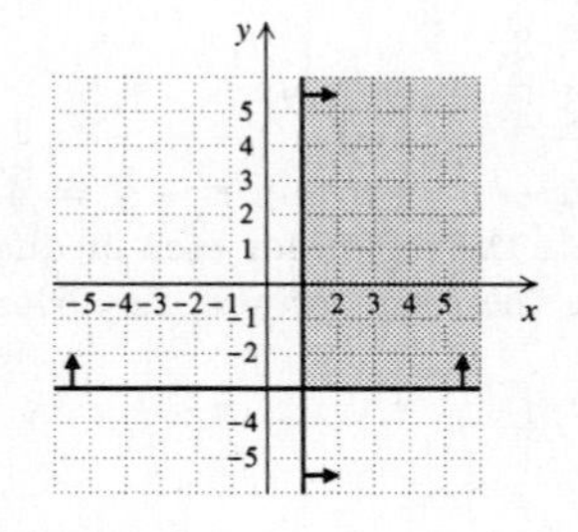

35. Graph: $x > -2$,

$y < -2x + 3$

Graph the lines $x = -2$ and $y = -2x + 3$, using dashed lines. Indicate the region for each inequality by arrows, and shade the region where they overlap.

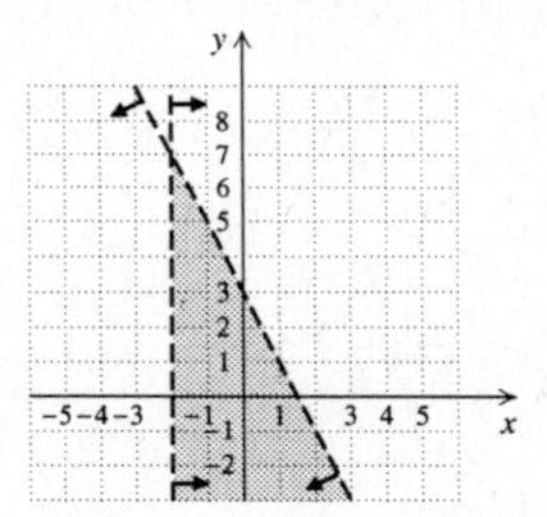

36.

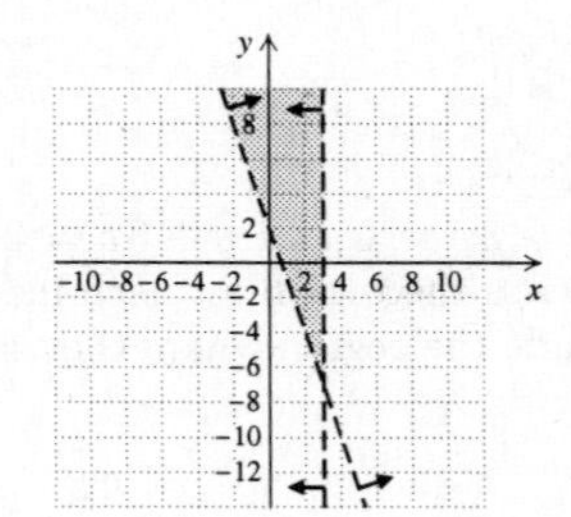

37. Graph: $y \leq 4$,

$y \geq -x + 2$

Graph the lines $y = 4$ and $y = -x + 2$, using solid lines. Indicate the region for each inequality by arrows, and shade the region where they overlap.

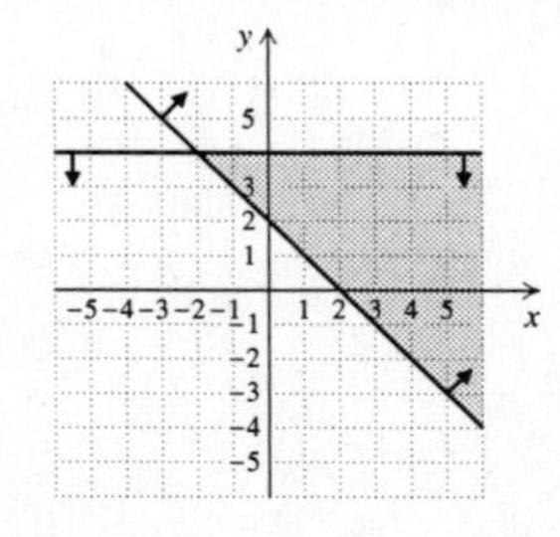

38.

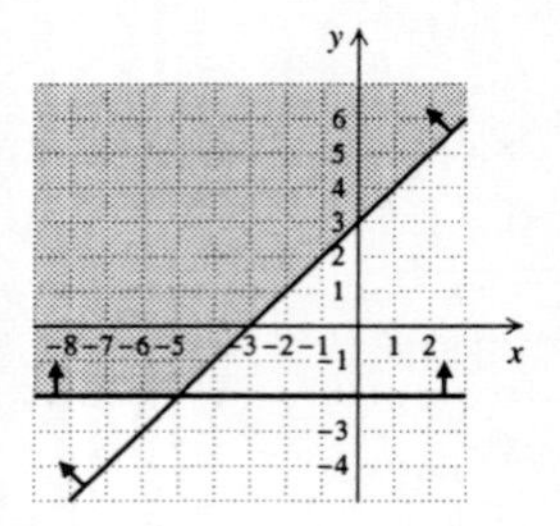

39. Graph: $x + y \leq 3,$

$x - y \leq 4$

Graph the lines $x + y = 3$ and $x - y = 4$, using solid lines. Indicate the region for each inequality by arrows, and shade the region where they overlap.

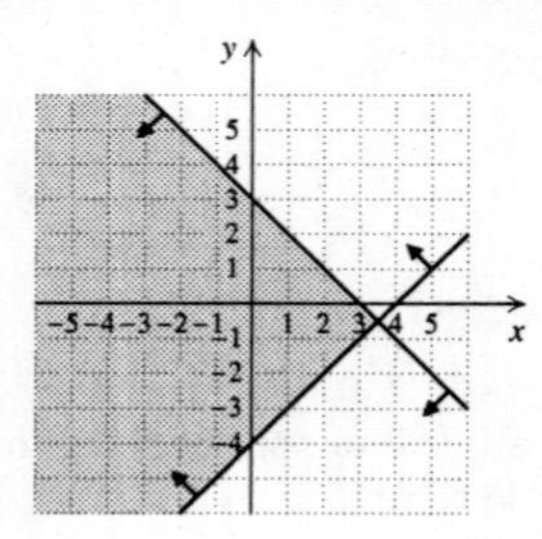

40.

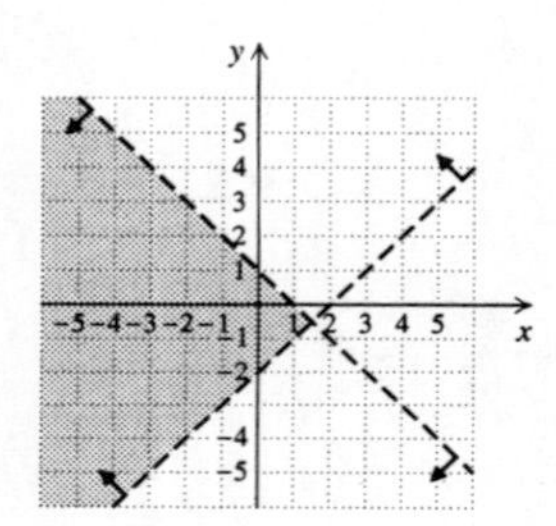

41. Graph: $y + 3x > 0,$

$y + 3x < 2$

Graph the lines $y + 3x = 0$ and $y + 3x = 2$, using dashed lines. Indicate the region for each inequality by arrows, and shade the region where they overlap.

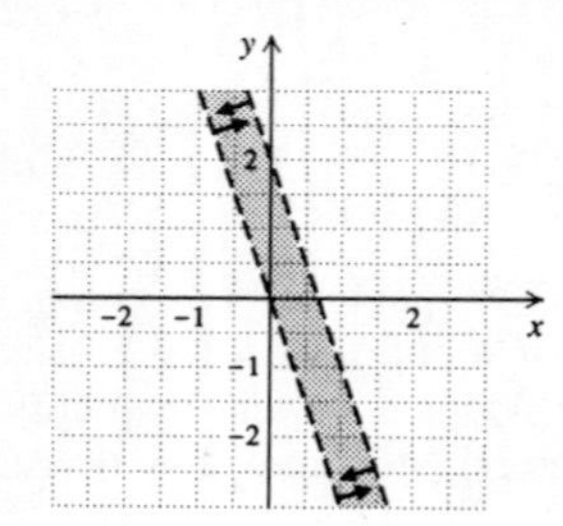

42.

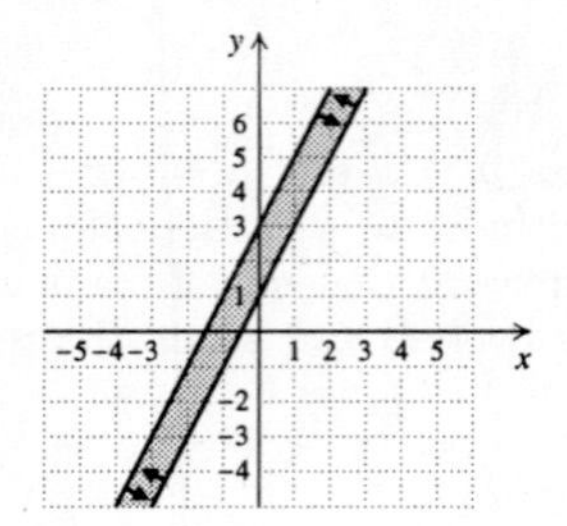

43. Graph:

$$y \leq 2x - 1, \quad (1)$$
$$y \geq -2x + 1, \quad (2)$$
$$x \leq 3 \quad (3)$$

Graph the lines $y = 2x - 1$, $y = -2x + 1$, and $x = 3$ using solid lines. Indicate the region for each inequality by arrows, and shade the region where they overlap.

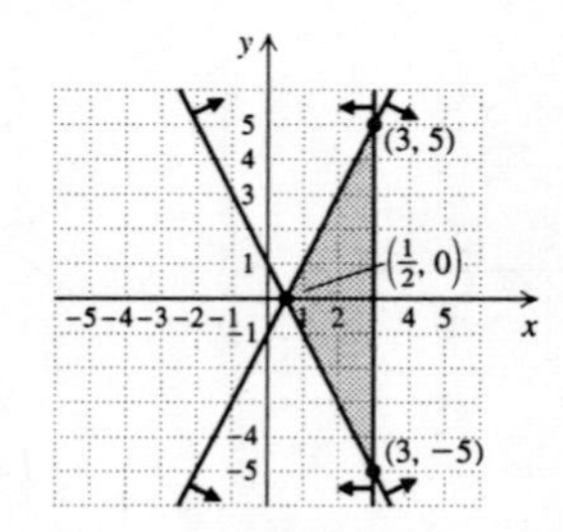

To find the vertex we solve three different systems of related equations.

From (1) and (2) we have $y = 2x - 1,$

$y = -2x + 1.$

Solving, we obtain the vertex $\left(\frac{1}{2}, 0\right)$.

From (1) and (3) we have $y = 2x - 1,$

$x = 3.$

Solving, we obtain the vertex $(3, 5)$.

From (2) and (3) we have $y = -2x + 1,$

$x = 3.$

Solving, we obtain the vertex $(3, -5)$.

44.

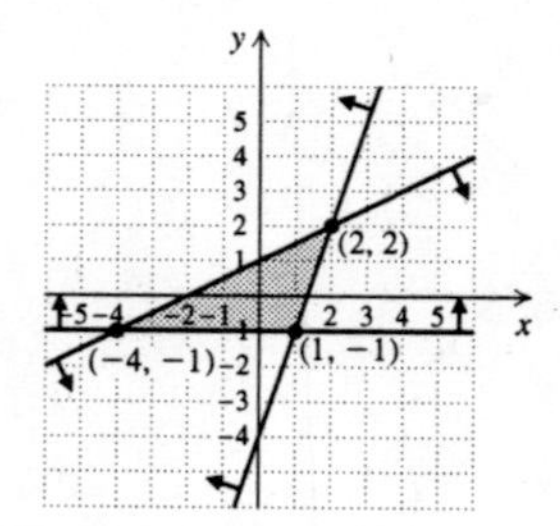

45. Graph:

$$x + 2y \leq 12, \quad (1)$$
$$2x + y \leq 12 \quad (2)$$
$$x \geq 0, \quad (3)$$
$$y \geq 0 \quad (4)$$

Graph the lines $x + 2y = 12$, $2x + y = 12$, $x = 0$, and $y = 0$ using solid lines. Indicate the region for each inequality by arrows, and shade the region where they overlap.

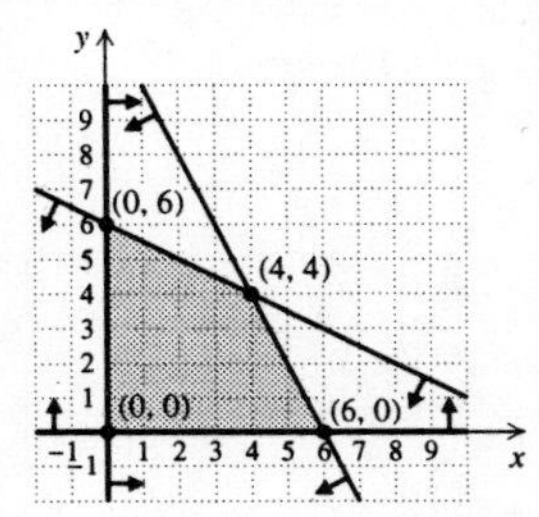

To find the vertices we solve four different systems of equations.

From (1) and (2) we have $x + 2y = 12$,
$2x + y = 12$.

Solving, we obtain the vertex $(4, 4)$.

From (1) and (3) we have $x + 2y = 12$,
$x = 0$.

Solving, we obtain the vertex $(0, 6)$.

From (2) and (4) we have $2x + y = 12$,
$y = 0$.

Solving, we obtain the vertex $(6, 0)$.

From (3) and (4) we have $x = 0$,
$y = 0$.

Solving, we obtain the vertex $(0, 0)$.

46.

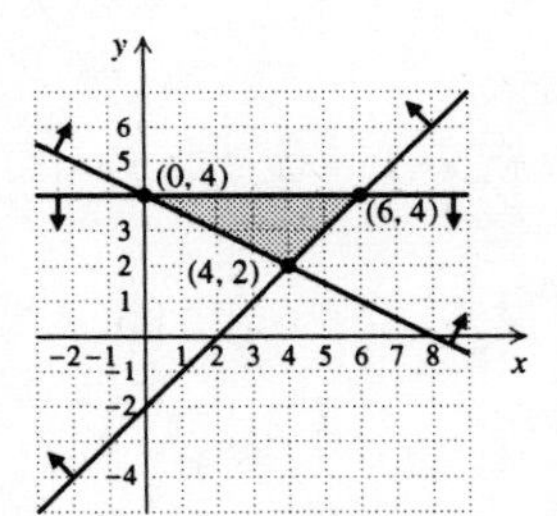

47. Graph: $8x + 5y \leq 40$, (1)
$x + 2y \leq 8$ (2)
$x \geq 0$, (3)
$y \geq 0$ (4)

Graph the lines $8x+5y = 40$, $x+2y = 8$, $x = 0$, and $y = 0$ using solid lines. Indicate the region for each inequality by arrows, and shade the region where they overlap.

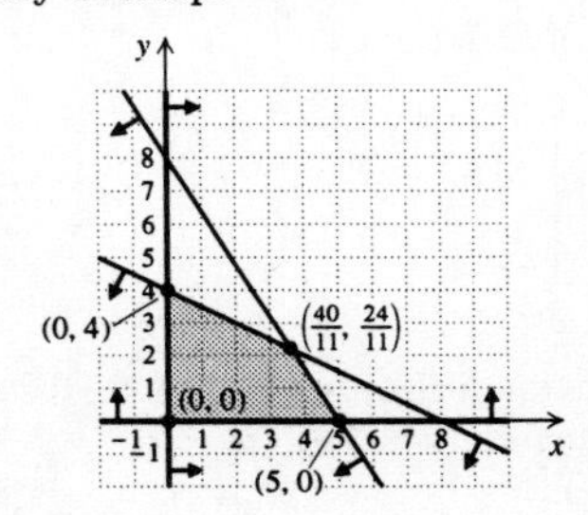

To find the vertices we solve four different systems of equations.

From (1) and (2) we have $8x + 5y = 40$,
$x + 2y = 8$.

Solving, we obtain the vertex $\left(\frac{40}{11}, \frac{24}{11}\right)$.

From (1) and (4) we have $8x + 5y = 40$,
$y = 0$.

Solving, we obtain the vertex $(5, 0)$.

From (2) and (3) we have $x + 2y = 8$,
$x = 0$.

Solving, we obtain the vertex $(0, 4)$.

From (3) and (4) we have $x = 0$,
$y = 0$.

Solving, we obtain the vertex $(0, 0)$.

48.

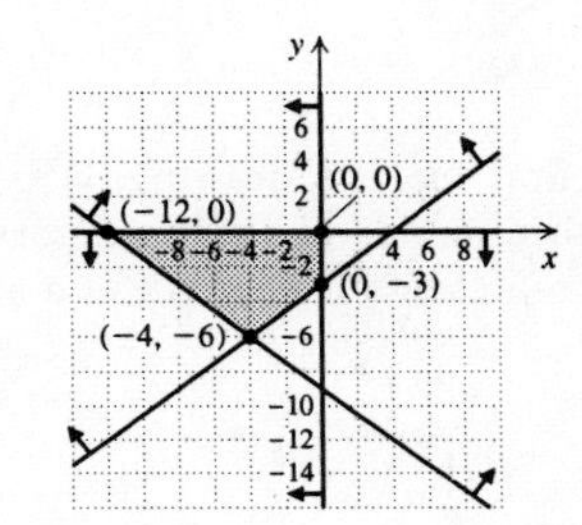

49. Graph: $y - x \geq 1$, (1)
$y - x \leq 3$, (2)
$2 \leq x \leq 5$ (3)

Think of (3) as two inequalities:

$2 \leq x$, (4)
$x \leq 5$ (5)

Graph the lines $y-x = 1$, $y-x = 3$, $x = 2$, and $x = 5$, using solid lines. Indicate the region for each inequality by arrows, and shade the region where they overlap.

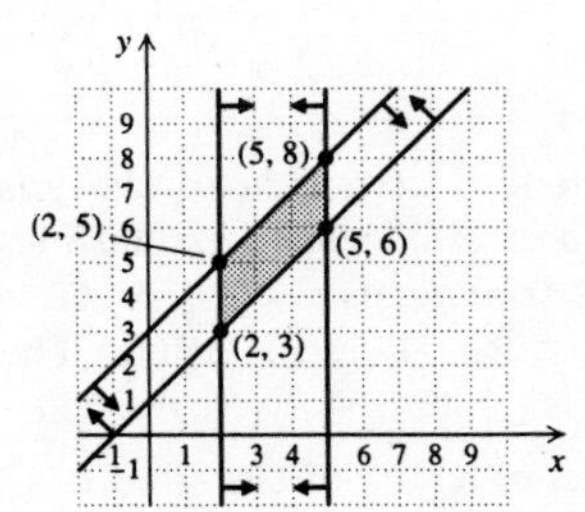

To find the vertices we solve four different systems of equations.

From (1) and (4) we have $y - x = 1$,
$x = 2$.

Solving, we obtain the vertex $(2, 3)$.

From (1) and (5) we have $y - x = 1,$

$x = 5.$

Solving, we obtain the vertex $(5, 6)$.

From (2) and (4) we have $y - x = 3,$

$x = 2.$

Solving, we obtain the vertex $(2, 5)$.

From (2) and (5) we have $y - x = 3,$

$x = 5.$

Solving, we obtain the vertex $(5, 8)$.

50.

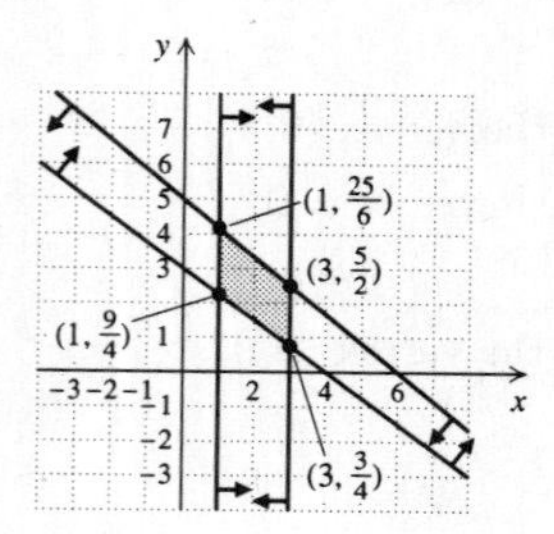

51. ***Familiarize.*** We first make a drawing. We let x represent the length of a side of the equilateral triangle. Then $x - 5$ represents the length of a side of the square.

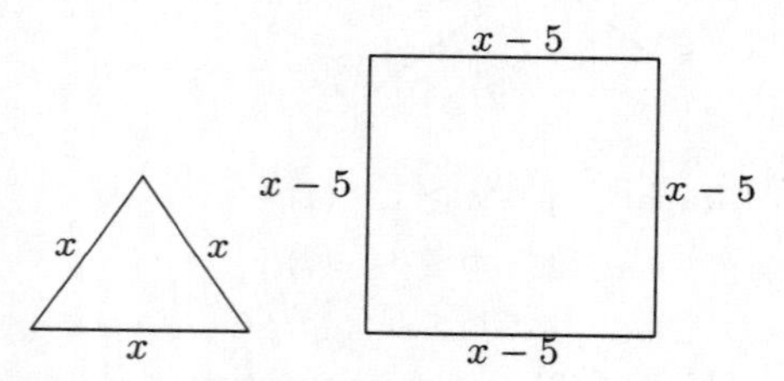

Translate.

Perimeter of triangle = Perimeter of square

$$3x \quad = \quad 4(x - 5)$$

Carry out.

$$3x = 4(x - 5)$$
$$3x = 4x - 20$$
$$20 = x$$

Then $x - 5 = 20 - 5 = 15$.

Check. If the length of a side of the triangle is 20 and the length of a side of the square is 15, the perimeter of the triangle is $3 \cdot 20$, or 60 and the perimeter of the square is $4 \cdot 15$, or 60. The values check.

State. The length of a side of the square is 15, and the length of a side of the triangle is 20.

52. $\left(-\frac{44}{23}, \frac{13}{23}\right)$

53.
$$5(3x - 4) = -2(x + 5)$$
$$15x - 20 = -2x - 10$$
$$17x - 20 = -10$$
$$17x = 10$$
$$x = \frac{10}{17}$$

The solution is $\frac{10}{17}$.

54. $-\frac{14}{13}$

55.

56.

57.

58.

59. Graph: $x + y > 8,$

$x + y \leq -2$

Graph the line $x + y = 8$ using a dashed line and graph $x + y = -2$, using a solid line. Indicate the region for each inequality by arrows. The regions do not overlap (the solution set is $\emptyset$), so we do not shade any portion of the graph.

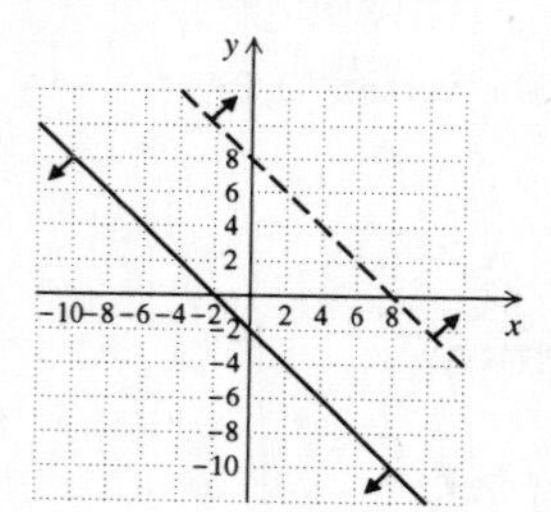

60.

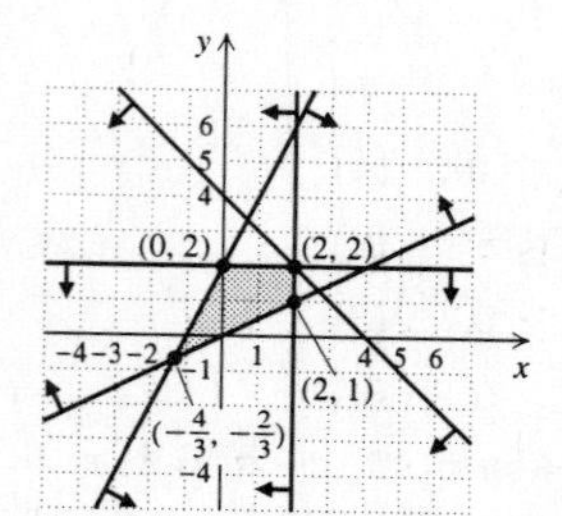

61. Graph: $x + y \geq 1,$

$-x + y \geq 2,$

$x \leq 4,$

$y \geq 0,$

$y \leq 4,$

$x \leq 2$

Graph the six inequalities above, and shade the region where they overlap.

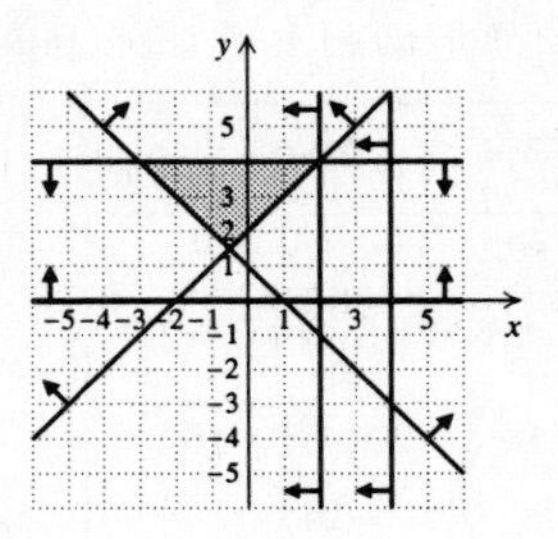

62. $x \geq -2,$
$y \leq 2,$
$x \leq 0,$
$y \geq 0;$ or
$x \geq 0,$
$y \leq 2,$
$x \leq 2,$
$y \geq 0;$ or
$x \geq 0,$
$y \leq 0,$
$x \leq 2,$
$y \geq -2;$ or
$x \geq -2,$
$y \leq 0,$
$x \leq 0,$
$y \geq -2$

63. Both the width and the height must be positive, but they must be less than 62 in. in order to be checked as luggage, so we have:

$$0 < w \leq 62,$$
$$0 < h \leq 62$$

The girth is represented by $2w + 2h$ and the length is 62 in. In order to meet postal regulations the sum of the girth and the length cannot exceed 108 in., so we have:

$$62 + 2w + 2h \leq 108, \text{ or}$$
$$2w + 2h \leq 46, \text{ or}$$
$$w + h \leq 23$$

Thus, have a system of inequalities:

$$0 < w \leq 62,$$
$$0 < h \leq 62,$$
$$w + h \leq 23$$

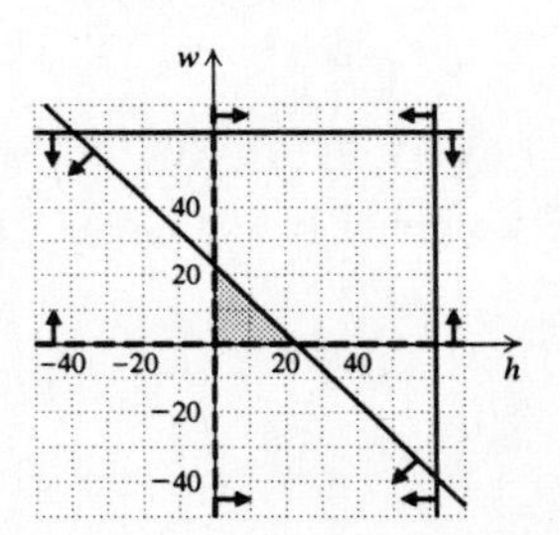

64.

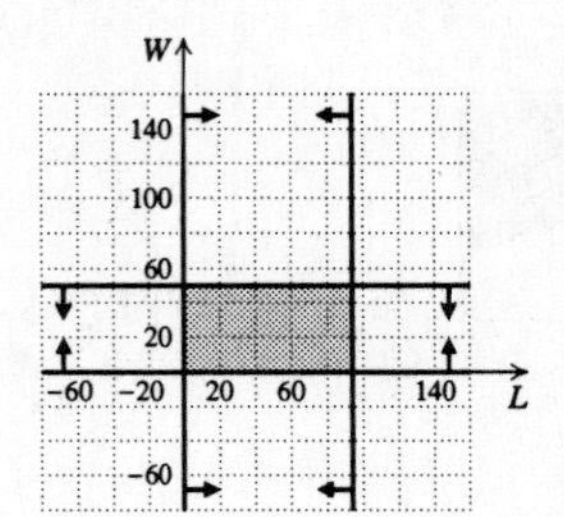

65. We graph the following system of inequalities:

$$2w + t \geq 60,$$
$$w \geq 0,$$
$$t \geq 0$$

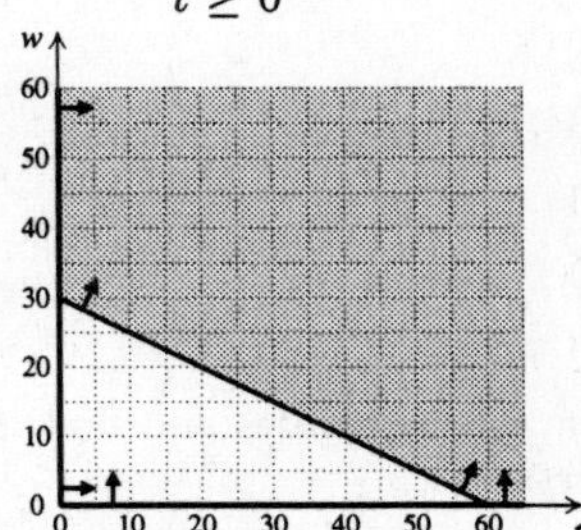

66.

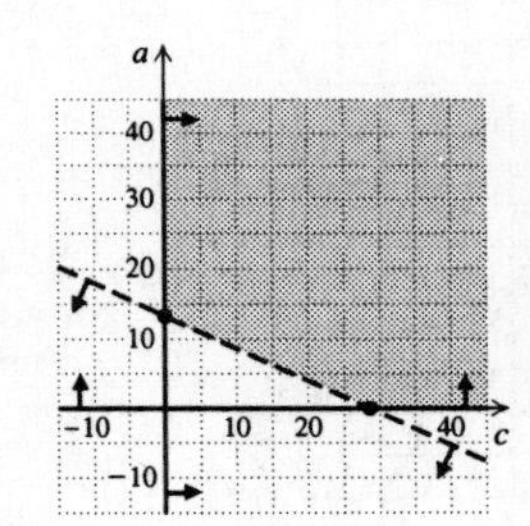

67. a)

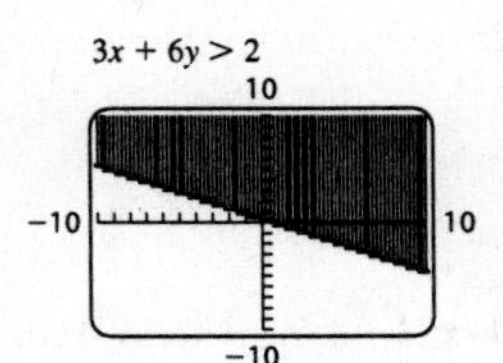

b)

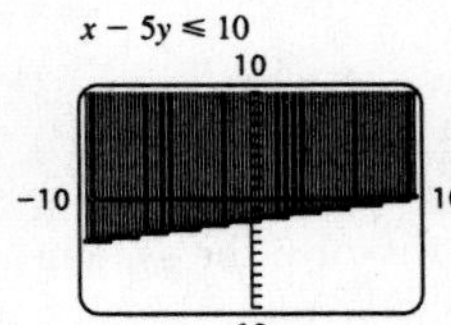

c)

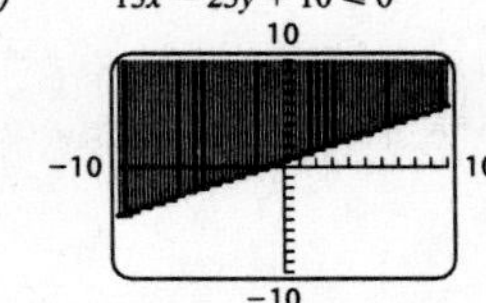

d)

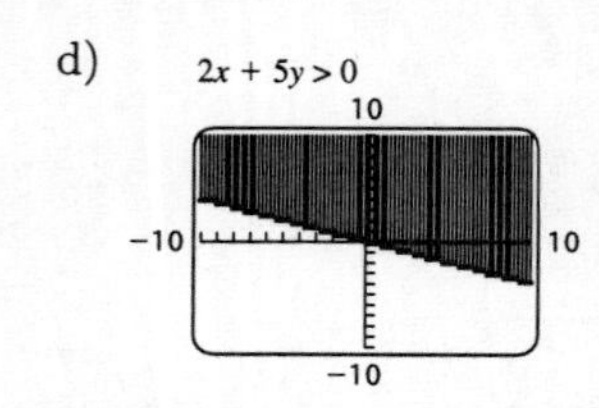

68.

Exercise Set 4.5

1. Find the maximum and minimum values of
$F = 4x + 28y$,

subject to

$$5x + 3y \leq 34, \quad (1)$$
$$3x + 5y \leq 30, \quad (2)$$
$$x \geq 0, \quad (3)$$
$$y \geq 0. \quad (4)$$

Graph the system of inequalities and find the coordinates of the vertices.

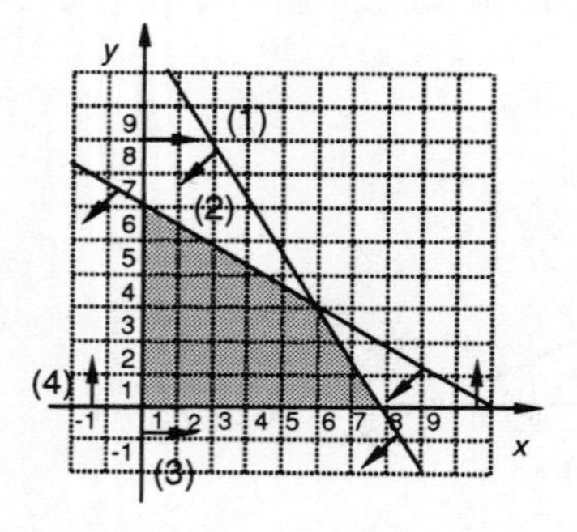

To find one vertex we solve the system

$x = 0$,

$y = 0$.

This vertex is $(0, 0)$.

To find a second vertex we solve the system

$5x + 3y = 34$,

$y = 0$.

This vertex is $\left(\frac{34}{5}, 0\right)$.

To find a third vertex we solve the system

$5x + 3y = 34$,

$3x + 5y = 30$.

This vertex is $(5, 3)$.

To find the fourth vertex we solve the system

$3x + 5y = 30$,

$x = 0$.

This vertex is $(0, 6)$.

Now find the value of F at each of these points.

Vertex (x, y)	$F = 4x + 28y$	
$(0, 0)$	$4 \cdot 0 + 28 \cdot 0 = 0 + 0 = 0$	← Minimum
$\left(\frac{34}{5}, 0\right)$	$4 \cdot \frac{34}{5} + 28 \cdot 0 = \frac{136}{5} + 0 = 27\frac{1}{5}$	
$(5, 3)$	$4 \cdot 5 + 28 \cdot 3 = 20 + 84 = 104$	
$(0, 6)$	$4 \cdot 0 + 28 \cdot 6 = 0 + 168 = 168$	← Maximum

The maximum value of F is 168 when $x = 0$ and $y = 6$. The minimum value of F is 0 when $x = 0$ and $y = 0$.

2. The maximum is 76 when $x = 2$ and $y = 3$. The minimum is 0 when $x = 0$ and $y = 0$.

3. Find the maximum and minimum values of
$P = 16x - 2y + 40$,

subject to

$$6x + 8y \leq 48, \quad (1)$$
$$0 \leq y \leq 4, \quad (2)$$
$$0 \leq x \leq 7. \quad (3)$$

Think of (2) as $0 \leq y$, (4)

$y \leq 4$. (5)

Think of (3) as $0 \leq x$, (6)

$x \leq 7$. (7)

Graph the system of inequalities.

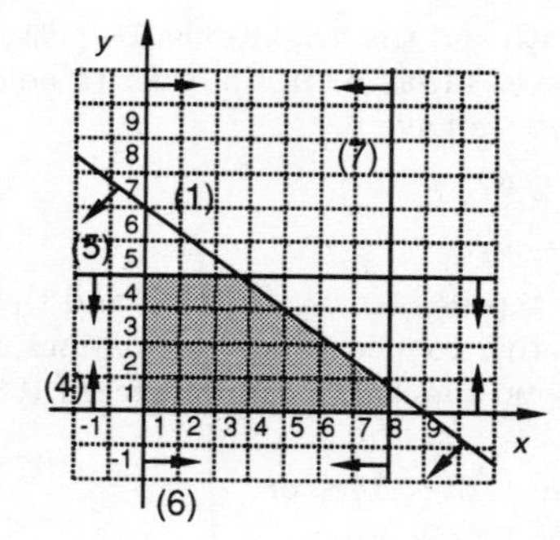

To determine the coordinates of the vertices, we solve the following systems:

$x = 0$, $y = 0$;

$x = 7$, $y = 0$;

$6x + 8y = 48$, $x = 7$;

$6x + 8y = 48$, $y = 4$;

$x = 0$, $y = 4$

The vertices are $(0, 0)$, $(7, 0)$, $\left(7, \frac{3}{4}\right)$, $\left(\frac{8}{3}, 4\right)$, and $(0, 4)$, respectively. Compute the value of P at each of these points.

Vertex (x, y)	$P = 16x - 2y + 40$
$(0, 0)$	$16 \cdot 0 - 2 \cdot 0 + 40 =$ $0 - 0 + 40 = 40$
$(7, 0)$	$16 \cdot 7 - 2 \cdot 0 + 40 =$ $112 - 0 + 40 =$ 152 ⟵ Maximum
$\left(7, \frac{3}{4}\right)$	$16 \cdot 7 - 2 \cdot \frac{3}{4} + 40 =$ $112 - \frac{3}{2} + 40 = 150\frac{1}{2}$
$\left(\frac{8}{3}, 4\right)$	$16 \cdot \frac{8}{3} - 2 \cdot 4 + 40 =$ $\frac{128}{3} - 8 + 40 = 74\frac{2}{3}$
$(0, 4)$	$16 \cdot 0 - 2 \cdot 4 + 40 =$ $0 - 8 + 40 =$ 32 ⟵ Minimum

The maximum is 152 when $x = 7$ and $y = 0$. The minimum is 32 when $x = 0$ and $y = 4$.

4. The maximum is 124 when $x = 3$ and $y = 0$. The minimum is 40 when $x = 0$ and $y = 4$.

5. Find the maximum and minimum values of
$F = 2y - 3x$,

subject to

$$y \leq 2x + 1, \quad (1)$$
$$y \geq -2x + 3, \quad (2)$$
$$x \leq 3 \quad (3)$$

Graph the system of inequalities and find the coordinates of the vertices.

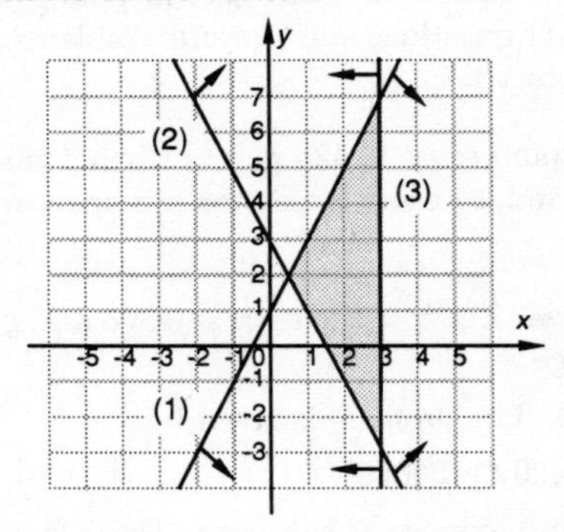

To determine the coordinates of the vertices, we solve the following systems:

$y = 2x + 1,$ $\quad y = -2x + 3;$

$y = 2x + 1,$ $\quad x = 3;$

$y = -2x + 3,$ $\quad x = 3$

The solutions of the systems are $\left(\frac{1}{2}, 2\right)$, $(3, 7)$, and $(3, -3)$, respectively. Now find the value of F at each of these points.

Vertex (x, y)	$F = 2y - 3x$
$\left(\frac{1}{2}, 2\right)$	$2 \cdot 2 - 3 \cdot \frac{1}{2} = \frac{5}{2}$
$(3, 7)$	$2 \cdot 7 - 3 \cdot 3 = 5$ ⟵ Maximum
$(3, -3)$	$2(-3) - 3 \cdot 3 = -15$ ⟵ Minimum

The maximum value is 5 when $x = 3$ and $y = 7$. The minimum value is -15 when $x = 3$ and $y = -3$.

6. The maximum is 51 when $x = 5$ and $y = 11$. The minimum is 12 when $x = \frac{2}{3}$ and $y = \frac{7}{3}$.

7. ***Familiarize.*** Let x = the number of orders of chili and y = the number of burritos sold each day.

Translate. The profit P is given by

$$P = \$1.65x + \$1.05y.$$

We wish to maximize P subject to these facts (constraints) about x and y:

$$10 \leq x \leq 40,$$
$$30 \leq y \leq 70,$$
$$x + y \leq 90.$$

Carry out. We graph the system of inequalities, determine the vertices, and evaluate P at each vertex.

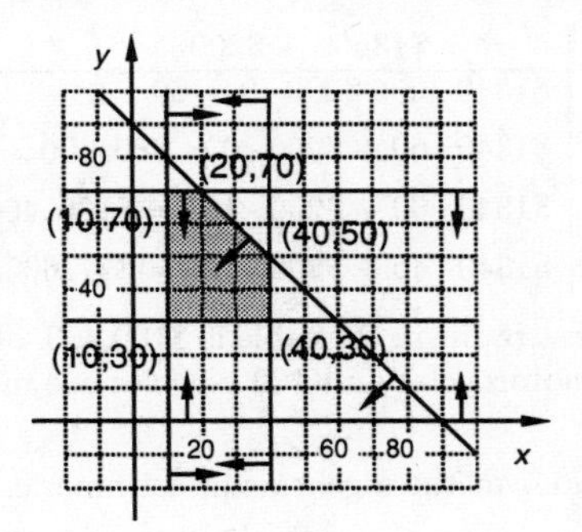

Vertex	$P = \$1.65x + \$1.05y$
$(10, 30)$	$\$1.65(10) + \$1.05(30) = \$48$
$(40, 30)$	$\$1.65(40) + \$1.05(30) = \$97.50$
$(40, 50)$	$\$1.65(40) + \$1.05(50) = \$118.50$
$(20, 70)$	$\$1.65(20) + \$1.05(70) = \$106.50$
$(10, 70)$	$\$1.65(10) + \$1.05(70) = \$90$

The greatest profit in the table is \$118.50, obtained when 40 orders of chili and 50 burritos are sold.

Check. Go over the algebra and arithmetic.

State. The maximum profit occurs when 40 orders of chili and 50 burritos are sold.

8. The maximum profit is achieved by producing 100 units of lumber and 300 units of plywood.

9. ***Familiarize.*** Let x = the number of motorcycles manufactured and y = the number of bicycles manufactured each month.

Translate. The profit P is given by

$$P = \$1340x + \$200y.$$

We wish to maximize P subject to these facts (constraints) about x and y.

$y \leq 3x$	The number of bicycles cannot exceed three times the number of motorcycles.
$0 \leq x \leq 60$	No more than 60 motorcycles will be produced.
$0 \leq y \leq 120$	No more than 120 bicycles can be produced.
$x + y \leq 160$	Total production cannot exceed 160.

Carry out. We graph the system of inequalities, determine the vertices, and evaluate P at each vertex.

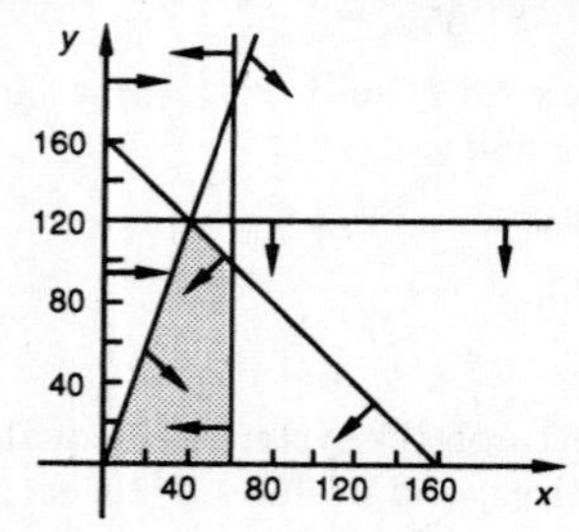

Vertex	$P = \$1340x + \$200y$
$(0, 0)$	$\$1340 \cdot 0 + \$200 \cdot 0 = \$0$
$(60, 0)$	$\$1340 \cdot 60 + \$200 \cdot 0 = \$80,400$
$(60, 100)$	$\$1340 \cdot 60 + \$200 \cdot 100 = \$100,400$
$(40, 120)$	$\$1340 \cdot 40 + \$200 \cdot 120 = \$77,600$

The greatest profit in the table is \$100,400, obtained when 60 motorcycles and 100 bicycles are manufactured.

Check. Go over the algebra and arithmetic.

State. The maximum profit occurs when 60 motorcycles and 100 bicycles are manufactured.

10. The maximum number of miles is 480 when the car uses 9 gal and the moped uses 3 gal.

11. ***Familiarize***. We organize the information in a table. Let x = the number of short-answer questions and y = the number of word problems answered.

Type	Number of points for each	Number answered	Total points
Short-answer	4	$5 \leq x \leq 10$	$4x$
Word problems	7	$3 \leq y \leq 10$	$7y$
Total		$x + y \leq 18$	$4x + 7y$

Since Edy can do no more than 18 questions in total, we have the inequality $x + y \leq 18$ in the "Number answered" column. The expression $4x + 7y$ in the "Total points" column gives the total score on the test.

Translate. The score S is given by

$$S = 4x + 7y.$$

We wish to maximize S subject to these facts (constraints) about x and y.

$$5 \leq x \leq 10,$$
$$3 \leq y \leq 10,$$
$$x + y \leq 18.$$

Carry out. We graph the system of inequalities, determine the vertices, and evaluate S at each vertex.

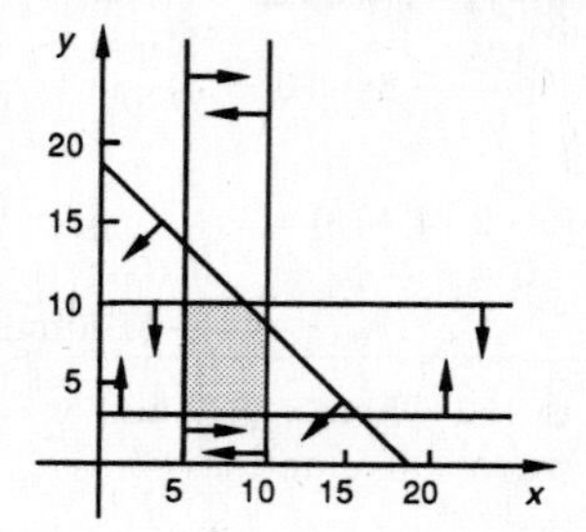

Vertex	$S = 4x + 7y$
$(5, 3)$	$4 \cdot 5 + 7 \cdot 3 = 41$
$(10, 3)$	$4 \cdot 10 + 7 \cdot 3 = 61$
$(10, 8)$	$4 \cdot 10 + 7 \cdot 8 = 96$
$(8, 10)$	$4 \cdot 8 + 7 \cdot 10 = 102$
$(5, 10)$	$4 \cdot 5 + 7 \cdot 10 = 90$

The greatest score in the table is 102, obtained when 8 short-answer questions and 10 word problems are answered correctly.

Check. Go over the algebra and arithmetic.

State. The maximum score is 102 points when 8 short-answer questions and 10 word problems are answered correctly.

12. The maximum score is 425 points when 5 matching questions and 15 essay questions are answered correctly.

13. ***Familiarize***. Let x = the corn acreage and y = the oats acreage.

Translate. The profit P is given by

$$P = \$400x + \$300y.$$

We wish to maximize P subject to these facts (constraints) about x and y.

$$x + y \leq 240,$$
$$2x + y \leq 320,$$
$$x \geq 0,$$
$$y \geq 0.$$

Carry out. We graph the system of inequalities, determine the vertices, and evaluate P at each vertex.

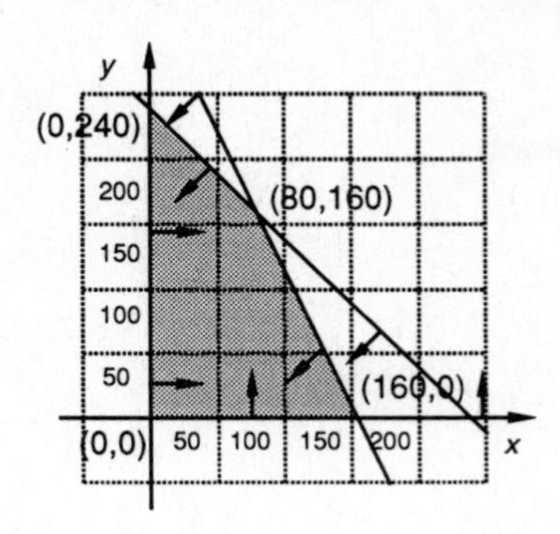

Vertex	$P = \$400x + \$300y$
$(0, 0)$	\$0
$(0, 240)$	\$72,000
$(80, 160)$	\$80,000
$(160, 0)$	\$64,000

Check. Go over the algebra and arithmetic.

State. The maximum profit occurs by planting 80 acres of corn and 160 acres of oats.

14. The maximum income of \$3110 occurs when \$22,000 is invested in corporate bonds and \$18,000 is invested in municipal bonds.

15. ***Familiarize***. Let x = the amount invested in City Bank and y = the amount invested in State Bank.

Translate. The income I is given by

$$I = 0.06x + 0.065y.$$

We wish to maximize I subject to these facts (constraints) about x and y.

$$x + y \leq \$22,000,$$
$$\$2000 \leq x \leq \$14,000,$$
$$\$0 \leq y \leq \$15,000$$

Carry out. We graph the system of inequalities, determine the vertices, and evaluate I at each vertex.

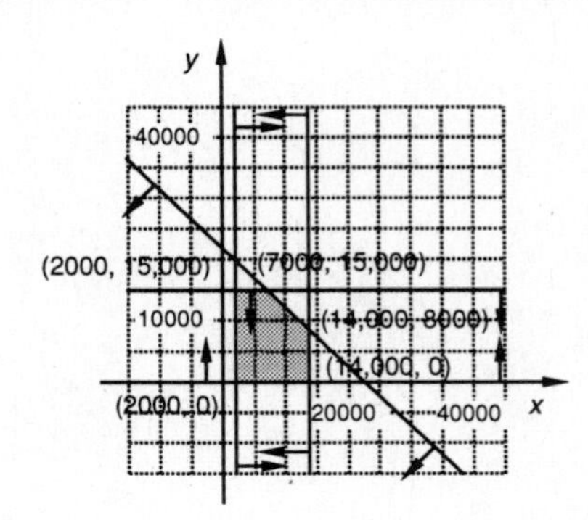

Vertex	$I = 0.06x + 0.065y$
$(\$2000, \$0)$	\$120
$(\$2000, \$15,000)$	\$1095
$(\$7000, \$15,000)$	\$1395
$(\$14,000, \$8000)$	\$1360
$(\$14,000, \$0)$	\$840

Check. Go over the algebra and arithmetic.

State. The maximum interest income is \$1395 when \$7000 is invested in City Bank and \$15,000 is invested in State Bank.

16. The maximum profit of \$8265 occurs when 68 batches of Hawaiian Blend and 39 batches of Classic Blend are made.

17. ***Familiarize***. Let x = the number of knit suits and y = the number of worsted suits made per day.

Translate. The profit P is given by

$$P = \$34x + \$31y.$$

We wish to maximize P subject to these facts (constraints) about x and y.

$$2x + 4y \leq 20,$$
$$4x + 2y \leq 16,$$
$$x \geq 0,$$
$$y \geq 0.$$

Carry out. Graph the system of inequalities, determine the vertices, the evaluate P at each vertex.

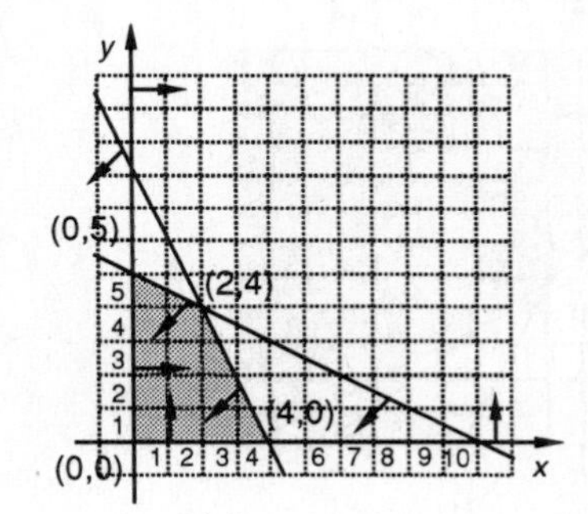

Vertex	$P = \$34x + \$31y$
$(0, 0)$	$\$34 \cdot 0 + \$31 \cdot 0 = \$0$
$(0, 5)$	$\$34 \cdot 0 + \$31 \cdot 5 = \$155$
$(2, 4)$	$\$34 \cdot 2 + \$31 \cdot 4 = \$192$
$(4, 0)$	$\$34 \cdot 4 + \$31 \cdot 0 = \$136$

Check. Go over the algebra and arithmetic.

State. The maximum profit per day is \$192 when 2 knit suits and 4 worsted suits are made.

18. The company will have a maximum income of \$180 when 100 of each type of biscuit is made.

19. ◈

20. 30 P-1's, 10 P-2's

21. ***Familiarize***. Let x represent the number of P-2 planes and y represent the number of P-3 planes. Organize the information in a table.

Plane	Number of planes	Passengers		
		First	Tourist	Economy
P-2	x	$80x$	$30x$	$40x$
P-3	y	$40y$	$40y$	$80y$

Plane	Cost per mile
P-2	$25x$
P-3	$37.50y$

Translate. Suppose C is the total cost per mile. Then $C = 25x + 37.50y$. We wish to minimize C subject to these facts (constraints) about x and y.

$$80x + 40y \geq 2000,$$
$$30x + 40y \geq 1500,$$
$$40x + 80y \geq 2400,$$
$$x \geq 0,$$
$$y \geq 0.$$

Carry out. Graph the system of inequalities, determine the vertices, and evaluate C at each vertex.

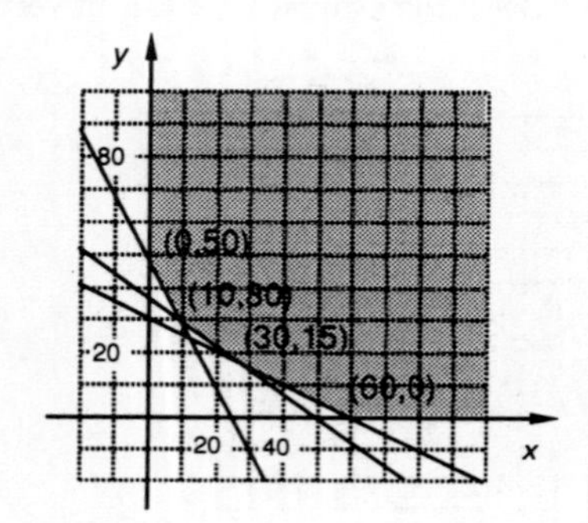

Vertex	$C = 25x + 37.50y$
$(0, 50)$	1875
$(10, 30)$	1375
$(30, 15)$	1312.50
$(60, 0)$	1500

Check. Go over the algebra and arithmetic.

State. The airline should use 30 P-2's and 15 P-3's.

22. 25 chairs, 9 sofas

Chapter 5

Polynomials and Polynomial Functions

Exercise Set 5.1

1. $-7x^5 - x^3 + x^2 + 3x - 9$

Term	$-7x^5$	$-x^3$	x^2	$3x$	-9
Degree	5	3	2	1	0
Degree of polynomial	5				

2. 3, 2, 1, 0; 3

3. $y^3 + 2y^7 + x^2y^4 - 8$

Term	y^3	$2y^7$	x^2y^4	-8
Degree	3	7	6	0
Degree of polynomial	7			

4. 2, 5, 7, 0; 7

5. $a^5 + 4a^2b^4 + 6ab + 4a - 3$

Term	a^5	$4a^2b^4$	$6ab$	$4a$	-3
Degree	5	6	2	1	0
Degree of polynomial	6				

6. 6, 8, 4, 2, 0; 8

7. $-4y^3 - 6y^2 + 7y + 15$; $-4y^3$; -4

8. $-18y^4 + 11y^3 + 6y^2 - 5y + 2$; $-18y^4$; -18

9. $3x^7 + 5x^2 - x + 12$; $3x^7$; 3

10. $-10x^4 + 7x^2 - 3x + 9$; $-10x^4$; -10

11. $-a^7 + 8a^5 + 5a^3 - 19a^2 + a$; $-a^7$; -1

12. $a^9 + 11a^4 + a^3 - 5a^2 - 7$; a^9; 1

13. $-9 + 7x - 5x^2 + 3x^4$

14. $9 - x + 2x^3 - 3x^4$

15. $3xy^3 + x^2y^2 - 9x^3y + 2x^4$

16. $-9xy + 5x^2y^2 + 8x^3y^2 - 5x^4$

17. $-7ab + 4ax - 7ax^2 + 4x^6$

18. $-12a + 5xy^8 + 4ax^3 - 3ax^5 + 5x^5$

19.
$$\begin{aligned} P(x) &= 3x^2 - 2x + 5 \\ P(4) &= 3 \cdot 4^2 - 2 \cdot 4 + 5 \\ &= 48 - 8 + 5 \\ &= 45 \\ P(0) &= 3 \cdot 0^2 - 2 \cdot 0 + 5 \\ &= 0 - 0 + 5 \\ &= 5 \end{aligned}$$

20. -46; 10

21.
$$\begin{aligned} P(y) &= 8y^3 - 12y - 5 \\ P(-2) &= 8(-2)^3 - 12(-2) - 5 \\ &= -64 + 24 - 5 \\ &= -45 \end{aligned}$$

$$\begin{aligned} P\left(\frac{1}{3}\right) &= 8\left(\frac{1}{3}\right)^3 - 12 \cdot \frac{1}{3} - 5 \\ &= 8 \cdot \frac{1}{27} - 4 - 5 \\ &= \frac{8}{27} - 9 \\ &= \frac{8}{27} - \frac{243}{27} \\ &= -\frac{235}{27}, \text{ or } -8\frac{19}{27} \end{aligned}$$

22. -168; -9

23. $-7x + 5 = -7 \cdot 4 + 5 = -28 + 5 = -23$

24. 3

25. $x^3 - 5x^2 + x = 4^3 - 5 \cdot 4^2 + 4 = 64 - 5 \cdot 16 + 4 = 64 - 80 + 4 = -12$

26. 51

27.
$$\begin{aligned} f(x) &= -5x^3 + 3x^2 - 4x - 3 \\ f(-1) &= -5(-1)^3 + 3(-1)^2 - 4(-1) - 3 \\ &= 5 + 3 + 4 - 3 \\ &= 9 \end{aligned}$$

28. -6

29.
$$\begin{aligned} p(n) &= n^3 - 3n^2 + 2n \\ p(12) &= 12^3 - 3 \cdot 12^2 + 2 \cdot 12 \\ &= 1728 - 432 + 24 \\ &= 1320 \end{aligned}$$

A president, vice president, and treasurer can be elected in 1320 ways.

30. 6840

31. $s(t) = 16t^2$

$s(3) = 16 \cdot 3^2 = 16 \cdot 9 = 144$

The scaffold is 144 ft high.

32. 1024 ft

33. Evaluate the polynomial function for $x = 75$:

$$\begin{aligned} R(x) &= 280x - 0.4x^2 \\ R(75) &= 280 \cdot 75 - 0.4(75)^2 \\ &= 21,000 - 0.4(5625) \\ &= 21,000 - 2250 = 18,750 \end{aligned}$$

The total revenue is $18,750.

34. $24,000

35. Evaluate the polynomial function for $x = 75$:

$$\begin{aligned} C(x) &= 5000 + 0.6x^2 \\ C(75) &= 5000 + 0.6(75)^2 \\ &= 5000 + 0.6(5625) \\ &= 5000 + 3375 \\ &= 8375 \end{aligned}$$

The total cost is $8375.

36. $11,000

37.

$$\begin{aligned} P(a) &= 0.4a^2 - 40a + 1039 \\ P(18) &= 0.4(18)^2 - 40(18) + 1039 \\ &= 0.4(324) - 720 + 1039 \\ &= 129.6 - 720 + 1039 \\ &= 448.6 \end{aligned}$$

There are approximately 449 accidents daily involving an 18-year-old driver.

38. 289

39. Locate 2 on the horizontal axis. From there move vertically to the graph and then horizontally to the $M(t)$-axis. This locates a value of about 340. Thus, about 340 mg of ibuprofen is in the the bloodstream 2 hr after 400 mg have been swallowed.

40. About 185 mg

41. Locate 5 on the horizontal axis. From there move vertically to the graph and then horizontally to the $M(t)$-axis. This locates a value of about 65. Thus, about 65 mg of ibuprofen is in the the bloodstream 5 hr after 400 mg have been swallowed.

42. $M(3) \approx 300$

43. We evaluate the polynomial for $h = 6.3$ and $r = 1.2$:

$2\pi rh + 2\pi r^2 = 2\pi(1.2)(6.3) + 2\pi(1.2)^2 \approx 56.5$

The surface area is about 56.5 in^2.

44. 44.5 in^2

45.

$$\begin{aligned} &4a + 7 - 4 + 2a^3 - 6a + 3 \\ &= 2a^3 + (4-6)a + (7-4+3) \\ &= 2a^3 - 2a + 6 \end{aligned}$$

46. $5x^2 + x + 14$

47.

$$\begin{aligned} &3a^2b + 4b^2 - 9a^2b - 6b^2 \\ &= (3-9)a^2b + (4-6)b^2 \\ &= -6a^2b - 2b^2 \end{aligned}$$

48. $-8x^3 - 3x^2y^2$

49.

$$\begin{aligned} &8x^2 - 3xy + 12y^2 + x^2 - y^2 + 5xy + 4y^2 \\ &= (8+1)x^2 + (-3+5)xy + (12-1+4)y^2 \\ &= 9x^2 + 2xy + 15y^2 \end{aligned}$$

50. $11a^2 + 3ab - 3b^2$

51.

$$\begin{aligned} &(5a + 6b - 3c) + (4a - 2b + 2c) \\ &= (5+4)a + (6-2)b + (-3+2)c \\ &= 9a + 4b - c \end{aligned}$$

52. $15x + 7y - 5z$

53.

$$\begin{aligned} &(a^2 - 3b^2 + 4c^2) + (-5a^2 + 2b^2 - c^2) \\ &= (1-5)a^2 + (-3+2)b^2 + (4-1)c^2 \\ &= -4a^2 - b^2 + 3c^2 \end{aligned}$$

54. $-5x^2 + 4y^2 - 11z^2$

55.

$$\begin{aligned} &(x^2 + 2x - 3xy - 7) + (-3x^2 - x + 2xy + 6) \\ &= (1-3)x^2 + (2-1)x + (-3+2)xy + (-7+6) \\ &= -2x^2 + x - xy - 1 \end{aligned}$$

56. $2a^2 + 3b - 4ab + 4$

57.

$$\begin{aligned} &(7x^2y - 3xy^2 + 4xy) + (-2x^2y - xy^2 + xy) \\ &= (7-2)x^2y + (-3-1)xy^2 + (4+1)xy \\ &= 5x^2y - 4xy^2 + 5xy \end{aligned}$$

58. $20ab - 18ac - 3bc$

59.

$$\begin{aligned} &(2r^2 + 12r - 11) + (6r^2 - 2r + 4) + (r^2 - r - 2) \\ &= (2+6+1)r^2 + (12-2-1)r + (-11+4-2) \\ &= 9r^2 + 9r - 9 \end{aligned}$$

60. $-3x^2 - x - 3$

61.

$$\begin{aligned} &\left(\frac{1}{8}xy - \frac{3}{5}x^3y^2 + 4.3y^3\right) + \\ &\quad \left(-\frac{1}{3}xy - \frac{3}{4}x^3y^2 - 2.9y^3\right) \\ &= \left(\frac{1}{8} - \frac{1}{3}\right)xy + \left(-\frac{3}{5} - \frac{3}{4}\right)x^3y^2 + (4.3 - 2.9)y^3 \\ &= \left(\frac{3}{24} - \frac{8}{24}\right)xy + \left(-\frac{12}{20} - \frac{15}{20}\right)x^3y^2 + 1.4y^3 \\ &= -\frac{5}{24}xy - \frac{27}{20}x^3y^2 + 1.4y^3 \end{aligned}$$

62. $-\frac{2}{15}xy + \frac{19}{12}xy^2 + 1.7x^2y$

63. $5x^3 - 7x^2 + 3x - 6$

a) $-(5x^3 - 7x^2 + 3x - 6)$ Writing the opposite of P as $-P$

b) $-5x^3 + 7x^2 - 3x + 6$ Changing the sign of every term

64. $-(-8y^4 - 18y^3 + 4y - 9)$, $8y^4 + 18y^3 - 4y + 9$

65. $-12y^5 + 4ay^4 - 7by^2$

a) $-(-12y^5 + 4ay^4 - 7by^2)$

b) $12y^5 - 4ay^4 + 7by^2$

66. $-(7ax^3y^2 - 8by^4 - 7abx - 12ay)$,
$-7ax^3y^2 + 8by^4 + 7abx + 12ay$

67.
$$\begin{aligned}&(8x - 4) - (-5x + 2)\\ &= (8x - 4) + (5x - 2)\\ &= 13x - 6\end{aligned}$$

68. $13y + 5$

69.
$$\begin{aligned}&(-3x^2 + 2x + 9) - (x^2 + 5x - 4)\\ &= (-3x^2 + 2x + 9) + (-x^2 - 5x + 4)\\ &= -4x^2 - 3x + 13\end{aligned}$$

70. $-13y^2 + 2y + 11$

71.
$$\begin{aligned}&(5a - 2b + c) - (3a + 2b - 2c)\\ &= (5a - 2b + c) + (-3a - 2b + 2c)\\ &= 2a - 4b + 3c\end{aligned}$$

72. $4x - 10y + 4z$

73.
$$\begin{aligned}&(3x^2 - 2x - x^3) - (5x^2 - 8x - x^3)\\ &= (3x^2 - 2x - x^3) + (-5x^2 + 8x + x^3)\\ &= -2x^2 + 6x\end{aligned}$$

74. $5y^2 + 6y + 3y^3$

75.
$$\begin{aligned}&(5a^2 + 4ab - 3b^2) - (9a^2 - 4ab + 2b^2)\\ &= (5a^2 + 4ab - 3b^2) + (-9a^2 + 4ab - 2b^2)\\ &= -4a^2 + 8ab - 5b^2\end{aligned}$$

76. $-3y^2 - 6yz - 12z^2$

77.
$$\begin{aligned}&(6ab - 4a^2b + 6ab^2) - (3ab^2 - 10ab - 12a^2b)\\ &= (6ab - 4a^2b + 6ab^2) + (-3ab^2 + 10ab + 12a^2b)\\ &= 8a^2b + 16ab + 3ab^2\end{aligned}$$

78. $17xy + 5x^2y^2 - 7y^3$

79.
$$\begin{aligned}&\left(\frac{5}{8}x^4 - \frac{1}{4}x^2 - \frac{1}{2}\right) - \left(-\frac{3}{8}x^4 + \frac{3}{4}x^2 + \frac{1}{2}\right)\\ &= \left(\frac{5}{8}x^4 - \frac{1}{4}x^2 - \frac{1}{2}\right) + \left(\frac{3}{8}x^4 - \frac{3}{4}x^2 - \frac{1}{2}\right)\\ &= x^4 - x^2 - 1\end{aligned}$$

80. $\frac{29}{24}y^4 - \frac{5}{4}y^2 - 11.2y + \frac{8}{15}$

81.
$$\begin{aligned}P(x) &= R(x) - C(x)\\ P(x) &= (280x - 0.4x^2) - (5000 + 0.6x^2)\\ P(x) &= (280x - 0.4x^2) + (-5000 - 0.6x^2)\\ P(x) &= 280x - x^2 - 5000\\ P(70) &= 280 \cdot 70 - 70^2 - 5000\\ &= 19,600 - 4900 - 5000 = 9700\end{aligned}$$

The profit is \$9700.

82. \$8000

83.

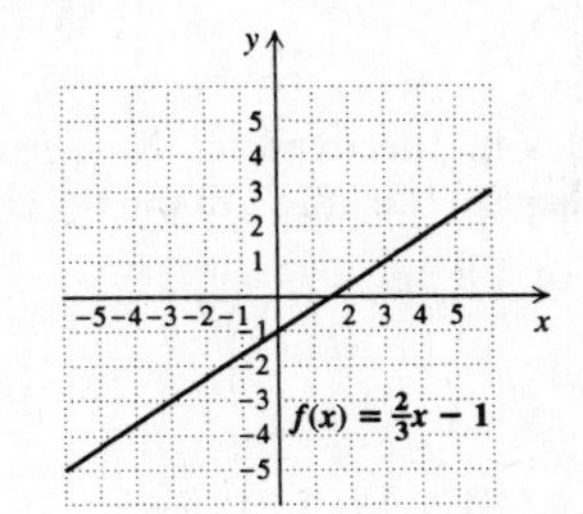

84. $\{-2, 5\}$

85. $3(y - 2) = 3y - 6$

86. $-2x - 92$

87.

88.

89.

90.

91.
$$\begin{aligned}&2[P(x)]\\ &= 2(13x^5 - 22x^4 - 36x^3 + 40x^2 - 16x + 75)\\ &= 26x^5 - 44x^4 - 72x^3 + 80x^2 - 32x + 150\end{aligned}$$

Use columns to add:

$$\begin{array}{rrrrrr} 26x^5 & -\ 44x^4 & -\ 72x^3 & +\ 80x^2 & -\ 32x & +\ 150\\ 42x^5 & -\ 37x^4 & +\ 50x^3 & -\ 28x^2 & +\ 34x & +\ 100\\ \hline 68x^5 & -\ 81x^4 & -\ 22x^3 & +\ 52x^2 & +\ 2x & +\ 250\end{array}$$

92. $-3x^5 - 29x^4 - 158x^3 + 148x^2 - 82x + 125$

93.
$$\begin{aligned}&2[Q(x)]\\ &= 2(42x^5 - 37x^4 + 50x^3 - 28x^2 + 34x + 100)\\ &= 84x^5 - 74x^4 + 100x^3 - 56x^2 + 68x + 200\end{aligned}$$

$$\begin{aligned}&3[P(x)]\\ &= 3(13x^5 - 22x^4 - 36x^3 + 40x^2 - 16x + 75)\\ &= 39x^5 - 66x^4 - 108x^3 + 120x^2 - 48x + 225\end{aligned}$$

Use columns to subtract, adding the opposite of $3[P(x)]$:

$$\begin{array}{rrrrrr} 84x^5 & -\ 74x^4 & +\ 100x^3 & -\ 56x^2 & +\ 68x & +\ 200\\ -\ 39x^5 & +\ 66x^4 & +\ 108x^3 & -\ 120x^2 & +\ 48x & -\ 225\\ \hline 45x^5 & -\ 8x^4 & +\ 208x^3 & -\ 176x^2 & +\ 116x & -\ 25\end{array}$$

94. $178x^5 - 199x^4 + 6x^3 + 76x^2 + 38x + 600$

95. First we find the number of truffles in the display.

$$N(x) = \frac{1}{6}x^3 + \frac{1}{2}x^2 + \frac{1}{3}x$$
$$N(5) = \frac{1}{6} \cdot 5^3 + \frac{1}{2} \cdot 5^2 + \frac{1}{3} \cdot 5$$
$$= \frac{1}{6} \cdot 125 + \frac{1}{2} \cdot 25 + \frac{5}{3}$$
$$= \frac{125}{6} + \frac{25}{2} + \frac{5}{3}$$
$$= \frac{125}{6} + \frac{75}{6} + \frac{10}{6}$$
$$= \frac{210}{6} = 35$$

There are 35 truffles in the display. Now find the volume of one truffle. Each truffle's diameter is 3 cm, so the radius is $\frac{3}{2}$, or 1.5 cm.

$$V(r) = \frac{4}{3}\pi r^3$$
$$V(1.5) \approx \frac{4}{3}(3.14)(1.5)^3 \approx 14.13 \text{ cm}^3$$

Finally, multiply the number of truffles and the volume of a truffle to find the total volume of chocolate.

$$35(14.13 \text{ cm}^3) = 494.55 \text{ cm}^3$$

The display contains about 494.55 cm^3 of chocolate.

96. $5x^2 - 8x$

97. To express the surface area in square centimeters, first convert r meters to centimeters:

$$r \text{ meters} \times \frac{100 \text{ centimeters}}{1 \text{ meters}} = 100r \text{ centimeters}$$

Then substitute h for h and $100r$ for r in the formula for Exercises 43 and 44:

$$2\pi(100r)(h) + 2\pi(100r)^2$$
$$= 200\pi rh + 20{,}000\pi r^2$$

Thus, a formula for the surface area, in square centimeters, of a right circular cylinder with height h centimeters and radius r meters is $200\pi rh + 20{,}000\pi r^2$.

To express the surface area in square meters, first convert h centimeters to meters:

$h \text{ centimeters} \times \frac{1 \text{ meter}}{100 \text{ centimeters}} = \frac{1}{100}h$ meters, or $0.01h$ meters

Then substitute $0.01h$ for h and r for r in the formula for Exercises 43 and 44:

$$2\pi r(0.01h) + 2\pi r^2 = 0.02\pi rh + 2\pi r^2$$

Thus, a formula for the surface area, in square meters, of a right circular cylinder with height h centimeters and radius r meters is $0.02\pi rh + 2\pi r^2$.

98. $8x^{2a} + 7x^a + 7$

99.
$$(3x^{6a} - 5x^{5a} - 4x^{3a} + 8) -$$
$$(2x^{6a} + 4x^{4a} + 3x^{3a} + 2x^{2a})$$
$$= (3-2)x^{6a} - 5x^{5a} - 4x^{4a} + (4-3)x^{3a} - 2x^{2a} + 8$$
$$= x^{6a} - 5x^{5a} - 4x^{4a} + x^{3a} - 2x^{2a} + 8$$

100. $x^{5b} + 4x^{4b} + x^{3b} - 6x^{2b} - 9x^b$

101.

102.

103. Observe that $p(0) = 5$ but that the y-intercept of the graph shown is not $(0, 5)$.

Exercise Set 5.2

1. $3a^2 \cdot 7a = (3 \cdot 7)(a^2 \cdot a) = 21a^3$

2. $-10x^4$

3. $5x(-4x^2y) = 5(-4)(x \cdot x^2)y = -20x^3y$

4. $-6a^3b^4$

5. $2x^3y^2(-5x^2y^4) = 2(-5)(x^3 \cdot x^2)(y^2 \cdot y^4) = -10x^5y^6$

6. $-56a^3b^4c^6$

7.
$$8x(2-x) = 8x \cdot 2 - 8x \cdot x$$
$$= 16x - 8x^2$$

8. $3a^3 - 12a^2$

9.
$$5cd(3c^2d - 5cd^2)$$
$$= 5cd \cdot 3c^2d - 5cd \cdot 5cd^2$$
$$= 15c^3d^2 - 25c^2d^3$$

10. $2a^4 - 5a^5$

11.
$$(2x+5)(3x-4)$$
$$= 6x^2 - 8x + 15x - 20 \quad \text{FOIL}$$
$$= 6x^2 + 7x - 20$$

12. $8a^2 + 10ab - 3b^2$

13.
$$(m+2n)(m-3n)$$
$$= m^2 - 3mn + 2mn - 6n^2 \quad \text{FOIL}$$
$$= m^2 - mn - 6n^2$$

14. $m^2 - 25$

15.
$$(y+8x)(2y-7x)$$
$$= 2y^2 - 7xy + 16xy - 56x^2 \quad \text{FOIL}$$
$$= 2y^2 + 9xy - 56x^2$$

16. $x^2 - xy - 2y^2$

17. $(a^2 - 2b^2)(a^2 - 3b^2)$
$= a^4 - 3a^2b^2 - 2a^2b^2 + 6b^4$ FOIL
$= a^4 - 5a^2b^2 + 6b^4$

18. $6m^4 - 13m^2n^2 + 5n^4$

19. $(x - 4)(x^2 + 4x + 16)$
$= (x - 4)(x^2) + (x - 4)(4x) + (x - 4)(16)$ Distributive law
$= x(x^2) - 4(x^2) + x(4x) - 4(4x) + x(16) - 4(16)$ Distributive law
$= x^3 - 4x^2 + 4x^2 - 16x + 16x - 64$ Multiplying monomials
$= x^3 - 64$ Collecting like terms

20. $y^3 + 27$

21. $(x + y)(x^2 - xy + y^2)$
$= (x + y)x^2 + (x + y)(-xy) + (x + y)(y^2)$
$= x(x^2) + y(x^2) + x(-xy) + y(-xy) + x(y^2) + y(y^2)$
$= x^3 + x^2y - x^2y - xy^2 + xy^2 + y^3$
$= x^3 + y^3$

22. $a^3 - b^3$

23.

$$\begin{array}{rrrrl} & & a^2 + & a - 1 & \\ & & a^2 + & 4a - 5 & \\ \hline & & -5a^2 - & 5a + 5 & \text{Multiplying by } -5 \\ & 4a^3 + & 4a^2 - & 4a & \text{Multiplying by } 4a \\ a^4 + & a^3 - & a^2 & & \text{Multiplying by } a^2 \\ \hline a^4 + & 5a^3 - & 2a^2 - & 9a + 5 & \text{Adding} \end{array}$$

24. $x^4 - x^3 + x^2 - 3x + 2$

25.

$$\begin{array}{rl} 4a^2b - 2ab + 3b^2 & \\ ab - 2b + a & \\ \hline 4a^3b - 2a^2b + 3ab^2 & (1) \\ -6b^3 \quad +4ab^2 - 8a^2b^2 & (2) \\ 3ab^3 \quad - 2a^2b^2 + 4a^3b^2 & (3) \\ \hline 3ab^3 - 6b^3 + 4a^3b - 2a^2b + 7ab^2 - 10a^2b^2 + 4a^3b^2 & (4) \end{array}$$

(1) Multiplying by a
(2) Multiplying by $-2b$
(3) Multiplying by ab
(4) Adding

26. $-4x^3y + 3xy^3 - x^2y^2 - 2y^4 + 2x^4$

27. $\left(x - \frac{1}{2}\right)\left(x - \frac{1}{4}\right)$
$= x^2 - \frac{1}{4}x - \frac{1}{2}x + \frac{1}{8}$ FOIL
$= x^2 - \frac{1}{4}x - \frac{2}{4}x + \frac{1}{8}$
$= x^2 - \frac{3}{4}x + \frac{1}{8}$

28. $b^2 - \frac{2}{3}b + \frac{1}{9}$

29. $(1.2x - 3y)(2.5x + 5y)$
$= 3x^2 + 6xy - 7.5xy - 15y^2$ FOIL
$= 3x^2 - 1.5xy - 15y^2$

30. $12a^2 + 399.928ab - 2.4b^2$

31. $P(x) \cdot Q(x) = (3x^2 - 5)(4x^2 - 7x + 2)$
$= (3x^2 - 5)(4x^2) + (3x^2 - 5)(-7x) + (3x^2 - 5)(2)$
$= 12x^4 - 20x^2 - 21x^3 + 35x + 6x^2 - 10$
$= 12x^4 - 21x^3 - 14x^2 + 35x - 10$

32. $x^5 + x^3 + 2x^2 - x + 2$

33. $(a + 8)(a + 5)$
$= a^2 + 5a + 8a + 40$ FOIL
$= a^2 + 13a + 40$

34. $x^2 + 5x + 6$

35. $(y - 4)(y + 3)$
$= y^2 + 3y - 4y - 12$
$= y^2 - y - 12$

36. $y^2 + 4y - 5$

37. $(x + 3)^2$
$= x^2 + 2 \cdot x \cdot 3 + 3^2$ $\quad (A + B)^2 = A^2 + 2AB + B^2$
$= x^2 + 6x + 9$

38. $y^2 - 14y + 49$

39. $(x - 2y)^2$
$= x^2 - 2(x)(2y) + (2y)^2$
$(A - B)^2 = A^2 - 2AB + B^2$
$= x^2 - 4xy + 4y^2$

40. $4s^2 + 12st + 9t^2$

41. $(2x + 9)(x + 2)$
$= 2x^2 + 4x + 9x + 18$ FOIL
$= 2x^2 + 13x + 18$

42. $6b^2 - 11b - 10$

43. $(10a - 0.12b)^2$
$= (10a)^2 - 2(10a)(0.12b) + (0.12b)^2$
$(A - B)^2 = A^2 - 2AB + B^2$
$= 100a^2 - 2.4ab + 0.0144b^2$

44. $100p^4 + 46p^2q + 5.29q^2$

45. $(2x - 3y)(2x + y)$
$= 4x^2 + 2xy - 6xy - 3y^2$ FOIL
$= 4x^2 - 4xy - 3y^2$

46. $4a^2 - 8ab + 3b^2$

47. $\left(2a+\frac{1}{3}\right)^2$

$= (2a)^2 + 2(2a)\left(\frac{1}{3}\right) + \left(\frac{1}{3}\right)^2$

$(A+B)^2 = A^2 + 2AB + B^2$

$= 4a^2 + \frac{4}{3}a + \frac{1}{9}$

48. $9c^2 - 3c + \frac{1}{4}$

49. $(2x^3 - 3y^2)^2$

$= (2x^3)^2 - 2(2x^3)(3y^2) + (3y^2)^2$

$(A-B)^2 = A^2 - 2AB + B^2$

$= 4x^6 - 12x^3y^2 + 9y^4$

50. $9s^4 + 24s^2t^3 + 16t^6$

51. $(a^2b^2 + 1)^2$

$= (a^2b^2)^2 + 2(a^2b^2)\cdot 1 + 1^2$

$(A+B)^2 = A^2 + 2AB + B^2$

$= a^4b^4 + 2a^2b^2 + 1$

52. $x^4y^2 - 2x^3y^3 + x^2y^4$

53. $P(x)\cdot P(x)$

$= (4x-1)(4x-1)$

$= (4x)^2 - 2(4x)(1) + 1^2$

$(A-B)^2 = A^2 - 2AB + B^2$

$= 16x^2 - 8x + 1$

54. $9x^4 + 6x^2 + 1$

55. $(c+2)(c-2)$

$= c^2 - 2^2 \quad (A+B)(A-B) = A^2 - B^2$

$= c^2 - 4$

56. $x^2 - 9$

57. $(2a+1)(2a-1)$

$= (2a)^2 - 1^2 \quad (A+B)(A-B) = A^2 - B^2$

$= 4a^2 - 1$

58. $9 - 4x^2$

59. $(3m-2n)(3m+2n)$

$= (3m)^2 - (2n)^2 \quad (A+B)(A-B) = A^2 - B^2$

$= 9m^2 - 4n^2$

60. $9x^2 - 25y^2$

61. $(x^3 + yz)(x^3 - yz)$

$= (x^3)^2 - (yz)^2 \quad (A+B)(A-B) = A^2 - B^2$

$= x^6 - y^2z^2$

62. $4a^6 - 25a^2b^2$

63. $(-mn + m^2)(mn + m^2)$

$= (m^2 - mn)(m^2 + mn)$

$= (m^2)^2 - (mn)^2 \quad (A+B)(A-B) = A^2 - B^2$

$= m^4 - m^2n^2$

64. $-9b^2 + a^4$, or $a^4 - 9b^2$

65. $(x+1)(x-1)(x^2+1)$

$= (x^2 - 1^2)(x^2+1)$

$= (x^2 - 1)(x^2 + 1)$

$= (x^2)^2 - 1^2$

$= x^4 - 1$

66. $y^4 - 16$

67. $(a-b)(a+b)(a^2-b^2)$

$= (a^2 - b^2)(a^2 - b^2)$

$= (a^2 - b^2)^2$

$= (a^2)^2 - 2(a^2)(b^2) + (b^2)^2$

$= a^4 - 2a^2b^2 + b^4$

68. $16x^4 - 8x^2y^2 + y^4$

69. $(a+b+1)(a+b-1)$

$= [(a+b)+1][(a+b)-1]$

$= (a+b)^2 - 1^2$

$= a^2 + 2ab + b^2 - 1$

70. $m^2 + 2mn + n^2 - 4$

71. $(2x+3y+4)(2x+3y-4)$

$= [(2x+3y)+4][(2x+3y)-4]$

$= (2x+3y)^2 - 4^2$

$= 4x^2 + 12xy + 9y^2 - 16$

72. $9a^2 - 12ab + 4b^2 - c^2$

73. $A = P(1+i)^2$

$A = P(1 + 2i + i^2)$

$A = P + 2Pi + Pi^2$

74. $A = P + Pi + \frac{Pi^2}{4}$

75. a) Replace each occurrence of x by $t-1$.

$f(t-1) = 5(t-1) + (t-1)^2$

$= 5t - 5 + t^2 - 2t + 1$

$= t^2 + 3t - 4$

b) $f(a+h) - f(a)$

$= [5(a+h) + (a+h)^2] - [5a + a^2]$

$= 5a + 5h + a^2 + 2ah + h^2 - 5a - a^2$

$= 5h + 2ah + h^2$

76. (a) $-p^2 + p + 6$; (b) $3h - 2ah - h^2$

77. ***Familiarize.*** We let x represent the number of days worked during the week and y represent the number of days working during the weekend. We organize the information in a table.

	Days worked	Pay per day	Total earned
During the week	x	\$50	$50x$
During the weekend	y	\$60	$60y$
Total	17		\$940

Translate.

$$\underbrace{\text{Total number of days worked}}_{x+y}\ \underbrace{\text{is}}_{=}\ \underbrace{17.}_{17}$$

$$\underbrace{\text{Total earned}}_{50x+60y}\ \underbrace{\text{is}}_{=}\ \underbrace{\$940.}_{940}$$

We now have a system of equations.

$$x + y = 17,$$
$$50x + 60y = 940$$

Carry out. Solving the system we get $(8, 9)$.

Check. The total number of days is $8 + 9$, or 17. The total earned is $\$50 \cdot 8 + \$60 \cdot 9$, or $\$400 + \$540$, or \$940. The values check.

State. Takako worked 8 weekdays.

78. 56, 58, 60

79. ***Familiarize.*** Let a, b, and c represent the daily production of machines A, B, and C, respectively.

Translate. Rewording, we have:

$$\underbrace{\text{Production of } A, B, \text{ and } C}_{a+b+c}\ \underbrace{\text{is}}_{=}\ \underbrace{222 \text{ per day.}}_{222}$$

$$\underbrace{\text{Production of } A \text{ and } B \text{ alone}}_{a+b}\ \underbrace{\text{is}}_{=}\ \underbrace{159 \text{ per day.}}_{159}$$

$$\underbrace{\text{Production of } B \text{ and } C \text{ alone}}_{b+c}\ \underbrace{\text{is}}_{=}\ \underbrace{147.}_{147}$$

We have a system of equations.

$$a + b + c = 222,$$
$$a + b = 159,$$
$$b + c = 147$$

Carry out. Solving the system we get $(75, 84, 63)$.

Check. The daily production of the three machines together is $75 + 84 + 63$, or 222. The daily production of A and B alone is $75 + 84$, or 159. The daily production of B and C alone is $84 + 63$, or 147. The numbers check.

State. The daily production of suitcases by machines A, B, and C is 75, 84, and 63, respectively.

80. 5 rolls of dimes, 2 rolls of nickels, 6 rolls of quarters

81.

82.

83.

84.

85. $(ab^{3n})^{2n} = a^{2n}(b^{3n})^{2n} = a^{2n}b^{6n^2}$

86. $x^{4a^2}y^{4ab}$

87.
$$\begin{aligned} &(z^{n^2})^{n^3}(z^{4n^3})^{n^2} \\ &= z^{n^5} \cdot z^{4n^5} \quad \text{Multiplying exponents} \\ &= z^{5n^5} \quad \text{Adding exponents} \end{aligned}$$

88. $\frac{1}{4}a^{7x}b^{2y+2}$

89.
$$\begin{aligned} &(a^x b^y)^{w+z} \\ &= a^{xw+xz}b^{yw+yz} \quad \text{Multiplying exponents} \end{aligned}$$

90. $y^{3n+3}z^{n+3} - 4y^4z^{3n}$

91.
$$\begin{aligned} &[(a+b)(a-b)][5-(a+b)][5+(a+b)] \\ &= [a^2-b^2][5^2-(a+b)^2] \\ &= [a^2-b^2][25-(a^2+2ab+b^2)] \\ &= [a^2-b^2][25-a^2-2ab-b^2] \\ &= 25a^2-a^4-2a^3b-a^2b^2-25b^2+a^2b^2+2ab^3+b^4 \\ &= -a^4-2a^3b+25a^2+2ab^3-25b^2+b^4 \end{aligned}$$

92. $x^3 + y^3 + 3y^2 + 3y + 1$

93.
$$\begin{aligned} &(y-1)^6(y+1)^6 \\ &= [(y-1)(y+1)]^6 \\ &= (y^2-1)^6 \\ &= [(y^2-1)^2]^3 \\ &= (y^4-2y^2+1)^3 \\ &= [(y^4-2y^2)+1]^2(y^4-2y^2+1) \\ &= (y^8-4y^6+4y^4+2y^4-4y^2+1)(y^4-2y^2+1) \\ &= (y^8-4y^6+6y^4-4y^2+1)(y^4-2y^2+1) \\ &= y^{12}-4y^{10}+6y^8-4y^6+y^4-2y^{10}+8y^8- \\ &\quad 12y^6+8y^4-2y^2+y^8-4y^6+6y^4-4y^2+1 \\ &= y^{12}-6y^{10}+15y^8-20y^6+15y^4-6y^2+1 \end{aligned}$$

94. $a^2 + 2ac + c^2 - b^2 - 2bd - d^2$

95. $\left(\frac{2}{3}x+\frac{1}{3}y+1\right)\left(\frac{2}{3}x-\frac{1}{3}y-1\right)$

$= \left[\frac{2}{3}x+\left(\frac{1}{3}y+1\right)\right]\left[\frac{2}{3}x-\left(\frac{1}{3}y+1\right)\right]$

$= \left(\frac{2}{3}x\right)^2-\left(\frac{1}{3}y+1\right)^2$

$= \frac{4}{9}x^2-\left(\frac{1}{9}y^2+\frac{2}{3}y+1\right)$

$= \frac{4}{9}x^2-\frac{1}{9}y^2-\frac{2}{3}y-1$

96. $x^3-\frac{1}{343}$

97. $(4x^2+2xy+y^2)(4x^2-2xy+y^2)$

$= [(4x^2+y^2)+2xy][(4x^2+y^2)-2xy]$

$= (4x^2+y^2)^2-(2xy)^2$

$= 16x^4+8x^2y^2+y^4-4x^2y^2$

$= 16x^4+4x^2y^2+y^4$

98. x^4-25x^2+144

99. $(x^a+y^b)(x^a-y^b)(x^{2a}+y^{2b})$

$= (x^{2a}-y^{2b})(x^{2a}+y^{2b})$

$= x^{4a}-y^{4b}$

100. x^6-1

101. $(x^{a-b})^{a+b}=x^{(a-b)(a+b)}=x^{a^2-b^2}$

102. $M^{x^2+2xy+y^2}$

103.

	x	5
2	$2x$	10
x	x^2	$5x$

Side: $x+2$ (= 2 + x); bottom: $x+5$ (= x + 5)

104.

105. One method is as follows. For each equation, let y_1 represent the left-hand side and y_2 represent the right-hand side, and let $y_3=y_2-y_1$. Then use a grapher to view the graph of y_3 and/or a table of values for y_3. If $y_3=0$, the equation is an identity. If $y_3\neq 0$, the equation is not an identity.

a) Not an identity

b) Identity

c) Identity

d) Not an identity

e) Not an identity

Exercise Set 5.3

1. $8t^2+2t$

$= 2t\cdot 4t+2t\cdot 1$

$= 2t(4t+1)$

2. $3y(2y+1)$

3. y^2-5y

$= y\cdot y-5\cdot y$

$= y(y-5)$

4. $x(x+9)$

5. y^3+9y^2

$= y\cdot y^2+9\cdot y^2$

$= y^2(y+9)$

6. $x^2(x+8)$

7. $5x^2-15x^4$

$= 5x^2\cdot 1-5x^2\cdot 3x^2$

$= 5x^2(1-3x^2)$

8. $4y^2(2+y^2)$

9. $4x^2y-12xy^2$

$= 4xy\cdot x-4xy\cdot 3y$

$= 4xy(x-3y)$

10. $5x^2y^2(y+3x)$

11. $3y^2-3y-9$

$= 3\cdot y^2-3\cdot y-3\cdot 3$

$= 3(y^2-y-3)$

12. $5(x^2-x+3)$

13. $6ab-4ad+12ac$

$= 2a\cdot 3b-2a\cdot 2d+2a\cdot 6c$

$= 2a(3b-2d+6c)$

14. $2x(4y+5z-7w)$

15. $9x^3y^6z^2-12x^4y^4z^4+15x^2y^5z^3$

$= 3x^2y^4z^2\cdot 3xy^2-3x^2y^4z^2\cdot 4x^2z^2+3x^2y^4z^2\cdot 5yz$

$= 3x^2y^4z^2(3xy^2-4x^2z^2+5yz)$

16. $7a^3b^3c^3(2ac^2+3b^2c-5ab)$

17. $-5x+15=-5(x-3)$

18. $-5(x+8)$

19. $-6y-72=-6(y+12)$

20. $-8(t-9)$

21. $-2x^2+4x-12=-2(x^2-2x+6)$

22. $-2(x^2-6x-20)$

23. $-3y^2+24x=-3(y^2-8x)$

24. $-7(x^2+8y)$

25. $-3y^3+12y^2-15y=-3y(y^2-4y+5)$

26. $-4m(m^3+8m^2-16)$

27. $-x^2+5x-9=-(x^2-5x+9)$

28. $-(p^3+4p^2-11)$

29. $-a^4+2a^3-13a=-a(a^3-2a^2+13)$

30. $-(m^3+m^2-m+2)$

31. a) $h(t)=-16t^2+96t$

$h(t)=-16t(t-6)$ Factoring out $-16t$

b) Using $h(t)=-16t^2+96t$:

$h(2)=-16\cdot 2^2+96\cdot 2=-16\cdot 4+192$

$=-64+192=128$

Using $h(t)=-16t(t-6)$:

$h(2)=-16(2)(2-6)=-16(2)(-4)=128$

The expressions have the same value for $t=2$, so the factorization is probably correct.

32. (a) $h(t)=-8t(2t-9)$; (b) $h(2)=80$

33. $N(x)=\frac{1}{6}x^3+\frac{1}{2}x^2+\frac{1}{3}x$

$N(x)=\frac{1}{6}x(x^2+3x+2)$ Factoring out $\frac{1}{6}x$

34. $f(n)=\frac{1}{2}n(n-1)$

35. $2\pi rh+\pi r^2=\pi r(2h+r)$

36. $P(n)=\frac{1}{2}n(n-3)$

37. $R(x)=280x-0.4x^2$

$R(x)=0.4x(700-x)$

38. $C(x)=0.6x(0.3+x)$

39. $a(b-2)+c(b-2)$

$=(b-2)(a+c)$

40. $(x^2-3)(a-2)$

41. $(x+7)(x-1)+(x+7)(x-2)$

$=(x+7)(x-1+x-2)$

$=(x+7)(2x-3)$

42. $(a+5)(2a-1)$

43. $a^2(x-y)+5(y-x)$

$=a^2(x-y)+5(-1)(x-y)$ Factoring out -1 to reverse the second subtraction

$=a^2(x-y)-5(x-y)$ Simplifying

$=(x-y)(a^2-5)$

44. $(x-6)(3x^2-2)$

45. $ac+ad+bc+bd$

$=a(c+d)+b(c+d)$

$=(c+d)(a+b)$

46. $(y+z)(x+w)$

47. b^3-b^2+2b-2

$=b^2(b-1)+2(b-1)$

$=(b-1)(b^2+2)$

48. $(y-1)(y^2+3)$

49. a^3-3a^2+6-2a

$=a^2(a-3)+2(3-a)$

$=a^2(a-3)+2(-1)(a-3)$ Factoring out -1 to reverse the second subtraction

$=a^2(a-3)-2(a-3)$

$=(a-3)(a^2-2)$

50. $(t+6)(t^2-2)$

51. $24x^3-36x^2+72x$

$=12x\cdot 2x^2-12x\cdot 3x+12x\cdot 6$

$=12x(2x^2-3x+6)$

52. $3a^2(4a^2-7a-3)$

53. $x^6-x^5-x^3+x^4$

$=x^3(x^3-x^2-1+x)$

$=x^3[x^2(x-1)+x-1]$ $(-1+x=x-1)$

$=x^3(x-1)(x^2+1)$

54. $y(y-1)(y^2+1)$

55. $2y^4+6y^2+5y^2+15$

$=2y^2(y^2+3)+5(y^2+3)$

$=(y^2+3)(2y^2+5)$

56. $(2-x)(xy-3)$

57.

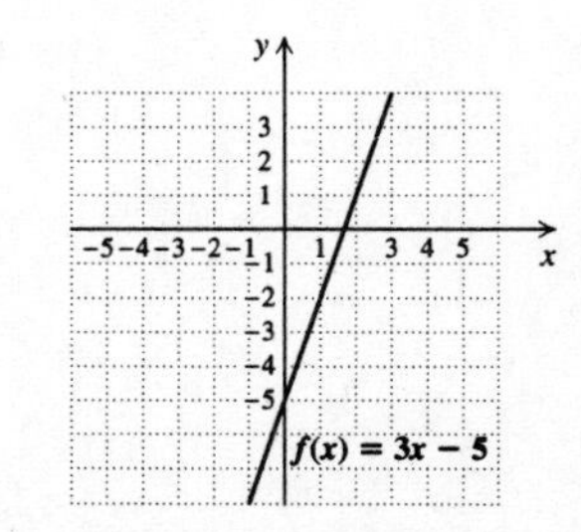

58. $6a - 2$

59. ***Familiarize.*** Let x = the number of pounds of Countryside rice and y = the number of pounds of Mystic rice to be used in the mixture.

Translate. We organize the information in a table.

	Countryside	Mystic	Mixture
Number of pounds	x	y	25
Percent of wild rice	10%	50%	35%
Pounds of wild rice	$0.1x$	$0.5x$	$0.35(25)$

From the "Number of pounds" row of the table we get one equation.

$$x + y = 25$$

We get a second equation from the last row of the table.

$$0.1x + 0.5x = 0.35(25), \text{ or}$$
$$0.1x + 0.5x = 8.75$$

Clearing decimals, we have the following system of equations:

$$x + y = 25, \quad (1)$$
$$10x + 50y = 875 \quad (2)$$

Carry out. We use the elimination method to solve the system of equations. We multiply Equation (1) by -10 and add.

$$\begin{aligned} -10x - 10y &= -250 \\ 10x + 50y &= 875 \\ \hline 40y &= 625 \\ y &= 15.625 \end{aligned}$$

Now substitute 15.625 for y in Equation (1) and solve for x.

$$x + 15.625 = 25$$
$$x = 9.375$$

Check. The total weight of the mixture is $9.375 + 15.625$, or 25 lb. The amount of wild rice in the mixture is $0.1(9.375) + 0.5(15.625) = 0.9375 + 7.8125 = 8.75$ lb. This is 35% of 25 lb. The numbers check.

State. 9.375 lb of Countryside Rice and 15.625 lb of Mystic Rice should be used.

60. 84°, 21°

61.

62.

63.

64.

65. $x^5y^4 + \_\_\_\_ = x^3y(\_\_\_\_ + xy^5)$

The term that goes in the first blank is the product of x^3y and xy^5, or x^4y^6.

The term that goes in the second blank is the expression that is multiplied with x^3y to obtain x^5y^4, or x^2y^3. Thus, we have

$$x^5y^4 + x^4y^6 = x^3y(x^2y^3 + xy^5).$$

66. $a^3b^7 - a^2b^3c^2 = a^2b^3(ab^4 - c^2)$

67.
$$\begin{aligned} &rx^2 - rx + 5r + sx^2 - sx + 5s \\ &= r(x^2 - x + 5) + s(x^2 - x + 5) \\ &= (x^2 - x + 5)(r + s) \end{aligned}$$

68. $(a^2 + 2a + 10)(3 + 7b)$

69.
$$\begin{aligned} &5x^2 - x^2y + 10x - 2xy + 15xz - 3xyz \\ &= x(5x - xy + 10 - 2y + 15z - 3yz) \\ &= x[x(5 - y) + 2(5 - y) + 3z(5 - y)] \\ &= x(5 - y)(x + 2 + 3z) \end{aligned}$$

70. $(x^4 + x^2 + 5)(a^4 + a^2 + 5)$

71.
$$\begin{aligned} &2x^{3a} + 8x^a + 4x^{2a} \\ &= 2x^a(x^{2a} + 4 + 2x^a) \end{aligned}$$

72. $3a^n(a + 2 - 5a^2)$

73.
$$\begin{aligned} &4x^{a+b} + 7x^{a-b} \\ &= 4 \cdot x^a \cdot x^b + 7 \cdot x^a \cdot x^{-b} \\ &= x^a(4x^b + 7x^{-b}) \end{aligned}$$

74. $y^{a+b}(7y^a - 5 + 3y^b)$

75.

76. One method is to let $y_1 = (x^2 - 3x + 2)^4$ and let $y_2 = x^8 + 81x^4 + 16$. Then use a table to show that $y_1 \neq y_2$ for all values of x.

Exercise Set 5.4

1. $x^2 + 8x + 12$

We look for two numbers whose product is 12 and whose sum is 8. Since 12 and 8 are both positive, we need only consider positive factors.

Pair of Factors	Sum of Factors
1,12	13
2, 6	8

The numbers we need are 2 and 6. The factorization is $(x + 2)(x + 6)$.

2. $(x + 1)(x + 5)$

3. $t^2 - 8t + 15$

Since the constant term is positive and the coefficient of the middle term is negative, we look for a factorization of 15 in which both factors are negative. Their sum must be -8.

Pair of Factors	Sum of Factors
$-1, -15$	-16
$-3, -5$	-8

The numbers we need are -3 and -5. The factorization is $(t-3)(t-5)$.

4. $(y-3)(y-9)$

5. $x^2 - 27 - 6x = x^2 - 6x - 27$

Since the constant term is negative, we look for a factorization of -27 in which one factor is positive and one factor is negative. Their sum must be -6, so the negative factor must have the larger absolute value. Thus we consider only pairs of factors in which the negative factor has the larger absolute value.

Pair of Factors	Sum of Factors
$-27, 1$	-26
$-9, 3$	-6

The numbers we need are -9 and 3. The factorization is $(x-9)(x+3)$.

6. $(t-5)(t+3)$

7. $2n^2 - 20n + 50$

$= 2(n^2 - 10n + 25)$ Removing the common factor

We now factor $n^2 - 10n + 25$. We look for two numbers whose product is 25 and whose sum is -10. Since the constant term is positive and the coefficient of the middle term is negative, we look for factorization of 25 in which both factors are negative.

Pair of Factors	Sum of Factors
$-1, -25$	-26
$-5, -5$	-10

The numbers we need are -5 and -5.

$$n^2 - 10n + 25 = (n-5)(n-5)$$

We must not forget to include the common factor 2.

$$2n^2 - 20n + 50 = 2(n-5)(n-5), \text{ or } 2(n-5)^2$$

8. $2(a-4)^2$

9. $a^3 + a^2 - 72a$

$= a(a^2 + a - 72)$ Removing the common factor

We now factor $a^2 + a - 72$. Since the constant term is negative, we look for a factorization of -72 in which one factor is positive and one factor is negative. We consider only pairs of factors in which the positive factor has the larger absolute value, since the sum of the factors, 1, is positive.

Pair of Factors	Sum of Factors
$72, -1$	71
$36, -2$	34
$18, -4$	14
$9, -8$	1

The numbers we need are 9 and -8.

$$a^2 + a - 72 = (a+9)(a-8)$$

We must not forget to include the common factor a.

$$a^3 + a^2 - 72a = a(a+9)(a-8)$$

10. $x(x+9)(x-6)$

11. $14x + x^2 + 45 = x^2 + 14x + 45$

Since the constant term and the middle term are both positive, we look for a factorization of 45 in which both factors are positive. Their sum must be 14.

Pair of Factors	Sum of Factors
45,1	46
15,3	18
9,5	14

The numbers we need are 9 and 5. The factorization is $(x+9)(x+5)$.

12. $(y+8)(y+4)$

13. $y^2 + 2y - 63$

Since the constant term is negative, we look for a factorization of -63 in which one factor is positive and one factor is negative. We consider only pairs of factors in which the positive factor has the larger absolute value, since the sum of the factors, 2, is positive.

Pair of Factors	Sum of Factors
$63, -1$	62
$21, -3$	18
$9, -7$	2

The numbers we need are 9 and -7. The factorization is $(y+9)(y-7)$.

14. $(p+8)(p-5)$

15. $t^2 - 14 + 45$

Since the constant term is positive and the coefficient of the middle term is negative, we look for a factorization of 45 in which both factors are negative. Their sum must be -14.

Pair of Factors	Sum of Factors
$-1, -45$	-46
$-3, -15$	-18
$-5, -9$	-14

The numbers we need are -5 and -9. The factorization is $(t-5)(t-9)$.

16. $(a-4)(a-7)$

17. $3x + x^2 - 10 = x^2 + 3x - 10$

Since the constant term is negative, we look for a factorization of -10 in which one factor is positive and one factor is negative. We consider only pairs of factors in which the positive factor has the larger absolute value, since the sum of the factors, 3, is positive.

Pair of Factors	Sum of Factors
10, −1	9
5, −2	3

The numbers we need are 5 and -2. The factorization is $(x + 5)(x - 2)$.

18. $(x + 3)(x - 2)$

19. $3x^2 + 15x + 18$

$= 3(x^2 + 5x + 6)$ Removing the common factor

We now factor $x^2 + 5x + 6$. We look for two numbers whose product is 6 and whose sum is 5. Since 6 and 5 are both positive, we need consider only positive factors.

Pair of Factors	Sum of Factors
1, 6	7
2, 3	5

The numbers we need are 2 and 3.

$$x^2 + 5x + 6 = (x + 2)(x + 3)$$

We must not forget to include the common factor 3.

$$3x^2 + 15x + 18 = 3(x + 2)(x + 3)$$

20. $5(y + 1)(y + 7)$

21. $56 + x - x^2 = -x^2 + x + 56 = -(x^2 - x - 56)$

We now factor $x^2 - x - 56$. Since the constant term is negative, we look for a factorization of -56 in which one factor is positive and one factor is negative. We consider only pairs of factors in which the negative factor has the larger absolute value, since the sum of the factors, -1, is negative.

Pair of Factors	Sum of Factors
−56, 1	−55
−28, 2	−26
−14, 4	−10
−8, 7	−1

The numbers we need are -8 and 7. Thus, $x^2 - x - 56 = (x - 8)(x + 7)$. We must not forget to include the factor that was factored out earlier:

$$56 + x - x^2 = -(x - 8)(x + 7), \text{ or}$$

$$(-x + 8)(x + 7), \text{ or } (8 - x)(7 + x)$$

22. $(8 - y)(4 + y)$

23. $32y + 4y^2 - y^3$

There is a common factor, y. We also factor out -1 in order to make the leading coefficient positive.

$$32y + 4y^2 - y^3 = -y(-32 - 4y + y^2)$$
$$= -y(y^2 - 4y - 32)$$

Now we factor $y^2 - 4y - 32$. Since the constant term is negative, we look for a factorization of -32 in which one factor is positive and one factor is negative. We consider only pairs of factors in which the negative factor has the larger absolute value, since the sum of the factors, -4, is negative.

Pair of Factors	Sum of Factors
−32, 1	−31
−16, 2	−14
−8, 4	−4

The numbers we need are -8 and 4. Thus, $y^2 - 4y - 32 = (y - 8)(y + 4)$. We must not forget to include the common factor:

$$32y + 4y^2 - y^3 = -y(y - 8)(y + 4), \text{ or}$$

$$y(-y + 8)(y + 4), \text{ or } y(8 - y)(4 + y)$$

24. $x(8 - x)(7 + x)$

25. $x^4 + 11x^2 - 80$

First make a substitution. We let $u = x^2$, so $u^2 = x^4$. Then we consider $u^2 + 11u - 80$. We look for pairs of factors of -80, one positive and one negative, such that the positive factor has the larger absolute value and the sum of the factors is 11.

Pair of Factors	Sum of Factors
80, −1	79
40, −2	38
20, −4	16
16, −5	11
10, −8	2

The numbers we need are 16 and -5. Then $u^2 + 11u - 80 = (u + 16)(u - 5)$. Replacing u by x^2 we obtain the factorization of the original trinomial: $(x^2 + 16)(x^2 - 5)$

26. $(y^2 + 12)(y^2 - 7)$

27. $x^2 + 12x + 13$

There are no factors of 13 whose sum is 12. This trinomial is not factorable into binomials with integer coefficients. The polynomial is prime.

28. Prime

29. $p^2 - 5pq - 24q^2$

We look for numbers r and s such that $p^2 - 5pq - 24q^2 = (p + rq)(p + sq)$. Our thinking is much the same as if we were factoring $p^2 - 5p - 24$. We look for factors of -24 whose sum is -5, one positive and one negative, such that the negative factor has the larger absolute value.

Pair of Factors	Sum of Factors
−24, 1	−23
−12, 2	−10
−8, 3	−5

The numbers we need are -8 and 3. The factorization is $(p - 8q)(p + 3q)$.

30. $(x+3y)(x+9y)$

31. $y^2+8yz+16z^2$

We look for numbers p and q such that $y^2+8yz+16z^2=(y+pz)(y+qz)$. Our thinking is much the same as if we factor $y^2+8y+16$. Since the constant term is positive and the coefficient of the middle term is negative, we look for a factorization of 16 in which both factors are positive. Their sum must be 8.

Pair of Factors	Sum of Factors
1, 16	17
2, 8	10
4, 4	8

The numbers we need are 4 and 4. The factorization is $(y+4z)(y+4z)$, or $(y+4z)^2$.

32. $(x-7y)(x-7y)$, or $(x-7y)^2$

33. p^4+80p^2+79

Substitute u for p^2 (and hence u^2 for p^4). Consider $u^2+80u+79$. We look for a pair of factors of 79 whose sum is 80. The only positive pair of factors is 1 and 79. These are the numbers we need. Then $u^2+80u+79=(u+1)(u+79)$. Replacing u by p^2 we have $p^4+80p^2+79=(p^2+1)(p^2+79)$.

34. $(x^2+1)(x^2+49)$

35. x^8-7x^4+10

Substitute u for x^4 (and hence u^2 for x^8). Consider $u^2-7u+10$. Since the constant term is positive and the coefficient of the middle term is negative, we look for a factorization of 10 in which both factors are negative. Their sum must be -7.

Pair of Factors	Sum of Factors
-1, -10	-11
-2, -5	-7

The numbers we need are -2 and -5. Then $u^2-7u+10=(u-2)(u-5)$. Replacing u by x^4 we have $x^8-7x^4+10=(x^4-2)(x^4-5)$.

36. $(x^3-7)(x^3+9)$

37. $6x^2-5x-25$

We will use the FOIL method.

1. There is no common factor (other than 1 or -1.)
2. Factor the first term, $6x^2$. The factors are $6x$, x and $3x$, $2x$. We have these possibilities:

   $(6x+\quad)(x+\quad)$ or $(3x+\quad)(2x+\quad)$
3. Factor the last term, -25. The possibilities are $25(-1)$, $-25\cdot 1$, and $-5\cdot 5$.
4. We need factors for which the sum of the products (the "outer" and "inner" parts of FOIL) is the middle term, $-5x$. Try some possibilities and check by multiplying.

   $(6x-5)(x+5)=6x^2+25x-25$

   We try again.

   $(3x+5)(2x-5)=6x^2-5x-25$

   The factorization is $(3x+5)(2x-5)$.

38. $(3x+2)(x-6)$

39. $10y^3-12y-7y^2=10y^3-7y^2-12y$

We will use the grouping method.

1. Look for a common factor. We factor out y:

   $y(10y^2-7y-12)$
2. Factor the trinomial $10y^2-7y-12$. Multiply the leading coefficient, 10, and the constant, -12.

   $10(-12)=-120$
3. Try to factor -120 so the sum of the factors is -7. We need only consider pairs of factors in which the negative factor has the larger absolute value, since their sum is negative.

Pair of Factors	Sum of Factors
-120, 1	-119
-30, 4	-26
-15, 8	-7

4. We split the middle term, $-12y$, using the results of step (3).

   $-7y=-15y+8y$
5. Factor by grouping:

$$\begin{aligned}10y^2-7y-12&=10y^2-15y+8y-12\\&=5y(2y-3)+4(2y-3)\\&=(2y-3)(5y+4)\end{aligned}$$

We must include the common factor to get a factorization of the original trinomial:

$$10y^3-12y-7y^2=y(2y-3)(5y+4)$$

40. $x(3x-5)(2x+3)$

41. $24a^2-14a+2$

We will use the FOIL method.

1. Factor out the common factor, 2:

   $2(12a^2-7a+1)$
2. Now we factor the trinomial $12a^2-7a+1$. Factor the first term, $12a^2$. The factors are $12a$, a and $6a$, $2a$ and $4a$, $3a$. We have these possibilities: $(12a+\quad)(a+\quad)$, $(6a+\quad)(2a+\quad)$, $(4a+\quad)(3a+\quad)$.
3. Factor the last term, 1. The possibilities are $1\cdot 1$ and $-1(-1)$.
4. Look for factors such that the sum of the products is the middle term, $-7a$. Trial and error leads us to the correct factorization:

   $12a^2-7a+1=(4a-1)(3a-1)$

We must include the common factor to get a factorization of the original trinomial:

$$24a^2-14a+2=2(4a-1)(3a-1)$$

42. $(3a-4)(a-2)$

43. $35y^2+34y+8$

We will use the grouping method.

1. There is no common factor (other than 1 or -1).

2. Multiply the leading coefficient, 35, and the constant, 8: $35(8)=280$

3. Try to factor 280 so the sum of the factors is 34. We need only consider pairs of positive factors since 280 and 34 are both positive.

Pair of Factors	Sum of Factors
280, 1	281
140, 2	142
70, 4	74
56, 5	61
40, 7	47
28, 10	38
20, 14	34

4. Split $34y$ using the results of step (3):

$$34y=20y+14y$$

5. Factor by grouping:

$$\begin{aligned}35y^2+34y+8&=35y^2+20y+14y+8\\&=5y(7y+4)+2(7y+4)\\&=(7y+4)(5y+2)\end{aligned}$$

44. $(3a+2)(3a+4)$

45. $4t+10t^2-6=10t^2+4t-6$

We will use the FOIL method.

1. Factor out the common factor, 2:

$$2(5t^2+2t-3)$$

2. Now we factor the trinomial $5t^2+2t-3$. Factor the first term, $5t^2$. The factors are $5t$ and t. We have this possibility: $(5t+\quad)(t+\quad)$

3. Factor the last term, -3. The possibilities are $(1)(-3)$ and $(-1)3$ as well as $(-3)(1)$ and $3(-1)$.

4. Look for factors such that the sum of the products is the middle term, $2t$. Trial and error leads us to the correct factorization:

$$5t^2+2t-3=(5t-3)(t+1)$$

We must include the common factor to get a factorization of the original trinomial:

$$4t+10t^2-6=2(5t-3)(t+1)$$

46. $2(5x+3)(3x-1)$

47. $8x^2-16-28x=8x^2-28x-16$

We will use the grouping method.

1. Factor out the common factor, 4:

$$4(2x^2-7x-4)$$

2. Now we factor the trinomial $2x^2-7x-4$. Multiply the leading coefficient, 2, and the constant, -4: $2(-4)=-8$

3. Factor -8 so the sum of the factors is -7. We need only consider pairs of factors in which the negative factor has the larger absolute value, since their sum is negative.

Pair of Factors	Sum of Factors
−4, 2	−2
−8, 1	−7

4. Split $-7x$ using the results of step (3):

$$-7x=-8x+x$$

5. Factor by grouping:

$$\begin{aligned}2x^2-7x-4&=2x^2-8x+x-4\\&=2x(x-4)+(x-4)\\&=(x-4)(2x+1)\end{aligned}$$

We must include the common factor to get a factorization of the original trinomial:

$$8x^2-16-28x=4(x-4)(2x+1)$$

48. $6(3x-4)(x+1)$

49. a^6-a^3-6

Substitute u for a^3 (and hence u^2 for a^6). Then factor u^2-u-6.

We look for a pair of factors of -6 whose sum is -1. The numbers we need are -3 and 2. Then $u^2-u-6=(u-3)(u+2)$. Replacing u by a^3 we have $a^6-a^3-6=(a^3-3)(a^3+2)$.

50. $(t^4-3)(t^4-2)$

51. $14x^4-19x^3-3x^2$

We will use the grouping method.

1. Factor out the common factor, x^2:

$$x^2(14x^2-19x-3)$$

2. Now we factor the trinomial $14x^2-19x-3$. Multiply the leading coefficient, 14, and the constant, -3: $14(-3)=-42$

3. Factor -42 so the sum of the factors is -19. We need only consider pairs of factors in which the negative factor has the larger absolute value, since the sum is negative.

Pair of Factors	Sum of Factors
−42, 1	−41
−21, 2	−19
−14, 3	−11
−7, 6	−1

4. Split $-19x$ using the results of step (3):

$$-19x=-21x+2x$$

5. Factor by grouping:

$$\begin{aligned}14x^2 - 19x - 3 &= 14x^2 - 21x + 2x - 3\\ &= 7x(2x - 3) + 2x - 3\\ &= (2x - 3)(7x + 1)\end{aligned}$$

We must include the common factor to get a factorization of the original trinomial:

$$14x^4 - 19x^3 - 3x^2 = x^2(2x - 3)(7x + 1)$$

52. $2x^2(5x - 2)(7x - 4)$

53. $12a^2 - 4a - 16$

We will use the FOIL method.

1. Factor out the common factor, 4:

$$4(3a^2 - a - 4)$$

2. We now factor the trinomial $3a^2 - a - 4$. Factor the first term, $3a^2$. The possibility is $(3a+\quad)(a+\quad)$.

3. Factor the last term, -4. The possibilities are $-4 \cdot 1$, $4(-1)$, and $-2 \cdot 2$.

4. We need factors for which the sum of the products is the middle term, $-a$. Trial and error leads us to the correct factorization:

$$3a^2 - a - 4 = (3a - 4)(a + 1)$$

We must include the common factor to get a factorization of the original trinomial:

$$12a^2 - 4a - 16 = 4(3a - 4)(a + 1)$$

54. $2(6a + 5)(a - 2)$

55. $9x^2 + 15x + 4$

We will use the grouping method.

1. There is no common factor (other than 1 or -1).

2. Multiply the leading coefficient and constant: $9(4) = 36$

3. Factor 36 so the sum of the factors is 15. We need only consider pairs of positive factors since 36 and 15 are both positive.

Pair of Factors	Sum of Factors
36, 1	37
18, 2	20
12, 3	15
9, 4	13
6, 6	12

4. Split $15x$ using the results of step (3):

$$15x = 12x + 3x$$

5. Factor by grouping:

$$\begin{aligned}9x^2 + 15x + 4 &= 9x^2 + 12x + 3x + 4\\ &= 3x(3x + 4) + 3x + 4\\ &= (3x + 4)(3x + 1)\end{aligned}$$

56. $(3y - 2)(2y + 1)$

57. $8 - 6z - 9z^2$

We will use the FOIL method.

1. There is no common factor (other than 1 or -1).

2. Factor the first term, 8. The possibilities are $(8+\quad)(1+\quad)$ and $(4+\quad)(2+\quad)$.

3. Factor the last term, $-9z^2$. The possibilities are $-9z \cdot z$, $-3z \cdot 3z$, and $9z(-z)$.

4. We need factors for which the sum of products is the middle term, $-6z$. Trial and error leads us to the correct factorization:

$$(4 + 3z)(2 - 3z)$$

58. $(3 - a)(1 + 12a)$

59. $-8t^2 - 8t + 30$

We will use the grouping method.

1. Factor out -2: $-2(4t^2 + 4t - 15)$

2. Now we factor the trinomial $4t^2 + 4t - 15$. Multiply the leading coefficient and the constant: $4(-15) = -60$

3. Factor -60 so the sum of the factors is 4. The desired factorization is $10(-6)$.

4. Split $4t$ using the results of step (3):

$$4t = 10t - 6t$$

5. Factor by grouping:

$$\begin{aligned}4t^2 + 4t - 15 &= 4t^2 + 10t - 6t - 15\\ &= 2t(2t + 5) - 3(2t + 5)\\ &= (2t + 5)(2t - 3)\end{aligned}$$

We must include the common factor to get a factorization of the original trinomial:

$$-8t^2 - 8t + 30 = -2(2t + 5)(2t - 3)$$

60. $-3(4a - 1)(3a - 1)$

61. $18xy^3 + 3xy^2 - 10xy$

We will use the FOIL method.

1. Factor out the common factor, xy.

$$xy(18y^2 + 3y - 10)$$

2. We now factor the trinomial $18y^2 + 3y - 10$. Factor the first term, $18y^2$. The possibilities are $(18y+\quad)(y+\quad)$, $(9y+\quad)(2y+\quad)$, and $(6y+\quad)(3y+\quad)$.

3. Factor the last term, -10. The possibilities are $-10 \cdot 1$, $-5 \cdot 2$, $10(-1)$ and $5(-2)$.

4. We need factors for which the sum of the products is the middle term, $3y$. Trial and error leads us to the correct factorization.

$$18y^2 + 3y - 10 = (6y + 5)(3y - 2)$$

We must include the common factor to get a factorization of the original trinomial:

$$18xy^3 + 3xy^2 - 10xy = xy(6y + 5)(3y - 2)$$

62. $xy^2(3x+1)(x-2)$

63. $24x^2 - 2 - 47x = 24x^2 - 47x - 2$

We will use the grouping method.

1. There is no common factor (other than 1 or -1).
2. Multiply the leading coefficient and the constant: $24(-2) = -48$
3. Factor -48 so the sum of the factors is -47. The desired factorization is $-48 \cdot 1$.
4. Split $-47x$ using the results of step (3):
   $-47x = -48x + x$
5. Factor by grouping:
$$\begin{aligned} 24x^2 - 47x - 2 &= 24x^2 - 48x + x - 2 \\ &= 24x(x-2) + (x-2) \\ &= (x-2)(24x+1) \end{aligned}$$

64. $(5y+1)(3y-10)$

65. $63x^3 + 111x^2 + 36x$

We will use the FOIL method.

1. Factor out the common factor, $3x$.
   $3x(21x^2 + 37x + 12)$
2. Now we will factor the trinomial $21x^2 + 37x + 12$. Factor the first term, $21x^2$. The factors are $21x$, x and $7x$, $3x$. We have these possibilities: $(21x+\quad)(x+\quad)$ and $(7x+\quad)(3x+\quad)$.
3. Factor the last term, 12. The possibilities are $12 \cdot 1$, $(-12)(-1)$, $6 \cdot 2$, $(-6)(-2)$, $4 \cdot 3$, and $(-4)(-3)$ as well as $1 \cdot 12$, $(-1)(-12)$, $2 \cdot 6$, $(-2)(-6)$, $3 \cdot 4$, and $(-3)(-4)$.
4. Look for factors such that the sum of the products is the middle term, $37x$. Trial and error leads us to the correct factorization:
   $(7x+3)(3x+4)$
   We must include the common factor to get a factorization of the original trinomial:
   $$63x^3 + 111x^2 + 36x = 3x(7x+3)(3x+4)$$

66. $5y(5y+4)(2y+3)$

67. $24x^4 + 2x^2 - 15$

We will use the grouping method. Substitute u for x^2 (and u^2 for x^4), and factor $24u^2 + 2u - 15$.

1. There is no common factor (other than 1 or -1).
2. Multiply the leading coefficient and the constant: $24(-15) = -360$
3. Factor -360 so the sum of the factors is 2. The desired factorization is $-18 \cdot 20$.
4. Split $2u$ using the results of step (3):
   $2u = -18u + 20u$
5. Factor by grouping:
$$\begin{aligned} 24u^2 + 2u - 15 &= 24u^2 - 18u + 20u - 15 \\ &= 6u(4u-3) + 5(4u-3) \\ &= (4u-3)(6u+5) \end{aligned}$$
   Replace u by x^2 to obtain the factorization of the original trinomial:
   $$24x^4 + 2x^2 - 15 = (4x^2-3)(6x^2+5)$$

68. $4(5y^2+3)(2y^2-1)$

69. $12a^2 - 17ab + 6b^2$

We will use the FOIL method. (Our thinking is much the same as if we were factoring $12a^2 - 17a + 6$.)

1. There is no common factor (other than 1 or -1).
2. Factor the first term, $12a^2$. The factors are $12a$, a and $6a$, $2a$ and $4a$, $3a$. We have these possibilities: $(12a+\quad)(a+\quad)$ and $(6a+\quad)(2a+\quad)$ and $(4a+\quad)(3a+\quad)$.
3. Factor the last term, $6b^2$. The possibilities are $6b \cdot b$, $(-6b)(-b)$, $3b \cdot 2b$, and $(-3b)(-2b)$ as well as $b \cdot 6b$, $(-b)(-6b)$, $2b \cdot 3b$, and $(-2b)(-3b)$.
4. Look for factors such that the sum of the products is the middle term, $-17ab$. Trial and error leads us to the correct factorization:
   $(4a-3b)(3a-2b)$

70. $(4p-3q)(5p-2q)$

71. $2x^2 + xy - 6y^2$

We will use the grouping method.

1. There is no common factor (other than 1 or -1).
2. Multiply the coefficients of the first and last terms: $2(-6) = -12$
3. Factor -12 so the sum of the factors is 1. The desired factorization is $4(-3)$.
4. Split xy using the results of step (3):
   $xy = 4xy - 3xy$
5. Factor by grouping:
$$\begin{aligned} 2x^2 + xy - 6y^2 &= 2x^2 + 4xy - 3xy - 6y^2 \\ &= 2x(x+2y) - 3y(x+2y) \\ &= (x+2y)(2x-3y) \end{aligned}$$

72. $(4m+3n)(2m-3n)$

73. $6x^2 - 29xy + 28y^2$

We will use the FOIL method.

1. There is no common factor (other than 1 or -1).
2. Factor the first term, $6x^2$. The factors are $6x$, x and $3x$, $2x$. We have these possibilities: $(6x+\quad)(x+\quad)$ and $(3x+\quad)(2x+\quad)$.

3. Factor the last term, $28y^2$. The possibilities are $28y\cdot y$, $(-28y)(-y)$, $14y\cdot 2y$, $(-14y)(-2y)$, $7y\cdot 4y$, and $(-7y)(-4y)$ as well as $y\cdot 28y$, $(-y)(-28y)$, $2y\cdot 14y$, $(-2y)(-14y)$, $4y\cdot 7y$, and $(-4y)(-7y)$.

4. Look for factors such that the sum of the products is the middle term, $-29xy$. Trial and error leads us to the correct factorization: $(3x-4y)(2x-7y)$

74. $(2p+3q)(5p-4q)$

75. $9x^2-30xy+25y^2$

We will use the grouping method.

1. There is no common factor (other than 1 or -1).
2. Multiply the coefficients of the first and last terms: $9(25)=225$
3. Factor 225 so the sum of the factors is -30. The desired factorization is $-15(-15)$.
4. Split $-30xy$ using the results of step (3):
   $-30xy=-15xy-15xy$
5. Factor by grouping:

$$\begin{aligned}9x^2-30xy+25y^2 &= 9x^2-15xy-15xy+25y^2\\ &= 3x(3x-5y)-5y(3x-5y)\\ &= (3x-5y)(3x-5y),\text{ or}\\ &\quad (3x-5y)^2\end{aligned}$$

76. $(2p+3q)(2p+3q)$, or $(2p+3q)^2$

77. $9x^2y^2+5xy-4$

Let $u=xy$ and $u^2=x^2y^2$. Factor $9u^2+5u-4$. We will use the FOIL method.

1. There is no common factor (other than 1 or -1).
2. Factor the first term, $9u^2$. The factors are $9u$, u and $3u$, $3u$. We have these possibilities: $(9u+\quad)(u+\quad)$ and $(3u+\quad)(3u+\quad)$.
3. Factor the last term, -4. The possibilities are $-4\cdot 1$, $-2\cdot 2$, and $-1\cdot 4$.
4. We need factors for which the sum of the products is the middle term, $5u$. Trial and error leads us to the factorization: $(9u-4)(u+1)$. Replace u by xy. We have $9x^2y^2+5xy-4=(9xy-4)(xy+1)$.

78. $(7ab+6)(ab+1)$

79.

$$\begin{aligned}h(t) &= -16t^2+80t+224\\ h(0) &= -16(0)^2+80(0)+224=224\text{ ft}\\ h(1) &= -16(1)^2+80(1)+224=288\text{ ft}\\ h(3) &= -16(3)^2+80(3)+224=320\text{ ft}\\ h(4) &= -16(4)^2+80(4)+224=288\text{ ft}\\ h(6) &= -16(6)^2+80(6)+224=128\text{ ft}\end{aligned}$$

80.

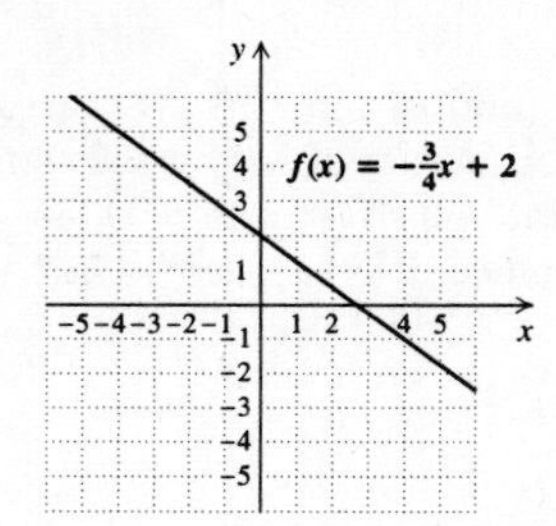

81.

$$\begin{aligned}g(x) &= -5x^2-7x\\ g(-3) &= -5(-3)^2-7(-3)=-5\cdot 9+21=\\ &\quad -45+21=-24\end{aligned}$$

82. 880 ft; 960 ft; 1024 ft; 624 ft; 240 ft

83. ◈

84. ◈

85. ◈

86. ◈

87. $2a^4b^6-3a^2b^3-20$

Let $u=a^2b^3$ (and $u^2=a^4b^6$). Factor $2u^2-3u-20$. We will use the FOIL method.

1. There is no common factor (other than 1 or -1).
2. Factor the first term, $2u^2$. The factors are $2u$, u. The possibility is $(2u+\quad)(u+\quad)$.
3. Factor the last term, -20. The possibilities are $-20\cdot 1$, $-10\cdot 2$, $-5\cdot 4$, $-4\cdot 5$, $-2\cdot 10$, and $-1\cdot 20$.
4. We need factors for which the sum of the products is the middle term, $-3u$. Trial and error leads us to the factorization: $(2u+5)(u-4)$. Replace u by a^2b^3. We have $(2a^2b^3+5)(a^2b^3-4)$.

88. $5(x^4y^3+4)(x^4y^3+3)$

89. $x^2-\frac{4}{25}+\frac{3}{5}x=x^2+\frac{3}{5}x-\frac{4}{25}$

We look for factors of $-\frac{4}{25}$ whose sum is $\frac{3}{5}$. The factors are $\frac{4}{5}$ and $-\frac{1}{5}$. The factorization is $\left(x+\frac{4}{5}\right)\left(x-\frac{1}{5}\right)$.

90. $\left(y+\frac{4}{7}\right)\left(y-\frac{2}{7}\right)$

91. $y^2+0.4y-0.05$

We look for factors of -0.05 whose sum is 0.4. The factors are -0.1 and 0.5. The factorization is $(y-0.1)(y+0.5)$.

92. $(2x^a+1)(2x^a-3)$

93. $x^{2a} + 5x^a - 24$

Substitute u for x^a (and u^2 for x^{2a}). We factor $u^2 + 5u - 24$. We look for factors of -24 whose sum is 5. The factors are 8 and -3. We have $u^2 + 5u - 24 = (u+8)(u-3)$. Replace u by x^a: $x^{2a} + 5x^a - 24 = (x^a + 8)(x^a - 3)$.

94. $(x+a)(x+b)$

95. $bdx^2 + adx + bcx + ac$

$= dx(bx+a) + c(bx+a)$ Factoring by grouping

$= (bx+a)(dx+c)$

96. $a(2r+s)(r+s)$

97. $a^2p^{2a} + a^2p^a - 2a^2 = a^2(p^{2a} + p^a - 2)$

Substitute u for p^a (and u^2 for p^{2a}). We factor $u^2 + u - 2$. Look for factors of -2 whose sum is 1. The factors are 2 and -1. We have $u^2 + u - 2 = (u+2)(u-1)$. Replace u by p^a: $p^{2a} + p^a - 2 = (p^a+2)(p^a-1)$. We must include the common factor a^2 to get a factorization of the original trinomial:

$$a^2p^{2a} + a^2p^a - 2a^2 = a^2(p^a+2)(p^a-1)$$

98. $(x-4)(x+8)$

99. $6(x-7)^2 + 13(x-7) - 5$

Let $u = x - 7$, make the substitutions and factor using the grouping method. We have $6u^2 + 13u - 5$. Multiply the leading coefficient and the constant: $6(-5) = -30$. Factor -30 so the sum of the factors is 13. The desired factorization is $15(-2)$.

Split the middle term and factor by grouping.

$$6u^2 + 13u - 5 = 6u^2 + 15u - 2u - 5$$
$$= 3u(2u+5) - (2u+5)$$
$$= (2u+5)(3u-1)$$

Replace u by $x - 7$. The factorization is $[2(x-7)+5][3(x-7)-1]$, or $(2x-9)(3x-22)$.

100. 76, -76, 28, -28, 20, -20

101. $x^2 + qx - 32$

All such q are the sums of the factors of -32.

Pair of Factors	Sum of Factors
32, −1	31
−32, 1	−31
16, −2	14
−16, 2	−14
8, −4	4
−8, 4	−4

q can be 31, -31, 14, -14, 4, or -4.

102. $(x - 365)$

103. See the answer section in the text.

104.

105.

106.

Exercise Set 5.5

1. $x^2 + 8x + 16 = (x+4)^2$

Find the square terms and write the quantities that were squared with a plus sign between them.

2. $(t+3)^2$

3. $a^2 - 16a + 64 = (a-8)^2$

Find the square terms and write the quantities that were squared with a minus sign between them.

4. $(a-7)^2$

5. $2a^2 + 8a + 8$

$= 2(a^2 + 4a + 4)$ Factoring out the common factor

$= 2(a+2)^2$ Factoring the perfect-square trinomial

6. $4(a-2)^2$

7. $y^2 + 36 - 12y$

$= y^2 - 12y + 36$ Changing order

$= (y-6)^2$ Factoring the perfect-square trinomial

8. $(y+6)^2$

9. $24a^2 + a^3 + 144a$

$= a^3 + 24a^2 + 144a$ Changing order

$= a(a^2 + 24a + 144)$ Factoring out the common factor

$= a(a+12)^2$ Factoring the perfect-square trinomial

10. $y(y-9)^2$

11. $32x^2 + 48x + 18$

$= 2(16x^2 + 24x + 9)$ Factoring out the common factor

$= 2(4x+3)^2$ Factoring the perfect-square trinomial

12. $2(x-10)^2$

13. $64 + 25y^2 - 80y$

$= 25y^2 - 80y + 64$ Changing order

$= (5y-8)^2$ Factoring the perfect-square trinomial

14. $(1-4d)^2$

15. $a^4 - 10a^2 + 25$
$= (a^2-5)^2$ Note that $a^4 = (a^2)^2$.

16. $(y^2+4)^2$

17. $0.25x^2 + 0.30x + 0.09 = (0.5x+0.3)^2$
Find the square terms and write the quantities that were squared with a plus sign between them.

18. $(0.2x-0.7)^2$

19. $p^2 - 2pq + q^2 = (p-q)^2$

20. $(m+n)^2$

21. $25a^2 - 30ab + 9b^2 = (5a-3b)^2$

22. $(7p-6q)^2$

23. $t^8 + 2t^4s^4 + s^8 = (t^4+s^4)^2$

24. $(a^2+b^2)^2$

25. $x^2 - 16 = x^2 - 4^2 = (x+4)(x-4)$

26. $(y+10)(y-10)$

27. $p^2 - 49 = p^2 - 7^2 = (p+7)(p-7)$

28. $(m+8)(m-8)$

29. $a^2b^2 - 81 = (ab)^2 - 9^2 = (ab+9)(ab-9)$

30. $(pq+5)(pq-5)$

31. $6x^2 - 6y^2$
$= 6(x^2-y^2)$ Factoring out the common factor
$= 6(x+y)(x-y)$ Factoring the difference of squares

32. $8(x+y)(x-y)$

33. $7xy^4 - 7xz^4$
$= 7x(y^4-z^4)$
$= 7x[(y^2)^2 - (z^2)^2]$
$= 7x(y^2+z^2)(y^2-z^2)$
$= 7x(y^2+z^2)(y+z)(y-z)$

34. $25a(b^2+z^2)(b+z)(b-z)$

35. $4a^3 - 49a = a(4a^2-49)$
$= a[(2a)^2 - 7^2]$
$= a(2a+7)(2a-7)$

36. $x^2(3x+5)(3x-5)$

37. $3x^8 - 3y^8$
$= 3(x^8-y^8)$
$= 3[(x^4)^2 - (y^4)^2]$
$= 3(x^4+y^4)(x^4-y^4)$
$= 3(x^4+y^4)[(x^2)^2 - (y^2)^2]$
$= 3(x^4+y^4)(x^2+y^2)(x^2-y^2)$
$= 3(x^4+y^4)(x^2+y^2)(x+y)(x-y)$

38. $a^2(3a+b)(3a-b)$

39. $9a^4 - 25a^2b^4 = a^2(9a^2 - 25b^4)$
$= a^2[(3a)^2 - (5b^2)^2]$
$= a^2(3a+5b^2)(3a-5b^2)$

40. $x^2(4x^2+11y^2)(4x^2-11y^2)$

41. $\frac{1}{25} - x^2 = \left(\frac{1}{5}\right)^2 - x^2$
$= \left(\frac{1}{5}+x\right)\left(\frac{1}{5}-x\right)$

42. $\left(\frac{1}{4}+y\right)\left(\frac{1}{4}-y\right)$

43. $(a+b)^2 - 9 = (a+b)^2 - 3^2$
$= [(a+b)+3][(a+b)-3]$
$= (a+b+3)(a+b-3)$

44. $(p+q+5)(p+q-5)$

45. $x^2 - 6x + 9 - y^2$
$= (x^2-6x+9) - y^2$ Grouping as a difference of squares
$= (x-3)^2 - y^2$
$= (x-3+y)(x-3-y)$

46. $(a-4+b)(a-4-b)$

47. $m^2 - 2mn + n^2 - 25$
$= (m^2-2mn+n^2) - 25$ Grouping as a difference of squares
$= (m-n)^2 - 5^2$
$= (m-n+5)(m-n-5)$

48. $(x+y+3)(x+y-3)$

49. $36 - (x+y)^2 = 6^2 - (x+y)^2$
$= [6+(x+y)][6-(x+y)]$
$= (6+x+y)(6-x-y)$

50. $(7+a+b)(7-a-b)$

51. $r^2 - 2r + 1 - 4s^2$
$= (r^2-2r+1) - 4s^2$ Grouping as a difference of squares
$= (r-1)^2 - (2s)^2$
$= (r-1+2s)(r-1-2s)$

52. $(c+2d+3p)(c+2d-3p)$

53. $16-a^2+2ab-b^2$

$= 16-(a^2-2ab+b^2)$ Grouping as a difference of squares

$= 4^2-(a-b)^2$

$= [4+(a-b)][4-(a-b)]$

$= (4+a-b)(4-a+b)$

54. $(3+x-y)(3-x+y)$

55. $m^3-7m^2-4m+28$

$= m^2(m-7)-4(m-7)$ Factoring by grouping

$= (m-7)(m^2-4)$

$= (m-7)(m+2)(m-2)$ Factoring the difference of squares

56. $(x+8)(x+1)(x-1)$

57. $a^3-ab^2-2a^2+2b^2$

$= a(a^2-b^2)-2(a^2-b^2)$ Factoring by grouping

$= (a^2-b^2)(a-2)$

$= (a+b)(a-b)(a-2)$ Factoring the difference of squares

58. $(p+5)(p-5)(q+3)$

59. $x-y+z=6,\ (1)$

$2x+y-z=0,\ (2)$

$x+2y+z=3\ \ (3)$

Add (1) and (2).

$$\begin{array}{rl} x-y+z=6 & (1)\\ 2x+y-z=0 & (2)\\ \hline 3x\qquad\quad =6 & \text{Adding}\\ x=2 & \end{array}$$

Add (2) and (3).

$$\begin{array}{rl} 2x+y-z=0 & (2)\\ x+2y+z=3 & (3)\\ \hline 3x+3y\quad =3 & (4) \end{array}$$

Substitute 2 for x in (4).

$3(2)+3y=3$

$6+3y=3$

$3y=-3$

$y=-1$

Substitute 2 for x and -1 for y in (1).

$2-(-1)+z=6$

$3+z=6$

$z=3$

The solution is $(2,-1,3)$.

60. $\left\{x \middle| x \le -\frac{4}{7} \text{ or } x \ge 2\right\}$, or $\left(-\infty, -\frac{4}{7}\right] \cup [2,\infty)$

61. $|5-7x| \le 9$

$-9 \le 5-7x \le 9$

$-14 \le -7x \le 4$

$2 \ge x \ge -\frac{4}{7}$

The solution set is $\left\{x \middle| -\frac{4}{7} \le x \le 2\right\}$, or $\left[-\frac{4}{7}, 2\right]$.

62. $\left\{x \middle| x < \frac{14}{19}\right\}$, or $\left(-\infty, \frac{14}{19}\right)$

63. ◈

64. ◈

65. ◈

66. ◈

67. $-\frac{3}{4}p^2+\frac{6}{5}p-\frac{12}{25} = -3\left(\frac{1}{4}p^2-\frac{2}{5}p+\frac{4}{25}\right)$

$= -3\left(\frac{1}{2}p-\frac{2}{5}\right)^2$

68. $-\frac{1}{54}(4r+3s)^2$

69. $\frac{1}{36}x^8+\frac{2}{9}x^4+\frac{4}{9} = \left(\frac{1}{6}x^4+\frac{2}{3}\right)^2$

70. $(0.3x^4+0.8)^2$, or $\frac{1}{100}(3x^4+8)^2$

71. $a^2+2ab+b^2-c^2+6c-9$

$= (a^2+2ab+b^2)-(c^2-6c+9)$

$= (a+b)^2-(c-3)^2$

$= [(a+b)+(c-3)][(a+b)-(c-3)]$

$= (a+b+c-3)(a+b-c+3)$

72. $(r+s+1)(r-s-9)$

73. $x^{2a}-y^2 = (x^a)^2-y^2 = (x^a+y)(x^a-y)$

74. $(x^{2a}+y^b)(x^{2a}-y^b)$

75. $4y^{4a}+20y^{2a}+20y^{2a}+100$

$= 4y^{4a}+40y^{2a}+100$ Collecting like terms

$= 4(y^{4a}+10y^{2a}+25)$ Note $y^{4a}=(y^{2a})^2$.

$= 4(y^{2a}+5)^2$

76. $(5y^a+x^b-1)(5y^a-x^b+1)$

77. $8(a-3)^2-64(a-3)+128$

$= 8[(a-3)^2-8(a-3)+16]$

$= 8[(a-3)-4]^2$

$= 8(a-7)^2$

78. $3(x+3)^2$

79. $$\begin{aligned}&5c^{100} - 80d^{100}\\ &= 5(c^{100} - 16d^{100})\\ &= 5(c^{50} + 4d^{50})(c^{50} - 4d^{50})\\ &= 5(c^{50} + 4d^{50})(c^{25} + 2d^{25})(c^{25} - 2d^{25})\end{aligned}$$

80. $(3x^n - 1)^2$

81. $$\begin{aligned}c^{2w+1} + 2c^{w+1} + c &= c(c^{2w} + 2c^w + 1)\\ &= c(c^w + 1)^2\end{aligned}$$

82. $h(2a+h)$

83. If $P(x) = x^4$, then
$$\begin{aligned}&P(a+h) - P(a)\\ &= (a+h)^4 - a^4\\ &= [(a+h)^2 + a^2][(a+h)^2 - a^2]\\ &= [(a+h)^2 + a^2][(a+h) + a][(a+h) - a]\\ &= (a^2 + 2ah + h^2 + a^2)(2a+h)(h)\\ &= h(2a+h)(2a^2 + 2ah + h^2)\end{aligned}$$

84. (a) $\pi h(R+r)(R-r)$; (b) $3,014,400\text{ cm}^3$

85.

86.

Exercise Set 5.6

1. $$\begin{aligned}t^3 - 8 &= t^3 - 2^3\\ &= (t-2)(t^2 + 2t + 4)\\ A^3 - B^3 &= (A-B)(A^2 + AB + B^2)\end{aligned}$$

2. $(x+4)(x^2 - 4x + 16)$

3. $$\begin{aligned}x^3 + 27 &= x^3 + 3^3\\ &= (x+3)(x^2 - 3x + 9)\\ A^3 + B^3 &= (A+B)(A^2 - AB + B^2)\end{aligned}$$

4. $(z-1)(z^2 + z + 1)$

5. $$\begin{aligned}m^3 - 64 &= m^3 - 4^3\\ &= (m-4)(m^2 + 4m + 16)\\ A^3 - B^3 &= (A-B)(A^2 + AB + B^2)\end{aligned}$$

6. $(x-3)(x^2 + 3x + 9)$

7. $$\begin{aligned}8a^3 + 1 &= (2a)^3 + 1^3\\ &= (2a+1)(4a^2 - 2a + 1)\\ A^3 + B^3 &= (A+B)(A^2 - AB + B^2)\end{aligned}$$

8. $(3x+1)(9x^2 - 3x + 1)$

9. $$\begin{aligned}8 - 27b^3 &= 2^3 - (3b)^3\\ &= (2-3b)(4 + 6b + 9b^2)\end{aligned}$$

10. $(4-5x)(16 + 20x + 25x^2)$

11. $$\begin{aligned}8x^3 + 27 &= (2x)^3 + 3^3\\ &= (2x+3)(4x^2 - 6x + 9)\end{aligned}$$

12. $(3y+4)(9y^2 - 12y + 16)$

13. $y^3 - z^3 = (y-z)(y^2 + yz + z^2)$

14. $(x-y)(x^2 + xy + y^2)$

15. $$\begin{aligned}x^3 + \frac{1}{27} &= x^3 + \left(\frac{1}{3}\right)^3\\ &= \left(x + \frac{1}{3}\right)\left(x^2 - \frac{1}{3}x + \frac{1}{9}\right)\end{aligned}$$

16. $\left(a + \frac{1}{2}\right)\left(a^2 - \frac{1}{2}a + \frac{1}{4}\right)$

17. $$\begin{aligned}2y^3 - 128 &= 2(y^3 - 64)\\ &= 2(y^3 - 4^3)\\ &= 2(y-4)(y^2 + 4y + 16)\end{aligned}$$

18. $8(t-1)(t^2 + t + 1)$

19. $$\begin{aligned}8a^3 + 1000 &= 8(a^3 + 125)\\ &= 8(a^3 + 5^3)\\ &= 8(a+5)(a^2 - 5a + 25)\end{aligned}$$

20. $2(3x+1)(9x^2 - 3x + 1)$

21. $$\begin{aligned}rs^3 + 64r &= r(s^3 + 64)\\ &= r(s^3 + 4^3)\\ &= r(s+4)(s^2 - 4s + 16)\end{aligned}$$

22. $a(b+5)(b^2 - 5b + 25)$

23. $$\begin{aligned}2y^3 - 54z^3 &= 2(y^3 - 27z^3)\\ &= 2[y^3 - (3z)^3]\\ &= 2(y - 3z)(y^2 + 3yz + 9z^2)\end{aligned}$$

24. $5(x-2z)(x^2 + 2xz + 4z^2)$

25. $$\begin{aligned}y^3 + 0.125 &= y^3 + (0.5)^3\\ &= (y + 0.5)(y^2 - 0.5y + 0.25)\end{aligned}$$

26. $(x + 0.1)(x^2 - 0.1x + 0.01)$

27. $$\begin{aligned}125c^6 - 8d^6 &= (5c^2)^3 - (2d^2)^3\\ &= (5c^2 - 2d^2)(25c^4 + 10c^2d^2 + 4d^4)\end{aligned}$$

28. $8(2x^2 - t^2)(4x^4 + 2x^2t^2 + t^4)$

29. $$\begin{aligned}3z^5 - 3z^2 &= 3z^2(z^3 - 1)\\ &= 3z^2(z^3 - 1^3)\\ &= 3z^2(z-1)(z^2 + z + 1)\end{aligned}$$

30. $2y(y-4)(y^2 + 4y + 16)$

31. $t^6 + 1 = (t^2)^3 + 1^3$
$= (t^2 + 1)(t^4 - t^2 + 1)$

32. $(z + 1)(z^2 - z + 1)(z - 1)(z^2 + z + 1)$

33. $p^6 - q^6$
$= (p^3)^2 - (q^3)^2$ Writing as a difference of squares
$= (p^3 + q^3)(p^3 - q^3)$ Factoring a difference of squares
$= (p + q)(p^2 - pq + q^2)(p - q)(p^2 + pq + q^2)$ Factoring a sum and a difference of cubes

34. $(t^2 + 4y^2)(t^4 - 4t^2y^2 + 16y^4)$

35. $a^9 + b^{12}c^{15}$
$= (a^3)^3 + (b^4c^5)^3$
$= (a^3 + b^4c^5)(a^6 - a^3b^4c^5 + b^8c^{10})$

36. $(x^4 - yz^4)(x^8 + x^4yx^4 + y^2z^8)$

37. ***Familiarize***. Let w represent the width and l represent the length of the rectangle. If the width is increased by 2 ft, the new width is $w + 2$. Also, recall the formulas for the perimeter and area of a rectangle:

$P = 2l + 2w$
$A = lw$

Translate.

Width is length less 7 feet.
$w = l - 7$

When the length is l and the width is $w + 2$, the perimeter is 66 ft, so we have

$66 = 2l + 2(w + 2)$.

We have a system of equations:

$w = l - 7$,
$66 = 2l + 2(w + 2)$

Carry out. Solving the system we get $(19, 12)$.

Check. The width, 12 ft, is 7 ft less than the length, 19 ft. When the width is increased by 2 ft it becomes 14 ft, and the perimeter of the rectangle is $2 \cdot 19 + 2 \cdot 14 = 38 + 28 = 66$ ft. The numbers check.

State. The area of the original rectangle is 19 ft $\cdot$ 12 ft, or 228 ft^2.

38. -2

39. Write the equation in slope-intercept form, $y = mx + b$ where m is the slope and b is the y-intercept.

$4x - 3y = 8$
$-3y = -4x + 8$
$y = \frac{4}{3}x - \frac{8}{3}$

The slope is $\frac{4}{3}$, and the y-intercept is $\left(0, -\frac{8}{3}\right)$.

40. $\{27, -27\}$

41. $|5x - 6| \leq 39$
$-39 \leq 5x - 6 \leq 39$
$-33 \leq 5x \leq 45$
$-\frac{33}{5} \leq x \leq 9$

The solution set is $\left\{x \middle| -\frac{33}{5} \leq x \leq 9\right\}$, or $\left[-\frac{33}{5}, 9\right]$.

42. $\left\{x \middle| x < -\frac{33}{5} \text{ or } x > 9\right\}$, or $\left(-\infty, -\frac{33}{5}\right) \cup (9, \infty)$

43. ◈

44. ◈

45. ◈

46. $(x^{2a} - y^b)(x^{4a} + x^{2a}y^b + y^{2b})$

47. $2x^{3a} + 16y^{3b} = 2(x^{3a} + 8y^{3b})$
$= 2[(x^a)^3 + (2y^b)^3]$
$= 2(x^a + 2y^b)(x^{2a} - 2x^ay^b + 4y^{2b})$

48. $(x + y)(x^2 + 15x - xy - 15y + y^2 + 75)$

49. $\frac{1}{16}x^{3a} + \frac{1}{2}y^{6a}z^{9b}$
$= \frac{1}{2}\left(\frac{1}{8}x^{3a} + y^{6a}z^{9b}\right)$
$= \frac{1}{2}\left[\left(\frac{1}{2}x^a\right)^3 + (y^{2a}z^{3b})^3\right]$
$= \frac{1}{2}\left(\frac{1}{2}x^a + y^{2a}z^{3b}\right)\left(\frac{1}{4}x^{2a} - \frac{1}{2}x^ay^{2a}z^{3b} + y^{4a}z^{6b}\right)$

This result can also be expressed as $\frac{1}{16}(x^a + 2y^{2a}z^{3b})(x^{2a} - 2x^ay^{2a}z^{3b} + 4y^{4a}z^{6b})$.

50. $5\left(xy^2 - \frac{1}{2}\right)\left(x^2y^4 + \frac{1}{2}xy^2 + \frac{1}{4}\right)$

51. $x^3 - (x + y)^3$
$= [x - (x + y)][x^2 + x(x + y) + (x + y)^2]$
$= (x - x - y)(x^2 + x^2 + xy + x^2 + 2xy + y^2)$
$= -y(3x^2 + 3xy + y^2)$

52. $-(3x^{4a} + 3x^{2a} + 1)$, or $-3x^{4a} - 3x^{2a} - 1$

53. $(x^{2a} - 1)^3 - x^{6a}$
$= (x^{2a} - 1)^3 - (x^{2a})^3$
$= (x^{2a} - 1 - x^{2a})[(x^{2a} - 1)^2 + (x^{2a} - 1)(x^{2a}) + (x^{2a})^2]$
$= (-1)(x^{4a} - 2x^{2a} + 1 + x^{4a} - x^{2a} + x^{4a})$
$= -(3x^{4a} - 3x^{2a} + 1)$
$= -3x^{4a} + 3x^{2a} - 1$

54. $(t-8)(t-1)(t^2+t+1)$

55. If $P(x) = x^3$, then

$$\begin{aligned}&P(a+h) - P(a)\\ &= (a+h)^3 - a^3\\ &= [(a+h) - a][(a+h)^2 + (a+h)(a) + a^2]\\ &= (a+h-a)(a^2+2ah+h^2+a^2+ah+a^2)\\ &= h(3a^2+3ah+h^2)\end{aligned}$$

56. $h(2a+h)(a^2+ah+h^2)(3a^2+3ah+h^2)$

57.

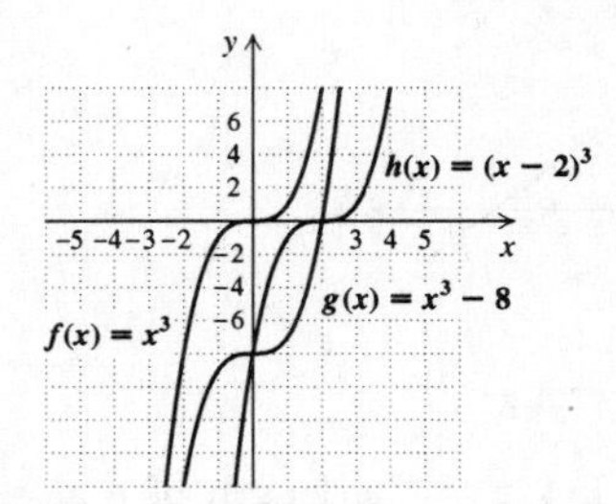

58.

59.

60.

Exercise Set 5.7

1.
$$\begin{aligned}&5m^4 - 20\\ &= 5(m^4 - 4)\\ &= 5[(m^2)^2 - 2^2] && \text{Difference of squares}\\ &= 5(m^2+2)(m^2-2)\end{aligned}$$

2. $(x+12)(x-12)$

3.
$$\begin{aligned}&a^2 - 81\\ &= a^2 - 9^2 && \text{Difference of squares}\\ &= (a+9)(a-9)\end{aligned}$$

4. $(2a-3)(a-4)$

5.
$$\begin{aligned}&8x^2 - 18x - 5\\ &= (4x+1)(2x-5) && \text{FOIL or grouping method}\end{aligned}$$

6. $2x(y+5)(y-5)$

7.
$$\begin{aligned}&a^2 + 25 + 10a\\ &= a^2 + 10a + 25 && \text{Perfect-square trinomial}\\ &= (a+5)^2\end{aligned}$$

8. $(p+8)^2$

9.
$$\begin{aligned}&3x^2 + 15x - 252\\ &= 3(x^2 + 5x - 84)\\ &= 3(x+12)(x-7) && \text{FOIL or grouping method}\end{aligned}$$

10. $2(y+11)(y-6)$

11.
$$\begin{aligned}&9x^2 - 25y^2\\ &= (3x)^2 - (5y)^2 && \text{Difference of squares}\\ &= (3x+5y)(3x-5y)\end{aligned}$$

12. $(4a+9b)(4a-9b)$

13.
$$\begin{aligned}&m^6 - 1\\ &= (m^3)^2 - 1^2 && \text{Difference of squares}\\ &= (m^3+1)(m^3-1) && \text{Sum and difference of cubes}\\ &= (m+1)(m^2-m+1)(m-1)(m^2+m+1)\end{aligned}$$

14. $(2t+1)(4t^2-2t+1)(2t-1)(4t^2+2t+1)$

15.
$$\begin{aligned}&x^2 + 6x - y^2 + 9\\ &= x^2 + 6x + 9 - y^2\\ &= (x+3)^2 - y^2 && \text{Difference of squares}\\ &= [(x+3)+y][(x+3)-y]\\ &= (x+y+3)(x-y+3)\end{aligned}$$

16. $(t+5+p)(t+5-p)$

17.
$$\begin{aligned}&343x^3 + 27y^3\\ &= (7x)^3 + (3y)^3 && \text{Sum of cubes}\\ &= (7x+3y)(49x^2 - 21xy + 9y^2)\end{aligned}$$

18. $2(4a+5b)(16a^2-20ab+25b^2)$

19.
$$\begin{aligned}&8m^3 + m^6 - 20\\ &= (m^3)^2 + 8m^3 - 20\\ &= (m^3-2)(m^3+10) && \text{Trial and error}\end{aligned}$$

20. $(x+6)(x-6)(x+1)(x-1)$

21.
$$\begin{aligned}&ac + cd - ab - bd\\ &= c(a+d) - b(a+d) && \text{Factoring by grouping}\\ &= (a+d)(c-b)\end{aligned}$$

22. $(x-y)(w+z)$

23.
$$\begin{aligned}&4c^2 - 4cd + d^2 && \text{Perfect-square trinomial}\\ &= (2c-d)^2\end{aligned}$$

24. $(10b+a)(7b-a)$

25.
$$\begin{aligned}&24 + 9t^2 + 8t + 3t^3\\ &= 3(8+3t^2) + t(8+3t^2) && \text{Factoring by grouping}\\ &= (8+3t^2)(3+t)\end{aligned}$$

26. $(2a-7)(2+a^2)$

27.
$$\begin{aligned}&2x^3 + 6x^2 - 8x - 24\\ &= 2(x^3 + 3x^2 - 4x - 12)\\ &= 2[x^2(x+3) - 4(x+3)] && \text{Factoring by grouping}\\ &= 2(x+3)(x^2-4) && \text{Difference of squares}\\ &= 2(x+3)(x+2)(x-2)\end{aligned}$$

28. $3(x+2)(x+3)(x-3)$

29.
$$\begin{aligned} &54a^3-16b^3 \\ &= 2(27a^3-8b^3) \\ &= 2[(3a)^3-(2b)^3] \quad \text{Difference of cubes} \\ &= 2(3a-2b)(9a^2+6ab+4b^2) \end{aligned}$$

30. $2(3x-5y)(9x^2+15xy+25y^2)$

31.
$$\begin{aligned} &36y^2-35+12y \\ &= 36y^2+12y-35 \\ &= (6y-5)(6y+7) \quad \text{FOIL or grouping method} \end{aligned}$$

32. $2b(1+7a)(1-2a)$, or $-2b(7a+1)(2a-1)$

33.
$$\begin{aligned} &a^8-b^8 \quad \text{Difference of squares} \\ &= (a^4+b^4)(a^4-b^4) \quad \text{Difference of squares} \\ &= (a^4+b^4)(a^2+b^2)(a^2-b^2) \quad \text{Difference of squares} \\ &= (a^4+b^4)(a^2+b^2)(a+b)(a-b) \end{aligned}$$

34. $2(x^2+4)(x+2)(x-2)$

35.
$$\begin{aligned} &a^3b-16ab^3 \\ &= ab(a^2-16b^2) \quad \text{Difference of squares} \\ &= ab(a+4b)(a-4b) \end{aligned}$$

36. $xy(x+5y)(x-5y)$

37.
$$\begin{aligned} &(a-3)(a+7)+(a-3)(a-1) \\ &= (a-3)(a+7+a-1) \\ &= (a-3)(2a+6) \\ &= (a-3)(2)(a+3) \\ &= 2(a-3)(a+3) \end{aligned}$$

38. $(x+3)(x+2)(x-2)$

39.
$$\begin{aligned} &7a^4-14a^3+21a^2-7a \\ &= 7a(a^3-2a^2+3a-1) \quad \text{Removing a common factor} \end{aligned}$$

40. $(a+b)^2(a-b)$

41.
$$\begin{aligned} &42ab+27a^2b^2+8 \\ &= 27a^2b^2+42ab+8 \\ &= (9ab+2)(3ab+4) \quad \text{FOIL or grouping method} \end{aligned}$$

42. $(4xy-3)(5xy-2)$

43.
$$\begin{aligned} &p+64p^4 \\ &= p(1+64p^3) \quad \text{Sum of cubes} \\ &= p(1+4p)(1-4p+16p^2) \end{aligned}$$

44. $a(5+2a)(25-10a+4a^2)$

45.
$$\begin{aligned} &a^2-b^2-6b-9 \\ &= a^2-(b^2+6b+9) \quad \text{Factoring out } -1 \\ &= a^2-(b+3)^2 \quad \text{Difference of squares} \\ &= [a+(b+3)][a-(b+3)] \\ &= (a+b+3)(a-b-3) \end{aligned}$$

46. $(m+n+4)(m-n-4)$

47.

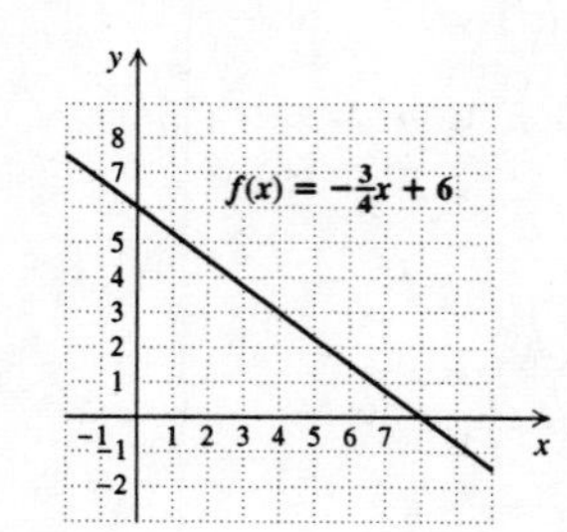

48. 55 correct, 20 wrong

49. ***Familiarize.*** Let x = the length of a side of the pentagon and y = the length of a side of the octagon. Then the perimeter of the pentagon is $5x$, and the perimeter of the octagon is $8y$.

Translate. We write two equations.

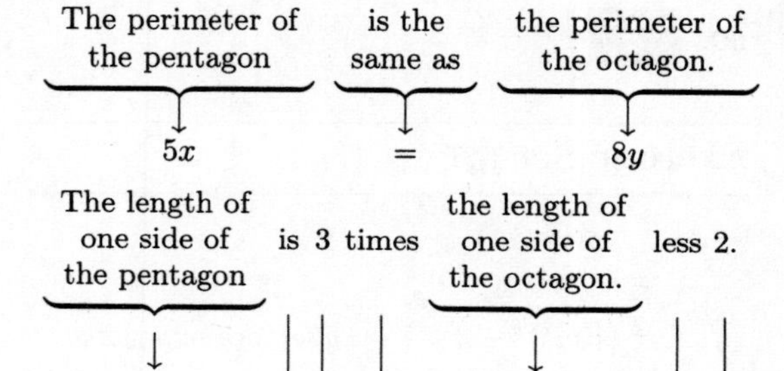

We have a system of equations.

$$5x = 8y,$$
$$x = 3y - 2$$

Carry out. Solving the system, we get $\left(\frac{16}{7}, \frac{10}{7}\right)$.

Check. The perimeter of the pentagon is $5 \cdot \frac{16}{7}$, or $\frac{80}{7}$. The perimeter of the octagon is $8 \cdot \frac{10}{7}$, or $\frac{80}{7}$. Thus, the perimeters are the same. Two less than 3 times the length of one side of the octagon is $3 \cdot \frac{10}{7} - 2 = \frac{30}{7} - \frac{14}{7} = \frac{16}{7}$, the length of one side of the pentagon. The numbers check.

State. The perimeters are $\frac{80}{7}$.

50. $\{x|4 < x < 10\}$, or $(4, 10)$

51. ◈

52. ◈

53. ◈

54. ◈

55. $60x^2 + 97xy^2 + 30y^4$
$= (12x + 5y^2)(5x + 6y^2)$ FOIL or grouping method

56. $a(7a + bc)(4a + 3bc)$

57. $-16 + 17(5 - y^2) - (5 - y^2)^2$
Substitute u for $5 - y^2$ (and u^2 for $(5 - y^2)^2$).
$-16 + 17u - u^2$
$= -(16 - 17u + u^2)$
$= -(16 - u)(1 - u)$
Now replace u by $5 - y^2$.
$-[16 - (5 - y^2)][1 - (5 - y^2)]$
$= -(16 - 5 + y^2)(1 - 5 + y^2)$
$= -(11 + y^2)(-4 + y^2)$
$= (11 + y^2)(-1)(-4 + y^2)$
$= (11 + y^2)(4 - y^2)$
$= (11 + y^2)(2 + y)(2 - y)$

58. $x(x - 2p)$

59. $a^4 - 50a^2b^2 + 49b^4$
$= (a^2 - b^2)(a^2 - 49b^2)$
$= (a + b)(a - b)(a + 7b)(a - 7b)$

60. $y(y - 1)^2(y - 2)$

61. $27x^{6s} + 64y^{3t}$
$= (3x^{2s})^3 + (4y^t)^3$ Sum of cubes
$= (3x^{2s} + 4y^t)(9x^{4s} - 12x^{2s}y^t + 16y^{2t})$

62. $(x - 1)^3(x^2 + 1)(x + 1)$

63. $4x^2 + 4xy + y^2 - r^2 + 6rs - 9s^2$
$= (4x^2 + 4xy + y^2) - (r^2 - 6rs + 9s^2)$ Grouping
$= (2x + y)^2 - (r - 3s)^2$ Difference of squares
$= [(2x + y) + (r - 3s)][(2x + y) - (r - 3s)]$
$= (2x + y + r - 3s)(2x + y - r + 3s)$

64. $-x(x - 1)^3(x^2 - 3x + 3)$, or $x(1 - x)^3(x^2 - 3x + 3)$

65. $24t^{2a} - 6$
$= 6(4t^{2a} - 1)$
$= 6[(2t^a)^2 - 1^2]$
$= 6(2t^a + 1)(2t^a - 1)$

66. $a(a^w + 1)^2$

67. $\dfrac{x^{27}}{1000} - 1$
$= \left(\dfrac{x^9}{10}\right)^3 - 1^3$
$= \left(\dfrac{x^9}{10} - 1\right)\left(\dfrac{x^{18}}{100} + \dfrac{x^9}{10} + 1\right)$

68. $(b + a)(1 + y^4)(1 + y^2)(1 + y)(1 - y)$

69. $3(x + 1)^2 - 9(x + 1) - 12$
Substitute u for $x + 1$ (and u^2 for $(x + 1)^2$.)
$3u^2 - 9u - 12$
$= 3(u^2 - 3u - 4)$
$= 3(u - 4)(u + 1)$
Now replace u by $x + 1$.
$3(x + 1 - 4)(x + 1 + 1)$
$= 3(x - 3)(x + 2)$

70. $3(a + b + c + d)(a + b - c - d)$

71. $3(a + 2)^2 + 30(a + 2) + 75$
Substitute u for $a + 2$ (and u^2 for $(a + 2)^2$.)
$3u^2 + 30u + 75$
$= 3(u^2 + 10u + 25)$
$= 3(u + 5)^2$
Now replace u by $a + 2$.
$3(a + 2 + 5)^2 = 3(a + 7)^2$

72. $-2(3m^2 + 1)$

73. $\left(x + \dfrac{2}{x}\right)^2 = 6$
$x^2 + 4 + \dfrac{4}{x^2} = 6$
$x^2 - 2 + \dfrac{4}{x^2} = 0$
Now consider $x^3 + \dfrac{8}{x^3}$.
$x^3 + \dfrac{8}{x^3}$ Sum of cubes
$= \left(x + \dfrac{2}{x}\right)\left(x^2 - 2 + \dfrac{4}{x^2}\right)$
$= \left(x + \dfrac{2}{x}\right)(0)$ Substituting 0 for $x^2 - 2 + \dfrac{4}{x^2}$
$= 0$

Exercise Set 5.8

1. $x^2 + 4x = 45$
$x^2 + 4x - 45 = 0$ Getting 0 on one side
$(x - 5)(x + 9) = 0$ Factoring
$x - 5 = 0$ *or* $x + 9 = 0$ Principle of zero products
$x = 5$ *or* $x = -9$

The solutions are 5 and -9. The solution set is $\{5, -9\}$.

2. $\{-4, 7\}$

3.
$$a^2 + 1 = 2a$$
$$a^2 - 2a + 1 = 0 \quad \text{Getting 0 on one side}$$
$$(a-1)(a-1) = 0 \quad \text{Factoring}$$
$$a - 1 = 0 \ \ or \ \ a - 1 = 0 \quad \text{Principle of zero products}$$
$$a = 1 \ \ or \ \ a = 1$$

There is only one solution, 1. The solution set is $\{1\}$.

4. $\{4\}$

5.
$$x^2 - 12x + 36 = 0$$
$$(x-6)(x-6) = 0 \quad \text{Factoring}$$
$$x - 6 = 0 \ \ or \ \ x - 6 = 0 \quad \text{Principle of zero products}$$
$$x = 6 \ \ or \ \ x = 6 \quad \text{We have a repeated root.}$$

The solution is 6. The solution set is $\{6\}$.

6. $\{-8\}$

7.
$$9x + x^2 + 20 = 0$$
$$x^2 + 9x + 20 = 0 \quad \text{Changing order}$$
$$(x+5)(x+4) = 0 \quad \text{Factoring}$$
$$x + 5 = 0 \ \ or \ \ x + 4 = 0 \quad \text{Principle of zero products}$$
$$x = -5 \ \ or \ \ x = -4$$

The solutions are -5 and -4. The solution set is $\{-5, -4\}$.

8. $\{-5, -3\}$

9.
$$x^2 - 8x = 0$$
$$x(x-8) = 0 \quad \text{Factoring}$$
$$x = 0 \ \ or \ \ x - 8 = 0 \quad \text{Principle of zero products}$$
$$x = 0 \ \ or \ \ x = 8$$

The solutions are 0 and 8. The solution set is $\{0, 8\}$.

10. $\{0, 9\}$

11.
$$a^3 - 3a^2 = 40a$$
$$a^3 - 3a^2 - 40a = 0 \quad \text{Getting 0 on one side}$$
$$a(a^2 - 3a - 40) = 0$$
$$a(a-8)(a+5) = 0$$
$$a = 0 \ \ or \ \ a - 8 = 0 \ \ or \ \ a + 5 = 0 \quad \text{Principle of zero products}$$
$$a = 0 \ \ or \ \ a = 8 \ \ or \ \ a = -5$$

The solutions are 0, 8, and -5. The solution set is $\{0, 8, -5\}$.

12. $\{-7, 0, 9\}$

13.
$$x^2 - 16 = 0$$
$$(x+4)(x-4) = 0$$
$$x + 4 = 0 \ \ or \ \ x - 4 = 0$$
$$x = -4 \ \ or \ \ x = 4$$

The solutions are -4 and 4. The solution set is $\{-4, 4\}$.

14. $\{-3, 3\}$

15.
$$t^2 = 81$$
$$t^2 - 81 = 0$$
$$(t+9)(t-9) = 0$$
$$t + 9 = 0 \ \ or \ \ t - 9 = 0$$
$$t = -9 \ \ or \ \ t = 9$$

The solutions are -9 and 9. The solution set is $\{-9, 9\}$.

16. $\{-6, 6\}$

17.
$$3x^2 - 8x + 4 = 0$$
$$(3x-2)(x-2) = 0$$
$$3x - 2 = 0 \ \ or \ \ x - 2 = 0$$
$$3x = 2 \ \ or \ \ x = 2$$
$$x = \frac{2}{3} \ \ or \ \ x = 2$$

The solutions are $\frac{2}{3}$ and 2. The solution set is $\left\{\frac{2}{3}, 2\right\}$.

18. $\left\{\frac{1}{3}, \frac{4}{3}\right\}$

19.
$$4t^3 + 11t^2 + 6t = 0$$
$$t(4t^2 + 11t + 6) = 0$$
$$t(4t+3)(t+2) = 0$$
$$t = 0 \ \ or \ \ 4t + 3 = 0 \ \ or \ \ t + 2 = 0$$
$$t = 0 \ \ or \ \ 4t = -3 \ \ or \ \ t = -2$$
$$t = 0 \ \ or \ \ t = -\frac{3}{4} \ \ or \ \ t = -2$$

The solutions are 0, $-\frac{3}{4}$, and -2. The solution set is $\left\{0, -\frac{3}{4}, -2\right\}$.

20. $\left\{-\frac{3}{4}, -\frac{1}{2}, 0\right\}$

21.
$$(y-3)(y+2) = 14$$
$$y^2 - y - 6 = 14 \quad \text{Multiplying}$$
$$y^2 - y - 20 = 0$$
$$(y-5)(y+4) = 0$$
$$y - 5 = 0 \ \ or \ \ y + 4 = 0$$
$$y = 5 \ \ or \ \ y = -4$$

The solutions are 5 and -4. The solution set is $\{5, -4\}$.

22. $\{-3, 1\}$

23.
$$x(5+12x) = 28$$
$$5x + 12x^2 = 28 \quad \text{Multiplying}$$
$$5x + 12x^2 - 28 = 0$$
$$12x^2 + 5x - 28 = 0 \quad \text{Rearranging}$$
$$(4x+7)(3x-4) = 0$$
$$4x+7 = 0 \quad or \quad 3x-4 = 0$$
$$4x = -7 \quad or \quad 3x = 4$$
$$x = -\frac{7}{4} \quad or \quad x = \frac{4}{3}$$

The solutions are $-\frac{7}{4}$ and $\frac{4}{3}$. The solution set is $\left\{-\frac{7}{4}, \frac{4}{3}\right\}$.

24. $\left\{-\frac{5}{7}, \frac{2}{3}\right\}$

25.
$$a^2 - \frac{1}{64} = 0$$
$$\left(a+\frac{1}{8}\right)\left(a-\frac{1}{8}\right) = 0$$
$$a + \frac{1}{8} = 0 \quad or \quad a - \frac{1}{8} = 0$$
$$a = -\frac{1}{8} \quad or \quad a = \frac{1}{8}$$

The solutions are $-\frac{1}{8}$ and $\frac{1}{8}$. The solution set is $\left\{-\frac{1}{8}, \frac{1}{8}\right\}$.

26. $\left\{-\frac{1}{5}, \frac{1}{5}\right\}$

27.
$$t^4 - 26t^2 + 25 = 0$$
$$(t^2-1)(t^2-25) = 0$$
$$(t+1)(t-1)(t+5)(t-5) = 0$$
$$t+1 = 0 \quad or \; t-1 = 0 \; or \; t+5 = 0 \quad or \; t-5 = 0$$
$$t = -1 \; or \quad t = 1 \; or \quad t = -5 \; or \quad t = 5$$

The solutions are -1, 1, -5, and 5. The solution set is $\{-1, 1, -5, 5\}$.

28. $\{-3, -2, 2, 3\}$.

29. We set $f(a)$ equal to 8.
$$a^2 + 12a + 40 = 8$$
$$a^2 + 12a + 32 = 0$$
$$(a+8)(a+4) = 0$$
$$a+8 = 0 \quad or \quad a+4 = 0$$
$$a = -8 \quad or \quad a = -4$$

The values of a for which $f(a) = 8$ are -8 and -4.

30. -9, -5

31. We set $g(a)$ equal to 12.
$$2a^2 + 5a = 12$$
$$2a^2 + 5a - 12 = 0$$
$$(2a-3)(a+4) = 0$$
$$2a-3 = 0 \quad or \quad a+4 = 0$$
$$2a = 3 \quad or \quad a = -4$$
$$a = \frac{3}{2} \quad or \quad a = -4$$

The values of a for which $g(a) = 12$ are $\frac{3}{2}$ and -4.

32. $\frac{1}{2}$, 7

33. We set $h(a)$ equal to -27.
$$12a + a^2 = -27$$
$$12a + a^2 + 27 = 0$$
$$a^2 + 12a + 27 = 0 \quad \text{Rearranging}$$
$$(a+3)(a+9) = 0$$
$$a+3 = 0 \quad or \quad a+9 = 0$$
$$a = -3 \quad or \quad a = -9$$

The values of a for which $h(a) = -27$ are -3 and -9.

34. -4, 8

35. $f(x) = \dfrac{3}{x^2 - 4x - 5}$

$f(x)$ cannot be calculated for any x-value for which the denominator, $x^2 - 4x - 5$, is 0. To find the excluded values, we solve:
$$x^2 - 4x - 5 = 0$$
$$(x-5)(x+1) = 0$$
$$x-5 = 0 \quad or \quad x+1 = 0$$
$$x = 5 \quad or \quad x = -1$$

The domain of f is $\{x|x$ is a real number and $x \neq 5$ and $x \neq -1\}$.

36. $\{x|x$ is a real number and $x \neq 1$ and $x \neq 6\}$

37. $f(x) = \dfrac{x}{6x^2 - 54}$

$f(x)$ cannot be calculated for any x-value for which the denominator, $6x^2 - 54$, is 0. To find the excluded values, we solve:
$$6x^2 - 54 = 0$$
$$6(x^2 - 9) = 0$$
$$6(x+3)(x-3) = 0$$
$$x+3 = 0 \quad or \quad x-3 = 0$$
$$x = -3 \quad or \quad x = 3$$

The domain of f is $\{x|x$ is a real number and $x \neq -3$ and $x \neq 3\}$.

38. $\{x|x$ is a real number and $x \neq -2$ and $x \neq 2\}$

39. $f(x) = \dfrac{x-5}{9x-18x^2}$

$f(x)$ cannot be calculated for any x-value for which the denominator, $9x - 18x^2$, is 0. To find the excluded values, we solve:

$$9x - 18x^2 = 0$$
$$9x(1-2x) = 0$$
$$9x = 0 \quad or \quad 1 - 2x = 0$$
$$x = 0 \quad or \quad -2x = -1$$
$$x = 0 \quad or \quad x = \frac{1}{2}$$

The domain of f is $\left\{x|x \text{ is a real number and } x \neq 0 \text{ and } x \neq \frac{1}{2}\right\}$.

40. $\left\{x|x \text{ is a real number and } x \neq 0 \text{ and } x \neq \frac{1}{5}\right\}$

41. $f(x) = \dfrac{7}{5x^3 - 35x^2 + 50x}$

$f(x)$ cannot be calculated for any x-value for which the denominator, $5x^3 - 35x^2 + 50x$, is 0. To find the excluded values, we solve:

$$5x^3 - 35x^2 + 50x = 0$$
$$5x(x^2 - 7x + 10) = 0$$
$$5x(x-2)(x-5) = 0$$
$$5x = 0 \quad or \quad x - 2 = 0 \quad or \quad x - 5 = 0$$
$$x = 0 \quad or \quad x = 2 \quad or \quad x = 5$$

The domain of f is $\{x|x \text{ is a real number and } x \neq 0 \text{ and } x \neq 2 \text{ and } x \neq 5\}$.

42. $\{x|x \text{ is a real number and } x \neq 0 \text{ and } x \neq -2 \text{ and } x \neq 3\}$

43. ***Familiarize***. Let x represent the number.

Translate.

$$\underbrace{\text{Square of number}}_{x^2} \quad \underbrace{\text{plus}}_{+} \quad \underbrace{\text{number}}_{x} \quad \underbrace{\text{is}}_{=} \quad \underbrace{156.}_{156}$$

Carry out. We solve the equation:

$$x^2 + x = 156$$
$$x^2 + x - 156 = 0$$
$$(x+13)(x-12) = 0$$
$$x + 13 = 0 \quad or \quad x - 12 = 0$$
$$x = -13 \quad or \quad x = 12$$

Check. The square of -13, which is 169, plus -13 is 156. The square of 12, which is 144, plus 12 is 156. Both numbers check.

State. The number is -13 or 12.

44. -12, 11

45. ***Familiarize***. We let w represent the width and $w+5$ represent the length. We make a drawing and label it.

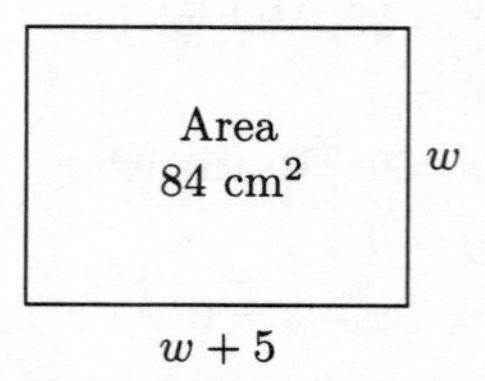

Recall that the formula for the area of a rectangle is $A =$ length $\times$ width.

Translate.

$$\underbrace{\text{Area}}_{w(w+5)} \quad \underbrace{\text{is}}_{=} \quad \underbrace{84\text{ cm}^2.}_{84}$$

Carry out. We solve the equation:

$$w(w+5) = 84$$
$$w^2 + 5w = 84$$
$$w^2 + 5w - 84 = 0$$
$$(w+12)(w-7) = 0$$
$$w + 12 = 0 \quad or \quad w - 7 = 0$$
$$w = -12 \quad or \quad w = 7$$

Check. The number -12 is not a solution, because width cannot be negative. If the width is 7 cm and the length is 5 cm more, or 12 cm, then the area is $12 \cdot 7$, or 84 cm^2. This is a solution.

State. The length is 12 cm, and the width is 7 cm.

46. Length: 12 cm, width: 8 cm

47. ***Familiarize***. We make a drawing and label it. We let x represent the length of a side of the original square.

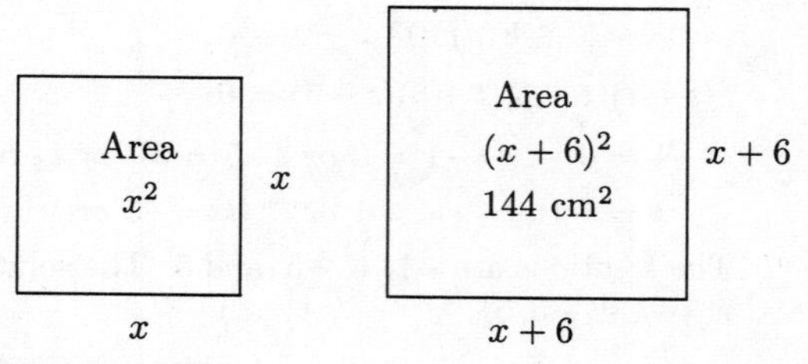

Translate.

$$\underbrace{\text{Area of new square}}_{(x+6)^2} \quad \underbrace{\text{is}}_{=} \quad \underbrace{144\text{ cm}^2.}_{144}$$

Carry out. We solve the equation:

$$(x+6)^2 = 144$$
$$x^2 + 12x + 36 = 144$$
$$x^2 + 12x - 108 = 0$$
$$(x-6)(x+18) = 0$$
$$x - 6 = 0 \quad or \quad x + 18 = 0$$
$$x = 6 \quad or \quad x = -18$$

$x - 6 = 0$ *or* $x + 18 = 0$

$x = 6$ *or* $x = -18$

Check. We only check 6 since the length of a side cannot be negative. If we increase the length by 6, the new length is $6+6$, or 12 cm. Then the new area is $12 \cdot 12$, or 144 cm^2. We have a solution.

State. The length of a side of the original square is 6 cm.

48. 3 m

49. ***Familiarize.*** We make a drawing and label it with both known and unknown information. We let x represent the width of the frame.

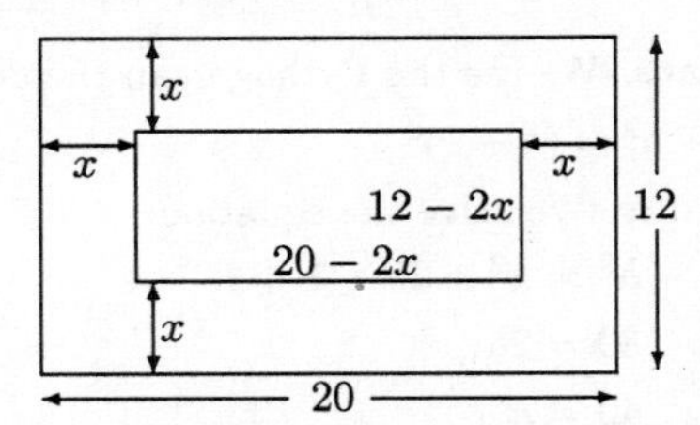

The length and width of the picture that shows are represented by $20 - 2x$ and $12 - 2x$. The area of the picture that shows is 84 cm^2.

Translate. Using the formula for the area of a rectangle, $A = l \cdot w$, we have

$$84 = (20 - 2x)(12 - 2x).$$

Carry out. We solve the equation:

$$84 = 240 - 64x + 4x^2$$
$$84 = 4(60 - 16x + x^2)$$
$$21 = 60 - 16x + x^2 \quad \text{Dividing by 4}$$
$$0 = x^2 - 16x + 39$$
$$0 = (x - 3)(x - 13)$$

$x - 3 = 0$ *or* $x - 13 = 0$

$x = 3$ *or* $x = 13$

Check. We see that 13 is not a solution because when $x = 13$, $20 - 2x = -6$ and $12 - 2x = -14$, and the length and width of the frame cannot be negative. We check 3. When $x = 3$, $20 - 2x = 14$ and $12 - 2x = 6$ and $14 \cdot 6 = 84$. The area is 84. The value checks.

State. The width of the frame is 3 cm.

50. 2 cm

51. ***Familiarize.*** We let x represent the width of the walkway. We make a drawing and label it with both the known and unknown information.

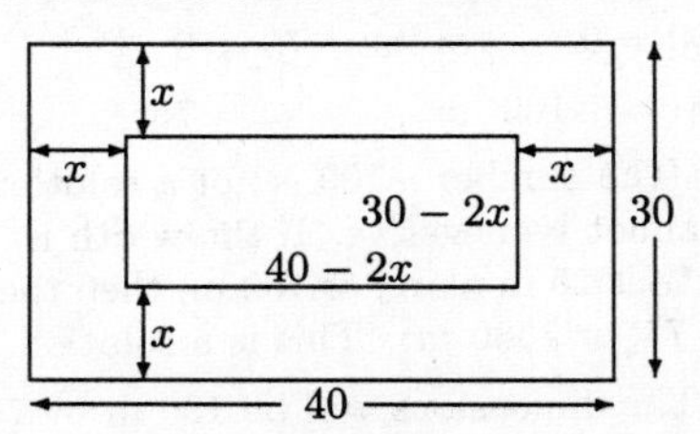

The area of the old garden is $40 \cdot 30$, or 1200 ft^2. The area of the new garden is $\frac{1}{2} \cdot 1200$, or 600 ft^2.

Translate. Rewording, we have

$$\underbrace{\text{Area of new garden}}_{(40 - 2x)(30 - 2x)} \;\; \underbrace{\text{is}}_{=} \;\; \underbrace{600 \text{ ft}^2}_{600}.$$

Carry out. We solve the equation:

$$(40 - 2x)(30 - 2x) = 600$$
$$1200 - 140x + 4x^2 = 600$$
$$4x^2 - 140x + 600 = 0$$
$$x^2 - 35x + 150 = 0 \quad \text{Dividing by 4}$$
$$(x - 5)(x - 30) = 0$$

$x - 5 = 0$ *or* $x - 30 = 0$

$x = 5$ *or* $x = 30$

Check. If the walkway is 5 ft wide, the length of the new garden will be $40 - 2 \cdot 5$, or 30 ft, and its width will be $30 - 2 \cdot 5$, or 20 ft. Then the area of the new garden will be $30 \cdot 20$, or 600 ft^2. This is $\frac{1}{2}$ of 1200 ft^2, the area of the old garden, so this answer checks.

If the walkway is 30 ft wide, the length of the new garden will be $40 - 2 \cdot 30$, or -20 ft. Since the length cannot be negative, 30 is not a solution.

State. The walkway is 5 ft wide.

52. 10 ft

53. ***Familiarize.*** Using the labels on the drawing in the text, we let x represent the base of the triangle and $x + 2$ represent the height. Recall that the formula for the area of the triangle with base b and height h is $\frac{1}{2}bh$.

Translate.

$$\underbrace{\text{The area}}_{\frac{1}{2}x(x + 2)} \;\; \underbrace{\text{is}}_{=} \;\; \underbrace{12 \text{ ft}^2}_{12}.$$

Carry out. We solve the equation:

$$\frac{1}{2}x(x + 2) = 12$$
$$x(x + 2) = 24 \quad \text{Multiplying by 2}$$
$$x^2 + 2x = 24$$
$$x^2 + 2x - 24 = 0$$
$$(x + 6)(x - 4) = 0$$

$x + 6 = 0$ *or* $x - 4 = 0$

$x = -6$ *or* $x = 4$

Check. We check only 4 since the length of the base cannot be negative. If the base is 4 ft, then the height is $4 + 2$, or 6 ft, and the area is $\frac{1}{2} \cdot 4 \cdot 6$, or 12 ft^2. The answer checks.

State. The height is 6 ft, and the base is 4 ft.

54. $-10, -8, -6$ or $6, 8, 10$

55. ***Familiarize.*** Let x represent the first integer, $x+2$ the second, and $x+4$ the third.

Translate.

$$\underbrace{\text{Square of the third}}_{(x+4)^2} \ \underbrace{\text{is}}_{=} \ \underbrace{76}_{76} \ \underbrace{\text{more than}}_{+} \ \underbrace{\text{square of the second.}}_{(x+2)^2}$$

Carry out. We solve the equation:

$$(x+4)^2 = 76 + (x+2)^2$$
$$x^2 + 8x + 16 = 76 + x^2 + 4x + 4$$
$$x^2 + 8x + 16 = x^2 + 4x + 80$$
$$4x = 64$$
$$x = 16$$

Check. We check the integers 16, 18, and 20. The square of 20, or 400, is 76 more than 324, the square of 18. The answer checks.

State. The integers are 16, 18, and 20.

56. Distance d: 12 ft, height of the tower: 16 ft

57. ***Familiarize.*** We make a drawing. Let w represent the width of the parking lot. Then $w+50$ represents the length.

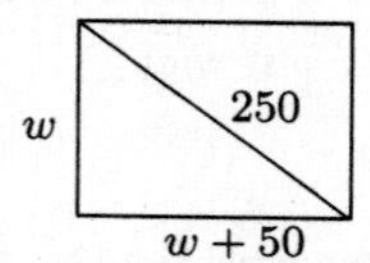

Since we have a right triangle, we can use the Pythagorean theorem:

$$a^2 + b^2 = c^2$$

Translate. Substituting, we have

$$w^2 + (w+50)^2 = 250^2$$

Carry out. We solve the equation:

$$w^2 + (w+50)^2 = 250^2$$
$$w^2 + w^2 + 100w + 2500 = 62,500$$
$$2w^2 + 100w - 60,000 = 0$$
$$w^2 + 50w - 30,000 = 0 \quad \text{Dividing by 2}$$
$$(w+200)(w-150) = 0$$
$$w + 200 = 0 \quad \textit{or} \quad w - 150 = 0$$
$$w = -200 \quad \textit{or} \quad w = 150$$

Check. We only check 150, since the width cannot be negative. If the width is 150 ft, then the length is 150 + 50, or 200 ft. The length of a diagonal is $150^2 + 200^2 = 62,500 = 250^2$. The answer checks.

State. The parking lot is 150 ft wide and 200 ft long.

58. Height: 16 m, base: 7 m

59. ***Familiarize.*** We make a drawing. Let h = the height the ladder reaches on the wall. Then the length of the ladder is $h+1$.

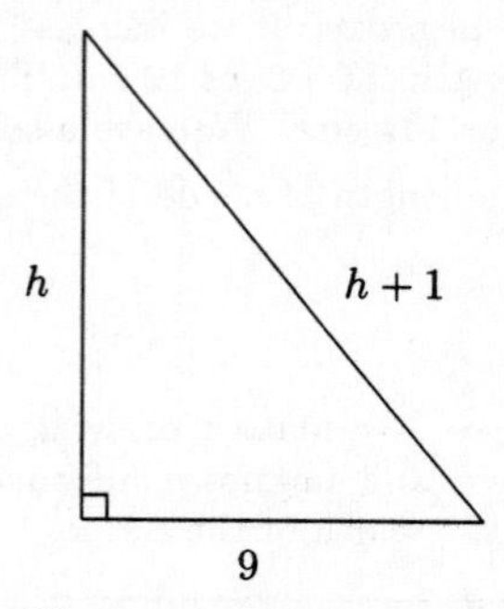

Translate. We use the Pythagorean theorem.

$$9^2 + h^2 = (h+1)^2$$

Carry out. We solve the equation:

$$81 + h^2 = h^2 + 2h + 1$$
$$80 = 2h$$
$$40 = h$$

Check. If $h = 40$, then $h + 1 = 41$; $9^2 + 40^2 = 81 + 1600 = 1681 = 41^2$, so the answer checks.

State. The ladder is 41 ft long.

60. 24 ft

61. ***Familiarize.*** Let w represent the width and $w+25$ represent the length. Make a drawing.

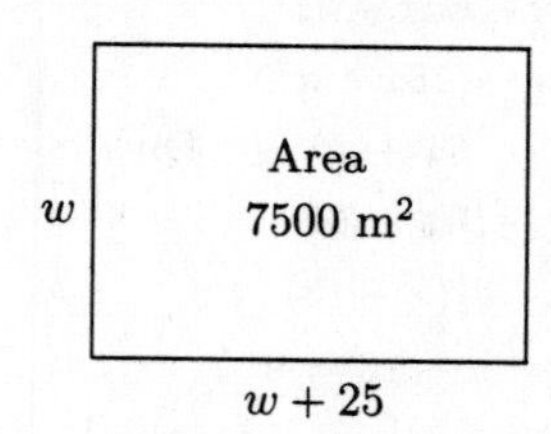

Recall that the formula for the area of a rectangle is $A =$ length $\times$ width.

Translate.

$$\underbrace{\text{Area}}_{w(w+25)} \ \underbrace{\text{is}}_{=} \ \underbrace{7500\text{ m}^2}_{7500}.$$

Carry out. We solve the equation:

$$w(w+25) = 7500$$
$$w^2 + 25w = 7500$$
$$w^2 + 25w - 7500 = 0$$
$$(w+100)(w-75) = 0$$
$$w + 100 = 0 \quad \textit{or} \quad w - 75 = 0$$
$$w = -100 \quad \textit{or} \quad w = 75$$

Check. The number -100 is not a solution because width cannot be negative. If the width is 75 m and the length is 25 m more, or 100 m, then the area will be $100 \cdot 75$, or 7500 m^2. This is a solution.

State. The dimensions will be 100 m by 75 m.

62. 9 m by 12 m

63. ***Familiarize.*** We will use the given function
$h(t) = -16t^2 + 96t + 880$.

Translate.

$$\underbrace{\text{Height}}_{-16t^2+96t+880} \quad \underbrace{\text{is}}_{=} \quad \underbrace{0 \text{ ft.}}_{0}$$

Carry out. We solve the equation:

$$-16t^2 + 96t + 880 = 0$$
$$t^2 - 6t - 55 = 0 \quad \text{Dividing by } -16$$
$$(t-11)(t+5) = 0$$
$$t - 11 = 0 \quad or \quad t + 5 = 0$$
$$t = 11 \quad or \quad t = -5$$

Check. We check only 11, since time cannot be negative in this application. When $t = 11$, $h(11) = -16 \cdot 11^2 + 96 \cdot 11 + 880 = 0$. The answer checks.

State. The rocket will reach the ground after 11 sec.

64. 7 sec

65. ***Familiarize.*** The firm breaks even when the cost and the revenue are the same. We use the functions given in the text.

Translate.

$$\underbrace{\text{Cost}}_{\frac{1}{9}x^2+2x+1} \quad \underbrace{\text{equals}}_{=} \quad \underbrace{\text{revenue.}}_{\frac{5}{36}x^2+2x}$$

Carry out. We solve the equation:

$$\frac{1}{9}x^2 + 2x + 1 = \frac{5}{36}x^2 + 2x$$
$$0 = \frac{1}{36}x^2 - 1$$
$$0 = \left(\frac{1}{6}x + 1\right)\left(\frac{1}{6}x - 1\right)$$
$$\frac{1}{6}x + 1 = 0 \quad or \quad \frac{1}{6}x - 1 = 0$$
$$\frac{1}{6}x = -1 \quad or \quad \frac{1}{6}x = 1$$
$$x = -6 \quad or \quad x = 6$$

Check. We check only 6 since the number of video cameras cannot be negative. If 6 cameras are produced, the cost is $C(6) = \frac{1}{9} \cdot 6^2 + 2 \cdot 6 + 1 = 4 + 12 + 1 = \$17$ thousand. If 6 cameras are sold, the revenue is $R(6) = \frac{5}{36} \cdot 6^2 + 2 \cdot 6 = 5 + 12 = \$17$ thousand. The answer checks.

State. The firm breaks even when 6 video cameras are produced and sold.

66. 2

67. ***Familiarize.*** Let r represent the speed of the faster car and d represent its distance. Then $r - 15$ and $651 - d$ represent the speed and distance of the slower car, respectively. We organize the information in a table.

	Speed	Time	Distance
Faster car	r	7	d
Slower car	$r-15$	7	$651-d$

Translate. We use the formula $rt = d$. Each row of the table gives us an equation.

$$7r = d$$
$$7(r - 15) = 651 - d$$

Carry out. We use the substitution method substituting $7r$ for d in the second equation and solving for r.

$$7(r-15) = 651 - 7r \quad \text{Substituting}$$
$$7r - 105 = 651 - 7r$$
$$14r = 756$$
$$r = 54$$

Check. If $r = 54$, then the speed of the faster car is 54 mph and the speed of the slower car is $54 - 15$, or 39 mph. The distance the faster car travels is $54 \cdot 7$, or 378 miles. The distance the slower car travels is $39 \cdot 7$, or 273 miles. The total of the two distances is $378 + 273$, or 651 miles. The result checks.

State. The speed of the faster car is 54 mph. The speed of the slower car is 39 mph.

68. Conventional: 36, surround-sound: 42

69.

$$2x - 14 + 9x > -8x + 16 + 10x$$
$$11x - 14 > 2x + 16 \quad \text{Collecting like terms}$$
$$9x - 14 > 16 \quad \text{Adding } -2x$$
$$9x > 30 \quad \text{Adding 14}$$
$$x > \frac{10}{3} \quad \text{Multiplying by } \frac{1}{9}$$

The solution set is $\left\{x \middle| x > \frac{10}{3}\right\}$, or $\left(\frac{10}{3}, \infty\right)$.

70. $(2, -2, -4)$

71. ◈

72. ◈

73. ◈

74. ◈

75.

$$(8x + 11)(12x^2 - 5x - 2) = 0$$
$$(8x + 11)(3x - 2)(4x + 1) = 0$$
$$8x + 11 = 0 \quad or \quad 3x - 2 = 0 \quad or \quad 4x + 1 = 0$$
$$8x = -11 \quad or \quad 3x = 2 \quad or \quad 4x = -1$$
$$x = -\frac{11}{8} \quad or \quad x = \frac{2}{3} \quad or \quad x = -\frac{1}{4}$$

The solution set is $\left\{-\frac{11}{8}, \frac{2}{3}, -\frac{1}{4}\right\}$.

76. $\{-2, 2\}$

77.
$$\begin{aligned}(x-2)^3 &= x^3-2\\ x^3-6x^2+12x-8 &= x^3-2\\ 0 &= 6x^2-12x+6\\ 0 &= 6(x^2-2x+1)\\ 0 &= 6(x-1)(x-1)\end{aligned}$$
$x-1=0$ *or* $x-1=0$

$x=1$ *or* $x=1$

The solution set is $\{1\}$.

78. $\{-1, 3\}$; $(-2, 4)$, or $\{x| -2 < x < 4\}$

79.

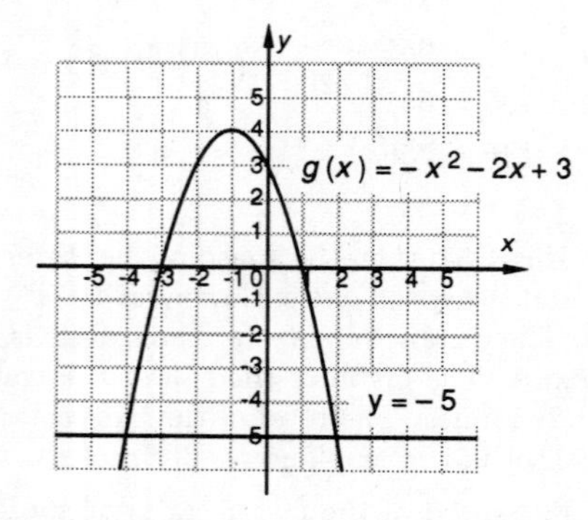

The solutions of $-x^2-2x+3=0$ are the first coordinates of the x-intercepts. From the graph we see that these are -3 and 1. The solution set is $\{-3, 1\}$.

To solve $-x^2-2x+3 \geq -5$ we find the x-values for which $g(x) \geq -5$. From the graph we see that these are the values in the interval $[-4, 2]$. The solution set can also be expressed as $\{x| -4 \leq x \leq 2\}$.

80. Answers may vary; $f(x) = 5x^3-20x^2+5x+30$.

81. Answers may vary. A polynomial function of lowest degree that meets the given criteria is of the form $g(x) = ax^3+bx^2+cx+d$. Substituting, we have

$a(-3)^3+b(-3)^2+c(-3)+d=0,$

$a\cdot 1^3+b\cdot 1^2+c\cdot 1+d=0,$

$a\cdot 5^3+b\cdot 5^2+c\cdot 5+d=0$

$a\cdot 0^3+b\cdot 0^2+c\cdot 0+d=45,$ or

$$\begin{aligned}-27a+9b-3c+d&=0,\\ a+b+c+d&=0,\\ 125a+25b+5c+d&=0,\\ d&=45.\end{aligned}$$

Solving the system of equations, we get $(3, -9, -39, 45)$, so the corresponding function is $g(x) = 3x^3-9x^2-39x+45$.

82. 3, 14

83. ***Familiarize.*** Using the labels on the drawing in the text, we let x represent the width of the piece of tin and $2x$ represent the length. Then the width and length of the base of the box are represented by $x-4$ and $2x-4$, respectively. Recall that the formula for the volume of a rectangular solid with length l, width w, and height h is $l\cdot w\cdot h$.

Translate.

$$\underbrace{\text{The volume}} \quad \text{is} \quad \underbrace{480 \text{ cm}^3}.$$
$$(2x-4)(x-4)(2) \;=\; 480$$

Carry out. We solve the equation:

$$\begin{aligned}(2x-4)(x-4)(2) &= 480\\ (2x-4)(x-4) &= 240 \quad \text{Dividing by 2}\\ 2x^2-12x+16 &= 240\\ 2x^2-12x-224 &= 0\\ x^2-6x-112 &= 0 \quad \text{Dividing by 2}\\ (x+8)(x-14) &= 0\end{aligned}$$

$x+8=0$ *or* $x-14=0$

$x=-8$ *or* $x=14$

Check. We check only 14 since the width cannot be negative. If the width of the piece of tin is 14 cm, then its length is $2\cdot 14$, or 28 cm, and the dimensions of the base of the box are $14-4$, or 10 cm by $28-4$, or 24 cm. The volume of the box is $24\cdot 10\cdot 2$, or 480 cm^3. The answer checks.

State. The dimensions of the piece of tin are 14 cm by 28 cm.

84. Tugboat: 15 km/h, freighter: 8 km/h

85. Graph $y_1 = 11.12(x+1)^2$ and $y_2 = 15.4x^2$ in a window that shows the point of intersection of the graphs. The window $[0, 10, 0, 500]$, Xscl = 1, Yscl = 100 is one good choice. Then find the first coordinate of the point of intersection. It is approximately 5.7, so it will take the camera about 5.7 sec to catch up to the skydiver.

86. Enter F on the Y = screen, as y_1, for example. Then, with the table set in ASK mode, enter -5 and 3 and observe that the grapher returns "ERROR" in the Y1-column.

87.

88. $\{-2.00, 4.00\}$

89. $\{-4.00, 1.00\}$

90. $\{6.90\}$

91. $\{-3.33, 5.15\}$

92. $\{3.48\}$

93. 

Chapter 6

Rational Expressions, Equations, and Functions

Exercise Set 6.1

1. $v(t) = \dfrac{4t^2 - 5t + 2}{t+3}$

$v(0) = \dfrac{4 \cdot 0^2 - 5 \cdot 0 + 2}{0+3} = \dfrac{0 - 0 + 2}{0+3} = \dfrac{2}{3}$

$v(3) = \dfrac{4 \cdot 3^2 - 5 \cdot 3 + 2}{3+3} = \dfrac{36 - 15 + 2}{3+3} = \dfrac{23}{6}$

$v(7) = \dfrac{4 \cdot 7^2 - 5 \cdot 7 + 2}{7+3} = \dfrac{196 - 35 + 2}{7+3} = \dfrac{163}{10}$

2. 2; $-\dfrac{11}{7}$; 15

3. $r(y) = \dfrac{3y^3 - 2y}{y-5}$

$r(0) = \dfrac{3 \cdot 0^3 - 2 \cdot 0}{0-5} = \dfrac{0-0}{0-5} = \dfrac{0}{-5} = 0$

$r(4) = \dfrac{3 \cdot 4^3 - 2 \cdot 4}{4-5} = \dfrac{192 - 8}{-1} = \dfrac{184}{-1} = -184$

$r(5) = \dfrac{3 \cdot 5^3 - 2 \cdot 5}{5-5} = \dfrac{375 - 10}{0}$

Since division by zero is not defined, $r(5)$ does not exist.

4. -6; 0; $-\dfrac{6}{5}$

5. $g(x) = \dfrac{2x^3 - 9}{x^2 - 4x + 4}$

$g(0) = \dfrac{2 \cdot 0^3 - 9}{0^2 - 4 \cdot 0 + 4} = \dfrac{0 - 9}{0 - 0 + 4} = -\dfrac{9}{4}$

$g(2) = \dfrac{2 \cdot 2^3 - 9}{2^2 - 4 \cdot 2 + 4} = \dfrac{16 - 9}{4 - 8 + 4} = \dfrac{7}{0}$

Since division by zero is not defined, $g(2)$ does not exist.

$g(-1) = \dfrac{2(-1)^3 - 9}{(-1)^2 - 4(-1) + 4} = \dfrac{-2 - 9}{1 + 4 + 4} = -\dfrac{11}{9}$

6. 0; $\dfrac{2}{5}$; does not exist

7. $\dfrac{5x}{5x} \cdot \dfrac{x-3}{x+2} = \dfrac{5x(x-3)}{5x(x+2)}$

8. $\dfrac{(3-a^2)(-1)}{(a-7)(-1)}$

9. $\dfrac{t-2}{t+3} \cdot \dfrac{-1}{-1} = \dfrac{(t-2)(-1)}{(t+3)(-1)}$

10. $\dfrac{(x-4)(x-5)}{(x+5)(x-5)}$

11. $\dfrac{15x}{5x^2}$

$= \dfrac{5x \cdot 3}{5x \cdot x}$ Factoring; the greatest common factor is $5x$.

$= \dfrac{5x}{5x} \cdot \dfrac{3}{x}$ Factoring the rational expression

$= 1 \cdot \dfrac{3}{x}$ $\dfrac{5x}{5x} = 1$

$= \dfrac{3}{x}$ Removing a factor equal to 1

12. $\dfrac{a^2}{3}$

13. $\dfrac{18t^3}{27t^7}$

$= \dfrac{9t^3 \cdot 2}{9t^3 \cdot 3t^4}$ Factoring the numerator and the denominator

$= \dfrac{9t^3}{9t^3} \cdot \dfrac{2}{3t^4}$ Factoring the rational expression

$= \dfrac{2}{3t^4}$ Removing a factor equal to 1

14. $\dfrac{2}{y^4}$

15. $\dfrac{2a - 10}{2} = \dfrac{2(a-5)}{2 \cdot 1} = \dfrac{2}{2} \cdot \dfrac{a-5}{1} = a - 5$

16. $a + 4$

17. $\dfrac{15}{25a - 30} = \dfrac{5 \cdot 3}{5(5a-6)} = \dfrac{5}{5} \cdot \dfrac{3}{5a-6} = \dfrac{3}{5a-6}$

18. $\dfrac{7}{2x-3}$

19. $\dfrac{3x - 12}{3x + 15} = \dfrac{3(x-4)}{3(x+5)} = \dfrac{3}{3} \cdot \dfrac{x-4}{x+5} = \dfrac{x-4}{x+5}$

20. $\dfrac{y-5}{y+3}$

21. $\dfrac{5x + 20}{x^2 + 4x} = \dfrac{5(x+4)}{x(x+4)} = \dfrac{5}{x} \cdot \dfrac{x+4}{x+4} = \dfrac{5}{x}$

22. $\dfrac{3}{x}$

23.

$$\frac{3a-1}{2-6a}$$

$$=\frac{3a-1}{2(1-3a)}$$

$$=\frac{-1(1-3a)}{2(1-3a)}$$

Factoring out -1 in the numerator reverses the subtraction.

$$=\frac{-1}{2}\cdot\frac{1-3a}{1-3a}$$

$$=-\frac{1}{2}$$

24. $-\frac{1}{2}$

25. $$\frac{8t-16}{t^2-4}=\frac{8(t-2)}{(t+2)(t-2)}=\frac{8}{t+2}\cdot\frac{t-2}{t-2}=\frac{8}{t+2}$$

26. $\frac{t-3}{5}$

27.

$$\frac{2t-1}{1-4t^2}$$

$$=\frac{2t-1}{(1+2t)(1-2t)}$$

$$=\frac{-1(1-2t)}{(1+2t)(1-2t)}$$

Factoring out -1 in the numerator reverses the subtraction

$$=\frac{-1}{1+2t}\cdot\frac{1-2t}{1-2t}$$

$$=-\frac{1}{1+2t}$$

28. $-\frac{1}{2+3a}$

29. $$\frac{12-6x}{5x-10}=\frac{-6(-2+x)}{5(x-2)}=\frac{-6(x-2)}{5(x-2)}=$$

$$\frac{-6}{5}\cdot\frac{x-2}{x-2}=-\frac{6}{5}$$

30. $-\frac{7}{3}$

31. $$\frac{a^2-25}{a^2+10a+25}=\frac{(a+5)(a-5)}{(a+5)(a+5)}=$$

$$\frac{a+5}{a+5}\cdot\frac{a-5}{a+5}=\frac{a-5}{a+5}$$

32. $\frac{a+4}{a-4}$

33. $$\frac{x^2+9x+8}{x^2-3x-4}=\frac{(x+1)(x+8)}{(x+1)(x-4)}=\frac{x+1}{x+1}\cdot\frac{x+8}{x-4}=$$

$$\frac{x+8}{x-4}$$

34. $\frac{t-9}{t+4}$

35. $$\frac{16-t^2}{t^2-8t+16}=\frac{16-t^2}{16-8t+t^2}=\frac{(4+t)(4-t)}{(4-t)(4-t)}=$$

$$\frac{4+t}{4-t}\cdot\frac{4-t}{4-t}=\frac{4+t}{4-t}$$

36. $\frac{5-p}{5+p}$

37.

$$\frac{5a^3}{3b}\cdot\frac{7b^3}{10a^7}$$

$$=\frac{5a^3\cdot 7b^3}{3b\cdot 10a^7}$$

Multiplying the numerators and also the denominators

$$=\frac{5\cdot a^3\cdot 7\cdot b\cdot b^2}{3\cdot b\cdot 2\cdot 5\cdot a^3\cdot a^4}$$

Factoring the numerator and the denominator

$$=\frac{\cancel{5}\cdot\cancel{a^3}\cdot 7\cdot\cancel{b}\cdot b^2}{3\cdot\cancel{b}\cdot 2\cdot\cancel{5}\cdot\cancel{a^3}\cdot a^4}$$

Removing a factor equal to 1

$$=\frac{7b^2}{6a^4}$$

38. $\frac{5}{3ab^3}$

39. $$\frac{3x-6}{5x}\cdot\frac{x^3}{5x-10}=\frac{(3x-6)(x^3)}{5x(5x-10)}$$

$$=\frac{3(x-2)(x)(x^2)}{5\cdot x\cdot 5(x-2)}$$

$$=\frac{3(\cancel{x-2})(\cancel{x})(x^2)}{5\cdot\cancel{x}\cdot 5(\cancel{x-2})}$$

$$=\frac{3x^2}{25}$$

40. $\frac{3t^2}{4}$

41. $$\frac{y^2-16}{2y+6}\cdot\frac{y+3}{y-4}=\frac{(y^2-16)(y+3)}{(2y+6)(y-4)}$$

$$=\frac{(y+4)(y-4)(y+3)}{2(y+3)(y-4)}$$

$$=\frac{(y+4)(\cancel{y-4})(\cancel{y+3})}{2(\cancel{y+3})(\cancel{y-4})}$$

$$=\frac{y+4}{2}$$

42. $\frac{m+n}{4}$

43. $$\frac{x^2-16}{x^2}\cdot\frac{x^2-4x}{x^2-x-12}=\frac{(x^2-16)(x^2-4x)}{x^2(x^2-x-12)}$$

$$=\frac{(x+4)(x-4)(x)(x-4)}{x\cdot x(x-4)(x+3)}$$

$$=\frac{(x+4)(\cancel{x-4})(\cancel{x})(x-4)}{\cancel{x}\cdot x(\cancel{x-4})(x+3)}$$

$$=\frac{(x+4)(x-4)}{x(x+3)}$$

44. $\frac{y(y+5)}{y-3}$

45. $$\frac{7a-14}{4-a^2}\cdot\frac{5a^2+6a+1}{35a+7}$$
$$=\frac{(7a-14)(5a^2+6a+1)}{(4-a^2)(35a+7)}$$
$$=\frac{7(a-2)(5a+1)(a+1)}{(2+a)(2-a)(7)(5a+1)}$$
$$=\frac{7(-1)(2-a)(5a+1)(a+1)}{(2+a)(2-a)(7)(5a+1)}$$
$$=\frac{7(-1)(2-a)(5a+1)(a+1)}{(2+a)(2-a)(7)(5a+1)}$$
$$=\frac{-1(a+1)}{2+a}$$
$$=\frac{-a-1}{2+a},\ \text{or}\ -\frac{a+1}{2+a}$$

46. $-\dfrac{3(a+1)}{a+6}$

47. $$\frac{6-2t}{t^2+4t+4}\cdot\frac{t^3+2t^2}{t^8-9t^6}$$
$$=\frac{(6-2t)(t^3+2t^2)}{(t^2+4t+4)(t^8-9t^6)}$$
$$=\frac{-2(-3+t)(t^2)(t+2)}{(t+2)(t+2)(t^6)(t+3)(t-3)}$$
$$=\frac{-2(t-3)(t^2)(t+2)}{(t+2)(t+2)(t^2)(t^4)(t+3)(t-3)}$$
$$=\frac{-2}{t^4(t+2)(t+3)}$$

48. $\dfrac{x^2(x+3)(x-3)}{-4}$

49. $$\frac{x^2-2x-35}{2x^3-3x^2}\cdot\frac{4x^3-9x}{7x-49}$$
$$=\frac{(x^2-2x-35)(4x^3-9x)}{(2x^3-3x^2)(7x-49)}$$
$$=\frac{(x-7)(x+5)(x)(2x+3)(2x-3)}{x^2(2x-3)(7)(x-7)}$$
$$=\frac{(x-7)(x+5)(x)(2x+3)(2x-3)}{x\cdot x(2x-3)(7)(x-7)}$$
$$=\frac{(x+5)(2x+3)}{7x}$$

50. $\dfrac{1}{y+1}$

51. $$\frac{c^3+8}{c^5-4c^3}\cdot\frac{c^6-4c^5+4c^4}{c^2-2c+4}$$
$$=\frac{(c^3+8)(c^6-4c^5+4c^4)}{(c^5-4c^3)(c^2-2c+4)}$$
$$=\frac{(c+2)(c^2-2c+4)(c^4)(c-2)(c-2)}{c^3(c+2)(c-2)(c^2-2c+4)}$$
$$=\frac{c^3(c+2)(c^2-2c+4)(c-2)}{c^3(c+2)(c^2-2c+4)(c-2)}\cdot\frac{c(c-2)}{1}$$
$$=c(c-2)$$

52. $\dfrac{x(x-3)^2}{x+3}$

53. $$\frac{a^3-b^3}{3a^2+9ab+6b^2}\cdot\frac{a^2+2ab+b^2}{a^2-b^2}$$
$$=\frac{(a^3-b^3)(a^2+2ab+b^2)}{(3a^2+9ab+6b^2)(a^2-b^2)}$$
$$=\frac{(a-b)(a^2+ab+b^2)(a+b)(a+b)}{3(a+b)(a+2b)(a+b)(a-b)}$$
$$=\frac{(a-b)(a^2+ab+b^2)(a+b)(a+b)}{3(a+b)(a+2b)(a+b)(a-b)}$$
$$=\frac{a^2+ab+b^2}{3(a+2b)}$$

54. $\dfrac{x^2-xy+y^2}{3(x+3y)}$

55. $$\frac{4x^2-9y^2}{8x^3-27y^3}\cdot\frac{4x^2+6xy+9y^2}{4x^2+12xy+9y^2}$$
$$=\frac{(4x^2-9y^2)(4x^2+6xy+9y^2)}{(8x^3-27y^3)(4x^2+12xy+9y^2)}$$
$$=\frac{(2x+3y)(2x-3y)(4x^2+6xy+9y^2)\cdot 1}{(2x-3y)(4x^2+6xy+9y^2)(2x+3y)(2x+3y)}$$
$$=\frac{(2x+3y)(2x-3y)(4x^2+6xy+9y^2)}{(2x+3y)(2x-3y)(4x^2+6xy+9y^2)}\cdot\frac{1}{2x+3y}$$
$$=\frac{1}{2x+3y}$$

56. $\dfrac{(x-y)(2x+3y)}{2(x+y)(9x^2+6xy+4y^2)}$

57. $$\frac{9x^5}{8y^2}\div\frac{3x}{16y^9}$$
$$=\frac{9x^5}{8y^2}\cdot\frac{16y^9}{3x}\qquad\text{Multiplying by the reciprocal of the divisor}$$
$$=\frac{9x^5(16y^9)}{8y^2(3x)}$$
$$=\frac{3\cdot 3\cdot x\cdot x^4\cdot 2\cdot 8\cdot y^2\cdot y^7}{8\cdot y^2\cdot 3\cdot x}$$
$$=\frac{3\cdot 3\cdot x\cdot x^4\cdot 2\cdot 8\cdot y^2\cdot y^7}{8\cdot y^2\cdot 3\cdot x\cdot 1}$$
$$=6x^4y^7$$

58. $\dfrac{4a^4}{b^4}$

59. $$\frac{6x+12}{x^8}\div\frac{x+2}{x^3}=\frac{6x+12}{x^8}\cdot\frac{x^3}{x+2}$$
$$=\frac{(6x+12)(x^3)}{x^8(x+2)}$$
$$=\frac{6(x+2)(x^3)}{x^3\cdot x^5(x+2)}$$
$$=\frac{6(x+2)(x^3)}{x^3\cdot x^5(x+2)}$$
$$=\frac{6}{x^5}$$

60. $\dfrac{3}{y^5}$

61.
$$\begin{aligned}\frac{x^2-4}{x^3} \div \frac{x^5-2x^4}{x+4} &= \frac{x^2-4}{x^3} \cdot \frac{x+4}{x^5-2x^4}\\ &= \frac{(x^2-4)(x+4)}{x^3(x^5-2x^4)}\\ &= \frac{(x+2)(x-2)(x+4)}{x^3(x^4)(x-2)}\\ &= \frac{(x+2)\cancel{(x-2)}(x+4)}{x^3(x^4)\cancel{(x-2)}}\\ &= \frac{(x+2)(x+4)}{x^7}\end{aligned}$$

62. $\dfrac{(y-3)(y+2)}{y^6}$

63.
$$\begin{aligned}\frac{25x^2-4}{x^2-9} \div \frac{5x-2}{x+3} &= \frac{25x^2-4}{x^2-9} \cdot \frac{x+3}{5x-2}\\ &= \frac{(25x^2-4)(x+3)}{(x^2-9)(5x-2)}\\ &= \frac{(5x+2)(5x-2)(x+3)}{(x+3)(x-3)(5x-2)}\\ &= \frac{(5x+2)\cancel{(5x-2)}\cancel{(x+3)}}{\cancel{(x+3)}(x-3)\cancel{(5x-2)}}\\ &= \frac{5x+2}{x-3}\end{aligned}$$

64. $\dfrac{-2a-1}{a+2}$

65.
$$\begin{aligned}\frac{5y-5x}{15y^3} \div \frac{x^2-y^2}{3x+3y} &= \frac{5y-5x}{15y^3} \cdot \frac{3x+3y}{x^2-y^2}\\ &= \frac{(5y-5x)(3x+3y)}{15y^3(x^2-y^2)}\\ &= \frac{5(y-x)(3)(x+y)}{5\cdot 3\cdot y^3(x+y)(x-y)}\\ &= \frac{5(-1)(x-y)(3)(x+y)}{5\cdot 3\cdot y^3(x+y)(x-y)}\\ &= \frac{\cancel{5}(-1)\cancel{(x-y)}\cancel{(3)}\cancel{(x+y)}}{\cancel{5}\cdot\cancel{3}\cdot y^3\cancel{(x+y)}\cancel{(x-y)}}\\ &= \frac{-1}{y^3},\ \text{or } -\frac{1}{y^3}\end{aligned}$$

66. $-x^2$

67.
$$\begin{aligned}&\frac{x^2-16}{x^2-10x+25} \div \frac{3x-12}{x^2-3x-10}\\ &= \frac{x^2-16}{x^2-10x+25} \cdot \frac{x^2-3x-10}{3x-12}\\ &= \frac{(x^2-16)(x^2-3x-10)}{(x^2-10x+25)(3x-12)}\\ &= \frac{(x+4)(x-4)(x-5)(x+2)}{(x-5)(x-5)(3)(x-4)}\\ &= \frac{(x+4)\cancel{(x-4)}\cancel{(x-5)}(x+2)}{\cancel{(x-5)}(x-5)(3)\cancel{(x-4)}}\\ &= \frac{(x+4)(x+2)}{3(x-5)}\end{aligned}$$

68. $\dfrac{(y+6)(y+3)}{3(y-4)}$

69.
$$\begin{aligned}&\frac{y^3+3y}{y^2-9} \div \frac{y^2+5y-14}{y^2+4y-21}\\ &= \frac{y^3+3y}{y^2-9} \cdot \frac{y^2+4y-21}{y^2+5y-14}\\ &= \frac{(y^3+3y)(y^2+4y-21)}{(y^2-9)(y^2+5y-14)}\\ &= \frac{y(y^2+3)(y+7)(y-3)}{(y+3)(y-3)(y+7)(y-2)}\\ &= \frac{y(y^2+3)\cancel{(y+7)}\cancel{(y-3)}}{(y+3)\cancel{(y-3)}\cancel{(y+7)}(y-2)}\\ &= \frac{y(y^2+3)}{(y+3)(y-2)}\end{aligned}$$

70. $\dfrac{a(a^2+4)}{(a+4)(a+3)}$

71.
$$\begin{aligned}&\frac{x^3-64}{x^3+64} \div \frac{x^2-16}{x^2-4x+16}\\ &= \frac{x^3-64}{x^3+64} \cdot \frac{x^2-4x+16}{x^2-16}\\ &= \frac{(x^3-64)(x^2-4x+16)}{(x^3+64)(x^2-16)}\\ &= \frac{(x-4)(x^2+4x+16)(x^2-4x+16)}{(x+4)(x^2-4x+16)(x+4)(x-4)}\\ &= \frac{(x-4)(x^2-4x+16)}{(x-4)(x^2-4x+16)} \cdot \frac{x^2+4x+16}{(x+4)(x+4)}\\ &= \frac{x^2+4x+16}{(x+4)(x+4)},\ \text{or } \frac{x^2+4x+16}{(x+4)^2}\end{aligned}$$

72. $\dfrac{4y^2+6y+9}{(4y-1)(2y+3)}$

73.
$$\begin{aligned}&\frac{8a^3+b^3}{2a^2+3ab+b^2} \div \frac{8a^2-4ab+2b^2}{4a^2+4ab+b^2}\\ &= \frac{8a^3+b^3}{2a^2+3ab+b^2} \cdot \frac{4a^2+4ab+b^2}{8a^2-4ab+2b^2}\\ &= \frac{(8a^3+b^3)(4a^2+4ab+b^2)}{(2a^2+3ab+b^2)(8a^2-4ab+2b^2)}\\ &= \frac{(2a+b)(4a^2-2ab+b^2)(2a+b)(2a+b)}{(2a+b)(a+b)(2)(4a^2-2ab+b^2)}\\ &= \frac{(2a+b)(4a^2-2ab+b^2)}{(2a+b)(4a^2-2ab+b^2)} \cdot \frac{(2a+b)(2a+b)}{(a+b)(2)}\\ &= \frac{(2a+b)(2a+b)}{2(a+b)},\ \text{or } \frac{(2a+b)^2}{2(a+b)}\end{aligned}$$

74. $\dfrac{2(2x-y)}{x}$

75. $3x + y = 13, \quad (1)$

$x = y + 1 \quad (2)$

$3(y+1) + y = 13$ Substituting $y+1$ for x in (1)

$$3y + 3 + y = 13$$
$$4y + 3 = 13$$
$$4y = 10$$
$$y = \frac{10}{4}, \text{ or } \frac{5}{2}$$

$x = \frac{5}{2} + 1$ Substituting $\frac{5}{2}$ for y in (2)

$$x = \frac{7}{2}$$

The solution is $\left(\frac{7}{2}, \frac{5}{2}\right)$.

76. 29

77. $\frac{2}{3}(3x - 4) = 8$

$\frac{3}{2} \cdot \frac{2}{3}(3x - 4) = \frac{3}{2} \cdot 8$ Multiplying by $\frac{3}{2}$

$$3x - 4 = 12$$
$$3x = 16$$
$$x = \frac{16}{3}$$

78. The full price is \$8.

79.

80.

81.

82. (a) $\frac{2x + 2h + 3}{4x + 4h - 1}$; (b) $\frac{2x + 3}{8x - 9}$; (c) $\frac{x + 5}{4x - 1}$

83. To find the domain of f we set

$$x - 3 = 0$$
$$x = 3.$$

The domain of $f = \{x | x \text{ is a real number and } x \neq 3\}$.

Simplify: $f(x) = \frac{x^2 - 9}{x - 3} = \frac{(x+3)(x-3)}{(x-3) \cdot 1}$

$$= \frac{x - 3}{x - 3} \cdot \frac{x + 3}{1} = x + 3$$

Graph $f(x) = x + 3$ using the domain found above.

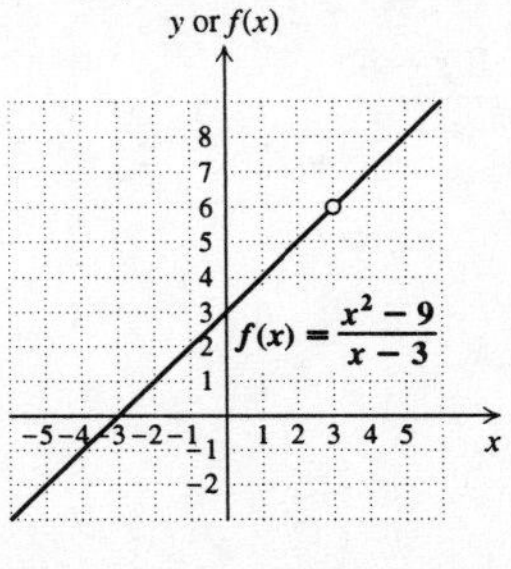

84. $\frac{2s}{r + 2s}$

85. $\left[\frac{d^2 - d}{d^2 - 6d + 8} \cdot \frac{d - 2}{d^2 + 5d}\right] \div \frac{5d}{d^2 - 9d + 20}$

$$= \left[\frac{d^2 - d}{d^2 - 6d + 8} \cdot \frac{d - 2}{d^2 + 5d}\right] \cdot \frac{d^2 - 9d + 20}{5d}$$
$$= \frac{(d^2 - d)(d - 2)(d^2 - 9d + 20)}{(d^2 - 6d + 8)(d^2 + 5d)(5d)}$$
$$= \frac{d(d-1)(d-2)(d-5)(d-4)}{(d-4)(d-2)(d)(d+5)(5d)}$$
$$= \frac{d(d-1)(d-2)(d-5)(d-4)}{(d-4)(d-2)(d)(d+5)(5d)}$$
$$= \frac{(d-1)(d-5)}{5d(d+5)}$$

86. $\frac{x - 3}{(x+1)(x+3)}$

87. $\frac{10x - 20 - 5x^2}{x^3 + 8}$

$$= \frac{-5(-2x + 4 + x^2)}{(x+2)(x^2 - 2x + 4)}$$
$$= \frac{-5(x^2 - 2x + 4)}{(x+2)(x^2 - 2x + 4)}$$
$$= \frac{-5}{x + 2} \cdot \frac{x^2 - 2x + 4}{x^2 - 2x + 4}$$
$$= \frac{-5}{x+2}, \text{ or } -\frac{5}{x+2}$$

88. $\frac{m - t}{m + t + 1}$

89. $\frac{a^3 - 2a^2 + 2a - 4}{a^3 - 2a^2 - 3a + 6}$

$$= \frac{a^2(a-2) + 2(a-2)}{a^2(a-2) - 3(a-2)}$$
$$= \frac{(a-2)(a^2+2)}{(a-2)(a^2-3)}$$
$$= \frac{(a-2)(a^2+2)}{(a-2)(a^2-3)}$$
$$= \frac{a^2 + 2}{a^2 - 3}$$

90. $\frac{x^2 + xy + y^2 + x + y}{x - y}$

91. $\dfrac{u^6+v^6+2u^3v^3}{u^3-v^3+u^2v-uv^2}$

$= \dfrac{u^6+2u^3v^3+v^6}{(u^3-v^3)+(u^2v-uv^2)}$

$= \dfrac{(u^3+v^3)^2}{(u-v)(u^2+uv+v^2)+uv(u-v)}$

$= \dfrac{[(u+v)(u^2-uv+v^2)]^2}{(u-v)(u^2+uv+v^2+uv)}$

$= \dfrac{(u+v)^2(u^2-uv+v^2)^2}{(u-v)(u^2+2uv+v^2)}$

$= \dfrac{(u+v)^2(u^2-uv+v^2)^2}{(u-v)(u+v)^2}$

$= \dfrac{\cancel{(u+v)^2}(u^2-uv+v^2)^2}{(u-v)\cancel{(u+v)^2}}$

$= \dfrac{(u^2-uv+v^2)^2}{u-v}$

92. $-\dfrac{2x}{x-1}$

93. a) $(f\cdot g)(x) = \dfrac{4}{x^2-1}\cdot\dfrac{4x^2+8x+4}{x^3-1}$

$= \dfrac{4(4x^2+8x+4)}{(x^2-1)(x^3-1)}$

$= \dfrac{4\cdot 4(x+1)(x+1)}{(x+1)(x-1)(x-1)(x^2+x+1)}$

$= \dfrac{4\cdot 4\cancel{(x+1)}(x+1)}{\cancel{(x+1)}(x-1)(x-1)(x^2+x+1)}$

$= \dfrac{16(x+1)}{(x-1)^2(x^2+x+1)}$

(Note that $x\neq -1$ is an additional restriction, since -1 is not in the domain of f.)

b) $(f/g)(x) = \dfrac{4}{x^2-1}\div\dfrac{4x^2+8x+4}{x^3-1}$

$= \dfrac{4}{x^2-1}\cdot\dfrac{x^3-1}{4x^2+8x+4}$

$= \dfrac{4(x^3-1)}{(x^2-1)(4x^2+8x+4)}$

$= \dfrac{4(x-1)(x^2+x+1)}{(x+1)(x-1)(4)(x+1)(x+1)}$

$= \dfrac{\cancel{4}\cancel{(x-1)}(x^2+x+1)}{(x+1)\cancel{(x-1)}\cancel{(4)}(x+1)(x+1)}$

$= \dfrac{x^2+x+1}{(x+1)^3}$

(Note that $x\neq 1$ is an additional restriction, since 1 is not in the domain of either f or g.)

c) $(g/f)(x) = \dfrac{1}{(f/g)(x)}$

$= \dfrac{(x+1)^3}{x^2+x+1}$ (See part (b) above.)

(Note that $x\neq -1$ and $x\neq 1$ are restrictions, since -1 is not in the domain of f and 1 is not in the domain of either f or g.)

94.

95.

96. One method is to graph $y_1 = \dfrac{x^2-16}{x+2}$ and $y_2 = x-8$ using DOT mode and observe that the graphs do not coincide. We could also check the table of values for y_1 and y_2 and observe that $y_1 \neq y_2$ for all values of x for which both functions are defined. We could also enter $y_1 = \dfrac{x^2-16}{x+2} - (x-8)$ and use a table or the TRACE feature to show that y_1 is not always 0.

97.

Exercise Set 6.2

1. $\dfrac{4}{3a}+\dfrac{8}{3a}$

$= \dfrac{12}{3a}$ Adding the numerators. The denominator is unchanged.

$= \dfrac{3\cdot 4}{3\cdot a}$

$= \dfrac{\cancel{3}\cdot 4}{\cancel{3}\cdot a}$

$= \dfrac{4}{a}$

2. $\dfrac{4}{y}$

3. $\dfrac{3}{4a^2b}-\dfrac{7}{4a^2b}=\dfrac{-4}{4a^2b}=\dfrac{-1\cdot 4}{4a^2b}=\dfrac{-1\cdot\cancel{4}}{\cancel{4}a^2b}=-\dfrac{1}{a^2b}$

4. $\dfrac{1}{3m^2n^2}$

5. $\dfrac{a-5b}{a+b}+\dfrac{a+7b}{a+b}=\dfrac{2a+2b}{a+b}$

$= \dfrac{2(a+b)}{a+b}$

$= \dfrac{2\cancel{(a+b)}}{1\cancel{(a+b)}}$

$= 2$

6. 2

7. $\dfrac{4y+2}{y-2}-\dfrac{y-3}{y-2}=\dfrac{4y+2-(y-3)}{y-2}$

$= \dfrac{4y+2-y+3}{y-2}$

$= \dfrac{3y+5}{y-2}$

8. $\dfrac{2t+4}{t-4}$

9. $$\begin{aligned}\frac{3x-4}{x^2-5x+4}+\frac{3-2x}{x^2-5x+4}&=\frac{3x-4+3-2x}{x^2-5x+4}\\&=\frac{x-1}{(x-4)(x-1)}\\&=\frac{1\cdot\cancel{(x-1)}}{(x-4)\cancel{(x-1)}}\\&=\frac{1}{x-4}\end{aligned}$$

10. $\dfrac{1}{x-7}$

11. $$\begin{aligned}\frac{3a-2}{a^2-25}-\frac{4a-7}{a^2-25}&=\frac{3a-2-(4a-7)}{a^2-25}\\&=\frac{3a-2-4a+7}{a^2-25}\\&=\frac{-a+5}{a^2-25}\\&=\frac{-1(a-5)}{(a+5)(a-5)}\\&=\frac{-1\cancel{(a-5)}}{(a+5)\cancel{(a-5)}}\\&=\frac{-1}{a+5},\text{ or }-\frac{1}{a+5}\end{aligned}$$

12. $-\dfrac{1}{a+3}$

13. $$\begin{aligned}\frac{a^2}{a-b}+\frac{b^2}{b-a}&=\frac{a^2}{a-b}+\frac{-1}{-1}\cdot\frac{b^2}{b-a}\\&=\frac{a^2}{a-b}+\frac{-b^2}{a-b}\\&=\frac{a^2-b^2}{a-b}=\frac{(a+b)(a-b)}{a-b}\\&=\frac{(a+b)\cancel{(a-b)}}{1\cdot\cancel{(a-b)}}=a+b\end{aligned}$$

14. $-(s+r)$

15. $\dfrac{3}{x}-\dfrac{8}{-x}=\dfrac{3}{x}+(-1)\cdot\dfrac{8}{-x}=\dfrac{3}{x}+\dfrac{1}{-1}\cdot\dfrac{8}{-x}=$
$\dfrac{3}{x}+\dfrac{8}{x}=\dfrac{11}{x}$

16. $\dfrac{7}{a}$

17. $$\begin{aligned}\frac{x-7}{x^2-16}-\frac{x-1}{16-x^2}&=\frac{x-7}{x^2-16}+(-1)\cdot\frac{x-1}{16-x^2}\\&=\frac{x-7}{x^2-16}+\frac{1}{-1}\cdot\frac{x-1}{16-x^2}\\&=\frac{x-7}{x^2-16}+\frac{x-1}{x^2-16}\\&=\frac{2x-8}{x^2-16}\\&=\frac{2(x-4)}{(x+4)(x-4)}\\&=\frac{2\cancel{(x-4)}}{(x+4)\cancel{(x-4)}}\\&=\frac{2}{x+4}\end{aligned}$$

18. $-\dfrac{1}{y+5}$

19. $$\begin{aligned}\frac{t^2+3}{t^4-16}+\frac{7}{16-t^4}&=\frac{t^2+3}{t^4-16}+\frac{-1}{-1}\cdot\frac{7}{16-t^4}\\&=\frac{t^2+3}{t^4-16}+\frac{-7}{t^4-16}\\&=\frac{t^2-4}{t^4-16}\\&=\frac{(t+2)(t-2)}{(t^2+4)(t+2)(t-2)}\\&=\frac{1\cdot\cancel{(t+2)}\cancel{(t-2)}}{(t^2+4)\cancel{(t+2)}\cancel{(t-2)}}\\&=\frac{1}{t^2+4}\end{aligned}$$

20. $\dfrac{1}{y^2+9}$

21. $$\begin{aligned}\frac{m-3n}{m^3-n^3}-\frac{2n}{n^3-m^3}&=\frac{m-3n}{m^3-n^3}+\frac{1}{-1}\cdot\frac{2n}{n^3-m^3}\\&=\frac{m-3n}{m^3-n^3}+\frac{2n}{m^3-n^3}\\&=\frac{m-n}{m^3-n^3}\\&=\frac{m-n}{(m-n)(m^2+mn+n^2)}\\&=\frac{1\cdot\cancel{(m-n)}}{\cancel{(m-n)}(m^2+mn+n^2)}\\&=\frac{1}{m^2+mn+n^2}\end{aligned}$$

22. $\dfrac{1}{r^2+rs+s^2}$

23. $$\frac{a+2}{a-4}+\frac{a-2}{a+3}$$
[LCD is $(a-4)(a+3)$.]
$$\begin{aligned}&=\frac{a+2}{a-4}\cdot\frac{a+3}{a+3}+\frac{a-2}{a+3}\cdot\frac{a-4}{a-4}\\&=\frac{(a^2+5a+6)+(a^2-6a+8)}{(a-4)(a+3)}\\&=\frac{2a^2-a+14}{(a-4)(a+3)}\end{aligned}$$

24. $\dfrac{2a^2+22}{(a-5)(a+4)}$

25. $$2+\frac{x-3}{x+1}=\frac{2}{1}+\frac{x-3}{x+1}$$
[LCD is $x+1$.]
$$\begin{aligned}&=\frac{2}{1}\cdot\frac{x+1}{x+1}+\frac{x-3}{x+1}\\&=\frac{(2x+2)+(x-3)}{x+1}\\&=\frac{3x-1}{x+1}\end{aligned}$$

26. $\dfrac{4y-13}{y-5}$

27. $\dfrac{4xy}{x^2-y^2}+\dfrac{x-y}{x+y}$

$=\dfrac{4xy}{(x+y)(x-y)}+\dfrac{x-y}{x+y}$

LCD is $(x+y)(x-y)$.]

$=\dfrac{4xy}{(x+y)(x-y)}+\dfrac{x-y}{x+y}\cdot\dfrac{x-y}{x-y}$

$=\dfrac{4xy+x^2-2xy+y^2}{(x+y)(x-y)}$

$=\dfrac{x^2+2xy+y^2}{(x+y)(x-y)}=\dfrac{(x+y)(x+y)}{(x+y)(x-y)}$

$=\dfrac{\cancel{(x+y)}(x+y)}{\cancel{(x+y)}(x-y)}=\dfrac{x+y}{x-y}$

28. $\dfrac{a^2+7ab+b^2}{(a+b)(a-b)}$

29. $\dfrac{8}{2x^2-7x+5}+\dfrac{3x+2}{2x^2-x-10}$

$=\dfrac{8}{(2x-5)(x-1)}+\dfrac{3x+2}{(2x-5)(x+2)}$

[LCD is $(2x-5)(x-1)(x+2)$.]

$=\dfrac{8}{(2x-5)(x-1)}\cdot\dfrac{x+2}{x+2}+\dfrac{3x+2}{(2x-5)(x+2)}\cdot\dfrac{x-1}{x-1}$

$=\dfrac{8x+16+3x^2-x-2}{(2x-5)(x-1)(x+2)}$

$=\dfrac{3x^2+7x+14}{(2x-5)(x-1)(x+2)}$

30. $\dfrac{3y-4}{(y-1)(y-2)}$

31. $\dfrac{4}{x+1}+\dfrac{x+2}{x^2-1}+\dfrac{3}{x-1}$

$=\dfrac{4}{x+1}+\dfrac{x+2}{(x+1)(x-1)}+\dfrac{3}{x-1}$

[LCD is $(x+1)(x-1)$.]

$=\dfrac{4}{x+1}\cdot\dfrac{x-1}{x-1}+\dfrac{x+2}{(x+1)(x-1)}+\dfrac{3}{x-1}\cdot\dfrac{x+1}{x+1}$

$=\dfrac{4x-4+x+2+3x+3}{(x+1)(x-1)}$

$=\dfrac{8x+1}{(x+1)(x-1)}$

32. $\dfrac{4y+17}{(y+2)(y-2)}$

33. $\dfrac{x+6}{5x+10}-\dfrac{x-2}{4x+8}$

$=\dfrac{x+6}{5(x+2)}-\dfrac{x-2}{4(x+2)}$

[LCD is $5\cdot 4(x+2)$.]

$=\dfrac{x+6}{5(x+2)}\cdot\dfrac{4}{4}-\dfrac{x-2}{4(x+2)}\cdot\dfrac{5}{5}$

$=\dfrac{4(x+6)-5(x-2)}{5\cdot 4(x+2)}$

$=\dfrac{4x+24-5x+10}{5\cdot 4(x+2)}$

$=\dfrac{-x+34}{5\cdot 4(x+2)}$, or $\dfrac{-x+34}{20(x+2)}$

34. $\dfrac{-2a+14}{15(a+5)}$

35. $\dfrac{5ab}{a^2-b^2}-\dfrac{a-b}{a+b}$

$=\dfrac{5ab}{(a+b)(a-b)}-\dfrac{a-b}{a+b}$

[LCD is $(a+b)(a-b)$.]

$=\dfrac{5ab}{(a+b)(a-b)}-\dfrac{a-b}{a+b}\cdot\dfrac{a-b}{a-b}$

$=\dfrac{5ab-(a^2-2ab+b^2)}{(a+b)(a-b)}$

$=\dfrac{5ab-a^2+2ab-b^2}{(a+b)(a-b)}$

$=\dfrac{-a^2+7ab-b^2}{(a+b)(a-b)}$

36. $\dfrac{-x^2+4xy-y^2}{(x+y)(x-y)}$

37. $\dfrac{x}{x^2+9x+20}-\dfrac{4}{x^2+7x+12}$

$=\dfrac{x}{(x+5)(x+4)}-\dfrac{4}{(x+3)(x+4)}$

[LCD is $(x+5)(x+4)(x+3)$.]

$=\dfrac{x}{(x+5)(x+4)}\cdot\dfrac{x+3}{x+3}-\dfrac{4}{(x+3)(x+4)}\cdot\dfrac{x+5}{x+5}$

$=\dfrac{x^2+3x-(4x+20)}{(x+5)(x+4)(x+3)}$

$=\dfrac{x^2+3x-4x-20}{(x+5)(x+4)(x+3)}$

$=\dfrac{x^2-x-20}{(x+5)(x+4)(x+3)}$

$=\dfrac{(x-5)(x+4)}{(x+5)(x+4)(x+3)}$

$=\dfrac{(x-5)\cancel{(x+4)}}{(x+5)\cancel{(x+4)}(x+3)}$

$=\dfrac{x-5}{(x+5)(x+3)}$

38. $\dfrac{x-6}{(x+6)(x+4)}$

39. $\dfrac{3y}{y^2-7y+10} - \dfrac{2y}{y^2-8y+15}$

$= \dfrac{3y}{(y-5)(y-2)} - \dfrac{2y}{(y-5)(y-3)}$

[LCD is $(y-5)(y-2)(y-3)$.]

$= \dfrac{3y}{(y-5)(y-2)} \cdot \dfrac{y-3}{y-3} - \dfrac{2y}{(y-5)(y-3)} \cdot \dfrac{y-2}{y-2}$

$= \dfrac{3y^2-9y-(2y^2-4y)}{(y-5)(y-2)(y-3)}$

$= \dfrac{3y^2-9y-2y^2+4y}{(y-5)(y-2)(y-3)}$

$= \dfrac{y^2-5y}{(y-5)(y-2)(y-3)} = \dfrac{y(y-5)}{(y-5)(y-2)(y-3)}$

$= \dfrac{y(\cancel{y-5})}{(\cancel{y-5})(y-2)(y-3)} = \dfrac{y}{(y-2)(y-3)}$

40. $\dfrac{2x^2+21x}{(x-4)(x-2)(x+3)}$

41. $\dfrac{2x+1}{x-y} + \dfrac{5x^2-5xy}{x^2-2xy+y^2}$

$= \dfrac{2x+1}{x-y} + \dfrac{5x(x-y)}{(x-y)(x-y)}$

$= \dfrac{2x+1}{x-y} + \dfrac{5x(\cancel{x-y})}{(\cancel{x-y})(x-y)}$

$= \dfrac{2x+1}{x-y} + \dfrac{5x}{x-y}$

$= \dfrac{7x+1}{x-y}$

42. $\dfrac{2}{a-b}$

43. $\dfrac{3y+2}{y^2+5y-24} + \dfrac{7}{y^2+4y-32}$

$= \dfrac{3y+2}{(y+8)(y-3)} + \dfrac{7}{(y+8)(y-4)}$

[LCD is $(y+8)(y-3)(y-4)$.]

$= \dfrac{3y+2}{(y+8)(y-3)} \cdot \dfrac{y-4}{y-4} + \dfrac{7}{(y+8)(y-4)} \cdot \dfrac{y-3}{y-3}$

$= \dfrac{3y^2-10y-8+7y-21}{(y+8)(y-3)(y-4)}$

$= \dfrac{3y^2-3y-29}{(y+8)(y-3)(y-4)}$

44. $\dfrac{5x^2-11x-6}{(x-5)(x-2)(x-3)}$

45. $\dfrac{a-3}{a^2-16} - \dfrac{3a-2}{a^2+2a-24}$

$= \dfrac{a-3}{(a+4)(a-4)} - \dfrac{3a-2}{(a+6)(a-4)}$

[LCD is $(a+4)(a-4)(a+6)$.]

$= \dfrac{a-3}{(a+4)(a-4)} \cdot \dfrac{a+6}{a+6} - \dfrac{3a-2}{(a+6)(a-4)} \cdot \dfrac{a+4}{a+4}$

$= \dfrac{(a-3)(a+6)-(3a-2)(a+4)}{(a+4)(a-4)(a+6)}$

$= \dfrac{a^2+3a-18-(3a^2+10a-8)}{(a+4)(a-4)(a+6)}$

$= \dfrac{a^2+3a-18-3a^2-10a+8}{(a+4)(a-4)(a+6)}$

$= \dfrac{-2a^2-7a-10}{(a+4)(a-4)(a+6)}$

46. $\dfrac{-2t^2+13t-7}{(t+3)(t-3)(t-1)}$

47. $\dfrac{2}{a^2-5a+4} + \dfrac{-2}{a^2-4}$

$= \dfrac{2}{(a-4)(a-1)} + \dfrac{-2}{(a+2)(a-2)}$

[LCD is $(a-4)(a-1)(a+2)(a-2)$.]

$= \dfrac{2}{(a-4)(a-1)} \cdot \dfrac{(a+2)(a-2)}{(a+2)(a-2)} +$

$\dfrac{-2}{(a+2)(a-2)} \cdot \dfrac{(a-4)(a-1)}{(a-4)(a-1)}$

$= \dfrac{2(a^2-4)-2(a^2-5a+4)}{(a-4)(a-1)(a+2)(a-2)}$

$= \dfrac{2a^2-8-2a^2+10a-8}{(a-4)(a-1)(a+2)(a-2)}$

$= \dfrac{10a-16}{(a-4)(a-1)(a+2)(a-2)}$, or

$\dfrac{10a-16}{(a^2-5a+4)(a^2-4)}$

48. $\dfrac{21a-45}{(a-6)(a-1)(a+3)(a-3)}$

49. $3 + \dfrac{t}{t+2} - \dfrac{2}{t^2-4} = \dfrac{3}{1} + \dfrac{t}{t+2} - \dfrac{2}{(t+2)(t-2)}$

[LCD is $(t+2)(t-2)$.]

$= \dfrac{3}{1} \cdot \dfrac{(t+2)(t-2)}{(t+2)(t-2)} + \dfrac{t}{t+2} \cdot \dfrac{t-2}{t-2} - \dfrac{2}{(t+2)(t-2)}$

$= \dfrac{3t^2-12+t^2-2t-2}{(t+2)(t-2)}$

$= \dfrac{4t^2-2t-14}{(t+2)(t-2)}$

50. $\dfrac{3t^2+3t-21}{(t+3)(t-3)}$

51. $\dfrac{2}{y+3} - \dfrac{y}{y-1} + \dfrac{y^2+2}{y^2+2y-3}$

$= \dfrac{2}{y+3} - \dfrac{y}{y-1} + \dfrac{y^2+2}{(y+3)(y-1)}$

[LCD is $(y+3)(y-1)$.]

$= \dfrac{2}{y+3}\cdot\dfrac{y-1}{y-1} - \dfrac{y}{y-1}\cdot\dfrac{y+3}{y+3} + \dfrac{y^2+2}{(y+3)(y-1)}$

$= \dfrac{2(y-1)-y(y+3)+y^2+2}{(y+3)(y-1)}$

$= \dfrac{2y-2-y^2-3y+y^2+2}{(y+3)(y-1)}$

$= \dfrac{-y}{(y+3)(y-1)}$, or $-\dfrac{y}{(y+3)(y-1)}$

52. 0

53. $\dfrac{5y}{1-2y} - \dfrac{2y}{2y+1} + \dfrac{3}{4y^2-1}$

$= \dfrac{-1}{-1}\cdot\dfrac{5y}{1-2y} - \dfrac{2y}{2y+1} + \dfrac{3}{(2y+1)(2y-1)}$

$= \dfrac{-5y}{2y-1} - \dfrac{2y}{2y+1} + \dfrac{3}{(2y+1)(2y-1)}$

[LCD is $(2y-1)(2y+1)$.]

$= \dfrac{-5y}{2y-1}\cdot\dfrac{2y+1}{2y+1} - \dfrac{2y}{2y+1}\cdot\dfrac{2y-1}{2y-1} + \dfrac{3}{(2y+1)(2y-1)}$

$= \dfrac{-5y(2y+1)-2y(2y-1)+3}{(2y+1)(2y-1)}$

$= \dfrac{-10y^2-5y-4y^2+2y+3}{(2y+1)(2y-1)}$

$= \dfrac{-14y^2-3y+3}{(2y+1)(2y-1)}$

54. $\dfrac{-3x^2-3x-4}{(x+1)(x-1)}$

55. $\dfrac{2}{x^2-5x+6} - \dfrac{4}{x^2-2x-3} + \dfrac{2}{x^2+4x+3}$

$= \dfrac{2}{(x-3)(x-2)} - \dfrac{4}{(x-3)(x+1)} + \dfrac{2}{(x+3)(x+1)}$

[LCD is $(x-3)(x-2)(x+1)(x+3)$.]

$= \dfrac{2}{(x-3)(x-2)}\cdot\dfrac{(x+1)(x+3)}{(x+1)(x+3)} - \dfrac{4}{(x-3)(x+1)}\cdot\dfrac{(x-2)(x+3)}{(x-2)(x+3)} + \dfrac{2}{(x+3)(x+1)}\cdot\dfrac{(x-3)(x-2)}{(x-3)(x-2)}$

$= \dfrac{2(x+1)(x+3)-4(x-2)(x+3)+2(x-3)(x-2)}{(x-3)(x-2)(x+1)(x+3)}$

$= \dfrac{2x^2+8x+6-4x^2-4x+24+2x^2-10x+12}{(x-3)(x-2)(x+1)(x+3)}$

$= \dfrac{-6x+42}{(x-3)(x-2)(x+1)(x+3)}$

56. $\dfrac{4t+26}{(t+3)(t+2)(t+1)(t-4)}$

57. $\dfrac{15x^{-7}y^{12}z^4}{35x^{-2}y^6z^{-3}} = \dfrac{15}{35}x^{-7-(-2)}y^{12-6}z^{4-(-3)}$

$= \dfrac{3}{7}x^{-5}y^6z^7 = \dfrac{3}{7}\cdot\dfrac{1}{x^5}\cdot y^6z^7$

$= \dfrac{3y^6z^7}{7x^5}$

58. $y = \dfrac{5}{4}x + \dfrac{11}{2}$

59. ***Familiarize***. We let x, y, and z represent the number of rolls of dimes, nickels, and quarters, respectively.

Coins	Number of rolls	Value per roll	Total Value
Dimes	x	50×0.10, or 5.00	$5x$
Nickels	y	40×0.05, or 2.00	$2y$
Quarters	z	40×0.25, or 10.00	$10z$
Total	12		\$70.00

Translate. The number of rolls of nickels is three more than the number of rolls of dimes. This gives us one equation.

$y = x + 3$

From the table we get two more equations.

$x + y + z = 12$

$5x + 2y + 10z = 70$

Carry out. Solving the system we get the ordered triple $(2, 5, 5)$.

Check.

2 rolls of dimes $= 2 \times \$5$, or \$10

5 rolls of nickels $= 5 \times \$2$, or \$10

5 rolls of quarters $= 5 \times \$10$, or \$50

The total value is $\$10 + \$10 + \$50 = \$70$.

The total number of rolls of coins is $2+5+5$, or 12, and the number of rolls of nickels, 5, is three more than the number of rolls of dimes, 2. The numbers check.

State. Robert has 2 rolls of dimes, 5 rolls of nickels, and 5 rolls of quarters.

60. 4 30-min, 8 60-min

61.

62.

63.

64.

65. The smallest number of beats possible is the least common multiple of 6 and 4.

$6 = 2 \cdot 3$

$4 = 2 \cdot 2$

$\text{LCM} = 2 \cdot 3 \cdot 2, \text{ or } 12$

A measure should be divided into 12 beats.

66. Every 420 years

67.
$$x^8 - x^4 = x^4(x^2+1)(x+1)(x-1)$$
$$x^5 - x^2 = x^2(x-1)(x^2+x+1)$$
$$x^5 - x^3 = x^3(x+1)(x-1)$$
$$x^5 + x^2 = x^2(x+1)(x^2-x+1)$$

The LCM is

$x^4(x^2+1)(x+1)(x-1)(x^2+x+1)(x^2-x+1)$.

68. $2ab(a^2+ab+b^2)(a+b)^2(a^2-ab+b^2)(a-b)(2b+3a)$

69. The LCM is $8a^4b^7$.

One expression is $2a^3b^7$.

Then the other expression must contain 8, a^4, and one of the following:

no factor of b, b, b^2, b^3, b^4, b^5, b^6, or b^7.

Thus, all the possibilities for the other expression are $8a^4$, $8a^4b$, $8a^4b^2$, $8a^4b^3$, $8a^4b^4$, $8a^4b^5$, $8a^4b^6$, $8a^4b^7$.

70. Domain: $\{x|x \text{ is a real number and } x \neq -2 \text{ and } x \neq 1\}$; range: $\{y|y \text{ is a real number and } y \neq 2 \text{ and } y \neq 3\}$

71.
$$\begin{aligned}(f+g)(x) &= \frac{x^3}{x^2-4} + \frac{x^2}{x^2+3x-10} \\ &= \frac{x^3}{(x+2)(x-2)} + \frac{x^2}{(x+5)(x-2)} \\ &= \frac{x^3(x+5)+x^2(x+2)}{(x+2)(x-2)(x+5)} \\ &= \frac{x^4+5x^3+x^3+2x^2}{(x+2)(x-2)(x+5)} \\ &= \frac{x^4+6x^3+2x^2}{(x+2)(x-2)(x+5)}\end{aligned}$$

72. $\dfrac{x^4+4x^3-2x^2}{(x+2)(x-2)(x+5)}$

73.
$$\begin{aligned}(f \cdot g)(x) &= \frac{x^3}{x^2-4} \cdot \frac{x^2}{x^2+3x-10} \\ &= \frac{x^5}{(x^2-4)(x^2+3x-10)}\end{aligned}$$

74. $\dfrac{x(x+5)}{x+2}$

(Note that $x \neq 0$, $x \neq -5$, and $x \neq 2$ are additional restrictions, since $g(0) = 0$, -5 is not in the domain of g, and 2 is not in the domain of either f or g.)

75.
$$\begin{aligned}&5(x-3)^{-1} + 4(x+3)^{-1} - 2(x+3)^{-2} \\ &= \frac{5}{x-3} + \frac{4}{x+3} - \frac{2}{(x+3)^2} \\ &[\text{LCD is } (x-3)(x+3)^2.] \\ &= \frac{5(x+3)^2 + 4(x-3)(x+3) - 2(x-3)}{(x-3)(x+3)^2} \\ &= \frac{5x^2+30x+45+4x^2-36-2x+6}{(x-3)(x+3)^2} \\ &= \frac{9x^2+28x+15}{(x-3)(x+3)^2}\end{aligned}$$

76. $\dfrac{-5y-23}{2y-5}$, or $\dfrac{5y+23}{5-2y}$

77.
$$\begin{aligned}&\frac{x+4}{6x^2-20x}\left(\frac{x}{x^2-x-20} + \frac{2}{x+4}\right) \\ &= \frac{x+4}{2x(3x-10)}\left(\frac{x}{(x-5)(x+4)} + \frac{2}{x+4}\right) \\ &= \frac{x+4}{2x(3x-10)}\left(\frac{x+2(x-5)}{(x-5)(x+4)}\right) \\ &= \frac{x+4}{2x(3x-10)}\left(\frac{x+2x-10}{(x-5)(x+4)}\right) \\ &= \frac{(x+4)(3x-10)}{2x(3x-10)(x-5)(x+4)} \\ &= \frac{\cancel{(x+4)}\cancel{(3x-10)}(1)}{2x\cancel{(3x-10)}(x-5)\cancel{(x+4)}} \\ &= \frac{1}{2x(x-5)}\end{aligned}$$

78. $\dfrac{3}{x+8}$

79.
$$\begin{aligned}&\frac{8t^5}{2t^2-10t+12} \div \left(\frac{2t}{t^2-8t+15} - \frac{3t}{t^2-7t+10}\right) \\ &= \frac{8t^5}{2t^2-10t+12} \div \left(\frac{2t}{(t-5)(t-3)} - \frac{3t}{(t-5)(t-2)}\right) \\ &= \frac{8t^5}{2t^2-10t+12} \div \left(\frac{2t(t-2)-3t(t-3)}{(t-5)(t-3)(t-2)}\right) \\ &= \frac{8t^5}{2t^2-10t+12} \div \left(\frac{2t^2-4t-3t^2+9t}{(t-5)(t-3)(t-2)}\right) \\ &= \frac{8t^5}{2t^2-10t+12} \div \frac{-t^2+5t}{(t-5)(t-3)(t-2)} \\ &= \frac{8t^5}{2(t-3)(t-2)} \cdot \frac{(t-5)(t-3)(t-2)}{-t(t-5)} \\ &= \frac{\cancel{2} \cdot 4 \cdot \cancel{t} \cdot t^4\cancel{(t-5)}\cancel{(t-3)}\cancel{(t-2)}}{\cancel{2}\cancel{(t-3)}\cancel{(t-2)}(-1)\cancel{(t)}\cancel{(t-5)}} \\ &= -4t^4\end{aligned}$$

80. $\dfrac{3t^2(t+3)}{-2t^2+13t-7}$

81. 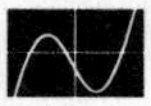

82. Domain: $\{x|x \text{ is a real number and } x \neq -1\}$; range: $\{y|y \text{ is a real number and } y \neq 3\}$

Exercise Set 6.3

1. $\dfrac{5+\frac{1}{a}}{\frac{1}{a}-2} = \dfrac{5+\frac{1}{a}}{\frac{1}{a}-2}\cdot\dfrac{a}{a}$ Multiplying by 1, using the LCD

$= \dfrac{\left(5+\frac{1}{a}\right)a}{\left(\frac{1}{a}-2\right)a}$ Multiplying the numerators and the denominator

$= \dfrac{5\cdot a+\frac{1}{a}\cdot a}{\frac{1}{a}\cdot a-2\cdot a}$

$= \dfrac{5a+\frac{\not a}{\not a}\cdot 1}{\frac{\not a}{\not a}\cdot 1-2a}$ Removing factors equal to 1

$= \dfrac{5a+1}{1-2a}$ Simplifying

2. $\dfrac{1+7y}{1-5y}$

3. $\dfrac{x-x^{-1}}{x+x^{-1}} = \dfrac{x-\frac{1}{x}}{x+\frac{1}{x}}$ Rewriting with positive exponents

$= \dfrac{x-\frac{1}{x}}{x+\frac{1}{x}}\cdot\dfrac{x}{x}$ Multiplying by 1, using the LCD

$= \dfrac{x\cdot x-\frac{1}{x}\cdot x}{x\cdot x+\frac{1}{x}\cdot x}$

$= \dfrac{x^2-1}{x^2+1}$

(Although the numerator can be factored, doing so does not lead to further simplification.)

4. $\dfrac{y^2+1}{y^2-1}$

5. $\dfrac{\frac{3}{x}+\frac{4}{y}}{\frac{4}{x}-\frac{3}{y}} = \dfrac{\frac{3}{x}+\frac{4}{y}}{\frac{4}{x}-\frac{3}{y}}\cdot\dfrac{xy}{xy}$ Multiplying by 1, using the LCD

$= \dfrac{\frac{3}{x}\cdot xy+\frac{4}{y}\cdot xy}{\frac{4}{x}\cdot xy-\frac{3}{y}\cdot xy}$

$= \dfrac{3y+4x}{4y-3x}$

6. $\dfrac{2z+5y}{z-4y}$

7. $\dfrac{\frac{x^2-y^2}{xy}}{\frac{x-y}{y}} = \dfrac{x^2-y^2}{xy}\cdot\dfrac{y}{x-y}$ Multiplying by the reciprocal of the divisor

$= \dfrac{(x+y)(x-y)\cdot y}{xy(x-y)}$

$= \dfrac{(x+y)\cancel{(x-y)}\cdot \cancel{y}}{x\cancel{y}\cancel{(x-y)}}$

$= \dfrac{x+y}{x}$

8. $\dfrac{a+b}{a}$

9. $\dfrac{\frac{3x}{y}-x}{2y-\frac{y}{x}} = \dfrac{\frac{3x}{y}-x}{2y-\frac{y}{x}}\cdot\dfrac{xy}{xy}$ Multiplying by 1, using the LCD

$= \dfrac{\frac{3x}{y}\cdot xy-x\cdot xy}{2y\cdot xy-\frac{y}{x}\cdot xy}$

$= \dfrac{3x^2-x^2y}{2xy^2-y^2}$

(Although both the numerator and the denominator can be factored, doing so does not lead to further simplification.)

10. $\dfrac{3}{3x+2}$

11. $\dfrac{a^{-1}+b^{-1}}{\frac{a^2-b^2}{ab}} = \dfrac{\frac{1}{a}+\frac{1}{b}}{\frac{a^2-b^2}{ab}}$

$= \dfrac{\frac{1}{a}+\frac{1}{b}}{\frac{a^2-b^2}{ab}}\cdot\dfrac{ab}{ab}$ Multiplying by 1, using the LCD

$= \dfrac{\frac{1}{a}\cdot ab+\frac{1}{b}\cdot ab}{\frac{a^2-b^2}{ab}\cdot ab}$

$= \dfrac{b+a}{a^2-b^2} = \dfrac{b+a}{(a+b)(a-b)}$

$= \dfrac{\cancel{(a+b)}\cdot(1)}{\cancel{(a+b)}(a-b)}$ $(b+a=a+b)$

$= \dfrac{1}{a-b}$

12. $\dfrac{1}{x-y}$

13. $\dfrac{\dfrac{1}{x+h}-\dfrac{1}{x}}{h} = \dfrac{\dfrac{1}{x+h}\cdot\dfrac{x}{x}-\dfrac{1}{x}\cdot\dfrac{x+h}{x+h}}{h}$ Adding in the numerator

$$= \dfrac{\dfrac{x-x-h}{x(x+h)}}{h} = \dfrac{\dfrac{-h}{x(x+h)}}{h}$$

$= \dfrac{-h}{x(x+h)}\cdot\dfrac{1}{h}$ Multiplying by the reciprocal of the divisor

$= \dfrac{-1\cdot \cancel{h}\cdot 1}{x(x+h)(\cancel{h})}$ $(-h=-1\cdot h)$

$$= -\dfrac{1}{x(x+h)}$$

14. $\dfrac{1}{a(a-h)}$

15. $\dfrac{\dfrac{a^2-4}{a^2+3a+2}}{\dfrac{a^2-5a-6}{a^2-6a-7}}$

$= \dfrac{a^2-4}{a^2+3a+2}\cdot\dfrac{a^2-6a-7}{a^2-5a-6}$ Multiplying by the reciprocal of the divisor

$$= \dfrac{(a+2)(a-2)}{(a+2)(a+1)}\cdot\dfrac{(a+1)(a-7)}{(a+1)(a-6)}$$

$$= \dfrac{(a+2)(a-2)(a+1)(a-7)}{(a+2)(a+1)(a+1)(a-6)}$$

$$= \dfrac{\cancel{(a+2)}(a-2)\cancel{(a+1)}(a-7)}{\cancel{(a+2)}\cancel{(a+1)}(a+1)(a-6)}$$

$$= \dfrac{(a-2)(a-7)}{(a+1)(a-6)}$$

16. $\dfrac{(x-4)(x-7)}{(x-5)(x+6)}$

17. $\dfrac{\dfrac{2}{y-3}+\dfrac{1}{y+1}}{\dfrac{3}{y+1}+\dfrac{4}{y-3}}$

$$= \dfrac{\dfrac{2}{y-3}+\dfrac{1}{y+1}}{\dfrac{3}{y+1}+\dfrac{4}{y-3}}\cdot\dfrac{(y-3)(y+1)}{(y-3)(y+1)}$$

$$= \dfrac{\dfrac{2}{y-3}\cdot(y-3)(y+1)+\dfrac{1}{y+1}\cdot(y-3)(y+1)}{\dfrac{3}{y+1}\cdot(y-3)(y+1)+\dfrac{4}{y-3}\cdot(y-3)(y+1)}$$

Multiplying by 1, using the LCD

$$= \dfrac{2(y+1)+(y-3)}{3(y-3)+4(y+1)}$$

$$= \dfrac{2y+2+y-3}{3y-9+4y+4}$$

$$= \dfrac{3y-1}{7y-5}$$

18. $\dfrac{4x-7}{7x-9}$

19. $\dfrac{a(a+3)^{-1}-2(a-1)^{-1}}{a(a+3)^{-1}-(a-1)^{-1}}$

$$= \dfrac{\dfrac{a}{a+3}-\dfrac{2}{a-1}}{\dfrac{a}{a+3}-\dfrac{1}{a-1}}$$

$$= \dfrac{\dfrac{a}{a+3}-\dfrac{2}{a-1}}{\dfrac{a}{a+3}-\dfrac{1}{a-1}}\cdot\dfrac{(a+3)(a-1)}{(a+3)(a-1)}$$

Multiplying by 1, using the LCD

$$= \dfrac{\dfrac{a}{a+3}\cdot(a+3)(a-1)-\dfrac{2}{a-1}\cdot(a+3)(a-1)}{\dfrac{a}{a+3}\cdot(a+3)(a-1)-\dfrac{1}{a-1}\cdot(a+3)(a-1)}$$

$$= \dfrac{a(a-1)-2(a+3)}{a(a-1)-(a+3)}$$

$$= \dfrac{a^2-a-2a-6}{a^2-a-a-3} = \dfrac{a^2-3a-6}{a^2-2a-3}$$

(Although the denominator can be factored, doing so does not lead to further simplification.)

20. $\dfrac{a^2-6a-6}{a^2-4a-2}$

21. $\dfrac{\dfrac{x}{x^2+3x-4}-\dfrac{1}{x^2+3x-4}}{\dfrac{x}{x^2+6x+8}+\dfrac{3}{x^2+6x+8}}$

$= \dfrac{\dfrac{x-1}{x^2+3x-4}}{\dfrac{x+3}{x^2+6x+8}}$ Adding in the numerator and the denominator

$$= \dfrac{x-1}{x^2+3x-4}\cdot\dfrac{x^2+6x+8}{x+3}$$

$$= \dfrac{(x-1)(x+4)(x+2)}{(x+4)(x-1)(x+3)}$$

$$= \dfrac{\cancel{(x-1)}\cancel{(x+4)}(x+2)}{\cancel{(x+4)}\cancel{(x-1)}(x+3)} = \dfrac{x+2}{x+3}$$

22. $\dfrac{x-4}{x-2}$

23. $$\frac{\frac{3}{a^2-9}+\frac{2}{a+3}}{\frac{4}{a^2-9}+\frac{1}{a+3}}$$

$$= \frac{\frac{3}{(a+3)(a-3)}+\frac{2}{a+3}}{\frac{4}{(a+3)(a-3)}+\frac{1}{a+3}}$$

$$= \frac{\frac{3}{(a+3)(a-3)}+\frac{2}{a+3}}{\frac{4}{(a+3)(a-3)}+\frac{1}{a+3}} \cdot \frac{(a+3)(a-3)}{(a+3)(a-3)}$$

Multiplying by 1, using the LCD

$$= \frac{\frac{3}{(a+3)(a-3)}\cdot(a+3)(a-3)+\frac{2}{a+3}\cdot(a+3)(a-3)}{\frac{4}{(a+3)(a-3)}\cdot(a+3)(a-3)+\frac{1}{a+3}\cdot(a+3)(a-3)}$$

$$= \frac{3+2(a-3)}{4+a-3}$$

$$= \frac{3+2a-6}{a+1} = \frac{2a-3}{a+1}$$

24. $\frac{a+1}{2a+5}$

25. $$\frac{\frac{4}{x^2-1}-\frac{3}{x+1}}{\frac{5}{x^2-1}-\frac{2}{x-1}}$$

$$= \frac{\frac{4}{(x+1)(x-1)}-\frac{3}{x+1}}{\frac{5}{(x+1)(x-1)}-\frac{2}{x-1}}$$

$$= \frac{\frac{4}{(x+1)(x-1)}-\frac{3}{x+1}}{\frac{5}{(x+1)(x-1)}-\frac{2}{x-1}} \cdot \frac{(x+1)(x-1)}{(x+1)(x-1)}$$

Multiplying by 1, using the LCD

$$= \frac{\frac{4}{(x+1)(x-1)}\cdot(x+1)(x-1)-\frac{3}{x+1}\cdot(x+1)(x-1)}{\frac{5}{(x+1)(x-1)}\cdot(x+1)(x-1)-\frac{2}{x-1}\cdot(x+1)(x-1)}$$

$$= \frac{4-3(x-1)}{5-2(x+1)}$$

$$= \frac{4-3x+3}{5-2x-2} = \frac{7-3x}{3-2x}$$

26. $\frac{-1-3x}{8-2x}$

27. $$\frac{\frac{y}{y^2-1}+\frac{3}{1-y^2}}{\frac{y^2}{y^2-1}+\frac{9}{1-y^2}}$$

$$= \frac{\frac{y}{y^2-1}+\frac{-1}{-1}\cdot\frac{3}{1-y^2}}{\frac{y^2}{y^2-1}+\frac{-1}{-1}\cdot\frac{9}{1-y^2}}$$

$$= \frac{\frac{y}{y^2-1}-\frac{3}{y^2-1}}{\frac{y^2}{y^2-1}-\frac{9}{y^2-1}}$$

$$= \frac{\frac{y-3}{y^2-1}}{\frac{y^2-9}{y^2-1}}$$ Adding in the numerator and the denominator

$$= \frac{y-3}{y^2-1}\cdot\frac{y^2-1}{y^2-9}$$ Multiplying by the reciprocal of the divisor

$$= \frac{(y-3)(y^2-1)}{(y^2-1)(y+3)(y-3)}$$

$$= \frac{\cancel{(y-3)}\cancel{(y^2-1)}(1)}{\cancel{(y^2-1)}(y+3)\cancel{(y-3)}}$$

$$= \frac{1}{y+3}$$

28. $\frac{1}{y+5}$

29. $$\frac{\frac{y^2}{y^2-9}-\frac{y}{y+3}}{\frac{y}{y^2-9}-\frac{1}{y-3}}$$

$$= \frac{\frac{y^2}{(y+3)(y-3)}-\frac{y}{y+3}}{\frac{y}{(y+3)(y-3)}-\frac{1}{y-3}}$$

$$= \frac{\frac{y^2}{(y+3)(y-3)}-\frac{y}{y+3}}{\frac{y}{(y+3)(y-3)}-\frac{1}{y-3}}\cdot\frac{(y+3)(y-3)}{(y+3)(y-3)}$$

Multiplying by 1, using the LCD

$$= \frac{\frac{y^2}{(y+3)(y-3)}\cdot(y+3)(y-3)-\frac{y}{y+3}\cdot(y+3)(y-3)}{\frac{y}{(y+3)(y-3)}\cdot(y+3)(y-3)-\frac{1}{y-3}\cdot(y+3)(y-3)}$$

$$= \frac{y^2-y(y-3)}{y-(y+3)} = \frac{y^2-y^2+3y}{y-y-3} = \frac{3y}{-3}$$

$$= \frac{\cancel{3}y}{-1\cdot\cancel{3}} = -y$$

30. $-y$

31. $\dfrac{\dfrac{a}{a+3}+\dfrac{4}{5a}}{\dfrac{a}{2a+6}+\dfrac{3}{a}}$

$=\dfrac{\dfrac{a}{a+3}+\dfrac{4}{5a}}{\dfrac{a}{2(a+3)}+\dfrac{3}{a}}$

$=\dfrac{\dfrac{a}{a+3}+\dfrac{4}{5a}}{\dfrac{a}{2(a+3)}\cdot\dfrac{3}{a}}\cdot\dfrac{10a(a+3)}{10a(a+3)}$

Multiplying by 1, using the LCD

$=\dfrac{\dfrac{a}{a+3}\cdot 10a(a+3)+\dfrac{4}{5a}\cdot 10a(a+3)}{\dfrac{a}{2(a+3)}\cdot 10a(a+3)+\dfrac{3}{a}\cdot 10a(a+3)}$

$=\dfrac{10a^2+8(a+3)}{5a^2+30(a+3)}=\dfrac{10a^2+8a+24}{5a^2+30a+90}$

$=\dfrac{2(5a^2+4a+12)}{5(a^2+6a+18)}$

32. $\dfrac{6a^2+30a+60}{3a^2+2a+4}$

33. $\dfrac{\dfrac{1}{x^2-3x+2}+\dfrac{1}{x^2-4}}{\dfrac{1}{x^2+4x+4}+\dfrac{1}{x^2-4}}$

$=\dfrac{\dfrac{1}{(x-1)(x-2)}+\dfrac{1}{(x+2)(x-2)}}{\dfrac{1}{(x+2)(x+2)}+\dfrac{1}{(x+2)(x-2)}}$

$=\dfrac{\dfrac{1}{(x-1)(x-2)}+\dfrac{1}{(x+2)(x-2)}}{\dfrac{1}{(x+2)(x+2)}+\dfrac{1}{(x+2)(x-2)}}\cdot\dfrac{(x-1)(x-2)(x+2)(x+2)}{(x-1)(x-2)(x+2)(x+2)}$

Multiplying by 1, using the LCD

$=\dfrac{(x+2)(x+2)+(x-1)(x+2)}{(x-1)(x-2)+(x-1)(x+2)}$

$=\dfrac{x^2+4x+4+x^2+x-2}{x^2-3x+2+x^2+x-2}$

$=\dfrac{2x^2+5x+2}{2x^2-2x}$

(Although both the numerator and the denominator can be factored, doing so will not lead to further simplification.)

34. $\dfrac{2x^2-5x-3}{2x^2+2x-4}$

35. $\dfrac{\dfrac{3}{a^2-4a+3}+\dfrac{3}{a^2-5a+6}}{\dfrac{3}{a^2-3a+2}+\dfrac{3}{a^2+3a-10}}$

$=\dfrac{\dfrac{3}{(a-1)(a-3)}+\dfrac{3}{(a-2)(a-3)}}{\dfrac{3}{(a-1)(a-2)}+\dfrac{3}{(a+5)(a-2)}}$

$=\dfrac{\dfrac{3}{(a-1)(a-3)}+\dfrac{3}{(a-2)(a-3)}}{\dfrac{3}{(a-1)(a-2)}+\dfrac{3}{(a+5)(a-2)}}\cdot\dfrac{(a-1)(a-3)(a-2)(a+5)}{(a-1)(a-3)(a-2)(a+5)}$

Multiplying by 1, using the LCD

$=\dfrac{3(a-2)(a+5)+3(a-1)(a+5)}{3(a-3)(a+5)+3(a-1)(a-3)}$

$=\dfrac{3[(a-2)(a+5)+(a-1)(a+5)]}{3[(a-3)(a+5)+(a-1)(a-3)]}$

$=\dfrac{\cancel{3}[(a-2)(a+5)+(a-1)(a+5)]}{\cancel{3}[(a-3)(a+5)+(a-1)(a-3)]}$

$=\dfrac{a^2+3a-10+a^2+4a-5}{a^2+2a-15+a^2-4a+3}$

$=\dfrac{2a^2+7a-15}{2a^2-2a-12}$

(Although both the numerator and the denominator can be factored, doing so will not lead to further simplification.)

36. $\dfrac{-a^2-21a-8}{a^2+3a-34}$

37. $\dfrac{\dfrac{y}{y^2-4}-\dfrac{y}{y^2+y-6}}{\dfrac{2y}{y^2-4}-\dfrac{2y}{y^2+5y+6}}$

$=\dfrac{\dfrac{y}{(y+2)(y-2)}-\dfrac{2y}{(y+3)(y-2)}}{\dfrac{2y}{(y+2)(y-2)}-\dfrac{y}{(y+3)(y+2)}}$

$=\dfrac{\dfrac{y}{(y+2)(y-2)}-\dfrac{2y}{(y+3)(y-2)}}{\dfrac{2y}{(y+2)(y-2)}-\dfrac{y}{(y+3)(y+2)}}\cdot\dfrac{(y+2)(y-2)(y+3)}{(y+2)(y-2)(y+3)}$

Multiplying by 1, using the LCD

$=\dfrac{y(y+3)-2y(y+2)}{2y(y+3)-y(y-2)}$

$=\dfrac{y^2+3y-2y^2-4y}{2y^2+6y-y^2+2y}=\dfrac{-y^2-y}{y^2+8y}$

$=\dfrac{y(-y-1)}{y(y+8)}=\dfrac{\cancel{y}(-y-1)}{\cancel{y}(y+8)}=\dfrac{-y-1}{y+8}$

38. $\dfrac{-2y^2+13y-21}{2(y^2-y-20)}$

39. $\dfrac{\dfrac{3}{x^2+2x-3}-\dfrac{1}{x^2-3x-10}}{\dfrac{3}{x^2-6x+5}-\dfrac{1}{x^2+5x+6}}$

$= \dfrac{\dfrac{3}{(x+3)(x-1)}-\dfrac{1}{(x-5)(x+2)}}{\dfrac{3}{(x-5)(x-1)}-\dfrac{1}{(x+3)(x+2)}}$

$= \dfrac{\dfrac{3}{(x+3)(x-1)}-\dfrac{1}{(x-5)(x+2)}}{\dfrac{3}{(x-5)(x-1)}-\dfrac{1}{(x+3)(x+2)}}\cdot \dfrac{(x+3)(x-1)(x-5)(x+2)}{(x+3)(x-1)(x-5)(x+2)}$

Multiplying by 1, using the LCD

$= \dfrac{3(x-5)(x+2)-(x+3)(x-1)}{3(x+3)(x+2)-(x-1)(x-5)}$

$= \dfrac{3(x^2-3x-10)-(x^2+2x-3)}{3(x^2+5x+6)-(x^2-6x+5)}$

$= \dfrac{3x^2-9x-30-x^2-2x+3}{3x^2+15x+18-x^2+6x-5}$

$= \dfrac{2x^2-11x-27}{2x^2+21x+13}$

40. $\dfrac{2a^2+4a+2}{2a^2+5a-3}$

41. $f(x) = x^2-3$

$f(-5) = (-5)^2-3 = 25-3 = 22$

42. $y = \dfrac{a-bx}{b}$

43.

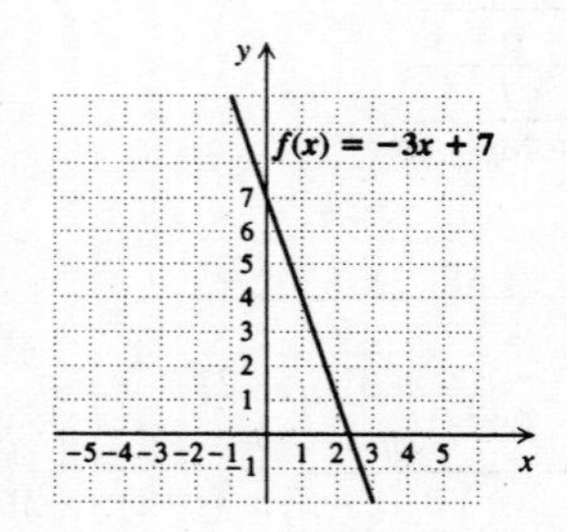

44. $\{-1, 6\}$

45. ***Familiarize***. Let $t =$ the amount Antonio received in tips on Thursday, in dollars. Then the average tip for the four days is given by $\dfrac{\$28+\$22+\$36+t}{4}$.

Translate.

The average tip is \$30.

$\dfrac{28+22+36+t}{4} = 30$

Carry out. We solve the equation.

$\dfrac{28+22+36+t}{4} = 30$

$28+22+36+t = 120$ Multiplying by 4

$86+t = 120$

$t = 34$

Check. If Antonio receives \$34 in tips on Thursday, then the average tip for the four days is $\dfrac{\$28+\$22+\$36+\$34}{4} = \dfrac{\$120}{4} = \$30$. The answer checks.

State. Antonio needs to earn \$34 in tips on Thursday in order for the average tip to be \$30.

46. First frame: 15 cm, second frame: 29 cm

47. ◈

48. ◈

49. $\dfrac{5x^{-2}+10x^{-1}y^{-1}+5y^{-2}}{3x^{-2}-3y^{-2}}$

$= \dfrac{\dfrac{5}{x^2}+\dfrac{10}{xy}+\dfrac{5}{y^2}}{\dfrac{3}{x^2}-\dfrac{3}{y^2}}$

$= \dfrac{\dfrac{5}{x^2}+\dfrac{10}{xy}+\dfrac{5}{y^2}}{\dfrac{3}{x^2}-\dfrac{3}{y^2}}\cdot\dfrac{x^2y^2}{x^2y^2}$

$= \dfrac{5y^2+10xy+5x^2}{3y^2-3x^2}$

$= \dfrac{5(y^2+2xy+x^2)}{3(y^2-x^2)}$

$= \dfrac{5(y+x)(y+x)}{3(y+x)(y-x)}$

$= \dfrac{5\cancel{(y+x)}(y+x)}{3\cancel{(y+x)}(y-x)}$

$= \dfrac{5(y+x)}{3(y-x)}$

50. $\dfrac{b-a}{ab}$

51. Substitute $\frac{c}{4}$ for both v_1 and v_2.

$$\frac{\frac{c}{4}+\frac{c}{4}}{1+\frac{\frac{c}{4}\cdot\frac{c}{4}}{c^2}}$$

$$=\frac{\frac{2c}{4}}{1+\frac{\frac{c^2}{16}}{c^2}}$$

$$=\frac{\frac{c}{2}}{1+\frac{c^2}{16}\cdot\frac{1}{c^2}}$$

$$=\frac{\frac{c}{2}}{1+\frac{1}{16}}$$

$$=\frac{\frac{c}{2}}{\frac{17}{16}}$$

$$=\frac{c}{2}\cdot\frac{16}{17}$$

$$=\frac{8c}{17}$$

The observed speed is $\frac{8c}{17}$, or $\frac{8}{17}$ the speed of light.

52. $\frac{2x+1}{3x+2}$

53. The reciprocal is $\frac{1}{x^2+x+1+\frac{1}{x}+\frac{1}{x^2}}$.

We simplify.

$$\frac{1}{x^2+x+1+\frac{1}{x}+\frac{1}{x^2}}$$

$$=\frac{1}{\frac{x^4+x^3+x^2+x+1}{x^2}} \quad \text{Adding in the denominator}$$

$$=1\cdot\frac{x^2}{x^4+x^3+x^2+x+1}$$

$$=\frac{x^2}{x^4+x^3+x^2+x+1}$$

54. $\frac{1+a}{2+a}$

55.

$$g(x)=\frac{x+1}{x-2}$$

$$g(a)=\frac{a+1}{a-2}$$

$$g(g(a))=\frac{\frac{a+1}{a-2}+1}{\frac{a+1}{a-2}-2}$$

$$=\frac{\frac{a+1}{a-2}+1}{\frac{a+1}{a-2}-2}\cdot\frac{a-2}{a-2}$$

$$=\frac{a+1+a-2}{a+1-2(a-2)}$$

$$=\frac{2a-1}{a+1-2a+4}$$

$$=\frac{2a-1}{-a+5}$$

56. $\frac{-3(2x+h)}{x^2(x+h)^2}$

57. $f(x)=\frac{5}{x}$, $f(x+h)=\frac{5}{x+h}$

$$\frac{f(x+h)-f(x)}{h}=\frac{\frac{5}{x+h}-\frac{5}{x}}{h}$$

$$=\frac{\frac{5x-5(x+h)}{x(x+h)}}{h}$$

$$=\frac{5x-5(x+h)}{x(x+h)}\cdot\frac{1}{h}$$

$$=\frac{5x-5x-5h}{xh(x+h)}$$

$$=\frac{-5h}{xh(x+h)}$$

$$=\frac{-5\not{h}}{x\not{h}(x+h)}$$

$$=\frac{-5}{x(x+h)}$$

58. $\frac{1}{(1-x-h)(1-x)}$

59. $f(x) = \dfrac{x}{1+x}$, $f(x+h) = \dfrac{x+h}{1+x+h}$

$$\frac{f(x+h)-f(x)}{h}$$
$$= \frac{\frac{x+h}{1+x+h} - \frac{x}{1+x}}{h}$$
$$= \frac{\frac{(x+h)(1+x) - x(1+x+h)}{(1+x+h)(1+x)}}{h}$$
$$= \frac{(x+h)(1+x) - x(1+x+h)}{(1+x+h)(1+x)} \cdot \frac{1}{h}$$
$$= \frac{x + x^2 + h + hx - x - x^2 - xh}{(1+x+h)(1+x)h}$$
$$= \frac{\cancel{h}}{(1+x+h)(1+x)\cancel{h}}$$
$$= \frac{1}{(1+x+h)(1+x)}$$

60. $\{x|x$ is a real number and $x \neq 0$ and $x \neq -2$ and $x \neq 2\}$

61. Division by zero occurs in $\dfrac{1}{x^2-1}$ when $x = 1$ or $x = -1$. Division by zero occurs in $\dfrac{1}{x^2-16}$ when $x = 4$ or $x = -4$. To avoid division in the complex fraction we solve:

$$\frac{1}{9} - \frac{1}{x^2-16} = 0$$
$$x^2 - 16 - 9 = 0 \quad \text{Multiplying by } 9(x^2-16)$$
$$x^2 - 25 = 0$$
$$(x+5)(x-5) = 0$$
$$x+5 = 0 \quad or \quad x-5 = 0$$
$$x = -5 \quad or \quad x = 5.$$

The domain of $G = \{x|x$ is a real number and $x \neq 1$ and $x \neq -1$ and $x \neq 4$ and $x \neq -4$ and $x \neq 5$ and $x \neq -5\}$.

62. $\dfrac{x^4}{81}$; $\{x|x$ is a real number and $x \neq 3\}$

63.

64.

65.

Exercise Set 6.4

1. $\dfrac{1}{3} + \dfrac{4}{5} = \dfrac{x}{9}$, LCD is 45

$$45\left(\frac{1}{3} + \frac{4}{5}\right) = 45 \cdot \frac{x}{9}$$
$$45 \cdot \frac{1}{3} + 45 \cdot \frac{4}{5} = 45 \cdot \frac{x}{9}$$
$$15 + 36 = 5x$$
$$51 = 5x$$
$$\frac{51}{5} = x$$

Check: $\dfrac{1}{3} + \dfrac{4}{5} = \dfrac{x}{9}$

$$\begin{array}{c|cl}
\frac{1}{3} + \frac{4}{5} \ ? & \frac{51/5}{9} & \\
\frac{5}{15} + \frac{12}{15} & \frac{51}{5} \cdot \frac{1}{9} & \\
\frac{17}{15} & \frac{17}{15} & \text{TRUE}
\end{array}$$

The solution is $\dfrac{51}{5}$.

2. $\dfrac{51}{2}$

3. $\dfrac{x}{3} - \dfrac{x}{4} = 12$, LCD is 12

$$12\left(\frac{x}{3} - \frac{x}{4}\right) = 12 \cdot 12$$
$$12 \cdot \frac{x}{3} - 12 \cdot \frac{x}{4} = 12 \cdot 12$$
$$4x - 3x = 144$$
$$x = 144$$

Check: $\dfrac{x}{3} - \dfrac{x}{4} = 12$

$$\begin{array}{c|cl}
\frac{144}{3} - \frac{144}{4} \ ? & 12 & \\
48 - 36 & & \\
12 & 12 & \text{TRUE}
\end{array}$$

The solution is 144.

4. $-\dfrac{225}{2}$

5. $\dfrac{5}{8} - \dfrac{1}{a} = \dfrac{2}{5}$

Because $\dfrac{1}{a}$ is undefined when a is 0, we note at the outset that $a \neq 0$. Then we multiply both sides by the LCD, $8 \cdot a \cdot 5$, or $40a$.

$$40a\left(\frac{5}{8}-\frac{1}{a}\right)=40a\cdot\frac{2}{5}$$
$$40a\cdot\frac{5}{8}-40a\cdot\frac{1}{a}=40a\cdot\frac{2}{5}$$
$$25a-40=16a$$
$$-40=-9a$$
$$\frac{40}{9}=a$$

Check: $\frac{5}{8}-\frac{1}{a}=\frac{2}{5}$

$$\frac{5}{8}-\frac{1}{40/9}\ ?\ \frac{2}{5}$$
$$\frac{5}{8}-1\cdot\frac{9}{40}$$
$$\frac{5}{8}-\frac{9}{40}$$
$$\frac{25}{40}-\frac{9}{40}$$
$$\frac{16}{40}$$
$$\frac{2}{5}\ \Big|\ \frac{2}{5}\quad \text{TRUE}$$

The solution is $\frac{40}{9}$.

6. -2

7. $\frac{2}{3}-\frac{1}{5}=\frac{7}{3x}$

Because $\frac{7}{3x}$ is undefined when x is 0, we note at the outset that $x\neq 0$. Then we multiply both sides by the LCD $3\cdot 5\cdot x$.

$$3\cdot 5\cdot x\left(\frac{2}{3}-\frac{1}{5}\right)=3\cdot 5\cdot x\cdot\frac{7}{3x}$$
$$3\cdot 5\cdot x\cdot\frac{2}{3}-3\cdot 5\cdot x\cdot\frac{1}{5}=3\cdot 5\cdot x\cdot\frac{7}{3x}$$
$$10x-3x=35$$
$$7x=35$$
$$x=5$$

Check: $\frac{2}{3}-\frac{1}{5}=\frac{7}{3x}$

$$\frac{2}{3}-\frac{1}{5}\ ?\ \frac{7}{3\cdot 5}$$
$$\frac{10}{15}-\frac{3}{15}\ \Big|\ \frac{7}{15}$$
$$\frac{7}{15}\ \Big|\ \frac{7}{15}\quad \text{TRUE}$$

The solution is 5.

8. 7

9. $\frac{2}{6}+\frac{1}{2x}=\frac{1}{3}$

Because $2x$ is undefined when x is 0, we note at the outset that $x\neq 0$. Then we multiply by the LCD, $2\cdot 3\cdot x$.

$$2\cdot 3\cdot x\left(\frac{2}{6}+\frac{1}{2x}\right)=2\cdot 3\cdot x\cdot\frac{1}{3}$$
$$2\cdot 3\cdot x\cdot\frac{2}{6}+2\cdot 3\cdot x\cdot\frac{1}{2x}=2\cdot 3\cdot x\cdot\frac{1}{3}$$
$$2x+3=2x$$
$$3=0$$

We get a false equation. The given equation has no solution.

10. No solution

11. $y+\frac{4}{y}=-5$

Because $\frac{4}{y}$ is undefined when y is 0, we note at the outset that $y\neq 0$. Then we multiply both sides by the LCD, y.

$$y\left(y+\frac{4}{y}\right)=y(-5)$$
$$y\cdot y+y\cdot\frac{4}{y}=-5y$$
$$y^2+4=-5y$$
$$y^2+5y+4=0$$
$$(y+1)(y+4)=0$$
$$y+1=0\quad or\quad y+4=0$$
$$y=-1\quad or\quad y=-4$$

Both values check. The solutions are -1 and -4.

12. $-\frac{1}{2}$

13. $\frac{y-1}{y-3}=\frac{2}{y-3}$

To assure that neither denominator is 0, we note at the outset that $y\neq 3$. Then we multiply on both sides by the LCD, $y-3$.

$$(y-3)\cdot\frac{y-1}{y-3}=(y-3)\cdot\frac{2}{y-3}$$
$$y-1=2$$
$$y=3$$

Recall that, because of the restriction above, 3 cannot be a solution. A check confirms this.

Check: $\frac{y-1}{y-3}=\frac{2}{y-3}$

$$\frac{3-1}{3-3}\ ?\ \frac{2}{3-3}$$
$$\frac{2}{0}\ \Big|\ \frac{2}{0}\quad \text{UNDEFINED}$$

The equation has no solution.

14. 4

15. $\dfrac{3}{x-2} = \dfrac{5}{x+4}$

To assure that neither denominator is 0, we note at the outset that $x \neq 2$ and $x \neq -4$. Then we multiply on both sides by the LCD, $(x-2)(x+4)$.

$$(x-2)(x+4) \cdot \frac{3}{x-2} = (x-2)(x+4) \cdot \frac{5}{x+4}$$
$$3(x+4) = 5(x-2)$$
$$3x + 12 = 5x - 10$$
$$22 = 2x$$
$$11 = x$$

This value checks. The solution is 11.

16. $-\dfrac{10}{3}$

17. $\dfrac{x^2-1}{x+2} = \dfrac{3}{x+2}$

To assure that neither denominator is 0. we note at the outset that $x \neq -2$. Then we multiply on both sides by the LCD, $x+2$.

$$(x+2) \cdot \frac{x^2-1}{x+2} = (x+2) \cdot \frac{3}{x+2}$$
$$x^2 - 1 = 3$$
$$x^2 - 4 = 0$$
$$(x+2)(x-2) = 0$$
$$x + 2 = 0 \quad or \quad x - 2 = 0$$
$$x = -2 \quad or \quad x = 2$$

Recall that, because of the restriction above, -2 cannot be a solution. The number 2 checks, so the solution is 2.

18. -1

19. $\dfrac{4}{a-7} = \dfrac{-2a}{a+3}$

To assure that neither denominator is 0, we note at the outset that $a \neq 7$ and $a \neq -3$. Then we multiply on both sides by the LCD, $(a-7)(a+3)$.

$$(a-7)(a+3) \cdot \frac{4}{a-7} = (a-7)(a+3) \cdot \frac{-2a}{a+3}$$
$$4(a+3) = -2a(a-7)$$
$$4a + 12 = -2a^2 + 14a$$
$$2a^2 - 10a + 12 = 0$$
$$2(a^2 - 5a + 6) = 0$$
$$2(a-2)(a-3) = 0$$
$$a - 2 = 0 \quad or \quad a - 3 = 0$$
$$a = 2 \quad or \quad a = 3$$

Both values check. The solutions are 2 and 3.

20. 2, 3

21. $\dfrac{50}{t-2} - \dfrac{16}{t} = \dfrac{30}{t}$

To assure that none of the denominators is 0, we note at the outset that $t \neq 2$ and $t \neq 0$. Then we multiply on both sides by the LCD, $t(t-2)$.

$$t(t-2)\left(\frac{50}{t-2} - \frac{16}{t}\right) = t(t-2) \cdot \frac{30}{t}$$
$$50t - 16(t-2) = 30(t-2)$$
$$50t - 16t + 32 = 30t - 60$$
$$34t + 32 = 30t - 60$$
$$4t = -92$$
$$t = -23$$

This value checks. The solution is -23.

22. -145

23. $\dfrac{3}{x} + \dfrac{x}{x+2} = \dfrac{4}{x^2+2x}$

$\dfrac{3}{x} + \dfrac{x}{x+2} = \dfrac{4}{x(x+2)}$

To assure that none of the denominators is 0, we note at the outset that $x \neq 0$ and $x \neq -2$. Then we multiply on both sides by the LCD, $x(x+2)$.

$$x(x+2)\left(\frac{3}{x} + \frac{x}{x+2}\right) = x(x+2) \cdot \frac{4}{x(x+2)}$$
$$3(x+2) + x \cdot x = 4$$
$$3x + 6 + x^2 = 4$$
$$x^2 + 3x + 2 = 0$$
$$(x+1)(x+2) = 0$$
$$x + 1 = 0 \quad or \quad x + 2 = 0$$
$$x = -1 \quad or \quad x = -2$$

Recall that, because of the restrictions above, -2 cannot be a solution. The number -1 checks. The solution is -1.

24. -4

25. We find all values of a for which $2a - \dfrac{15}{a} = 1$. First note that $a \neq 0$. Then multiply on both sides by the LCD, a.

$$a\left(2a - \frac{15}{a}\right) = a \cdot 1$$
$$a \cdot 2a - a \cdot \frac{15}{a} = a$$
$$2a^2 - 15 = a$$
$$2a^2 - a - 15 = 0$$
$$(2a+5)(a-3) = 0$$
$$a = -\frac{5}{2} \quad or \quad a = 3$$

Both values check. The solutions are $-\dfrac{5}{2}$ and 3.

26. $-\dfrac{3}{2}$, 2

27. We find all values of a for which $\dfrac{a-5}{a+1} = \dfrac{3}{5}$. First note that $a \neq -1$. Then multiply on both sides by the LCD, $5(a+1)$.

$$5(a+1)\cdot\frac{a-5}{a+1} = 5(a+1)\cdot\frac{3}{5}$$
$$5(a-5) = 3(a+1)$$
$$5a-25 = 3a+3$$
$$2a = 28$$
$$a = 14$$

This value checks. The solution is 14.

28. $\dfrac{17}{4}$

29. We find all values of a for which $\dfrac{12}{a}-\dfrac{12}{2a}=8$. First note that $a\neq 0$. Then multiply on both sides by the LCD, $2a$.

$$2a\left(\frac{12}{a}-\frac{12}{2a}\right) = 2a\cdot 8$$
$$2a\cdot\frac{12}{a}-2a\cdot\frac{12}{2a} = 16a$$
$$24-12 = 16a$$
$$12 = 16a$$
$$\frac{3}{4} = a$$

This value checks. The solution is $\dfrac{3}{4}$.

30. $\dfrac{3}{5}$

31. $$\frac{5}{x+2}-\frac{3}{x-2} = \frac{2x}{4-x^2}$$
$$\frac{5}{x+2}-\frac{3}{x-2} = \frac{2x}{(2+x)(2-x)}$$
$$\frac{5}{x+2}+\frac{3}{2-x} = \frac{2x}{(2+x)(2-x)} \quad \left(-\frac{3}{x-2}=\frac{3}{2-x}\right)$$

First note that $x\neq -2$ and $x\neq 2$. Then multiply on both sides by the LCD, $(2+x)(2-x)$.

$$(2+x)(2-x)\left(\frac{5}{x+2}+\frac{3}{2-x}\right) = (2+x)(2-x)\cdot\frac{2x}{(2+x)(2-x)}$$
$$5(2-x)+3(2+x) = 2x$$
$$10-5x+6+3x = 2x$$
$$16-2x = 2x$$
$$16 = 4x$$
$$4 = x$$

This value checks. The solution is 4.

32. -3

33. $$\frac{2}{a+4}+\frac{2a-1}{a^2+2a-8} = \frac{1}{a-2}$$
$$\frac{2}{a+4}+\frac{2a-1}{(a+4)(a-2)} = \frac{1}{a-2}$$

First note that $a\neq -4$ and $a\neq 2$. Then multiply on both sides by the LCD, $(a+4)(a-2)$.

$$(a+4)(a-2)\left(\frac{2}{a+4}+\frac{2a-1}{(a+4)(a-2)}\right) = (a+4)(a-2)\cdot\frac{1}{a-2}$$
$$2(a-2)+2a-1 = a+4$$
$$2a-4+2a-1 = a+4$$
$$4a-5 = a+4$$
$$3a = 9$$
$$a = 3$$

This value checks. The solution is 3.

34. -6, 5

35. $$\frac{2}{x+3}-\frac{3x+5}{x^2+4x+3} = \frac{5}{x+1}$$
$$\frac{2}{x+3}-\frac{3x+5}{(x+3)(x+1)} = \frac{5}{x+1}$$

Note that $x\neq -3$ and $x\neq -1$. Then multiply on both sides by the LCD, $(x+3)(x+1)$.

$$(x+3)(x+1)\left(\frac{2}{x+3}-\frac{3x+5}{(x+3)(x+1)}\right) = (x+3)(x+1)\cdot\frac{5}{x+1}$$
$$2(x+1)-(3x+5) = 5(x+3)$$
$$2x+2-3x-5 = 5x+15$$
$$-x-3 = 5x+15$$
$$-18 = 6x$$
$$-3 = x$$

Recall that, because of the restriction above, -3 is not a solution. Thus, the equation has no solution.

36. No solution

37. $$\frac{x-1}{x^2-2x-3}+\frac{x+2}{x^2-9} = \frac{2x+5}{x^2+4x+3}$$
$$\frac{x-1}{(x-3)(x+1)}+\frac{x+2}{(x+3)(x-3)} = \frac{2x+5}{(x+3)(x+1)}$$

Note that $x\neq 3$ and $x\neq -1$ and $x\neq -3$. Then multiply on both sides by the LCD, $(x-3)(x+1)(x+3)$.

$$(x-3)(x+1)(x+3)\left(\frac{x-1}{(x-3)(x+1)}+\frac{x+2}{(x+3)(x-3)}\right) = (x-3)(x+1)(x+3)\cdot\frac{2x+5}{(x+3)(x+1)}$$
$$(x+3)(x-1)+(x+1)(x+2) = (x-3)(2x+5)$$
$$x^2+2x-3+x^2+3x+2 = 2x^2-x-15$$
$$2x^2+5x-1 = 2x^2-x-15$$
$$5x-1 = -x-15$$
$$6x = -14$$
$$x = -\frac{7}{3}$$

This value checks. The solution is $-\dfrac{7}{3}$.

38. $\frac{5}{14}$

39. $$\frac{3}{x^2-x-12}+\frac{1}{x^2+x-6}=\frac{4}{x^2+3x-10}$$
$$\frac{3}{(x-4)(x+3)}+\frac{1}{(x+3)(x-2)}=\frac{4}{(x+5)(x-2)}$$

Note that $x \neq 4$ and $x \neq -3$ and $x \neq 2$ and $x \neq -5$. Then multiply on both sides by the LCD, $(x-4)(x+3)(x-2)(x+5)$.

$$(x-4)(x+3)(x-2)(x+5)\left(\frac{3}{(x-4)(x+3)}+\frac{1}{(x+3)(x-2)}\right)=$$
$$(x-4)(x+3)(x-2)(x+5)\cdot\frac{4}{(x+5)(x-2)}$$
$$3(x-2)(x+5)+(x-4)(x+5)=4(x-4)(x+3)$$
$$3(x^2+3x-10)+x^2+x-20=4(x^2-x-12)$$
$$3x^2+9x-30+x^2+x-20=4x^2-4x-48$$
$$4x^2+10x-50=4x^2-4x-48$$
$$10x-50=-4x-48$$
$$14x=2$$
$$x=\frac{1}{7}$$

This value checks. The solution is $\frac{1}{7}$.

40. $\frac{3}{5}$

41. $81x^4-y^4=(9x^2+y^2)(9x^2-y^2)$
$=(9x^2+y^2)(3x+y)(3x-y)$

42. (a) Inconsistent; (b) consistent

43. $|x-2|>3$

$x-2<-3$ *or* $x-2>3$

$x<-1$ *or* $x>5$

The solution set is $\{x|x<-1 \text{ or } x>5\}$, or $(-\infty,-1)\cup(5,\infty)$.

44. 25 multiple-choice, 30 true-false, 15 fill-ins

45. ***Familiarize***. Let x = the first number. Then $x+2$ = the second number.

Translate.

The product of two consecutive even integers is 288.

$$x(x+2) = 288$$

Carry out. We solve the equation.

$$x(x+2)=288$$
$$x^2+2x=288$$
$$x^2+2x-288=0$$
$$(x-16)(x+18)=0$$
$$x-16=0 \quad or \quad x+18=0$$
$$x=16 \quad or \quad x=-18$$

Check. If the first number is 16, then the next number is 16+2, or 18, and their product is 16·18, or 288. If the first number is −18, then the next number is −18 + 2, or −16, and their product is −18(−16), or 288. Both solutions check.

State. The numbers are 16 and 18 or −18 and −16.

46. $a^{22}b^9$

47. ◈

48. ◈

49. ◈

50. ◈

51. Set $f(a)$ equal to $g(a)$ and solve for a.

$$\frac{2+\frac{a}{2}}{2-\frac{a}{4}}=\frac{2}{\frac{a}{4}-2}$$
$$\frac{2+\frac{a}{2}}{2-\frac{a}{4}}=\frac{-1}{-1}\cdot\frac{2}{\frac{a}{4}-2}$$
$$\frac{2+\frac{a}{2}}{2-\frac{a}{4}}=\frac{-2}{2-\frac{a}{4}}$$
$$\left(2-\frac{a}{4}\right)\left[\frac{2+\frac{a}{2}}{2-\frac{a}{4}}\right]=\left(2-\frac{a}{4}\right)\left[\frac{-2}{2-\frac{a}{4}}\right]$$
$$2+\frac{a}{2}=-2$$
$$2\left(2+\frac{a}{2}\right)=2(-2)$$
$$4+a=-4$$
$$a=-8$$

This value checks. Thus, when $a=-8$, $f(a)=g(a)$.

52. $\frac{1}{5}$

53. Set $f(a)$ equal to $g(a)$ and solve for a.

$$\frac{1}{1+a}+\frac{a}{1-a}=\frac{1}{1-a}-\frac{a}{1+a}$$

Note that $a \neq -1$ and $a \neq 1$. Then multiply by the LCD, $(1+a)(1-a)$.

$$(1+a)(1-a)\left(\frac{1}{1+a}+\frac{a}{1-a}\right)=$$
$$(1+a)(1-a)\left(\frac{1}{1-a}-\frac{a}{1+a}\right)$$
$$1-a+a(1+a)=1+a-a(1-a)$$
$$1-a+a+a^2=1+a-a+a^2$$
$$1+a^2=1+a^2$$
$$1=1$$

We get an equation that is true for all real numbers. Recall that, because of the restrictions above, -1 and 1 are not solutions. Thus, for all values of a except -1 and 1, $f(a) = g(a)$.

54. $-\dfrac{7}{2}$

55. Set $f(a)$ equal to $g(a)$ and solve for a.

$$\frac{0.793}{a} + 18.15 = \frac{6.034}{a} - 43.17$$

Note that $a \neq 0$. Then multiply on both sides by the LCD, a.

$$a\left(\frac{0.793}{a} + 18.15\right) = a\left(\frac{6.034}{a} - 43.17\right)$$
$$0.793 + 18.15a = 6.034 - 43.17a$$
$$61.32a = 5.241$$
$$a \approx 0.0854697$$

This value checks. When $a \approx 0.0854697$, $f(a) = g(a)$.

56. -2.955341202

57.
$$\frac{x^2+6x-16}{x-2} = x+8, x \neq 2$$
$$\frac{(x+8)(x-2)}{x-2} = x+8$$
$$\frac{(x+8)\cancel{(x-2)}}{\cancel{x-2}} = x+8$$
$$x+8 = x+8$$
$$8 = 8$$

Since $8 = 8$ is true for all values of x, the original equation is true for any possible replacements of the variable. It is an identity.

58. Yes

59.

60.

61. Let $y_1 = \dfrac{x^2-4}{x-2}$ and observe that for $x = 2$ the entry in the Y1-column of the table is "ERROR."

62. 

Exercise Set 6.5

1. ***Familiarize.*** Let x = the number.

Translate.

The reciprocal of 3	plus	the reciprocal of 6	is	the reciprocal of the number.
↓	↓	↓	↓	↓
$\frac{1}{3}$	$+$	$\frac{1}{6}$	$=$	$\frac{1}{x}$

Carry out. We solve the equation.

$$\frac{1}{3} + \frac{1}{6} = \frac{1}{x}, \text{ LCD is } 6x$$
$$6x\left(\frac{1}{3} + \frac{1}{6}\right) = 6x \cdot \frac{1}{x}$$
$$2x + x = 6$$
$$3x = 6$$
$$x = 2$$

Check. $\dfrac{1}{3} + \dfrac{1}{6} = \dfrac{2}{6} + \dfrac{1}{6} = \dfrac{3}{6} = \dfrac{1}{2}$. This is the reciprocal of 2, so the result checks.

State. The number is 2.

2. $\dfrac{35}{12}$

3. ***Familiarize.*** We let x = the number.

Translate.

A number	plus	6	times	its reciprocal	is	-5.
↓	↓	↓	↓	↓	↓	↓
x	$+$	6	$\cdot$	$\frac{1}{x}$	$=$	-5

Carry out. We solve the equation.

$$x + \frac{6}{x} = -5, \text{ LCD is } x$$
$$x\left(x + \frac{6}{x}\right) = x(-5)$$
$$x^2 + 6 = -5x$$
$$x^2 + 5x + 6 = 0$$
$$(x+3)(x+2) = 0$$
$$x = -3 \text{ or } x = -2$$

Check. The possible solutions are -3 and -2. We check -3 in the conditions of the problem.

Number:	-3
6 times the reciprocal of the number:	$6\left(-\frac{1}{3}\right) = -2$
Sum of the number and 6 times its reciprocal:	$-3 + (-2) = -5$

The number -3 checks.

Now we check -2:

Number:	-2
6 times the reciprocal of the number:	$6\left(-\frac{1}{2}\right) = -3$
Sum of the number and 6 times its reciprocal:	$-2 + (-3) = -5$

The number -2 also checks.

State. The number is -3 or -2.

4. $-3, -7$

5. ***Familiarize.*** We let x = the first integer. Then $x+1$ = the second, and their product $= x(x+1)$.

Translate.

$$\underbrace{\text{Reciprocal of the product}} \text{ is } \frac{1}{42}.$$

$$\frac{1}{x(x+1)} = \frac{1}{42}$$

Carry out. We solve the equation.

$$\frac{1}{x(x+1)} = \frac{1}{42}, \text{ LCD is } 42x(x+1)$$

$$42x(x+1)\cdot\frac{1}{x(x+1)} = 42x(x+1)\cdot\frac{1}{42}$$

$$42 = x(x+1)$$

$$42 = x^2 + x$$

$$0 = x^2 + x - 42$$

$$0 = (x+7)(x-6)$$

$$x = -7 \text{ or } x = 6$$

Check. When $x = -7$, then $x + 1 = -6$ and $-7(-6) = 42$. The reciprocal of this product is $\frac{1}{42}$. When $x = 6$, then $x + 1 = 7$ and $6 \cdot 7 = 42$. The reciprocal of this product is also $\frac{1}{42}$. Both possible solutions check.

State. The integers are -7 and -6 or 6 and 7.

6. -9 and -8, 8 and 9

7. ***Familiarize.*** The job takes Otto 4 hours working alone and Sally 3 hours working alone. Then in 1 hour, Otto does $\frac{1}{4}$ of the job and Sally does $\frac{1}{3}$ of the job. Working together, they can do $\frac{1}{4}+\frac{1}{3}$ of the job in

1 hour. Let t represent the number of hours required for Otto and Sally, working together, to do the job.

Translate. We want to find t such that

$$t\left(\frac{1}{4}\right) + t\left(\frac{1}{3}\right) = 1, \text{ or } \frac{t}{4} + \frac{t}{3} = 1,$$

where 1 represents one entire job.

Carry out. We solve the equation.

$$\frac{t}{4} + \frac{t}{3} = 1, \text{ LCD is } 12$$

$$12\left(\frac{t}{4} + \frac{t}{3}\right) = 12 \cdot 1$$

$$3t + 4t = 12$$

$$7t = 12$$

$$t = \frac{12}{7}$$

Check. In $\frac{12}{7}$ hours, Otto will do $\frac{1}{4}\cdot\frac{12}{7}$, or $\frac{3}{7}$ of the job and Sally will do $\frac{1}{3}\cdot\frac{12}{7}$, or $\frac{4}{7}$ of the job. Together, they do $\frac{3}{7}+\frac{4}{7}$, or 1 entire job. The answer checks.

State. It will take $\frac{12}{7}$ hr, or $1\frac{5}{7}$ hr, for Otto and Sally, together, to paint the room.

8. $3\frac{3}{14}$ hr

9. ***Familiarize.*** The pool can be filled in 12 hours with only the pipe and in 30 hours with only the hose. Then in 1 hour, the pipe fills $\frac{1}{12}$ of the pool, and the hose fills $\frac{1}{30}$ of the pool. Using both the pipe and the hose, $\frac{1}{12}+\frac{1}{30}$ of the pool can be filled in 1 hour. Suppose that it takes t hours to fill the pool using both the pipe and hose.

Translate. We want to find t such that

$$t\left(\frac{1}{12}\right) + t\left(\frac{1}{30}\right) = 1, \text{ or } \frac{t}{12} + \frac{t}{30} = 1,$$

where 1 represents one entire job.

Carry out. We solve the equation. We multiply on both sides by the LCD, $60t$.

$$60\left(\frac{t}{12} + \frac{t}{30}\right) = 60 \cdot 1$$

$$5t + 2t = 60$$

$$7t = 60$$

$$t = \frac{60}{7}$$

Check. The possible solution is $\frac{60}{7}$ hours. If the pipe is used $\frac{60}{7}$ hours, it fills $\frac{1}{12}\cdot\frac{60}{7}$, or $\frac{5}{7}$ of the pool. If the hose is used $\frac{60}{7}$ hours, it fills $\frac{1}{30}\cdot\frac{60}{7}$, or $\frac{2}{7}$ of the pool. Using both, $\frac{5}{7}+\frac{2}{7}$ of the pool, or all of it, will be filled in $\frac{60}{7}$ hours.

State. Using both the pipe and the hose, it will take $\frac{60}{7}$, or $8\frac{4}{7}$ hours, to fill the pool.

10. 9.9 hr

11. ***Familiarize.*** In 1 hour Pronto Press does $\frac{1}{4.5}$ of the job and Red Dot Printers does $\frac{1}{5.5}$ of the job. Working together, they can do $\frac{1}{4.5}+\frac{1}{5.5}$ of the job in 1 hour. Suppose it takes them t hours working together.

Translate. We want to find t such that

$$t\left(\frac{1}{4.5}\right) + t\left(\frac{1}{5.5}\right) = 1, \text{ or } \frac{t}{4.5} + \frac{t}{5.5} = 1.$$

Carry out. We solve the equation.

$$4.5(5.5)\left(\frac{t}{4.5} + \frac{t}{5.5}\right) = 4.5(5.5)(1)$$

$$5.5t + 4.5t = 24.75$$

$$10t = 24.75$$

$$t = 2.475$$

Check. In 2.475 hr Pronto Press will do $\frac{2.475}{4.5}$, or 0.55 of the job and Red Dot will do $\frac{2.475}{5.5}$, or 0.45 of the job. Together they will do $0.55+0.45$, or 1 entire job.

State. Working together, it will take them 2.475 hours.

12. $3\frac{9}{52}$ hr

13. ***Familiarize***. Let t represent the number of hours it takes Henri, working alone, to sand the floor. Then in 1 hr, Mavis does $\frac{1}{3}$ of the job, and Henri does $\frac{1}{t}$ of the job.

Translate. Working together, they can do the entire job in 2 hr, so we want to find t such that

$$2\left(\frac{1}{3}\right)+2\left(\frac{1}{t}\right)=1, \text{ or } \frac{2}{3}+\frac{2}{t}=1.$$

Carry out. We solve the equation.

$$\begin{aligned} \frac{2}{3}+\frac{2}{t} &= 1, \text{ LCD is } 3t \\ 3t\left(\frac{2}{3}+\frac{2}{t}\right) &= 3t\cdot 1 \\ 2t+6 &= 3t \\ 6 &= t \end{aligned}$$

Check. In 2 hr, Mavis will do $2\cdot\frac{1}{3}$, or $\frac{2}{3}$ of the job, and Henri will do $2\cdot\frac{1}{6}$, or $\frac{1}{3}$ of the job. Together they will do $\frac{2}{3}+\frac{1}{3}$, or 1 entire job. The answer checks.

State. It would take Henri 6 hours, working by himself, to sand the floor.

14. Skyler: 12 hour, Jake: 6 hours

15. ***Familiarize***. Let t represent the number of hours it takes Kate to paint the floor. Then $t+3$ represents the time it takes Sara to paint the floor. In 1 hour, Kate does $\frac{1}{t}$ of the job and Sara does $\frac{1}{t+3}$.

Translate. Working together, it takes them 2 hr to do the job, so we want to find t such that

$$2\left(\frac{1}{t}\right)+2\left(\frac{1}{t+3}\right)=1, \text{ or } \frac{2}{t}+\frac{2}{t+3}=1.$$

Carry out. We solve the equation. We multiply by the LCD, $t(t+3)$.

$$\begin{aligned} t(t+3)\left(\frac{2}{t}+\frac{2}{t+3}\right) &= t(t+3)(1) \\ 2(t+3)+2t &= t^2+3t \\ 2t+6+2t &= t^2+3t \\ 4t+6 &= t^2+3t \\ 0 &= t^2-t-6 \\ 0 &= (t-3)(t+2) \\ t=3 \textit{ or } t &= -2 \end{aligned}$$

Check. We check only 3, since the time cannot be negative. If Kate does the job in 3 hr, then in 2 hr she does $2\left(\frac{1}{3}\right)$, or $\frac{2}{3}$ of the job. If Sara does the job in $3+3$ or 6 hr, then in 2 hr she does $2\left(\frac{1}{6}\right)$, or $\frac{1}{3}$ of the job. Together they do $\frac{2}{3}+\frac{1}{3}$, or 1 entire job in 2 hr. The result checks.

State. It would take Kate 3 hours to do the job and it would take Sara 6 hours to do the job working alone.

16. Claudia: 10 days, Jan: 40 days

17. ***Familiarize***. Let t represent the number of hours it takes Zsuzanna to deliver the papers alone. Then $3t$ represents the number of hours it takes Stan to deliver the papers alone.

Translate. In 1 hr Zsuzanna and Stan will do one entire job, so we have

$$1\left(\frac{1}{t}\right)+1\left(\frac{1}{3t}\right)=1, \text{ or } \frac{1}{t}+\frac{1}{3t}=1.$$

Carry out. We solve the equation. Multiply on both sides by the LCD, $3t$.

$$\begin{aligned} 3t\left(\frac{1}{t}+\frac{1}{3t}\right) &= 3t\cdot 1 \\ 3+1 &= 3t \\ 4 &= 3t \\ \frac{4}{3} &= t \end{aligned}$$

Check. If Zsuzanna does the job alone in $\frac{4}{3}$ hr, then in 1 hr she does $\frac{1}{4/3}$, or $\frac{3}{4}$ of the job. If Stan does the job alone in $3\cdot\frac{4}{3}$, or 4 hr, then in 1 hr he does $\frac{1}{4}$ of the job. Together, they do $\frac{3}{4}+\frac{1}{4}$, or 1 entire job, in 1 hr. The result checks.

State. It would take Zsuzanna $\frac{4}{3}$ hours and it would take Stan 4 hours to deliver the papers alone.

18. $1\frac{1}{5}$ hr

19. ***Familiarize***. We will convert hours to minutes:

2 hr = $2\cdot 60$ min = 120 min

2 hr 55 min = 120 min + 55 min = 175 min

Let t = the number of minutes it takes Deb to do the job alone. Then $t+120$ = the number of minutes it takes John alone. In 1 hour (60 minutes) Deb does $\frac{1}{t}$ and John does $\frac{1}{t+120}$ of the job.

Translate. In 175 min John and Deb will complete one entire job, so we have

$$175\left(\frac{1}{t}\right)+175\left(\frac{1}{t+120}\right)=1, \text{ or}$$

$$\frac{175}{t}+\frac{175}{t+120}=1.$$

Carry out. We solve the equation. Multiply on both sides by the LCD, $t(t+120)$.

$$\begin{aligned} t(t+120)\left(\frac{175}{t}+\frac{175}{t+120}\right) &= t(t+120)(1) \\ 175(t+120)+175t &= t^2+120t \\ 175t+21,000+175t &= t^2+120t \\ 0 &= t^2-230t-21,000 \\ 0 &= (t-300)(t+70) \\ t=300 \text{ or } t &= -70 \end{aligned}$$

Check. Since negative time has no meaning in this problem, -70 is not a solution of the original problem. If the job takes Deb 300 min and it takes John $300+120=420$ min, then in 175 min they would complete

$$175\left(\frac{1}{300}\right)+175\left(\frac{1}{420}\right)=\frac{7}{12}+\frac{5}{12}=1 \text{ job.}$$

The result checks.

State. It would take Deb 300 min, or 5 hr, to do the job alone.

20. 8 hr

21. ***Familiarize.*** Let t = the number of hours it takes the new machine to do the job alone. Then $2t$ = the number of hours it takes the old machine to do the job alone. In 1 hr the new machine does $\frac{1}{t}$ of the job and the old machine does $\frac{1}{2t}$ of the job.

Translate. In 15 hr the two machines will complete one entire job, so we have

$$15\left(\frac{1}{t}\right)+15\left(\frac{1}{2t}\right)=1, \text{ or } \frac{15}{t}+\frac{15}{2t}=1.$$

Carry out. We solve the equation. Multiply on both sides by the LCD, $2t$.

$$\begin{aligned} 2t\left(\frac{15}{t}+\frac{15}{2t}\right) &= 2t\cdot 1 \\ 30+15 &= 2t \\ 45 &= 2t \\ \frac{45}{2} &= t \end{aligned}$$

Check. In 1 hr, the new machine will do $\dfrac{1}{\frac{45}{2}}$, or $\dfrac{2}{45}$ of the job and the old machine will do $\dfrac{1}{2\cdot\frac{45}{2}}$, or $\dfrac{1}{45}$ of the job. Together they will do $\frac{2}{45}+\frac{1}{45}=\frac{3}{45}=\frac{1}{15}$ of the job. Then in 15 hr they will do $15\cdot\frac{1}{15}=1$ job. The answer checks.

State. Working alone the job would take the new machine $\frac{45}{2}$, or $22\frac{1}{2}$ hr, and it would take the old machine $2\cdot\frac{45}{2}$, or 45 hr.

22. 2 hr

23. ***Familiarize.*** We first make a drawing. Let r = the boat's speed in still water in mph. Then $r-4$ = the speed upstream and $r+4$ = the speed downstream.

Upstream 6 miles $r-4$ mph →

← 12 miles $r+4$ mph Downstream

We organize the information in a table. The one is the same both upstream and downstream so we use t for each time.

	Distance	Speed	Time
Upstream	6	$r-4$	t
Downstream	12	$r+4$	t

Translate. Using the formula Time = Distance/Rate in each row of the table and the fact that the times are the same, we can write an equation.

$$\frac{6}{r-4}=\frac{12}{r+4}$$

Carry out. We solve the equation.

$$\frac{6}{r-4}=\frac{12}{r+4},$$

LCD is $(r-4)(r+4)$

$$\begin{aligned} (r-4)(r+4)\cdot\frac{6}{r-4} &= (r-4)(r+4)\cdot\frac{12}{r+4} \\ 6(r+4) &= 12(r-4) \\ 6r+24 &= 12r-48 \\ 72 &= 6r \\ 12 &= r \end{aligned}$$

Check. If the boat's speed in still water is 12 mph, then its speed upstream is $12-4$, or 8 mph, and its speed downstream is $12+4$, or 16 mph. Traveling 6 mi at 8 mph takes the boat $\frac{6}{8}=\frac{3}{4}$ hr. Traveling 12 mi at 16 mph takes the boat $\frac{12}{16}=\frac{3}{4}$ hr. Since the times are the same, the answer checks.

State. The boat's speed in still water is 12 mph.

24. 7 mph

25. ***Familiarize.*** We first make a drawing. Let r = Camille's speed on a nonmoving sidewalk in ft/sec. Then her speed moving forward on the moving sidewalk is $r+1.8$, and her speed in the opposite direction is $r-1.8$.

Forward $r+1.8$ 105 ft →

← 51 ft $r-1.8$ Opposite direction

We organize the information in a table. The time is the same both forward and in the opposite direction so we use t for each time.

	Distance	Speed	Time
Forward	105	$r+1.8$	t
Opposite direction	51	$r-1.8$	t

Translate. Using the formula Time = Distance/Rate in each row of the table and the fact that the times are the same, we can write an equation.

$$\frac{105}{r+1.8}=\frac{51}{r-1.8}$$

Carry out. We solve the equation.

$$\frac{105}{r+1.8}=\frac{51}{r-1.8},$$

$$\text{LCD is } (r+1.8)(r-1.8)$$

$$(r+1.8)(r-1.8)\cdot\frac{105}{r+1.8}=(r+1.8)(r-1.8)\cdot\frac{51}{r-1.8}$$

$$105(r-1.8)=51(r+1.8)$$

$$105r-189=51r+91.8$$

$$54r=280.8$$

$$r=5.2$$

Check. If Camille's speed on a nonmoving sidewalk is 5.2 ft/sec, then her speed moving forward on the moving sidewalk is 5.2+1.8, or 7 ft/sec, and her speed moving in the opposite direction on the sidewalk is 5.2 − 1.8, or 3.4 ft/sec. Moving 105 ft at 7 ft/sec takes $\frac{105}{7}=15$ sec. Moving 51 ft at 3.4 ft/sec takes 15 sec. Since the times are the same, the answer checks.

State. Camille would be walking 5.2 ft/sec on a nonmoving sidewalk.

26. 4.3 ft/sec

27. ***Familiarize***. Let r = the speed of the passenger train in mph. Then $r-14$ = the speed of the freight train in mph. We organize the information in a table. The time is the same for both trains so we use t for each time.

	Distance	Speed	Time
Passenger train	400	r	t
Freight train	330	$r-14$	t

Translate. Using the formula Time = Distance/Rate in each row of the table and the fact that the times are the same, we can write an equation.

$$\frac{400}{r}=\frac{330}{r-14}$$

Carry out. We solve the equation.

$$\frac{400}{r}=\frac{330}{r-14},\quad \text{LCD is } r(r-14)$$

$$r(r-14)\cdot\frac{400}{r}=r(r-14)\cdot\frac{330}{r-14}$$

$$400(r-14)=330r$$

$$400r-5600=330r$$

$$-5600=-70r$$

$$80=r$$

Check. If the passenger train's speed is 80 mph, then the freight train's speed is 80 − 14, or 66 mph. Traveling 400 mi at 80 mph takes $\frac{400}{80}=5$ hr. Traveling 330 mi at 66 mph takes $\frac{330}{66}=5$ hr. Since the times are the same, the answer checks.

State. The speed of the passenger train is 80 mph; the speed of the freight train is 66 mph.

28. Rosanna: $3\frac{1}{3}$ mph, Simone: $5\frac{1}{3}$ mph

29. ***Familiarize***. Let r = the speed of the express bus in mph. Then $r-7$ = the speed of the local bus in mph. We organize the information in a table. The time is the same for both buses so we use t for each time.

	Distance	Speed	Time
Express	90	r	t
Local	75	$r-7$	t

Translate. Using the formula Time = Distance/Rate in each row of the table and the fact that the times are the same, we can write an equation.

$$\frac{90}{r}=\frac{75}{r-7}$$

Carry out. We solve the equation.

$$\frac{90}{r}=\frac{75}{r-7},\quad \text{LCD is } r(r-7)$$

$$r(r-7)\cdot\frac{90}{r}=r(r-7)\cdot\frac{75}{r-7}$$

$$90(r-7)=75r$$

$$90r-630=75r$$

$$-630=-15r$$

$$42=r$$

Check. If the speed of the express bus is 42 mph, then the speed of the local bus is 42 − 7, or 35 mph. Traveling 90 mi at 42 mph takes $\frac{90}{42}=\frac{15}{7}$ hr. Traveling 75 mi at 35 mph takes $\frac{75}{35}=\frac{15}{7}$ hr. Since the times are the same, the answer checks.

State. The speed of the express bus is 42 mph; the speed of the local bus is 35 mph.

30. A: 46 mph, B: 58 mph

31. ***Familiarize.*** We let r = the speed of the river. Then $15+r$ = Suzie's speed downstream in km/h and $15-r$ = her speed upstream in km/h. The times are the same. Let t represent the time. We organize the information in a table.

	Distance	Speed	Time
Downstream	140	$15+r$	t
Upstream	35	$15-r$	t

Translate. Using the formula Time = Distance/Rate in each row of the table and the fact that the times are the same, we can write an equation.

$$\frac{140}{15+r}=\frac{35}{15-r}$$

Carry out. We solve the equation.

$$\frac{140}{15+r}=\frac{35}{15-r},$$

LCD is $(15+r)(15-r)$

$$(15+r)(15-r)\cdot\frac{140}{15+r}=(15+r)(15-r)\cdot\frac{35}{15-r}$$
$$140(15-r)=35(15+r)$$
$$2100-140r=525+35r$$
$$1575=175r$$
$$9=r$$

Check. If $r=9$, then the speed downstream is $15+9$, or 24 km/h and the speed upstream is $15-9$, or 6 km/h. The time for the trip is downstream is $\frac{140}{24}$, or $5\frac{5}{6}$ hours. The time for the trip upstream is $\frac{35}{6}$, or $5\frac{5}{6}$ hours. The times are the same. The values check.

State. The speed of the river is 9 km/h.

32. $1\frac{1}{5}$ km/h

33. ***Familiarize.*** Let r = the speed of Mara's moped in km/h. Then $r+8$ = the speed of Jaime's moped in km/h. We organize the information in a table. The time is the same for both mopeds so we use t for each time.

	Distance	Speed	Time
Mara	45	r	t
Jaime	69	$r+8$	t

Translate. Using the formula Time = Distance/Rate in each row of the table and the fact that the times are the same, we can write an equation.

$$\frac{45}{r}=\frac{69}{r+8}$$

Carry out. We solve the equation.

$$\frac{45}{r}=\frac{69}{r+8},\ \text{LCD is } r(r+8)$$
$$r(r+8)\cdot\frac{45}{r}=r(r+8)\cdot\frac{69}{r+8}$$
$$45(r+8)=69r$$
$$45r+360=69r$$
$$360=24r$$
$$15=r$$

Check. If the speed of Mara's moped is 15 km/h, then the speed of Jaime's moped is $15+8$, or 23 mph. Traveling 45 km at 15 km/h takes $\frac{45}{15}=3$ hr. Traveling 69 km at 23 km/h takes $\frac{69}{23}=3$ hr. Since the times are the same, the answer checks.

State. The speed of Mara's moped is 15 km/h; the speed of Jaime's moped is 23 mph.

34. 2 km/h

35. ***Familiarize.*** Let r = the speed of the current in m per minute. Then Al's speed upstream is $55-r$, and his speed downstream is $55+r$. We organize the information in a table.

	Distance	Speed	Time
Upstream	150	$55-r$	t_1
Downstream	150	$55+r$	t_2

Translate. Using the formula Time = Distance/Rate we see that $t_1=\frac{150}{55-r}$ and $t_2=\frac{150}{55+r}$. The total time upstream and back is 5.5 min, so $t_1+t_2=5.5$, or

$$\frac{150}{55-r}+\frac{150}{55+r}=5.5.$$

Carry out. We solve the equation. Multiply on both sides by the LCD, $(55-r)(55+r)$.

$$(55-r)(55+r)\left(\frac{150}{55-r}+\frac{150}{55+r}\right)=$$
$$(55-r)(55+r)(5.5)$$
$$150(55+r)+150(55-r)=$$
$$5.5(3025-r^2)$$
$$8250+150r+8250-150r=$$
$$16,637.5-5.5r^2$$
$$16,500=$$
$$16,637.5-5.5r^2$$
$$5.5r^2-137.5=0$$
$$5.5(r^2-25)=0$$
$$5.5(r+5)(r-5)=0$$
$$r=-5\ \text{or}\ r=5$$

Check. We check only 5 since the speed cannot be negative. If the speed of the current is 5 m per minute, then Al's speed upstream is $55-5$, or 50 m per minute, and his speed downstream is $55+5$, or

60 m per minute. Swimming 150 m at 50 m per minute takes $\frac{150}{50}$ or 3 min. Swimming 150 m at 60 m per minute takes $\frac{150}{60}$, or 2.5 min. The total time is $3+2.5$, or 5.5 min. The answer checks.

State. The speed of the current is 5 m per minute.

36. 20 mph

37. ***Familiarize***. Let $r =$ the speed at which the van actually traveled in mph, and let $t =$ the actual travel time in hours. We organize the information in a table.

	Distance	Speed	Time
Actual speed	120	r	t
Faster speed	120	$r+10$	$t-2$

Translate. From the first row of the table we have $120 = rt$, and from the second row we have $120 = (r+10)(t-2)$. Solving the first equation for t, we have $t = \frac{120}{r}$. Substituting for t in the second equation, we have

$$120 = (r+10)\left(\frac{120}{r} - 2\right).$$

Carry out. We solve the equation.

$$\begin{aligned}
120 &= (r+10)\left(\frac{120}{r} - 2\right)\\
120 &= 120 - 2r + \frac{1200}{r} - 20\\
20 &= -2r + \frac{1200}{r}\\
r\cdot 20 &= r\left(-2r + \frac{1200}{r}\right)\\
20r &= -2r^2 + 1200
\end{aligned}$$

$$\begin{aligned}
2r^2 + 20r - 1200 &= 0\\
2(r^2 + 10r - 600) &= 0\\
2(r+30)(r-20) &= 0\\
r = -30 \text{ or } r &= 20
\end{aligned}$$

Check. Since speed cannot be negative in this problem, -30 cannot be a solution of the original problem. If the speed is 20 mph, it takes $\frac{120}{20}$, or 6 hr, to travel 120 mi. If the speed is 10 mph faster, or 30 mph, it takes $\frac{120}{30}$, or 4 hr, to travel 120 mi. Since 4 hr is 2 hr less time than 6 hr, the answer checks.

State. The speed was 20 mph.

38. 12 mph

39. We find the values of x for which the denominator is 0.

$$\begin{aligned}
x^2 - 4x - 5 &= 0\\
(x-5)(x+1) &= 0\\
x - 5 = 0 \quad or \quad x+1 &= 0\\
x = 5 \quad or \quad\quad x &= -1.
\end{aligned}$$

The domain of $f = \{x|x \text{ is a real number and } x \neq 5 \text{ and } x \neq -1\}$.

40. $\{-7, 11\}$

41.
$$\begin{aligned}
&4y - 5xy^2 + 6xy - 3xy^2 - 2y\\
&= (4-2)y + (-5-3)xy^2 + 6xy\\
&= 2y - 8xy^2 + 6xy
\end{aligned}$$

42. 1500

43. ◈

44. ◈

45. ◈

46. ◈

47. ***Familiarize***. Let $t =$ the time, in minutes, it will take to empty a full tub if the water is left on. In 1 minute, $\frac{1}{8}$ of the tub is drained and $\frac{1}{10}$ of the tub is filled for a total change of $\frac{1}{8} - \frac{1}{10}$ of the tub.

Translate. We want to find t such that

$$t\left(\frac{1}{8} - \frac{1}{10}\right) = 1, \text{ or } \frac{t}{8} - \frac{t}{10} = 1.$$

Carry out. We solve the equation.

$$\begin{aligned}
\frac{t}{8} - \frac{t}{10} &= 1, \text{ LCD is } 40\\
40\left(\frac{t}{8} - \frac{t}{10}\right) &= 40\cdot 1\\
5t - 4t &= 40\\
t &= 40
\end{aligned}$$

Check. In 40 min, $40\cdot\frac{1}{8}$ or 5 tubs of water are drained and $40\cdot\frac{1}{10}$ or 4 tubs of water are added. Since the tub was full to begin with, 5 tubs of water needed to be drained in order to empty the tub. The answer checks.

State. It will take 40 min to empty a full tub if the water is left on.

48. $49\frac{1}{2}$ hr

49. ***Familiarize***. Let $p =$ the number of people per hour moved by the 60 cm-wide escalator. Then $2p =$ the number of people per hour moved by the 100 cm-wide escalator. We convert 1575 people per 14 minutes to people per hour:

$$\frac{1575 \text{ people}}{14 \cancel{\text{min}}}\cdot\frac{60 \cancel{\text{min}}}{1 \text{ hr}} = 6750 \text{ people/hr}$$

Translate. We use the information that together the escalators move 6750 people per hour to write an equation.

$$p + 2p = 6750$$

Carry out. We solve the equation.

$$p + 2p = 6750$$
$$3p = 6750$$
$$p = 2250$$

Check. If the 60 cm-wide escalator moves 2250 people per hour, then the 100 cm-wide escalator moves $2 \cdot 2250$, or 4500 people per hour. Together, they move $2250 + 4500$, or 6750 people per hour. The answer checks.

State. The 60 cm-wide escalator moves 2250 people per hour.

50. 700 mi from the airport

51. ***Familiarize***. Let d = the distance, in miles, the paddleboat can cruise upriver before it is time to turn around. The boat's speed upriver is $12 - 5$, or 7 mph, and its speed downriver is $12 + 5$, or 17 mph. We organize the information in a table.

	Distance	Speed	Time
Upriver	d	7	t_1
Downriver	d	17	t_2

Translate. Using the formula Time = Distance/Rate we see that $t_1 = \frac{d}{7}$ and $t_2 = \frac{d}{17}$. The time upriver and back is 3 hr, so $t_1 + t_2 = 3$, or

$$\frac{d}{7} + \frac{d}{17} = 3.$$

Carry out. We solve the equation.

$$7 \cdot 17\left(\frac{d}{7} + \frac{d}{17}\right) = 7 \cdot 17 \cdot 3$$
$$17d + 7d = 357$$
$$24d = 357$$
$$d = \frac{119}{8}$$

Check. Traveling $\frac{119}{8}$ mi upriver at a speed of 7 mph takes $\frac{119/8}{7} = \frac{17}{8}$ hr. Traveling $\frac{119}{8}$ mi downriver at a speed of 17 mph takes $\frac{119/8}{17} = \frac{7}{8}$ hr. The total time is $\frac{17}{8} + \frac{7}{8} = \frac{24}{8} = 3$ hr. The answer checks.

State. The pilot can go $\frac{119}{8}$, or $14\frac{7}{8}$ mi upriver before it is time to turn around.

52. $3\frac{3}{4}$ km/h

53. ***Familiarize***. Let d = the distance, in miles, Melissa lives from work. Also let t = the travel time in hours, when Melissa arrives on time. Note that 1 min $= \frac{1}{60}$ hr and 5 min $= \frac{5}{60}$, or $\frac{1}{12}$ hr.

Translate. Melissa's travel time at 50 mph is $\frac{d}{50}$. This is $\frac{1}{60}$ hr more than t, so we write an equation using this information:

$$\frac{d}{50} = t + \frac{1}{60}$$

Her travel time at 60 mph, $\frac{d}{60}$, is $\frac{1}{12}$ hr less than t, so we write a second equation:

$$\frac{d}{60} = t - \frac{1}{12}$$

We have a system of equations:

$$\frac{d}{50} = t + \frac{1}{60},$$
$$\frac{d}{60} = t - \frac{1}{12}$$

Carry out. Solving the system of equations, we get $\left(30, \frac{7}{12}\right)$.

Check. Traveling 30 mi at 50 mph takes $\frac{30}{50}$, or $\frac{3}{5}$ hr. Since $\frac{7}{12} + \frac{1}{60} = \frac{36}{60} = \frac{3}{5}$, this time makes Melissa $\frac{1}{60}$ hr, or 1 min late. Traveling 30 mi at 60 mph takes $\frac{30}{60}$, or $\frac{1}{2}$ hr. Since $\frac{7}{12} - \frac{1}{12} = \frac{6}{12} = \frac{1}{2}$, this time makes Melissa $\frac{1}{12}$ hr, or 5 min early. The answer checks.

State. Melissa lives 30 mi from work.

54. $21\frac{9}{11}$ min after 4:00

55. ***Familiarize*** Express the position of the hands in terms of minute units on the face of the clock. At 10:30 the hour hand is at $\frac{10.5}{12}\text{ hr} \times \frac{60\text{ min}}{1\text{ hr}}$, or 52.5 minutes, and the minute hand is at 30 minutes. The rate of the minute hand is 12 times the rate of the hour hand. (When the minute hand moves 60 minutes, the hour hand moves 5 minutes.) Let t = the number of minutes after 10:30 that the hands will first be perpendicular. After t minutes the minute hand has moved t units, and the hour hand has moved $\frac{t}{12}$ units. The position of the hour hand will be 15 units "ahead" of the position of the minute hand when they are first perpendicular.

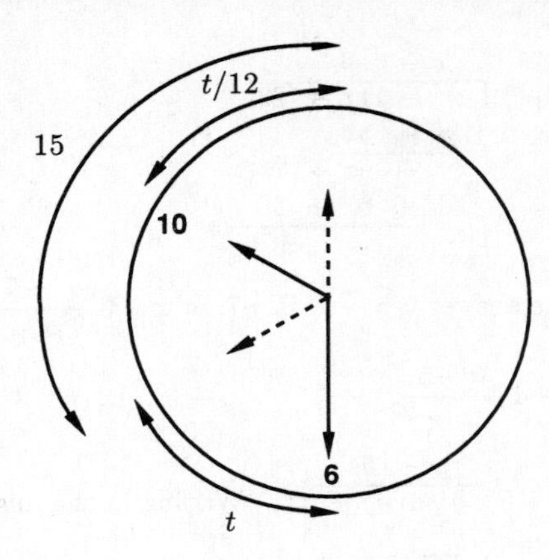

Translate.

Position of hour hand after t min is position of minute hand after t min plus 15 min.

$$52.5+\frac{t}{12} = 30+t + 15$$

Solve. We solve the equation.

$$52.5+\frac{t}{12}=30+t+15$$
$$52.5+\frac{t}{12}=45+t, \text{ LCM is } 12$$
$$12\left(52.5+\frac{t}{12}\right)=12(45+t)$$
$$630+t=540+12t$$
$$90=11t$$
$$\frac{90}{11}=t, \text{ or}$$
$$8\frac{2}{11}=t$$

Check. At $\frac{90}{11}$ min after 10:30, the position of the hour hand is at $52.5+\frac{90/11}{12}$, or $53\frac{2}{11}$ min. The minute hand is at $30+\frac{90}{11}$, or $38\frac{2}{11}$ min. The hour hand is 15 minutes ahead of the minute hand so the hands are perpendicular. The answer checks.

State. After 10:30 the hands of a clock will first be perpendicular in $8\frac{2}{11}$ min. The time is $10{:}38\frac{2}{11}$, or $21\frac{9}{11}$ min before 11:00.

56. 48 km/h

57. ***Familiarize.*** Let r = the speed in mph Chip would have to travel for the last half of the trip in order to average a speed of 45 mph for the entire trip. We organize the information in a table.

	Distance	Speed	Time
First half	50	40	t_1
Last half	50	r	t_2

The total distance is 50 + 50, or 100 mi.

The total time is t_1+t_2, or $\frac{50}{40}+\frac{50}{r}$, or $\frac{5}{4}+\frac{50}{r}$. The average speed is 45 mph.

Translate.

$$\text{Average speed} = \frac{\text{Total distance}}{\text{Total time}}$$
$$45=\frac{100}{\frac{5}{4}+\frac{50}{r}}$$

Carry out. We solve the equation.

$$45=\frac{100}{\frac{5}{4}+\frac{50}{r}}$$
$$45=\frac{100}{\frac{5r+200}{4r}}$$
$$45=100\cdot\frac{4r}{5r+200}$$
$$45=\frac{400r}{5r+200}$$
$$(5r+200)(45)=(5r+200)\cdot\frac{400r}{5r+200}$$
$$225r+9000=400r$$
$$9000=175r$$
$$\frac{360}{7}=r$$

Check. Traveling 50 mi at 40 mph takes $\frac{50}{40}$, or $\frac{5}{4}$ hr. Traveling 50 mi at $\frac{360}{7}$ mph takes $\frac{50}{360/7}$, or $\frac{35}{36}$ hr. Then the total time is $\frac{5}{4}+\frac{35}{36}=\frac{80}{36}=\frac{20}{9}$ hr. The average speed when traveling 100 mi for $\frac{20}{9}$ hr is $\frac{100}{20/9}=45$ mph. The answer checks.

State. Chip would have to travel at a speed of $\frac{360}{7}$, or $51\frac{3}{7}$ mph for the last half of the trip so that the average speed for the entire trip would be 45 mph.

Exercise Set 6.6

1. $$\frac{24x^6+18x^5-36x^2}{6x^2}$$
$$=\frac{24x^6}{6x^2}+\frac{18x^5}{6x^2}-\frac{36x^2}{6x^2}$$
$$=4x^4+3x^3-6$$

2. $6y^4-3y^2+8$

3. $$\frac{28a^3+7a^2-3a-14}{7a}$$
$$=\frac{28a^3}{7a}+\frac{7a^2}{7a}-\frac{3a}{7a}-\frac{14}{7a}$$
$$=4a^2+a-\frac{3}{7}-\frac{2}{a}$$

4. $-8x^2 + 4x - \frac{3}{5} + \frac{7}{5x}$

5. $$\frac{26y^3 - 9y^2 - 8y}{2y^2}$$
$$= \frac{26y^3}{2y^2} - \frac{9y^2}{2y^2} - \frac{8y}{2y^2}$$
$$= 13y - \frac{9}{2} - \frac{4}{y}$$

6. $3a^3 + \frac{9a}{2} - \frac{4}{a}$

7. $$\frac{18x^7 - 27x^4 - 3x^2}{-3x^2}$$
$$= \frac{18x^7}{-3x^2} + \frac{-27x^4}{-3x^2} + \frac{-3x^2}{-3x^2}$$
$$= -6x^5 + 9x^2 + 1$$

8. $-6y^5 + 3y^3 + 2y$

9. $$(a^2b - a^3b^3 - a^5b^5) \div (a^2b)$$
$$= \frac{a^2b}{a^2b} - \frac{a^3b^3}{a^2b} - \frac{a^5b^5}{a^2b}$$
$$= 1 - ab^2 - a^3b^4$$

10. $x - xy - x^2$

11. $$(6p^2q^2 - 9p^2q + 12pq^2) \div -3pq$$
$$= \frac{6p^2q^2}{-3pq} + \frac{-9p^2q}{-3pq} + \frac{12pq^2}{-3pq}$$
$$= -2pq + 3p - 4q$$

12. $4z - 2y^2z^3 + 3y^4z^2$

13.
$$\begin{array}{rl}
x + 7 & \\
x+3\,\overline{)\,x^2 + 10x + 21} & \\
\underline{x^2 + 3x} & \\
7x + 21 & (x^2+10x)-(x^2+3x)=7x \\
\underline{7x + 21} & \\
0 &
\end{array}$$

The answer is $x + 7$.

14. $y - 4$

15.
$$\begin{array}{rl}
a - 12 & \\
a+4\,\overline{)\,a^2 - 8a - 16} & \\
\underline{a^2 + 4a} & \\
-12a - 16 & (a^2-8a)-(a^2+4a)=-12a \\
\underline{-12a - 48} & \\
32 & (-12a-16)-(-12a-48)=32
\end{array}$$

The answer is $a - 12$, R 32, or $a - 12 + \frac{32}{a+4}$.

16. $y - 5 + \frac{-50}{y-5}$

17.
$$\begin{array}{r}
x - 6 \\
x-5\,\overline{)\,x^2 - 11x + 23} \\
\underline{x^2 - 5x} \\
-6x + 23 \\
\underline{-6x + 30} \\
-7
\end{array}$$

The answer is $x - 6$, R -7, or $x - 6 + \frac{-7}{x-5}$.

18. $x - 4 + \frac{-5}{x-7}$

19.
$$\begin{array}{rl}
y - 5 & \\
y+5\,\overline{)\,y^2 + 0y - 25} & \text{Writing in the missing term} \\
\underline{y^2 + 5y} & \\
-5y - 25 & \\
\underline{-5y - 25} & \\
0 &
\end{array}$$

The answer is $y - 5$.

20. $a + 9$

21.
$$\begin{array}{r}
y^2 - 2y - 1 \\
y-2\,\overline{)\,y^3 - 4y^2 + 3y - 6} \\
\underline{y^3 - 2y^2} \\
-2y^2 + 3y \\
\underline{-2y^2 + 4y} \\
-y - 6 \\
\underline{-y + 2} \\
-8
\end{array}$$

The answer is $y^2 - 2y - 1$, R -8, or $y^2 - 2y - 1 + \frac{-8}{y-2}$.

22. $x^2 - 2x - 2 + \frac{-13}{x-3}$

23.
$$\begin{array}{r}
2x^2 - x + 1 \\
x+2\,\overline{)\,2x^3 + 3x^2 - x - 3} \\
\underline{2x^3 + 4x^2} \\
-x^2 - x \\
\underline{-x^2 - 2x} \\
x - 3 \\
\underline{x + 2} \\
-5
\end{array}$$

The answer is $2x^2 - x + 1$, R -5, or $2x^2 - x + 1 + \frac{-5}{x+2}$.

24. $3x^2 + x - 1 + \frac{-4}{x-2}$

25.
$$\begin{array}{r}
a^2 + 4a + 15 \\
a-4\,\overline{)\,a^3 + 0a^2 - a + 12} \\
\underline{a^3 - 4a^2} \\
4a^2 - a \\
\underline{4a^2 - 16a} \\
15a + 12 \\
\underline{15a - 60} \\
72
\end{array}$$

The answer is $a^2 + 4a + 15$, R 72, or $a^2 + 4a + 15 + \frac{72}{a-4}$.

26. $x^2 - 2x + 3$

27.

$$\begin{array}{r}
2y^2 + 2y - 1 \\
5y - 2\,\overline{)\,10y^3 + 6y^2 - 9y + 10} \\
\underline{10y^3 - 4y^2} \qquad\qquad \\
10y^2 - 9y \qquad \\
\underline{10y^2 - 4y} \qquad \\
-5y + 10 \\
\underline{-5y + 2} \\
8
\end{array}$$

The answer is $2y^2 + 2y - 1$, R 8, or

$2y^2 + 2y - 1 + \dfrac{8}{5y - 2}$.

28. $3x^2 - x + 4 + \dfrac{10}{2x - 3}$

29.

$$\begin{array}{r}
2x^2 - x - 9 \\
x^2 + 2\,\overline{)\,2x^4 - x^3 - 5x^2 + x - 6} \\
\underline{2x^4 \qquad + 4x^2} \qquad\qquad\quad \\
-x^3 - 9x^2 + x \qquad \\
\underline{-x^3 \qquad\quad - 2x} \qquad \\
-9x^2 + 3x - 6 \\
\underline{-9x^2 \qquad - 18} \\
3x + 12
\end{array}$$

The answer is $2x^2 - x - 9$, R $3x + 12$, or

$2x^2 - x - 9 + \dfrac{3x + 12}{x^2 + 2}$.

30. $3x^2 + 2x - 5 + \dfrac{2x - 5}{x^2 - 2}$

31. $F(x) = \dfrac{f(x)}{g(x)} = \dfrac{64x^3 - 8}{4x - 2}$

$$\begin{array}{r}
16x^2 + 8x + 4 \\
4x - 2\,\overline{)\,64x^3 \qquad\qquad\qquad - 8} \\
\underline{64x^3 - 32x^2} \qquad\qquad\quad \\
32x^2 + 0x \qquad \\
\underline{32x^2 - 16x} \qquad \\
16x - 8 \\
\underline{16x - 8} \\
0
\end{array}$$

Since $g(x)$ is 0 for $x = \dfrac{1}{2}$, we have

$F(x) = 16x^2 + 8x + 4$, provided $x \neq \dfrac{1}{2}$.

32. $4x^2 - 6x + 9$, $x \neq -\dfrac{3}{2}$

33. $F(x) = \dfrac{f(x)}{g(x)} = \dfrac{6x^2 - 11x - 10}{3x + 2}$

$$\begin{array}{r}
2x - 5 \\
3x + 2\,\overline{)\,6x^2 - 11x - 10} \\
\underline{6x^2 + 4x} \qquad\quad \\
-15x - 10 \\
\underline{-15x - 10} \\
0
\end{array}$$

Since $g(x)$ is 0 for $x = -\dfrac{2}{3}$, we have

$F(x) = 2x - 5$, provided $x \neq -\dfrac{2}{3}$.

34. $4x + 3$, $x \neq \dfrac{7}{2}$

35. $F(x) = \dfrac{f(x)}{g(x)} = \dfrac{x^4 - 3x^2 - 54}{x^2 - 9}$

$$\begin{array}{r}
x^2 + 6 \\
x^2 - 9\,\overline{)\,x^4 - 3x^2 - 54} \\
\underline{x^4 - 9x^2} \qquad\quad \\
6x^2 - 54 \\
\underline{6x^2 - 54} \\
0
\end{array}$$

Since $g(x)$ is 0 for $x = -3$ or $x = 3$, we have $F(x) = x^2 + 6$, provided $x \neq -3$ and $x \neq 3$.

36. $x^2 + 1$, $x \neq -5$, $x \neq 5$

37. $F(x) = \dfrac{f(x)}{g(x)} = \dfrac{2x^5 - 3x^4 - 2x^3 + 8x^2 - 5}{x^2 - 1}$

$$\begin{array}{r}
2x^3 - 3x^2 + 5 \\
x^2 - 1\,\overline{)\,2x^5 - 3x^4 - 2x^3 + 8x^2 - 5} \\
\underline{2x^5 \qquad\quad - 2x^3} \qquad\qquad\quad \\
-3x^4 \qquad\quad + 8x^2 \qquad \\
\underline{-3x^4 \qquad\quad + 3x^2} \qquad \\
5x^2 - 5 \\
\underline{5x^2 - 5} \\
0
\end{array}$$

Since $g(x)$ is 0 for $x = -1$ or $x = 1$, we have $F(x) = 2x^3 - 3x^2 + 5$, provided $x \neq -1$ and $x \neq 1$.

38. $3x^2 - x + 2$, $x \neq -2$, $x \neq 2$

39.

$$\begin{aligned}
x^2 - 5x &= 0 \\
x(x - 5) &= 0 \\
x = 0 \text{ or } x - 5 &= 0 \quad \text{Principle of zero products} \\
x = 0 \text{ or } \quad x &= 5
\end{aligned}$$

The solutions are 0 and 5.

40. $-\dfrac{8}{5}, \dfrac{8}{5}$

41. ***Familiarize.*** Let x, $x + 1$, and $x + 2$ represent the three consecutive positive integers.

Translate. Rewording, we write an equation.

Product of first and second	is	product of second and third	less	26.
$\downarrow$	$\downarrow$	$\downarrow$	$\downarrow$	$\downarrow$
$x(x + 1)$	$=$	$(x + 1)(x + 2)$	$-$	26

Carry out. We solve the equation.

$$\begin{aligned}
x^2 + x &= x^2 + 3x + 2 - 26 \\
x &= 3x - 24 \\
24 &= 2x \\
12 &= x
\end{aligned}$$

If the first integer is 12, the next two are 13 and 14.

Check. The product of 12 and 13 is 156. The product of 13 and 14 is 182, and $182 - 26 = 156$. The numbers check.

State. The three consecutive positive integers are 12, 13, and 14.

42. $-54a^3$

43. $|2x-3| = 7$

$$2x - 3 = 7 \quad or \quad 2x - 3 = -7$$
$$2x = 10 \quad or \quad 2x = -4$$
$$x = 5 \quad or \quad x = -2$$

The solution set is $\{5, -2\}$.

44. $\left\{x \middle| -\frac{7}{3} < x < 3\right\}$, or $\left(-\frac{7}{3}, 3\right)$

45.

46.

47.

48.

49.
$$\begin{array}{r} x^2 + 2y \\ x^2 - xy + y^2 \overline{\big)\, x^4 - x^3y + x^2y^2 + 2x^2y - 2xy^2 + 2y^3} \\ \underline{x^4 - x^3y + x^2y^2 \qquad\qquad\qquad\qquad} \\ 0 + 2x^2y - 2xy^2 + 2y^3 \\ \underline{2x^2y - 2xy^2 + 2y^3} \\ 0 \end{array}$$

The answer is $x^2 + 2y$.

50. $a^2 + ab$

51.
$$\begin{array}{r} a^6 - a^5b + a^4b^2 - a^3b^3 + a^2b^4 - ab^5 + b^6 \\ a + b \overline{\big)\, a^7 \qquad\qquad\qquad\qquad\qquad\qquad + b^7} \\ \underline{a^7 + a^6b} \\ -a^6b \\ \underline{-a^6b - a^5b^2} \\ a^5b^2 \\ \underline{a^5b^2 + a^4b^3} \\ -a^4b^3 \\ \underline{-a^4b^3 - a^3b^4} \\ a^3b^4 \\ \underline{a^3b^4 + a^2b^5} \\ -a^2b^5 \\ \underline{-a^2b^5 - ab^6} \\ ab^6 + b^7 \\ \underline{ab^6 + b^7} \\ 0 \end{array}$$

The answer is $a^6 - a^5b + a^4b^2 - a^3b^3 + a^2b^4 - ab^5 + b^6$.

52. $\frac{14}{3}$

53.
$$\begin{array}{r} x - 5 \\ x + 2 \overline{\big)\, x^2 - 3x + 2k} \\ \underline{x^2 + 2x \qquad} \\ -5x + 2k \\ \underline{-5x - 10} \\ 2k + 10 \end{array}$$

The remainder is 7. Thus, we solve the following equation for k.

$$2k + 10 = 7$$
$$2k = -3$$
$$k = -\frac{3}{2}$$

54. (a), (b) $3 + \frac{1}{x+2}$

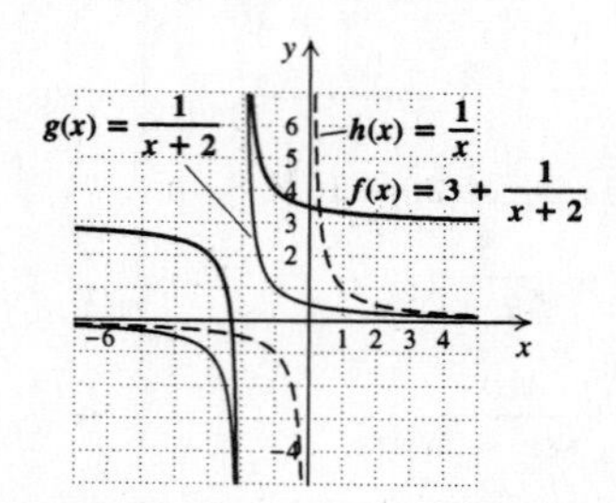

c) The graph of f looks like the graph of g, shifted up 3 units. The graph of g looks like the graph of h, shifted to the left 2 units.

55.

56.

57.

Exercise Set 6.7

1. $(x^3 - 2x^2 + 2x - 5) \div (x - 1)$

$$\begin{array}{r|rrrr} 1 & 1 & -2 & 2 & -5 \\ & & 1 & -1 & 1 \\ \hline & 1 & -1 & 1 & -4 \end{array}$$

The answer is $x^2 - x + 1$, R -4, or $x^2 - x + 1 + \frac{-4}{x-1}$.

2. $x^2 - 3x + 5$, R -10, or $x^2 - 3x + 5 + \frac{-10}{x+1}$

3. $(a^2 + 11a - 19) \div (a + 4) =$

$(a^2 + 11a - 19) \div [a - (-4)]$

$$\begin{array}{r|rrr} -4 & 1 & 11 & -19 \\ & & -4 & -28 \\ \hline & 1 & 7 & -47 \end{array}$$

The answer is $a + 7$, R -47, or $a + 7 + \frac{-47}{a+4}$.

4. $a + 15$, R 41, or $a + 15 + \frac{41}{a-4}$

5. $(x^3 - 7x^2 - 13x + 3) \div (x - 2)$

$$\begin{array}{r|rrrr} 2 & 1 & -7 & -13 & 3 \\ & & 2 & -10 & -46 \\ \hline & 1 & -5 & -23 & -43 \end{array}$$

The answer is $x^2 - 5x - 23$, R -43, or

$x^2 - 5x - 23 + \frac{-43}{x-2}$.

6. $x^2 - 9x + 5$, R -7, or $x^2 - 9x + 5 + \dfrac{-7}{x+2}$

7. $(3x^3 + 7x^2 - 4x + 3) \div (x + 3) =$

$(3x^3 + 7x^2 - 4x + 3) \div [x - (-3)]$

$$\begin{array}{r|rrrr} -3 & 3 & 7 & -4 & 3 \\ & & -9 & 6 & -6 \\ \hline & 3 & -2 & 2 & \mid -3 \end{array}$$

The answer is $3x^2 - 2x + 2$, R -3, or

$3x^2 - 2x + 2 + \dfrac{-3}{x+3}$.

8. $3x^2 + 16x + 44$, R 135, or $3x^2 + 16x + 44 + \dfrac{135}{x-3}$

9. $(y^3 - 3y + 10) \div (y - 2) =$

$(y^3 + 0y^2 - 3y + 10) \div (y - 2)$

$$\begin{array}{r|rrrr} 2 & 1 & 0 & -3 & 10 \\ & & 2 & 4 & 2 \\ \hline & 1 & 2 & 1 & \mid 12 \end{array}$$

The answer is $y^2 + 2y + 1$, R 12, or

$y^2 + 2y + 1 + \dfrac{12}{y-2}$.

10. $x^2 - 4x + 8$, R -8, or $x^2 - 4x + 8 + \dfrac{-8}{x+2}$

11. $(x^5 - 32) \div (x - 2) =$

$(x^5 + 0x^4 + 0x^3 + 0x^2 + 0x - 32) \div (x - 2)$

$$\begin{array}{r|rrrrrr} 2 & 1 & 0 & 0 & 0 & 0 & -32 \\ & & 2 & 4 & 8 & 16 & 32 \\ \hline & 1 & 2 & 4 & 8 & 16 & \mid 0 \end{array}$$

The answer is $x^4 + 2x^3 + 4x^2 + 8x + 16$.

12. $y^4 + y^3 + y^2 + y + 1$

13. $(3x^3 + 1 - x + 7x^2) \div \left(x + \dfrac{1}{3}\right) =$

$(3x^3 + 7x^2 - x + 1) \div \left[x - \left(-\dfrac{1}{3}\right)\right]$

$$\begin{array}{r|rrrr} -\frac{1}{3} & 3 & 7 & -1 & 1 \\ & & -1 & -2 & 1 \\ \hline & 3 & 6 & -3 & \mid 2 \end{array}$$

The answer is $3x^2 + 6x - 3$ R 2, or

$3x^2 + 6x - 3 + \dfrac{2}{x + \frac{1}{3}}$.

14. $8x^2 - 2x + 6$, R 2, or $8x^2 - 2x + 6 + \dfrac{2}{x - \frac{1}{2}}$

15.

$$\begin{array}{r|rrrrr} -3 & 6 & 15 & 0 & 28 & 6 \\ & & -18 & 9 & -27 & -3 \\ \hline & 6 & -3 & 9 & 1 & \mid 3 \end{array}$$

The remainder tells us that $f(-3) = 3$.

16. 0

17.

$$\begin{array}{r|rrrrr} -1 & 2 & -1 & -5 & 1 & 7 \\ & & -2 & 3 & 2 & -3 \\ \hline & 2 & -3 & -2 & 3 & \mid 4 \end{array}$$

The remainder tells us that $P(-1) = 4$.

18. 2

19.

$$\begin{array}{r|rrrrr} 4 & 1 & -1 & -19 & 49 & -30 \\ & & 4 & 12 & -28 & 84 \\ \hline & 1 & 3 & -7 & 21 & \mid 54 \end{array}$$

The remainder tells us that $f(4) = 54$.

20. 90

21. Graph: $2x - 3y < 6$

First graph the line $2x - 3y = 6$. The intercepts are $(0, -2)$ and $(3, 0)$. We draw the line dashed since the inequality is $<$. Since the ordered pair $(0, 0)$ is a solution of the inequality ($2 \cdot 0 - 3 \cdot 0 < 6$ is true), we shade the upper half-plane.

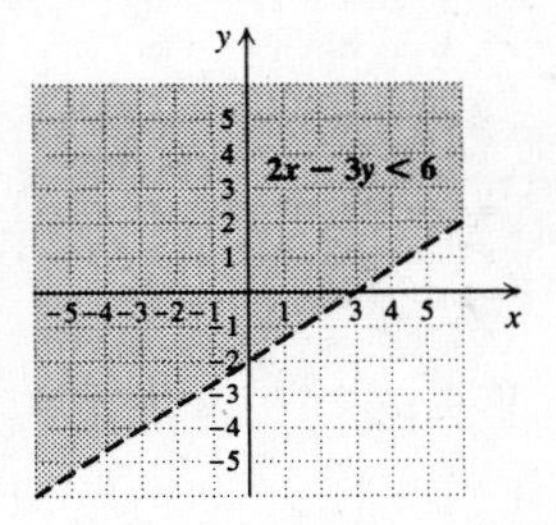

22.

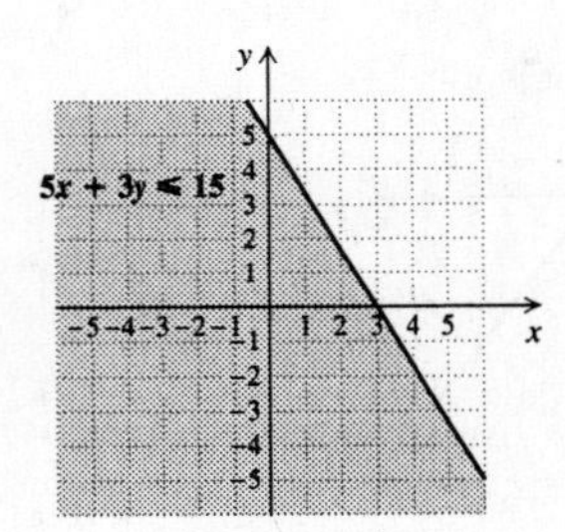

23. Graph: $y > 4$

First graph the line $y = 4$. The line is parallel to the x-axis with y-intercept $(0, 4)$. We draw the line dashed since the inequality is $>$. Since the ordered pair $(0, 0)$ is not a solution of the inequality ($0 > 4$ is false), we shade the upper half-plane.

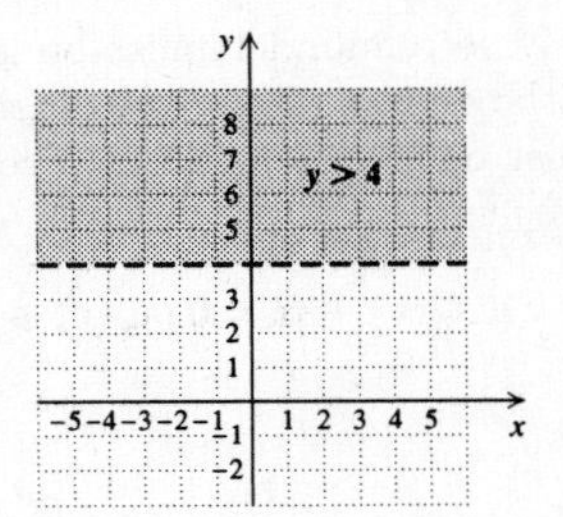

24.

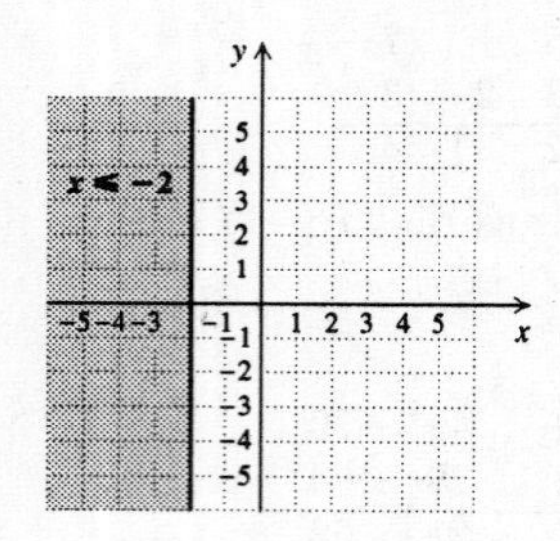

25. Graph: $y - 2 = \frac{3}{4}(x + 1)$

This is the equation of the line with slope $\frac{3}{4}$ and passing through $(-1, 2)$. To graph this equation, start at $(-1, 2)$ and count off a slope of $\frac{3}{4}$ by going up 3 units and to the right 4 units (or down 3 units and to the left 4 units). Then draw the line.

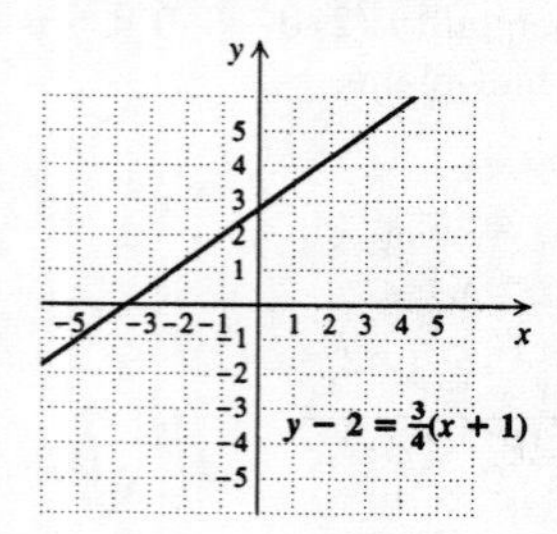

26.

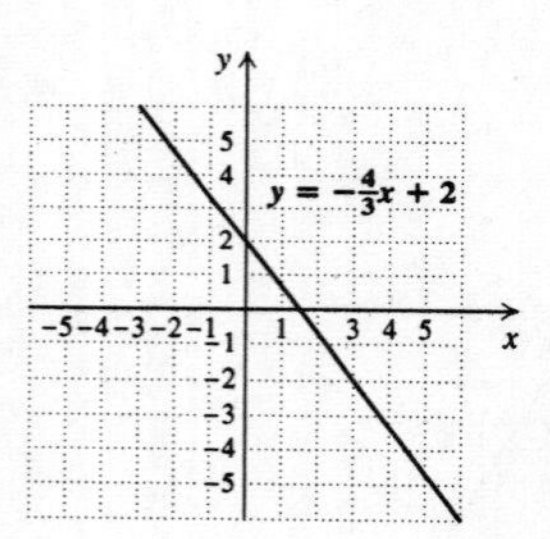

27.

28.

29.

30.

31. a) The degree of the remainder must be less than the degree of the divisor. Thus, the degree of the remainder must be 0, so R must be a constant.

b) $P(x) = (x - r) \cdot Q(x) + R$

$P(r) = (r - r) \cdot Q(r) + R = 0 \cdot Q(r) + R = R$

32. $0;\ -3,\ -\frac{5}{2},\ \frac{3}{2}$

33.
$$\begin{array}{r|rrrr} 4 & 6 & -13 & -79 & 140 \\ & & 24 & 44 & -140 \\ \hline & 6 & 11 & -35 & 0 \end{array}$$

The remainder tells us that $f(4) = 0$.

$f(x) = (x-4)(6x^2+11x-35) = (x-4)(2x+7)(3x-5)$

Solve $f(x) = 0$:

$(x - 4)(2x + 7)(3x - 5) = 0$

$x - 4 = 0 \quad or \quad 2x + 7 = 0 \quad or \quad 3x - 5 = 0$

$x = 4 \quad or \quad 2x = -7 \quad or \quad 3x = 5$

$x = 4 \quad or \quad x = -\frac{7}{2} \quad or \quad x = \frac{5}{3}$

The solutions are 4, $-\frac{7}{2}$, and $\frac{5}{3}$.

34.

35.

36. 0

37.
$$\begin{aligned} f(x) &= 6x^3 - 13x^2 - 79x + 140 \\ &= x(6x^2 - 13x - 79) + 140 \\ &= x(x(6x - 13) - 79) + 140 \\ f(4) &= 4(4(6 \cdot 4 - 13) - 79) + 140 \\ &= 4(4(24 - 13) - 79) + 140 \\ &= 4(4 \cdot 11 - 79) + 140 \\ &= 4(44 - 79) + 140 \\ &= 4(-35) + 140 \\ &= -140 + 140 \\ &= 0 \end{aligned}$$

Exercise Set 6.8

1.
$$\frac{W_1}{W_2} = \frac{d_1}{d_2}$$

$$W_1 = \frac{d_1 W_2}{d_2} \quad \text{Multiplying by } W_2$$

2. $d_1 = \frac{d_2 W_1}{W_2}$

3.
$$s = \frac{(v_1 + v_2)t}{2}$$

$$2s = (v_1 + v_2)t \quad \text{Multiplying by 2}$$

$$\frac{2s}{t} = v_1 + v_2 \quad \text{Dividing by } t$$

$$\frac{2s}{t} - v_2 = v_1$$

This result can also be expressed as $v_1 = \frac{2s - tv_2}{t}$.

4. $t = \frac{2s}{v_1 + v_2}$

5. $$\frac{1}{R} = \frac{1}{r_1} + \frac{1}{r_2}$$

$$Rr_1r_2 \cdot \frac{1}{R} = Rr_1r_2\left(\frac{1}{r_1} + \frac{1}{r_2}\right) \quad \text{Multiplying by the LCD}$$

$$Rr_1r_2 \cdot \frac{1}{R} = Rr_1r_2 \cdot \frac{1}{r_1} + Rr_1r_2 \cdot \frac{1}{r_2}$$

$$r_1r_2 = Rr_2 + Rr_1$$

$$r_1r_2 = R(r_2 + r_1) \quad \text{Factoring out } R$$

$$\frac{r_1r_2}{r_2 + r_1} = R \quad \text{Multiplying by } \frac{1}{r_2 + r_1}$$

6. $r_1 = \dfrac{Rr_2}{r_2 - R}$

7. $$R = \frac{gs}{g+s}$$

$$(g+s) \cdot R = (g+s) \cdot \frac{gs}{g+s} \quad \text{Multiplying by the LCD}$$

$$Rg + Rs = gs$$

$$Rs = gs - Rg$$

$$Rs = g(s - R) \quad \text{Factoring out } g$$

$$\frac{Rs}{s - R} = g \quad \text{Multiplying by } \frac{1}{s - R}$$

8. $t = \dfrac{Kr}{r + K}$

9. $$I = \frac{2V}{R + 2r}$$

$$I(R + 2r) = \frac{2V}{R + 2r} \cdot (R + 2r) \quad \text{Multiplying by the LCD}$$

$$I(R + 2r) = 2V$$

$$R + 2r = \frac{2V}{I}$$

$$R = \frac{2V}{I} - 2r, \text{ or } \frac{2V - 2Ir}{I}$$

10. $r = \dfrac{2V - IR}{2I}$

11. $$\frac{1}{p} + \frac{1}{q} = \frac{1}{f}$$

$$pqf\left(\frac{1}{p} + \frac{1}{q}\right) = pqf \cdot \frac{1}{f} \quad \text{Multiplying by the LCD}$$

$$qf + pf = pq$$

$$qf = pq - pf$$

$$qf = p(q - f)$$

$$\frac{qf}{q - f} = p$$

12. $q = \dfrac{pf}{p - f}$

13. $$I = \frac{nE}{R + nr}$$

$$I(R + nr) = \frac{nE}{R + nr} \cdot (R + nr) \quad \text{Multiplying by the LCD}$$

$$IR + Inr = nE$$

$$IR = nE - Inr$$

$$IR = n(E - Ir)$$

$$\frac{IR}{E - Ir} = n$$

14. $r = \dfrac{nE - IR}{In}$

15. $$S = \frac{H}{m(t_1 - t_2)}$$

$$(t_1 - t_2)S = \frac{H}{m} \quad \text{Multiplying by } t_1 - t_2$$

$$t_1 - t_2 = \frac{H}{Sm} \quad \text{Dividing by } S$$

$$t_1 = \frac{H}{Sm} + t_2, \text{ or } \frac{H + Smt_2}{Sm}$$

16. $H = m(t_1 - t_2)S$

17. $$\frac{E}{e} = \frac{R + r}{r}$$

$$er \cdot \frac{E}{e} = er \cdot \frac{R + r}{r} \quad \text{Multiplying by the LCD}$$

$$Er = e(R + r)$$

$$Er = eR + er$$

$$Er - er = eR$$

$$r(E - e) = eR$$

$$r = \frac{er}{E - e}$$

18. $e = \dfrac{rE}{R + r}$

19. $$S = \frac{a}{1 - r}$$

$$(1 - r)S = a \quad \text{Multiplying by the LCD, } 1 - r$$

$$1 - r = \frac{a}{S} \quad \text{Dividing by } S$$

$$1 - \frac{a}{S} = r \quad \text{Adding } r \text{ and } -\frac{a}{S}$$

This result can also be expressed as $r = \dfrac{S - a}{S}$.

20. $a = \dfrac{S - Sr}{1 - r^n}$

21. $P = \dfrac{A}{1+r}$

$P(1+r) = \dfrac{A}{1+r} \cdot (1+r)$ Multiplying by the LCD

$P(1+r) = A$

$1+r = \dfrac{A}{P}$ Dividing by P

$r = \dfrac{A}{P} - 1$, or $\dfrac{A-P}{P}$

22. $t_2 = \dfrac{d_2 - d_1}{v} + t_1$, or $\dfrac{d_2 - d_1 + t_1 v}{v}$

23. From Exercise 22 we know that

$$t_2 = \frac{d_2 - d_1}{v} + t_1.$$

We use this result, replacing v with 60, d_1 with 0, d_2 with 105, and t_1 with 0.

$$t_2 = \frac{105 - 0}{60} + 0$$

$t_2 = 1.75$ hr, or 1 hr 45 min

The arrival time is 1 hr 45 min after 2:00 A.M., or 3:45 A.M.

24. 7%

25. First we solve the formula for R.

$$A = \frac{9R}{I}$$

$$\frac{I}{9} \cdot A = \frac{I}{9} \cdot \frac{9R}{I}$$

$$\frac{AI}{9} = R$$

Then substitute 2.4 for A and 45 for I.

$$\frac{2.4(45)}{9} = R$$

$$12 = R$$

Thus, 12 earned runs were given up.

26. $5\frac{5}{23}$ ohms

27. Use the result of Example 4, replacing R with 5 and r_1 with 50.

$$r_2 = \frac{Rr_1}{r_1 - R}$$

$$r_2 = \frac{5 \cdot 50}{50 - 5}$$

$$= \frac{250}{45}$$

$$= \frac{50}{9}, \text{ or } 5\frac{5}{9}$$

A resistor with a resistance of $5\frac{5}{9}$ ohms should be used.

28. $t = \dfrac{ab}{b+a}$

29. First solve the formula for the area of the trapezoid for one of the bases, say b_2.

$$A = \frac{1}{2}h(b_1 + b_2)$$

$\dfrac{2A}{h} = b_1 + b_2$ Multiplying by $\dfrac{2}{h}$

$$\frac{2A}{h} - b_1 = b_2$$

Then substitute 25 for A, 5 for h, and 4 for b_1.

$$\frac{2 \cdot 25}{5} - 4 = b_2$$

$$10 - 4 = b_2$$

$$6 = b_2$$

The length of the other base is 6 cm.

30. $T = -\dfrac{I_f}{I_t} + 1$, or $1 - \dfrac{I_f}{I_t}$, or $\dfrac{I_t - I_f}{I_t}$

31. $\dfrac{V^2}{R^2} = \dfrac{2g}{R+h}$

$(R+h) \cdot \dfrac{V^2}{R^2} = (R+h) \cdot \dfrac{2g}{R+h}$ Multiplying by $(R+h)$

$\dfrac{(R+h)V^2}{R^2} = 2g$

$R + h = \dfrac{2gR^2}{V^2}$ Multiplying by $\dfrac{R^2}{V^2}$

$h = \dfrac{2gR^2}{V^2} - R$ Adding $-R$

The result can also be expressed as

$$h = \frac{2gR^2 - RV^2}{V^2}.$$

32. $d = \dfrac{LD}{R+L}$

33. $A = \dfrac{2Tt + Qq}{2T + Q}$

$(2T + Q) \cdot A = (2T + Q) \cdot \dfrac{2Tt + Qq}{2T + Q}$

$2AT + AQ = 2Tt + Qq$

$AQ - Qq = 2Tt - 2AT$ Adding $-2AT$ and $-Qq$

$Q(A - q) = 2Tt - 2AT$

$Q = \dfrac{2Tt - 2AT}{A - q}$

34. $t_1 = t_2 - \dfrac{v_2 - v_1}{a}$

35. We use the formula for slope:

$$m = \frac{y_1 - y_2}{x_1 - x_2}$$

Substitute $-\frac{2}{5}$ for m, 2 for y_1, 8 for y_2, x_1 for x_1 and $2x_1$ for x_2 and solve for x_1.

$$-\frac{2}{5} = \frac{2-8}{x_1 - 2x_1}$$
$$-\frac{2}{5} = \frac{-6}{-x_1}$$
$$-\frac{2}{5} = \frac{6}{x_1}$$
$$-2x_1 = 30 \qquad \text{Multiplying by } 5x_1$$
$$x_1 = -15$$

If $x_1 = -15$, then $2x_1 = 2(-15) = -30$. Thus, the coordinates of the points are $(-15, 2)$ and $(-30, 8)$.

36. 3

37. Graph: $6x - y < 6$

First graph the line $6x - y = 6$. The intercepts are $(0, -6)$ and $(1, 0)$. We draw the line dashed since the inequality is $<$. Since the ordered pair $(0, 0)$ is a solution of the inequality ($6 \cdot 0 - 0 < 6$ is true), we shade the half-plane containing $(0, 0)$.

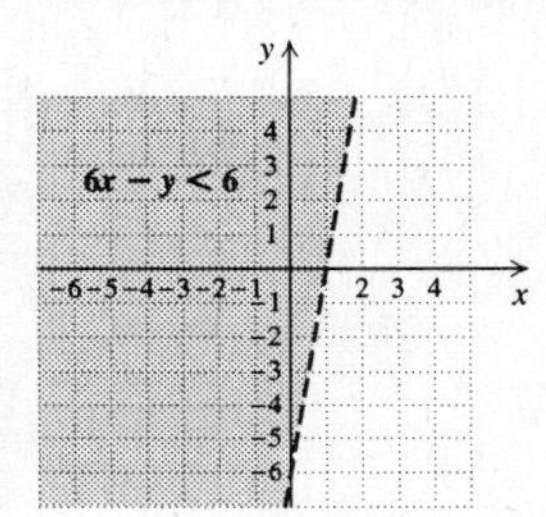

38. $8a^3 - 2a$

39. $t^3 + 8b^3 = t^3 + (2b)^3 = (t + 2b)(t^2 - 2tb + 4b^2)$

40. $-\frac{5}{3}, \frac{7}{2}$

41.

42. 567 mi

43.

$$\frac{1}{M} = \frac{\frac{1}{a} + \frac{1}{b}}{2}$$
$$2M \cdot \frac{1}{M} = 2M \cdot \frac{\frac{1}{a} + \frac{1}{b}}{2}$$
$$2 = M\left(\frac{1}{a} + \frac{1}{b}\right)$$
$$2 = \frac{M}{a} + \frac{M}{b}$$
$$ab \cdot 2 = ab\left(\frac{M}{a} + \frac{M}{b}\right)$$
$$2ab = bM + aM$$
$$2ab = M(b + a)$$
$$\frac{2ab}{b + a} = M$$

44. $pq,\ 2pq$

45.

$$a = \frac{\frac{d_4 - d_3}{t_4 - t_3} - \frac{d_2 - d_1}{t_2 - t_1}}{t_4 - t_2}$$
$$a(t_4 - t_2) = \frac{d_4 - d_3}{t_4 - t_3} - \frac{d_2 - d_1}{t_2 - t_1} \qquad \text{Multiplying by } t_4 - t_2$$
$$a(t_4 - t_2)(t_4 - t_3)(t_2 - t_1) = (d_4 - d_3)(t_2 - t_1) - (d_2 - d_1)(t_4 - t_3)$$
Multiplying by $(t_4 - t_3)(t_2 - t_1)$
$$a(t_4 - t_2)(t_4 - t_3)(t_2 - t_1) - (d_4 - d_3)(t_2 - t_1) = -(d_2 - d_1)(t_4 - t_3)$$
$$(t_2 - t_1)[a(t_4 - t_2)(t_4 - t_3) - (d_4 - d_3)] = -(d_2 - d_1)(t_4 - t_3)$$
$$t_2 - t_1 = \frac{-(d_2 - d_1)(t_4 - t_3)}{a(t_4 - t_2)(t_4 - t_3) - (d_4 - d_3)}$$
$$t_2 + \frac{(d_2 - d_1)(t_4 - t_3)}{a(t_4 - t_2)(t_4 - t_3) + d_3 - d_4} = t_1$$

Chapter 7

Exponents and Radicals

Exercise Set 7.1

1. The square roots of 16 are 4 and -4, because $4^2 = 16$ and $(-4)^2 = 16$.

2. 15, -15

3. The square roots of 144 are 12 and -12, because $12^2 = 144$ and $(-12)^2 = 144$.

4. 3, -3

5. The square roots of 400 are 20 and -20, because $20^2 = 400$ and $(-20)^2 = 400$.

6. 9, -9

7. The square roots of 49 are 7 and -7, because $7^2 = 49$ and $(-7)^2 = 49$.

8. 30, -30

9. $-\sqrt{\frac{49}{36}} = -\frac{7}{6}$ Since $\sqrt{\frac{49}{36}} = \frac{7}{6}$, $-\sqrt{\frac{49}{36}} = -\frac{7}{6}$.

10. $-\frac{19}{3}$

11. $\sqrt{196} = 14$ Remember, $\sqrt{\ }$ indicates the principle square root.

12. 21

13. $-\sqrt{\frac{16}{81}} = -\frac{4}{9}$ Since $\sqrt{\frac{16}{81}} = \frac{4}{9}$, $-\sqrt{\frac{16}{81}} = -\frac{4}{9}$.

14. $-\frac{9}{12}$, or $-\frac{3}{4}$

15. $\sqrt{0.09} = 0.3$

16. 0.6

17. $-\sqrt{0.0049} = -0.07$

18. 0.12

19. $5\sqrt{p^2+4}$

The radicand is the expression written under the radical sign, $p^2 + 4$.

Since the index is not written, we know it is 2.

20. $y^2 - 8$; 2

21. $x^2y^2\sqrt[3]{\frac{x}{y+4}}$

The radicand is the expression written under the radical sign, $\frac{x}{y+4}$.

The index is 3.

22. $\frac{a}{a^2-b}$; 3

23.
$$\begin{aligned} f(y) &= \sqrt{5y-10} \\ f(6) &= \sqrt{5\cdot 6-10} = \sqrt{20} \\ f(2) &= \sqrt{5\cdot 2-10} = \sqrt{0} = 0 \\ f(1) &= \sqrt{5\cdot 1-10} = \sqrt{-5} \end{aligned}$$

Since negative numbers do not have real-number square roots, 1 is not in the domain of f.

$$f(-1) = \sqrt{5(-1)-10} = \sqrt{-15}$$

Since negative numbers do not have real-number square roots, -1 is not in the domain of f.

24. $\sqrt{11}$; does not exist; $\sqrt{11}$, 12

25.
$$\begin{aligned} t(x) &= -\sqrt{2x+1} \\ t(4) &= -\sqrt{2\cdot 4+1} = -\sqrt{9} = -3 \\ t(-1) &= -\sqrt{2(-1)+1} = -\sqrt{-1}; \end{aligned}$$

-1 is not in the domain of t.

$$t\left(-\frac{1}{2}\right) = -\sqrt{2\left(-\frac{1}{2}\right)+1} = -\sqrt{0} = 0$$

26. $\sqrt{12}$; does not exist; $\sqrt{30}$; does not exist

27.
$$\begin{aligned} f(t) &= \sqrt{t^2+1} \\ f(0) &= \sqrt{0^2+1} = \sqrt{1} = 1 \\ f(-1) &= \sqrt{(-1)^2+1} = \sqrt{2} \\ f(-10) &= \sqrt{(-10)^2+1} = \sqrt{101} \end{aligned}$$

28. -2; -5; -4

29.
$$\begin{aligned} g(x) &= \sqrt{x^3+9} \\ g(-2) &= \sqrt{(-2)^3+9} = \sqrt{1} = 1 \\ g(-3) &= \sqrt{(-3)^3+9} = \sqrt{-18}; \end{aligned}$$

-3 is not in the domain of g.

$$g(3) = \sqrt{3^3+9} = \sqrt{36} = 6$$

30. Does not exist; $\sqrt{17}$; $\sqrt{54}$

31. $\sqrt{25t^2} = \sqrt{(5t)^2} = |5t| = 5|t|$

Since t might be negative, absolute-value notation is necessary.

32. $4|x|$

33. $\sqrt{(-6b)^2} = |-6b| = |-6| \cdot |b| = 6|b|$

Since b might be negative, absolute-value notation is necessary.

34. $7|c|$

35. $\sqrt{(5-b)^2} = |5-b|$

Since $5-b$ might be negative, absolute-value notation is necessary.

36. $|a+1|$

37. $\sqrt{y^2+16y+64} = \sqrt{(y+8)^2} = |y+8|$

Since $y+8$ might be negative, absolute-value notation is necessary.

38. $|x-2|$

39. $\sqrt{9x^2-30x+25} = \sqrt{(3x-5)^2} = |3x-5|$

Since $3x-5$ might be negative, absolute-value notation is necessary.

40. $|2x+7|$

41. $-\sqrt[4]{256} = -4$ Since $4^4 = 256$

42. 5

43. $-\sqrt[5]{7^5} = -7$

44. -1

45. $\sqrt[5]{-\frac{1}{32}} = -\frac{1}{2}$ Since $\left(-\frac{1}{2}\right)^5 = -\frac{1}{32}$

46. $-\frac{2}{3}$

47. $\sqrt[8]{y^8} = |y|$

The index is even. Use absolute-value notation since y could have a negative value.

48. $|x|$

49. $\sqrt[4]{(7b)^4} = |7b| = 7|b|$

The index is even. Use absolute-value notation since b could have a negative value.

50. $5|a|$

51. $\sqrt[12]{(-10)^{12}} = |-10| = 10$

52. 6

53. $\sqrt[1976]{(2a+b)^{1976}} = |2a+b|$

The index is even. Use absolute-value notation since $2a+b$ could have a negative value.

54. $|a+b|$

55. $\sqrt{x^{12}} = x^6$ Note that $(x^6)^2 = x^{12}$; x^6 is nonnegative regardless of the value of x.

56. $|a^{11}|$

57. $\sqrt{a^{14}} = |a^7|$ Note that $(a^7)^2 = a^{14}$; a^7 could have a negative value.

58. x^8

59. $\sqrt{25t^2} = \sqrt{(5t)^2} = 5t$ Assuming t is nonnegative

60. $4x$

61. $\sqrt{(7c)^2} = 7c$ Assuming c is nonnegative

62. $6b$

63. $\sqrt{(5+b)^2} = 5+b$ Assuming $5+b$ is nonnegative

64. $a+1$

65. $\sqrt{9x^2+36x+36} = \sqrt{9(x^2+4x+4)} = \sqrt{[3(x+2)]^2} = 3(x+2)$, or $3x+6$

66. $2(x+1)$, or $2x+2$

67. $\sqrt{25t^2-20t+4} = \sqrt{(5t-2)^2} = 5t-2$

68. $3t-2$

69. $-\sqrt[3]{64} = -4$ $(4^3 = 64)$

70. 3

71. $\sqrt[4]{81x^4} = \sqrt[4]{(3x)^4} = 3x$

72. $2x$

73. $-\sqrt[5]{-100,000} = -(-10) = 10$ $[(-10)^5 = -100,000]$

74. -6

75. $-\sqrt[3]{-64x^3} = -(-4x) = 4x$ $[(-4x)^3 = -64x^3]$

76. $5y$

77. $\sqrt{a^{14}} = \sqrt{(a^7)^2} = a^7$

78. a^{11}

79. $\sqrt{(x+3)^{10}} = \sqrt{[(x+3)^5]^2} = (x+3)^5$

80. $(x-2)^4$

81.

$$
\begin{aligned}
f(x) &= \sqrt[3]{x+1} \\
f(7) &= \sqrt[3]{7+1} = \sqrt[3]{8} = 2 \\
f(26) &= \sqrt[3]{26+1} = \sqrt[3]{27} = 3 \\
f(-9) &= \sqrt[3]{-9+1} = \sqrt[3]{-8} = -2 \\
f(-65) &= \sqrt[3]{-65+1} = \sqrt[3]{-64} = -4
\end{aligned}
$$

82. 1; 5; 3; -5

83.
$$\begin{aligned} g(t) &= \sqrt[4]{t-3} \\ g(19) &= \sqrt[4]{19-3} = \sqrt[4]{16} = 2 \\ g(-13) &= \sqrt[4]{-13-3} = \sqrt[4]{-16}; \\ &\quad -13 \text{ is not in the domain of } g. \\ g(1) &= \sqrt[4]{1-3} = \sqrt[4]{-2}; \\ &\quad 1 \text{ is not in the domain of } g. \\ g(84) &= \sqrt[4]{84-3} = \sqrt[4]{81} = 3 \end{aligned}$$

84. 1; 2; does not exist; 3

85. $f(x) = \sqrt{x-5}$

Since the index is even, the radicand, $x-5$, must be nonnegative. We solve the inequality:

$$\begin{aligned} x - 5 &\geq 0 \\ x &\geq 5 \end{aligned}$$

Domain of $f = \{x|x \geq 5\}$, or $[5, \infty)$

86. $\{x|x \geq -8\}$, or $[-8, \infty)$

87. $g(t) = \sqrt[4]{t+3}$

Since the index is even, the radicand, $t+3$, must be nonnegative. We solve the inequality:

$$\begin{aligned} t + 3 &\geq 0 \\ t &\geq -3 \end{aligned}$$

Domain of $g = \{t|t \geq -3\}$, or $[-3, \infty)$

88. $\{x|x \geq 7\}$, or $[7, \infty)$

89. $g(x) = \sqrt[4]{5-x}$

Since the index is even, the radicand, $5-x$, must be nonnegative. We solve the inequality:

$$\begin{aligned} 5 - x &\geq 0 \\ 5 &\geq x \end{aligned}$$

Domain of $g = \{x|x \leq 5\}$, or $(-\infty, 5]$

90. $\{t|t \text{ is a real number}\}$, or $(-\infty, \infty)$

91. $f(t) = \sqrt[5]{2t+9}$

Since the index is odd, the radicand can be any real number.

Domain of $f = \{t|t \text{ is a real number}\}$, or $(-\infty, \infty)$

92. $\left\{t\middle|t \geq -\frac{5}{2}\right\}$, or $\left[-\frac{5}{2}, \infty\right)$

93. $h(z) = -\sqrt[6]{5z+3}$

Since the index is even, the radicand, $5z+3$, must be nonnegative. We solve the inequality:

$$\begin{aligned} 5z + 3 &\geq 0 \\ 5z &\geq -3 \\ z &\geq -\frac{3}{5} \end{aligned}$$

Domain of $h = \left\{z\middle|z \geq -\frac{3}{5}\right\}$, or $\left[-\frac{3}{5}, \infty\right)$

94. $\left\{x\middle|x \geq \frac{5}{7}\right\}$, or $\left[\frac{5}{7}, \infty\right)$

95. $f(t) = 4 + 2\sqrt[8]{3t-7}$

Since the index is even, the radicand, $3t-7$, must be nonnegative. We solve the inequality:

$$\begin{aligned} 3t - 7 &\geq 0 \\ 3t &\geq 7 \\ t &\geq \frac{7}{3} \end{aligned}$$

Domain of $f = \left\{t\middle|t \geq \frac{7}{3}\right\}$, or $\left[\frac{7}{3}, \infty\right)$

96. $\left\{t\middle|t \geq \frac{4}{5}\right\}$, or $\left[\frac{4}{5}, \infty\right)$

97. $(a^3b^2c^5)^3 = a^{3\cdot 3}b^{2\cdot 3}c^{5\cdot 3} = a^9b^6c^{15}$

98. $10a^{10}b^9$

99. $(x-3)(x+3) = x^2 - 3^2 = x^2 - 9$

100. $a^2 - b^2x^2$

101.
$$\begin{aligned} &(2x+1)(x^2-3x+1) \\ &= 2x^3 - 6x^2 + 2x + x^2 - 3x + 1 \\ &= 2x^3 - 5x^2 - x + 1 \end{aligned}$$

102. $x^3 - 3x^2 + 7x - 5$

103.

104.

105.

106.

107. $N = 2.5\sqrt{A}$

a) $N = 2.5\sqrt{25} = 2.5(5) = 12.5 \approx 13$

b) $N = 2.5\sqrt{36} = 2.5(6) = 15$

c) $N = 2.5\sqrt{49} = 2.5(7) = 17.5 \approx 18$

d) $N = 2.5\sqrt{64} = 2.5(8) = 20$

108. $\{x|x \geq -5\}$, or $[-5, \infty)$

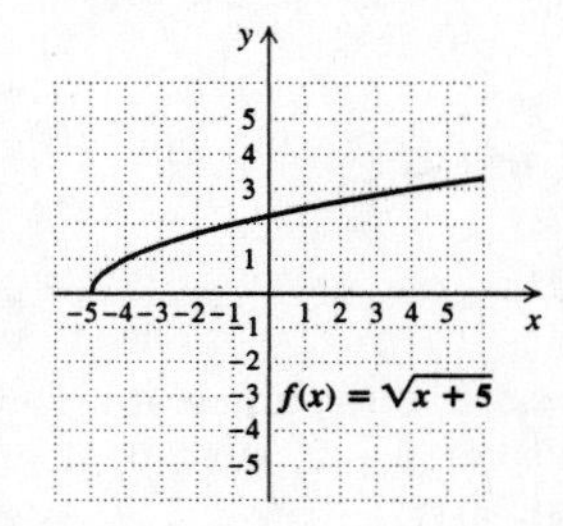

109. $g(x) = \sqrt{x} + 5$

Since the index is even, th radicand, x, must be nonnegative, so we have $x \geq 0$.

Domain of $g = \{x|x \geq 0\}$, or $[0, \infty)$

Make a table of values, keeping in mind that x must be nonnegative. Plot these points and draw the graph.

x	y
0	5
1	6
2	6.4
3	6.7
4	7
5	7.2

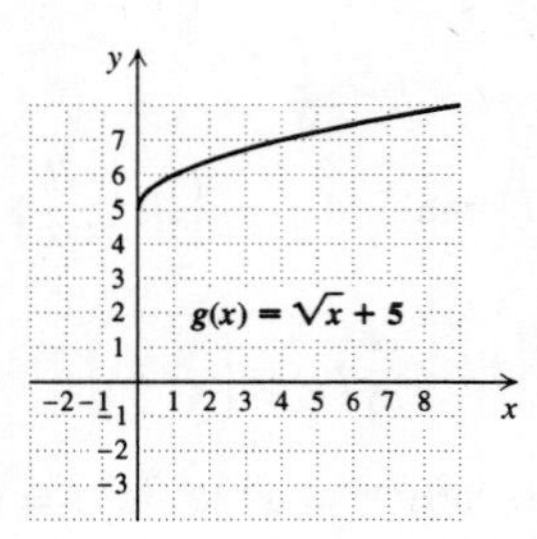

110. $\{x|x \geq 0\}$, or $[0, \infty)$

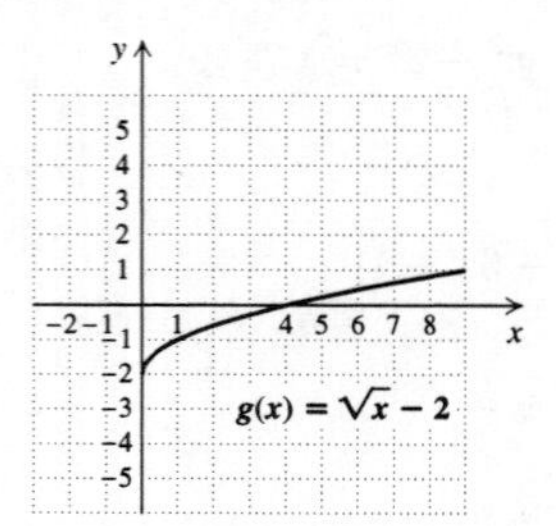

111. $f(x) = \sqrt{x-2}$

Since the index is even, the radicand, $x - 2$, must be nonnegative. We solve the inequality.

$$x - 2 \geq 0$$
$$x \geq 2$$

Domain of $f = \{x|x \geq 2\}$, or $[2, \infty)$

Make a table of values, keeping in mind that x must be 2 or greater. Plot these points and draw the graph.

x	y
2	0
3	1
4	1.4
6	2
8	2.4

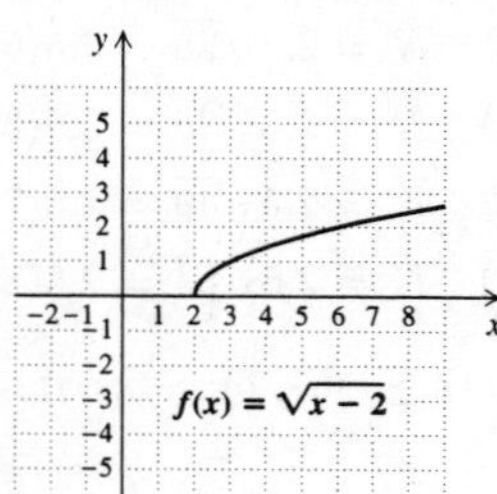

112. $\{x| -3 \leq x < 2\}$, or $[-3, 2)$.

113. $g(x) = \dfrac{\sqrt[4]{5-x}}{\sqrt[6]{x+4}}$

The radical expression in the numerator has an even index, so the radicand, $5 - x$, must be nonnegative. We solve the inequality:

$$5 - x \geq 0$$
$$5 \geq x$$

The radical expression in the denominator also has an even index, so the radicand, $x + 4$, must be nonnegative in order for $\sqrt[6]{x+4}$ to exist. In addition, the denominator cannot be zero, so the radicand must be positive. We solve the inequality:

$$x + 4 > 0$$
$$x > -4$$

We have $x \leq 5$ *and* $x > -4$ so

Domain of $g = \{x| -4 < x \leq 5\}$, or $(-4, 5]$.

114.

115.

Exercise Set 7.2

1. $x^{1/4} = \sqrt[4]{x}$

2. $\sqrt[5]{y}$

3. $(16)^{1/2} = \sqrt{16} = 4$

4. 2

5. $81^{1/4} = \sqrt[4]{81} = 3$

6. 2

7. $9^{1/2} = \sqrt{9} = 3$

8. 5

9. $(xyz)^{1/3} = \sqrt[3]{xyz}$

10. $\sqrt[4]{ab}$

11. $(a^2b^2)^{1/5} = \sqrt[5]{a^2b^2}$

12. $\sqrt[4]{x^3y^3}$

13. $a^{2/3} = \sqrt[3]{a^2}$

14. $\sqrt{b^3}$

15. $16^{3/4} = \sqrt[4]{16^3} = (\sqrt[4]{16})^3 = 2^3 = 8$

16. 128

17. $49^{3/2} = \sqrt{49^3} = (\sqrt{49})^3 = 7^3 = 343$

18. 81

19. $9^{5/2} = \sqrt{9^5} = (\sqrt{9})^5 = 3^5 = 243$

20. 729

21. $(81x)^{3/4} = \sqrt[4]{(81x)^3} = \sqrt[4]{81^3x^3}$, or $\sqrt[4]{81^3} \cdot \sqrt[4]{x^3} = (\sqrt[4]{81})^3 \cdot \sqrt[4]{x^3} = 3^3\sqrt[4]{x^3} = 27\sqrt[4]{x^3}$

22. $25\sqrt[3]{a^2}$

23. $(25x^4)^{3/2} = \sqrt{(25x^4)^3} = \sqrt{25^3 \cdot x^{12}} = \sqrt{25^3} \cdot \sqrt{x^{12}} =$
$(\sqrt{25})^3 x^6 = 5^3 x^6 = 125x^6$

24. $27y^9$

25. $\sqrt[3]{20} = 20^{1/3}$

26. $19^{1/3}$

27. $\sqrt{17} = 17^{1/2}$

28. $6^{1/2}$

29. $\sqrt{x^3} = x^{3/2}$

30. $a^{5/2}$

31. $\sqrt[5]{m^2} = m^{2/5}$

32. $n^{4/5}$

33. $\sqrt[4]{cd} = (cd)^{1/4}$ Parentheses are required.

34. $(xy)^{1/5}$

35. $\sqrt[5]{xy^2z} = (xy^2z)^{1/5}$

36. $(x^3y^2z^2)^{1/7}$

37. $(\sqrt{3mn})^3 = (3mn)^{3/2}$

38. $(7xy)^{4/3}$

39. $(\sqrt[7]{8x^2y})^5 = (8x^2y)^{5/7}$

40. $(2a^5b)^{7/6}$

41. $\dfrac{2x}{\sqrt[3]{z^2}} = \dfrac{2x}{z^{2/3}}$

42. $\dfrac{3a}{c^{2/5}}$

43. $x^{-1/3} = \dfrac{1}{x^{1/3}}$

44. $\dfrac{1}{y^{1/4}}$

45. $(2rs)^{-3/4} = \dfrac{1}{(2rs)^{3/4}}$

46. $\dfrac{1}{(5xy)^{5/6}}$

47. $\left(\dfrac{1}{10}\right)^{-2/3} = \left(\dfrac{10}{1}\right)^{2/3}$

$= 10^{2/3}$ Finding the reciprocal of the base and changing the sign of the exponent

48. $8^{3/4}$

49. $\dfrac{1}{a^{-5/7}} = a^{5/7}$

50. $a^{3/5}$

51. $2a^{3/4}b^{-1/2}c^{2/3} = 2 \cdot a^{3/4} \cdot \dfrac{1}{b^{1/2}} \cdot c^{2/3} = \dfrac{2a^{3/4}c^{2/3}}{b^{1/2}}$

52. $\dfrac{5y^{4/5}z}{x^{2/3}}$

53. $2^{-1/3}x^4y^{-2/7} = \dfrac{1}{2^{1/3}} \cdot x^4 \cdot \dfrac{1}{y^{2/7}} = \dfrac{x^4}{2^{1/3}y^{2/7}}$

54. $\dfrac{a^3}{3^{5/2}b^{7/3}}$

55. $\left(\dfrac{7x}{8yz}\right)^{-3/5} = \left(\dfrac{8yz}{7x}\right)^{3/5}$ Finding the reciprocal of the base and changing the sign of the exponent

56. $\left(\dfrac{3c}{2ab}\right)^{5/6}$

57. $\dfrac{7x}{\sqrt[3]{z}} = \dfrac{7x}{z^{1/3}}$

58. $\dfrac{6a}{b^{1/4}}$

59. $\dfrac{5a}{3c^{-1/2}} = \dfrac{5a}{3} \cdot c^{1/2} = \dfrac{5ac^{1/2}}{3}$

60. $\dfrac{2x^{1/3}z}{5}$

61. $5^{3/4} \cdot 5^{1/8} = 5^{3/4+1/8} = 5^{6/8+1/8} = 5^{7/8}$

We added exponents after finding a common denominator.

62. $11^{7/6}$

63. $\dfrac{3^{5/8}}{3^{-1/8}} = 3^{5/8-(-1/8)} = 3^{5/8+1/8} = 3^{6/8} = 3^{3/4}$

We subtracted exponents and simplified.

64. $8^{9/11}$

65. $\dfrac{4.1^{-1/6}}{4.1^{-2/3}} = 4.1^{-1/6-(-2/3)} = 4.1^{-1/6+2/3} =$
$4.1^{-1/6+4/6} = 4.1^{3/6} = 4.1^{1/2}$

We subtracted exponents after finding a common denominator. Then we simplified.

66. $\dfrac{1}{2.3^{1/10}}$

67. $(10^{3/5})^{2/5} = 10^{3/5 \cdot 2/5} = 10^{6/25}$

We multiplied exponents.

68. $5^{15/28}$

69. $a^{2/3} \cdot a^{5/4} = a^{2/3+5/4} = a^{8/12+15/12} = a^{23/12}$

We added exponents after finding a common denominator.

70. $x^{17/12}$

71. $(x^{2/3})^{-3/7} = x^{2/3(-3/7)} = x^{-2/7} = \dfrac{1}{x^{2/7}}$

We multiplied exponents, simplified, and wrote the result without a negative exponent.

72. $\dfrac{1}{a^{1/3}}$

73. $(m^{2/3}n^{-1/4})^{1/2} = m^{2/3\cdot 1/2}n^{-1/4\cdot 1/2} = m^{1/3}n^{-1/8} = m^{1/3}\cdot\dfrac{1}{n^{1/8}} = \dfrac{m^{1/3}}{n^{1/8}}$

74. $\dfrac{y^{1/10}}{x^{1/12}}$

75.
$\sqrt[6]{a^2} = a^{2/6}$ Converting to exponential notation
$= a^{1/3}$ Simplifying the exponent
$= \sqrt[3]{a}$ Returning to radical notation

76. $\sqrt[3]{t^2}$

77.
$\sqrt[3]{x^{15}} = x^{15/3}$ Converting to exponential notation
$= x^5$ Simplifying

78. a^3

79.
$\sqrt[6]{x^{18}} = x^{18/6}$ Converting to exponential notation
$= x^3$ Simplifying

80. a^2

81.
$(\sqrt[3]{ab})^{15} = (ab)^{15/3}$ Converting to exponential notation
$= (ab)^5$ Simplifying the exponent
$= a^5b^5$ Using the law of exponents

82. x^2y^2

83.
$\sqrt[8]{(3x)^2} = (3x)^{2/8}$ Converting to exponential notation
$= (3x)^{1/4}$ Simplifying the exponent
$= \sqrt[4]{3x}$ Returning to radical notation

84. $\sqrt{7a}$

85.
$(\sqrt[10]{3a})^5 = (3a)^{5/10}$ Converting to exponential notation
$= (3a)^{1/2}$ Simplifying the exponent
$= \sqrt{3a}$ Returning to radical notation

86. $\sqrt[4]{8x^3}$

87.
$\sqrt[4]{\sqrt{x}} = \sqrt[4]{x^{1/2}}$ Converting to
$= (x^{1/2})^{1/4}$ exponential notation
$= x^{1/8}$ Using a law of exponents
$= \sqrt[8]{x}$ Returning to radical notation

88. $\sqrt[18]{m}$

89.
$\sqrt{(ab)^6} = (ab)^{6/2}$ Converting to exponential notation
$= (ab)^3$ Using the laws
$= a^3b^3$ of exponents

90. x^3y^3

91.
$(\sqrt[3]{x^2y^5})^{12} = (x^2y^5)^{12/3}$ Converting to exponential notation
$= (x^2y^5)^4$ Simplifying the exponent
$= x^8y^{20}$ Using the laws of exponents

92. a^6b^{12}

93.
$\sqrt[3]{\sqrt[4]{xy}} = \sqrt[3]{(xy)^{1/4}}$ Converting to
$= [(xy)^{1/4}]^{1/3}$ exponential notation
$= (xy)^{1/12}$ Using a law of exponents
$= \sqrt[12]{xy}$ Returning to radical notation

94. $\sqrt[10]{2a}$

95.
$$x^2 - 1 = 8$$
$$x^2 - 9 = 0$$
$$(x+3)(x-3) = 0$$
$$x + 3 = 0 \quad \textit{or} \quad x - 3 = 0$$
$$x = -3 \quad \textit{or} \quad x = 3$$

Both values check. The solutions are -3 and 3.

96. $-\dfrac{11}{2}$

97.
$$\frac{1}{x} + 2 = 5$$
$$\frac{1}{x} = 3$$
$$x\cdot\frac{1}{x} = x\cdot 3$$
$$1 = 3x$$
$$\frac{1}{3} = x$$

This value checks. The solution is $\dfrac{1}{3}$.

98. -7, 7

99. ***Familiarize.*** Let $p =$ the selling price of the home.

Translate.

$$\underbrace{0.5\% \text{ of the selling price}}_{\downarrow} \ \underset{\downarrow}{\text{is}} \ \underset{\downarrow}{\$467.50}$$
$$0.005p \quad = \quad 467.50$$

Carry out. We solve the equation.

$$0.005p = 467.50$$
$$p = 93{,}500 \quad \text{Dividing by 0.005}$$

Check. 0.5% of \$93,500 is $0.005(\$93,500)$, or \$467.50. The answer checks.

State. The selling price of the home was \$93,500.

100. 0, 1

101. ◈

102. ◈

103. ◈

104. ◈

105. $\sqrt[5]{x^2y\sqrt{xy}} = \sqrt[5]{x^2y(xy)^{1/2}} = \sqrt[5]{x^2yx^{1/2}y^{1/2}} = \sqrt[5]{x^{5/2}y^{3/2}} = (x^{5/2}y^{3/2})^{1/5} = x^{5/10}y^{3/10} = (x^5y^3)^{1/10} = \sqrt[10]{x^5y^3}$

106. $\sqrt[6]{x^5}$

107. $\sqrt[4]{\sqrt[3]{8x^3y^6}} = \sqrt[4]{(2^3x^3y^6)^{1/3}} = \sqrt[4]{2^{3/3}x^{3/3}y^{6/3}} = \sqrt[4]{2xy^2}$

108. $\sqrt[6]{p+q}$

109. a) $L = \dfrac{(0.000169)60^{2.27}}{1} \approx 1.8 \text{ m}$

b) $L = \dfrac{(0.000169)75^{2.27}}{0.9906} \approx 3.1 \text{ m}$

c) $L = \dfrac{(0.000169)80^{2.27}}{2.4} \approx 1.5 \text{ m}$

d) $L = \dfrac{(0.000169)100^{2.27}}{1.1} \approx 5.3 \text{ m}$

110. About 7.937×10^{-13} to 1

111. $m = m_0(1 - v^2c^{-2})^{1/2}$

$$m = 8\left[1 - \left(\frac{9}{5} \times 10^8\right)^2 (3 \times 10^8)^{-2}\right]^{1/2}$$
$$= 8\left[1 - \frac{\left(\frac{9}{5} \times 10^8\right)^2}{(3 \times 10^8)^2}\right]^{1/2}$$
$$= 8\left[1 - \frac{\frac{81}{25} \times 10^{16}}{9 \times 10^6}\right]^{1/2}$$
$$= 8\left[1 - \frac{81}{25} \cdot \frac{1}{9}\right]^{1/2}$$
$$= 8\left[1 - \frac{9}{25}\right]^{1/2}$$
$$= 8\left(\frac{16}{25}\right)^{1/2}$$
$$= 8 \cdot \frac{4}{5} = \frac{32}{5}$$
$$= 6.4$$

The particle's new mass is 6.4 mg.

112. $y_1 = x^{1/2}$, $y_2 = 3x^{2/5}$,
$y_3 = x^{4/7}$, $y_4 = \frac{1}{5}x^{3/4}$

Exercise Set 7.3

1. $\sqrt{6}\sqrt{7} = \sqrt{6 \cdot 7} = \sqrt{42}$

2. $\sqrt{35}$

3. $\sqrt[3]{2}\sqrt[3]{5} = \sqrt[3]{2 \cdot 5} = \sqrt[3]{10}$

4. $\sqrt[3]{14}$

5. $\sqrt[4]{8}\sqrt[4]{9} = \sqrt[4]{8 \cdot 9} = \sqrt[4]{72}$

6. $\sqrt[4]{18}$

7. $\sqrt{5a}\sqrt{3b} = \sqrt{5a \cdot 3b} = \sqrt{15ab}$

8. $\sqrt{26xy}$

9. $\sqrt[5]{9t^2}\sqrt[5]{2t} = \sqrt[5]{9t^2 \cdot 2t} = \sqrt[5]{18t^3}$

10. $\sqrt[5]{80y^4}$

11. $\sqrt{x-a}\sqrt{x+a} = \sqrt{(x-a)(x+a)} = \sqrt{x^2 - a^2}$

12. $\sqrt{y^2 - b^2}$

13. $\sqrt[3]{0.5x}\sqrt[3]{0.2x} = \sqrt[3]{0.5x \cdot 0.2x} = \sqrt[3]{0.1x^2}$

14. $\sqrt[3]{0.21y^2}$

15. $\sqrt[4]{x-1}\sqrt[4]{x^2+x+1} = \sqrt[4]{(x-1)(x^2+x+1)} = \sqrt[4]{x^3 - 1}$

16. $\sqrt[5]{(x-2)^3}$

17. $\sqrt{\dfrac{x}{5}}\sqrt{\dfrac{3}{y}} = \sqrt{\dfrac{x}{5} \cdot \dfrac{3}{y}} = \sqrt{\dfrac{3x}{5y}}$

18. $\sqrt{\dfrac{7s}{11t}}$

19. $\sqrt[7]{\dfrac{x-3}{4}}\sqrt[7]{\dfrac{5}{x+2}} = \sqrt[7]{\dfrac{x-3}{4} \cdot \dfrac{5}{x+2}} = \sqrt[7]{\dfrac{5x-15}{4x+8}}$

20. $\sqrt[6]{\dfrac{3a}{b^2-4}}$

21. $\sqrt[3]{5}\cdot\sqrt{6}$

$= 5^{1/3}\cdot 6^{1/2}$ Converting to exponential notation

$= 5^{2/6}\cdot 6^{3/6}$ Rewriting so that exponents have a common denominator

$= (5^2\cdot 6^3)^{1/6}$ Using the laws of exponents

$= \sqrt[6]{25\cdot 216}$ Squaring 5, cubing 6, and returning to radical notation

$= \sqrt[6]{5400}$ Multiplying under the radical

22. $\sqrt[12]{300,125}$

23. $\sqrt{x}\sqrt[3]{7y} = x^{1/2}(7y)^{1/3} = x^{3/6}(7y)^{2/6} = [x^3(7y)^2]^{1/6} = \sqrt[6]{x^3\cdot 49y^2} = \sqrt[6]{49x^3y^2}$

24. $\sqrt[15]{27y^5z^3}$

25. $\sqrt{x}\sqrt[3]{x-2} = x^{1/2}\cdot(x-2)^{1/3} = x^{3/6}\cdot(x-2)^{2/6} = [x^3(x-2)^2]^{1/6} = \sqrt[6]{x^3(x-2)^2} = \sqrt[6]{x^3(x^2-4x+4)} = \sqrt[6]{x^5-4x^4+4x^3}$

26. $\sqrt[4]{3xy^2+24xy+48x}$

27. $\sqrt[5]{yx^2}\sqrt{xy} = (yx^2)^{1/5}(xy)^{1/2} = y^{1/5}x^{2/5}x^{1/2}y^{1/2} = x^{2/5+1/2}y^{1/5+1/2} = x^{4/10+5/10}y^{2/10+5/10} = x^{9/10}y^{7/10} = (x^9y^7)^{1/10} = \sqrt[10]{x^9y^7}$

28. $\sqrt[10]{4a^9b^9}$

29. $\sqrt[4]{xy^2}\sqrt[3]{x^2y} = (xy^2)^{1/4}(x^2y)^{1/3} = (xy^2)^{3/12}(x^2y)^{4/12} = [(xy^2)^3(x^2y)^4]^{1/12} = \sqrt[12]{x^3y^6\cdot x^8y^4} = \sqrt[12]{x^{11}y^{10}}$

30. $\sqrt[20]{a^{18}b^{17}}$

31. $\sqrt[4]{a^2bc^2}\sqrt[5]{a^2b^3c} = (a^2bc^2)^{1/4}(a^2b^3c)^{1/5} = (a^2bc^2)^{5/20}(a^2b^3c)^{4/20} = [(a^2bc^2)^5(a^2b^3c)^4]^{1/20} = \sqrt[20]{a^{10}b^5c^{10}\cdot a^8b^{12}c^4} = \sqrt[20]{a^{18}b^{17}c^{14}}$

32. $\sqrt[30]{x^{22}y^{11}z^{27}}$

33. $\sqrt{27}$

$= \sqrt{9\cdot 3}$ 9 is the largest perfect square factor of 27.

$= \sqrt{9}\cdot\sqrt{3}$

$= 3\sqrt{3}$

34. $2\sqrt{7}$

35. $\sqrt{12}$

$= \sqrt{4\cdot 3}$ 4 is the largest perfect square factor of 12.

$= \sqrt{4}\cdot\sqrt{3}$

$= 2\sqrt{3}$

36. $3\sqrt{5}$

37. $\sqrt{8} = \sqrt{4\cdot 2} = \sqrt{4}\cdot\sqrt{2} = 2\sqrt{2}$

38. $3\sqrt{2}$

39. $\sqrt{44} = \sqrt{4\cdot 11} = \sqrt{4}\cdot\sqrt{11} = 2\sqrt{11}$

40. $2\sqrt{6}$

41. $\sqrt{36a^4b}$

$= \sqrt{36a^4\cdot b}$ $36a^4$ is a perfect square.

$= \sqrt{36a^4}\cdot\sqrt{b}$ Factoring into two radicals

$= 6a^2\sqrt{b}$ Taking the square root of $36a^4$

42. $5y^4\sqrt{7}$

43. $\sqrt[3]{8x^3y^2}$

$= \sqrt[3]{8x^3\cdot y^2}$ $8x^3$ is a perfect cube.

$= \sqrt[3]{8x^3}\cdot\sqrt[3]{y^2}$ Factoring into two radicals

$= 2x\sqrt[3]{y^2}$ Taking the cube root of $8x^3$

44. $3b^2\sqrt[3]{a}$

45. $\sqrt[3]{-16x^6}$

$= \sqrt[3]{-8x^6\cdot 2}$ $-8x^6$ is a perfect cube.

$= \sqrt[3]{-8x^6}\cdot\sqrt[3]{2}$

$= -2x^2\sqrt[3]{2}$ Taking the cube root of $-8x^6$

46. $-2a^2\sqrt[3]{4}$

47. $f(x) = \sqrt[3]{125x^5}$

$= \sqrt[3]{125x^3\cdot x^2}$

$= \sqrt[3]{125x^3}\cdot\sqrt[3]{x^2}$

$= 5x\sqrt[3]{x^2}$

48. $2x^2\sqrt[3]{2}$

49. $f(x) = \sqrt{49(x+5)^2}$ $49(x+5)^2$ is a perfect square.

$= |7(x+5)|$, or $7|x+5|$

50. $|9(x-1)|$, or $9|x-1|$

51. $f(x) = \sqrt{5x^2-10x+5}$

$= \sqrt{5(x^2-2x+1)}$

$= \sqrt{5(x-1)^2}$

$= \sqrt{(x-1)^2}\cdot\sqrt{5}$

$= |x-1|\sqrt{5}$

52. $|x+2|\sqrt{2}$

53.
$$\begin{aligned} &\sqrt{a^3b^4} \\ &= \sqrt{a^2\cdot a\cdot b^4} && \text{Identifying the largest even powers of } a \text{ and } b \\ &= \sqrt{a^2}\sqrt{b^4}\sqrt{a} && \text{Factoring into several radicals} \\ &= ab^2\sqrt{a} \end{aligned}$$

54. $x^3y^4\sqrt{y}$

55.
$$\begin{aligned} &\sqrt[3]{x^5y^6z^{10}} \\ &= \sqrt[3]{x^3\cdot x^2\cdot y^6\cdot z^9\cdot z} && \text{Identifying the largest perfect-cube powers of } x,\ y, \text{ and } z \\ &= \sqrt[3]{x^3}\cdot\sqrt[3]{y^6}\cdot\sqrt[3]{z^9}\cdot\sqrt[3]{x^2z} && \text{Factoring into several radicals} \\ &= xy^2z^3\sqrt[3]{x^2z} \end{aligned}$$

56. $a^2b^2c^4\sqrt[3]{bc}$

57. $\sqrt[5]{-32a^7b^{11}} = \sqrt[5]{-32\cdot a^5\cdot a^2\cdot b^{10}\cdot b} =$
$\sqrt[5]{-32}\sqrt[5]{a^5}\sqrt[5]{b^{10}}\sqrt[5]{a^2b} = -2ab^2\sqrt[5]{a^2b}$

58. $2xy^2\sqrt[4]{xy^3}$

59. $\sqrt[5]{a^6b^{12}c^7} = \sqrt[5]{a^5\cdot a\cdot b^{10}\cdot b^2\cdot c^5\cdot c^2} =$
$\sqrt[5]{a^5}\cdot\sqrt[5]{b^{10}}\cdot\sqrt[5]{c^5}\sqrt[5]{ab^2c^2} =$
$ab^2c\sqrt[5]{ab^2c^2}$

60. $x^2yz^3\sqrt[5]{x^3y^3z^2}$

61. $\sqrt[4]{810x^9} = \sqrt[4]{81\cdot 10\cdot x^8\cdot x} =$
$\sqrt[4]{81}\cdot\sqrt[4]{x^8}\cdot\sqrt[4]{10x} = 3x^2\sqrt[4]{10x}$

62. $-2a^4\sqrt[3]{10a^2}$

63. $3\sqrt{7}+2\sqrt{7} = (3+2)\sqrt{7} = 5\sqrt{7}$

64. $17\sqrt{5}$

65. $9\sqrt[3]{5}-6\sqrt[3]{5} = (9-6)\sqrt[3]{5} = 3\sqrt[3]{5}$

66. $8\sqrt[5]{2}$

67. $4\sqrt[3]{y}+9\sqrt[3]{y} = (4+9)\sqrt[3]{y} = 13\sqrt[3]{y}$

68. $6\sqrt[4]{t}$

69. $8\sqrt{2}-6\sqrt{2}+5\sqrt{2} = (8-6+5)\sqrt{2} = 7\sqrt{2}$

70. $7\sqrt{6}$

71. $9\sqrt[3]{7}-\sqrt{3}+4\sqrt[3]{7}+2\sqrt{3} =$
$(9+4)\sqrt[3]{7}+(-1+2)\sqrt{3} = 13\sqrt[3]{7}+\sqrt{3}$

72. $6\sqrt{7}+\sqrt[4]{11}$

73.
$$\begin{aligned} &8\sqrt{27}-3\sqrt{3} \\ &= 8\sqrt{9\cdot 3}-3\sqrt{3} && \text{Factoring the} \\ &= 8\sqrt{9}\cdot\sqrt{3}-3\sqrt{3} && \text{first radical} \\ &= 8\cdot 3\sqrt{3}-3\sqrt{3} && \text{Taking the square root of 9} \\ &= 24\sqrt{3}-3\sqrt{3} \\ &= 21\sqrt{3} && \text{Combining like radicals} \end{aligned}$$

74. $41\sqrt{2}$

75.
$$\begin{aligned} &3\sqrt{45}+7\sqrt{20} \\ &= 3\sqrt{9\cdot 5}+7\sqrt{4\cdot 5} && \text{Factoring the} \\ &= 3\sqrt{9}\cdot\sqrt{5}+7\sqrt{4}\cdot\sqrt{5} && \text{radicals} \\ &= 3\cdot 3\sqrt{5}+7\cdot 2\sqrt{5} && \text{Taking the square roots} \\ &= 9\sqrt{5}+14\sqrt{5} \\ &= 23\sqrt{5} && \text{Combining like radicals} \end{aligned}$$

76. $58\sqrt{3}$

77. $3\sqrt[3]{16}+\sqrt[3]{54} = 3\sqrt[3]{8\cdot 2}+\sqrt[3]{27\cdot 2} =$
$3\sqrt[3]{8}\cdot\sqrt[3]{2}+\sqrt[3]{27}\cdot\sqrt[3]{2} = 3\cdot 2\sqrt[3]{2}+3\sqrt[3]{2} =$
$6\sqrt[3]{2}+3\sqrt[3]{2} = 9\sqrt[3]{2}$

78. -7

79. $\sqrt{5a}+2\sqrt{45a^3} = \sqrt{5a}+2\sqrt{9a^2\cdot 5a} =$
$\sqrt{5a}+2\sqrt{9a^2}\cdot\sqrt{5a} = \sqrt{5a}+2\cdot 3a\sqrt{5a} =$
$\sqrt{5a}+6a\sqrt{5a} = (1+6a)\sqrt{5a}$

80. $(4x-2)\sqrt{3x}$

81. $\sqrt[3]{6x^4}+\sqrt[3]{48x} = \sqrt[3]{x^3\cdot 6x}+\sqrt[3]{8\cdot 6x} =$
$\sqrt[3]{x^3}\cdot\sqrt[3]{6x}+\sqrt[3]{8}\cdot\sqrt[3]{6x} = x\sqrt[3]{6x}+2\sqrt[3]{6x} =$
$(x+2)\sqrt[3]{6x}$

82. $(3-x)\sqrt[3]{2x}$

83. $\sqrt{4a-4}+\sqrt{a-1} = \sqrt{4(a-4)}+\sqrt{a-1} =$
$\sqrt{4}\sqrt{a-1}+\sqrt{a-1} = 2\sqrt{a-1}+\sqrt{a-1} = 3\sqrt{a-1}$

84. $4\sqrt{y+3}$

85. $\sqrt{x^3-x^2}+\sqrt{9x-9} = \sqrt{x^2(x-1)}+\sqrt{9(x-1)} =$
$\sqrt{x^2}\cdot\sqrt{x-1}+\sqrt{9}\cdot\sqrt{x-1} =$
$x\sqrt{x-1}+3\sqrt{x-1} = (x+3)\sqrt{x-1}$

86. $(2-x)\sqrt{x-1}$

87.
$$\begin{aligned} f(x) &= \sqrt{20x^2+4x^3}-3x\sqrt{45+9x}+\sqrt{5x^2+x^3} \\ &= \sqrt{4x^2(5+x)}-3x\sqrt{9(5+x)}+\sqrt{x^2(5+x)} \\ &= \sqrt{4x^2}\sqrt{5+x}-3x\sqrt{9}\sqrt{5+x}+\sqrt{x^2}\sqrt{5+x} \\ &= 2x\sqrt{5+x}-3x\cdot 3\sqrt{5+x}+x\sqrt{5+x} \\ &= 2x\sqrt{5+x}-9x\sqrt{5+x}+x\sqrt{5+x} \\ &= -6x\sqrt{5+x} \end{aligned}$$

88. $2x\sqrt{x-1}$

89.
$$\begin{aligned} f(x) &= \sqrt[4]{x^5-x^4}+3\sqrt[4]{x^9-x^8} \\ &= \sqrt[4]{x^4(x-1)}+3\sqrt[4]{x^8(x-1)} \\ &= \sqrt[4]{x^4}\cdot\sqrt[4]{x-1}+3\sqrt[4]{x^8}\sqrt[4]{x-1} \\ &= x\sqrt[4]{x-1}+3x^2\sqrt[4]{x-1} \\ &= (x+3x^2)\sqrt[4]{x-1} \end{aligned}$$

90. $(2x-2x^2)\sqrt[4]{1+x}$

91. ***Familiarize***. Let x and y represent the number of 30-sec and 60-sec commercials, respectively. Then the total number of minutes of commercial time during the show is $\dfrac{30x+60y}{60}$, or $\dfrac{x}{2}+y$. (We divide by 60 to convert seconds to minutes.)

Translate. Rewording when necessary, we write two equations.

Total number of commercials is 12.
$$x+y=12$$

Number of 30-sec commercials is total minutes of commercial time less 6.
$$x=\frac{x}{2}+y-6$$

Carry out. Solving the system of equations, we get (4,8).

Check. If there are 4 30-sec and 8 60-sec commercials, the total number of commercials is 12. The total amount of commercial time is $4\cdot 30$ sec + $8\cdot 60$ sec = 600 sec, or 10 min. Then the number of 30-sec commercials is 6 less than the total number of minutes of commercial time. The values check.

State. 8 60-sec commercials were used.

92. $5x^3+5x^2-3x-16$

93.
$$\begin{aligned} &(7a^3b^2+5a^2b^2-a^2b)+(2a^3b^2-7a^2b+3ab) \\ &= (7+2)a^3b^2+5a^2b^2+(-1-7)a^2b+3ab \\ &= 9a^3b^2+5a^2b^2-8a^2b+3ab \end{aligned}$$

94. $4x^2-9$

95. $4x^2-49=(2x)^2-7^2=(2x+7)(2x-7)$

96. $2(x-9)(x-4)$

97.

98.

99.

100. (a) 20 mph; (b) 37.4 mph; (c) 42.4 mph

101. a) $T_w = 33-\dfrac{(10.45+10\sqrt{8}-8)(33-7)}{22}$
$\approx -3.3°$ C

b) $T_w = 33-\dfrac{(10.45+10\sqrt{12}-12)(33-0)}{22}$
$\approx -16.6°$ C

c) $T_w = 33-\dfrac{(10.45+10\sqrt{14}-14)[33-(-5)]}{22}$
$\approx -25.5°$ C

d) $T_w = 33-\dfrac{(10.45+10\sqrt{15}-15)[33-(-23)]}{22}$
$\approx -54.0°$ C

102. $r^{10}t^3\sqrt{rt}$

103. $(\sqrt[3]{25x^4})^4=\sqrt[3]{(25x^4)^4}=\sqrt[3]{25^4x^{16}}=$
$\sqrt[3]{25^3\cdot 25\cdot x^{15}\cdot x}=\sqrt[3]{25^3}\sqrt[3]{x^{15}}\sqrt[3]{25x}=$
$25x^5\sqrt[3]{25x}$

104. $ac^2[(3a+2c)\sqrt{ab}-2\sqrt[3]{ab}]$

105.
$$\begin{aligned} &7x\sqrt{(x+y)^3}-5xy\sqrt{x+y}-2y\sqrt{(x+y)^3} \\ &= 7x\sqrt{(x+y)^2(x+y)}-5xy\sqrt{x+y}- \\ &\qquad 2y\sqrt{(x+y)^2(x+y)} \\ &= 7x(x+y)\sqrt{x+y}-5xy\sqrt{x+y}-2y(x+y)\sqrt{x+y} \\ &= [7x(x+y)-5xy-2y(x+y)]\sqrt{x+y} \\ &= (7x^2+7xy-5xy-2xy-2y^2)\sqrt{x+y} \\ &= (7x^2-2y^2)\sqrt{x+y} \end{aligned}$$

106.

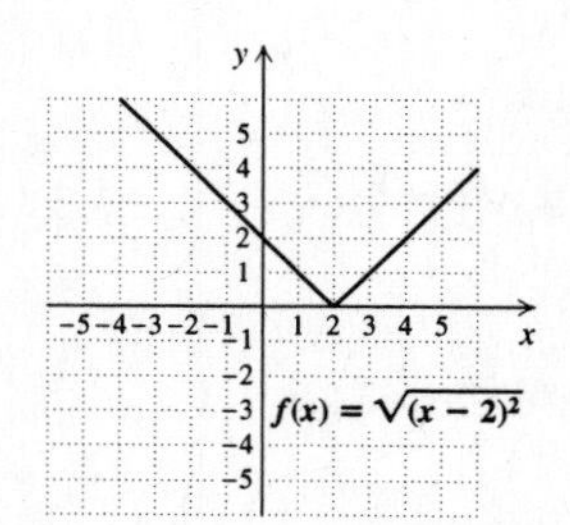

$\{x|x$ is a real number$\}$, or $(-\infty,\infty)$

107.

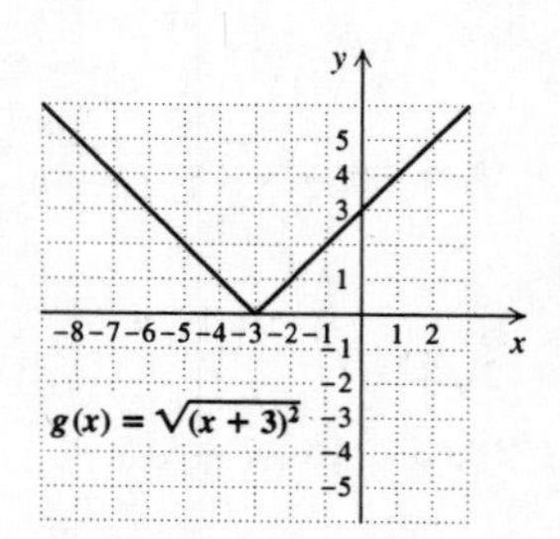

Since $(x+3)^2$ is nonnegative for all values of x, the domain of g is $\{x|x$ is a real number$\}$, or $(-\infty,\infty)$.

108.

109. 

Exercise Set 7.4

1. $\sqrt{10}\sqrt{5} = \sqrt{10 \cdot 5} = \sqrt{50} = \sqrt{25 \cdot 2} = 5\sqrt{2}$

2. $3\sqrt{2}$

3. $\sqrt{6}\sqrt{14} = \sqrt{6 \cdot 14} = \sqrt{84} = \sqrt{4 \cdot 21} = 2\sqrt{21}$

4. $3\sqrt{35}$

5. $\sqrt[3]{2}\sqrt[3]{4} = \sqrt[3]{2 \cdot 4} = \sqrt[3]{8} = 2$

6. 3

7. $\sqrt[3]{5a^2}\sqrt[3]{2a} = \sqrt[3]{5a^2 \cdot 2a} = \sqrt[3]{10a^3} = \sqrt[3]{a^3 \cdot 10} = a\sqrt[3]{10}$

8. $x\sqrt[3]{21}$

9. $\sqrt{3x^3}\sqrt{6x^5} = \sqrt{18x^8} = \sqrt{9x^8 \cdot 2} = 3x^4\sqrt{2}$

10. $5a^5\sqrt{3}$

11. $\sqrt[3]{s^2t^4}\sqrt[3]{s^4t^6} = \sqrt[3]{s^6t^{10}} = \sqrt[3]{s^6t^9 \cdot t} = s^2t^3\sqrt[3]{t}$

12. $xy^3\sqrt[3]{xy}$

13. $\sqrt[3]{(x+5)^2}\sqrt[3]{(x+5)^4} = \sqrt[3]{(x+5)^6} = (x+5)^2$

14. $(a-b)^4$

15. $\sqrt[4]{12a^3b^7}\sqrt[4]{4a^2b^5} = \sqrt[4]{48a^5b^{12}} = \sqrt[4]{16a^4b^{12} \cdot 3a} =$
$2ab^3\sqrt[4]{3a}$

16. $3x^2y^2\sqrt[4]{xy^3}$

17. $\sqrt[5]{x^3(y+z)^4}\sqrt[5]{x^3(y+z)^6} = \sqrt[5]{x^6(y+z)^{10}} =$
$\sqrt[5]{x^5(y+z)^{10} \cdot x} = x(y+z)^2\sqrt[5]{x}$

18. $a^2(b-c)\sqrt[5]{(b-c)^3}$

19. $\sqrt{a}\sqrt[4]{a^3}$

$= a^{1/2} \cdot a^{3/4}$ Converting to exponential notation

$= a^{5/4}$ Adding exponents

$= a^{1+1/4}$ Writing 5/4 as a mixed number

$= a \cdot a^{1/4}$ Factoring

$= a\sqrt[4]{a}$ Returning to radical notation

20. $x\sqrt{x}$

21. $\sqrt[5]{b^2}\sqrt{b^3}$

$= b^{2/5} \cdot b^{3/2}$ Converting to exponential notation

$= b^{19/10}$ Adding exponents

$= b^{1+9/10}$ Writing 19/10 as a mixed number

$= b \cdot b^{9/10}$ Factoring

$= b\sqrt[10]{b^9}$ Returning to radical notation

22. $a\sqrt[12]{a^5}$

23.
$$\begin{aligned}\sqrt{xy^3}\sqrt[3]{x^2y} &= (xy^3)^{1/2}(x^2y)^{1/3}\\ &= (xy^3)^{3/6}(x^2y)^{2/6}\\ &= [(xy^3)^3(x^2y)^2]^{1/6}\\ &= \sqrt[6]{x^3y^9 \cdot x^4y^2}\\ &= \sqrt[6]{x^7y^{11}}\\ &= \sqrt[6]{x^6y^6 \cdot xy^5}\\ &= xy\sqrt[6]{xy^5}\end{aligned}$$

24. $a\sqrt[10]{ab^7}$

25.
$$\begin{aligned}\sqrt[4]{9ab^3}\sqrt{3a^4b} &= (9ab^3)^{1/4}(3a^4b)^{1/2}\\ &= (9ab^3)^{1/4}(3a^4b)^{2/4}\\ &= [(9ab^3)(3a^4b)^2]^{1/4}\\ &= \sqrt[4]{9ab^3 \cdot 9a^8b^2}\\ &= \sqrt[4]{81a^9b^5}\\ &= \sqrt[4]{81a^8b^4 \cdot ab}\\ &= 3a^2b\sqrt[4]{ab}\end{aligned}$$

26. $2xy^2\sqrt[6]{2x^5y}$

27.
$$\begin{aligned}\sqrt[3]{xy^2z}\sqrt{x^3yz^2} &= (xy^2z)^{1/3}(x^3yz^2)^{1/2}\\ &= (xy^2z)^{2/6}(x^3yz^2)^{3/6}\\ &= [(xy^2z)^2(x^3yz^2)^3]^{1/6}\\ &= \sqrt[6]{x^2y^4z^2 \cdot x^9y^3z^6}\\ &= \sqrt[6]{x^{11}y^7z^8}\\ &= \sqrt[6]{x^6y^6z^6 \cdot x^5yz^2}\\ &= xyz\sqrt[6]{x^5yz^2}\end{aligned}$$

28. $a^2b^2c^2\sqrt[6]{a^2bc^2}$

29.
$$\begin{aligned}&\sqrt{27a^5(b+1)}\sqrt[3]{81a(b+1)^4}\\ &= [27a^5(b+1)]^{1/2}[81a(b+1)^4]^{1/3}\\ &= [27a^5(b+1)]^{3/6}[81a(b+1)^4]^{2/6}\\ &= \{[3^3a^5(b+1)]^3[3^4a(b+1)^4]^2\}^{1/6}\\ &= \sqrt[6]{3^9a^{15}(b+1)^3 \cdot 3^8a^2(b+1)^8}\\ &= \sqrt[6]{3^{17}a^{17}(b+1)^{11}}\\ &= \sqrt[6]{3^{12}a^{12}(b+1)^6 \cdot 3^5a^5(b+1)^5}\\ &= 3^2a^2(b+1)\sqrt[6]{3^5a^5(b+1)^5}, \text{ or}\\ &\quad 9a^2(b+1)\sqrt[6]{243a^5(b+1)^5}\end{aligned}$$

30. $4x(y+z)^3\sqrt[6]{2x(y+z)}$

31. $\sqrt{\dfrac{25}{36}} = \dfrac{\sqrt{25}}{\sqrt{36}} = \dfrac{5}{6}$

32. $\dfrac{10}{9}$

33. $\sqrt[3]{\dfrac{64}{27}} = \dfrac{\sqrt[3]{64}}{\sqrt[3]{27}} = \dfrac{4}{3}$

34. $\frac{7}{10}$

35. $\sqrt{\frac{49}{y^2}} = \frac{\sqrt{49}}{\sqrt{y^2}} = \frac{7}{y}$

36. $\frac{11}{x}$

37. $\sqrt{\frac{25y^3}{x^4}} = \frac{\sqrt{25y^3}}{\sqrt{x^4}} = \frac{\sqrt{25y^2 \cdot y}}{\sqrt{x^4}} = \frac{\sqrt{25y^2}\,\sqrt{y}}{\sqrt{x^4}} =$
$\frac{5y\sqrt{y}}{x^2}$

38. $\frac{6a^2\sqrt{a}}{b^3}$

39. $\sqrt[3]{\frac{27a^4}{8b^3}} = \frac{\sqrt[3]{27a^4}}{\sqrt[3]{8b^3}} = \frac{\sqrt[3]{27a^3 \cdot a}}{\sqrt[3]{8b^3}} = \frac{\sqrt[3]{27a^3}\,\sqrt[3]{a}}{\sqrt[3]{8b^3}} =$
$\frac{3a\sqrt[3]{a}}{2b}$

40. $\frac{2x^2\sqrt[3]{x}}{3y^2}$

41. $\sqrt[4]{\frac{16a^4}{b^4c^8}} = \frac{\sqrt[4]{16a^4}}{\sqrt[4]{b^4c^8}} = \frac{2a}{bc^2}$

42. $\frac{3x}{y^2z}$

43. $\sqrt[4]{\frac{a^5b^8}{c^{10}}} = \frac{\sqrt[4]{a^5b^8}}{\sqrt[4]{c^{10}}} = \frac{\sqrt[4]{a^4b^8 \cdot a}}{\sqrt[4]{c^8 \cdot c^2}} = \frac{\sqrt[4]{a^4b^8}\,\sqrt[4]{a}}{\sqrt[4]{c^8}\,\sqrt[4]{c^2}} =$
$\frac{ab^2\sqrt[4]{a}}{c^2\sqrt[4]{c^2}}$, or $\frac{ab^2}{c^2}\sqrt[4]{\frac{a}{c^2}}$

44. $\frac{x^2y^3}{z}\sqrt[4]{\frac{x}{z^2}}$

45. $\sqrt[5]{\frac{32x^6}{y^{11}}} = \frac{\sqrt[5]{32x^6}}{\sqrt[5]{y^{11}}} = \frac{\sqrt[5]{32x^5 \cdot x}}{\sqrt[5]{y^{10} \cdot y}} = \frac{\sqrt[5]{32x^5} \cdot \sqrt[5]{x}}{\sqrt[5]{y^{10}}\sqrt[5]{y}} =$
$\frac{2x\sqrt[5]{x}}{y^2\sqrt[5]{y}}$, or $\frac{2x}{y^2}\sqrt[5]{\frac{x}{y}}$

46. $\frac{3a}{b^2}\sqrt[5]{\frac{a^4}{b^3}}$

47. $\sqrt[6]{\frac{x^6y^8}{z^{15}}} = \frac{\sqrt[6]{x^6y^8}}{\sqrt[6]{z^{15}}} = \frac{\sqrt[6]{x^6y^6 \cdot y^2}}{\sqrt[6]{z^{12} \cdot z^3}} = \frac{\sqrt[6]{x^6y^6}\,\sqrt[6]{y^2}}{\sqrt[6]{z^{12}}\,\sqrt[6]{z^3}} =$
$\frac{xy\sqrt[6]{y^2}}{z^2\sqrt[6]{z^3}}$, or $\frac{xy}{z^2}\sqrt[6]{\frac{y^2}{z^3}}$

48. $\frac{ab^2}{c^2}\sqrt[6]{\frac{a^3}{c}}$

49. $\frac{\sqrt{35x}}{\sqrt{7x}} = \sqrt{\frac{35x}{7x}} = \sqrt{5}$

50. $\sqrt{7}$

51. $\frac{\sqrt[3]{270}}{\sqrt[3]{10}} = \sqrt[3]{\frac{270}{10}} = \sqrt[3]{27} = 3$

52. 2

53. $\frac{\sqrt{40xy^3}}{\sqrt{8x}} = \sqrt{\frac{40xy^3}{8x}} = \sqrt{5y^3} = \sqrt{y^2 \cdot 5y} =$
$\sqrt{y^2}\,\sqrt{5y} = y\sqrt{5y}$

54. $2b\sqrt{2b}$

55. $\frac{\sqrt[3]{96a^4b^2}}{\sqrt[3]{12a^2b}} = \sqrt[3]{\frac{96a^4b^2}{12a^2b}} = \sqrt[3]{8a^2b} = \sqrt[3]{8}\,\sqrt[3]{a^2b} =$
$2\sqrt[3]{a^2b}$

56. $3xy\sqrt[3]{y^2}$

57. $\frac{\sqrt{100ab}}{5\sqrt{2}} = \frac{1}{5}\,\frac{\sqrt{100ab}}{\sqrt{2}} = \frac{1}{5}\sqrt{\frac{100ab}{2}} = \frac{1}{5}\sqrt{50ab} =$
$\frac{1}{5}\sqrt{25 \cdot 2ab} = \frac{1}{5} \cdot 5\sqrt{2ab} = \sqrt{2ab}$

58. $\frac{5}{3}\sqrt{ab}$

59. $\frac{\sqrt[4]{48x^9y^{13}}}{\sqrt[4]{3xy^{-2}}} = \sqrt[4]{\frac{48x^9y^{13}}{3xy^{-2}}} = \sqrt[4]{16x^8y^{15}} =$
$\sqrt[4]{16x^8y^{12}}\,\sqrt[4]{y^3} = 2x^2y^3\sqrt[4]{y^3}$

60. $2a^2b^6$

61. $\frac{\sqrt[3]{x^3 - y^3}}{\sqrt[3]{x - y}} = \sqrt[3]{\frac{x^3 - y^3}{x - y}} =$
$\sqrt[3]{\frac{(x-y)(x^2+xy+y^2)}{x-y}} =$
$\sqrt[3]{\frac{\cancel{(x-y)}(x^2+xy+y^2)}{\cancel{x-y}}} = \sqrt[3]{x^2+xy+y^2}$

62. $\sqrt[3]{r^2 - rs + s^2}$

63. $\frac{\sqrt[3]{a^2}}{\sqrt[4]{a}}$

$= \frac{a^{2/3}}{a^{1/4}}$ Converting to exponential notation

$= a^{2/3-1/4}$ Subtracting exponents

$= a^{5/12}$ Converting back

$= \sqrt[12]{a^5}$ to radical notation

64. $\sqrt[15]{x^7}$

65. $\dfrac{\sqrt[4]{x^2y^3}}{\sqrt[3]{xy}}$

$= \dfrac{(x^2y^3)^{1/4}}{(xy)^{1/3}}$ Converting to exponential notation

$= \dfrac{x^{2/4}y^{3/4}}{x^{1/3}y^{1/3}}$ Using the power and product rules

$= x^{2/4-1/3}y^{3/4-1/3}$ Subtracting exponents

$= x^{2/12}y^{5/12}$

$= (x^2y^5)^{1/2}$ Converting back to

$= \sqrt[12]{x^2y^5}$ radical notation

66. $\sqrt[15]{\dfrac{a^7}{b^4}}$

67. $\dfrac{\sqrt{ab^3c}}{\sqrt[5]{a^2b^3c^{-1}}} = \dfrac{(ab^3c)^{1/2}}{(a^2b^3c^{-1})^{1/5}}$

$= \dfrac{a^{1/2}b^{3/2}c^{1/2}}{a^{2/5}b^{3/5}c^{-1/5}}$

$= a^{1/10}b^{9/10}c^{7/10}$ Subtracting exponents

$= (ab^9c^7)^{1/10}$

$= \sqrt[10]{ab^9c^7}$

68. $yz\sqrt[10]{xy^8z^3}$

69. $\dfrac{\sqrt[4]{(3x-1)^3}}{\sqrt[5]{(3x-1)^3}}$

$= \dfrac{(3x-1)^{3/4}}{(3x-1)^{3/5}}$ Converting to exponential notation

$= (3x-1)^{3/4-3/5}$ Subtracting exponents

$= (3x-1)^{3/20}$ Converting back

$= \sqrt[20]{(3x-1)^3}$ to radical notation

70. $\sqrt[12]{(2+5x)^5}$

71. $\dfrac{\sqrt[3]{(2x+1)^2}}{\sqrt[5]{(2x+1)^2}}$

$= \dfrac{(2x+1)^{2/3}}{(2x+1)^{2/5}}$ Converting to exponential notation

$= (2x+1)^{2/3-2/5}$ Subtracting exponents

$= (2x+1)^{4/15}$ Converting back to

$= \sqrt[15]{(2x+1)^4}$ radical notation

72. $\sqrt[12]{5-3x}$

73. $\dfrac{12x}{x-4} - \dfrac{3x^2}{x+4} = \dfrac{384}{x^2-16}$

$\dfrac{12x}{x-4} - \dfrac{3x^2}{x+4} = \dfrac{384}{(x+4)(x-4)}$,

Note that $x \neq -4$ and $x \neq 4$

$$(x+4)(x-4)\left[\frac{12x}{x-4} - \frac{3x^2}{x+4}\right] = (x+4)(x-4)\cdot\frac{384}{(x+4)(x-4)}$$

$$12x(x+4) - 3x^2(x-4) = 384$$
$$12x^2 + 48x - 3x^3 + 12x^2 = 384$$
$$-3x^3 + 24x^2 + 48x - 384 = 0$$
$$-3(x^3 - 8x^2 - 16x + 128) = 0$$
$$-3[x^2(x-8) - 16(x-8)] = 0$$
$$-3(x-8)(x^2-16) = 0$$
$$-3(x-8)(x+4)(x-4) = 0$$

$x - 8 = 0$ or $x + 4 = 0$ or $x - 4 = 0$

$x = 8$ or $x = -4$ or $x = 4$

Check: For 8:

$$\begin{array}{c|c} \dfrac{12x}{x-4} - \dfrac{3x^2}{x+4} & \dfrac{384}{x^2-16} \\ \hline \dfrac{12\cdot 8}{8-4} - \dfrac{3\cdot 8^2}{8+4} & \dfrac{384}{8^2-16} \\ \dfrac{96}{4} - \dfrac{192}{12} & \dfrac{384}{48} \\ 24 - 16 & 8 \\ 8 & \end{array} \quad \text{TRUE}$$

8 is a solution.

For -4:

$$\begin{array}{c|c} \dfrac{12x}{x-4} - \dfrac{3x^2}{x+4} & \dfrac{384}{x^2-16} \\ \hline \dfrac{12(-4)}{-4-4} - \dfrac{3(-4)^2}{-4+4} & \dfrac{384}{(-4)^2-16} \\ \dfrac{-48}{-8} - \dfrac{48}{0} & \dfrac{384}{16-16} \end{array} \quad \text{UNDEFINED}$$

-4 is not a solution.

For 4:

$$\begin{array}{c|c} \dfrac{12x}{x-4} - \dfrac{3x^2}{x+4} & \dfrac{384}{x^2-16} \\ \hline \dfrac{12\cdot 4}{4-4} - \dfrac{3\cdot 4^2}{4+4} & \dfrac{384}{4^2-16} \\ \dfrac{48}{0} - \dfrac{48}{8} & \dfrac{384}{16-16} \end{array} \quad \text{UNDEFINED}$$

4 is not a solution.

The checks confirm that -4 and 4 are not solutions. The solution is 8.

74. $\dfrac{15}{2}$

75. ***Familiarize***. Let x and y represent the width and length of the rectangle, respectively.

Translate. We write two equations.

$\underbrace{\text{The width}}_{\downarrow}\ \underbrace{\text{is}}_{\downarrow}\ \underbrace{\text{one-fourth}}_{\downarrow}\ \underbrace{\text{the length.}}_{\downarrow}$

$x \quad = \quad \frac{1}{4}\cdot \quad y$

$\underbrace{\text{The area}}_{\downarrow}\ \underbrace{\text{is}}_{\downarrow}\ \underbrace{\text{twice}}_{\downarrow}\ \underbrace{\text{the perimeter.}}_{\downarrow}$

$xy \quad = \quad 2\cdot \quad (2x+2y)$

Carry out. Solving the system of equations we get (5,20).

Check. The width, 5, is one-fourth the length, 20. The area is $5\cdot 20$, or 100. The perimeter is $2\cdot 5 + 2\cdot 20$, or 50. Since $100 = 2\cdot 50$, the area is twice the perimeter. The values check.

State. The width is 5, and the length is 20.

76. -5, 4

77.
$$\begin{aligned} A &= \frac{m}{a_2-a_1} \\ A(a_2-a_1) &= m \qquad \text{Multiplying by } a_2-a_1 \\ Aa_2 - Aa_1 &= m \\ -Aa_1 &= m - Aa_2 \\ a_1 &= \frac{m-Aa_2}{-A}, \text{ or } \frac{Aa_2-m}{A}, \text{ or} \\ & a_2 - \frac{m}{A} \end{aligned}$$

78. $n = \dfrac{4m-P}{7}$

79. ◈

80. ◈

81. ◈

82. ◈

83. a) $T = 2\pi\sqrt{\frac{65}{980}} \approx 1.62$ sec

b) $T = 2\pi\sqrt{\frac{98}{980}} \approx 1.99$ sec

c) $T = 2\pi\sqrt{\frac{120}{980}} \approx 2.20$ sec

84. a^3bxy^2

85.
$$\begin{aligned} \frac{(\sqrt[3]{81mn^2})^2}{(\sqrt[3]{mn})^2} &= \frac{\sqrt[3]{(81mn^2)^2}}{\sqrt[3]{(mn)^2}} \\ &= \frac{\sqrt[3]{6561m^2n^4}}{\sqrt[3]{m^2n^2}} \\ &= \sqrt[3]{\frac{6561m^2n^4}{m^2n^2}} \\ &= \sqrt[3]{6561n^2} \\ &= \sqrt[3]{729\cdot 9n^2} \\ &= \sqrt[3]{729}\ \sqrt[3]{9n^2} \\ &= 9\sqrt[3]{9n^2} \end{aligned}$$

86. $2yz\sqrt{2z}$

87.
$$\begin{aligned} &\frac{\sqrt{x^5-2x^4y} - \sqrt{xy^4-2y^5}}{\sqrt{xy^2-2y^3} + \sqrt{x^3-2x^2y}} \\ &= \frac{\sqrt{x^4(x-2y)} - \sqrt{y^4(x-2y)}}{\sqrt{y^2(x-2y)} + \sqrt{x^2(x-2y)}} \\ &= \frac{x^2\sqrt{x-2y} - y^2\sqrt{x-2y}}{y\sqrt{x-2y} + x\sqrt{x-2y}} \\ &= \frac{(x^2-y^2)\sqrt{x-2y}}{(y+x)\sqrt{x-2y}} \\ &= \frac{(x+y)(x-y)\sqrt{x-2y}}{(y+x)\sqrt{x-2y}} \\ &= (x-y)\cdot\frac{(x+y)\sqrt{x-2y}}{(y+x)\sqrt{x-2y}} \\ &= x-y \end{aligned}$$

88. 10

89.
$$\begin{aligned} \sqrt[5]{4a^{3k+2}}\sqrt[5]{8a^{6-k}} &= 2a^4 \\ \sqrt[5]{32a^{2k+8}} &= 2a^4 \\ 2\sqrt[5]{a^{2k+8}} &= 2a^4 \\ \sqrt[5]{a^{2k+8}} &= a^4 \\ a^{\frac{2k+8}{5}} &= a^4 \end{aligned}$$

Since the base is the same, the exponents must be equal. We have:

$$\begin{aligned} \frac{2k+8}{5} &= 4 \\ 2k+8 &= 20 \\ 2k &= 12 \\ k &= 6 \end{aligned}$$

90. 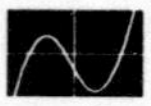

91. 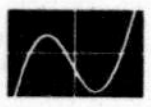

Exercise Set 7.5

1. $\sqrt{7}(3-\sqrt{7}) = \sqrt{7}\cdot 3 - \sqrt{7}\cdot\sqrt{7} = 3\sqrt{7} - 7$

2. $4\sqrt{3}+3$

3. $\sqrt{2}(\sqrt{3}-\sqrt{5}) = \sqrt{2}\cdot\sqrt{3} - \sqrt{2}\cdot\sqrt{5} = \sqrt{6} - \sqrt{10}$

4. $5-\sqrt{10}$

5. $\sqrt{3}(2\sqrt{5}-3\sqrt{4}) = \sqrt{3}(2\sqrt{5}-3\cdot 2) =$
$\sqrt{3}\cdot 2\sqrt{5} - \sqrt{3}\cdot 6 = 2\sqrt{15} - 6\sqrt{3}$

6. $6\sqrt{5}-4$

7. $\sqrt[3]{2}(\sqrt[3]{4} - 2\sqrt[3]{32}) = \sqrt[3]{2}\cdot\sqrt[3]{4} - \sqrt[3]{2}\cdot 2\sqrt[3]{32} =$
$\sqrt[3]{8} - 2\sqrt[3]{64} = 2 - 2\cdot 4 = 2 - 8 = -6$

8. $3 - 4\sqrt[3]{63}$

9. $\sqrt[3]{a}(\sqrt[3]{a^2} + \sqrt[3]{24a^2}) = \sqrt[3]{a} \cdot \sqrt[3]{a^2} + \sqrt[3]{a}\sqrt[3]{24a^2} =$
$\sqrt[3]{a^3} + \sqrt[3]{24a^3} = \sqrt[3]{a^3} + \sqrt[3]{8a^3 \cdot 3} =$
$a + 2a\sqrt[3]{3}$

10. $-2x\sqrt[3]{3}$

11. $(5+\sqrt{6})(5-\sqrt{6}) = 5^2 - (\sqrt{6})^2 = 25 - 6 = 19$

12. -1

13. $(3-2\sqrt{7})(3+2\sqrt{7}) = 3^2 - (2\sqrt{7})^2 = 9 - 4 \cdot 7 =$
$9 - 28 = -19$

14. -2

15. $(5+\sqrt[3]{10})(3-\sqrt[3]{10})$
$= 15 - 5\sqrt[3]{10} + 3\sqrt[3]{10} - \sqrt[3]{10}\sqrt[3]{10}$ Using FOIL
$= 15 - 5\sqrt[3]{10} + 3\sqrt[3]{10} - \sqrt[3]{100}$ Multiplying radicals
$= 15 - 2\sqrt[3]{10} - \sqrt[3]{100}$ Simplifying

16. $\sqrt[3]{49} + \sqrt[3]{7} - 20$

17. $(2\sqrt{7} - 4\sqrt{2})(3\sqrt{7} + 6\sqrt{2}) =$
$2\sqrt{7} \cdot 3\sqrt{7} + 2\sqrt{7} \cdot 6\sqrt{2} - 4\sqrt{2} \cdot 3\sqrt{7} - 4\sqrt{2} \cdot 6\sqrt{2} =$
$6 \cdot 7 + 12\sqrt{14} - 12\sqrt{14} - 24 \cdot 2 =$
$42 + 12\sqrt{14} - 12\sqrt{14} - 48 = -6$

18. $24 - 7\sqrt{15}$

19. $(2\sqrt[3]{3} - \sqrt[3]{2})(\sqrt[3]{3} + 2\sqrt[3]{2}) =$
$2\sqrt[3]{3} \cdot \sqrt[3]{3} + 2\sqrt[3]{3} \cdot 2\sqrt[3]{2} - \sqrt[3]{2} \cdot \sqrt[3]{3} - \sqrt[3]{2} \cdot 2\sqrt[3]{2} =$
$2\sqrt[3]{9} + 4\sqrt[3]{6} - \sqrt[3]{6} - 2\sqrt[3]{4} = 2\sqrt[3]{9} + 3\sqrt[3]{6} - 2\sqrt[3]{4}$

20. $6\sqrt[4]{63} - 9\sqrt[4]{42} + 2\sqrt[4]{54} - 3\sqrt[4]{36}$

21. $(\sqrt{3x} + \sqrt{y})^2$
$= (\sqrt{3x})^2 + 2 \cdot \sqrt{3x} \cdot \sqrt{y} + (\sqrt{y})^2$ Squaring a binomial
$= 3x + 2\sqrt{3xy} + y$

22. $t - 2\sqrt{2rt} + 2r$

23. $\sqrt[3]{x^2y}(\sqrt{xy} - \sqrt[5]{xy^3})$
$= (x^2y)^{1/3}[(xy)^{1/2} - (xy^3)^{1/5}]$
$= x^{2/3}y^{1/3}(x^{1/2}y^{1/2} - x^{1/5}y^{3/5})$
$= x^{2/3}y^{1/3}x^{1/2}y^{1/2} - x^{2/3}y^{1/3}x^{1/5}y^{3/5}$
$= x^{2/3+1/2}y^{1/3+1/2} - x^{2/3+1/5}y^{1/3+3/5}$
$= x^{7/6}y^{5/6} - x^{13/15}y^{14/15}$
$= x^{1\frac{1}{6}}y^{\frac{5}{6}} - x^{13/15}y^{14/15}$
Writing a mixed numeral
$= x \cdot x^{1/6}y^{5/6} - x^{13/15}y^{14/15}$
$= x(xy^5)^{1/6} - (x^{13}y^{14})^{1/15}$
$= x\sqrt[6]{xy^5} - \sqrt[15]{x^{13}y^{14}}$

24. $a\sqrt[12]{a^2b^7} - \sqrt[20]{a^{18}b^{13}}$

25. $(m + \sqrt[3]{n^2})(2m + \sqrt[4]{n})$
$= (m + n^{2/3})(2m + n^{1/4})$ Converting to exponential notation
$= 2m^2 + mn^{1/4} + 2mn^{2/3} + n^{2/3}n^{1/4}$ Using FOIL
$= 2m^2 + mn^{1/4} + 2mn^{2/3} + n^{2/3+1/4}$ Adding exponents
$= 2m^2 + mn^{1/4} + 2mn^{2/3} + n^{11/12}$
$= 2m^2 + m\sqrt[4]{n} + 2m\sqrt[3]{n^2} + \sqrt[12]{n^{11}}$ Converting back to radical notation

26. $3r^2 - r\sqrt[5]{s} - 3r\sqrt[4]{s^3} + \sqrt[20]{s^{19}}$

27. $f(x) = \sqrt[4]{x},\ g(x) = \sqrt[4]{2x} - \sqrt[4]{x^{11}}$
$(f \cdot g)(x) = \sqrt[4]{x}(\sqrt[4]{2x} - \sqrt[4]{x^{11}})$
$= \sqrt[4]{2x^2} - \sqrt[4]{x^{12}}$
$= \sqrt[4]{2x^2} - x^3$

28. $x^2 + \sqrt[4]{3x^3}$

29. $f(x) = x + \sqrt{7},\ g(x) = x - \sqrt{7}$
$(f \cdot g)(x) = (x + \sqrt{7})(x - \sqrt{7})$
$= x^2 - (\sqrt{7})^2$
$= x^2 - 7$

30. $x^2 + x\sqrt{6} - x\sqrt{2} - 2\sqrt{3}$

31. $f(x) = 2 - \sqrt{x},\ g(x) = 1 - \sqrt{x}$
$(f \cdot g)(x) = (2 - \sqrt{x})(1 - \sqrt{x})$
$= 2 - 2\sqrt{x} - \sqrt{x} + \sqrt{x^2}$
$= 2 - 3\sqrt{x} + x$

32. $x + \sqrt{2x} + \sqrt{3x} + \sqrt{6}$

33. $f(x) = x^2$
$f(5 - \sqrt{2}) = (5 - \sqrt{2})^2 = 25 - 10\sqrt{2} + (\sqrt{2})^2 =$
$25 - 10\sqrt{2} + 2 = 27 - 10\sqrt{2}$

34. $52 + 14\sqrt{3}$

35. $f(x) = x^2$
$f(\sqrt{3} + \sqrt{5}) = (\sqrt{3} + \sqrt{5})^2 =$
$(\sqrt{3})^2 + 2 \cdot \sqrt{3} \cdot \sqrt{5} + (\sqrt{5})^2 =$
$3 + 2\sqrt{15} + 5 = 8 + 2\sqrt{15}$

36. $9 - 6\sqrt{2}$

37. $f(x) = x^2$
$f(\sqrt{10} - \sqrt{5}) = (\sqrt{10} - \sqrt{5})^2 =$
$(\sqrt{10})^2 - 2 \cdot \sqrt{10} \cdot \sqrt{5} + (\sqrt{5})^2 =$
$10 - 2\sqrt{50} + 5 = 15 - 2\sqrt{25 \cdot 2} =$
$15 - 2 \cdot 5\sqrt{2} = 15 - 10\sqrt{2}$

38. $15+4\sqrt{14}$

39. $\sqrt{\frac{5}{7}}=\sqrt{\frac{5}{7}\cdot\frac{7}{7}}=\sqrt{\frac{35}{49}}=\frac{\sqrt{35}}{\sqrt{49}}=\frac{\sqrt{35}}{7}$

40. $\frac{\sqrt{66}}{6}$

41. $\frac{6\sqrt{5}}{5\sqrt{3}}=\frac{6\sqrt{5}}{5\sqrt{3}}\cdot\frac{\sqrt{3}}{\sqrt{3}}=\frac{6\sqrt{15}}{5\cdot 3}=\frac{2\sqrt{15}}{5}$

42. $\frac{2\sqrt{10}}{3}$

43. $\sqrt[3]{\frac{16}{9}}=\sqrt[3]{\frac{16}{9}\cdot\frac{3}{3}}=\sqrt[3]{\frac{48}{27}}=\frac{\sqrt[3]{8\cdot 6}}{\sqrt[3]{27}}=\frac{2\sqrt[3]{6}}{3}$

44. $\frac{\sqrt[3]{6}}{3}$

45. $\frac{\sqrt[3]{3a}}{\sqrt[3]{5c}}=\frac{\sqrt[3]{3a}}{\sqrt[3]{5c}}\cdot\frac{\sqrt[3]{5^2c^2}}{\sqrt[3]{5^2c^2}}=\frac{\sqrt[3]{75ac^2}}{\sqrt[3]{5^3c^3}}=\frac{\sqrt[3]{75ac^2}}{5c}$

46. $\frac{\sqrt[3]{63xy^2}}{3y}$

47. $\frac{\sqrt[3]{5y^4}}{\sqrt[3]{6x^4}}=\frac{\sqrt[3]{5y^4}}{\sqrt[3]{6x^4}}\cdot\frac{\sqrt[3]{36x^2}}{\sqrt[3]{36x^2}}=\frac{\sqrt[3]{y^3\cdot 180x^2y}}{\sqrt[3]{216x^6}}=$
$\frac{y\sqrt[3]{180x^2y}}{6x^2}$

48. $\frac{a\sqrt[3]{147ab}}{7b}$

49. $\sqrt[3]{\frac{2}{x^2y}}=\sqrt[3]{\frac{2}{x^2y}\cdot\frac{xy^2}{xy^2}}=\sqrt[3]{\frac{2xy^2}{x^3y^3}}=\frac{\sqrt[3]{2xy^2}}{\sqrt[3]{x^3y^3}}=$
$\frac{\sqrt[3]{2xy^2}}{xy}$

50. $\frac{\sqrt[3]{5a^2b}}{ab}$

51. $\sqrt{\frac{7a}{18}}=\sqrt{\frac{7a}{18}\cdot\frac{2}{2}}=\sqrt{\frac{14a}{36}}=\frac{\sqrt{14a}}{\sqrt{36}}=\frac{\sqrt{14a}}{6}$

52. $\frac{\sqrt{30x}}{10}$

53. $\sqrt{\frac{9}{20x^2y}}=\sqrt{\frac{9}{20x^2y}\cdot\frac{5y}{5y}}=\sqrt{\frac{9\cdot 5y}{100x^2y^2}}=$
$\frac{\sqrt{9\cdot 5y}}{\sqrt{100x^2y^2}}=\frac{3\sqrt{5y}}{10xy}$

54. $\frac{\sqrt{10a}}{8ab}$

55. $\frac{5}{7-\sqrt{2}}=\frac{5}{7-\sqrt{2}}\cdot\frac{7+\sqrt{2}}{7+\sqrt{2}}=\frac{5(7+\sqrt{2})}{7^2-(\sqrt{2})^2}=$
$\frac{35+5\sqrt{2}}{49-2}=\frac{35+5\sqrt{2}}{47}$

56. $\frac{15-3\sqrt{6}}{19}$

57. $\frac{\sqrt{x}}{\sqrt{x}+\sqrt{y}}=\frac{\sqrt{x}}{\sqrt{x}+\sqrt{y}}\cdot\frac{\sqrt{x}-\sqrt{y}}{\sqrt{x}-\sqrt{y}}=$
$\frac{\sqrt{x}(\sqrt{x}-\sqrt{y})}{(\sqrt{x})^2-(\sqrt{y})^2}=\frac{x-\sqrt{xy}}{x-y}$

58. $\frac{\sqrt{ab}+b}{a-b}$

59. $\frac{\sqrt{3}+4\sqrt{5}}{\sqrt{3}-2\sqrt{6}}=\frac{\sqrt{3}+4\sqrt{5}}{\sqrt{3}-2\sqrt{6}}\cdot\frac{\sqrt{3}+2\sqrt{6}}{\sqrt{3}+2\sqrt{6}}=$
$\frac{(\sqrt{3}+4\sqrt{5})(\sqrt{3}+2\sqrt{6})}{(\sqrt{3}-2\sqrt{6})(\sqrt{3}+2\sqrt{6})}=$
$\frac{(\sqrt{3})^2+\sqrt{3}\cdot 2\sqrt{6}+4\sqrt{5}\cdot\sqrt{3}+4\sqrt{5}\cdot 2\sqrt{6}}{(\sqrt{3})^2-(2\sqrt{6})^2}=$
$\frac{3+2\sqrt{18}+4\sqrt{15}+8\sqrt{30}}{3-4\cdot 6}=$
$\frac{3+2\cdot 3\sqrt{2}+4\sqrt{15}+8\sqrt{30}}{3-24}=$
$\frac{3+6\sqrt{2}+4\sqrt{15}+8\sqrt{30}}{-21}$

60. $\frac{5\sqrt{5}-15\sqrt{3}-\sqrt{15}+9}{-11}$

61. $\frac{5\sqrt{3}-3\sqrt{2}}{3\sqrt{2}-2\sqrt{3}}=\frac{5\sqrt{3}-3\sqrt{2}}{3\sqrt{2}-2\sqrt{3}}\cdot\frac{3\sqrt{2}+2\sqrt{3}}{3\sqrt{2}+2\sqrt{3}}=$
$\frac{15\sqrt{6}+10\cdot 3-9\cdot 2-6\sqrt{6}}{9\cdot 2-4\cdot 3}=$
$\frac{15\sqrt{6}+30-18-6\sqrt{6}}{18-12}=\frac{9\sqrt{6}+12}{6}=$
$\frac{3(3\sqrt{6}+4)}{3\cdot 2}=\frac{3\sqrt{6}+4}{2}$

62. $\frac{4\sqrt{6}+9}{3}$

63. $\frac{\sqrt{5}}{\sqrt{7x}}=\frac{\sqrt{5}}{\sqrt{7x}}\cdot\frac{\sqrt{5}}{\sqrt{5}}=\frac{\sqrt{25}}{\sqrt{35x}}=\frac{5}{\sqrt{35x}}$

64. $\frac{10}{\sqrt{30x}}$

65. $\sqrt{\frac{14}{21}}=\sqrt{\frac{2}{3}}=\sqrt{\frac{2}{3}\cdot\frac{2}{2}}=\sqrt{\frac{4}{6}}=\frac{\sqrt{4}}{\sqrt{6}}=\frac{2}{\sqrt{6}}$

66. $\frac{2}{\sqrt{5}}$

67. $\frac{4\sqrt{13}}{3\sqrt{7}}=\frac{4\sqrt{13}}{3\sqrt{7}}\cdot\frac{\sqrt{13}}{\sqrt{13}}=\frac{4\sqrt{169}}{3\sqrt{91}}=\frac{4\cdot 13}{3\sqrt{91}}=\frac{52}{3\sqrt{91}}$

68. $\frac{105}{2\sqrt{105}}$

69. $\dfrac{\sqrt[3]{7}}{\sqrt[3]{2}} = \dfrac{\sqrt[3]{7}}{\sqrt[3]{2}} \cdot \dfrac{\sqrt[3]{7^2}}{\sqrt[3]{7^2}} = \dfrac{\sqrt[3]{7^3}}{\sqrt[3]{98}} = \dfrac{7}{\sqrt[3]{98}}$

70. $\dfrac{5}{\sqrt[3]{100}}$

71. $\sqrt{\dfrac{7x}{3y}} = \sqrt{\dfrac{7x}{3y} \cdot \dfrac{7x}{7x}} = \dfrac{\sqrt{(7x)^2}}{\sqrt{21xy}} = \dfrac{7x}{\sqrt{21xy}}$

72. $\dfrac{6a}{\sqrt{30ab}}$

73. $\sqrt[3]{\dfrac{2a^5}{5b}} = \sqrt[3]{\dfrac{2a^5}{5b} \cdot \dfrac{4a}{4a}} = \sqrt[3]{\dfrac{8a^6}{20ab}} = \dfrac{2a^2}{\sqrt[3]{20ab}}$

74. $\dfrac{2a^2}{\sqrt[3]{28a^2b}}$

75. $\sqrt{\dfrac{x^3y}{2}} = \sqrt{\dfrac{x^3y}{2} \cdot \dfrac{xy}{xy}} = \sqrt{\dfrac{x^4y^2}{2xy}} = \dfrac{\sqrt{x^4y^2}}{\sqrt{2xy}} = \dfrac{x^2y}{\sqrt{2xy}}$

76. $\dfrac{ab^3}{\sqrt{3ab}}$

77. $\dfrac{\sqrt{5}+2}{6} = \dfrac{\sqrt{5}+2}{6} \cdot \dfrac{\sqrt{5}-2}{\sqrt{5}-2} = \dfrac{(\sqrt{5})^2 - 2^2}{6(\sqrt{5}-2)} =$

$\dfrac{5-4}{6\sqrt{5}-12} = \dfrac{1}{6\sqrt{5}-12}$

78. $\dfrac{23}{14+2\sqrt{3}}$

79. $\dfrac{\sqrt{3}-5}{\sqrt{2}+5} = \dfrac{\sqrt{3}-5}{\sqrt{2}+5} \cdot \dfrac{\sqrt{3}+5}{\sqrt{3}+5} =$

$\dfrac{3-25}{\sqrt{6}+5\sqrt{2}+5\sqrt{3}+25} = \dfrac{-22}{\sqrt{6}+5\sqrt{2}+5\sqrt{3}+25}$

80. $\dfrac{-3}{3\sqrt{2}+3\sqrt{3}+7\sqrt{6}+21}$

81. $\dfrac{\sqrt{x}+\sqrt{y}}{\sqrt{x}-\sqrt{y}} = \dfrac{\sqrt{x}+\sqrt{y}}{\sqrt{x}-\sqrt{y}} \cdot \dfrac{\sqrt{x}-\sqrt{y}}{\sqrt{x}-\sqrt{y}} =$

$\dfrac{x-y}{x-\sqrt{xy}-\sqrt{xy}+y} = \dfrac{x-y}{x-2\sqrt{xy}+y}$

82. $\dfrac{x-y}{x+2\sqrt{xy}+y}$

83. $\dfrac{a\sqrt{b}-\sqrt{c}}{\sqrt{b}+\sqrt{c}} = \dfrac{a\sqrt{b}-\sqrt{c}}{\sqrt{b}+\sqrt{c}} \cdot \dfrac{a\sqrt{b}+\sqrt{c}}{a\sqrt{b}+\sqrt{c}} =$

$\dfrac{a^2b-c}{ab+\sqrt{bc}+a\sqrt{bc}+c}$, or $\dfrac{a^2b-c}{ab+(1+a)\sqrt{bc}+c}$

84. $\dfrac{a-b^2c}{a-b\sqrt{ac}-\sqrt{ac}+bc}$, or $\dfrac{a-b^2c}{a+(-b-1)\sqrt{ac}+bc}$

85. $\dfrac{1}{2} - \dfrac{1}{3} = \dfrac{1}{t}$, LCD is $6t$

Note that $t \neq 0$.

$$6t\left(\frac{1}{2}-\frac{1}{3}\right) = 6t\left(\frac{1}{t}\right)$$
$$3t - 2t = 6$$
$$t = 6$$

Check:

$\frac{1}{2} - \frac{1}{3} = \frac{1}{t}$	
$\frac{1}{2} - \frac{1}{3}$	$\frac{1}{6}$
$\frac{3}{6} - \frac{2}{6}$	
$\frac{1}{6}$	TRUE

The solution is 6.

86. $-\dfrac{19}{5}$

87. $\dfrac{2x^2-x-6}{x^2+4x+3} \div \dfrac{2x^2+x-3}{x^2-1}$

$= \dfrac{2x^2-x-6}{x^2+4x+3} \cdot \dfrac{x^2-1}{2x^2+x-3}$

$= \dfrac{(2x^2-x-6)(x^2-1)}{(x^2+4x+3)(2x^2+x-3)}$

$= \dfrac{(2x+3)(x-2)(x+1)(x-1)}{(x+3)(x+1)(2x+3)(x-1)}$

$= \dfrac{\cancel{(2x+3)}(x-2)\cancel{(x+1)}\cancel{(x-1)}}{(x+3)\cancel{(x+1)}\cancel{(2x+3)}\cancel{(x-1)}}$

$= \dfrac{x-2}{x+3}$

88. 1

89.

90.

91. $x-5 = (\sqrt{x})^2 - (\sqrt{5})^2 = (\sqrt{x}+\sqrt{5})(\sqrt{x}-\sqrt{5})$

92. $(\sqrt{y}+\sqrt{7})(\sqrt{y}-\sqrt{7})$

93. $x-a = (\sqrt{x})^2 - (\sqrt{a})^2 = (\sqrt{x}+\sqrt{a})(\sqrt{x}-\sqrt{a})$

94. 6

95. $(\sqrt{x+2}-\sqrt{x-2})^2 =$

$x+2-2\sqrt{(x+2)(x-2)}+x-2 =$

$x+2-2\sqrt{x^2-4}+x-2 = 2x-2\sqrt{x^2-4}$

96. $\dfrac{ab+(a-b)\sqrt{a+b}-a-b}{a+b-b^2}$

97. $\dfrac{b+\sqrt{b}}{1+b+\sqrt{b}} = \dfrac{b+\sqrt{b}}{(1+b)+\sqrt{b}}\cdot\dfrac{(1+b)-\sqrt{b}}{(1+b)-\sqrt{b}}$

$= \dfrac{(b+\sqrt{b})(1+b-\sqrt{b})}{(1+b)^2-(\sqrt{b})^2}$

$= \dfrac{b+b^2-b\sqrt{b}+\sqrt{b}+b\sqrt{b}-b}{1+2b+b^2-b}$

$= \dfrac{b^2+\sqrt{b}}{1+b+b^2}$

98. $\dfrac{1}{\sqrt{y+18}+\sqrt{y}}$

99. $\dfrac{\sqrt{x+6}-5}{\sqrt{x+6}+5} = \dfrac{\sqrt{x+6}-5}{\sqrt{x+6}+5}\cdot\dfrac{\sqrt{x+6}+5}{\sqrt{x+6}+5}$

$= \dfrac{(x+6)-25}{(x+6)+10\sqrt{x+6}+25}$

$= \dfrac{x-19}{x+10\sqrt{x+6}+31}$

100. $\dfrac{-3\sqrt{a^2-3}}{a^2-3}$, or $\dfrac{-3}{\sqrt{a^2-3}}$

101. $5\sqrt{\dfrac{x}{y}}+4\sqrt{\dfrac{y}{x}}-\dfrac{3}{\sqrt{xy}} = \dfrac{5\sqrt{x}}{\sqrt{y}}+\dfrac{4\sqrt{y}}{\sqrt{x}}-\dfrac{3}{\sqrt{xy}} =$

$\dfrac{5\sqrt{x}}{\sqrt{y}}\cdot\dfrac{\sqrt{x}}{\sqrt{x}}+\dfrac{4\sqrt{y}}{\sqrt{x}}\cdot\dfrac{\sqrt{y}}{\sqrt{y}}-\dfrac{3}{\sqrt{xy}} = \dfrac{5x}{\sqrt{xy}}+\dfrac{4y}{\sqrt{xy}}-\dfrac{3}{\sqrt{xy}} =$

$\dfrac{5x+4y-3}{\sqrt{xy}} = \dfrac{5x+4y-3}{\sqrt{xy}}\cdot\dfrac{\sqrt{xy}}{\sqrt{xy}} =$

$\dfrac{(5x+4y-3)\sqrt{xy}}{xy}$

102. $1-\sqrt{w}$

103. $\dfrac{1}{4+\sqrt{3}}+\dfrac{1}{\sqrt{3}}+\dfrac{1}{\sqrt{3}-4} =$

$\dfrac{1}{4+\sqrt{3}}\cdot\dfrac{\sqrt{3}(\sqrt{3}-4)}{\sqrt{3}(\sqrt{3}-4)}+\dfrac{1}{\sqrt{3}}\cdot\dfrac{(4+\sqrt{3})(\sqrt{3}-4)}{(4+\sqrt{3})(\sqrt{3}-4)}+$

$\dfrac{1}{\sqrt{3}-4}\cdot\dfrac{\sqrt{3}(4+\sqrt{3})}{\sqrt{3}(4+\sqrt{3})} =$

$\dfrac{3-4\sqrt{3}-16+3+4\sqrt{3}+3}{\sqrt{3}(4+\sqrt{3})(\sqrt{3}-4)} = \dfrac{-7}{\sqrt{3}(-16+3)} =$

$\dfrac{-7}{-13\sqrt{3}}\cdot\dfrac{\sqrt{3}}{\sqrt{3}} = \dfrac{7\sqrt{3}}{39}$

104. ◈

105. [graphing calculator icon]

Exercise Set 7.6

1. $\sqrt{5x+1} = 6$

$(\sqrt{5x+1})^2 = 6^2$ Principle of powers (squaring)

$5x+1 = 36$

$5x = 35$

$x = 7$

Check: $\sqrt{5x+1} = 6$

$\sqrt{5\cdot 7+1}$? 6

$\sqrt{36}$ |

6 | 6 TRUE

The solution is 7.

2. 33

3. $\sqrt{3x}+1 = 7$

$\sqrt{3x} = 6$ Adding to isolate the radical

$(\sqrt{3x})^2 = 6^2$ Principle of powers (squaring)

$3x = 36$

$x = 12$

Check: $\sqrt{3x}+1 = 7$

$\sqrt{3\cdot 12}+1$? 7

$6+1$ |

7 | 7 TRUE

The solution is 12.

4. 32

5. $\sqrt{y+1}-5 = 8$

$\sqrt{y+1} = 13$ Adding to isolate the radical

$(\sqrt{y+1})^2 = 13^2$ Principle of powers (squaring)

$y+1 = 169$

$y = 168$

Check: $\sqrt{y+1}-5 = 8$

$\sqrt{168+1}-5$? 8

$13-5$ |

8 | 8 TRUE

The solution is 168.

6. 11

7. $\sqrt{x-7}+3 = 10$

$\sqrt{x-7} = 7$ Adding to isolate the radical

$(\sqrt{x-7})^2 = 7^2$ Principle of powers (squaring)

$x-7 = 49$

$x = 56$

Check:
$$\begin{array}{c|c} \sqrt{x-7}+3 = 10 & \\ \hline \sqrt{56-7}+3 \ ? & 10 \\ \sqrt{49}+3 & \\ 7+3 & \\ 10 & 10 \quad \text{TRUE} \end{array}$$

The solution is 56.

8. -3

9.
$$\sqrt[3]{x+5} = 2$$
$$(\sqrt[3]{x+5})^3 = 2^3$$
$$x+5 = 8$$
$$x = 3$$

Check:
$$\begin{array}{c|c} \sqrt[3]{x+5} = 2 & \\ \hline \sqrt[3]{3+5} \ ? & 2 \\ \sqrt[3]{8} & \\ 2 & 2 \quad \text{TRUE} \end{array}$$

The solution is 3.

10. 29

11.
$$\sqrt[4]{y-3} = 2$$
$$(\sqrt[4]{y-3})^4 = 2^4$$
$$y-3 = 16$$
$$y = 19$$

Check:
$$\begin{array}{c|c} \sqrt[4]{y-3} = 2 & \\ \hline \sqrt[4]{19-3} \ ? & 2 \\ \sqrt[4]{16} & \\ 2 & 2 \quad \text{TRUE} \end{array}$$

The solution is 19.

12. 78

13.
$$3\sqrt{x} = x$$
$$(3\sqrt{x})^2 = x^2$$
$$9x = x^2$$
$$0 = x^2 - 9x$$
$$0 = x(x-9)$$
$$x = 0 \ \ or \ \ x = 9$$

Check:

For 0:
$$\begin{array}{c|c} 3\sqrt{x} = x & \\ \hline 3\sqrt{0} \ ? & 0 \\ 3 \cdot 0 & \\ 0 & 0 \quad \text{TRUE} \end{array}$$

For 9:
$$\begin{array}{c|c} 3\sqrt{x} = x & \\ \hline 3\sqrt{9} \ ? & 9 \\ 3 \cdot 3 & \\ 9 & 9 \quad \text{TRUE} \end{array}$$

The solutions are 0 and 9.

14. 0, 64

15.
$$2y^{1/2} - 7 = 9$$
$$2\sqrt{y} - 7 = 9$$
$$2\sqrt{y} = 16$$
$$\sqrt{y} = 8$$
$$(\sqrt{y})^2 = 8^2$$
$$y = 64$$

Check:
$$\begin{array}{c|c} 2y^{1/2} - 7 = 9 & \\ \hline 2 \cdot 64^{1/2} - 7 \ ? & 9 \\ 2 \cdot 8 - 7 & \\ 9 & 9 \quad \text{TRUE} \end{array}$$

The solution is 64.

16. No solution

17.
$$\sqrt[3]{x} = -3$$
$$(\sqrt[3]{x})^3 = (-3)^3$$
$$x = -27$$

Check:
$$\begin{array}{c|c} \sqrt[3]{x} = -3 & \\ \hline \sqrt[3]{-27} \ ? & -3 \\ -3 & -3 \quad \text{TRUE} \end{array}$$

The solution is -27.

18. -64

19.
$$t^{1/3} - 2 = 3$$
$$t^{1/3} = 5$$
$$(t^{1/3})^3 = 5^3 \quad \text{Principle of powers}$$
$$t = 125$$

Check:
$$\begin{array}{c|c} t^{1/3} - 2 = 3 & \\ \hline 125^{1/3} - 2 \ ? & 3 \\ 5 - 2 & \\ 3 & 3 \quad \text{TRUE} \end{array}$$

The solution is 125.

20. 81

21.
$$(x+2)^{1/2} = -4$$
$$\sqrt{x+2} = -4$$

We might observe that this equation has no real-number solution, since the principal square root of a number is never negative. However, we will go through the solution process.

$$(\sqrt{x+2})^2 = (-4)^2$$
$$x+2 = 16$$
$$x = 14$$

Check:
$$\begin{array}{c|c} (x+2)^{1/2} = -4 & \\ \hline (14+2)^{1/2} \ ? & -4 \\ 16^{1/2} & \\ 4 & -4 \quad \text{FALSE} \end{array}$$

The number 14 does not check. The equation has no solution.

22. No solution

23.
$$\sqrt[4]{2x+3}-5=-2$$
$$\sqrt[4]{2x+3}=3$$
$$(\sqrt[4]{2x+3})^4=3^4$$
$$2x+3=81$$
$$2x=78$$
$$x=39$$

Check: $\sqrt[4]{2x+3}-5=-2$

$\sqrt[4]{2\cdot 39+3}-5$?	-2	
$\sqrt[4]{81}-5$		
$3-5$		
-2	-2	TRUE

The solution is 39.

24. $\frac{80}{3}$

25.
$$(y-7)^{1/4}=3$$
$$[(y-7)^{1/4}]^4=3^4$$
$$y-7=81$$
$$y=88$$

Check: $(y-7)^{1/4}=3$

$(88-7)^{1/4}$?	3	
$81^{1/4}$		
3	3	TRUE

The solution is 88.

26. 59

27.
$$\sqrt[3]{6x+9}+8=5$$
$$\sqrt[3]{6x+9}=-3$$
$$(\sqrt[3]{6x+9})^3=(-3)^3$$
$$6x+9=-27$$
$$6x=-36$$
$$x=-6$$

Check: $\sqrt[3]{6x+9}+8=5$

$\sqrt[3]{6(-6)+9}+8$?	5	
$\sqrt[3]{-27}+8$		
5	5	TRUE

The solution is -6.

28. $-\frac{5}{3}$

29.
$$\sqrt{2t-7}=\sqrt{3t-12}$$
$$(\sqrt{2t-7})^2=(\sqrt{3t-12})^2$$
$$2t-7=3t-12$$
$$-7=t-12$$
$$5=t$$

Check: $\sqrt{2t-7}=\sqrt{3t-12}$

$\sqrt{2\cdot 5-7}$?	$\sqrt{3\cdot 5-12}$	
$\sqrt{3}$	$\sqrt{3}$	TRUE

The solution is 5.

30. 1

31.
$$2(1-x)^{1/3}=4^{1/3}$$
$$[2(1-x)^{1/3}]^3=(4^{1/3})^3$$
$$8(1-x)=4$$
$$8-8x=4$$
$$-8x=-4$$
$$x=\frac{1}{2}$$

The number $\frac{1}{2}$ checks and is the solution.

32. $\frac{106}{27}$

33.
$$x=\sqrt{x-1}+3$$
$$x-3=\sqrt{x-1}$$
$$(x-3)^2=(\sqrt{x-1})^2$$
$$x^2-6x+9=x-1$$
$$x^2-7x+10=0$$
$$(x-2)(x-5)=0$$
$$x=2 \quad or \quad x=5$$

Check:

For 2: $x=\sqrt{x-1}+3$

2 ?	$\sqrt{2-1}+3$	
	$\sqrt{1}+3$	
	$1+3$	
2	4	FALSE

For 5: $x=\sqrt{x-1}+3$

5 ?	$\sqrt{5-1}+3$	
	$\sqrt{4}+3$	
	$2+3$	
5	5	TRUE

Since 5 checks but 2 does not, the solution is 5.

34. 4

35.
$$3+\sqrt{z-6}=\sqrt{z+9}$$ One radical is already isolated.
$$(3+\sqrt{z-6})^2=(\sqrt{z+9})^2$$ Squaring both sides
$$9+6\sqrt{z-6}+z-6=z+9$$
$$6\sqrt{z-6}=6$$
$$\sqrt{z-6}=1$$ Multiplying by $\frac{1}{6}$
$$(\sqrt{z-6})^2=1^2$$
$$z-6=1$$
$$z=7$$

The number 7 checks and is the solution.

36. 3, 7

37.
$$\begin{aligned}
\sqrt{20-x}+8 &= \sqrt{9-x}+11 \\
\sqrt{20-x} &= \sqrt{9-x}+3 \quad \text{Isolating one radical} \\
(\sqrt{20-x})^2 &= (\sqrt{9-x}+3)^2 \quad \text{Squaring both sides} \\
20-x &= 9-x+6\sqrt{9-x}+9 \\
2 &= 6\sqrt{9-x} \quad \text{Isolating the remaining radical} \\
1 &= 3\sqrt{9-x} \quad \text{Multiplying by } \frac{1}{2} \\
1^2 &= (3\sqrt{9-x})^2 \quad \text{Squaring both sides} \\
1 &= 9(9-x) \\
1 &= 81-9x \\
-80 &= -9x \\
\frac{80}{9} &= x
\end{aligned}$$

The number $\frac{80}{9}$ checks and is the solution.

38. $\frac{15}{4}$

39.
$$\begin{aligned}
\sqrt{x+2}+\sqrt{3x+4} &= 2 \\
\sqrt{x+2} &= 2-\sqrt{3x+4} \quad \text{Isolating one radical} \\
(\sqrt{x+2})^2 &= (2-\sqrt{3x+4})^2 \\
x+2 &= 4-4\sqrt{3x+4}+3x+4 \\
-2x-6 &= -4\sqrt{3x+4} \quad \text{Isolating the remaining radical} \\
x+3 &= 2\sqrt{3x+4} \quad \text{Multiplying by } -\frac{1}{2} \\
(x+3)^2 &= (2\sqrt{3x+4})^2 \\
x^2+6x+9 &= 4(3x+4) \\
x^2+6x+9 &= 12x+16 \\
x^2-6x-7 &= 0 \\
(x-7)(x+1) &= 0 \\
x-7=0 \quad &\text{or} \quad x+1=0 \\
x=7 \quad &\text{or} \quad x=-1
\end{aligned}$$

Check:

For 7:

$$\begin{array}{r|l}
\multicolumn{2}{c}{\sqrt{x+2}+\sqrt{3x+4}=2} \\
\hline
\sqrt{7+2}+\sqrt{3\cdot 7+4} & ?\ 2 \\
\sqrt{9}+\sqrt{25} & \\
8 & 2 \quad \text{FALSE}
\end{array}$$

For -1:

$$\begin{array}{r|l}
\multicolumn{2}{c}{\sqrt{x+2}+\sqrt{3x+4}=2} \\
\hline
\sqrt{-1+2}+\sqrt{3\cdot(-1)+4} & ?\ 2 \\
\sqrt{1}+\sqrt{1} & \\
2 & 2 \quad \text{TRUE}
\end{array}$$

Since -1 checks but 7 does not, the solution is -1.

40. -1, $\frac{1}{3}$

41. We must have $f(x)=2$, or $\sqrt{x}+\sqrt{x-9}=1$.

$$\begin{aligned}
\sqrt{x}+\sqrt{x-9} &= 1 \\
\sqrt{x-9} &= 1-\sqrt{x} \quad \text{Isolating one radical term} \\
(\sqrt{x-9})^2 &= (1-\sqrt{x})^2 \\
x-9 &= 1-2\sqrt{x}+x \\
-10 &= -2\sqrt{x} \quad \text{Isolating the remaining radical term} \\
5 &= \sqrt{x} \\
25 &= x
\end{aligned}$$

This value does not check. There is no solution, so there is no value of x for which $f(x)=1$.

42. 9

43. We must have $g(a)=-1$, or $\sqrt{2a+7}-\sqrt{a+15}=-1$.

$$\begin{aligned}
\sqrt{2a+7}-\sqrt{a+15} &= -1 \\
\sqrt{2a+7} &= \sqrt{a+15}-1 \quad \text{Isolating one radical term} \\
(\sqrt{2a+7})^2 &= (\sqrt{a+15}-1)^2 \\
2a+7 &= a+15-2\sqrt{a+15}+1 \\
a-9 &= -2\sqrt{a+15} \quad \text{Isolating the remaining radical} \\
(a-9)^2 &= (-2\sqrt{a+15})^2 \\
a^2-18a+81 &= 4(a+15) \\
a^2-18a+81 &= 4a+60 \\
a^2-22a+21 &= 0 \\
(a-1)(a-21) &= 0 \\
a=1 \quad &\textit{or} \quad a=21
\end{aligned}$$

Since 1 checks but 21 does not, we have $g(a)=-1$ when $a=1$.

44. 2, 6

45. We must have $\sqrt{2x-3}=\sqrt{x+7}-2$.

$$\begin{aligned}
\sqrt{2x-3} &= \sqrt{x+7}-2 \\
(\sqrt{2x-3})^2 &= (\sqrt{x+7}-2)^2 \\
2x-3 &= x+7-4\sqrt{x+7}+4 \\
x-14 &= -4\sqrt{x+7} \\
(x-14)^2 &= (-4\sqrt{x+7})^2 \\
x^2-28x+196 &= 16(x+7) \\
x^2-28x+196 &= 16x+112 \\
x^2-44x+84 &= 0 \\
(x-2)(x-42) &= 0 \\
x=2 \quad &\textit{or} \quad x=42
\end{aligned}$$

Since 2 checks but 42 does not, we have $f(x)=g(x)$ when $x=2$.

46. 10

47. We must have $4-\sqrt{a-3}=(a+5)^{1/2}$.

$$\begin{aligned} 4-\sqrt{a-3} &= (a+5)^{1/2} \\ (4-\sqrt{a-3})^2 &= [(a+5)^{1/2}]^2 \\ 16-8\sqrt{a-3}+a-3 &= a+5 \\ -8\sqrt{a-3} &= -8 \\ \sqrt{a-3} &= 1 \\ (\sqrt{a-3})^2 &= 1^2 \\ a-3 &= 1 \\ a &= 4 \end{aligned}$$

The number 4 checks, so we have $f(a)=g(a)$ when $a=4$.

48. 15

49. $$\frac{3}{2x}+\frac{1}{x}=\frac{2x+3.5}{3x} \qquad \text{LCD is } 6x$$

Note that $x \neq 0$.

$$\begin{aligned} 6x\left(\frac{3}{2x}+\frac{1}{x}\right) &= 6x\left(\frac{2x+3.5}{3x}\right) \\ 9+6 &= 4x+7 \\ 8 &= 4x \\ 2 &= x \end{aligned}$$

The number 2 checks and is the solution.

50. Height: 7 in., base: 9 in.

51. Graph: $f(x)=\frac{2}{5}x-7$

The y-intercept is $(0,-7)$, and the slope is $\frac{2}{5}$. From the y-intercept we go up 2 units and to the right 5 units to the point $(5,-5)$. Knowing two points, we can draw the graph.

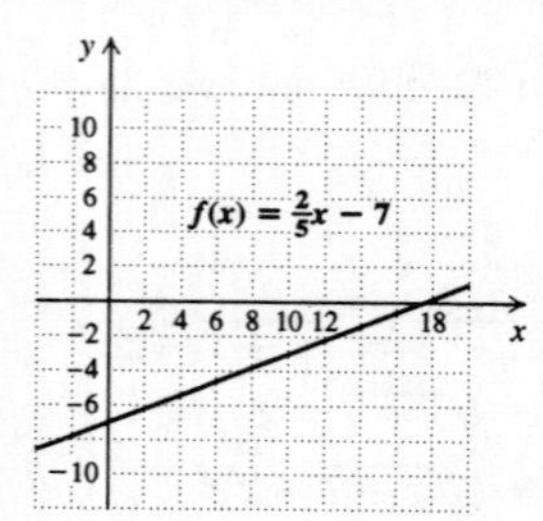

52.

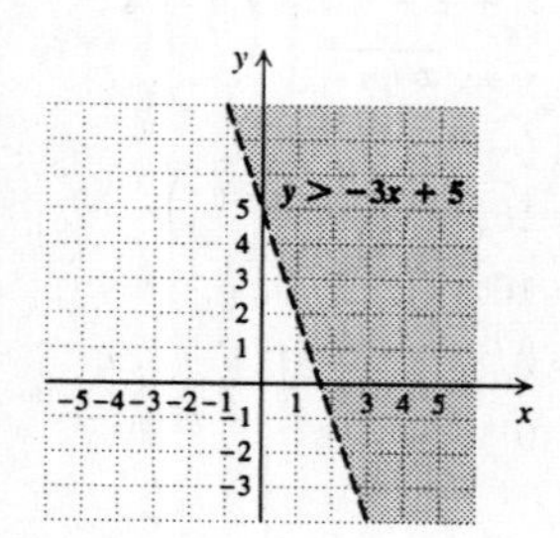

53. ◈

54. ◈

55. ◈

56. ◈

57.

$$\begin{aligned} v &= \sqrt{2gr}\sqrt{\frac{h}{r+h}} \\ v^2 &= 2gr\cdot\frac{h}{r+h} && \text{Squaring both sides} \\ v^2(r+h) &= 2grh && \text{Multiplying by } r+h \\ v^2r+v^2h &= 2grh \\ v^2r &= 2grh-v^2h \\ v^2r &= h(2gr-v^2) \\ \frac{v^2r}{2gr-v^2} &= h \end{aligned}$$

58. $r=\dfrac{v^2h}{2gh-v^2}$

59.

$$\begin{aligned} D(h) &= 1.2\sqrt{h} \\ 180 &= 1.2\sqrt{h} \\ (180)^2 &= (1.2\sqrt{h})^2 \\ 32,400 &= 1.44h \\ 22,500 &= h \end{aligned}$$

The pilot must fly 22,500 ft above sea level.

60. 72.25 ft

61.

$$\begin{aligned} \frac{x+\sqrt{x+1}}{x-\sqrt{x+1}} &= \frac{5}{11} \\ 11(x+\sqrt{x+1}) &= 5(x-\sqrt{x+1}) \\ 11x+11\sqrt{x+1} &= 5x-5\sqrt{x+1} \\ 16\sqrt{x+1} &= -6x \\ 8\sqrt{x+1} &= -3x \\ (8\sqrt{x+1})^2 &= (-3x)^2 \\ 64(x+1) &= 9x^2 \\ 64x+64 &= 9x^2 \\ 0 &= 9x^2-64x-64 \\ 0 &= (9x+8)(x-8) \end{aligned}$$

$$\begin{aligned} 9x+8=0 \quad &\text{or} \quad x-8=0 \\ 9x=-8 \quad &\text{or} \quad x=8 \\ x=-\frac{8}{9} \quad &\text{or} \quad x=8 \end{aligned}$$

Since $-\frac{8}{9}$ checks but 8 does not, the solution is $-\frac{8}{9}$.

62. $\dfrac{2504}{125}, \dfrac{2496}{125}$

63.
$$(z^2+17)^{3/4} = 27$$
$$[(z^2+17)^{3/4}]^{4/3} = (3^3)^{4/3}$$
$$z^2+17 = 3^4$$
$$z^2+17 = 81$$
$$z^2-64 = 0$$
$$(z+8)(z-8) = 0$$
$$z=-8 \ \ or \ \ z=8$$

Both -8 and 8 check. They are the solutions.

64. 0

65.
$$x^2-5x-\sqrt{x^2-5x-2} = 4$$
$$x^2-5x-\sqrt{x^2-5x-2} = 2+2$$
$$x^2-5x-2-\sqrt{x^2-5x-2}-2 = 0$$

Let $u=\sqrt{x^2-5x-2}$.
$$u^2-u-2=0$$
$$(u+1)(u-2)=0$$
$$u=-1 \ \ or \ \ u=2$$

Now we replace u with $\sqrt{x^2-5x-2}$.
$$\sqrt{x^2-5x-2}=-1 \ \ or \ \ \sqrt{x^2-5x-2}=2$$
$$\text{No solution} \qquad (\sqrt{x^2-5x-2})^2 = 2^2$$
$$x^2-5x-2 = 4$$
$$x^2-5x-6 = 0$$
$$(x-6)(x+1) = 0$$
$$x=6 \ \ or \ \ x=-1$$

Both 6 and -1 check. They are the solutions.

66. 1, 8

67. We find the values of x for which $f(x)=0$.
$$\sqrt{x-2}-\sqrt{x+2}+2 = 0$$
$$\sqrt{x-2}+2 = \sqrt{x+2}$$
$$(\sqrt{x-2}+2)^2 = (\sqrt{x+2})^2$$
$$x-2+4\sqrt{x-2}+4 = x+2$$
$$4\sqrt{x-2} = 0$$
$$\sqrt{x-2} = 0$$
$$(\sqrt{x-2})^2 = 0^2$$
$$x-2 = 0$$
$$x = 2$$

The number 2 checks. The x-intercept is $(2,0)$.

68. $\left(\dfrac{1}{36},0\right)$, $(36,0)$

69. We find the values of x for which $f(x)=0$.
$$(x^2+30x)^{1/2}-x-(5x)^{1/2} = 0$$
$$\sqrt{x^2+30x}-x-\sqrt{5x} = 0$$
$$\sqrt{x^2+30x}-x = \sqrt{5x}$$
$$(\sqrt{x^2+30x}-x)^2 = (\sqrt{5x})^2$$
$$x^2+30x-2x\sqrt{x^2+30x}+x^2 = 5x$$
$$2x^2+25x = 2x\sqrt{x^2+30x}$$
$$(2x^2+25x)^2 = (2x\sqrt{x^2+30x})^2$$
$$4x^4+100x^3+625x^2 = 4x^2(x^2+30x)$$
$$4x^4+100x^3+625x^2 = 4x^4+120x^3$$
$$-20x^3+625x^2 = 0$$
$$-5x^2(4x-125) = 0$$
$$x=0 \text{ or } x=\frac{125}{4}$$

Both 0 and $\dfrac{125}{4}$ check. The x-intercepts are $(0,0)$ and $\left(\dfrac{125}{4},0\right)$.

70.

71.

72.

Exercise Set 7.7

1. $a=5, \ \ b=3$

Find c.

$c^2 = a^2+b^2$	Pythagorean equation
$c^2 = 5^2+3^2$	Substituting
$c^2 = 25+9$	
$c^2 = 34$	
$c = \sqrt{34}$	Exact answer
$c \approx 5.831$	Approximation

2. $\sqrt{164}$; 12.806

3. $a=7, \ \ b=7$

Find c.

$c^2 = a^2+b^2$	Pythagorean equation
$c^2 = 7^2+7^2$	Substituting
$c^2 = 49+49$	
$c^2 = 98$	
$c = \sqrt{98}$	Exact answer
$c \approx 9.899$	Approximation

4. $\sqrt{200}$; 14.142

5. $b = 12, \quad c = 13$

Find a.

$$\begin{aligned} a^2 + b^2 &= c^2 && \text{Pythagorean equation} \\ a^2 + 12^2 &= 13^2 && \text{Substituting} \\ a^2 + 144 &= 169 \\ a^2 &= 25 \\ a &= 5 \end{aligned}$$

6. $\sqrt{119}$; 10.909

7. $c = 6, \quad a = \sqrt{5}$

Find b.

$$\begin{aligned} c^2 &= a^2 + b^2 \\ (\sqrt{5})^2 + b^2 &= 6^2 \\ 5 + b^2 &= 36 \\ b^2 &= 31 \\ b &= \sqrt{31} && \text{Exact answer} \\ b &\approx 5.568 && \text{Approximation} \end{aligned}$$

8. 4

9. $b = 1, \quad c = \sqrt{13}$

Find a.

$$\begin{aligned} a^2 + b^2 &= c^2 && \text{Pythagorean equation} \\ a^2 + 1^2 &= (\sqrt{13})^2 && \text{Substituting} \\ a^2 + 1 &= 13 \\ a^2 &= 12 \\ a &= \sqrt{12} && \text{Exact answer} \\ a &\approx 3.464 && \text{Approximation} \end{aligned}$$

10. $\sqrt{19}$; 4.359

11. $a = 1, \quad c = \sqrt{n}$

Find b.

$$\begin{aligned} a^2 + b^2 &= c^2 \\ 1^2 + b^2 &= (\sqrt{n})^2 \\ 1 + b^2 &= n \\ b^2 &= n - 1 \\ b &= \sqrt{n-1} \end{aligned}$$

12. $\sqrt{4-n}$

13. We make a drawing and let d = the length of the guy wire.

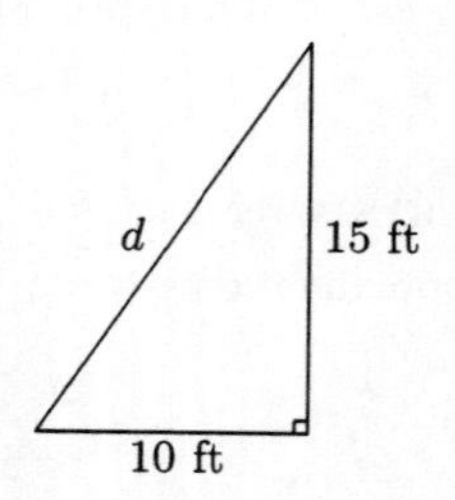

We use the Pythagorean equation to find d.

$$\begin{aligned} d^2 &= 10^2 + 15^2 \\ d^2 &= 100 + 225 \\ d^2 &= 325 \\ d &= \sqrt{325} \\ d &\approx 18.028 \end{aligned}$$

The wire is $\sqrt{325}$, or about 18.028 ft long.

14. $\sqrt{8450}$ ft; 91.924 ft

15. We first make a drawing and let d = the distance, in feet, to second base. A right triangle is formed in which the length of the leg from second base to third base is 90 ft. The length of the leg from third base to where the catcher fields the ball is 90 − 10, or 80 ft.

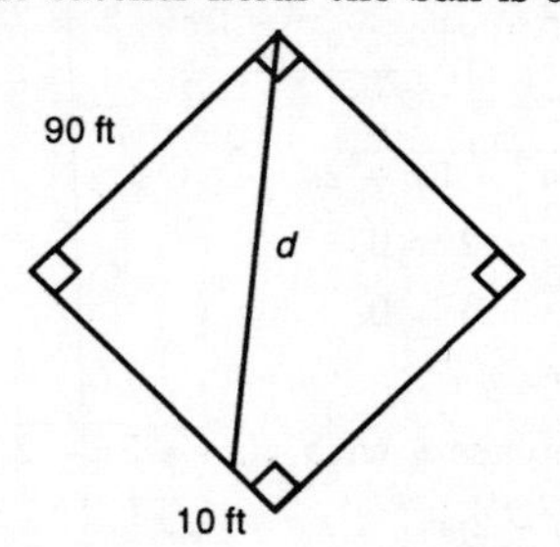

We substitute these values into the Pythagorean equation to find d.

$$\begin{aligned} d^2 &= 90^2 + 80^2 \\ d^2 &= 8100 + 6400 \\ d^2 &= 14,500 \\ d &= \sqrt{14,500} \end{aligned}$$

Exact answer: $d = \sqrt{14,500}$ ft

Approximation: $d \approx 120.416$ ft

16. 12 in.

17. We make a drawing.

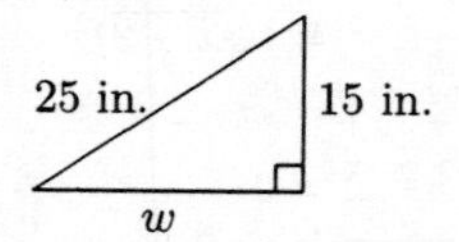

We use the Pythagorean equation to find w.

$$\begin{aligned} w^2 + 15^2 &= 25^2 \\ w^2 + 225 &= 625 \\ w^2 &= 400 \\ w &= 20 \end{aligned}$$

The width is 20 in.

18. $\sqrt{340} + 8$ ft; 26.439 ft

19.

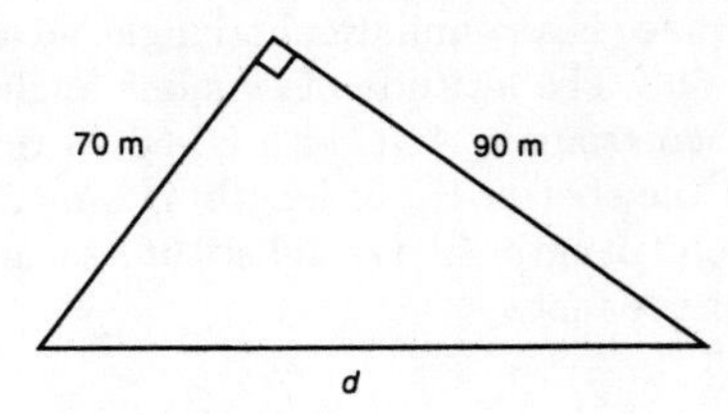

$$d^2 = 70^2 + 90^2$$
$$d^2 = 4900 + 8100$$
$$d^2 = 13,000$$
$$d = \sqrt{13,000} \text{ m} \quad \text{Exact answer}$$
$$d \approx 114.018 \text{ m} \quad \text{Approximation}$$

20. 50 ft

21. Since one acute angle is 45°, this is an isosceles right triangle with $b = 5$. Then $a = 5$ also. We substitute to find c.

$$c = a\sqrt{2}$$
$$c = 5\sqrt{2}$$

Exact answer: $a = 5,\ c = 5\sqrt{2}$

Approximation: $c \approx 7.071$

22. $a = 14;\ c = 14\sqrt{2} \approx 19.799$

23. This is a 30-60-90 right triangle with $c = 14$. We substitute to find a and b.

$$c = 2a$$
$$14 = 2a$$
$$7 = a$$

$$b = a\sqrt{3}$$
$$b = 7\sqrt{3}$$

Exact answer: $a = 7,\ b = 7\sqrt{3}$

Approximation: $b \approx 12.124$

24. $a = 9;\ b = 9\sqrt{3} \approx 15.588$

25. This is a 30-60-90 right triangle with $b = 15$. We substitute to find a and c.

$$b = a\sqrt{3}$$
$$15 = a\sqrt{3}$$
$$\frac{15}{\sqrt{3}} = a$$
$$\frac{15\sqrt{3}}{3} = a \quad \text{Rationalizing the denominator}$$
$$5\sqrt{3} = a \quad \text{Simplifying}$$
$$c = 2a$$
$$c = 2 \cdot 5\sqrt{3}$$
$$c = 10\sqrt{3}$$

Exact answer: $a = 5\sqrt{3},\ c = 10\sqrt{3}$

Approximations: $a \approx 8.660,\ c \approx 17.321$

26. $a = 4\sqrt{2} \approx 5.657;\ b = 4\sqrt{2} \approx 5.657$

27. This is an isosceles right triangle with $c = 13$. We substitute to find a.

$$a = \frac{c\sqrt{2}}{2}$$
$$a = \frac{13\sqrt{2}}{2}$$

Since $a = b$, we have $b = \dfrac{13\sqrt{2}}{2}$ also.

Exact answer: $a = \dfrac{13\sqrt{2}}{2},\ b = \dfrac{13\sqrt{2}}{2}$

Approximations: $a \approx 9.192,\ b \approx 9.192$

28. $a = \dfrac{7}{\sqrt{3}} \approx 4.041;\ c = \dfrac{14\sqrt{3}}{3} \approx 8.083$

29. This is a 30-60-90 triangle with $a = 14$. We substitute to find b and c.

$$b = a\sqrt{3} \qquad c = 2a$$
$$b = 14\sqrt{3} \qquad c = 2 \cdot 14$$
$$c = 28$$

Exact answer: $b = 14\sqrt{3},\ c = 28$

Approximation: $b \approx 24.249$

30. $b = 9\sqrt{3} \approx 15.588;\ c = 18$

31.

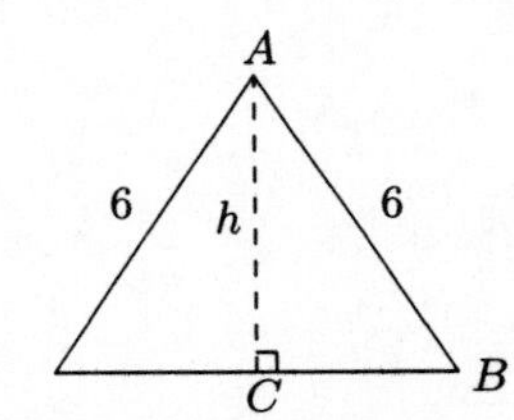

This is an equilateral triangle, so all the angles are 60°. The altitude bisects one angle and one side. Then triangle ABC is a 30-60-90 right triangle with the shorter leg of length 6/2, or 3, and hypotenuse of length 6. We substitute to find the length of the other leg.

$$b = a\sqrt{3}$$
$$h = 3\sqrt{3} \quad \text{Substituting } h \text{ for } b \text{ and 3 for } a$$

Exact answer: $h = 3\sqrt{3}$

Approximation: $h \approx 5.196$

32. $5\sqrt{3} \approx 8.660$

33.

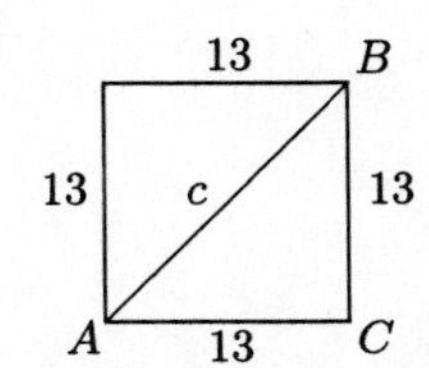

Triangle ABC is an isosceles right triangle with $a = 13$. We substitute to find c.

$$c = a\sqrt{2}$$
$$c = 13\sqrt{2}$$

Exact answer: $c = 13\sqrt{2}$

Approximation: $c \approx 18.385$

34. $7\sqrt{2} \approx 9.899$

35.

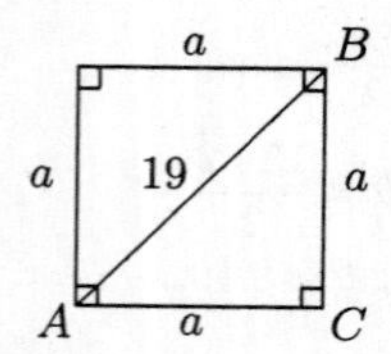

Triangle ABC is an isosceles right triangle with $c = 19$. We substitute to find a.

$$a = \frac{c\sqrt{2}}{2}$$
$$a = \frac{19\sqrt{2}}{2}$$

Exact answer: $a = \dfrac{19\sqrt{2}}{2}$

Approximation: $a \approx 13.435$

36. $\dfrac{15\sqrt{2}}{2} \approx 10.607$

37. We will express all distances in feet. Recall that 1 mi = 5280 ft.

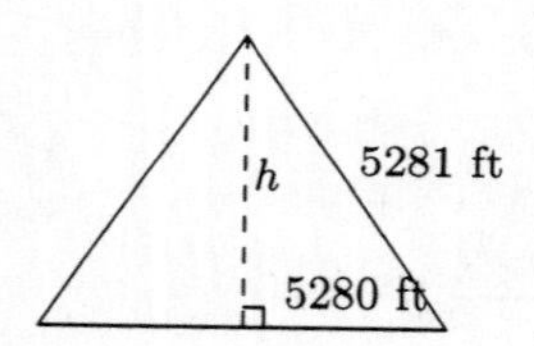

We use the Pythagorean equation to find h.

$$h^2 + (5280)^2 = (5281)^2$$
$$h^2 + 27,878,400 = 27,888,961$$
$$h^2 = 10,561$$
$$h = \sqrt{10,561}$$
$$h \approx 102.767$$

The height of the bulge is $\sqrt{10,561}$ ft, or about 102.767 ft.

38. Neither; they have the same area, 300 ft^2.

39.

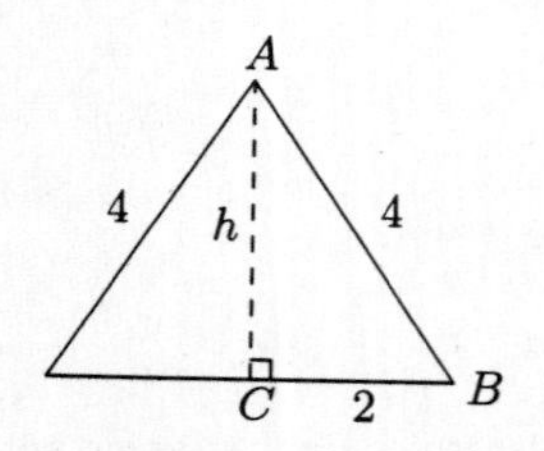

The entrance is an equilateral triangle, so all the angles are 60°. The altitude bisects one angle and one side. Then triangle ABC is a 30-60-90 right triangle with the shorter leg of length 4/2, or 2, and hypotenuse of length 4. We substitute to find h, the height of the tent.

$$b = a\sqrt{3}$$
$$h = 2\sqrt{3} \quad \text{Substituting } h \text{ for } b \text{ and 2 for } a$$

Exact answer: $h = 2\sqrt{3}$ ft

Approximation: $h \approx 3.464$ ft

40. $d = s + s\sqrt{2}$

41.

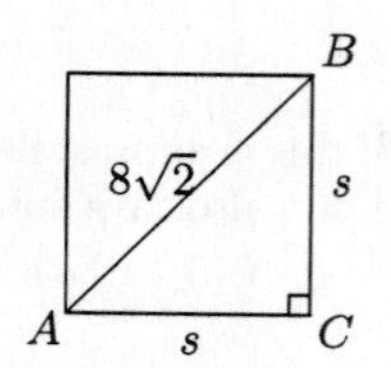

Triangle ABC is an isosceles right triangle with $c = 8\sqrt{2}$. We substitute to find a.

$$a = \frac{c\sqrt{2}}{2} = \frac{8\sqrt{2} \cdot \sqrt{2}}{2} = \frac{8 \cdot 2}{2} = 8$$

The length of a side of the square is 8 ft.

42. $\sqrt{181}$, cm; 13.454 cm

43.

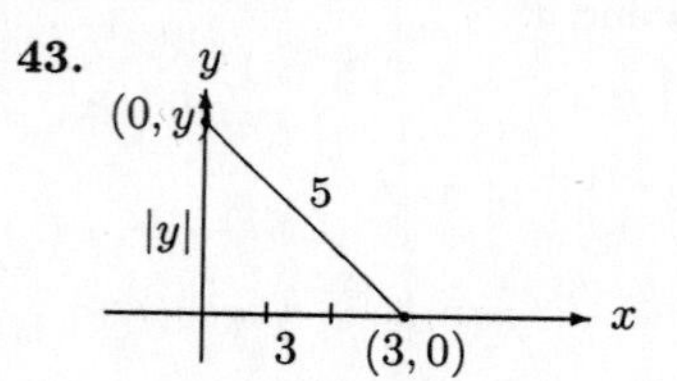

$$|y|^2 + 3^2 = 5^2$$
$$y^2 + 9 = 25$$
$$y^2 = 16$$
$$y = \pm 4$$

The points are $(0, -4)$ and $(0, 4)$.

44. $(-3, 0)$ and $(3, 0)$

45.
$$x^2 - 11x + 24 = 0$$
$$(x - 8)(x - 3) = 0$$
$$x - 8 = 0 \quad \text{or} \quad x - 3 = 0$$
$$x = 8 \quad \text{or} \quad x = 3$$

The solutions are 8 and 3.

46. $-7, \dfrac{3}{2}$

47. $|3x-5|=7$

$$3x-5=7 \quad or \quad 3x-5=-7$$
$$3x=12 \quad or \quad 3x=-2$$
$$x=4 \quad or \quad x=-\frac{2}{3}$$

The solution set is $\left\{4,-\frac{2}{3}\right\}$.

48. $\left\{-\frac{4}{3},10\right\}$.

49. ◈

50. ◈

51.

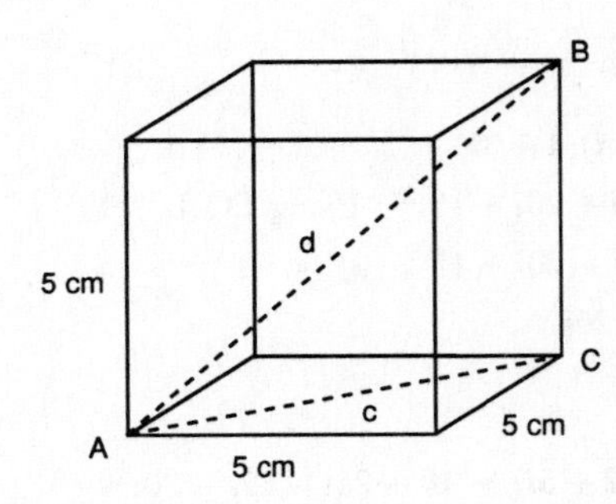

First find the length of a diagonal of the base of the cube. It is the hypotenuse of an isosceles right triangle with $a=5$ cm. Then $c=a\sqrt{2}=5\sqrt{2}$ cm.

Triangle ABC is a right triangle with legs of $5\sqrt{2}$ cm and 5 cm and hypotenuse d. Use the Pythagorean equation to find d, the length of the diagonal that connects two opposite corners of the cube.

$$d^2=(5\sqrt{2})^2+5^2$$
$$d^2=25\cdot 2+25$$
$$d^2=50+25$$
$$d^2=75$$
$$d=\sqrt{75}$$

Exact answer: $d=\sqrt{75}$ cm

52. 9

53.

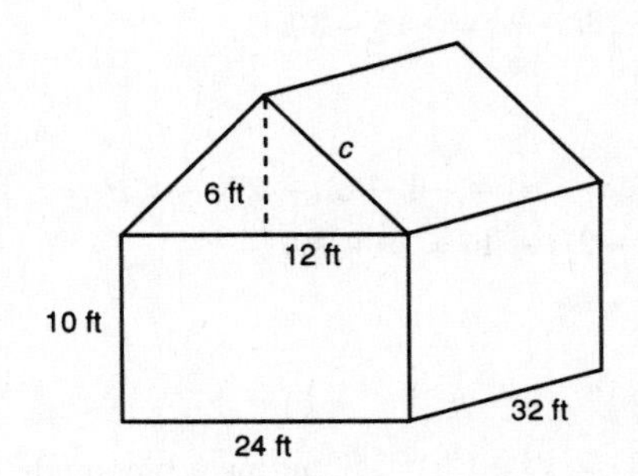

The area to be painted consists of two 10 ft by 24 ft rectangles, two 10 ft by 32 ft rectangles, and two triangles with height 6 ft and base 24 ft. The area of the two 10 ft by 24 ft rectangle is $2\cdot 10\text{ ft}\cdot 24\text{ ft}=480\text{ ft}^2$.

The area of the two 10 ft by 32 ft rectangles is $2\cdot 10\text{ ft}\cdot 32\text{ ft}=640\text{ ft}^2$. The area of the two triangles is $2\cdot\frac{1}{2}\cdot 24\text{ ft}\cdot 6\text{ ft}=144\text{ ft}^2$. Thus, the total area to be painted is $480\text{ ft}^2+640\text{ ft}^2+144\text{ ft}^2=1264\text{ ft}^2$.

One gallon of paint covers 275 ft^2, so we divide to determine how many gallons of paint are required: $\frac{1264}{275}\approx 4.6$. Thus, 4 gallons of paint should be bought to paint the house. This answer assumes that the total area of the doors and windows is 164 ft^2 ($4\cdot 275=1100$ and $1264=1100+164$).

Exercise Set 7.8

1. $\sqrt{-25}=\sqrt{-1\cdot 25}=\sqrt{-1}\cdot\sqrt{25}=i\cdot 5=5i$

2. $6i$

3. $\sqrt{-13}=\sqrt{-1\cdot 13}=\sqrt{-1}\cdot\sqrt{13}=i\sqrt{13}$, or $\sqrt{13}i$

4. $i\sqrt{19}$, or $\sqrt{19}i$

5. $\sqrt{-18}=\sqrt{-1}\cdot\sqrt{9}\cdot\sqrt{2}=i\cdot 3\cdot\sqrt{2}=3i\sqrt{2}$, or $3\sqrt{2}i$

6. $7i\sqrt{2}$, or $7\sqrt{2}i$

7. $\sqrt{-3}=\sqrt{-1\cdot 3}=\sqrt{-1}\cdot\sqrt{3}=i\sqrt{3}$, or $\sqrt{3}i$

8. $2i$

9. $\sqrt{-81}=\sqrt{-1\cdot 81}=\sqrt{-1}\cdot\sqrt{81}=i\cdot 9=9i$

10. $3i\sqrt{3}$, or $3\sqrt{3}i$

11. $\sqrt{-300}=\sqrt{-1}\cdot\sqrt{100}\cdot\sqrt{3}=i\cdot 10\cdot\sqrt{3}=10i\sqrt{3}$, or $10\sqrt{3}i$

12. $-5i\sqrt{3}$, or $-5\sqrt{3}i$

13. $-\sqrt{-49}=-\sqrt{-1\cdot 49}=-\sqrt{-1}\cdot\sqrt{49}=-i\cdot 7=-7i$

14. $-5i\sqrt{5}$, or $-5\sqrt{5}i$

15. $4-\sqrt{-60}=4-\sqrt{-1\cdot 60}=4-\sqrt{-1}\cdot\sqrt{60}=$ $4-i\cdot 2\sqrt{15}=4-2\sqrt{15}i$, or $4-2i\sqrt{15}$

16. $6-2i\sqrt{21}$, or $6-2\sqrt{21}i$

17. $\sqrt{-4}+\sqrt{-12}=\sqrt{-1\cdot 4}+\sqrt{-1\cdot 12}=$ $\sqrt{-1}\cdot\sqrt{4}+\sqrt{-1}\cdot\sqrt{12}=i\cdot 2+i\cdot 2\sqrt{3}=$ $(2+2\sqrt{3})i$

18. $(-2\sqrt{19}+5\sqrt{5})i$

19. $(4+7i)+(5-2i)$

$=(4+5)+(7-2)i$ Combining the real and the imaginary parts

$=9+5i$

20. $12+i$

21. $(-2+8i)+(5+3i)$
$= (-2+5)+(8+3)i$ Combining the real and the imaginary parts
$= 3+11i$

22. $7+4i$

23. $(9+8i)-(5+3i) = (9-5)+(8-3)i$
$= 4+5i$

24. $7+3i$

25. $(8-3i)-(9+2i) = (8-9)+(-3-2)i$
$= -1-5i$

26. $2-i$

27. $(-2+6i)-(-7+i) = -2-(-7)+(6-1)i$
$= 5+5i$

28. $-12-5i$

29. $6i\cdot 5i = 30\cdot i^2$
$= 30\cdot(-1)$ $i^2=-1$
$= -30$

30. -42

31. $7i\cdot(-9i) = -63\cdot i^2$
$= -63\cdot(-1)$ $i^2=-1$
$= 63$

32. -24

33. $\sqrt{-49}\sqrt{-25} = \sqrt{-1}\cdot\sqrt{49}\cdot\sqrt{-1}\cdot\sqrt{25}$
$= i\cdot 7\cdot i\cdot 5$
$= i^2\cdot 35$
$= -1\cdot 35$
$= -35$

34. -18

35. $\sqrt{-6}\sqrt{-7} = \sqrt{-1}\cdot\sqrt{6}\cdot\sqrt{-1}\cdot\sqrt{7}$
$= i\cdot\sqrt{6}\cdot i\cdot\sqrt{7}$
$= i^2\cdot\sqrt{42}$
$= -1\cdot\sqrt{42}$
$= -\sqrt{42}$

36. $-\sqrt{10}$

37. $\sqrt{-15}\sqrt{-10} = \sqrt{-1}\cdot\sqrt{15}\cdot\sqrt{-1}\cdot\sqrt{10}$
$= i\cdot\sqrt{15}\cdot i\cdot\sqrt{10}$
$= i^2\cdot\sqrt{150}$
$= -\sqrt{25\cdot 6}$
$= -5\sqrt{6}$

38. $-3\sqrt{14}$

39. $2i(7+3i)$
$= 2i\cdot 7+2i\cdot 3i$ Using the distributive law
$= 14i+6i^2$
$= 14i-6$ $i^2=-1$
$= -6+14i$

40. $-30+10i$

41. $-4i(6-5i) = -4i\cdot 6-4i(-5i)$
$= -24i+20i^2$
$= -24i-20$
$= -20-24i$

42. $-28-21i$

43. $(2+5i)(4+3i)$
$= 8+6i+20i+15i^2$ Using FOIL
$= 8+6i+20i-15$ $i^2=-1$
$= -7+26i$

44. $1+5i$

45. $(5-6i)(2+5i) = 10+25i-12i-30i^2$
$= 10+25i-12i+30$
$= 40+13i$

46. $38+9i$

47. $(-4+5i)(3-4i) = -12+16i+15i-20i^2$
$= -12+16i+15i+20$
$= 8+31i$

48. $2-46i$

49. $(7-3i)(4-7i) = 28-49i-12i+21i^2 =$
$28-49i-12i-21 = 7-61i$

50. $5-37i$

51. $(-3+6i)(-3+4i) = 9-12i-18i+24i^2 =$
$9-12i-18i-24 = -15-30i$

52. $(-11-16i$

53. $(2+9i)(-3-5i) = -6-10i-27i-45i^2 =$
$-6-10i-27i+45 = 39-37i$

54. $13-47i$

55. $(5-2i)^2$
$= 5^2-2\cdot 5\cdot 2i+(2i)^2$ Squaring a binomial
$= 25-20i+4i^2$
$= 25-20i-4$ $i^2=-1$
$= 21-20i$

56. $5 - 12i$

57. $(4+2i)^2$

$= 4^2 + 2\cdot 4\cdot 2i + (2i)^2$ Squaring a binomial

$= 16 + 16i + 4i^2$

$= 16 + 16i - 4$ $\quad i^2 = -1$

$= 12 + 16i$

58. $-5 + 12i$

59. $(-5-2i)^2 = 25 + 20i + 4i^2 = 25 + 20i - 4 = 21 + 20i$

60. $-5 - 12i$

61. $\frac{7}{2-i}$

$= \frac{7}{2-i}\cdot\frac{2+i}{2+i}$ Multiplying by 1, using the conjugate

$= \frac{14+7i}{4-i^2}$ Multiplying

$= \frac{14+7i}{4-(-1)}$ $\quad i^2 = -1$

$= \frac{14+7i}{5}$

$= \frac{14}{5} + \frac{7}{5}i$

62. $\frac{6}{5} - \frac{2}{5}i$

63. $\frac{3i}{5+2i}$

$= \frac{3i}{5+2i}\cdot\frac{5-2i}{5-2i}$ Multiplying by 1, using the conjugate

$= \frac{15i - 6i^2}{25 - 4i^2}$ Multiplying

$= \frac{15i+6}{25+4}$

$= \frac{15i+6}{29}$

$= \frac{6}{29} + \frac{15}{29}i$

64. $-\frac{6}{17} + \frac{10}{17}i$

65. $\frac{8}{9i} = \frac{8}{9i}\cdot\frac{-9i}{-9i} = \frac{-72i}{-81i^2} = \frac{-72i}{81} = -\frac{8}{9}i$

66. $-\frac{5}{8}i$

67. $\frac{7-2i}{6i} = \frac{7-2i}{6i}\cdot\frac{-6i}{-6i} = \frac{-42i+12i^2}{-36i^2} = \frac{-42i-12}{36} = -\frac{12}{36} - \frac{42}{36}i = -\frac{1}{3} - \frac{7}{6}i$

68. $\frac{8}{9} - \frac{1}{3}i$

69. $\frac{4+5i}{3-7i} = \frac{4+5i}{3-7i}\cdot\frac{3+7i}{3+7i} = \frac{12+28i+15i+35i^2}{9-49i^2} = \frac{12+28i+15i-35}{9+49} = \frac{-23+43i}{58} = -\frac{23}{58} + \frac{43}{58}i$

70. $\frac{23}{65} + \frac{41}{65}i$

71. $\frac{3-2i}{4+3i} = \frac{3-2i}{4+3i}\cdot\frac{4-3i}{4-3i} = \frac{12-9i-8i+6i^2}{16-9i^2} = \frac{12-9i-8i-6}{16+9} = \frac{6-17i}{25} = \frac{6}{25} - \frac{17}{25}i$

72. $\frac{1}{15} - \frac{4}{5}i$

73. $i^7 = i^6\cdot i = (i^2)^3\cdot i = (-1)^3\cdot i = -1\cdot i = -i$

74. $-i$

75. $i^{24} = (i^2)^{12} = (-1)^{12} = 1$

76. i

77. $i^{42} = (i^2)^{21} = (-1)^{21} = -1$

78. 1

79. $i^9 = (i^2)^4\cdot i = (-1)^4\cdot i = 1\cdot i = i$

80. i

81. $i^6 = (i^2)^3 = (-1)^3 = -1$

82. 1

83. $(5i)^3 = 5^3\cdot i^3 = 125\cdot i^2\cdot i = 125(-1)(i) = -125i$

84. $-243i$

85. $i^2 + i^4 = -1 + (i^2)^2 = -1 + (-1)^2 = -1 + 1 = 0$

86. i

87. $i^5 + i^7 = i^4\cdot i + i^6\cdot i = (i^2)^2\cdot i + (i^2)^3\cdot i = (-1)^2\cdot i + (-1)^3\cdot i = 1\cdot i + (-1)i = i - i = 0$

88. 0

89. $\dfrac{196}{x^2-7x+49} - \dfrac{2x}{x+7} = \dfrac{2058}{x^3+343}$

Note: $x^3+343 = (x+7)(x^2-7x+49)$.

The LCD $= (x+7)(x^2-7x+49)$.

Note that $x \neq -7$.

$$(x+7)(x^2-7x+49)\left(\frac{196}{x^2-7x+49} - \frac{2x}{x+7}\right) = (x+7)(x^2-7x+49)\cdot\frac{2058}{x^3+343}$$

$$196(x+7) - 2x(x^2-7x+49) = 2058$$

$$196x + 1372 - 2x^3 + 14x^2 - 98x = 2058$$

$$98x - 686 - 2x^3 + 14x^2 = 0$$

$$49x - 343 - x^3 + 7x^2 = 0 \quad \text{Dividing by 2}$$

$$49(x-7) - x^2(x-7) = 0$$

$$(49-x^2)(x-7) = 0$$

$$(7-x)(7+x)(x-7) = 0$$

$7-x=0$ or $7+x=0$ or $x-7=0$

$7=x$ or $x=-7$ or $x=7$

As noted above, -7 cannot be a solution. The number 7 checks. It is the solution.

90. $\dfrac{70}{29}$

91. $28 = 3x^2 - 17x$

$0 = 3x^2 - 17x - 28$

$0 = (3x+4)(x-7)$

$3x+4=0$ or $x-7=0$

$3x=-4$ or $x=7$

$x=-\dfrac{4}{3}$ or $x=7$

Both values check. The solutions are $-\dfrac{4}{3}$ and 7.

92. $\left\{x \middle| -\dfrac{29}{3} < x < 5\right\}$, or $\left(-\dfrac{29}{3}, 5\right)$

93. ◈

94. ◈

95. ◈

96. ◈

97. $g(2i) = \dfrac{(2i)^4-(2i)^2}{2i-1} = \dfrac{16i^4-4i^2}{-1+2i} = \dfrac{20}{-1+2i} = \dfrac{20}{-1+2i}\cdot\dfrac{-1-2i}{-1-2i} = \dfrac{-20-40i}{5} = -4-8i;$

$g(i+1) = \dfrac{(i+1)^4-(i+1)^2}{(i+1)-1} = \dfrac{(i+1)^2[(i+1)^2-1]}{i} = \dfrac{2i(2i-1)}{i} = 2(2i-1) = -2+4i;$

$g(2i-1) = \dfrac{(2i-1)^4-(2i-1)^2}{(2i-1)-1} = \dfrac{(2i-1)^2[(2i-1)^2-1]}{2i-2} = \dfrac{(-3-4i)(-4-4i)}{-2+2i} = \dfrac{(-3-4i)(-2-2i)}{-1+i} = \dfrac{-2+14i}{-1+i} = \dfrac{-2+14i}{-1+i}\cdot\dfrac{-1-i}{-1-i} = \dfrac{16-12i}{2} = 8-6i$

98. $\dfrac{250}{41} + \dfrac{200}{41}i$

99. $\dfrac{i^5+i^6+i^7+i^8}{(1-i)^4} = \dfrac{(i^2)^2\cdot i + (i^2)^3 + (i^2)^3\cdot i + (i^2)^4}{(1-i)^2(1-i)^2} = \dfrac{(-1)^2\cdot i + (-1)^3 + (-1)^3\cdot i + (-1)^4}{-2i(-2i)}$

$\dfrac{i-1-i+1}{-4} = 0$

100. 8

101. $\dfrac{5-\sqrt{5}i}{\sqrt{5}i} = \dfrac{5-\sqrt{5}i}{\sqrt{5}i}\cdot\dfrac{-\sqrt{5}i}{-\sqrt{5}i} = \dfrac{-5\sqrt{5}i-5}{5} = -\dfrac{5}{5} - \dfrac{5\sqrt{5}}{5}i = -1-\sqrt{5}i$

102. $\dfrac{3}{5} + \dfrac{9}{5}i$

103. $\left(\dfrac{1}{2}-\dfrac{1}{3}i\right)^2 - \left(\dfrac{1}{2}+\dfrac{1}{3}i\right)^2 = \dfrac{1}{4} - \dfrac{1}{3}i - \dfrac{1}{9} - \left(\dfrac{1}{4}+\dfrac{1}{3}i-\dfrac{1}{9}\right) = \dfrac{1}{4} - \dfrac{1}{3}i - \dfrac{1}{9} - \dfrac{1}{4} - \dfrac{1}{3}i + \dfrac{1}{9} = -\dfrac{2}{3}i$

104. 1

Chapter 8

Quadratic Functions and Equations

Exercise Set 8.1

1. $5x^2 = 15$

$x^2 = 3$ Multiplying by $\frac{1}{5}$

$x = \sqrt{3}$ *or* $x = -\sqrt{3}$ Using the principle of square roots

The solutions are $\sqrt{3}$ and $-\sqrt{3}$, or $\pm\sqrt{3}$.

2. $\pm\sqrt{5}$

3. $25x^2 + 4 = 0$

$x^2 = -\frac{4}{25}$ Isolating x^2

$x = \sqrt{-\frac{4}{25}}$ *or* $x = -\sqrt{-\frac{4}{25}}$ Principle of square roots

$x = \sqrt{\frac{4}{25}}\sqrt{-1}$ *or* $x = -\sqrt{\frac{4}{25}}\sqrt{-1}$

$x = \frac{2}{5}i$ *or* $x = -\frac{2}{5}i$

The solutions are $\frac{2}{5}i$ and $-\frac{2}{5}i$, or $\pm\frac{2}{5}i$.

4. $\pm\frac{4}{3}i$

5. $2x^2 - 3 = 0$

$x^2 = \frac{3}{2}$

$x = \sqrt{\frac{3}{2}}$ *or* $x = -\sqrt{\frac{3}{2}}$ Principle of square roots

$x = \sqrt{\frac{3}{2}\cdot\frac{2}{2}}$ *or* $x = -\sqrt{\frac{3}{2}\cdot\frac{2}{2}}$ Rationalizing denominators

$x = \frac{\sqrt{6}}{2}$ *or* $x = -\frac{\sqrt{6}}{2}$

The solutions are $\sqrt{\frac{3}{2}}$ and $-\sqrt{\frac{3}{2}}$. This can also be written as $\pm\sqrt{\frac{3}{2}}$ or, if we rationalize the denominator, $\pm\frac{\sqrt{6}}{2}$.

6. $\pm\sqrt{\frac{7}{3}}$, or $\pm\frac{\sqrt{21}}{3}$

7. $(x+2)^2 = 49$

$x + 2 = 7$ *or* $x + 2 = -7$ Principle of square roots

$x = 5$ *or* $x = -9$

The solutions are 5 and -9.

8. $1 \pm \sqrt{6}$

9. $(a+5)^2 = 8$

$a + 5 = \sqrt{8}$ *or* $a + 5 = -\sqrt{8}$ Principle of square roots

$a + 5 = 2\sqrt{2}$ *or* $a + 5 = -2\sqrt{2}$ $(\sqrt{8} = \sqrt{4\cdot 2} = 2\sqrt{2})$

$a = -5 + 2\sqrt{2}$ *or* $a = -5 - 2\sqrt{2}$

The solutions are $-5 + 2\sqrt{2}$ and $-5 - 2\sqrt{2}$, or $-5 \pm 2\sqrt{2}$.

10. 5, 21

11. $(x-7)^2 = -4$

$x - 7 = \sqrt{-4}$ *or* $x - 7 = -\sqrt{-4}$

$x - 7 = 2i$ *or* $x - 7 = -2i$

$x = 7 + 2i$ *or* $x = 7 - 2i$

The solutions are $7 + 2i$ and $7 - 2i$, or $7 \pm 2i$.

12. $-1 \pm 3i$

13. $\left(x + \frac{3}{2}\right)^2 = \frac{7}{2}$

$x + \frac{3}{2} = \sqrt{\frac{7}{2}}$ *or* $x + \frac{3}{2} = -\sqrt{\frac{7}{2}}$

$x + \frac{3}{2} = \sqrt{\frac{7}{2}\cdot\frac{2}{2}}$ *or* $x + \frac{3}{2} = -\sqrt{\frac{7}{2}\cdot\frac{2}{2}}$

$x + \frac{3}{2} = \frac{\sqrt{14}}{2}$ *or* $x + \frac{3}{2} = -\frac{\sqrt{14}}{2}$

$x = -\frac{3}{2} + \frac{\sqrt{14}}{2}$ *or* $x = -\frac{3}{2} - \frac{\sqrt{14}}{2}$

$x = \frac{-3 + \sqrt{14}}{2}$ *or* $x = \frac{-3 - \sqrt{14}}{2}$

The solutions are $\frac{-3 + \sqrt{14}}{2}$ and $\frac{-3 - \sqrt{14}}{2}$, or $\frac{-3 \pm \sqrt{14}}{2}$.

14. $\frac{-3 \pm \sqrt{17}}{4}$

15. $x^2 - 6x + 9 = 100$

$(x-3)^2 = 100$

$x - 3 = 10$ *or* $x - 3 = -10$

$x = 13$ *or* $x = -7$

The solutions are 13 and -7.

16. -3, 13

17. $f(x) = 16$

$(x-7)^2 = 16$ Substituting

$x-7=4$ *or* $x-7=-4$

$x = 11$ *or* $x = 3$

The solutions are 11 and 3.

18. $-3, 7$

19. $F(x) = 13$

$(x-3)^2 = 13$ Substituting

$x-3=\sqrt{13}$ *or* $x-3=-\sqrt{13}$

$x = 3+\sqrt{13}$ *or* $x = 3-\sqrt{13}$

The solutions are $3+\sqrt{13}$ and $3-\sqrt{13}$, or $3\pm\sqrt{13}$.

20. $-3\pm\sqrt{17}$

21. $g(x) = 36$

$x^2+14x+49 = 36$ Substituting

$(x+7)^2 = 36$

$x+7=6$ *or* $x+7=-6$

$x=-1$ *or* $x=-13$

The solutions are -1 and -13.

22. -7, -1

23. x^2+10x

We take half the coefficient of x and square it:

Half of 10 is 5, and $5^2 = 25$. We add 25.

$x^2+10x+25$, $(x+5)^2$

24. $x^2+16x+64$, $(x+8)^2$

25. x^2-6x

We take half the coefficient of x and square it:

Half of -6 is -3, and $(-3)^2 = 9$. We add 9.

x^2-6x+9, $(x-3)^2$

26. $x^2-8x+16$, $(x-4)^2$

27. x^2-24x

We take half the coefficient of x and square it:

$\frac{1}{2}(-24) = -12$ and $(-12)^2 = 144$. We add 144.

$x^2-24x+144$, $(x-12)^2$

28. $x^2-18x+81$, $(x-9)^2$

29. x^2+9x

$\frac{1}{2}\cdot 9 = \frac{9}{2}$, and $\left(\frac{9}{2}\right)^2 = \frac{81}{4}$. We add $\frac{81}{4}$.

$x^2+9x+\frac{81}{4}$, $\left(x+\frac{9}{2}\right)^2$

30. $x^2+3x+\frac{9}{4}$, $\left(x+\frac{3}{2}\right)^2$

31. x^2-3x

We take half the coefficient of x and square it:

$\frac{1}{2}(-3) = -\frac{3}{2}$ and $\left(-\frac{3}{2}\right)^2 = \frac{9}{4}$. We add $\frac{9}{4}$.

$x^2-3x+\frac{9}{4}$, $\left(x-\frac{3}{2}\right)^2$

32. $x^2-7x+\frac{49}{4}$, $\left(x-\frac{7}{2}\right)^2$

33. $x^2+\frac{2}{3}x$

$\frac{1}{2}\cdot\frac{2}{3} = \frac{1}{3}$, and $\left(\frac{1}{3}\right)^2 = \frac{1}{9}$. We add $\frac{1}{9}$.

$x^2+\frac{2}{3}x+\frac{1}{9}$, $\left(x+\frac{1}{3}\right)^2$

34. $x^2+\frac{2}{5}x+\frac{1}{25}$, $\left(x+\frac{1}{5}\right)^2$

35. $x^2-\frac{5}{6}x$

$\frac{1}{2}\left(-\frac{5}{6}\right) = -\frac{5}{12}$, and $\left(-\frac{5}{12}\right)^2 = \frac{25}{144}$. We add $\frac{25}{144}$.

$x^2-\frac{5}{6}x+\frac{25}{144}$, $\left(x-\frac{5}{12}\right)^2$

36. $x^2-\frac{5}{3}x+\frac{25}{36}$, $\left(x-\frac{5}{6}\right)^2$

37. $x^2+\frac{9}{5}x$

$\frac{1}{2}\cdot\frac{9}{5} = \frac{9}{10}$, and $\left(\frac{9}{10}\right)^2 = \frac{81}{100}$. We add $\frac{81}{100}$.

$x^2+\frac{9}{5}x+\frac{81}{100}$, $\left(x+\frac{9}{10}\right)^2$

38. $x^2+\frac{9}{4}x+\frac{81}{64}$, $\left(x+\frac{9}{8}\right)^2$

39. $x^2+6x = 7$

$x^2+6x+9 = 7+9$ Adding 9 on both sides to complete the square

$(x+3)^2 = 16$ Factoring

$x+3 = \pm 4$ Principle of square roots

$x = -3\pm 4$

$x = -3+4$ *or* $x = -3-4$

$x = 1$ *or* $x = -7$

The solutions are 1 and -7.

40. -3, -2

41. $x^2 - 10x = 22$

$x^2 - 10x + 25 = 22 + 25$ Adding 25 on both sides to complete the square

$(x-5)^2 = 47$

$x - 5 = \pm\sqrt{47}$ Principle of square roots

$x = 5 \pm \sqrt{47}$

The solutions are $5 \pm \sqrt{47}$.

42. $4 \pm \sqrt{7}$

43. $x^2 + 6x + 5 = 0$

$x^2 + 6x = -5$ Adding -5 on both sides

$x^2 + 6x + 9 = -5 + 9$ Completing the square

$(x+3)^2 = 4$

$x + 3 = \pm 2$

$x = -3 \pm 2$

$x = -3 - 2$ *or* $x = -3 + 2$

$x = -5$ *or* $x = -1$

The solutions are -5 and -1.

44. -9, -1

45. $x^2 - 10x + 21 = 0$

$x^2 - 10x = -21$

$x^2 - 10x + 25 = -21 + 25$

$(x-5)^2 = 4$

$x - 5 = \pm 2$

$x = 5 \pm 2$

$x = 5 - 2$ *or* $x = 5 + 2$

$x = 3$ *or* $x = 7$

The solutions are 3 and 7.

46. 4, 6

47. $x^2 + 4x + 1 = 0$

$x^2 + 4x = -1$

$x^2 + 4x + 4 = -1 + 4$

$(x+2)^2 = 3$

$x + 2 = \pm\sqrt{3}$

$x = -2 \pm \sqrt{3}$

The solutions are $-2 \pm \sqrt{3}$.

48. $-3 \pm \sqrt{2}$

49. $x^2 + 4 = 6x$

$x^2 - 6x = -4$

$x^2 - 6x + 9 = -4 + 9$

$(x-3)^2 = 5$

$x - 3 = \pm\sqrt{5}$

$x = 3 \pm \sqrt{5}$

The solutions are $3 \pm \sqrt{5}$.

50. $5 \pm \sqrt{2}$

51. $x^2 + 6x + 13 = 0$

$x^2 + 6x = -13$

$x^2 + 6x + 9 = -13 + 9$

$(x+3)^2 = -4$

$x + 3 = \pm 2i$

$x = -3 \pm 2i$

The solutions are $-3 \pm 2i$.

52. $-4 \pm 3i$

53. $2x^2 - 5x - 3 = 0$

$2x^2 - 5x = 3$

$x^2 - \frac{5}{2}x = \frac{3}{2}$ Dividing by 2 on both sides

$x^2 - \frac{5}{2}x + \frac{25}{16} = \frac{3}{2} + \frac{25}{16}$

$\left(x - \frac{5}{4}\right)^2 = \frac{49}{16}$

$x - \frac{5}{4} = \pm\frac{7}{4}$

$x = \frac{5}{4} \pm \frac{7}{4}$

$x = \frac{5}{4} - \frac{7}{4}$ *or* $x = \frac{5}{4} + \frac{7}{4}$

$x = -\frac{1}{2}$ *or* $x = 3$

The solutions are $-\frac{1}{2}$ and 3.

54. $-2, \frac{1}{3}$

55. $4x^2 + 8x + 3 = 0$

$4x^2 + 8x = -3$

$x^2 + 2x = -\frac{3}{4}$

$x^2 + 2x + 1 = -\frac{3}{4} + 1$

$(x+1)^2 = \frac{1}{4}$

$x + 1 = \pm\frac{1}{2}$

$x = -1 \pm \frac{1}{2}$

$x = -1 - \frac{1}{2}$ *or* $x = -1 + \frac{1}{2}$

$x = -\frac{3}{2}$ *or* $x = -\frac{1}{2}$

The solutions are $-\frac{3}{2}$ and $-\frac{1}{2}$.

56. $-\frac{4}{3}, -\frac{2}{3}$

57.
$$6x^2 - x = 15$$
$$x^2 - \frac{1}{6}x = \frac{5}{2}$$
$$x^2 - \frac{1}{6}x + \frac{1}{144} = \frac{5}{2} + \frac{1}{144}$$
$$\left(x - \frac{1}{12}\right)^2 = \frac{361}{144}$$
$$x - \frac{1}{12} = \pm\frac{19}{12}$$
$$x = \frac{1}{12} \pm \frac{19}{12}$$
$$x = \frac{1}{12} + \frac{19}{12} \quad or \quad x = \frac{1}{12} - \frac{19}{12}$$
$$x = \frac{20}{12} \quad or \quad x = -\frac{18}{12}$$
$$x = \frac{5}{3} \quad or \quad x = -\frac{3}{2}$$

The solutions are $\frac{5}{3}$ and $-\frac{3}{2}$.

58. $-\frac{1}{2}, \frac{2}{3}$

59. $2x^2 + 4x + 1 = 0$
$$2x^2 + 4x = -1$$
$$x^2 + 2x = -\frac{1}{2}$$
$$x^2 + 2x + 1 = -\frac{1}{2} + 1$$
$$(x+1)^2 = \frac{1}{2}$$
$$x + 1 = \pm\sqrt{\frac{1}{2}}$$
$$x + 1 = \pm\frac{\sqrt{2}}{2} \qquad \text{Rationalizing the denominator}$$
$$x = -1 \pm \frac{\sqrt{2}}{2}$$

The solutions are $-1 \pm \frac{\sqrt{2}}{2}$, or $\frac{-2 \pm \sqrt{2}}{2}$.

60. $-2, -\frac{1}{2}$

61.
$$3x^2 - 5x - 3 = 0$$
$$3x^2 - 5x = 3$$
$$x^2 - \frac{5}{3}x = 1$$
$$x^2 - \frac{5}{3}x + \frac{25}{36} = 1 + \frac{25}{36}$$
$$\left(x - \frac{5}{6}\right)^2 = \frac{61}{36}$$
$$x - \frac{5}{6} = \pm\frac{\sqrt{61}}{6}$$
$$x = \frac{5 \pm \sqrt{61}}{6}$$

The solutions are $\frac{5 \pm \sqrt{61}}{6}$.

62. $\frac{3 \pm \sqrt{13}}{4}$

63. ***Familiarize.*** We are already familiar with the compound-interest formula.

Translate. We substitute into the formula.
$$A = P(1+r)^t$$
$$2420 = 2000(1+r)^2$$

Carry out. We solve for r.
$$2420 = 2000(1+r)^2$$
$$\frac{2420}{2000} = (1+r)^2$$
$$\frac{121}{100} = (1+r)^2$$
$$\pm\sqrt{\frac{121}{100}} = 1 + r$$
$$\pm\frac{11}{10} = 1 + r$$
$$-\frac{10}{10} + \frac{11}{10} = r$$
$$\frac{1}{10} = r \quad or \quad -\frac{21}{10} = r$$

Check. Since the interest rate cannot be negative, we need only check $\frac{1}{10}$, or 10%. If \$2000 were invested at 10% interest, compounded annually, then in 2 years it would grow to $\$2000(1.1)^2$, or \$2420. The number 10% checks.

State. The interest rate is 10%.

64. 6.25%

65. ***Familiarize.*** We are already familiar with the compound-interest formula.

Translate. We substitute into the formula.
$$A = P(1+r)^t$$
$$1805 = 1280(1+r)^2$$

Carry out. We solve for r.
$$1805 = 1280(1+r)^2$$
$$\frac{1805}{1280} = (1+r)^2$$
$$\frac{361}{256} = (1+r)^2$$
$$\pm\frac{19}{16} = 1 + r$$
$$-\frac{16}{16} \pm \frac{19}{16} = r$$
$$\frac{3}{16} = r \quad or \quad -\frac{35}{16} = r$$

Check. Since the interest rate cannot be negative, we need only check $\frac{3}{16}$ or 18.75%. If \$1280 were invested at 18.75% interest, compounded annually, then in 2 years it would grow to $\$1280(1.1875)^2$, or \$1805. The number 18.75% checks.

State. The interest rate is 18.75%.

66. 20%

67. ***Familiarize.*** We are already familiar with the compound-interest formula.

Translate. We substitute into the formula.

$$A = P(1+r)^t$$
$$6760 = 6250(1+r)^2$$

Carry out. We solve for r.

$$\frac{6760}{6250} = (1+r)^2$$
$$\frac{676}{625} = (1+r)^2$$
$$\pm\frac{26}{25} = 1+r$$
$$-\frac{25}{25} \pm \frac{26}{25} = r$$
$$\frac{1}{25} = r \text{ or } -\frac{51}{25} = r$$

Check. Since the interest rate cannot be negative, we need only check $\frac{1}{25}$, or 4%. If \$6250 were invested at 4% interest, compounded annually, then in 2 years it would grow to $\$6250(1.04)^2$, or \$6760. The number 4% checks.

State. The interest rate is 4%.

68. 8%

69. ***Familiarize.*** We will use the formula $s = 16t^2$.

Translate. We substitute into the formula.

$$s = 16t^2$$
$$1815 = 16t^2$$

Carry out. We solve for t.

$$1815 = 16t^2$$
$$\frac{1815}{16} = t^2$$
$$\sqrt{\frac{1815}{16}} = t \quad \text{Principle of square roots; rejecting the negative square root}$$
$$10.7 \approx t$$

Check. Since $16(10.7)^2 = 1831.84 \approx 1815$, our answer checks.

State. It would take an object about 10.7 sec to fall freely from the top of the CN Tower.

70. About 6.8 sec

71. ***Familiarize.*** We will use the formula $s = 16t^2$.

Translate. We substitute into the formula.

$$s = 16t^2$$
$$640 = 16t^2$$

Carry out. We solve for t.

$$640 = 16t^2$$
$$40 = t^2$$
$$\sqrt{40} = t \quad \text{Principle of square roots; rejecting the negative square root}$$
$$6.3 \approx t$$

Check. Since $16(6.3)^2 = 635.04 \approx 640$, our answer checks.

State. It would take an object about 6.3 sec to fall freely from the top of the Gateway Arch.

72. About 9.5 sec

73. Graph: $f(x) = 5 - 2x$

Select some x-values and find the corresponding values of $f(x)$. Then plot these ordered pairs and draw the graph.

x	$f(x)$, or y
-1	7
0	5
2	1
4	-3

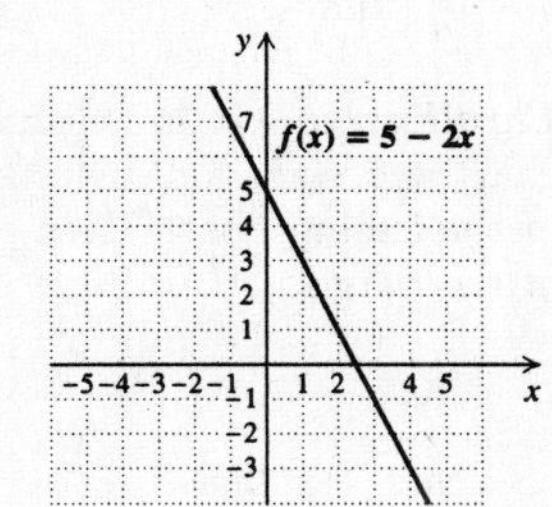

74.

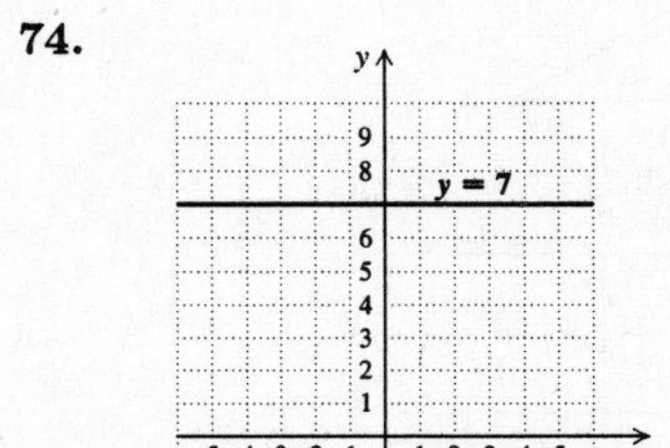

75. $\sqrt[3]{270} = \sqrt[3]{27 \cdot 10} = \sqrt[3]{27}\sqrt[3]{10} = 3\sqrt[3]{10}$

76. $4\sqrt{5}$

77.
$$f(x) = \sqrt{3x-5}$$
$$f(10) = \sqrt{3 \cdot 10 - 5} = \sqrt{30-5} = \sqrt{25} = 5$$

78. 7

79.

80.

81.

82.

83. In order for $x^2 + bx + 81$ to be a square, the following must be true:

$$\left(\frac{b}{2}\right)^2 = 81$$
$$\frac{b^2}{4} = 81$$
$$b^2 = 324$$
$$b = 18 \text{ or } b = -18$$

84. ± 14

85. $x(2x^2 - 9x - 56)(x^2 - 5) = 0$

$x(2x + 7)(x - 8)(x^2 - 5) = 0$

$x=0 \;\; or \;\; 2x+7=0 \;\; or \;\; x-8=0 \;\; or \;\; x^2-5=0$

$x=0 \;\; or \;\; 2x=-7 \;\; or \;\; x=8 \;\; or \;\; x^2=5$

$x=0 \;\; or \;\; x=-\frac{7}{2} \;\; or \;\; x=8 \;\; or \;\; x=\pm\sqrt{5}$

The solutions are 0, $-\frac{7}{2}$, 8, $\sqrt{5}$, and $-\sqrt{5}$.

86. $-\frac{1}{2}, \frac{1}{3}$, and $\frac{1}{2}$

87. ***Familiarize.*** It is helpful to list information in a chart and make a drawing. Let r represent the speed of the fishing boat. Then $r - 7$ represents the speed of the barge.

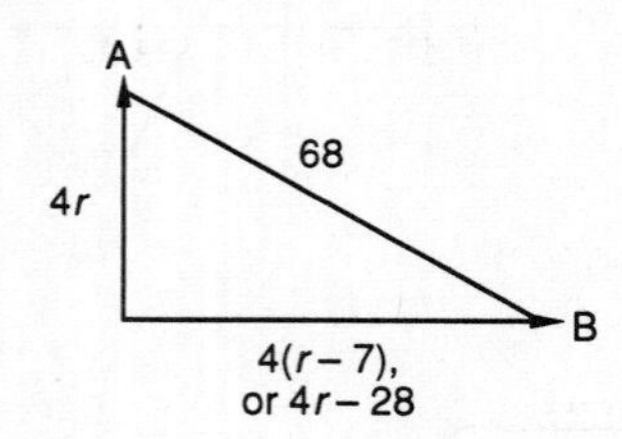

Boat	r	t	d
Fishing	r	4	$4r$
Barge	$r - 7$	4	$4(r - 7)$

Translate. We use the Pythagorean equation:

$a^2 + b^2 = c^2$

$(4r - 28)^2 + (4r)^2 = 68^2$

Carry out.

$$(4r - 28)^2 + (4r)^2 = 68^2$$
$$16r^2 - 224r + 784 + 16r^2 = 4624$$
$$32r^2 - 224r - 3840 = 0$$
$$r^2 - 7r - 120 = 0$$
$$(r + 8)(r - 15) = 0$$

$r + 8 = 0 \;\; or \;\; r - 15 = 0$

$r = -8 \;\; or \;\; r = 15$

Check. We check only 15 since the speeds of the boats cannot be negative. If the speed of the fishing boat is 15 km/h, then the speed of the barge is $15-7$, or 8 km/h, and the distances they travel are $4 \cdot 15$ (or 60) and $4 \cdot 8$ (or 32).

$60^2 + 32^2 = 3600 + 1024 = 4624 = 68^2$

The values check.

State. The speed of the fishing boat is 15 km/h, and the speed of the barge is 8 km/h.

88. 5, 6, 7

89.

90.

91. 

Exercise Set 8.2

1. $x^2 + 7x + 4 = 0$

$a = 1,\ b = 7,\ c = 4$

$x = \frac{-b \pm \sqrt{b^2 - 4ac}}{2a}$

$x = \frac{-7 \pm \sqrt{7^2 - 4 \cdot 1 \cdot 4}}{2 \cdot 1} = \frac{-7 \pm \sqrt{49 - 16}}{2}$

$x = \frac{-7 \pm \sqrt{33}}{2}$

The solutions are $\frac{-7 + \sqrt{33}}{2}$ and $\frac{-7 - \sqrt{33}}{2}$.

2. $\frac{7 \pm \sqrt{61}}{2}$

3. $3p^2 = -8p - 5$

$3p^2 + 8p + 5 = 0$

$(3p + 5)(p + 1) = 0$

$3p + 5 = 0 \;\; or \;\; p + 1 = 0$

$p = -\frac{5}{3} \;\; or \;\; p = -1$

The solutions are $-\frac{5}{3}$ and -1.

4. $3 \pm \sqrt{7}$

5. $x^2 - x + 2 = 0$

$a = 1,\ b = -1,\ c = 2$

$x = \frac{-(-1) \pm \sqrt{(-1)^2 - 4 \cdot 1 \cdot 2}}{2 \cdot 1} = \frac{1 \pm \sqrt{1 - 8}}{2}$

$x = \frac{1 \pm \sqrt{-7}}{2} = \frac{1 \pm i\sqrt{7}}{2}$

The solutions are $\frac{1 + i\sqrt{7}}{2}$ and $\frac{1 - i\sqrt{7}}{2}$.

6. $\frac{1 \pm i\sqrt{3}}{2}$

7. $x^2 + 13 = 6x$

$x^2 - 6x + 13 = 0$

$a = 1,\ b = -6,\ c = 13$

$x = \frac{-(-6) \pm \sqrt{(-6)^2 - 4 \cdot 1 \cdot 13}}{2 \cdot 1} = \frac{6 \pm \sqrt{36 - 52}}{2}$

$x = \frac{6 \pm \sqrt{-16}}{2} = \frac{6 \pm 4i}{2} = 3 \pm 2i$

The solutions are $3 + 2i$ and $3 - 2i$.

8. $2 \pm 3i$

9. $h^2 + 4 = 6h$

$h^2 - 6h + 4 = 0$

$a = 1,\ b = -6,\ c = 4$

$$x = \frac{-(-6) \pm \sqrt{(-6)^2 - 4 \cdot 1 \cdot 4}}{2 \cdot 1} = \frac{6 \pm \sqrt{36 - 16}}{2}$$

$$x = \frac{6 \pm \sqrt{20}}{2} = \frac{6 \pm \sqrt{4 \cdot 5}}{2} = \frac{6 \pm 2\sqrt{5}}{2}$$

$x = 3 \pm \sqrt{5}$

The solutions are $3 + \sqrt{5}$ and $3 - \sqrt{5}$.

10. $\dfrac{-3 \pm \sqrt{41}}{2}$

11. $3 + \dfrac{8}{x} = \dfrac{1}{x^2}$, LCD is x^2

$$x^2\left(3 + \frac{8}{x}\right) = x^2 \cdot \frac{1}{x^2}$$

$3x^2 + 8x = 1$

$3x^2 + 8x - 1 = 0$

$a = 3,\ b = 8,\ c = -1$

$$x = \frac{-8 \pm \sqrt{8^2 - 4 \cdot 3 \cdot (-1)}}{2 \cdot 3} = \frac{-8 \pm \sqrt{64 + 12}}{6}$$

$$x = \frac{-8 \pm \sqrt{76}}{6} = \frac{-8 \pm \sqrt{4 \cdot 19}}{6} = \frac{-8 \pm 2\sqrt{19}}{6}$$

$$x = \frac{-4 \pm \sqrt{19}}{3}$$

The solutions are $\dfrac{-4 + \sqrt{19}}{3}$ and $\dfrac{-4 - \sqrt{19}}{3}$.

12. $\dfrac{9 \pm \sqrt{41}}{4}$

13. $3x + x(x - 2) = 0$

$3x + x^2 - 2x = 0$

$x^2 + x = 0$

$x(x + 1) = 0$

$x = 0$ *or* $x + 1 = 0$

$x = 0$ *or* $x = -1$

The solutions are 0 and -1.

14. -1, 0

15. $14x^2 + 9x = 0$

$x(14x + 9) = 0$

$x = 0$ *or* $14x + 9 = 0$

$x = 0$ *or* $14x = -9$

$x = 0$ *or* $x = -\dfrac{9}{14}$

The solutions are 0 and $-\dfrac{9}{14}$.

16. $-\dfrac{8}{19}$, 0

17. $25x^2 - 20x + 4 = 0$

$(5x - 2)(5x - 2) = 0$

$5x - 2 = 0$ *or* $5x - 2 = 0$

$5x = 2$ *or* $5x = 2$

$x = \dfrac{2}{5}$ *or* $x = \dfrac{2}{5}$

The solution is $\dfrac{2}{5}$.

18. $-\dfrac{7}{6}$

19. $7x(x + 2) + 6 = 3x(x + 1)$

$7x^2 + 14x + 6 = 3x^2 + 3x$ Removing parentheses

$4x^2 + 11x + 6 = 0$

$(4x + 3)(x + 2) = 0$

$4x + 3 = 0$ *or* $x + 2 = 0$

$4x = -3$ *or* $x = -2$

$x = -\dfrac{3}{4}$ *or* $x = -2$

The solutions are $-\dfrac{3}{4}$ and -2.

20. -2, -1

21. $14(x - 4) - (x + 2) = (x + 2)(x - 4)$

$14x - 56 - x - 2 = x^2 - 2x - 8$ Removing parentheses

$13x - 58 = x^2 - 2x - 8$

$0 = x^2 - 15x + 50$

$0 = (x - 10)(x - 5)$

$x - 10 = 0$ *or* $x - 5 = 0$

$x = 10$ *or* $x = 5$

The solutions are 10 and 5.

22. 1, 15

23. $5x^2 = 13x + 17$

$5x^2 - 13x - 17 = 0$

$a = 5,\ b = -13,\ c = -17$

$$x = \frac{-(-13) \pm \sqrt{(-13)^2 - 4(5)(-17)}}{2 \cdot 5}$$

$$x = \frac{13 \pm \sqrt{169 + 340}}{10} = \frac{13 \pm \sqrt{509}}{10}$$

The solutions are $\dfrac{13 + \sqrt{509}}{10}$ and $\dfrac{13 - \sqrt{509}}{10}$.

24. $\dfrac{4}{3}$, 7

25. $x^2 + 9 = 4x$

$x^2 - 4x + 9 = 0$

$a = 1,\ b = -4,\ c = 9$

$$x = \frac{-(-4) \pm \sqrt{(-4)^2 - 4 \cdot 1 \cdot 9}}{2 \cdot 1} = \frac{4 \pm \sqrt{16 - 36}}{2}$$

$$x = \frac{4 \pm \sqrt{-20}}{2} = \frac{4 \pm \sqrt{-4 \cdot 5}}{2}$$

$$x = \frac{4 \pm 2i\sqrt{5}}{2} = 2 \pm i\sqrt{5}$$

The solutions are $2 + i\sqrt{5}$ and $2 - i\sqrt{5}$.

26. $\frac{3 \pm i\sqrt{19}}{2}$

27.
$$x + \frac{1}{x} = \frac{13}{6}, \text{ LCD is } 6x$$
$$6x\left(x + \frac{1}{x}\right) = 6x \cdot \frac{13}{6}$$
$$6x^2 + 6 = 13x$$
$$6x^2 - 13x + 6 = 0$$
$$(2x - 3)(3x - 2) = 0$$
$$2x - 3 = 0 \quad or \quad 3x - 2 = 0$$
$$2x = 3 \quad or \quad 3x = 2$$
$$x = \frac{3}{2} \quad or \quad x = \frac{2}{3}$$

The solutions are $\frac{3}{2}$ and $\frac{2}{3}$.

28. $\frac{3}{2}$, 6

29.
$$x^3 - 8 = 0$$
$$x^3 - 2^3 = 0$$
$$(x - 2)(x^2 + 2x + 4) = 0$$
$$x - 2 = 0 \quad or \quad x^2 + 2x + 4 = 0$$
$$x = 2 \quad or \quad x = \frac{-2 \pm \sqrt{2^2 - 4 \cdot 1 \cdot 4}}{2 \cdot 1}$$
$$x = 2 \quad or \quad x = \frac{-2 \pm \sqrt{-12}}{2} = \frac{-2 \pm 2i\sqrt{3}}{2}$$
$$x = 2 \quad or \quad x = -1 \pm i\sqrt{3}$$

The solutions are 2, $-1 + i\sqrt{3}$, and $-1 - i\sqrt{3}$.

30. -1, $\frac{1 \pm i\sqrt{3}}{2}$

31.
$$f(x) = 0$$
$$3x^2 - 5x - 1 = 0 \qquad \text{Substituting}$$
$$a = 3,\ b = -5,\ c = -1$$
$$x = \frac{-(-5) \pm \sqrt{(-5)^2 - 4 \cdot 3 \cdot (-1)}}{2 \cdot 3}$$
$$x = \frac{5 \pm \sqrt{25 + 12}}{6} = \frac{5 \pm \sqrt{37}}{6}$$

The solutions are $\frac{5 + \sqrt{37}}{6}$ and $\frac{5 - \sqrt{37}}{6}$.

32. $\frac{1 \pm \sqrt{13}}{4}$

33.
$$f(x) = 1$$
$$\frac{7}{x} + \frac{7}{x + 4} = 1 \qquad \text{Substituting}$$
$$x(x + 4)\left(\frac{7}{x} + \frac{7}{x + 4}\right) = x(x + 4) \cdot 1$$
Multiplying by the LCD
$$7(x + 4) + 7x = x^2 + 4x$$
$$7x + 28 + 7x = x^2 + 4x$$
$$14x + 28 = x^2 + 4x$$
$$0 = x^2 - 10x - 28$$
$$a = 1,\ b = -10,\ c = -28$$
$$x = \frac{-(-10) \pm \sqrt{(-10)^2 - 4 \cdot 1 \cdot (-28)}}{2 \cdot 1}$$
$$x = \frac{10 \pm \sqrt{100 + 112}}{2} = \frac{10 \pm \sqrt{212}}{2}$$
$$x = \frac{10 \pm \sqrt{4 \cdot 53}}{2} = \frac{10 \pm 2\sqrt{53}}{2}$$
$$x = 5 \pm \sqrt{53}$$

The solutions are $5 + \sqrt{53}$ and $5 - \sqrt{53}$.

34. -2, 3

35.
$$F(x) = G(x)$$
$$\frac{x + 3}{x} = \frac{x - 4}{3} \qquad \text{Substituting}$$
$$3x\left(\frac{x + 3}{x}\right) = 3x\left(\frac{x - 4}{3}\right) \qquad \text{Multiplying by the LCD}$$
$$3x + 9 = x^2 - 4x$$
$$0 = x^2 - 7x - 9$$
$$a = 1,\ b = -7,\ c = -9$$
$$x = \frac{-(-7) \pm \sqrt{(-7)^2 - 4 \cdot 1 \cdot (-9)}}{2 \cdot 1}$$
$$x = \frac{7 \pm \sqrt{49 + 36}}{2} = \frac{7 \pm \sqrt{85}}{2}$$

The solutions are $\frac{7 + \sqrt{85}}{2}$ and $\frac{7 - \sqrt{85}}{2}$.

36. $\pm 2\sqrt{7}$

37. $x^2 + 4x - 7 = 0$
$$a = 1,\ b = 4,\ c = -7$$
$$x = \frac{-4 \pm \sqrt{4^2 - 4 \cdot 1 \cdot (-7)}}{2 \cdot 1} = \frac{-4 \pm \sqrt{16 + 28}}{2}$$
$$x = \frac{-4 \pm \sqrt{44}}{2}$$

Using a calculator we find that $\frac{-4 + \sqrt{44}}{2} \approx 1.31662479$ and $\frac{-4 - \sqrt{44}}{2} \approx -5.31662479$.

The solutions are approximately 1.31662479 and -5.31662479.

38. -5.236067978, -0.7639320225

39. $x^2 - 6x + 4 = 0$

$a = 1$, $b = -6$, $c = 4$

$$x = \frac{-(-6) \pm \sqrt{(-6)^2 - 4 \cdot 1 \cdot 4}}{2 \cdot 1} = \frac{6 \pm \sqrt{36 - 16}}{2}$$

$$x = \frac{6 \pm \sqrt{20}}{2}$$

Using a calculator we find that $\frac{6+\sqrt{20}}{2} \approx$ 5.236067978 and $\frac{6-\sqrt{20}}{2} \approx 0.7639320225$.

The solutions are approximately 5.236067978 and 0.7639320225.

40. 0.2679491924, 3.732050808

41. $2x^2 - 3x - 7 = 0$

$a = 2$, $b = -3$, $c = -7$

$$x = \frac{-(-3) \pm \sqrt{(-3)^2 - 4 \cdot 2 \cdot (-7)}}{2 \cdot 2}$$

$$x = \frac{3 \pm \sqrt{9 + 56}}{4} = \frac{3 \pm \sqrt{65}}{4}$$

Using a calculator we find that $\frac{3+\sqrt{65}}{4} \approx$ 2.765564437 and $\frac{3-\sqrt{65}}{4} \approx -1.265564437$.

The solutions are approximately 2.765564437 and -1.265564437.

42. -0.4574271078, 1.457427108

43. ***Familiarize.*** Let x = the number of pounds of Kenyan coffee and y = the number of pounds of Peruvian coffee in the mixture. We organize the information in a table.

Type of Coffee	Kenyan	Peruvian	Mixture
Price per pound	\$6.75	\$11.25	\$8.55
Number of pounds	x	y	50
Total cost	\$6.75x	\$11.25y	\$8.55 × 50, or \$427.50

Translate. From the last two rows of the table we get a system of equations.

$x + y = 50$,

$6.75x + 11.25y = 427.50$

Solve. Solving the system of equations, we get $(30, 20)$.

Check. The total number of pounds in the mixture is $30 + 20$, or 50. The total cost of the mixture is $\$6.75(30) + \$11.25(20) = \$427.50$. The values check.

State. The mixture should consist of 30 lb of Kenyan coffee and 20 lb of Peruvian coffee.

44. $4a^2b^3\sqrt{6}$

45. $$\frac{\frac{3}{x-1}}{\frac{1}{x+1} + \frac{2}{x-1}}$$

$$= \frac{\frac{3}{x-1}}{\frac{1}{x+1} + \frac{2}{x-1}} \cdot \frac{(x-1)(x+1)}{(x-1)(x+1)}$$

$$= \frac{3(x+1)}{x - 1 + 2(x+1)}$$

$$= \frac{3x+3}{x - 1 + 2x + 2}$$

$$= \frac{3x+3}{3x+1}, \text{ or } \frac{3(x+1)}{3x+1}$$

46. $\dfrac{4b}{3ab^2 - 4a^2}$

47. ◈

48. ◈

49. ◈

50. ◈

51. $g(x) = \dfrac{4x-2}{x-2} + \dfrac{x+4}{2}$

To find the x-coordinates of the x-intercepts of the graph of g, we solve $g(x) = 0$.

$$\frac{4x-2}{x-2} + \frac{x+4}{2} = 0, \text{ LCD is } 2(x-2)$$

$$2(x-2)\left(\frac{4x-2}{x-2} + \frac{x+4}{2}\right) = 2(x-2) \cdot 0$$

$$2(4x-2) + (x-2)(x+4) = 0$$

$$8x - 4 + x^2 + 2x - 8 = 0$$

$$x^2 + 10x - 12 = 0$$

$a = 1$, $b = 10$, $c = -12$

$$x = \frac{-10 \pm \sqrt{10^2 - 4 \cdot 1 \cdot (-12)}}{2 \cdot 1}$$

$$x = \frac{-10 \pm \sqrt{100 + 48}}{2} = \frac{-10 \pm \sqrt{148}}{2}$$

$$x = \frac{-10 \pm \sqrt{4 \cdot 37}}{2} = \frac{-10 \pm 2\sqrt{37}}{2}$$

$$x = -5 \pm \sqrt{37}$$

The x-intercepts are $(-5-\sqrt{37}, 0)$ and $(-5+\sqrt{37}, 0)$.

52. $(-2, 0)$, $(1, 0)$

53.
$$f(x) = g(x)$$
$$\frac{x^2}{x-2}+1 = \frac{4x-2}{x-2}+\frac{x+4}{2}$$
Substituting
$$2(x-2)\left(\frac{x^2}{x-2}+1\right) = 2(x-2)\left(\frac{4x-2}{x-2}+\frac{x+4}{2}\right)$$
Multiplying by the LCD
$$2x^2+2(x-2) = 2(4x-2)+(x-2)(x+4)$$
$$2x^2+2x-4 = 8x-4+x^2+2x-8$$
$$2x^2+2x-4 = x^2+10x-12$$
$$x^2-8x+8 = 0$$
$a = 1$, $b = -8$, $c = 8$
$$x = \frac{-(-8)\pm\sqrt{(-8)^2-4\cdot 1\cdot 8}}{2\cdot 1} = \frac{8\pm\sqrt{64-32}}{2}$$
$$x = \frac{8\pm\sqrt{32}}{2} = \frac{8\pm\sqrt{16\cdot 2}}{2} = \frac{8\pm 4\sqrt{2}}{2}$$
$$x = 4\pm 2\sqrt{2}$$
The solutions are $4+2\sqrt{2}$ and $4-2\sqrt{2}$.

54. -0.4253905297, 1.17539053

55. $z^2+0.84z-0.4 = 0$
$a = 1$, $b = 0.84$, $c = -0.4$
$$z = \frac{-0.84\pm\sqrt{(0.84)^2-4\cdot 1\cdot(-0.4)}}{2\cdot 1}$$
$$z = \frac{-0.84\pm\sqrt{2.3056}}{2}$$
$$z = \frac{-0.84+\sqrt{2.3056}}{2} \approx 0.3392101158$$
$$z = \frac{-0.84-\sqrt{2.3056}}{2} \approx -1.179210116$$
The solutions are approximately 0.3392101158 and −1.179210116.

56. $\dfrac{-5\sqrt{2}\pm\sqrt{34}}{4}$

57. $(1+\sqrt{3})x^2-(3+2\sqrt{3})x+3 = 0$
$a = 1+\sqrt{3}$, $b = -(3+2\sqrt{3})$, $c = 3$
$$x = \frac{-[-(3+2\sqrt{3})]\pm\sqrt{[-(3+2\sqrt{3})]^2-4\cdot(1+\sqrt{3})\cdot 3}}{2(1+\sqrt{3})}$$
$$x = \frac{3+2\sqrt{3}\pm\sqrt{9}}{2+2\sqrt{3}} = \frac{3+2\sqrt{3}\pm 3}{2+2\sqrt{3}}$$
$$x = \frac{3+2\sqrt{3}+3}{2+2\sqrt{3}} = \frac{6+2\sqrt{3}}{2+2\sqrt{3}} = \frac{3+\sqrt{3}}{1+\sqrt{3}}$$
$$= \frac{3+\sqrt{3}}{1+\sqrt{3}}\cdot\frac{1-\sqrt{3}}{1-\sqrt{3}} = \frac{-2\sqrt{3}}{-2} = \sqrt{3}$$
$$x = \frac{3+2\sqrt{3}-3}{2+2\sqrt{3}} = \frac{2\sqrt{3}}{2+2\sqrt{3}} = \frac{\sqrt{3}}{1+\sqrt{3}}$$
$$= \frac{\sqrt{3}}{1+\sqrt{3}}\cdot\frac{1-\sqrt{3}}{1-\sqrt{3}} = \frac{\sqrt{3}-3}{-2} = \frac{3-\sqrt{3}}{2}$$

The solutions are $\dfrac{3+\sqrt{3}}{1+\sqrt{3}}$ and $\dfrac{\sqrt{3}}{1+\sqrt{3}}$ or, if we rationalize denominators, $\sqrt{3}$ and $\dfrac{3-\sqrt{3}}{2}$.

58. $-i\pm i\sqrt{1-i}$

59.
$$kx^2+3x-k = 0$$
$$k(-2)^2+3(-2)-k = 0 \quad \text{Substituting } -2 \text{ for } x$$
$$4k-6-k = 0$$
$$3k = 6$$
$$k = 2$$
$$2x^2+3x-2 = 0 \quad \text{Substituting 2 for } k$$
$$(2x-1)(x+2) = 0$$
$$2x-1 = 0 \quad or \quad x+2 = 0$$
$$x = \frac{1}{2} \quad or \quad x = -2$$
The other solution is $\dfrac{1}{2}$.

60.

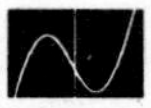

61.

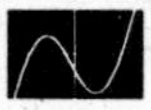

62.

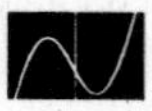

Exercise Set 8.3

1. ***Familiarize***. We first make a drawing, labeling it with the known and unknown information. We can also organize the information in a table. We let r represent the speed and t the time for the first part of the trip.

r mph t hr — 120 mi; $r-10$ mph $4-t$ hr — 100 mi

Trip	Distance	Speed	Time
1st part	120	r	t
2nd part	100	$r-10$	$4-t$

Translate. Using $r = \dfrac{d}{t}$, we get two equations from the table, $r = \dfrac{120}{t}$ and $r-10 = \dfrac{100}{4-t}$.

Carry out. We substitute $\dfrac{120}{t}$ for r in the second equation and solve for t.

$$\frac{120}{t} - 10 = \frac{100}{4-t}, \text{ LCD is } t(4-t)$$

$$t(4-t)\left(\frac{120}{t} - 10\right) = t(4-t)\cdot\frac{100}{4-t}$$

$$120(4-t) - 10t(4-t) = 100t$$

$$480 - 120t - 40t + 10t^2 = 100t$$

$$10t^2 - 260t + 480 = 0 \quad \text{Standard form}$$

$$t^2 - 26t + 48 = 0 \quad \text{Multiplying by } \frac{1}{10}$$

$$(t-2)(t-24) = 0$$

$t = 2$ or $t = 24$

Check. Since the time cannot be negative (If $t = 24$, $4 - t = -20$.), we check only 2 hr. If $t = 2$, then $4 - t = 2$. The speed of the first part is $\frac{120}{2}$, or 60 mph. The speed of the second part is $\frac{100}{2}$, or 50 mph. The speed of the second part is 10 mph slower than the first part. The value checks.

State. The speed of the first part was 60 mph, and the speed of the second part was 50 mph.

2. First part: 12 km/h, second part: 8 km/h

3. ***Familiarize.*** We first make a drawing. We also organize the information in a table. We let r = the speed and t = the time of the slower trip.

200 mi — r mph — t hr

200 mi — $r+10$ mph — $t-1$ hr

Trip	Distance	Speed	Time
Slower	200	r	t
Faster	200	$r+10$	$t-1$

Translate. Using $t = d/r$, we get two equations from the table:

$$t = \frac{200}{r} \text{ and } t - 1 = \frac{200}{r+10}$$

Carry out. We substitute $\frac{200}{r}$ for t in the second equation and solve for r.

$$\frac{200}{r} - 1 = \frac{200}{r+10}, \text{ LCD is } r(r+10)$$

$$r(r+10)\left(\frac{200}{r} - 1\right) = r(r+10)\cdot\frac{200}{r+10}$$

$$200(r+10) - r(r+10) = 200r$$

$$200r + 2000 - r^2 - 10r = 200r$$

$$0 = r^2 + 10r - 2000$$

$$0 = (r+50)(r-40)$$

$r = -50$ *or* $r = 40$

Check. Since negative speed has no meaning in this problem, we check only 40. If $r = 40$, then the time for the slower trip is $\frac{200}{40}$, or 5 hours. If $r = 40$, then $r + 10 = 50$ and the time for the faster trip is $\frac{200}{50}$, or 4 hours. This is 1 hour less time than the slower trip took, so we have an answer to the problem.

State. The speed is 40 mph.

4. 35 mph

5. ***Familiarize.*** We make a drawing and then organize the information in a table. We let r = the speed and t = the time of the Cessna.

600 mi — r mph — t hr

1000 mi — $r+50$ mph — $t+1$ hr

Plane	Distance	Speed	Time
Cessna	600	r	t
Beechcraft	1000	$r+50$	$t+1$

Translate. Using $t = d/r$, we get two equations from the table:

$$t = \frac{600}{r} \text{ and } t + 1 = \frac{1000}{r+50}$$

Carry out. We substitute $\frac{600}{r}$ for t in the second equation and solve for r.

$$\frac{600}{r} + 1 = \frac{1000}{r+50},$$

LCD is $r(r+50)$

$$r(r+50)\left(\frac{600}{r} + 1\right) = r(r+50)\cdot\frac{1000}{r+50}$$

$$600(r+50) + r(r+50) = 1000r$$

$$600r + 30,000 + r^2 + 50r = 1000r$$

$$r^2 - 350r + 30,000 = 0$$

$$(r-150)(r-200) = 0$$

$r = 150$ *or* $r = 200$

Check. If $r = 150$, then the Cessna's time is $\frac{600}{150}$, or 4 hr and the Beechcraft's time is $\frac{1000}{150+50}$, or $\frac{1000}{200}$, or 5 hr. If $r = 200$, then the Cessna's time is $\frac{600}{200}$, or 3 hr and the Beechcraft's time is $\frac{1000}{200+50}$, or $\frac{1000}{250}$, or 4 hr. Since the Beechcraft's time is 1 hr longer in each case, both values check. There are two solutions.

State. The speed of the Cessna is 150 mph and the speed of the Beechcraft is 200 mph; or the speed of the Cessna is 200 mph and the speed of the Beechcraft is 250 mph.

6. Super-prop: 350 mph, turbo-jet: 400 mph

7. ***Familiarize.*** We make a drawing and then organize the information in a table. We let r represent the speed and t the time of the trip to Hillsboro.

Hillsboro

40 mi r mph t hr →

← 40 mi $r-6$ mph $14-t$ hr

Trip	Distance	Speed	Time
To Hillsboro	40	r	t
Return	40	$r-6$	$14-t$

Translate. Using $t=\frac{d}{r}$, we get two equations from the table,

$$t=\frac{40}{r} \text{ and } 14-t=\frac{40}{r-6}.$$

Carry out. We substitute $\frac{40}{r}$ for t in the second equation and solve for r.

$$14-\frac{40}{r}=\frac{40}{r-6},$$

LCD is $r(r-6)$

$$r(r-6)\left(14-\frac{40}{r}\right)=r(r-6)\cdot\frac{40}{r-6}$$
$$14r(r-6)-40(r-6)=40r$$
$$14r^2-84r-40r+240=40r$$
$$14r^2-164r+240=0$$
$$7r^2-82r+120=0$$
$$(7r-12)(r-10)=0$$
$$r=\frac{12}{7} \quad or \quad r=10$$

Check. Since negative speed has no meaning in this problem (If $r=\frac{12}{7}$, then $r-6=-\frac{30}{7}$.), we check only 10 mph. If $r=10$, then the time of the trip to Hillsboro is $\frac{40}{10}$, or 4 hr. The speed of the return trip is $10-6$, or 4 mph, and the time is $\frac{40}{4}$, or 10 hr. The total time for the round trip is 4 hr + 10 hr, or 14 hr. The value checks.

State. Naoki's speed on the trip to Hillsboro was 10 mph and it was 4 mph on the return trip.

8. Speed to Richmond: 60 mph, speed returning: 50 mph

9. ***Familiarize.*** We make a drawing and organize the information in a table. Let r represent the speed of the barge in still water, and let t represent the time of the trip upriver.

24 mi $r-4$ mph t hr → Upriver

Downriver ← 24 mi $r+4$ mph $5-t$ hr

Trip	Distance	Speed	Time
Upriver	24	$r-4$	t
Downriver	24	$r+4$	$5-t$

Translate. Using $t=\frac{d}{r}$, we get two equations from the table,

$$t=\frac{24}{r-4} \text{ and } 5-t=\frac{24}{r+4}.$$

Carry out. We substitute $\frac{24}{r-4}$ for t in the second equation and solve for r.

$$5-\frac{24}{r-4}=\frac{24}{r+4},$$

LCD is $(r-4)(r+4)$

$$(r-4)(r+4)\left(5-\frac{24}{r-4}\right)=(r-4)(r+4)\cdot\frac{24}{r+4}$$
$$5(r-4)(r+4)-24(r+4)=24(r-4)$$
$$5r^2-80-24r-96=24r-96$$
$$5r^2-48r-80=0$$

We use the quadratic formula.

$$r=\frac{-(-48)\pm\sqrt{(-48)^2-4\cdot5\cdot(-80)}}{2\cdot5}$$
$$r=\frac{48\pm\sqrt{3904}}{10}$$
$$r\approx 11 \quad or \quad r\approx -1.5$$

Check. Since negative speed has no meaning in this problem, we check only 11 mph. If $r\approx 11$, then the speed upriver is about $11-4$, or 7 mph, and the time is about $\frac{24}{7}$, or 3.4 hr. The speed downriver is about $11+4$, or 15 mph, and the time is about $\frac{24}{15}$, or 1.6 hr. The total time of the round trip is $3.4+1.6$, or 5 hr. The value checks.

State. The barge must be able to travel about 11 mph in still water.

10. About 14 mph

11. ***Familiarize.*** Let x represent the time it takes the smaller hose to fill the pool. Then $x-6$ represents the time it takes the larger hose to fill the pool. It takes them 4 hr to fill the pool when both hoses are working together, so they can fill $\frac{1}{4}$ of the pool in 1 hr. The smaller hose will fill $\frac{1}{x}$ of the pool in 1 hr, and the larger hose will fill $\frac{1}{x-6}$ of the pool in 1 hr.

Translate. We have an equation.

$$\frac{1}{x}+\frac{1}{x-6}=\frac{1}{4}$$

Carry out. We solve the equation.

We multiply by the LCD, $4x(x-6)$.

$$4x(x-6)\left(\frac{1}{x}+\frac{1}{x-6}\right) = 4x(x-6)\cdot\frac{1}{4}$$
$$4(x-6)+4x = x(x-6)$$
$$4x-24+4x = x^2-6x$$
$$0 = x^2-14x+24$$
$$0 = (x-2)(x-12)$$

$x=2$ *or* $x=12$

Check. Since negative time has no meaning in this problem, 2 is not a solution $(2-6=-4)$. We check only 12 hr. This is the time it would take the smaller hose working alone. Then the larger hose would take $12-6$, or 6 hr working alone. The larger hose would fill $4\left(\frac{1}{6}\right)$, or $\frac{2}{3}$, of the pool in 4 hr, and the smaller hose would fill $4\left(\frac{1}{12}\right)$, or $\frac{1}{3}$, of the pool in 4 hr. Thus in 4 hr they would fill $\frac{2}{3}+\frac{1}{3}$ of the pool. This is all of it, so the numbers check.

State. It takes the smaller hose, working alone, 12 hr to fill the pool.

12. 6 hr

13. ***Familiarize***. We make a drawing and then organize the information in a table. We let r represent Dan's speed in still water. Then $r-5$ is the speed upstream and $r+5$ is the speed downstream. Using $t=\frac{d}{r}$, we let $\frac{10}{r-5}$ represent the time upstream and $\frac{10}{r+5}$ represent the time downstream.

10 km $\quad r-5$ km/h → Upstream

Downstream ← 10 km $\quad r+5$ km/h

Trip	Distance	Speed	Time
Upstream	10	$r-5$	$\frac{10}{r-5}$
Downstream	10	$r+5$	$\frac{10}{r+5}$

Translate. The time for the round trip is 3 hours. We now have an equation.

$$\frac{10}{r-5}+\frac{10}{r+5}=3$$

Carry out. We solve the equation. We multiply by the LCD, $(r-5)(r+5)$.

$$(r-5)(r+5)\left(\frac{10}{r-5}+\frac{10}{r+5}\right) = (r-5)(r+5)\cdot 3$$
$$10(r+5)+10(r-5) = 3(r^2-25)$$
$$10r+50+10r-50 = 3r^2-75$$
$$0 = 3r^2-20r-75$$

We use the quadratic formula

$$r=\frac{-(-20)\pm\sqrt{(-20)^2-4\cdot3\cdot(-75)}}{2\cdot3}$$
$$r=\frac{20\pm\sqrt{400+900}}{6}=\frac{20\pm\sqrt{1300}}{6}$$

$r\approx 9.34$ *or* $r\approx -2.68$

Check. Since negative speed has no meaning in this problem, we check only 9.34. If $r\approx 9.34$, then $r-5\approx 4.34$ and $r+5\approx 14.34$. The time it takes to travel upstream is approximately $\frac{10}{4.34}$, or 2.3 hr, and the time it takes to travel downstream is approximately $\frac{10}{14.34}$, or 0.7 hr. The total time is approximately $2.3+0.7$ or approximately 3 hr. The value checks.

State. Dan's speed in still water is approximately 9.34 km/h.

14. About 3.24 mph

15.
$$A=4\pi r^2$$
$$\frac{A}{4\pi}=r^2 \quad \text{Dividing by } 4\pi$$
$$\frac{1}{2}\sqrt{\frac{A}{\pi}}=r \quad \text{Taking the positive square root}$$

16. $s=\sqrt{\frac{A}{6}}$

17. $A=2\pi r^2+2\pi rh$

$$0=2\pi r^2+2\pi rh-A \quad \text{Standard form}$$

$a=2\pi,\ b=2\pi h,\ c=-A$

$$r=\frac{-2\pi h\pm\sqrt{(2\pi h)^2-4\cdot2\pi\cdot(-A)}}{2\cdot2\pi} \quad \text{Using the quadratic formula}$$
$$r=\frac{-2\pi h\pm\sqrt{4\pi^2h^2+8\pi A}}{4\pi}$$
$$r=\frac{-2\pi h\pm2\sqrt{\pi^2h^2+2\pi A}}{4\pi}$$
$$r=\frac{-\pi h\pm\sqrt{\pi^2h^2+2\pi A}}{2\pi}$$

Since taking the negative square root would result in a negative answer, we take the positive one.

$$r=\frac{-\pi h+\sqrt{\pi^2h^2+2\pi A}}{2\pi}$$

18. $r=\sqrt{\frac{Gm_1m_2}{F}}$

19.
$$N=\frac{kQ_1Q_2}{s^2}$$
$$Ns^2=kQ_1Q_2 \quad \text{Multiplying by } s^2$$
$$s^2=\frac{kQ_1Q_2}{N} \quad \text{Dividing by } N$$
$$s=\sqrt{\frac{kQ_1Q_2}{N}} \quad \text{Taking the positive square root}$$

20. $r = \sqrt{\dfrac{A}{\pi}}$

21. $T = 2\pi\sqrt{\dfrac{l}{g}}$

$\dfrac{T}{2\pi} = \sqrt{\dfrac{l}{g}}$ Multiplying by $\dfrac{1}{2\pi}$

$\dfrac{T^2}{4\pi^2} = \dfrac{l}{g}$ Squaring

$gT^2 = 4\pi^2 l$ Multiplying by $4\pi^2 g$

$g = \dfrac{4\pi^2 l}{T^2}$ Multiplying by $\dfrac{1}{T^2}$

22. $b = \sqrt{c^2 - a^2}$

23. $a^2 + b^2 + c^2 = d^2$

$c^2 = d^2 - a^2 - b^2$ Subtracting a^2 and b^2

$c = \sqrt{d^2 - a^2 - b^2}$ Taking the positive square root

24. $k = \dfrac{3 + \sqrt{9 + 8N}}{2}$

25. $s = v_0 t + \dfrac{gt^2}{2}$

$0 = \dfrac{gt^2}{2} + v_0 t - s$ Standard form

$a = \dfrac{g}{2},\ b = v_0,\ c = -s$

$$t = \frac{-v_0 \pm \sqrt{v_0^2 - 4\left(\frac{g}{2}\right)(-s)}}{2\left(\frac{g}{2}\right)}$$

$$t = \frac{-v_0 \pm \sqrt{v_0^2 + 2gs}}{g}$$

Since taking the negative square root would result in a negative answer, we take the positive one.

$$t = \frac{-v_0 + \sqrt{v_0^2 + 2gs}}{g}$$

26. $r = \dfrac{-\pi s + \sqrt{\pi^2 s^2 + 4\pi A}}{2\pi}$

27. $N = \dfrac{1}{2}(n^2 - n)$

$N = \dfrac{1}{2}n^2 - \dfrac{1}{2}n$

$0 = \dfrac{1}{2}n^2 - \dfrac{1}{2}n - N$

$a = \dfrac{1}{2},\ b = -\dfrac{1}{2},\ c = -N$

$$n = \frac{-\left(-\frac{1}{2}\right) \pm \sqrt{\left(-\frac{1}{2}\right)^2 - 4\cdot\frac{1}{2}\cdot(-N)}}{2\left(\frac{1}{2}\right)}$$

$$n = \frac{1}{2} \pm \sqrt{\frac{1}{4} + 2N}$$

$$n = \frac{1}{2} \pm \sqrt{\frac{1 + 8N}{4}}$$

$$n = \frac{1}{2} \pm \frac{1}{2}\sqrt{1 + 8N}$$

Since taking the negative square root would result in a negative answer, we take the positive one.

$$n = \frac{1}{2} + \frac{1}{2}\sqrt{1 + 8N}, \text{ or } \frac{1 + \sqrt{1 + 8N}}{2}$$

28. $r = 1 - \sqrt{\dfrac{A}{A_0}}$

29. $V = 3.5\sqrt{h}$

$V = 12.25h$ Squaring

$\dfrac{V^2}{12.25} = h$

30. $L = \dfrac{1}{W^2 C}$

31. $A = P_1(1 + r)^2 + P_2(1 + r)$

$0 = P_1(1 + r)^2 + P_2(1 + r) - A$

Let $u = 1 + r$.

$0 = P_1 u^2 + P_2 u - A$ Substituting

$$u = \frac{-P_2 \pm \sqrt{P_2^2 - 4(P_1)(-A)}}{2P_1}$$

$u = \dfrac{-P_2 + \sqrt{P_2^2 + 4AP_1}}{2P_1}$ Simplifying and taking the positive square root

$1 + r = \dfrac{-P_2 + \sqrt{P_2^2 + 4AP_1}}{2P_1}$ Substituting $1 + r$ for u

$$r = -1 + \frac{-P_2 + \sqrt{P_2^2 + 4AP_1}}{2P_1}$$

32. $r = -2 + \dfrac{-P_2 + \sqrt{P_2^2 + 4AP_1}}{P_1}$

33. a) ***Familiarize and Translate.*** From Example 4, we know

$$t = \frac{-v_0 + \sqrt{{v_0}^2 + 19.6s}}{9.8}.$$

Carry out. Substituting 500 for s and 0 for v_0, we have

$$t = \frac{0 + \sqrt{0^2 + 19.6(500)}}{9.8}$$

$t \approx 10.1$

Check. Substitute 10.1 for t and 0 for v_0 in the original formula. (See Example 4.)

$$s = 4.9t^2 + v_0t = 4.9(10.1)^2 + 0 \cdot (10.1)^2$$
$$\approx 500$$

The answer checks.

State. It takes about 10.1 sec to reach the ground.

b) ***Familiarize and Translate.*** From Example 4, we know

$$t = \frac{-v_0 + \sqrt{v_0^2 + 19.6s}}{9.8}.$$

Carry out. Substitute 500 for s and 30 for v_0.

$$t = \frac{-30 + \sqrt{30^2 + 19.6(500)}}{9.8}$$
$$t \approx 7.49$$

Check. Substitute 30 for v_0 and 7.49 for t in the original formula. (See Example 4.)

$$s = 4.9t^2 + v_0t = 4.9(7.49)^2 + (30)(7.49)$$
$$\approx 500$$

The answer checks.

State. It takes about 7.49 sec to reach the ground.

c) ***Familiarize and Translate.*** We will use the formula in Example 4, $s = 4.9t^2 + v_0t$.

Carry out. Substitute 5 for t and 30 for v_0.

$$s = 4.9(5)^2 + 30(5) = 272.5$$

Check. We can substitute 30 for v_0 and 272.5 for s in the form of the formula we used in part (b).

$$t = \frac{-v_0 + \sqrt{v_0^2 + 19.6s}}{9.8}$$
$$= \frac{-30 + \sqrt{(30)^2 + 19.6(272.5)}}{9.8} = 5$$

The answer checks.

State. The object will fall 272.5 m.

34. (a) 3.9 sec; (b) 1.9 sec; (c) 79.6 m

35. ***Familiarize and Translate.*** From Example 3, we know

$$T = \frac{\sqrt{3V}}{12}.$$

Carry out. Substituting 36 for V, we have

$$T = \frac{\sqrt{3 \cdot 36}}{12}$$
$$T \approx 0.87$$

Check. Substitute 0.87 for T in the original formula. (See Example 3.)

$$48T^2 = V$$
$$48(0.87)^2 = V$$
$$36 \approx V$$

The answer checks.

State. Anfernee Hardaway's hang time is about 0.87 sec.

36. 30.625 m

37. ***Familiarize and Translate.*** From Example 4, we know

$$t = \frac{-v_0 + \sqrt{v_0^2 + 19.6s}}{9.8}.$$

Carry out. Substituting 40 for s and 0 for v_0 we have

$$t = \frac{0 + \sqrt{0^2 + 19.6(40)}}{9.8}$$
$$t \approx 2.9$$

Check. Substitute 2.9 for t and 0 for v_0 in the original formula. (See Example 4.)

$$s = 4.9t^2 + v_0t = 4.9(2.9)^2 + 0(2.9)$$
$$\approx 40$$

The answer checks.

State. He will be falling for about 2.9 sec.

38. 12

39. ***Familiarize and Translate.*** We will use the formula in Example 4, $s = 4.9t^2 + v_0t$.

Carry out. Solve the formula for v_0.

$$s - 4.9t^2 = v_0t$$
$$\frac{s - 4.9t^2}{t} = v_0$$

Now substitute 51.6 for s and 3 for t.

$$\frac{51.6 - 4.9(3)^2}{3} = v_0$$
$$2.5 = v_0$$

Check. Substitute 3 for t and 2.5 for v_0 in the original formula.

$$s = 4.9(3)^2 + 2.5(3) = 51.6$$

The solution checks.

State. The initial velocity is 2.5 m/sec.

40. 3.2 m/sec

41. ***Familiarize and Translate.*** From Exercise 31 we know that

$$r = -1 + \frac{-P_2 + \sqrt{P_2^2 + 4P_1A}}{2P_1},$$

where A is the total amount in the account after two years, P_1 is the amount of the original deposit, P_2 is deposited at the begining of the second year, and r is the annual interest rate.

Carry out. Substitute 3000 for P_1, 1700 for P_2, and 5253.70 for A.

$$r = -1 + \frac{-1700 + \sqrt{(1700)^2 + 4(3000)(5253.70)}}{2(3000)}$$

Using a calculator, we have $r = 0.07$.

Check. Substitute in the original formula in Exercise 31.

$$\begin{aligned} P_1(1+r)^2 + P_2(1+r) &= A \\ 3000(1.07)^2 + 1700(1.07) &= A \\ 5253.70 &= A \end{aligned}$$

The answer checks.

State. The annual interest rate is 0.07, or 7%.

42. 8.5%

43.

$$\begin{aligned} \sqrt{3x+1} &= \sqrt{2x-1}+1 \\ 3x+1 &= 2x-1+2\sqrt{2x+1}+1 \quad \text{Squaring both sides} \\ x+1 &= 2\sqrt{2x-1} \\ x^2+2x+1 &= 4(2x-1) \quad \text{Squaring both sides again} \\ x^2+2x+1 &= 8x-4 \\ x^2-6x+5 &= 0 \\ (x-1)(x-5) &= 0 \\ x=1 \quad &\text{or} \quad x=5 \end{aligned}$$

Both numbers check. The solutions are 1 and 5.

44. $\dfrac{1}{x-2}$

45. $\sqrt[3]{18y^3}\,\sqrt[3]{4x^2} = \sqrt[3]{72x^2y^3} = \sqrt[3]{8y^3 \cdot 9x^2} = 2y\sqrt[3]{9x^2}$

46. $\dfrac{1}{7}$

47.

48.

49.

50. $\pm\sqrt{2}$

51. ***Familiarize***. Let x = the number of beach towels purchased for $250.

Then $\dfrac{250}{x}$ = cost per towel,

$\dfrac{250}{x} + 3.50$ = amount received per towel,

$x - 15$ = number of towels sold,

$\left(\dfrac{250}{x} + 3.50\right)(x-15)$ = total amount received from the sale of $x-15$ towels,

$2x+4$ = number of towels in new purchase,

and $\dfrac{250}{x}(2x+4)$ = total amount spent on new purchase.

Translate. The total amount received from the sale of $x-15$ towels is the same as the total amount spent on the new purchase. We now have an equation.

$$\left(\frac{250}{x} + 3.50\right)(x-15) = \frac{250}{x}(2x+4)$$

Carry out. After multiplying and simplifying we have:

$$\begin{aligned} &7x^2 - 605x - 9500 = 0 \\ &(7x+95)(x-100) = 0 \\ &x = -\frac{95}{7} \quad \text{or} \quad x = 100 \end{aligned}$$

Check. Since the cost per towel cannot be negative, we check only 100. When 100 towels are purchased for $250, the cost per towel is $250/100, or $2.50. The amount received per towel is $2.50 + $3.50, or $6. When all but 15 towels, or 100 − 15, or 85, towels are sold, the total amount received is $6 · 85, or $510. If 4 more than twice as many towels as before were purchased, then 2 · 100 + 4, or 204 towels would be purchased. At $2.50 per towel, this would cost $510. Since this is the same as the amount received from the sale of the original towels, the answer checks.

State. The cost per towel was $\dfrac{\$250}{100}$, or $2.50.

52. $n = \pm\sqrt{\dfrac{r^2 \pm \sqrt{r^4 + 4m^4r^2p - 4mp}}{2m}}$

53.

$$\begin{aligned} \frac{w}{l} &= \frac{l}{w+l} \\ l(w+l)\cdot\frac{w}{l} &= l(w+l)\cdot\frac{l}{w+l} \\ w(w+l) &= l^2 \\ w^2 + lw &= l^2 \\ 0 &= l^2 - lw - w^2 \end{aligned}$$

Use the quadratic formula with $a = 1$, $b = -w$, and $c = -w^2$.

$$l = \frac{-(-w) \pm \sqrt{(-w)^2 - 4\cdot 1(-w^2)}}{2\cdot 1}$$

$$l = \frac{w \pm \sqrt{w^2+4w^2}}{2} = \frac{w \pm \sqrt{5w^2}}{2}$$

$$l = \frac{w \pm w\sqrt{5}}{2}$$

Since $\dfrac{w - w\sqrt{5}}{2}$ is negative we use the positive square root:

$$l = \frac{w + w\sqrt{5}}{2}$$

54. $L(A) = \sqrt{\dfrac{A}{2}}$

55.
$$m = \frac{m_0}{\sqrt{1 - \frac{v^2}{c^2}}}$$
$$m\sqrt{1 - \frac{v^2}{c^2}} = m_0$$
$$\sqrt{1 - \frac{v^2}{c^2}} = \frac{m_0}{m}$$
$$1 - \frac{v^2}{c^2} = \frac{m_0^2}{m^2} \quad \text{Squaring}$$
$$c^2m^2 - m^2v^2 = c^2m_0^2 \quad \text{Multiplying by } c^2m^2$$
$$c^2m^2 - c^2m_0^2 = m^2v^2$$
$$c^2(m^2 - m_0^2) = m^2v^2$$
$$c^2 = \frac{m^2v^2}{m^2 - m_0^2}$$
$$c = \frac{mv}{\sqrt{m^2 - m_0^2}} \quad \text{Taking the positive square root}$$

56. $d = \dfrac{-\pi h + \sqrt{\pi^2h^2 + 2\pi A}}{\pi}$

57. Let s represent a length of a side of the cube, let S represent the surface area of the cube, and let A represent the surface area of the sphere. Then the diameter of the sphere is s, so the radius r is $s/2$. From Exercise 15, we know, $A = 4\pi r^2$, so when $r = s/2$ we have $A = 4\pi\left(\frac{s}{2}\right)^2 = 4\pi \cdot \frac{s^2}{4} = \pi s^2$. From the formula for the surface area of a cube (See Exercise 16.) we know that $S = 6s^2$, so $\frac{S}{6} = s^2$ and then $A = \pi \cdot \frac{S}{6}$, or $A(S) = \frac{\pi S}{6}$.

58.

Exercise Set 8.4

1. $x^2 - 4x + 3 = 0$

$a = 1,\ b = -4,\ c = 3$

We substitute and compute the discriminant.

$$\begin{aligned} b^2 - 4ac &= (-4)^2 - 4 \cdot 1 \cdot 3 \\ &= 16 - 12 \\ &= 4 \end{aligned}$$

Since the discriminant is positive and a perfect square, there are two rational solutions.

2. Two rational

3. $x^2 + 5 = 0$

$a = 1,\ b = 0,\ c = 5$

We substitute and compute the discriminant.

$$\begin{aligned} b^2 - 4ac &= 0^2 - 4 \cdot 1 \cdot 5 \\ &= -20 \end{aligned}$$

Since the discriminant is negative, there are two imaginary-number solutions.

4. Two imaginary

5. $x^2 - 2 = 0$

$a = 1,\ b = 0,\ c = -2$

We substitute and compute the discriminant.

$$\begin{aligned} b^2 - 4ac &= 0^2 - 4 \cdot 1 \cdot (-2) \\ &= 8 \end{aligned}$$

Since the discriminant is a positive number that is not a perfect square, there are two irrational solutions.

6. Two irrational

7. $4x^2 - 12x + 9 = 0$

$a = 4,\ b = -12,\ c = 9$

We substitute and compute the discriminant.

$$\begin{aligned} b^2 - 4ac &= (-12)^2 - 4 \cdot 4 \cdot 9 \\ &= 144 - 144 \\ &= 0 \end{aligned}$$

Since the discriminant is 0, there is just one solution, and it is a rational number.

8. Two rational

9. $x^2 - 2x + 4 = 0$

$a = 1,\ b = -2,\ c = 4$

We substitute and compute the discriminant.

$$\begin{aligned} b^2 - 4ac &= (-2)^2 - 4 \cdot 1 \cdot 4 \\ &= 4 - 16 \\ &= -12 \end{aligned}$$

Since the discriminant is negative, there are two imaginary-number solutions.

10. Two imaginary

11. $a^2 + 11a + 28 = 0$

$a = 1,\ b = 11,\ c = 28$

We substitute and compute the discriminant.

$$\begin{aligned} b^2 - 4ac &= 11^2 - 4 \cdot 1 \cdot 28 \\ &= 121 - 112 = 9 \end{aligned}$$

Since the discriminant is a positive number and a perfect square, there are two rational solutions.

12. One rational

13. $6x^2 + 5x - 4 = 0$

$a = 6,\ b = 5,\ c = -4$

We substitute and compute the discriminant.

$$\begin{aligned} b^2 - 4ac &= 5^2 - 4 \cdot 6 \cdot (-4) \\ &= 25 + 96 = 121 \end{aligned}$$

Since the discriminant is a positive number and a perfect square, there are two rational solutions.

14. Two rational

15. $9t^2 - 3t = 0$

$a = 9,\ b = -3,\ c = 0$

We substitute and compute the discriminant.

$$\begin{aligned} b^2 - 4ac &= (-3)^2 - 4 \cdot 9 \cdot 0 \\ &= 9 - 0 \\ &= 9 \end{aligned}$$

Since the discriminant is a positive number and a perfect square, there are two rational solutions.

16. Two rational

17. $x^2 + 5x = 7$

$x^2 + 5x - 7 = 0$ Standard form

$a = 1,\ b = 5,\ c = -7$

We substitute and compute the discriminant.

$$\begin{aligned} b^2 - 4ac &= 5^2 - 4 \cdot 1 \cdot (-7) \\ &= 25 + 28 = 53 \end{aligned}$$

Since the discriminant is a positive number that is not a perfect square, there are two irrational solutions.

18. Two irrational

19. $2a^2 - 3a = -5$

$2a^2 - 3a + 5 = 0$ Standard form

$a = 2,\ b = -3,\ c = 5$

We substitute and compute the discriminant.

$$\begin{aligned} b^2 - 4ac &= (-3)^2 - 4 \cdot 2 \cdot 5 \\ &= 9 - 40 \\ &= -31 \end{aligned}$$

Since the discriminant is negative, there are two imaginary-number solutions.

20. Two imaginary

21. $y^2 + \frac{9}{4} = 4y$

$y^2 - 4y + \frac{9}{4} = 0$ Standard form

$a = 1,\ b = -4,\ c = \frac{9}{4}$

We substitute and compute the discriminant.

$$\begin{aligned} b^2 - 4ac &= (-4)^2 - 4 \cdot 1 \cdot \frac{9}{4} \\ &= 16 - 9 \\ &= 7 \end{aligned}$$

The discriminant is a positive number that is not a perfect square. There are two irrational solutions.

22. Two imaginary

23. The solutions are -7 and 3.

$x = -7$ *or* $x = 3$

$x + 7 = 0$ *or* $x - 3 = 0$

$(x + 7)(x - 3) = 0$ Principle of zero products

$x^2 + 4x - 21 = 0$ FOIL

24. $x^2 + 2x - 24 = 0$

25. The only solution is 3. It must be a repeated solution.

$x = 3$ *or* $x = 3$

$x - 3 = 0$ *or* $x - 3 = 0$

$(x - 3)(x - 3) = 0$ Principle of zero products

$x^2 - 6x + 9 = 0$ FOIL

26. $x^2 + 10x + 25 = 0$

27. The solutions are -2 and -5.

$x = -2$ *or* $x = -5$

$x + 2 = 0$ *or* $x + 5 = 0$

$(x + 2)(x + 5) = 0$

$x^2 + 7x + 10 = 0$

28. $x^2 + 4x + 3 = 0$

29. The solutions are 4 and $\frac{2}{3}$.

$x = 4$ *or* $x = \frac{2}{3}$

$x - 4 = 0$ *or* $x - \frac{2}{3} = 0$

$$\begin{aligned} (x - 4)\left(x - \frac{2}{3}\right) &= 0 \\ x^2 - \frac{2}{3}x - 4x + \frac{8}{3} &= 0 \\ x^2 - \frac{14}{3}x + \frac{8}{3} &= 0 \\ 3x^2 - 14x + 8 &= 0 \quad \text{Multiplying by 3} \end{aligned}$$

30. $4x^2 - 23x + 15 = 0$

31. The solutions are $\frac{1}{2}$ and $\frac{1}{3}$.

$x = \frac{1}{2}$ *or* $x = \frac{1}{3}$

$x - \frac{1}{2} = 0$ *or* $x - \frac{1}{3} = 0$

$$\begin{aligned} \left(x - \frac{1}{2}\right)\left(x - \frac{1}{3}\right) &= 0 \\ x^2 - \frac{1}{3}x - \frac{1}{2}x + \frac{1}{6} &= 0 \\ x^2 - \frac{5}{6}x + \frac{1}{6} &= 0 \\ 6x^2 - 5x + 1 &= 0 \quad \text{Multiplying by 6} \end{aligned}$$

32. $8x^2 + 6x + 1 = 0$

33. The solutions are -0.6 and 1.4.

$x = -0.6$ *or* $x = 1.4$

$x + 0.6 = 0$ *or* $x - 1.4 = 0$

$$\begin{aligned} (x + 0.6)(x - 1.4) &= 0 \\ x^2 - 1.4x + 0.6x - 0.84 &= 0 \\ x^2 - 0.8x - 0.84 &= 0 \end{aligned}$$

34. $x^2 - 2.1x - 1 = 0$

35. The solutions are $-\sqrt{7}$ and $\sqrt{7}$.

$$x = -\sqrt{7} \quad or \quad x = \sqrt{7}$$
$$x + \sqrt{7} = 0 \quad or \quad x - \sqrt{7} = 0$$
$$(x + \sqrt{7})(x - \sqrt{7}) = 0$$
$$x^2 - 7 = 0$$

36. $x^2 - 3 = 0$

37. The solutions are $3\sqrt{2}$ and $-3\sqrt{2}$.

$$x = 3\sqrt{2} \quad or \quad x = -3\sqrt{2}$$
$$x - 3\sqrt{2} = 0 \quad or \quad x + 3\sqrt{2} = 0$$
$$(x - 3\sqrt{2})(x + 3\sqrt{2}) = 0$$
$$x^2 - (3\sqrt{2})^2 = 0$$
$$x^2 - 9 \cdot 2 = 0$$
$$x^2 - 18 = 0$$

38. $x^2 - 20 = 0$

39. The solutions are $3i$ and $-3i$.

$$x = 3i \quad or \quad x = -3i$$
$$x - 3i = 0 \quad or \quad x + 3i = 0$$
$$(x - 3i)(x + 3i) = 0$$
$$x^2 - (3i)^2 = 0$$
$$x^2 + 9 = 0$$

40. $x^2 + 16 = 0$

41. The solutions are $5 - 2i$ and $5 + 2i$.

$$x = 5 - 2i \quad or \quad x = 5 + 2i$$
$$x - 5 + 2i = 0 \quad or \quad x - 5 - 2i = 0$$
$$[x + (-5 + 2i)][x + (-5 - 2i)] = 0$$
$$x^2 + x(-5-2i) + x(-5+2i) + (-5+2i)(-5-2i) = 0$$
$$x^2 - 5x - 2xi - 5x + 2xi + 25 - 4i^2 = 0$$
$$x^2 - 10x + 29 = 0$$
$$(i^2 = -1)$$

42. $x^2 - 4x + 53 = 0$

43. The solutions are $2 - \sqrt{10}$ and $2 + \sqrt{10}$.

$$x = 2 - \sqrt{10} \quad or \quad x = 2 + \sqrt{10}$$
$$x - (2 - \sqrt{10}) = 0 \quad or \quad x - (2 + \sqrt{10}) = 0$$
$$[x - (2 - \sqrt{10})][x - (2 + \sqrt{10})] = 0$$
$$x^2 - x(2 + \sqrt{10}) - x(2 - \sqrt{10}) + (2 - \sqrt{10})(2 + \sqrt{10}) = 0$$
$$x^2 - 2x - x\sqrt{10} - 2x + x\sqrt{10} + 4 - 10 = 0$$
$$x^2 - 4x - 6 = 0$$

44. $x^2 - 6x - 5 = 0$

45. The solutions are -3, 0, and 4.

$$x = -3 \quad or \quad x = 0 \quad or \quad x = 4$$
$$x + 3 = 0 \quad or \quad x = 0 \quad or \quad x - 4 = 0$$
$$(x + 3)(x)(x - 4) = 0$$
$$(x^2 + 3x)(x - 4) = 0$$
$$x^3 - 4x^2 + 3x^2 - 12x = 0$$
$$x^3 - x^2 - 12x = 0$$

46. $x^3 + 3x^2 - 10x = 0$

47. The solutions are -1, 1, and 2.

$$x = -1 \quad or \quad x = 1 \quad or \quad x = 2$$
$$x + 1 = 0 \quad or \quad x - 1 = 0 \quad or \quad x - 2 = 0$$
$$(x + 1)(x - 1)(x - 2) = 0$$
$$(x^2 - 1)(x - 2) = 0$$
$$x^3 - 2x^2 - x + 2 = 0$$

48. $x^3 - 3x^2 - 4x + 12 = 0$

49. ***Familiarize***. Let x and y represent the number of 30-sec and 60-sec commercials, respectively. Then the amount of time for the 30-sec commercials was $30x$ sec, or $\frac{30x}{60} = \frac{x}{2}$ min. The amount of time for the 60-sec commercials was $60x$ sec, or $\frac{60x}{60} = x$ min.

Translate. Rewording, we write two equations. We will express time in minutes.

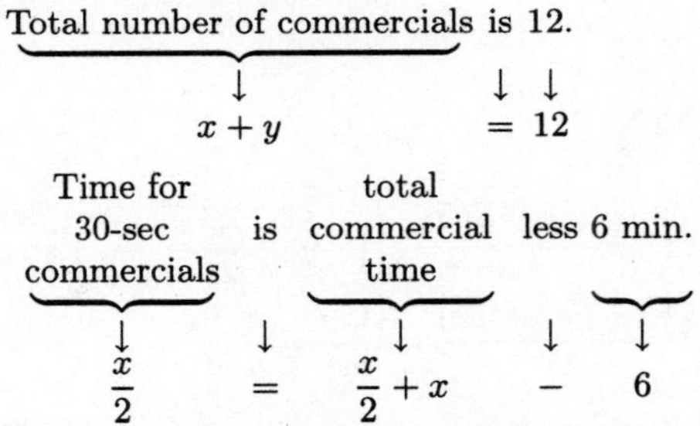

Carry out. Solving the system of equations we get $(6, 6)$.

Check. If there are six 30-sec and six 60-sec commercials, the total number of commercials is 12. The amount of time for six 30-sec commercials is 180 sec, or 3 min, and for six 60-sec commercials is 360 sec, or 6 min. The total commercial time is 9 min, and the amount of time for 30-sec commercials is 6 min less than this. The numbers check.

State. There were six 30-sec commercials.

50.

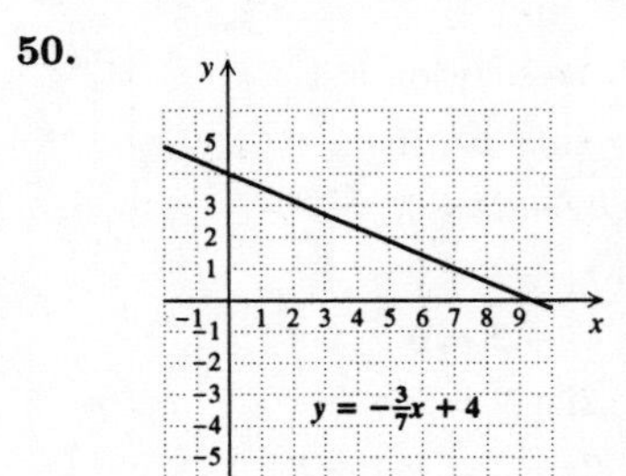

51. Graph: $f(x) = -x - 3$.

Select some x-values and find the corresponding values of $f(x)$. Then plot these ordered pairs and draw the graph.

x	$f(x)$
-5	2
0	-3
3	-6

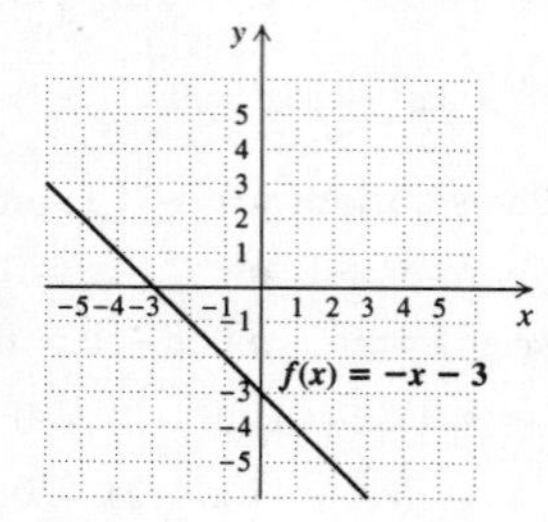

52.

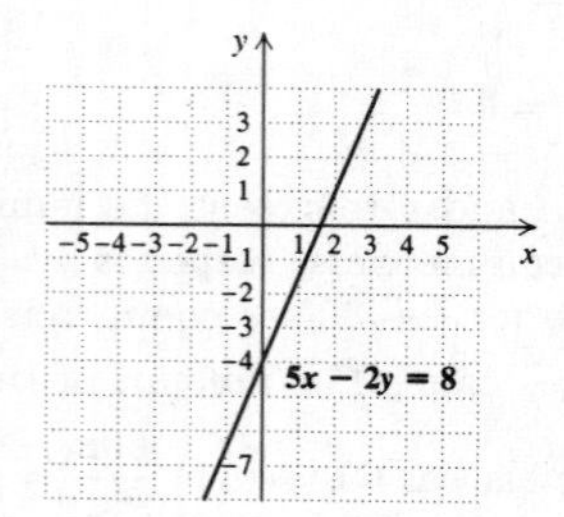

53.

54.

55.

56. $\left(\dfrac{-b+\sqrt{b^2-4ac}}{2a}\right)\left(\dfrac{-b-\sqrt{b^2-4ac}}{2a}\right) =$

$$\frac{b^2-(b^2-4ac)}{4a^2} = \frac{4ac}{4a^2} = \frac{c}{a}$$

57. The graph includes the points $(-3,0)$, $(0,-3)$, and $(1,0)$. Substituting in $y = ax^2+bx+c$, we have three equations.

$$\begin{aligned} 0 &= 9a - 3b + c, \\ -3 &= \phantom{9a - 3b + {}} c, \\ 0 &= a + b + c \end{aligned}$$

The solution of this system of equations is $a = 1$, $b = 2$, $c = -3$.

58. (a) $-\dfrac{3}{5}$; (b) $-\dfrac{1}{3}$

59. a) $x^2 - kx + 2 = 0$; one solution is $1 + i$

We first find k by substituting $1 + i$ for x.

$$\begin{aligned} x^2 - kx + 2 &= 0 \\ (1+i)^2 - k(1+i) + 2 &= 0 \\ 1 + 2i - 1 - k - ki + 2 &= 0 \\ 2i + 2 &= k + ki \\ 2i + 2 &= k(1+i) \\ \frac{2(i+1)}{1+i} &= k \\ 2 &= k \end{aligned}$$

b) Now substitute 2 for k in the original equation.

$$\begin{aligned} x^2 - 2x + 2 &= 0 \\ x &= \frac{2 \pm \sqrt{4-8}}{2} = \frac{2 \pm 2i}{2} \\ x &= 1 \pm i \end{aligned}$$

The other solution is $1 - i$.

60. (a) $9 + 9i$; (b) $3 + 3i$

61. Consider a quadratic equation in standard form, $ax^2 + bx + c = 0$. The solutions are

$$\frac{-b \pm \sqrt{b^2-4ac}}{2a}.$$

The sum of the solutions is

$$\frac{-b+\sqrt{b^2-4ac}}{2a} + \frac{-b-\sqrt{b^2-4ac}}{2a} =$$

$$\frac{-2b}{2a} = -\frac{b}{a}.$$

62. -1

63. $3x^2 - hx + 4k = 0$

$a = 3$, $b = -h$, $c = 4k$

The sum of the solutions is -12.

$-\dfrac{b}{a} = -\dfrac{-h}{3} = \dfrac{h}{3}$ (See Exercise 61.)

$\dfrac{h}{3} = -12$, so $h = -36$.

The product of the solutions is 20.

$\dfrac{c}{a} = \dfrac{4k}{3}$ (See Exercise 56.)

$\dfrac{4k}{3} = 20$, so $4k = 60$ and $k = 15$.

Thus, $h = -36$ and $k = 15$.

64. $a = 8$, $b = 20$, $c = -12$

65. We have $x = 2 - \sqrt{3}$ *or* $x = 2 + \sqrt{3}$ *or* $x = 5 - 2i$ *or* $x = 5 + 2i$.

Then $x - (2 - \sqrt{3}) = 0$ *or* $x - (2 + \sqrt{3}) = 0$ *or* $x - (5 - 2i) = 0$ *or* $x - (5 + 2i) = 0$.

$$\begin{aligned} [x-(2-\sqrt{3})][x-(2+\sqrt{3})][x-(5-2i)][x-(5+2i)] &= 0 \\ (x^2 - 4x + 1)(x^2 - 10x + 29) &= 0 \\ x^4 - 14x^3 + 70x^2 - 126x + 29 &= 0 \end{aligned}$$

66. $x^4 - 8x^3 + 21x^2 - 2x - 52 = 0$

67.

68.

69.

Exercise Set 8.5

1. $x^4 - 5x^2 + 4 = 0$

Let $u = x^2$ and think of x^4 as $(x^2)^2$.

$u^2 - 5u + 4 = 0$ Substituting u for x^2

$(u-1)(u-4) = 0$

$u - 1 = 0$ or $u - 4 = 0$

$u = 1$ or $u = 4$

Now replace u with x^2 and solve these equations:

$x^2 = 1$ *or* $x^2 = 4$

$x = \pm 1$ *or* $x = \pm 2$

The numbers 1, -1, 2, and -2 check. They are the solutions.

2. ± 1, ± 3

3. $x^4 - 12x^2 + 27 = 0$

Let $u = x^2$.

$u^2 - 12u + 27 = 0$ Substituting u for x^2

$(u-9)(u-3) = 0$

$u = 9$ or $u = 3$

Now replace u with x^2 and solve these equations:

$x^2 = 9$ *or* $x^2 = 3$

$x = \pm 3$ *or* $x = \pm\sqrt{3}$

The numbers 3, -3, $\sqrt{3}$, and $-\sqrt{3}$ check. They are the solutions.

4. ± 2, $\pm\sqrt{5}$

5. $4x^4 - 19x^2 + 12 = 0$

Let $u = x^2$.

$4u^2 - 19u + 12 = 0$ Substituting u for x^2

$(4u-3)(u-4) = 0$

$4u - 3 = 0$ *or* $u - 4 = 0$

$u = \frac{3}{4}$ *or* $u = 4$

Now replace u with x^2 and solve these equations:

$x^2 = \frac{3}{4}$ *or* $x^2 = 4$

$x = \pm\frac{\sqrt{3}}{2}$ *or* $x = \pm 2$

The numbers $\frac{\sqrt{3}}{2}$, $-\frac{\sqrt{3}}{2}$, 2, and -2 check. They are the solutions.

6. $\pm\frac{\sqrt{5}}{3}$, ± 1

7. $x - 4\sqrt{x} - 1 = 0$

Let $u = \sqrt{x}$ and view x as $(\sqrt{x})^2$.

$u^2 - 4u - 1 = 0$ Substituting u for $\sqrt{x}$

$$u = \frac{-(-4) \pm \sqrt{(-4)^2 - 4\cdot 1\cdot(-1)}}{2\cdot 1}$$

$$u = \frac{4 \pm \sqrt{20}}{2} = \frac{2\cdot 2 \pm 2\sqrt{5}}{2}$$

$$u = 2 \pm \sqrt{5}$$

$u = 2 + \sqrt{5}$ *or* $u = 2 - \sqrt{5}$

Replace u with $\sqrt{x}$ and solve these equations.

$\sqrt{x} = 2 + \sqrt{5}$ *or* $\sqrt{x} = 2 - \sqrt{5}$

$(\sqrt{x})^2 = (2+\sqrt{5})^2$ No solution: $2 - \sqrt{5}$ is negative

$x = 4 + 4\sqrt{5} + 5$

$x = 9 + 4\sqrt{5}$

The number $9 + 4\sqrt{5}$ checks. It is the solution.

8. $8 + 2\sqrt{7}$

9. $(x^2-7)^2 - 3(x^2-7) + 2 = 0$

Let $u = x^2 - 7$.

$u^2 - 3u + 2 = 0$ Substituting u for $x^2 - 7$

$(u-1)(u-2) = 0$

$u = 1$ *or* $u = 2$

$x^2 - 7 = 1$ *or* $x^2 - 7 = 2$ Replacing u with $x^2 - 7$

$x^2 = 8$ *or* $x^2 = 9$

$x = \pm\sqrt{8}$ *or* $x = \pm 3$

$x = \pm 2\sqrt{2}$ *or* $x = \pm 3$

The numbers $2\sqrt{2}$, $-2\sqrt{3}$, 3, and -3 check. They are the solutions.

10. $\pm\sqrt{3}$, ± 2

11. $(3+\sqrt{x})^2 + 3(3+\sqrt{x}) - 10 = 0$

Let $u = 3 + \sqrt{x}$.

$u^2 + 3u - 10 = 0$ Substituting u for $3 + \sqrt{x}$

$(u+5)(u-2) = 0$

$u = -5$ *or* $u = 2$

$3 + \sqrt{x} = -5$ *or* $3 + \sqrt{x} = 2$ Replacing u with $3 + \sqrt{x}$

$\sqrt{x} = -8$ *or* $\sqrt{x} = -1$

Since the principal square root cannot be negative, this equation has no solution.

12. No solution

13. $x^{-2} - x^{-1} - 6 = 0$

Let $u = x^{-1}$ and think of x^{-2} as $(x^{-1})^2$.

$u^2 - u - 6 = 0$ Substituting u for x^{-1}

$(u-3)(u+2) = 0$

$u = 3$ or $u = -2$

Now we replace u with x^{-1} and solve these equations:

$x^{-1} = 3$ *or* $x^{-1} = -2$

$\frac{1}{x} = 3$ *or* $\frac{1}{x} = -2$

$\frac{1}{3} = x$ *or* $-\frac{1}{2} = x$

Both $\frac{1}{3}$ and $-\frac{1}{2}$ check. They are the solutions.

14. $-2, 1$

15. $4x^{-2} + x^{-1} - 5 = 0$

Let $u = x^{-1}$.

$4u^2 + u - 5 = 0$ Substituting u for x^{-1}

$(4u + 5)(u - 1) = 0$

$u = -\frac{5}{4}$ or $u = 1$

Now we replace u with x^{-1} and solve these equations:

$x^{-1} = -\frac{5}{4}$ *or* $x^{-1} = 1$

$\frac{1}{x} = -\frac{5}{4}$ *or* $\frac{1}{x} = 1$

$4 = -5x$ *or* $1 = x$

$-\frac{4}{5} = x$ *or* $1 = x$

The numbers $-\frac{4}{5}$ and 1 check. They are the solutions.

16. $-\frac{1}{10}, 1$

17. $t^{2/3} + t^{1/3} - 6 = 0$

Let $u = t^{1/3}$ and think of $t^{2/3}$ as $(t^{1/3})^2$.

$u^2 + u - 6 = 0$ Substituting u for $t^{1/3}$

$(u + 3)(u - 2) = 0$

$u = -3$ or $u = 2$

Now we replace u with $t^{1/3}$ and solve these equations:

$t^{1/3} = -3$ *or* $t^{1/3} = 2$

$t = (-3)^3$ *or* $t = 2^3$ Raising to the third power

$t = -27$ *or* $t = 8$

Both -27 and 8 check. They are the solutions.

18. $-8, 64$

19. $y^{1/3} - y^{1/6} - 6 = 0$

Let $u = y^{1/6}$.

$u^2 - u - 6 = 0$ Substituting u for $y^{1/6}$

$(u - 3)(u + 2) = 0$

$u = 3$ or $u = -2$

Now we replace u with $y^{1/6}$ and solve these equations:

$y^{1/6} = 3$ *or* $y^{1/6} = -2$

$\sqrt[6]{y} = 3$ *or* $\sqrt[6]{y} = -2$

$y = 3^6$ This equation has no solution since principal sixth roots are never negative.

$y = 729$

The number 729 checks and is the solution.

20. No solution

21. $t^{1/3} + 2t^{1/6} = 3$

$t^{1/3} + 2t^{1/6} - 3 = 0$

Let $u = t^{1/6}$.

$u^2 + 2u - 3 = 0$ Substituting u for $t^{1/6}$

$(u + 3)(u - 1) = 0$

$u = -3$ *or* $u = 1$

$t^{1/6} = -3$ *or* $t^{1/6} = 1$ Substituting $t^{1/6}$ for u

No solution $\quad t = 1$

The number 1 checks and is the solution.

22. $16, 81$

23. $(3 - \sqrt{x})^2 - 10(3 - \sqrt{x}) + 23 = 0$

Let $u = 3 - \sqrt{x}$.

$u^2 - 10u + 23 = 0$ Substituting u for $3 - \sqrt{x}$

$$u = \frac{-(-10) \pm \sqrt{(-10)^2 - 4 \cdot 1 \cdot 23}}{2 \cdot 1}$$

$$u = \frac{10 \pm \sqrt{8}}{2} = \frac{2 \cdot 5 \pm 2\sqrt{2}}{2}$$

$$u = 5 \pm \sqrt{2}$$

$u = 5 + \sqrt{2}$ or $u = 5 - \sqrt{2}$

Now we replace u with $3 - \sqrt{x}$ and solve these equations:

$3 - \sqrt{x} = 5 + \sqrt{2}$ *or* $3 - \sqrt{x} = 5 - \sqrt{2}$

$-\sqrt{x} = 2 + \sqrt{2}$ *or* $-\sqrt{x} = 2 - \sqrt{2}$

$\sqrt{x} = -2 - \sqrt{2}$ *or* $\sqrt{x} = -2 + \sqrt{2}$

Since both $-2 - \sqrt{2}$ and $-2 + \sqrt{2}$ are negative and principal square roots are never negative, the equation has no solution.

24. $4 + 2\sqrt{3}$

25. $16\left(\frac{x - 1}{x - 8}\right)^2 + 8\left(\frac{x - 1}{x - 8}\right) + 1 = 0$

Let $u = \frac{x - 1}{x - 8}$.

$16u^2 + 8u + 1 = 0$ Substituting u for $\frac{x - 1}{x - 8}$

$(4u + 1)(4u + 1) = 0$

$u = -\frac{1}{4}$

Now we replace u with $\frac{x - 1}{x - 8}$ and solve this equation:

$\frac{x - 1}{x - 8} = -\frac{1}{4}$

$4x - 4 = -x + 8$ Multiplying by $4(x - 8)$

$5x = 12$

$x = \frac{12}{5}$

The number $\frac{12}{5}$ checks and is the solution.

26. $-\dfrac{3}{2}$

27. The x-intercepts occur where $f(x) = 0$. Thus, we must have $5x + 13\sqrt{x} - 6 = 0$.

Let $u = \sqrt{x}$.

$5u^2 + 13u - 6 = 0$ Substituting

$(5u - 2)(u + 3) = 0$

$u = \dfrac{2}{5}$ *or* $u = -3$

Now replace u with $\sqrt{x}$ and solve these equations:

$\sqrt{x} = \dfrac{2}{5}$ *or* $\sqrt{x} = -3$

$x = \dfrac{4}{25}$ No solution

The number $\dfrac{4}{25}$ checks. Thus, the x-intercept is $\left(\dfrac{4}{25}, 0\right)$.

28. $\left(\dfrac{4}{9}, 0\right)$

29. The x-intercepts occur where $f(x) = 0$. Thus, we must have $(x^2 - 3x)^2 - 10(x^2 - 3x) + 24 = 0$.

Let $u = x^2 - 3x$.

$u^2 - 10u + 24 = 0$ Substituting

$(u - 6)(u - 4) = 0$

$u = 6$ *or* $u = 4$

Now replace u with $x^2 - 3x$ and solve these equations:

$x^2 - 3x = 6$ *or* $x^2 - 3x = 4$

$x^2 - 3x - 6 = 0$ *or* $x^2 - 3x - 4 = 0$

$x = \dfrac{-(-3) \pm \sqrt{(-3)^2 - 4(1)(-6)}}{2 \cdot 1}$ *or*

$(x - 4)(x + 1) = 0$

$x = \dfrac{3 \pm \sqrt{33}}{2}$ *or* $x = 4$ *or* $x = -1$

All four numbers check. Thus, the x-intercepts are $\left(\dfrac{3 + \sqrt{33}}{2}, 0\right)$, $\left(\dfrac{3 - \sqrt{33}}{2}, 0\right)$, $(4, 0)$, and $(-1, 0)$.

30. $(-1, 0)$, $(1, 0)$, $(5, 0)$, and $(7, 0)$

31. The x-intercepts occur where $f(x) = 0$. Thus, we must have $x^{2/5} + x^{1/5} - 6 = 0$.

Let $u = x^{1/5}$.

$u^2 + u - 6 = 0$ Substituting u for $x^{1/5}$

$(u + 3)(u - 2) = 0$

$u = -3$ *or* $u = 2$

$x^{1/5} = -3$ *or* $x^{1/5} = 2$ Replacing u with $x^{1/5}$

$x = -243$ *or* $x = 32$ Raising to the fifth power

Both -243 and 32 check. Thus, the x-intercepts are $(-243, 0)$ and $(32, 0)$.

32. $(81, 0)$

33. The x-intercepts occur where $f(x) = 0$. Thus, we must have $\left(\dfrac{x^2 - 2}{x}\right)^2 - 7\left(\dfrac{x^2 - 2}{x}\right) - 18 = 0$.

Let $u = \dfrac{x^2 - 2}{x}$.

$u^2 - 7u - 18 = 0$ Substituting

$(u - 9)(u + 2) = 0$

$u = 9$ *or* $u = -2$

$\dfrac{x^2 - 2}{x} = 9$ *or* $\dfrac{x^2 - 2}{x} = -2$

Replacing u with $\dfrac{x^2 - 2}{x}$

$x^2 - 2 = 9x$ *or* $x^2 - 2 = -2x$

$x^2 - 9x - 2 = 0$ *or* $x^2 + 2x - 2 = 0$

$x = \dfrac{-(-9) \pm \sqrt{(-9)^2 - 4 \cdot 1 \cdot (-2)}}{2 \cdot 1}$ *or*

$x = \dfrac{-2 \pm \sqrt{2^2 - 4 \cdot 1 \cdot (-2)}}{2 \cdot 1}$

$x = \dfrac{9 \pm \sqrt{89}}{2}$ *or* $x = \dfrac{-2 \pm \sqrt{12}}{2}$

$x = \dfrac{9 \pm \sqrt{89}}{2}$ *or* $x = \dfrac{-2 \pm 2\sqrt{3}}{2}$

$x = \dfrac{9 \pm \sqrt{89}}{2}$ *or* $x = -1 \pm \sqrt{3}$

All four numbers check. Thus, the x-intercepts are $\left(\dfrac{9 + \sqrt{89}}{2}, 0\right)$, $\left(\dfrac{9 - \sqrt{89}}{2}, 0\right)$, $(-1 + \sqrt{3}, 0)$, and $(-1 - \sqrt{3}, 0)$.

34. $(3 + \sqrt{10}, 0)$, $(3 - \sqrt{10}, 0)$, $(-1 + \sqrt{2}, 0)$, and $(-1 - \sqrt{2}, 0)$

35. $\sqrt{3x^2}\sqrt{3x^3} = \sqrt{3x^2 \cdot 3x^3} = \sqrt{9x^5} = \sqrt{9x^4 \cdot x} = 3x^2\sqrt{x}$

36. 4 L of A, 8 L of B

37. $\dfrac{x + 1}{x - 1} - \dfrac{x + 1}{x^2 + x + 1}$, LCD $= (x - 1)(x^2 + x + 1)$

$= \dfrac{x + 1}{x - 1} \cdot \dfrac{x^2 + x + 1}{x^2 + x + 1} - \dfrac{x + 1}{x^2 + x + 1} \cdot \dfrac{x - 1}{x - 1}$

$= \dfrac{(x^3 + 2x^2 + 2x + 1) - (x^2 - 1)}{(x - 1)(x^2 + x + 1)}$

$= \dfrac{x^3 + x^2 + 2x + 2}{x^3 - 1}$

38. $a^2 + a$

39. ◈

40. ◈

41. $3x^4 + 5x^2 - 1 = 0$

Let $u = x^2$.

$3u^2 + 5u - 1 = 0$ Substituting

$u = \frac{-5 \pm \sqrt{5^2 - 4 \cdot 3 \cdot (-1)}}{2 \cdot 3}$

$u = \frac{-5 \pm \sqrt{37}}{6}$

$x^2 = \frac{-5 \pm \sqrt{37}}{6}$ Replacing u with x^2

$x = \pm\sqrt{\frac{-5 \pm \sqrt{37}}{6}}$

All four numbers check and are the solutions.

42. $\pm\sqrt{\frac{7 \pm \sqrt{29}}{10}}$

43. $(x^2 - 5x - 1)^2 - 18(x^2 - 5x - 1) + 65 = 0$

Let $u = x^2 - 5x - 1$.

$u^2 - 18u + 65 = 0$ Substituting

$(u - 5)(u - 13) = 0$

$u = 5$ *or* $u = 13$

$x^2 - 5x - 1 = 5$ *or* $x^2 - 5x - 1 = 13$

Replacing u with $x^2 - 5x - 1$

$x^2 - 5x - 6 = 0$ *or* $x^2 - 5x - 14 = 0$

$(x - 6)(x + 1) = 0$ *or* $(x - 7)(x + 2) = 0$

$x = 6$ *or* $x = -1$ *or* $x = 7$ *or* $x = -2$

The numbers 6, -1, 7, and -2 check and are the solutions.

44. $-2, -1, 5, 6$

45. $\frac{x}{x-1} - 6\sqrt{\frac{x}{x-1}} - 40 = 0$

Let $u = \sqrt{\frac{x}{x-1}}$.

$u^2 - 6u - 40 = 0$ Substituting for $\sqrt{\frac{x}{x-1}}$

$(u - 10)(u + 4) = 0$

$u = 10$ *or* $u = -4$

$\sqrt{\frac{x}{x-1}} = 10$ *or* $\sqrt{\frac{x}{x-1}} = -4$

$\frac{x}{x-1} = 100$ *or* No solution

$x = 100x - 100$ Multiplying by $(x - 1)$

$100 = 99x$

$\frac{100}{99} = x$

The number $\frac{100}{99}$ checks. It is the solution.

46. $\frac{432}{143}$

47. $a^5(a^2 - 25) + 13a^3(25 - a^2) + 36a(a^2 - 25) = 0$

$a^5(a^2 - 25) - 13a^3(a^2 - 25) + 36a(a^2 - 25) = 0$

$a(a^2 - 25)(a^4 - 13a^2 + 36) = 0$

$a(a^2 - 25)(a^2 - 4)(a^2 - 9) = 0$

$a=0$ *or* $a^2 - 25=0$ *or* $a^2 - 4=0$ *or* $a^2 - 9 = 0$

$a=0$ *or* $a^2=25$ *or* $a^2=4$ *or* $a^2 = 9$

$a=0$ *or* $a=\pm 5$ *or* $a=\pm 2$ *or* $a = \pm 3$

All seven numbers check. The solutions are 0, 5, -5, 2, -2, 3, and -3.

48. 9

49. $x^6 - 28x^3 + 27 = 0$

Let $u = x^3$.

$u^2 - 28u + 27 = 0$

$(u - 27)(u - 1) = 0$

$u = 27$ *or* $u = 1$

$x^3 = 27$ *or* $x^3 = 1$

$x = 3$ *or* $x = 1$

Both 3 and 1 check. They are the solutions.

50. $-2, 1$

51.

52. $-3, -1, 1, 4$

53.

Exercise Set 8.6

1. $y = kx$

$28 = k \cdot 7$ Substituting

$4 = k$

The variation constant is 4.

The equation of variation is $y = 4x$.

2. $k = \frac{5}{12}$; $y = \frac{5}{12}x$

3. $y = kx$

$3.4 = k \cdot 2$ Substituting

$1.7 = k$

The variation constant is 1.7.

The equation of variation is $y = 1.7x$.

4. $k = \frac{2}{5}$; $y = \frac{2}{5}x$

5. $y = kx$

$30 = k \cdot 8$ Substituting

$\frac{30}{8} = k$

$\frac{15}{4} = k$

The variation constant is $\frac{15}{4}$.

The equation of variation is $y = \frac{15}{4}x$.

6. $k = 3; y = 3x$

7. $y = kx$

$0.8 = k(0.5)$ Substituting

$8 = k \cdot 5$ Clearing decimals

$\frac{8}{5} = k$

$1.6 = k$

The variation constant is 1.6.

The equation of variation is $y = 1.6x$.

8. $k = 1.5;\ y = 1.5x$

9. ***Familiarize***. Because of the phrase "I ... varies directly as ... V," we express the current as a function of the voltage. Thus we have $I(V) = kV$. We know that $I(15) = 5$.

Translate. We find the variation constant and then find the equation of variation.

$I(V) = kV$

$I(15) = k \cdot 15$ Replacing V with 15

$5 = k \cdot 15$ Replacing $I(15)$ with 5

$\frac{5}{15} = k$

$\frac{1}{3} = k$ Variation constant

The equation of variation is $I(V) = \frac{1}{3}V$.

Carry out. We compute $I(18)$.

$I(V) = \frac{1}{3}V$

$I(18) = \frac{1}{3} \cdot 18$ Replacing V with 18

$= 6$

Check. Reexamine the calculations. Note that the answer seems reasonable since $15/5 = 18/6$.

State. The current is 6 amperes when 18 volts is applied.

10. $33\frac{1}{3}$ cm

11. ***Familiarize***. Because N varies directly as the number of people P using the cans, we write N as a function of P: $N(P) = kP$. We know that $N(250) = 60,000$.

Translate.

$N(P) = kP$

$N(250) = k \cdot 250$ Replacing P with 250

$60,000 = k \cdot 250$ Replacing $N(250)$ with 60,000

$\frac{60,000}{250} = k$

$240 = k$ Variation constant

$N(P) = 240P$ Equation of variation

Carry out. Find $N(1,008,000)$.

$N(P) = 240P$

$N(1,008,000) = 240 \cdot 1,008,000$

$= 241,920,000$

Check. Reexamine the calculation.

State. 241,920,000 aluminum cans are used each year in Dallas.

12. \$4.29

13. ***Familiarize***. Because W varies directly as the total mass, we write $W(m) = km$. We know that $W(96) = 64$.

Translate.

$W(m) = km$

$W(96) = k \cdot 96$ Replacing m with 96

$64 = k \cdot 96$ Replacing $W(96)$ with 64

$\frac{2}{3} = k$ Variation constant

$W(m) = \frac{2}{3}m$ Equation of variation

Carry out. Find $W(60)$.

$W(m) = \frac{2}{3}m$

$W(60) = \frac{2}{3} \cdot 60$

$= 40$

Check. Reexamine the calculations.

State. There are 40 kg of water in a 60 kg person.

14. 40 lb

15. ***Familiarize***. The amount A of lead released varies directly as the population P. We write A as a function of P: $A(P) = kP$. We know that $A(12,500) = 385$.

Translate.

$A(P) = kP$

$A(12,500) = k \cdot 12,500$ Replacing P with 12,500

$385 = k \cdot 12,500$ Replacing $A(12,500)$ with 385

$0.0308 = k$ Variation constant

$A(P) = 0.0308P$ Equation of variation

Carry out. Find $A(250,000,000)$.

$$A(P) = 0.0308P$$
$$A(250,000,000) = 0.0308(250,000,000) = 7,700,000$$

Check. Reexamine the calculations.

State. 7,700,000 tons of lead were released nationally.

16. 3.36

17. $y = \frac{k}{x}$

$6 = \frac{k}{10}$ Substituting

$60 = k$

The variation constant is 60.

The equation of variation is $y = \frac{60}{x}$.

18. $k = 64$; $y = \frac{64}{x}$

19. $y = \frac{k}{x}$

$4 = \frac{k}{3}$ Substituting

$12 = k$

The variation constant is 12.

The equation of variation is $y = \frac{12}{x}$.

20. $k = 36$; $y = \frac{36}{x}$

21. $y = \frac{k}{x}$

$12 = \frac{k}{3}$ Substituting

$36 = k$

The variation constant is 36.

The equation of variation is $y = \frac{36}{x}$.

22. $k = 45$; $y = \frac{45}{x}$

23. $y = \frac{k}{x}$

$27 = \frac{k}{\frac{1}{3}}$ Substituting

$9 = k$

The variation constant is 9.

The equation of variation is $y = \frac{9}{x}$.

24. $k = 9$; $y = \frac{9}{x}$

25. ***Familiarize***. Because t varies inversely as r, we express t as a function of r. Thus we write $t(r) = k/r$. We know that $t(600) = 45$.

Translate.

$t(r) = \frac{k}{r}$

$t(600) = \frac{k}{600}$ Replacing r with 600

$45 = \frac{k}{600}$ Replacing $t(600)$ with 45

$27,000 = k$ Variation constant

$t(r) = \frac{27,000}{r}$ Equation of variation

Carry out. Find $t(1000)$.

$$t(1000) = \frac{27,000}{1000} = 27$$

Check. Reexamine the calculation. Note that, as expected, when the rate increases the time decreases.

State. It will take the pump 27 min to empty the tank at the rate of 1000 kL/min.

26. $\frac{2}{9}$ ampere

27. ***Familiarize***. Because V varies inversely as P, we write $V(P) = k/P$. We know that $V(32) = 200$.

Translate.

$V(P) = \frac{k}{P}$

$V(32) = \frac{k}{32}$ Replacing P with 32

$200 = \frac{k}{32}$ Replacing $V(32)$ with 200

$6400 = k$ Variation constant

$V(P) = \frac{6400}{P}$ Equation of variation

Carry out. Find $V(40)$.

$$V(40) = \frac{6400}{40} = 160$$

Check. Reexamine the calculations.

State. The volume will be 160 cm^3.

28. 3.5 hr

29. ***Familiarize***. Because W varies inversely as F, we write $W(F) = k/F$. We know that $W(1200) = 300$.

Translate.

$$W(F) = \frac{k}{F}$$

$$W(1200) = \frac{k}{1200} \quad \text{Replacing } F \text{ with } 1200$$

$$300 = \frac{k}{1200} \quad \text{Replacing } W(1200) \text{ with } 300$$

$$360,000 = k \quad \text{Variation constant}$$

$$W(F) = \frac{360,000}{F} \quad \text{Equation of variation}$$

Carry out. Find $W(800)$.

$$W(800) = \frac{360,000}{800} = 450$$

Check. Reexamine the calculations.

State. The length of the wave is 450 meters.

30. $5\frac{5}{7}$ hr

31.
$$y = kx^2$$
$$6 = k \cdot 3^2 \quad \text{Substituting}$$
$$6 = 9k$$
$$\frac{6}{9} = k$$
$$\frac{2}{3} = k \quad \text{Variation constant}$$

The equation of variation is $y = \frac{2}{3}x^2$.

32. $y = 15x^2$

33.
$$y = \frac{k}{x^2}$$
$$6 = \frac{k}{3^2} \quad \text{Substituting}$$
$$6 = \frac{k}{9}$$
$$6 \cdot 9 = k$$
$$54 = k \quad \text{Variation constant}$$

The equation of variation is $y = \frac{54}{x^2}$.

34. $y = \frac{0.0015}{x^2}$

35.
$$y = kxz$$
$$56 = k \cdot 14 \cdot 8 \quad \text{Substituting 56 for } y, \text{ 14 for } x, \text{ and 8 for } z$$
$$56 = 112k$$
$$0.5 = k \quad \text{Variation constant}$$

The equation of variation is $y = 0.5xz$.

36. $y = \frac{5x}{z}$

37.
$$y = kxz^2$$
$$105 = k \cdot 14 \cdot 5^2 \quad \text{Substituting 105 for } y, \text{ 14 for } x, \text{ and 5 for } z$$
$$105 = 350k$$
$$\frac{105}{350} = k$$
$$0.3 = k$$

The equation of variation is $y = 0.3xz^2$.

38. $y = \frac{xz}{w}$

39.
$$y = k \cdot \frac{wx^2}{z}$$
$$49 = k \cdot \frac{3 \cdot 7^2}{12} \quad \text{Substituting}$$
$$4 = k \quad \text{Variation constant}$$

The equation of variation is $y = \frac{4wx^2}{z}$.

40. $y = \frac{6x}{wz^2}$

41.
$$y = k \cdot \frac{xz}{wp}$$
$$\frac{3}{28} = k \cdot \frac{3 \cdot 10}{7 \cdot 8} \quad \text{Substituting}$$
$$\frac{3}{28} = k \cdot \frac{30}{56}$$
$$\frac{3}{28} \cdot \frac{56}{30} = k$$
$$\frac{1}{5} = k \quad \text{Variation constant}$$

The equation of variation is $y = \frac{xz}{5wp}$.

42. $y = \frac{5xz}{4w^2}$

43. ***Familiarize.*** Because d varies directly as the square of r, we write $d = kr^2$. We know that $d = 200$ when $r = 60$.

Translate.
$$d = kr^2$$
$$200 = k(60)^2$$
$$\frac{1}{18} = k$$
$$d = \frac{1}{18}r^2 \quad \text{Equation of variation}$$

Carry out. Substitute 72 for d and solve for r.
$$72 = \frac{1}{18}r^2$$
$$1296 = r^2$$
$$36 = r$$

Check. Recheck the calculations and perhaps make an estimate to see if the answer seems reasonable.

State. The car can go 36 mph.

44. 220 cm^3

45. ***Familiarize.*** I varies inversely as d^2, so we write $I = k/d^2$. We know that $I = 90$ when $d = 5$.

Translate. Find k.

$$I = \frac{k}{d^2}$$
$$90 = \frac{k}{5^2}$$
$$2250 = k$$

$$I = \frac{2250}{d^2} \quad \text{Equation of variation}$$

Carry out. Substitute 40 for I and solve for d.

$$40 = \frac{2250}{d^2}$$
$$d^2 = \frac{2250}{40}$$
$$d = 7.5$$

We subtract to find how much farther this is:

$7.5 - 5 = 2.5$

Check. Reexamine the calculations.

State. It would be 2.5 m farther.

46. 6.25 km

47. ***Familiarize.*** W varies inversely as d^2, so we write $W = k/d^2$. We know that $W = 100$ when $d = 6400$.

Translate. Find k.

$$W = \frac{k}{d^2}$$
$$100 = \frac{k}{(6400)^2}$$
$$4,096,000,000 = k$$

$$W = \frac{4,096,000,000}{d^2} \quad \text{Equation of variation}$$

Carry out. Substitute 64 for w and solve for d.

$$64 = \frac{4,096,000,000}{d^2}$$
$$d^2 = \frac{4,096,000,000}{64}$$
$$d = 8000$$

Note that a distance of 8000 km from the center of the earth is 8000−6400, or 1600 km, above the earth.

Check. Recheck the calculations.

State. The astronaut must be 1600 km above the earth in order to weigh 64 lb.

48. 2 mm

49. ***Familiarize.*** The drag D varies jointly as the surface area A and velocity v, so we write $D = kAv$. We know that $D = 222$ when $A = 37.8$ and $v = 40$.

Translate. Find k.

$$D = kAv$$
$$222 = k(37.8)(40)$$
$$\frac{222}{37.8(40)} = k$$
$$\frac{37}{252} = k$$
$$D = \frac{37}{252}Av \quad \text{Equation of variation}$$

Carry out. Substitute 51 for A and 430 for D and solve for v.

$$430 = \frac{37}{252} \cdot 51 \cdot v$$
$$57.42 \text{ mph} \approx v$$

(If we had used the rounded value 0.1468 for k, the resulting speed would have been approximately 57.43 mph.)

Check. Reexamine the calculations.

State. The car must travel about 57.42 mph.

50. About 8.2 mph

51. Use the slope-interest form, $y = mx + b$, where m is the slope and b is the y-intercept.

$$y = -\frac{2}{3}x - 5$$

52. $y - 7 = -\frac{2}{7}(x - 4)$

53. $$\frac{\frac{1}{ab} - \frac{2}{bc}}{\frac{3}{ab} + \frac{4}{bc}} = \frac{\left(\frac{1}{ab} - \frac{2}{bc}\right)abc}{\left(\frac{3}{ab} + \frac{4}{bc}\right)abc} = \frac{c - 2a}{3c + 4a}$$

54. $\dfrac{18}{-x - 10}$

55. $f(x) = x^3 - 2x^2$

$f(3) = 3^3 - 2 \cdot 3^2 = 27 - 18 = 9$

56. $9x^2 - 12xy + 4y^2$

57. ◈

58. ◈

59. ◈

60. ◈

61. Write y as a function of x, and then substitute $0.5x$ for x.

$$y(x) = \frac{k}{x^3}$$
$$y(0.5x) = \frac{k}{(0.5x)^3} = \frac{k}{0.125x^3} = \frac{1}{0.125} \cdot \frac{k}{x^3} = 8 \cdot y(x)$$

y is multiplied by 8.

62. Q varies directly as the square of p and inversely as the cube of q.

63. $W = \dfrac{km_1M_1}{d^2}$

W varies jointly as m_1 and M_1 and inversely as the square of d.

64. About 1.697 m

65. a) ***Familiarize.*** We write $N = \dfrac{kP_1P_2}{d^2}$. We let P_1 = the population of Indianapolis and P_2 = the population of Cincinnati. We know that $N = 11,153$ when $P_1 = 752,279$, $P_2 = 358,170$, and $d = 174$.

Translate. We substitute.

$$11,153 = \frac{k(752,279)(358,170)}{(174)^2}$$

Carry out. We solve for k.

$$11,153 = \frac{k(752,279)(358,170)}{(174)^2}$$
$$11,153 = \frac{k(752,279)(358,170)}{30,276}$$
$$337,668,228 = k(752,279)(358,170)$$
$$0.001 \approx k$$

Check. Reexamine the calculations.

State. The value of k is approximately 0.001. The equation of variation is $N = \dfrac{0.001P_1P_2}{d^2}$.

b) ***Familiarize.*** We will use the equation of variation found in part (a): $N = \dfrac{0.001P_1P_2}{d^2}$. We let P_1 = the population of Indianapolis and P_2 = the population of New York. We know that $N = 4270$ when $P_1 = 752,279$ and $P_2 = 7,333,153$.

Translate. We substitute.

$$4270 = \frac{0.001(752,279)(7,333,153)}{d^2}$$

Carry out. We solve for d.

$$4270 = \frac{0.001(752,279)(7,333,153)}{d^2}$$
$$d^2 = \frac{0.001(752,279)(7,333,153)}{4270}$$
$$d^2 \approx 1,291,938$$
$$d \approx 1137$$

Check. Reexamine the calculations.

State. The distance between Indianapolis and New York is approximately 1137 km.

66. \$7.20

67. ***Familiarize.*** Because d varies inversely as s, we write $d(s) = k/s$. We know that $d(0.56) = 50$.

Translate.

$$d(s) = \frac{k}{s}$$
$$d(0.56) = \frac{k}{0.56} \quad \text{Replacing } s \text{ with } 0.56$$
$$50 = \frac{k}{0.56} \quad \text{Replacing } d(0.56) \text{ with } 50$$
$$28 = k$$
$$d(s) = \frac{28}{s} \quad \text{Equation of variation}$$

Carry out. Find $d(0.40)$.

$$d(0.40) = \frac{28}{0.40} = 70$$

Check. Reexamine the calculations. Also observe that, as expected, when d decreases, then s increases.

State. The distance is 70 yd.

Exercise Set 8.7

1. $f(x) = x^2$

See Example 1 in the text.

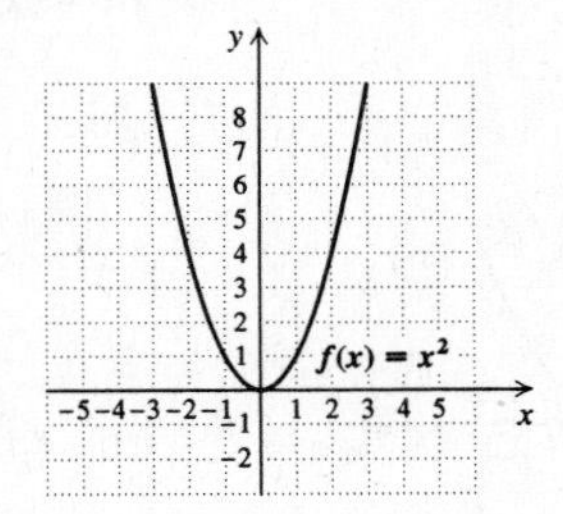

2.

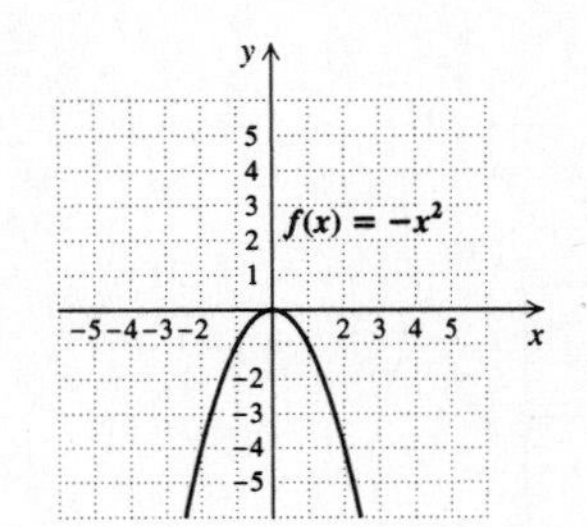

3. $f(x) = -2x^2$

We choose some numbers for x and compute $f(x)$ for each one. Then we plot the ordered pairs $(x, f(x))$ and connect them with a smooth curve.

x	$f(x) = -4x^2$
0	0
1	−2
2	−8
−1	−2
−2	−8

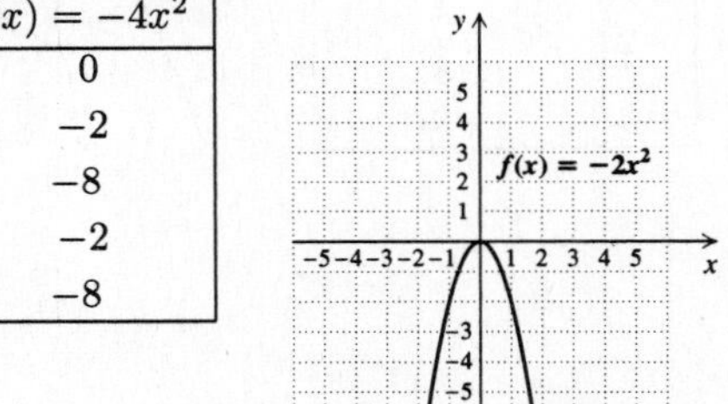

4.

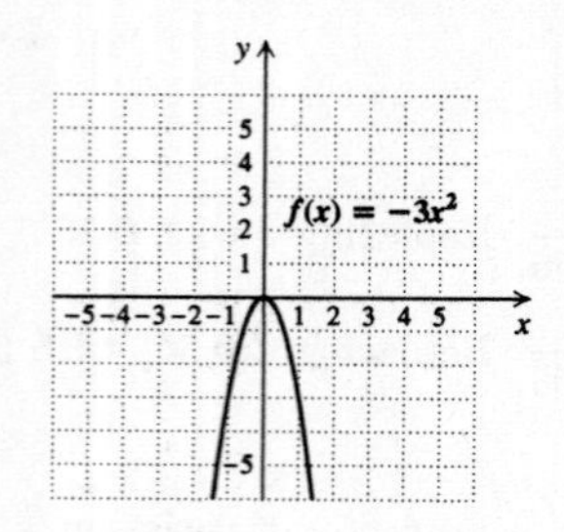

5. $g(x) = \frac{1}{4}x^2$

x	$g(x) = \frac{1}{4}x^2$
0	0
1	$\frac{1}{4}$
2	1
3	$\frac{9}{4}$
-1	$\frac{1}{4}$
-2	1
-3	$\frac{9}{4}$

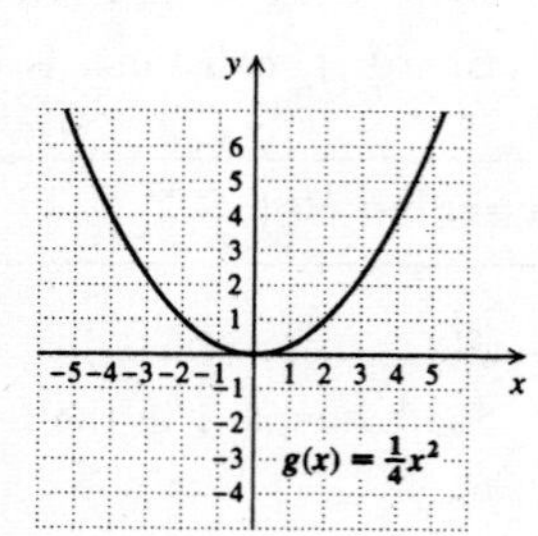

6.

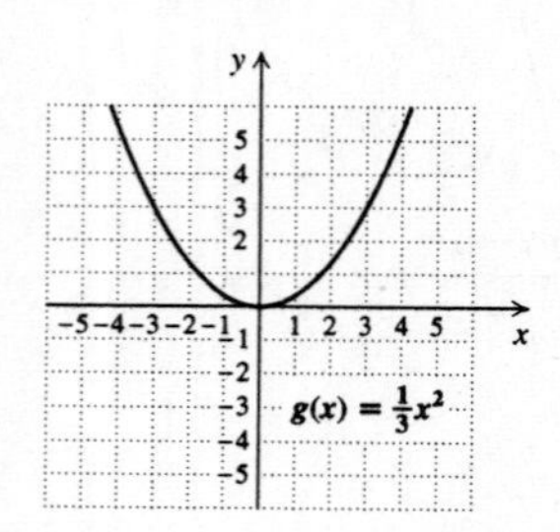

7. $h(x) = -\frac{1}{3}x^2$

x	$h(x) = -\frac{1}{3}x^2$
0	0
1	$-\frac{1}{3}$
2	$-\frac{4}{3}$
3	-3
-1	$-\frac{1}{3}$
-2	$-\frac{4}{3}$
-3	-3

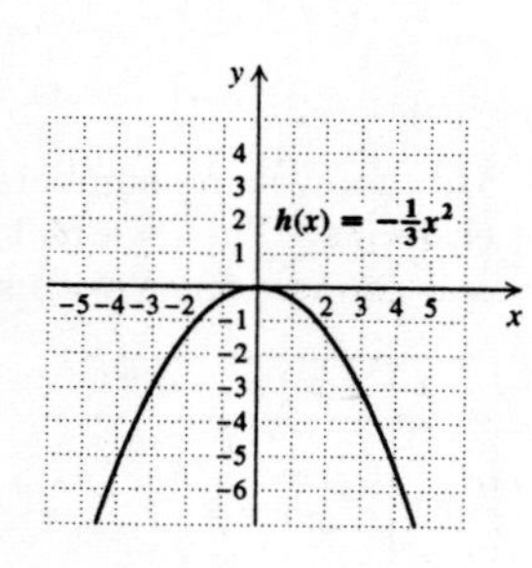

8.

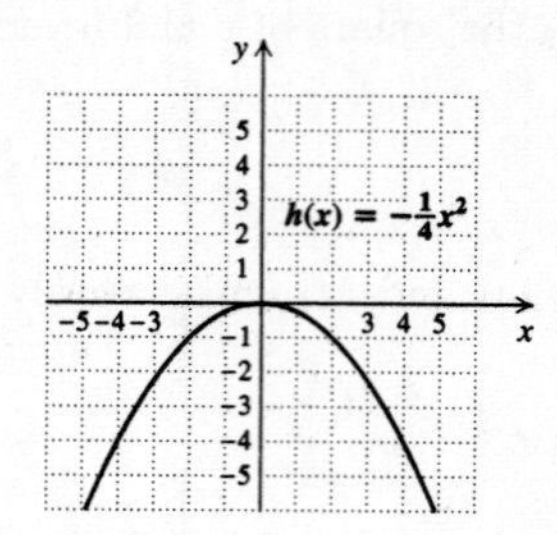

9. $f(x) = \frac{3}{2}x^2$

x	$f(x) = \frac{3}{2}x^2$
0	0
1	$\frac{3}{2}$
2	6
-1	$\frac{3}{2}$
-2	6

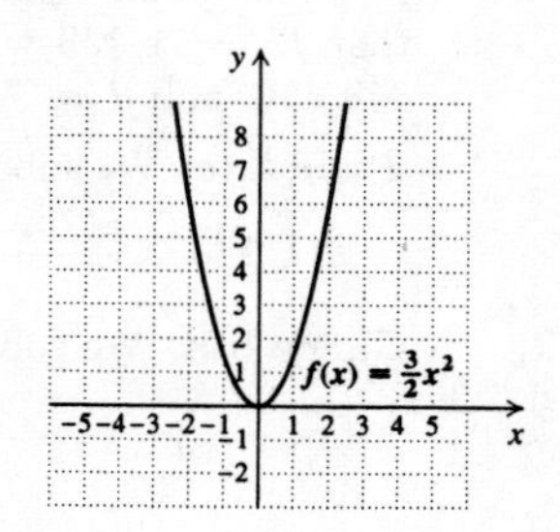

10.

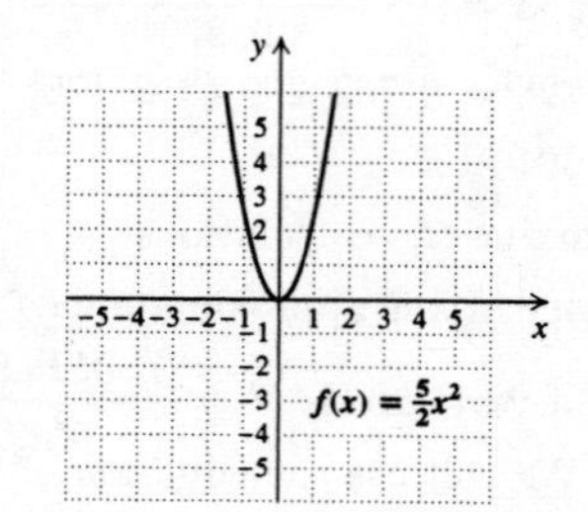

11. $g(x) = (x+1)^2 = [x-(-1)]^2$

We know that the graph of $g(x) = (x+1)^2$ looks like the graph of $f(x) = x^2$ (see Exercise 1) but moved to the left 1 unit.

Vertex: $(-1, 0)$, axis of symmetry: $x = -1$

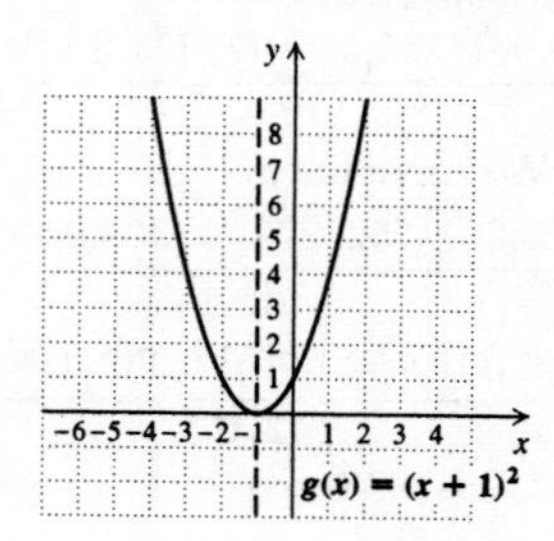

12. Vertex: $(-4, 0)$, axis of symmetry: $x = -4$

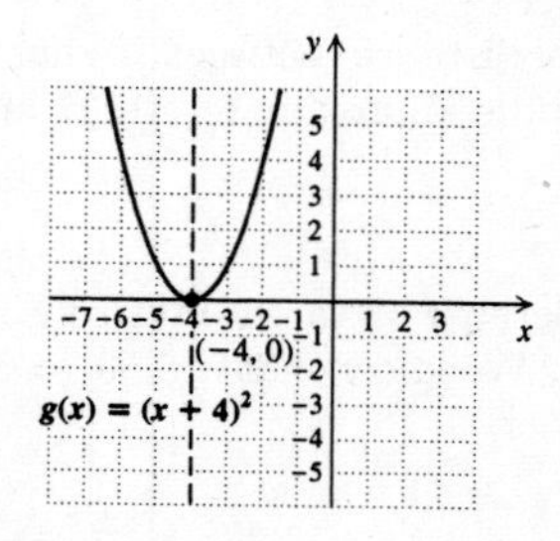

13. $f(x) = (x-2)^2$

The graph of $f(x) = (x-2)^2$ looks like the graph of $f(x) = x^2$ (see Exercise 1) but moved to the right 2 units.

Vertex: $(2, 0)$, axis of symmetry: $x = 2$

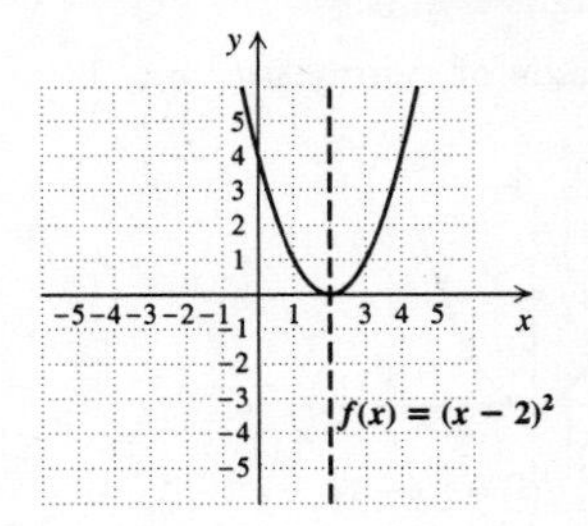

14. Vertex: $(1, 0)$, axis of symmetry: $x = 1$

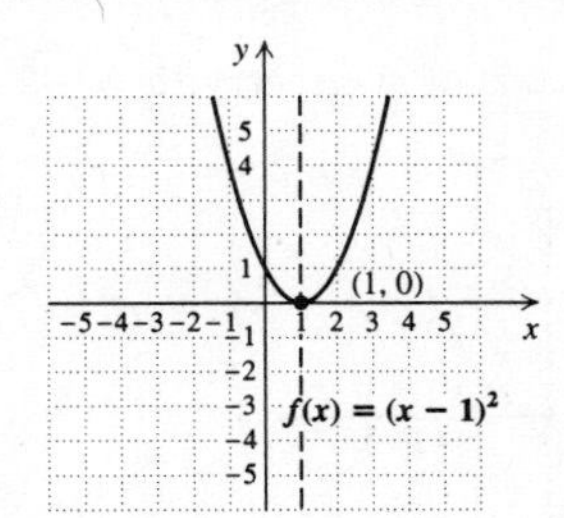

15. $h(x) = (x-3)^2$

The graph of $h(x) = (x-3)^2$ looks like the graph of $f(x) = x^2$ (see Exercise 1) but moved to the right 3 units.

Vertex: $(3, 0)$, axis of symmetry: $x = 3$

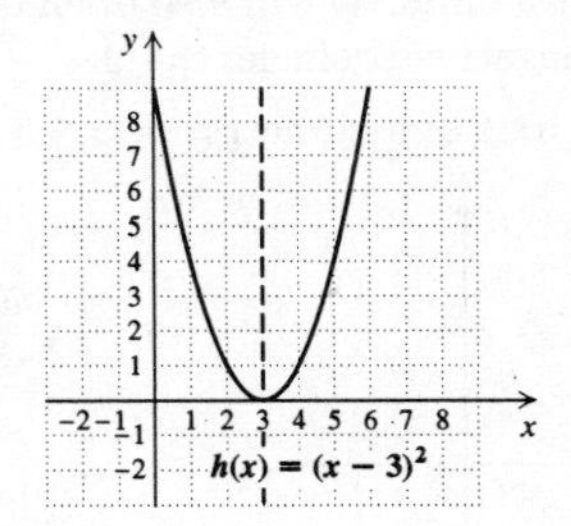

16. Vertex: $(4, 0)$, axis of symmetry: $x = 4$

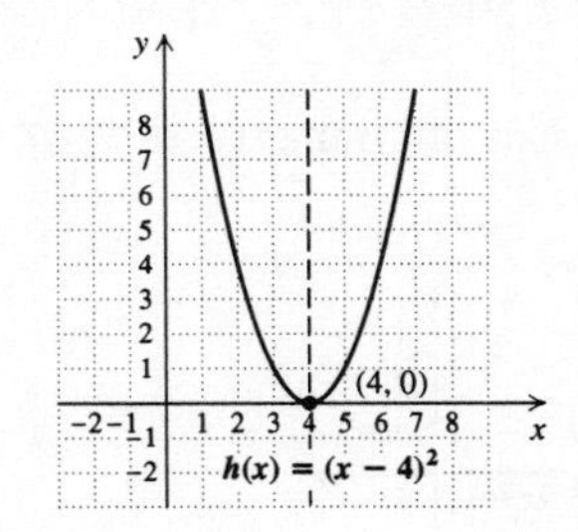

17. $f(x) = -(x+4)^2 = -[x-(-4)]^2$

The graph of $f(x) = -(x+4)^2$ looks like the graph of $f(x) = x^2$ (see Exercise 1) but moved to the left 4 units. It will also open downward because of the negative coefficient, -1.

Vertex: $(-4, 0)$, axis of symmetry: $x = -4$

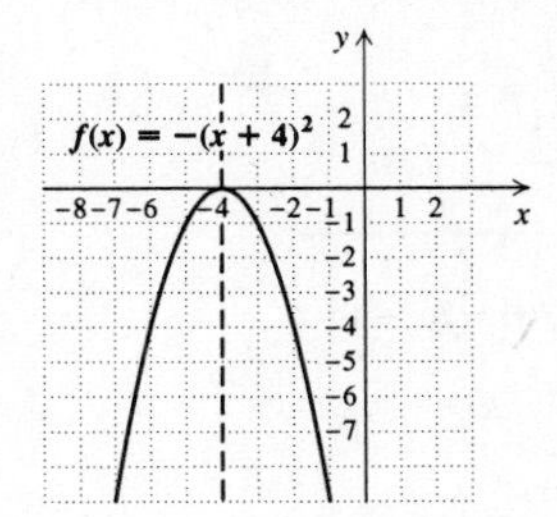

18. Vertex: $(2, 0)$, axis of symmetry: $x = 2$

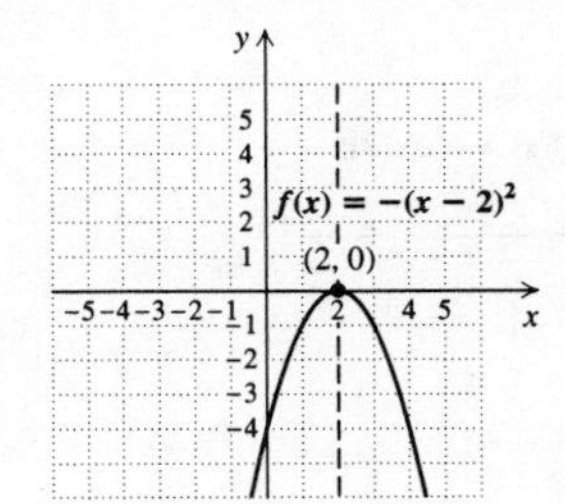

19. $g(x) = -(x-1)^2$

The graph of $g(x) = -(x-1)^2$ looks like the graph of $f(x) = x^2$ (see Exercise 1) but moved to the right 1 unit. It will also open downward because of the negative coefficient, -1.

Vertex: $(1, 0)$, axis of symmetry: $x = 1$

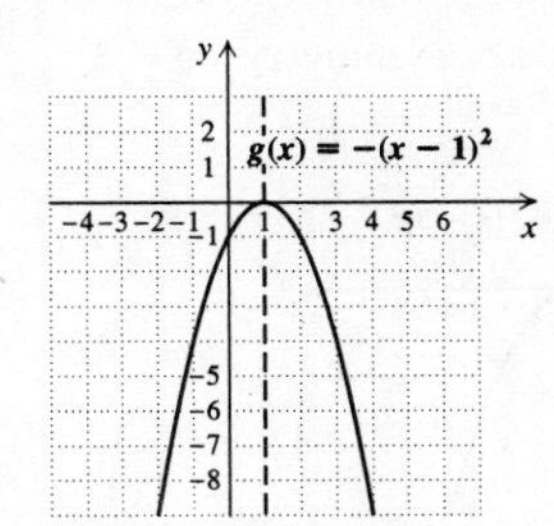

20. Vertex: $(-5, 0)$, axis of symmetry: $x = -5$

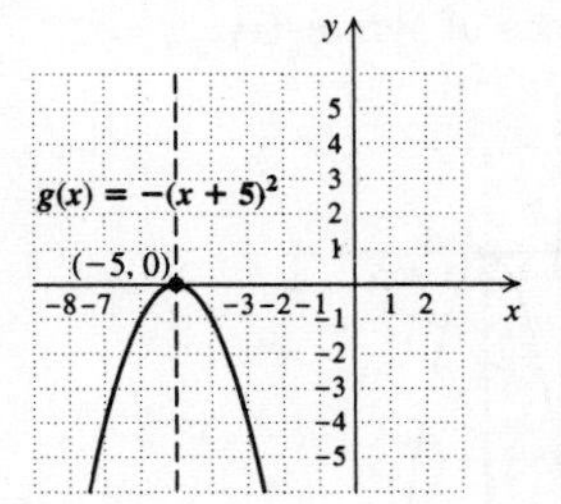

21. $f(x) = 2(x+1)^2$

The graph of $f(x) = 2(x+1)^2$ looks like the graph of $h(x) = 2x^2$ (see graph following Example 1) but moved to the left 1 unit.

Vertex: $(-1, 0)$, axis of symmetry: $x = -1$

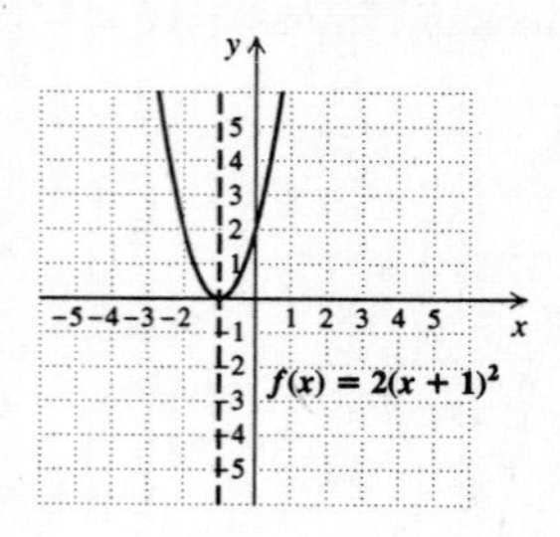

22. Vertex: $(-4, 0)$, axis of symmetry: $x = -4$

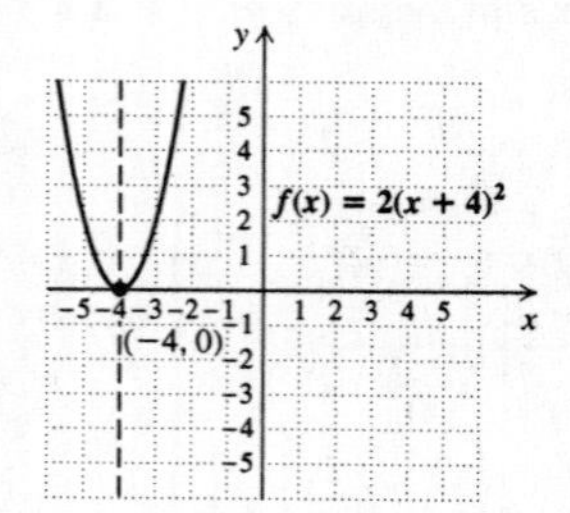

23. $h(x) = -\frac{1}{2}(x-3)^2$

The graph of $h(x) = -\frac{1}{2}(x-3)^2$ looks like the graph of $g(x) = \frac{1}{2}x^2$ (see graph following Example 1) but moved to the right 3 units. It will also open downward because of the negative coefficient, $-\frac{1}{2}$.

Vertex: $(3, 0)$, axis of symmetry: $x = 3$

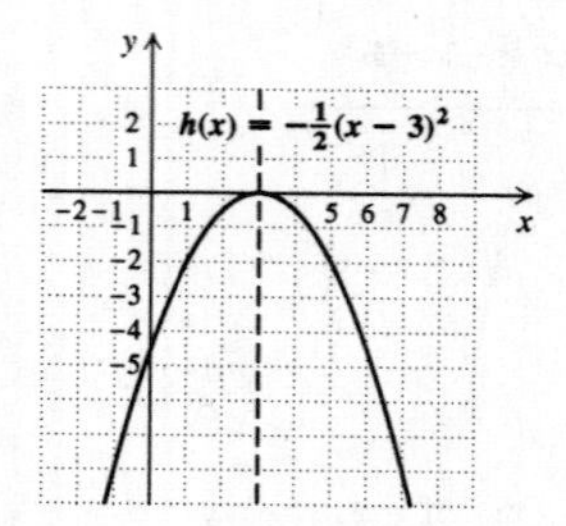

24. Vertex: $(2, 0)$, axis of symmetry: $x = 2$

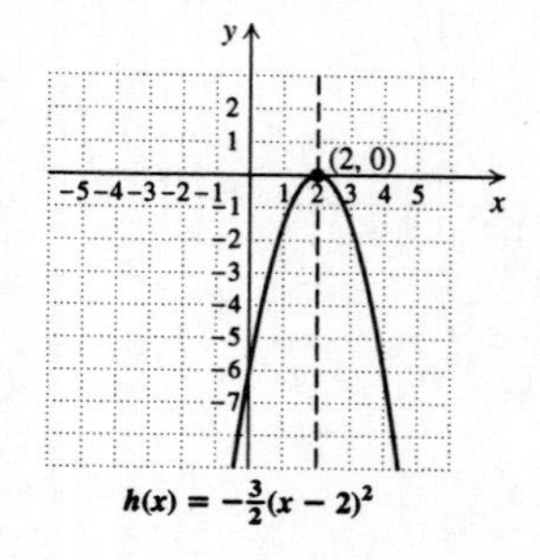

25. $f(x) = \frac{1}{2}(x-1)^2$

The graph of $f(x) = \frac{1}{2}(x-1)^2$ looks like the graph of $g(x) = \frac{1}{2}x^2$ (see graph following Example 1) but moved to the right 1 unit.

Vertex: $(1, 0)$, axis of symmetry: $x = 1$

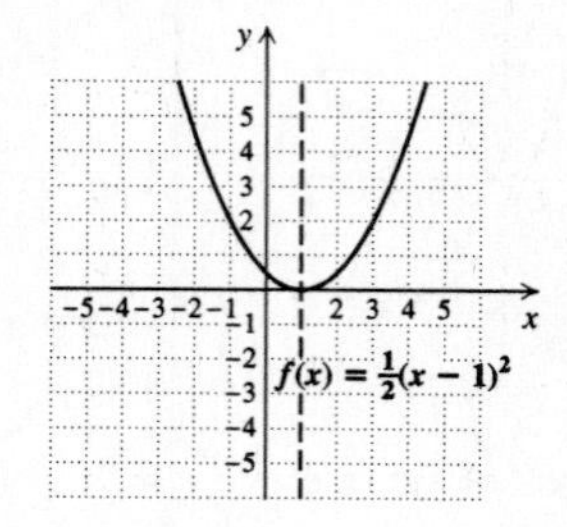

26. Vertex: $(-2, 0)$, axis of symmetry: $x = -2$

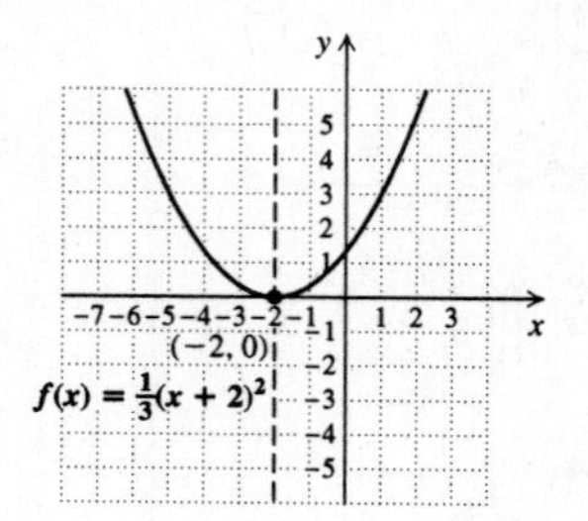

27. $f(x) = -2(x+5)^2 = -2[x-(-5)]^2$

The graph of $f(x) = -2(x+5)^2$ looks like the graph of $h(x) = 2x^2$ (see graph following Example 1) but moved to the left 5 units. It will also open downward because of the negative coefficient, -2.

Vertex: $(-5, 0)$, axis of symmetry: $x = -5$

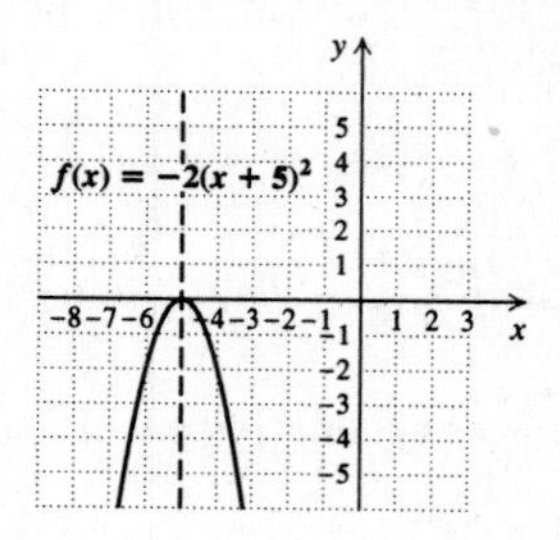

28. Vertex: $(-7, 0)$, axis of symmetry: $x = -7$

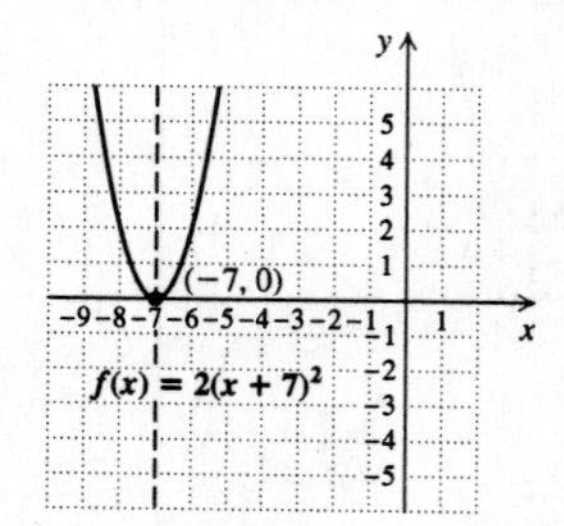

29. $h(x) = -3\left(x - \frac{1}{2}\right)^2$

The graph of $h(x) = -3\left(x - \frac{1}{2}\right)^2$ looks like the graph of $f(x) = -3x^2$ (see Exercise 4) but moved to the right $\frac{1}{2}$ unit.

Vertex: $\left(\frac{1}{2}, 0\right)$, axis of symmetry: $x = \frac{1}{2}$

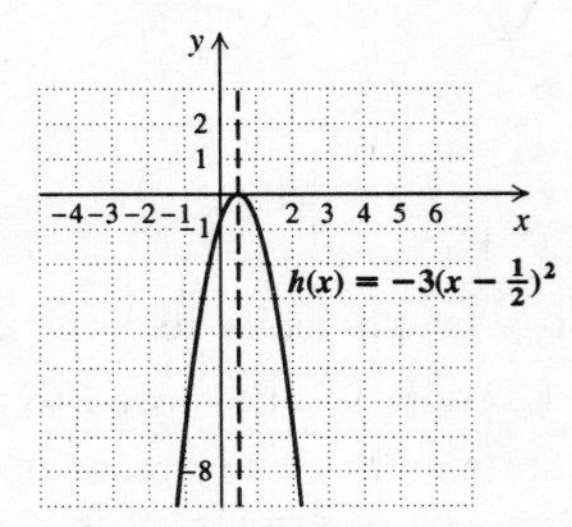

30. Vertex: $\left(-\frac{1}{2}, 0\right)$, axis of symmetry: $x = -\frac{1}{2}$

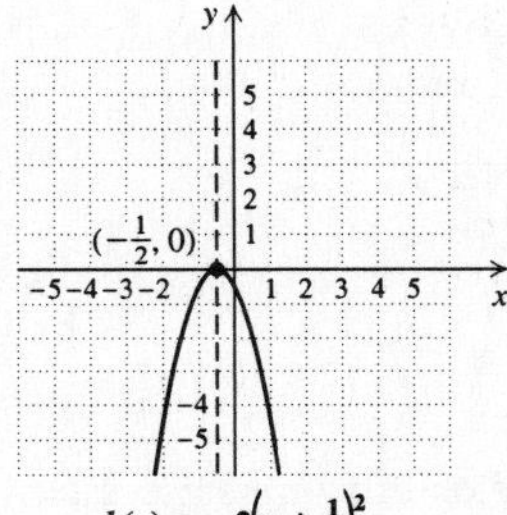

31. $f(x) = (x - 5)^2 + 1$

We know that the graph looks like the graph of $f(x) = x^2$ (see Example 1) but moved to the right 5 units and up 1 unit. The vertex is $(5, 1)$, and the axis of symmetry is $x = 5$. Since the coefficient of $(x - 5)^2$ is positive $(1 > 0)$, there is a minimum function value, 1.

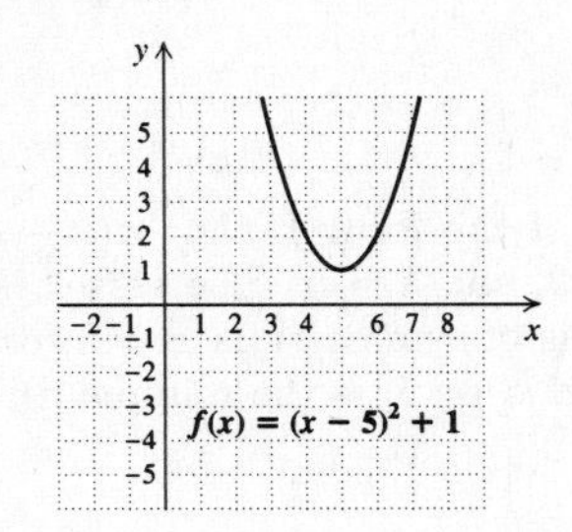

32. Vertex: $(-3, -2)$, axis of symmetry: $x = -3$

Minimum: -2

$f(x) = (x + 3)^2 - 2$

33. $f(x) = (x + 1)^2 - 2$

We know that the graph looks like the graph of $f(x) = x^2$ (see Example 1) but moved to the left 1 unit and down 2 units. The vertex is $(-1, -2)$, and the axis of symmetry is $x = -1$. Since the coefficient of $(x + 1)^2$ is positive $(1 > 0)$, there is a minimum function value, -2.

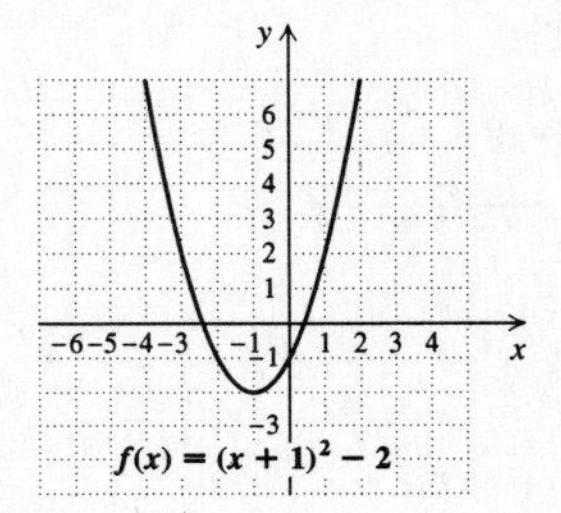

34. Vertex: $(1, 2)$, axis of symmetry: $x = 1$

Minimum: 2

$f(x) = (x - 1)^2 + 2$

35. $g(x) = (x + 4)^2 + 1$

We know that the graph looks like the graph of $f(x) = x^2$ (see Example 1) but moved to the left 4 units and up 1 unit. The vertex is $(-4, 1)$, and the axis of symmetry is $x = -4$. Since the coefficient of $(x + 4)^2$ is positive $(1 > 0)$, there is a minimum function value, 1.

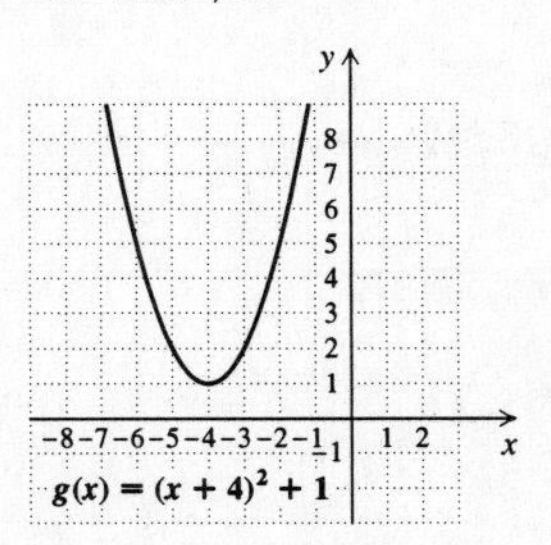

36. Vertex: $(2, -4)$, axis of symmetry: $x = 2$

Maximum: -4

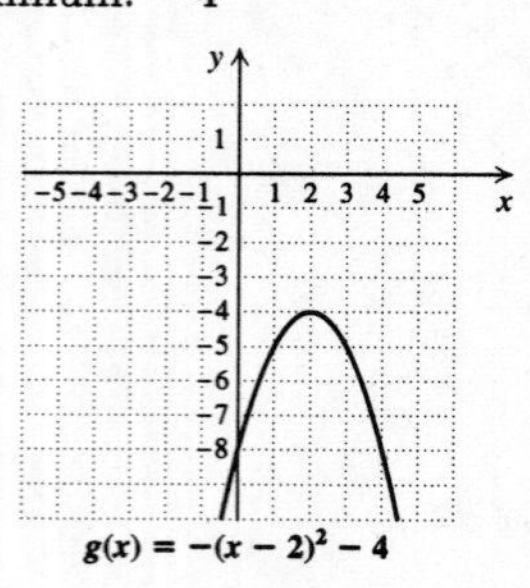

37. $h(x) = -2(x-1)^2 - 3$

We know that the graph looks like the graph of $h(x) = 2x^2$ (see graph following Example 1) but moved to the right 1 unit and down 3 units and turned upside down. The vertex is $(1, -3)$, and the axis of symmetry is $x = 1$. The maximum function value is -3.

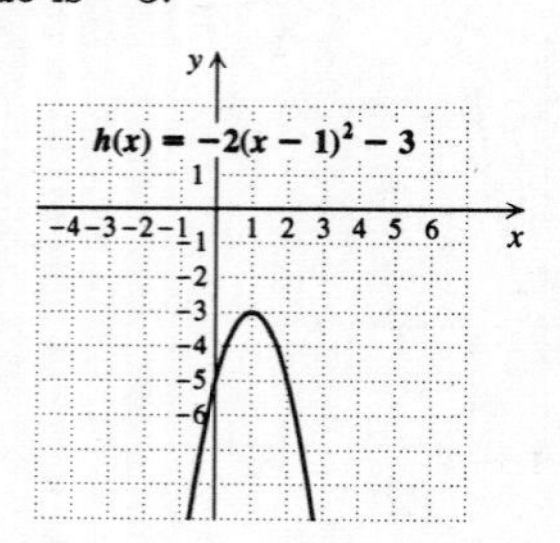

38. Vertex: $(-1, 4)$, axis of symmetry: $x = -1$

Maximum: 4

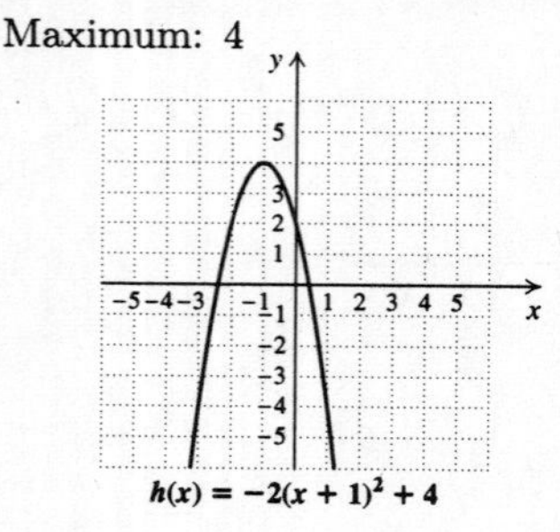

39. $f(x) = 2(x+4)^2 + 1$

We know that the graph looks like the graph of $f(x) = 2x^2$ (see graph following Example 1) but moved to the left 4 units and up 1 unit. The vertex is $(-4, 1)$, the axis of symmetry is $x = -4$, and the minimum function value is 1.

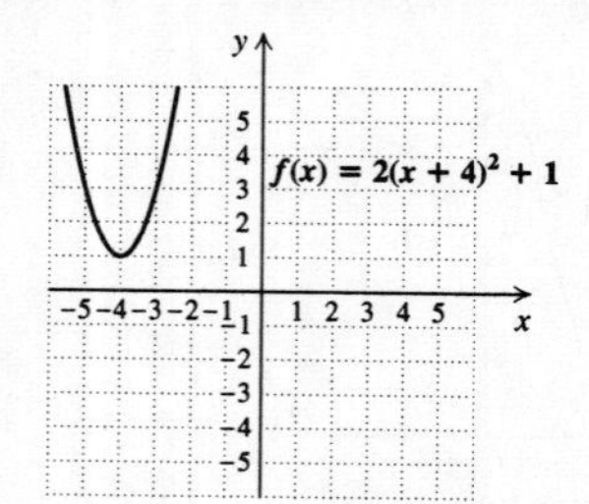

40. Vertex: $(5, -3)$, axis of symmetry: $x = 5$

Minimum: -3

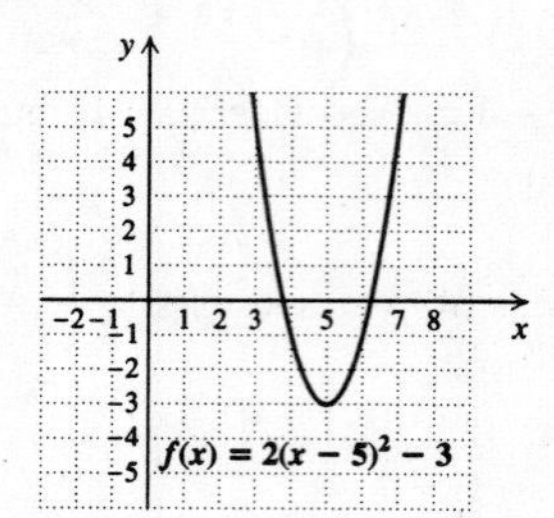

41. $g(x) = -\frac{3}{2}(x-1)^2 + 2$

We know that the graph looks like the graph of $f(x) = \frac{3}{2}x^2$ (see Exercise 9) but moved to the right 1 unit and up 2 units and turned upside down. The vertex is $(1, 2)$, the axis of symmetry is $x = 1$, and the maximum function value is 2.

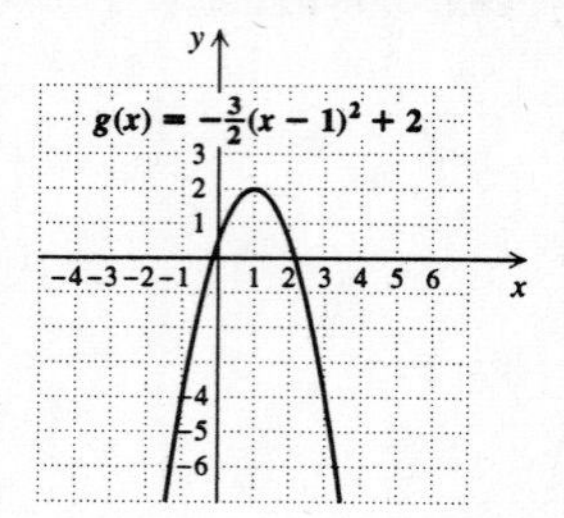

42. Vertex: $(-2, -1)$, axis of symmetry: $x = -2$

Minimum: -1

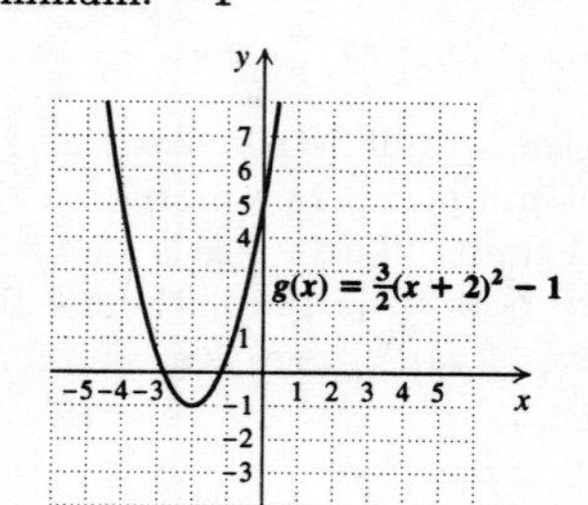

43. $f(x) = 8(x-9)^2 + 5$

This function is of the form $f(x) = a(x-h)^2 + k$ with $a = 8$, $h = 9$, and $k = 5$. The vertex is (h, k), or $(9, 5)$. The axis of symmetry is $x = h$, or $x = 9$. Since $a > 0$, then k, or 5, is the minimum function value.

44. Vertex: $(-5, -8)$

Axis of symmetry: $x = -5$

Minimum: -8

45. $h(x) = -\frac{2}{7}(x+6)^2 + 11$

This function is of the form $f(x) = a(x-h)^2 + k$ with $a = -\frac{2}{7}$, $h = -6$, and $k = 11$. The vertex is (h, k), or $(-6, 11)$. The axis of symmetry is $x = h$, or $x = -6$. Since $a < 0$, then k, or 11, is the maximum function value.

46. Vertex: $(7, -9)$

Axis of symmetry: $x = 7$

Maximum: -9

47. $f(x) = 5\left(x + \frac{1}{4}\right)^2 - 13$

This function is of the form $f(x) = a(x - h)^2 + k$ with $a = 5$, $h = -\frac{1}{4}$, and $k = -13$. The vertex is (h, k), or $\left(-\frac{1}{4}, -13\right)$. The axis of symmetry is $x = h$, or $x = -\frac{1}{4}$. Since $a > 0$, then k, or -13, is the minimum function value.

48. Vertex: $\left(\frac{1}{4}, 19\right)$

Axis of symmetry: $x = \frac{1}{4}$

Minimum: 19

49. $f(x) = \sqrt{2}(x + 4.58)^2 + 65\pi$

This function is of the form $f(x) = a(x - h)^2 + k$ with $a = \sqrt{2}$, $h = -4.58$, and $k = 65\pi$. The vertex is (h, k), or $(-4.58, 65\pi)$. The axis of symmetry is $x = h$, or $x = -4.58$. Since $a > 0$, then k, or 65π, is the minimum function value.

50. Vertex: $(38.2, -\sqrt{34})$

Axis of symmetry: $x = 38.2$

Minimum: $-\sqrt{34}$

51. $3x + 4y = -19$, (1)

$7x - 6y = -29$ (2)

Multiply Equation (1) by 3 and multiply Equation (2) by 2. Then add the equations to eliminate the y-term.

$$\begin{aligned} 9x + 12y &= -57 \\ 14x - 12y &= -58 \\ \hline 23x \qquad &= -115 \\ x &= -5 \end{aligned}$$

Now substitute -5 for x in one of the original equations and solve for y. We use Equation (1).

$$\begin{aligned} 3(-5) + 4y &= -19 \\ -15 + 4y &= -19 \\ 4y &= -4 \\ y &= -1 \end{aligned}$$

The pair $(-5, -1)$ checks and it is the solution.

52. $(-1, 2)$

53. $x^2 + 5x$

We take half the coefficient of x and square it.

$\frac{1}{2} \cdot 5 = \frac{5}{2}$, $\left(\frac{5}{2}\right)^2 = \frac{25}{4}$

Then we have $x^2 + 5x + \frac{25}{4}$.

54. $x^2 - 9x + \frac{81}{4}$

55. ◈

56. ◈

57. ◈

58. ◈

59. Since there is a minimum at $(5, 0)$, the parabola will have the same shape as $f(x) = 2x^2$. It will be of the form $f(x) = 2(x - h)^2 + k$ with $h = 5$ and $k = 0$: $f(x) = 2(x - 5)^2$

60. $f(x) = 2(x - 2)^2$

61. Since there is a maximum at $(-4, 0)$, the parabola will have the same shape as $f(x) = -2x^2$. It will be of the form $f(x) = -2(x - h)^2 + k$ with $h = -4$ and $k = 0$: $g(x) = -2[x - (-4)]^2$, or $f(x) = -2(x + 4)^2$

62. $g(x) = -2x^2 + 3$

63. Since there is a maximum at $(3, 8)$, the parabola will have the same shape as $f(x) = -2x^2$. It will be of the form $f(x) = -2(x - h)^2 + k$ with $h = 3$ and $k = 8$: $g(x) = -2(x - 3)^2 + 8$

64. $f(x) = 2(x + 2)^2 + 3$

65. The maximum value of $g(x)$ is 1 and occurs at the point $(5, 1)$, so for $F(x)$ we have $h = 5$ and $k = 1$. $F(x)$ has the same shape as $f(x)$ and has a minimum, so $a = 3$. Thus, $F(x) = 3(x - 5)^2 + 1$.

66. $F(x) = -\frac{1}{3}(x + 4)^2 - 6$

67. The graph of $y = f(x - 1)$ looks like the graph of $y = f(x)$ moved horizontally 1 unit to the right.

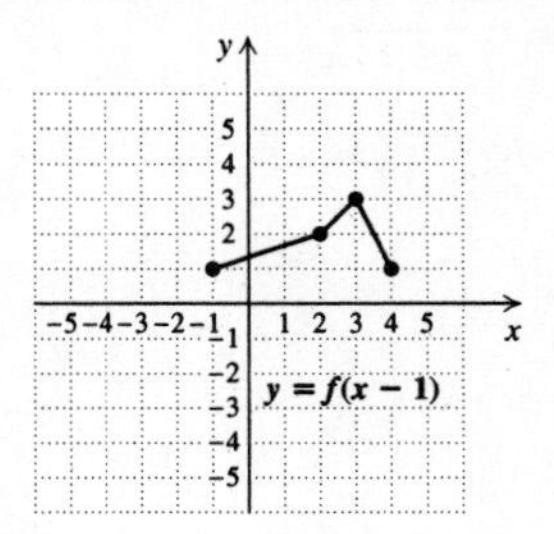

68.

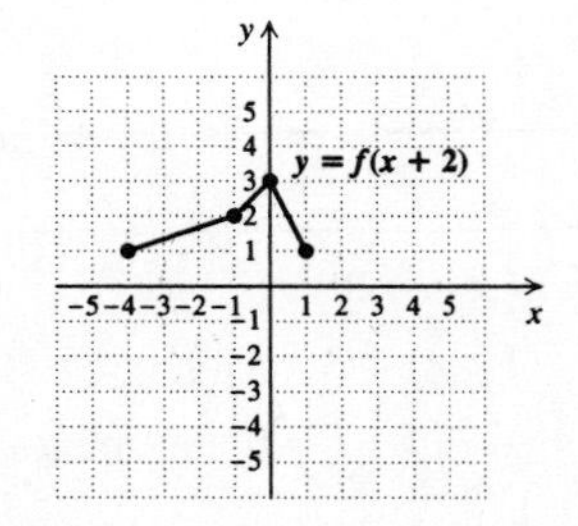

69. The graph of $y = f(x) + 2$ looks like the graph of $y = f(x)$

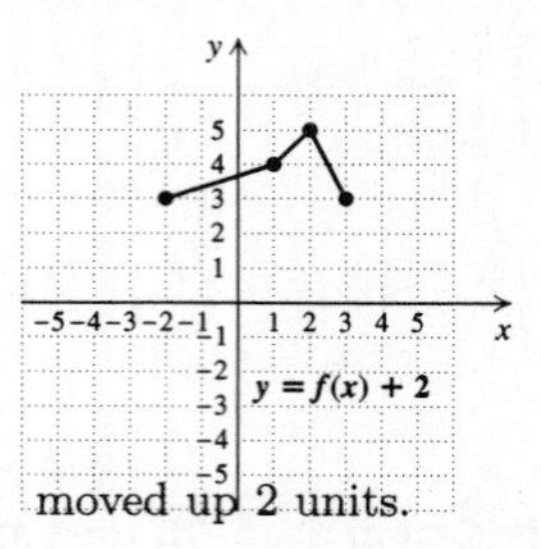

moved up 2 units.

70.

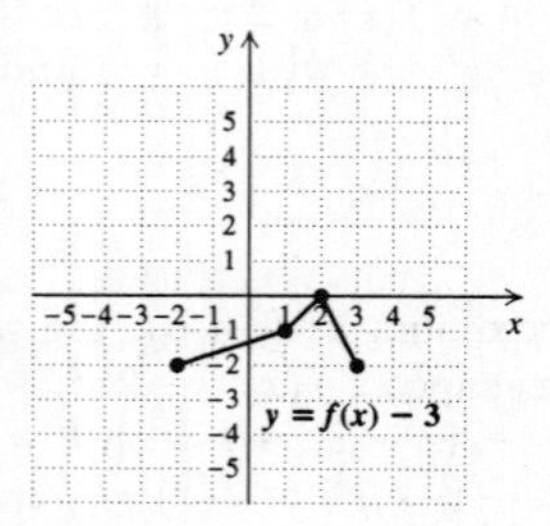

71. The graph of $y = f(x+3) - 2$ looks like the graph of $y = f(x)$ moved horizontally 3 units to the left and also moved down 2 units.

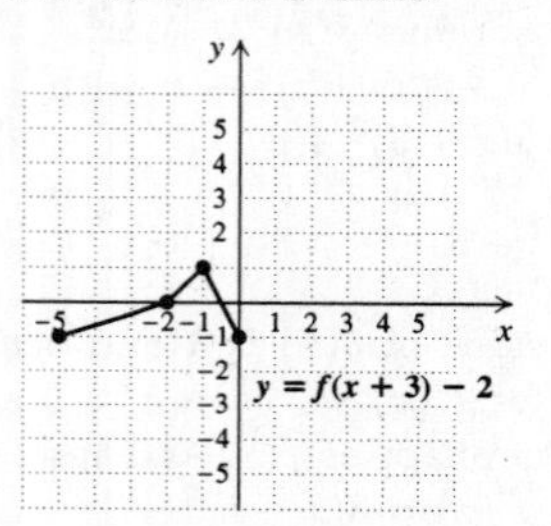

72.

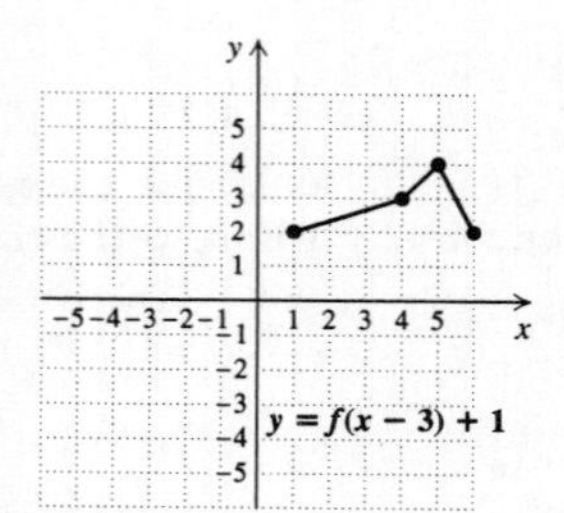

73.

74.

75.

Exercise Set 8.8

1.

$$\begin{aligned} f(x) &= x^2 - 4x + 5 \\ &= (x^2 - 4x + 4 - 4) + 5 \quad \text{Adding } 4 - 4 \\ &= (x^2 - 4x + 4) - 4 + 5 \quad \text{Regrouping} \\ &= (x-2)^2 + 1 \end{aligned}$$

The vertex is $(2, 1)$, the axis of symmetry is $x = 2$, and the graph opens upward since the coefficient 1 is positive. We plot a few points as a check and draw the curve.

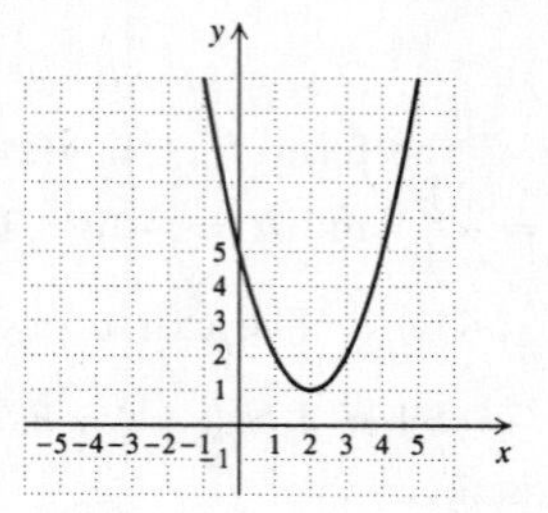

2. (a) Vertex: $(-1, -6)$, axis of symmetry: $x = -1$

(b)

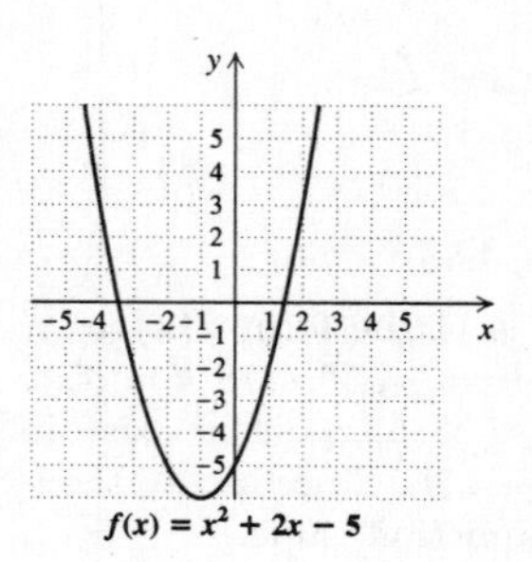

3.

$$\begin{aligned} g(x) &= x^2 + 6x + 13 \\ &= (x^2 + 6x + 9 - 9) + 13 \quad \text{Adding } 9 - 9 \\ &= (x^2 + 6x + 9) - 9 + 13 \quad \text{Regrouping} \\ &= (x+3)^2 + 4 \end{aligned}$$

The vertex is $(-3, 4)$, the axis of symmetry is $x = -3$, and the graph opens upward since the coefficient 1 is positive. We plot a few points as a check and draw the curve.

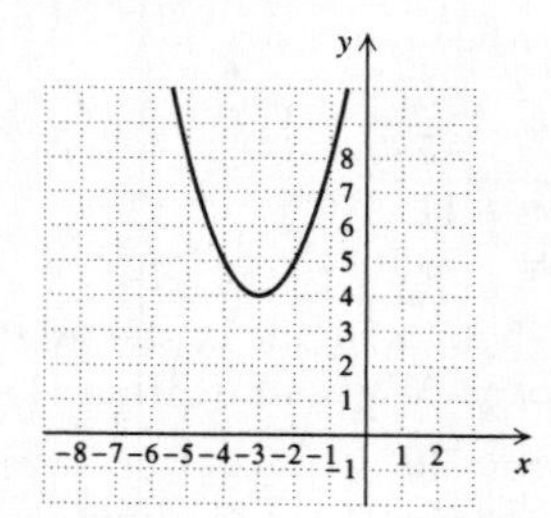

4. (a) Vertex: $(-2, 1)$, axis of symmetry: $x = -2$

(b)

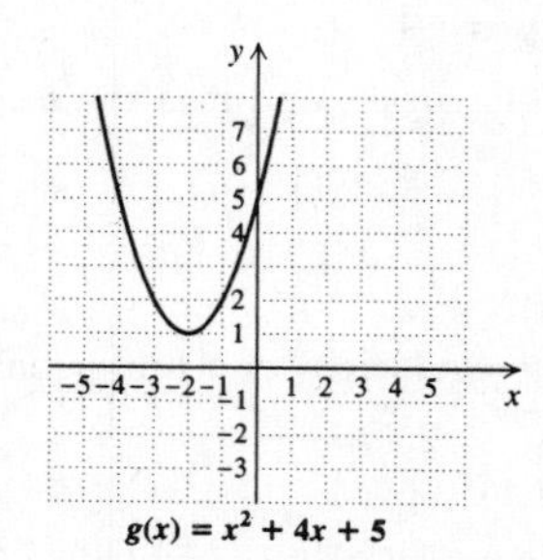

5. $f(x) = x^2 + 8x + 20$

$= (x^2 + 8x + 16 - 16) + 20$ Adding 16−16

$= (x^2 + 8x + 16) - 16 + 20$ Regrouping

$= (x + 4)^2 + 4$

The vertex is $(-4, 4)$, the axis of symmetry is $x = -4$, and the graph opens upward since the coefficient 1 is positive.

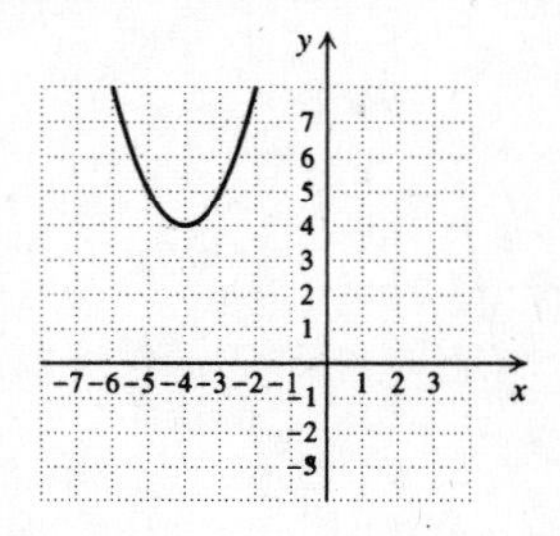

6. (a) Vertex: $(5, -4)$, axis of symmetry: $x = 5$

(b)

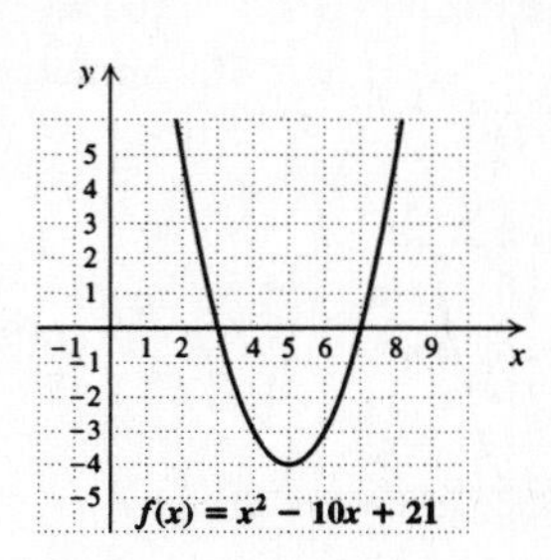

7. $h(x) = 2x^2 + 16x + 25$

$= 2(x^2 + 8x) + 25$ Factoring 2 from the first two terms

$= 2(x^2 + 8x + 16 - 16) + 25$ Adding 16–16 inside the parentheses

$= 2(x^2 + 8x + 16) + 2(-16) + 25$ Distributing to obtain a trinomial square

$= 2(x + 4)^2 - 7$

The vertex is $(-4, -7)$, the axis of symmetry is $x = -4$, and the graph opens upward since the coefficient 2 is positive.

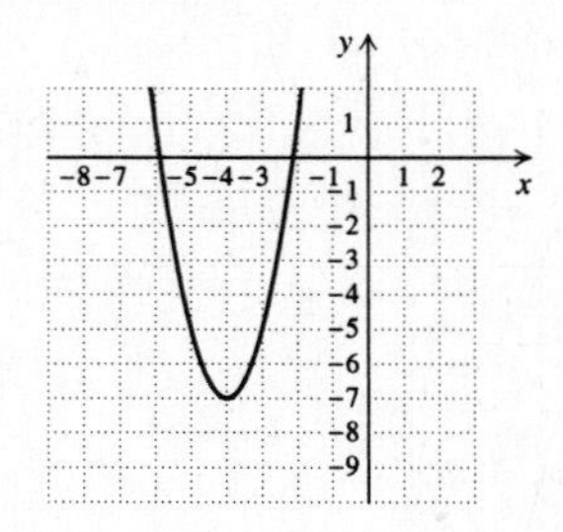

8. (a) Vertex: $(4, -9)$, axis of symmetry: $x = 4$

(b)

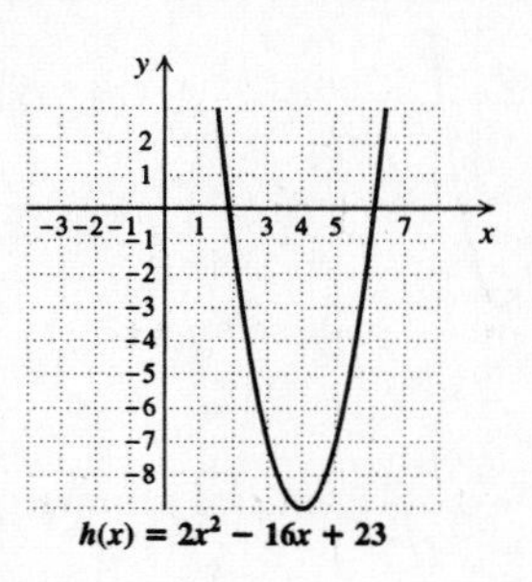

9. $f(x) = -x^2 + 2x + 5$

$= -(x^2 - 2x) + 5$ Factoring −1 from the first two terms

$= -(x^2 - 2x + 1 - 1) + 5$ Adding 1 − 1 inside the parentheses

$= -(x^2 - 2x + 1) - (-1) + 5$

$= -(x - 1)^2 + 6$

The vertex is $(1, 6)$, the axis of symmetry is $x = 1$, and the graph opens downward since the coefficient −1 is negative.

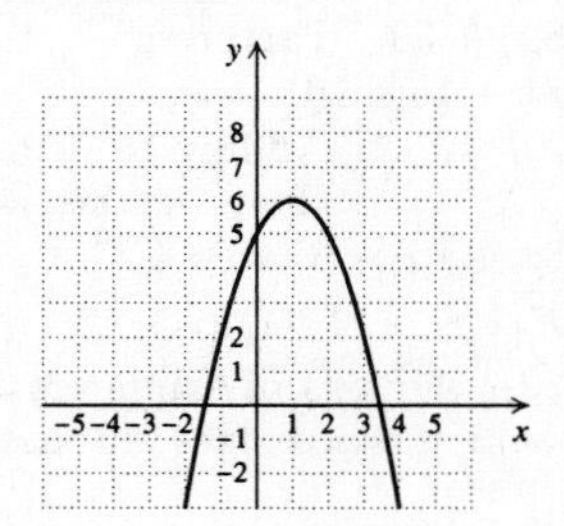

10. (a) Vertex: $(-1, 8)$, axis of symmetry: $x = -1$

(b)

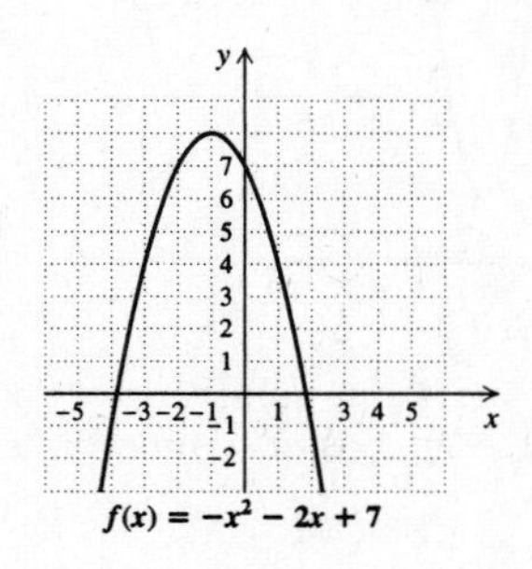

11. $g(x) = x^2 + 3x - 10$

$= \left(x^2 + 3x + \frac{9}{4} - \frac{9}{4}\right) - 10$

$= \left(x^2 + 3x + \frac{9}{4}\right) - \frac{9}{4} - 10$

$= \left(x + \frac{3}{2}\right)^2 - \frac{49}{4}$

The vertex is $\left(-\frac{3}{2}, -\frac{49}{4}\right)$, the axis of symmetry is $x = -\frac{3}{2}$, and the graph opens upward since the coefficient 1 is positive.

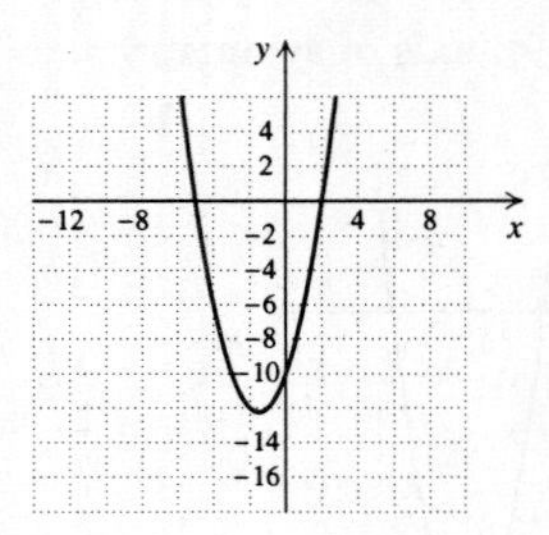

12. (a) Vertex: $\left(-\frac{5}{2}, -\frac{9}{4}\right)$, axis of symmetry: $x = -\frac{5}{2}$

(b)

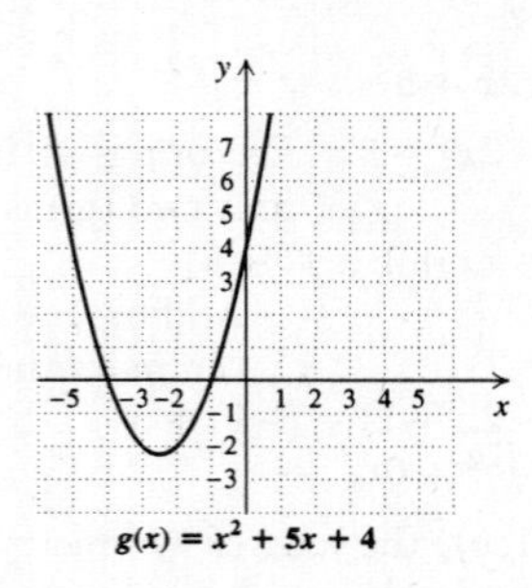

13. $f(x) = 3x^2 - 24x + 50$

$= 3(x^2 - 8x) + 50$ Factoring

$= 3(x^2 - 8x + 16 - 16) + 50$ Adding $16 - 16$ inside the parentheses

$= 3(x^2 - 8x + 16) - 3 \cdot 16 + 50$

$= 3(x - 4)^2 + 2$

The vertex is $(4, 2)$, the axis of symmetry is $x = 4$, and the graph opens upward since the coefficient 3 is positive.

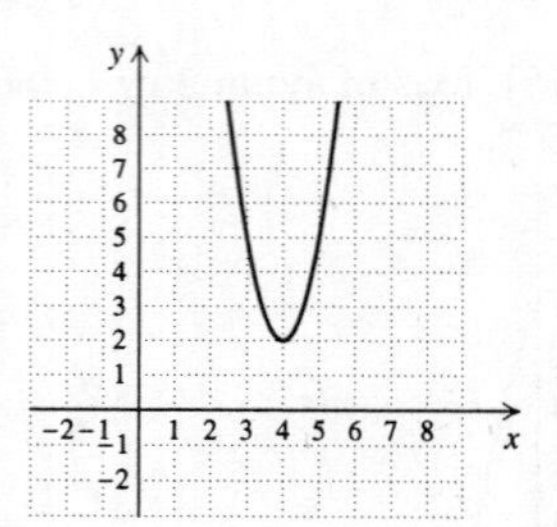

14. (a) Vertex: $(-1, -7)$, axis of symmetry: $x = -1$

(b)

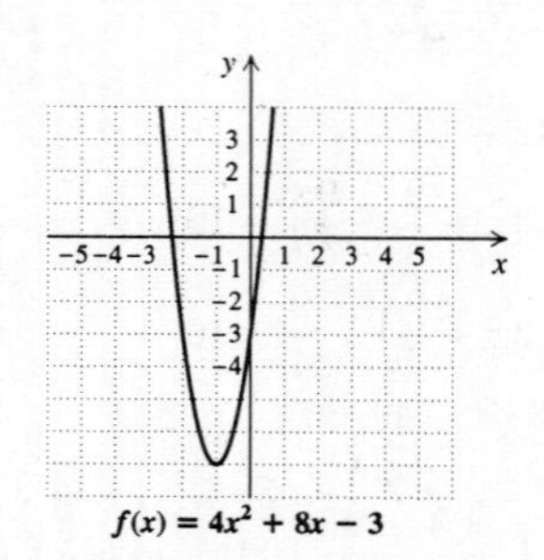

15. $h(x) = x^2 - 9x$

$= \left(x^2 - 9x + \frac{81}{4}\right) - \frac{81}{4}$

$= \left(x - \frac{9}{2}\right)^2 - \frac{81}{4}$

The vertex is $\left(\frac{9}{2}, -\frac{81}{4}\right)$, the axis of symmetry is $x = \frac{9}{2}$, and the graph opens upward since the coefficient 1 is positive.

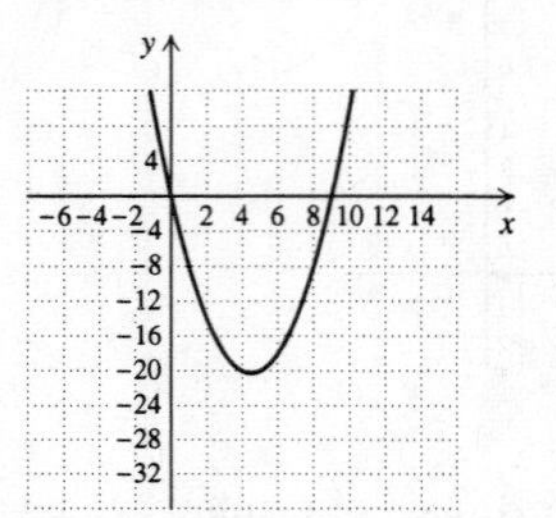

16. (a) Vertex: $\left(-\frac{1}{2}, -\frac{1}{4}\right)$, axis of symmetry: $x = -\frac{1}{2}$

(b)

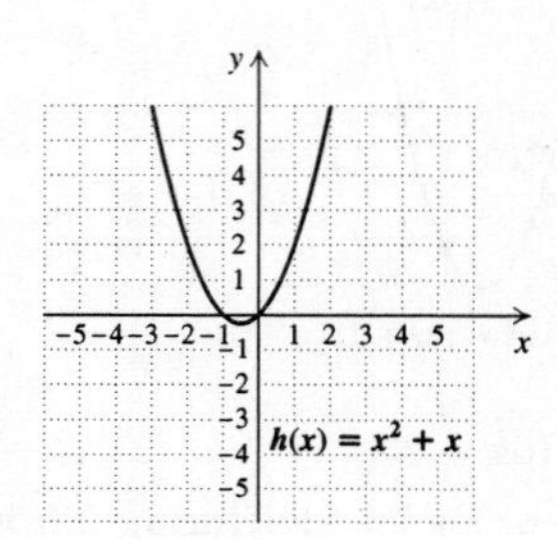

17. $f(x) = -2x^2 - 4x - 6$

$= -2(x^2 + 2x) - 6$ Factoring

$= -2(x^2 + 2x + 1 - 1) - 6$ Adding $1 - 1$ inside the parentheses

$= -2(x^2 + 2x + 1) - 2(-1) - 6$

$= -2(x + 1)^2 - 4$

The vertex is $(-1, -4)$, the axis of symmetry is $x = -1$, and the graph opens downward since the coefficient -2 is negative.

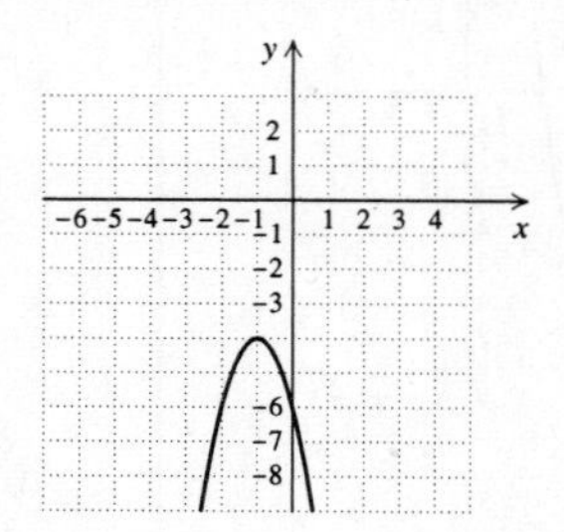

18. (a) Vertex: $(1, 5)$, axis of symmetry: $x = 1$

(b)

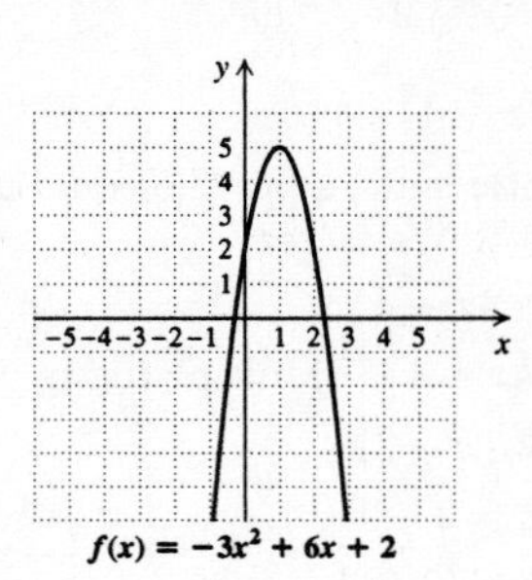

$f(x) = -3x^2 + 6x + 2$

19.
$$\begin{aligned} g(x) &= 2x^2 - 7x + 1 \\ &= 2(x^2 - \frac{7}{2}x) + 1 \qquad \text{Factoring} \\ &= 2\left(x^2 - \frac{7}{2}x + \frac{49}{16} - \frac{49}{16}\right) + 1 \\ &\qquad \text{Adding } \frac{49}{16} - \frac{49}{16} \text{ inside the parentheses} \\ &= 2\left(x^2 - \frac{7}{2}x + \frac{49}{16}\right) + 2\left(-\frac{49}{16}\right) + 1 \\ &= 2\left(x - \frac{7}{4}\right)^2 - \frac{41}{8} \end{aligned}$$

The vertex is $\left(\frac{7}{4}, -\frac{41}{8}\right)$, the axis of symmetry is $x = \frac{7}{4}$, and the graph opens upward since the coefficient 2 is positive.

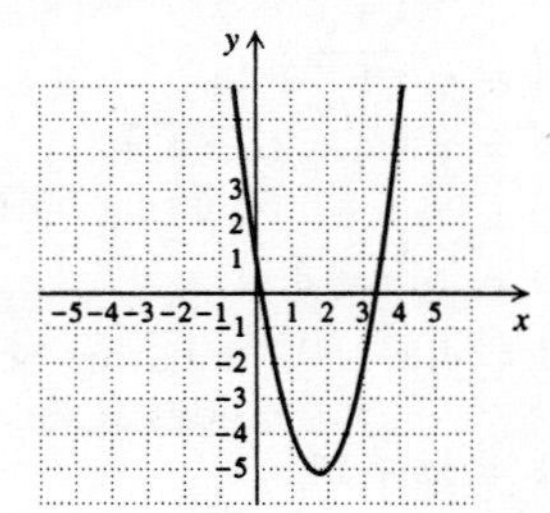

20. (a) Vertex: $\left(-\frac{5}{4}, -\frac{33}{8}\right)$, axis of symmetry: $x = -\frac{5}{4}$

(b)

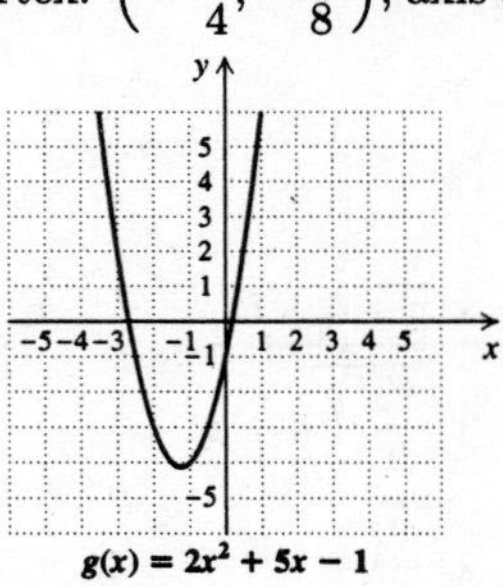

$g(x) = 2x^2 + 5x - 1$

21.
$$\begin{aligned} f(x) &= -3x^2 + 5x - 2 \\ &= -3\left(x^2 - \frac{5}{3}x\right) - 2 \qquad \text{Factoring} \\ &= -3\left(x^2 - \frac{5}{3}x + \frac{25}{36} - \frac{25}{36}\right) - 2 \\ &\qquad \text{Adding } \frac{25}{36} - \frac{25}{36} \text{ inside the parentheses} \\ &= -3\left(x^2 - \frac{5}{3}x + \frac{25}{36}\right) - 3\left(-\frac{25}{36}\right) - 2 \\ &= -3\left(x - \frac{5}{6}\right)^2 + \frac{1}{12} \end{aligned}$$

The vertex is $\left(\frac{5}{6}, \frac{1}{12}\right)$, the axis of symmetry is $x = \frac{5}{6}$, and the graph opens downward since the coefficient -3 is negative.

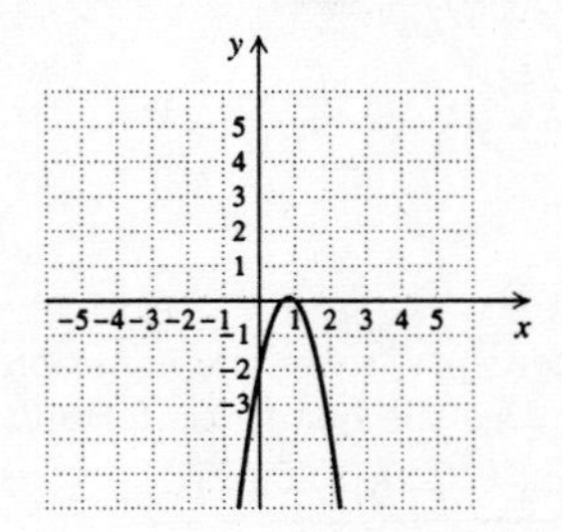

22. (a) Vertex: $\left(-\frac{7}{6}, \frac{73}{12}\right)$, axis of symmetry: $x = -\frac{7}{6}$

(b)

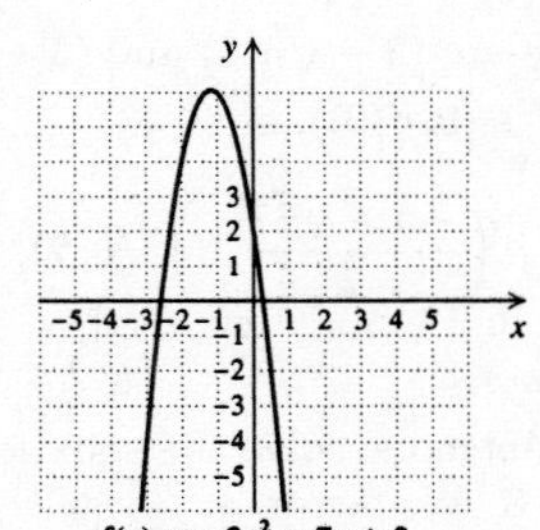

$f(x) = -3x^2 - 7x + 2$

23.
$$\begin{aligned} h(x) &= \frac{1}{2}x^2 + 4x + \frac{19}{3} \\ &= \frac{1}{2}(x^2 + 8x) + \frac{19}{3} \qquad \text{Factoring} \\ &= \frac{1}{2}(x^2 + 8x + 16 - 16) + \frac{19}{3} \\ &\qquad \text{Adding } 16 - 16 \text{ inside the parentheses} \\ &= \frac{1}{2}(x^2 + 8x + 16) + \frac{1}{2}(-16) + \frac{19}{3} \\ &= \frac{1}{2}(x + 4)^2 - \frac{5}{3} \end{aligned}$$

The vertex is $\left(-4, -\frac{5}{3}\right)$, the axis of symmetry is $x = -4$, and the graph opens upward since the coefficient $\frac{1}{2}$ is positive.

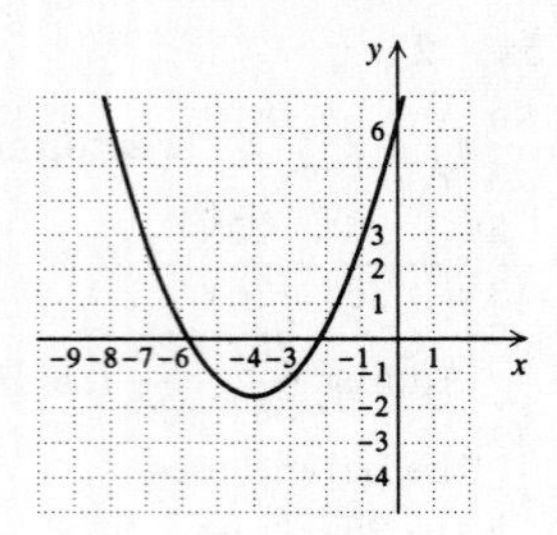

24. (a) Vertex: $\left(3, -\dfrac{5}{2}\right)$, axis of symmetry: $x = 3$

(b)

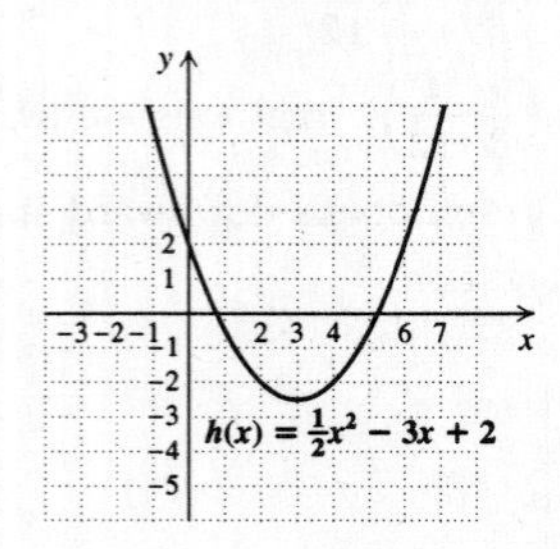

25. $f(x) = x^2 - 6x + 3$

To find the x-intercepts, solve the equation $0 = x^2 - 6x + 3$. Use the quadratic formula.

$$x = \frac{-(-6) \pm \sqrt{(-6)^2 - 4 \cdot 1 \cdot 3}}{2 \cdot 1}$$

$$x = \frac{6 \pm \sqrt{24}}{2} = \frac{6 \pm 2\sqrt{6}}{2} = 3 \pm \sqrt{6}$$

The x-intercepts are $(3 - \sqrt{6}, 0)$ and $(3 + \sqrt{6}, 0)$.

The y-intercept is $(0, f(0))$, or $(0, 3)$.

26. $\left(\dfrac{-5 - \sqrt{17}}{2}, 0\right)$, $\left(\dfrac{-5 + \sqrt{17}}{2}, 0\right)$; $(0, 2)$

27. $g(x) = -x^2 + 2x + 3$

To find the x-intercepts, solve the equation $0 = -x^2 + 2x + 3$. We factor.

$$\begin{aligned} 0 &= -x^2 + 2x + 3 \\ 0 &= x^2 - 2x - 3 \quad \text{Multiplying by } -1 \\ 0 &= (x - 3)(x + 1) \\ x &= 3 \text{ or } x = -1 \end{aligned}$$

The x-intercepts are $(-1, 0)$ and $(3, 0)$.

The y-intercept is $(0, g(0))$, or $(0, 3)$.

28. $(3, 0)$; $(0, 9)$

29. $f(x) = x^2 - 3x + 4$

To find the x-intercepts, solve the equation $0 = x^2 - 3x + 4$. We use the quadratic formula.

$$x = \frac{-(-3) \pm \sqrt{(-3)^2 - 4 \cdot 1 \cdot 4}}{2 \cdot 1}$$

$$x = \frac{3 \pm \sqrt{-7}}{2} = \frac{3 \pm i\sqrt{7}}{2}$$

The equation has no real solutions, so there is no x-intercept.

The y-intercept is $(0, f(0))$, or $(0, 4)$.

30. $\left(\dfrac{7 - \sqrt{57}}{2}, 0\right)$, $\left(\dfrac{7 + \sqrt{57}}{2}, 0\right)$; $(0, -2)$

31. $h(x) = -x^2 + 4x - 4$

To find the x-intercepts, solve the equation $0 = -x^2 + 4x - 4$. We factor.

$$\begin{aligned} 0 &= -x^2 + 4x - 4 \\ 0 &= x^2 - 4x + 4 \quad \text{Multiplying by } -1 \\ 0 &= (x - 2)(x - 2) \\ x &= 2 \text{ or } x = 2 \end{aligned}$$

The x-intercept is $(2, 0)$.

The y-intercept is $(0, h(0))$, or $(0, -4)$.

32. No x-intercept; $(0, 6)$

33. $f(x) = 4x^2 - 12x + 3$

To find the x-intercepts, solve the equation $0 = 4x^2 - 12x + 3$. We use the quadratic formula.

$$x = \frac{-(-12) \pm \sqrt{(-12)^2 - 4 \cdot 4 \cdot 3}}{2 \cdot 4}$$

$$x = \frac{12 \pm \sqrt{96}}{8} = \frac{12 \pm 4\sqrt{6}}{8} = \frac{3 \pm \sqrt{6}}{2}$$

The x-intercepts are $\left(\dfrac{3 - \sqrt{6}}{2}, 0\right)$ and $\left(\dfrac{3 + \sqrt{6}}{2}, 0\right)$.

The y-intercept is $(0, f(0))$, or $(0, 3)$.

34. No x-intercept; $(0, 2)$

35.

$$\begin{aligned} \sqrt{4x - 4} &= \sqrt{x + 4} + 1 \\ 4x - 4 &= x + 4 + 2\sqrt{x + 4} + 1 \quad \text{Squaring both sides} \\ 3x - 9 &= 2\sqrt{x + 4} \\ 9x^2 - 54x + 81 &= 4(x + 4) \quad \text{Squaring both sides again} \\ 9x^2 - 54x + 81 &= 4x + 16 \\ 9x^2 - 58x + 65 &= 0 \\ (9x - 13)(x - 5) &= 0 \\ x = \frac{13}{9} \quad &\text{or} \quad x = 5 \end{aligned}$$

Check: For $x = \dfrac{13}{9}$:

$$\begin{array}{c|l} \sqrt{4x - 4} & = \sqrt{x + 4} + 1 \\ \hline \sqrt{4\left(\frac{13}{9}\right) - 4} & ? \ \sqrt{\frac{13}{9} + 4} + 1 \\ \sqrt{\frac{16}{9}} & \sqrt{\frac{49}{9}} + 1 \\ \frac{4}{3} & \frac{7}{3} + 1 \\ \frac{4}{3} & \frac{10}{3} \end{array} \quad \text{FALSE}$$

For $x = 5$:

$$\begin{array}{c|c} \sqrt{4x-4} & \sqrt{x+4}+1 \\ \hline \sqrt{4\cdot 5-4} \;?& \sqrt{5+4}+1 \\ \sqrt{16} & \sqrt{9}+1 \\ 4 & 3+1 \\ 4 & 4 \quad \text{TRUE} \end{array}$$

5 checks, but $\frac{13}{9}$ does not. The solution is 5.

36. 4

37.

$$A = \frac{3Q + nT}{n}$$

$$\begin{aligned} An &= 3Q + nT && \text{Multiplying by } n \\ An - nT &= 3Q && \text{Adding } -nT \\ n(A-T) &= 3Q && \text{Factoring} \\ n &= \frac{3Q}{A-T} && \text{Dividing by } A - T \end{aligned}$$

38. $\frac{x^2 - 2x - 3}{5}$

39. ◈

40. ◈

41. ◈

42. (a) Minimum value: -6.953660714;
(b) $(-1.056433682, 0)$, $(2.413576539, 0)$; $(0, -5.89)$

43. a)
$$\begin{aligned} f(x) &= -18.8x^2 + 7.92x + 6.18 \\ &= -18.8(x^2 - 0.421276595x + \\ &\quad 0.044368492) + 0.834127659 + 6.18 \\ &= -18.8(x - 0.210638297)^2 + 7.01412766 \end{aligned}$$

Maximum: 7.01412766

b) To find the x-intercepts, solve the equation $0 = -18.8x^2 + 7.92x + 6.18$.

$$x = \frac{-7.92 \pm \sqrt{(7.92)^2 - 4(-18.8)(6.18)}}{2(-18.8)}$$

$x \approx -0.400174191$ or $x \approx 0.821450786$

The x-intercepts are $(-0.400174191, 0)$ and $(0.821450787, 0)$.

The y-intercept is $(0, f(0))$, or $(0, 6.18)$.

44. (a) -2.4, 3.4; (b) -1.3, 2.3

45. 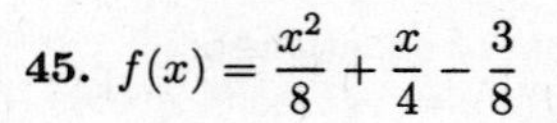
$f(x) = \frac{x^2}{8} + \frac{x}{4} - \frac{3}{8}$

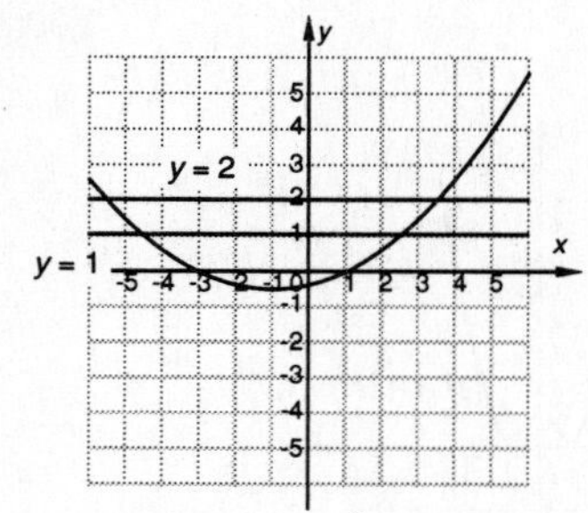

a) The solutions of $\frac{x^2}{8} + \frac{x}{4} - \frac{3}{8} = 0$ are the first coordinates of the x-intercepts of the graph of $f(x) = \frac{x^2}{8} + \frac{x}{4} - \frac{3}{8}$. From the graph we see that the solutions are -3 and 1.

b) The solutions of $\frac{x^2}{8} + \frac{x}{4} - \frac{3}{8} = 1$ are the first coordinates of the points of intersection of the graph of $f(x) = \frac{x^2}{8} + \frac{x}{4} - \frac{3}{8}$ and $y = 1$. From the graph we see that they are approximately -4.5 and 2.5.

c) The solutions of $\frac{x^2}{8} + \frac{x}{4} - \frac{3}{8} = 2$ are the first coordinates of the points of intersection of the graphs of $f(x) = \frac{x^2}{8} + \frac{x}{4} - \frac{3}{8}$ and $y = 2$. From the graph we see that they are approximately -5.5 and 3.5.

46. $f(x) = m\left(x - \frac{n}{2m}\right)^2 + \frac{4mp - n^2}{4m}$

47.
$$\begin{aligned} f(x) &= 3x^2 + mx + m^2 \\ &= 3\left(x^2 + \frac{m}{3}x\right) + m^2 \\ &= 3\left(x^2 + \frac{m}{3}x + \frac{m^2}{36} - \frac{m^2}{36}\right) + m^2 \\ &= 3\left(x + \frac{m}{6}\right)^2 - \frac{m^2}{12} + m^2 \\ &= 3\left[x - \left(-\frac{m}{6}\right)\right]^2 + \frac{11m^2}{12} \end{aligned}$$

48. $f(x) = \frac{5}{16}x^2 - \frac{15}{8}x - \frac{35}{16}$

49. The horizontal distance from $(4, 0)$ to $(-1, 7)$ is $|-1-4|$, or 5, so by symmetry the other x-intercept is $(-1-5, 0)$, or $(-6, 0)$. Substituting the three ordered pairs $(4, 0)$, $(-1, 7)$, and $(-6, 0)$ in the equation $f(x) = ax^2 + bx + c$ yields a system of equations:

$$\begin{aligned} 0 &= a \cdot 4^2 + b \cdot 4 + c, \\ 7 &= a(-1)^2 + b(-1) + c, \\ 0 &= a(-6)^2 + b(-6) + c \end{aligned}$$

or

$$\begin{aligned} 0 &= 16a + 4b + c, \\ 7 &= a - b + c, \\ 0 &= 36a - 6b + c \end{aligned}$$

The solution of this system of equations is $(-0.28, -0.56, 6.72)$, so $f(x) = -0.28x^2 - 0.56x + 6.72$.

50.

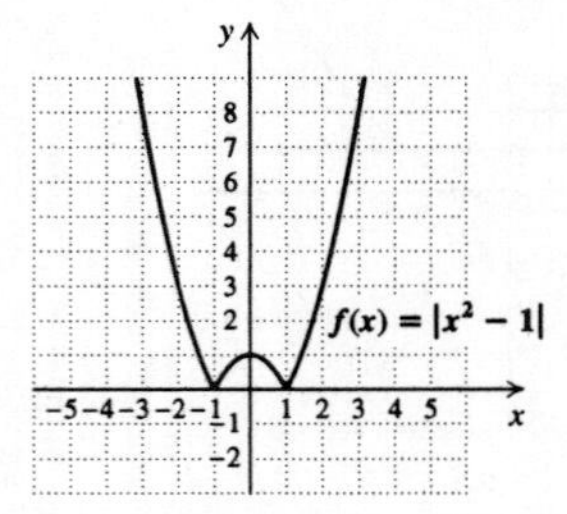

51. $f(x) = |x^2 - 3x - 4|$

We plot some points and draw the curve. Note that it will lie entirely on or above the x-axis since absolute value is never negative.

x	$f(x)$
-4	24
-3	14
-2	6
-1	0
0	4
1	6
2	6
3	4
4	0
5	6
6	14

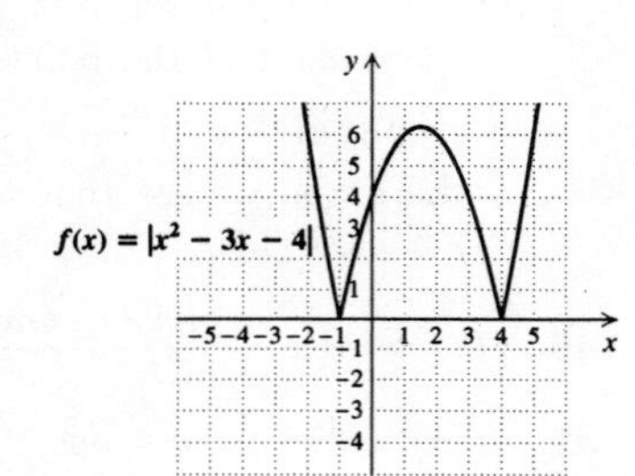

52.

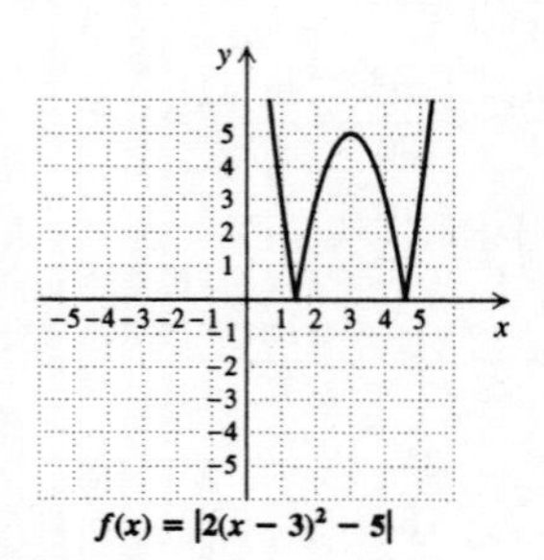

53.

Exercise Set 8.9

1. ***Familiarize***. We make a drawing and label it.

l

w w

l

Perimeter: $2l + 2w = 56$ ft

Area: $A = l \cdot w$

Translate. We have a system of equations.

$$2l + 2w = 56$$
$$A = lw$$

Carry out. Solving the first equation for l, we get $l = 28 - w$. Substituting for l in the second equation we get a quadratic function A:

$$A = (28 - w)w$$
$$A = -w^2 + 28w$$

Completing the square, we get

$$A = -(w - 14)^2 + 196.$$

The maximum function value is 196. It occurs when w is 14. When $w = 14$, $l = 28 - 14$, or 14.

Check. We check a function value for w less than 14 and for w greater than 14.

$$A(13) = -(13)^2 + 28 \cdot 13 = 195$$
$$A(15) = -(15)^2 + 28 \cdot 15 = 195$$

Since 196 is greater than these numbers, it looks as though we have a maximum.

State. The maximum area of 196 ft^2 occurs when the dimensions are 14 ft by 14 ft.

2. 21 in. by 21 in.

3. ***Familiarize***. We let x and y represent the two numbers, and we let P represent their product.

Translate. We have two equations.

$$x - y = 6,$$
$$P = xy$$

Carry out. Solve the first equation for x.

$$x = 6 + y$$

Substitute for x in the second equation.

$$P = (6 + y)y$$
$$P = y^2 + 6y$$

Completing the square, we get

$$P = (y + 3)^2 - 9.$$

The minimum function value is -9. It occurs when $y = -3$. When $y = -3$, $x = 6 + (-3)$, or 3.

Check. Check a function value for y less than -3 and for y greater than -3.

$$P(-4) = (-4)^2 + 6(-4) = -8$$
$$P(-2) = (-2)^2 + 6(-2) = -8$$

Since -9 is less than these numbers, it looks as though we have a minimum.

State. The minimum product of -9 occurs for the numbers 3 and -3.

4. 81; 9 and 9

5. ***Familiarize.*** We let x and y represent the numbers, and we let P represent their product.

   Translate. We have two equations.

   $x + y = 26,$

   $P = xy$

   Carry out. Solving the first equation for y, we get $y = 26 - x$. Substituting for y in the second equation we get a quadratic function P:

   $P = x(26 - x)$

   $P = -x^2 + 26x$

   Completing the square, we get

   $P = -(x - 13)^2 + 169.$

   The maximum function value is 169. It occurs when $x = 13$. When $x = 13$, $y = 26 - 13$, or 13.

   Check. We can check a function value for x less than 13 and for x greater than 13.

   $P(12) = -(12)^2 + 26 \cdot 12 = 168$

   $P(14) = -(14)^2 + 26 \cdot 14 = 168$

   Since 169 is greater than these numbers, it looks as though we have a maximum.

   State. The maximum product of 169 occurs for the numbers 13 and 13.

6. -16; 4 and -4

7. ***Familiarize.*** We let x and y represent the two numbers, and we let P represent their product.

   Translate. We have two equations.

   $x - y = 7,$

   $P = xy$

   Carry out. Solve the first equation for x.

   $x = y + 7$

   Substitute for x in the second equation.

   $P = (y + 7)y$

   $P = y^2 + 7y$

   Completing the square, we get

   $P = \left(y + \frac{7}{2}\right)^2 - \frac{49}{4}.$

   The minimum function value is $-\frac{49}{4}$. It occurs when $y = -\frac{7}{2}$. When $y = -\frac{7}{2}$, $x = -\frac{7}{2} + 7 = \frac{7}{2}$.

   Check. Check a function value for y less than $-\frac{7}{2}$ and for y greater than $-\frac{7}{2}$.

   $P(-4) = (-4)^2 + 7(-4) = -12$

   $P(-3) = (-3)^2 + 7(-3) = -12$

   Since $-\frac{49}{4}$ is less than these numbers, it looks as though we have a minimum.

   State. The minimum product of $-\frac{49}{4}$ occurs for the numbers $\frac{7}{2}$ and $-\frac{7}{2}$.

8. 25; -5 and -5

9. ***Familiarize.*** We let x and y represent the two numbers, and we let P represent their product.

   Translate. We have two equations.

   $x + y = -12,$

   $P = xy$

   Carry out. Solve the first equation for y.

   $y = -12 - x$

   Substitute for y in the second equation.

   $P = x(-12 - x)$

   $P = -x^2 - 12x$

   Completing the square, we get

   $P = -(x + 6)^2 + 36$

   The maximum function value is 36. It occurs when $x = -6$. When $x = -6$, $y = -12 - (-6)$, or -6.

   Check. Check a function value for x less than -6 and for x greater than -6.

   $P(-7) = -(-7)^2 - 12(-7) = 35$

   $P(-5) = -(-5)^2 - 12(-5) = 35$

   Since 36 is greater than these numbers, it looks as though we have a maximum.

   State. The maximum product of 36 occurs for the numbers -6 and -6.

10. $-\frac{81}{4}$; $\frac{9}{2}$ and $-\frac{9}{2}$

11. ***Familiarize.*** We make a drawing and label it.

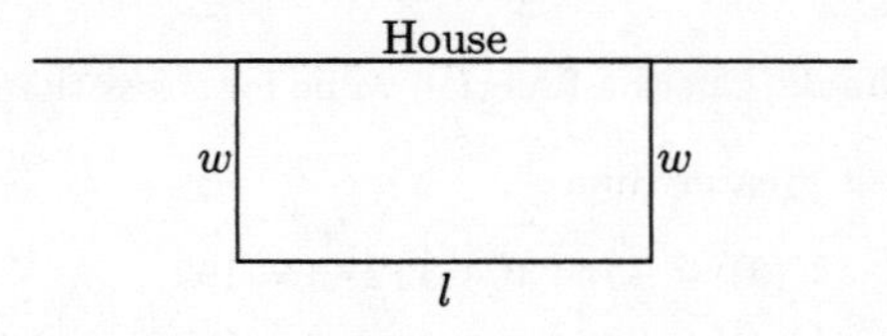

   Translate. We have two equations.

   $l + 2w = 60,$

   $A = lw$

   Carry out. Solve the first equation for l.

   $l = 60 - 2w$

   Substitute for l in the second equation.

   $A = (60 - 2w)w$

   $A = -2w^2 + 60w$

   Completing the square, we get

   $A = -2(w - 15)^2 + 450.$

   The maximum function value of 450 occurs when $w = 15$. When $w = 15$, $l = 60 - 2 \cdot 15 = 30$.

   Check. Check a function value for w less than 15 and for w greater than 15.

   $A(14) = -2 \cdot 14^2 + 60 \cdot 14 = 448$

   $A(16) = -2 \cdot 16^2 + 60 \cdot 16 = 448$

Since 450 is greater than these numbers, it looks as though we have a maximum.

State. The maximum area of 450 ft^2 will occur when the dimensions are 15 ft by 30 ft.

12. 200 ft^2; 10 ft by 20 ft

13. ***Familiarize.*** Let x represent the height of the file and y represent the width. We make a drawing.

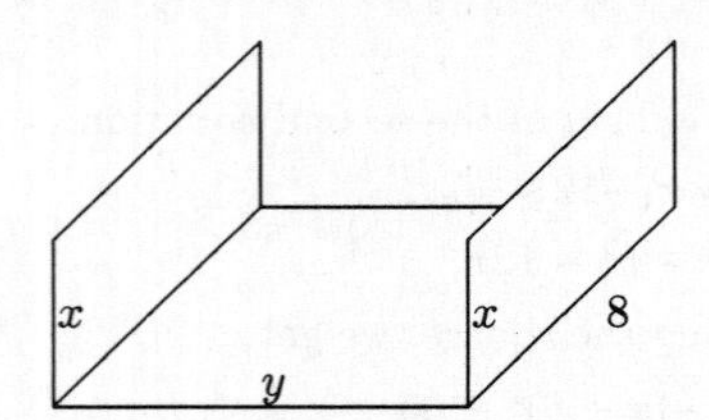

Translate. We have two equations.

$2x + y = 14$

$V = 8xy$

Carry out. Solve the first equation for y.

$y = 14 - 2x$

Substitute for y in the second equation.

$V = 8x(14 - 2x)$

$V = -16x^2 + 112x$

Completing the square, we get

$$V = -16\left(x - \frac{7}{2}\right)^2 + 196.$$

The maximum function value of 196 occurs when $x = \frac{7}{2}$. When $x = \frac{7}{2}$, $y = 14 - 2 \cdot \frac{7}{2} = 7$.

Check. Check a function value for x less than $\frac{7}{2}$ and for x greater than $\frac{7}{2}$.

$$V(3) = -16 \cdot 3^2 + 112 \cdot 3 = 192$$

$$V(4) = -16 \cdot 4^2 + 112 \cdot 4 = 192$$

Since 196 is greater than these numbers, it looks as though we have a maximum.

State. The file should be $\frac{7}{2}$ in., or 3.5 in., tall.

14. 4 ft by 4 ft

15. ***Familiarize and Translate.*** We want to find the value of x for which $C(x) = 0.1x^2 - 0.7x + 2.425$ is a minimum.

Carry out. We complete the square.

$$C(x) = 0.1(x^2 - 7x + 12.25) + 2.425 - 1.225$$

$$C(x) = 0.1(x - 3.5)^2 + 1.2$$

The minimum function value of 1.2 occurs when $x = 3.5$.

Check. Check a function value for x less than 3.5 and for x greater than 3.5.

$$C(3) = 0.1(3)^2 - 0.7(3) + 2.425 = 1.225$$

$$C(4) = 0.1(4)^2 - 0.7(4) + 2.425 = 1.225$$

Since 1.2 is less than these numbers, it looks as though we have a minimum.

State. The shop should build 3.5 hundred, or 350 bicycles.

16. 2700 yd^2

17. Find the total profit:

$$P(x) = R(x) - C(x)$$

$$P(x) = (1000x - x^2) - (3000 + 20x)$$

$$P(x) = -x^2 + 980x - 3000$$

To find the maximum value of the total profit and the value of x at which it occurs we complete the square:

$$\begin{aligned} P(x) &= -(x^2 - 980x) - 3000 \\ &= -(x^2 - 980x + 240,100 - 240,100) - 3000 \\ &= -(x^2 - 980x + 240,100) - (-240,100) - 3000 \\ &= -(x - 490)^2 + 237,100 \end{aligned}$$

The maximum profit of \$237,100 occurs at $x = 490$.

18. $P(x) = -x^2 + 192x - 5000$; \$4216 at $x = 96$

19. We look for a function of the form $f(x) = ax^2 + bx + c$. Substituting the data points, we get

$$\begin{aligned} 4 &= a(1)^2 + b(1) + c, \\ -2 &= a(-1)^2 + b(-1) + c, \\ 13 &= a(2)^2 + b(2) + c, \end{aligned}$$

or

$$\begin{aligned} 4 &= a + b + c, \\ -2 &= a - b + c, \\ 13 &= 4a + 2b + c. \end{aligned}$$

Solving this system, we get

$$a = 2,\ b = 3, \text{ and } c = -1.$$

Therefore the function we are looking for is

$$f(x) = 2x^2 + 3x - 1.$$

20. $f(x) = 3x^2 - x + 2$

21. We look for a function of the form $f(x) = ax^2 + bx + c$. Substituting the data points, we get

$$\begin{aligned} 0 &= a(2)^2 + b(2) + c, \\ 3 &= a(4)^2 + b(4) + c, \\ -5 &= a(12)^2 + b(12) + c, \end{aligned}$$

or

$$\begin{aligned} 0 &= 4a + 2b + c, \\ 3 &= 16a + 4b + c, \\ -5 &= 144a + 12b + c. \end{aligned}$$

Solving this system, we get

$$a = -\frac{1}{4},\ b = 3,\ c = -5.$$

Therefore the function we are looking for is

$$f(x) = -\frac{1}{4}x^2 + 3x - 5.$$

22. $f(x) = -\frac{1}{3}x^2 + 5x - 12$

23. a) ***Familiarize.*** We look for a function of the form $A(s) = as^2+bs+c$, where $A(s)$ represents the number of nighttime accidents (for every 200 million km) and s represents the travel speed (in km/h).

Translate. We substitute the given values of s and $A(s)$.

$$400 = a(60)^2 + b(60) + c,$$
$$250 = a(80)^2 + b(80) + c,$$
$$250 = a(100)^2 + b(100) + c,$$

or

$$400 = 3600a + 60b + c,$$
$$250 = 6400a + 80b + c,$$
$$250 = 10,000a + 100b + c.$$

Carry out. Solving the system of equations, we get

$a = \frac{3}{16}$, $b = -\frac{135}{4}$, $c = 1750$.

Check. Recheck the calculations.

State. The function

$A(s) = \frac{3}{16}s^2 - \frac{135}{4}s + 1750$ fits the data.

b) Find $A(50)$.

$$A(50) = \frac{3}{16}(50)^2 - \frac{135}{4}(50) + 1750 = 531.25$$

About 531 accidents occur at 50 km/h.

24. (a) $A(s) = 0.05x^2 - 5.5x + 250$; (b) 100

25. ***Familiarize.*** Think of a coordinate system placed on the drawing in the text with the origin at the point where the arrow is released. Then three points on the arrow's parabolic path are $(0, 0)$, $(63, 27)$, and $(126, 0)$. We look for a function of the form $h(d) = ad^2+bd+c$, where $h(d)$ represents the arrow's height and d represents the distance the arrow has traveled horizontally.

Translate. We substitute the values given above for d and $h(d)$.

$$0 = a \cdot 0^2 + b \cdot 0 + c,$$
$$27 = a \cdot 63^2 + b \cdot 63 + c,$$
$$0 = a \cdot 126^2 + b \cdot 126 + c$$

or

$$0 = c,$$
$$27 = 3969a + 63b + c,$$
$$0 = 15,876a + 126b + c$$

Carry out. Solving the system of equations, we get $a \approx -0.0068$, $b \approx 0.8571$, and $c = 0$.

Check. Recheck the calculations.

State. The function $h(d) = -0.0068d^2 + 0.8571d$ expresses the arrow's height as a function of the distance it has traveled horizontally.

26. (a) $P(d) = \frac{1}{64}d^2 + \frac{5}{16}d + \frac{5}{2}$; (b) \$9.94

27.
$$\frac{x}{x^2+17x+72} - \frac{8}{x^2+15x+56}$$
$$= \frac{x}{(x+8)(x+9)} - \frac{8}{(x+8)(x+7)}$$
$$= \frac{x}{(x+8)(x+9)} \cdot \frac{x+7}{x+7} - \frac{8}{(x+8)(x+7)} \cdot \frac{x+9}{x+9}$$
$$= \frac{x(x+7) - 8(x+9)}{(x+8)(x+9)(x+7)}$$
$$= \frac{x^2+7x-8x-72}{(x+8)(x+9)(x+7)}$$
$$= \frac{x^2-x-72}{(x+8)(x+9)(x+7)} = \frac{(x-9)\cancel{(x+8)}}{\cancel{(x+8)}(x+9)(x+7)}$$
$$= \frac{x-9}{(x+9)(x+7)}$$

28. $\dfrac{(x-3)(x+1)}{(x-7)(x+3)}$

29. ◈

30. ◈

31. ***Familiarize.*** We add labels to the drawing in the text.

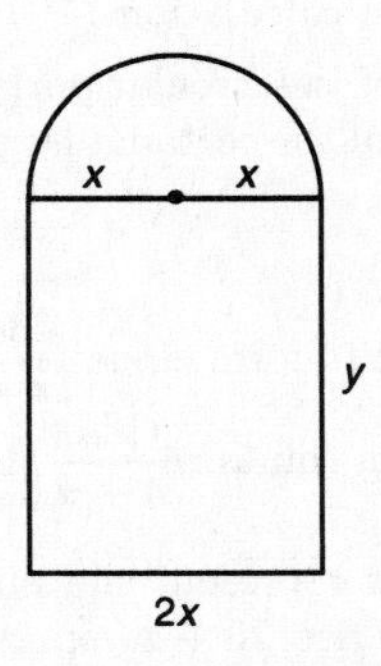

The perimeter of the semicircular portion of the window is $\frac{1}{2} \cdot 2\pi x$, or πx. The perimeter of the rectanglar portion is $y + 2x + y$, or $2x + 2y$. The area of the semicircular portion of the window is $\frac{1}{2} \cdot \pi x^2$, or $\frac{\pi}{2}x^2$. The area of the rectangular portion is $2xy$.

Translate. We have two equations, one giving the perimeter of the window and the other giving the area.

$$\pi x + 2x + 2y = 24,$$
$$A = \frac{\pi}{2}x^2 + 2xy$$

Carry out. Solve the first equation for y.

$$\pi x + 2x + 2y = 24$$
$$2y = 24 - \pi x - 2x$$
$$y = 12 - \frac{\pi x}{2} - x$$

Substitute for y in the second equation.

$$A = \frac{\pi}{2}x^2 + 2x\left(12 - \frac{\pi x}{2} - x\right)$$
$$A = \frac{\pi}{2}x^2 + 24x - \pi x^2 - 2x^2$$
$$A = -2x^2 - \frac{\pi}{2}x^2 + 24x$$
$$A = -\left(2x + \frac{\pi}{2}\right)x^2 + 24x$$

Completing the square, we get

$$A = -\left(2 + \frac{\pi}{2}\right)\left(x^2 + \frac{24}{-\left(2 + \frac{\pi}{2}\right)}x\right)$$
$$A = -\left(2 + \frac{\pi}{2}\right)\left(x^2 - \frac{48}{4+\pi}x\right)$$
$$A = -\left(2 + \frac{\pi}{2}\right)\left(x - \frac{24}{4+\pi}\right)^2 + \left(\frac{24}{4+\pi}\right)^2$$

The maximum function value occurs when $x = \frac{24}{4+\pi}$. When $x = \frac{24}{4+\pi}$,

$$y = 12 - \frac{\pi}{2}\left(\frac{24}{4+\pi}\right) - \frac{24}{4+\pi} =$$
$$\frac{48+12\pi}{4+\pi} - \frac{12\pi}{4+\pi} - \frac{24}{4+\pi} = \frac{24}{4+\pi}.$$

Check. Recheck the calculations.

State. The radius of the circular portion of the window and the height of the rectangular portion should each be $\frac{24}{4+\pi}$ ft.

32. Length of piece used to form circle: $\frac{36\pi}{4+\pi}$ in., length of piece used to form square: $\frac{144}{4+\pi}$ in.

33. ***Familiarize***. Let x represent the number of trees added to an acre. Then $20 + x$ represents the total number of trees per acre and $40 - x$ represents the corresponding yield per tree. Let T represent the total yield per acre.

Translate. Since total yield is number of trees times yield per tree we have the following function for total yield per acre.

$$T(x) = (20 + x)(40 - x)$$
$$T(x) = -x^2 + 20x + 800$$

Carry out. Completing the square, we get

$$T(x) = -(x - 10)^2 + 900.$$

The maximum function value of 900 occurs when $x = 10$. When $x = 10$, the number of trees per acre is $20 + 10$, or 30.

Check. We check a function value for x less than 10 and for x greater than 10.

$$T(9) = (20 + 9)(40 - 9) = 899$$
$$T(11) = (20 + 11)(40 - 11) = 899$$

Since 900 is greater than these numbers, it looks as though we have a maximum.

State. The grower should plant 30 trees per acre.

34. \$15

35. ***Familiarize***. We want to find the maximum value of a function of the form $h(t) = at^2 + bt + c$ that fits the following data.

Time (sec)	Height (ft)
0	64
3	64
3+2, or 5	0

Translate. Substitute the given values for t and $h(t)$.

$$64 = a(0)^2 + b(0) + c,$$
$$64 = a(3)^2 + b(3) + c,$$
$$0 = a(5)^2 + b(5) + c,$$

or

$$64 = c,$$
$$64 = 9a + 3b + c,$$
$$0 = 25a + 5b + c.$$

Carry out. Solving the system of equations, we get $a = -6.4$, $b = 19.2$, $c = 64$. The function $h(t) = -6.4t^2 + 19.2t + 64$ fits the data.

Completing the square, we get

$$h(t) = -6.4(t - 1.5)^2 + 78.4.$$

The maximum funtion value of 78.4 occurs at $t = 1.5$.

Check. Recheck the calculations. Also check a function value for t less than 1.5 and for t greater than 1.5.

$$h(1) = -6.4(1)^2 + 19.2(1) + 64 = 76.8$$
$$h(2) = -6.4(2)^2 + 19.2(2) + 64 = 76.8$$

Since 78.4 is greater than these numbers, it looks as though we have a maximum.

State. The maximum height is 78.4 ft.

36. 158 ft

37.

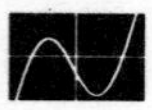

Exercise Set 8.10

1. $(x + 4)(x - 3) > 0$

The solutions of $(x + 4)(x - 3) = 0$ are -4 and 3. They are not solutions of the inequality, but they divide the real-number line in a natural way. The product $(x + 4)(x - 3)$ is positive or negative, for values other than -4 and 3, depending on the signs of the factors $x + 4$ and $x - 3$.

$x + 4 > 0$ when $x > -4$ and $x + 4 < 0$ when $x < -4$.

$x - 3 > 0$ when $x > 3$ and $x - 3 < 0$ when $x < 3$.

We make a diagram.

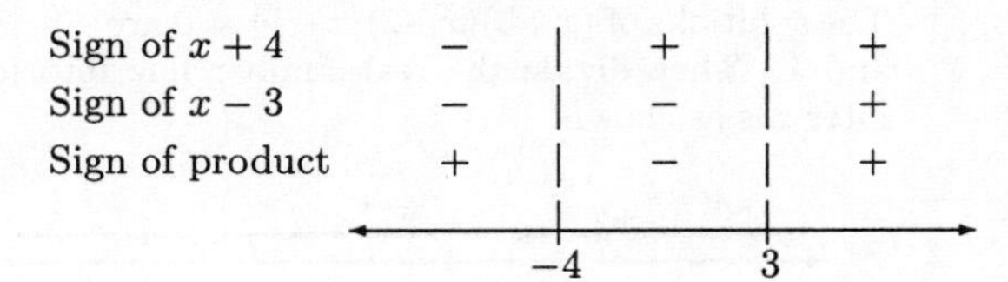

For the product $(x+4)(x-3)$ to be positive, both factors must be positive or both factors must be negative. We see from the diagram that numbers satisfying $x < -4$ or $x > 3$ are solutions. The solution set of the inequality is $(-\infty, -4) \cup (3, \infty)$, or $\{x|x < -4 \text{ or } x > 3\}$.

2. $(-\infty, -2) \cup (5, \infty)$, or $\{x|x < -2 \text{ or } x > 5\}$

3. $(x+7)(x-2) \le 0$

The solutions of $(x+7)(x-2) = 0$ are -7 and 2. They divide the number line into three intervals as shown:

A B C
−7 2

We try test numbers in each interval.

A: Test -8, $f(-8) = (-8+7)(-8-2) = 10$

B: Test 0, $f(0) = (0+7)(0-2) = -14$

C: Test 3, $f(3) = (3+7)(3-2) = 10$

Since $f(0)$ is negative, the function value will be negative for all numbers in the interval containing 0. The inequality symbol is $\le$, so we need to include the intercepts.

The solution set is $[-7, 2]$, or $\{x|-7 \le x \le 2\}$.

4. $[-4, 1]$, or $\{x|-4 \le x \le 1\}$

5. $x^2 - x - 2 < 0$

$(x+1)(x-2) < 0$ Factoring

The solutions of $(x+1)(x-2) = 0$ are -1 and 2. They divide the number line into three intervals as shown:

A B C
−1 2

We try test numbers in each interval.

A: Test -2, $f(-2) = (-2+1)(-2-2) = 4$

B: Test 0, $f(0) = (0+1)(0-2) = -2$

C: Test 3, $f(3) = (3+1)(3-2) = 4$

Since $f(0)$ is negative, the function value will be negative for all numbers in the interval containing 0. The solution set is $(-1, 2)$, or $\{x|-1 < x < 2\}$.

6. $(-2, 1)$, or $\{x|-2 < x < 1\}$

7. $9 - x^2 \le 0$

$(3-x)(3+x) \le 0$

The solutions of $(3-x)(3+x) = 0$ are 3 and -3. They divide the real-number line in a natural way. The product $(3-x)(3+x)$ is positive or negative, for values other than 3 and -3, depending on the signs of the factors $3-x$ and $3+x$.

$3 - x > 0$ when $x < 3$ and $3 - x < 0$ when $x > 3$.

$3 + x > 0$ when $x > -3$ and $3 + x < 0$ when $x < -3$.

We make a diagram.

Sign of $3-x$	+	+	−
Sign of $3+x$	−	+	+
Sign of product	−	+	−

−3 3

For the product $(3-x)(3+x)$ to be negative, one factor must be positive and the other negative. We see from the diagram that numbers satisfying $x < -3$ or $x > 3$ are solutions. The intercepts are also solutions, because the inequality symbol is $\le$. The solution set of the inequality is $(-\infty, -3] \cup [3, \infty)$, or $\{x|x \le -3 \text{ or } x \ge 3\}$.

8. $[-2, 2]$, or $\{x|-2 \le x \le 2\}$

9. $x^2 - 2x + 1 \ge 0$

$(x-1)^2 \ge 0$

The solution of $(x-1)^2 = 0$ is 1. For all real-number values of x except 1, $(x-1)^2$ will be positive. Thus the solution set is $(-\infty, \infty)$, or $\{x|x \text{ is a real number}\}$.

10. $\emptyset$

11. $x^2 - 4x < 12$

$x^2 - 4x - 12 < 0$

$(x-6)(x+2) < 0$

The solutions of $(x-6)(x+2) = 0$ are 6 and -2. They are not solutions of the inequality, but they divide the real-number line in a natural way. The product $(x-6)(x+2)$ is positive or negative, for values other than 6 and -2, depending on the signs of the factors $x-6$ and $x+2$.

$x - 6 > 0$ when $x > 6$ and $x - 6 < 0$ when $x < 6$.

$x + 2 > 0$ when $x > -2$ and $x + 2 < 0$ when $x < -2$.

We make a diagram.

Sign of $x-6$	−	−	+
Sign of $x+2$	−	+	+
Sign of product	+	−	+

−2 6

For the product $(x-6)(x+2)$ to be negative, one factor must be positive and the other negative. The only situation in the diagram for which this happens is when $-2 < x < 6$. The solution set of the inequality is $(-2, 6)$, or $\{x| -2 < x < 6\}$.

12. $(-\infty, -4) \cup (-2, \infty)$, or $\{x|x < -4 \text{ or } x > -2\}$

13. $3x(x+2)(x-2) < 0$

The solutions of $3x(x+2)(x-2) = 0$ are 0, -2, and 2. They divide the real-number line into four intervals as shown:

A B C D
-2 0 2

We try test numbers in each interval.

A: Test -3, $f(-3) = 3(-3)(-3+2)(-3-2) = -45$

B: Test -1, $f(-1) = 3(-1)(-1+2)(-1-2) = 9$

C: Test 1, $f(1) = 3(1)(1+2)(1-2) = -9$

D: Test 3, $f(3) = 3(3)(3+2)(3-2) = 45$

Since $f(-3)$ amd $f(1)$ are negative, the function value will be negative for all numbers in the intervals containing -3 and 1. The solution set is $(-\infty, -2) \cup (0, 2)$, or $\{x|x < -2 \text{ or } 0 < x < 2\}$.

14. $(-1, 0) \cup (1, \infty)$, or $\{x| -1 < x < 0 \text{ or } x > 1\}$

15. $(x+3)(x-2)(x+1) > 0$

The solutions of $(x+3)(x-2)(x+1) = 0$ are -3, 2, and -1. They are not solutions of the inequality, but they divide the real-number line in a natural way. The product $(x+3)(x-2)(x+1)$ is positive or negative, for values other than -3, 2, and -1, depending on the signs of the factors $x+3$, $x-2$, and $x+1$.

$x+3 > 0$ when $x > -3$ and $x+3 < 0$ when $x < -3$.

$x-2 > 0$ when $x > 2$ and $x-2 < 0$ when $x < 2$.

$x+1 > 0$ when $x > -1$ and $x+1 < 0$ when $x < -1$.

We make a diagram.

Sign of $x+3$	$-$	$+$	$+$	$+$
Sign of $x-2$	$-$	$-$	$-$	$+$
Sign of $x+1$	$-$	$-$	$+$	$+$
Sign of product	$-$	$+$	$-$	$+$

-3 -1 2

The product of three numbers is positive when all three are positive or two are negative and one is positive. We see from the diagram that numbers satisfying $-3 < x < -1$ or $x > 2$ are solutions. The solution set of the inequality is
$(-3, -1) \cup (2, \infty)$, or $\{x| -3 < x < -1 \text{ or } x > 2\}$

16. $(-\infty, -2) \cup (1, 4)$, or $\{x|x < -2 \text{ or } 1 < x < 4\}$

17. $(x+3)(x+2)(x-1) < 0$

The solutions of $(x+3)(x+2)(x-1) = 0$ are -3, -2, and 1. They divide the real-number line into four intervals as shown:

A B C D
-3 -2 1

We try test numbers in each interval.

A: Test -4, $f(-4) = (-4+3)(-4+2)(-4-1) = -10$

B: Test $-\frac{5}{2}$, $f\left(-\frac{5}{2}\right) = \left(-\frac{5}{2}+3\right)\left(-\frac{5}{2}+2\right)\left(-\frac{5}{2}-1\right) = \frac{7}{8}$

C: Test 0, $f(0) = (0+3)(0+2)(0-1) = -6$

D: Test 2, $f(2) = (2+3)(2+2)(2-1) = 20$

The function value will be negative for all numbers in intervals A and C. The solution set is
$(-\infty, -3) \cup (-2, 1)$, or $\{x|x < -3 \text{ or } -2 < x < 1\}$.

18. $(-\infty, -1) \cup (2, 3)$, or $\{x|x < -1 \text{ or } 2 < x < 3\}$

19. $\dfrac{1}{x+7} < 0$

We write the related equation by changing the $<$ symbol to $=$:

$$\frac{1}{x+7} = 0$$

We solve the related equation.

$$(x+7) \cdot \frac{1}{x+7} = (x+7) \cdot 0$$
$$1 = 0$$

The related equation has no solution.

Next we find the values that make the denominator 0 by setting the denominate equal to 0 and solving:

$$x + 7 = 0$$
$$x = -7$$

We use -7 to divide the number line into two intervals as shown:

A B
-7

We try test numbers in each interval.

A: Test -8, $\dfrac{1}{-8+7} = \dfrac{1}{-1} = -1 < 0$

The number -8 is a solution of the inequality, so the interval A is part of the solution set.

B: Test 0, $\dfrac{1}{0+7} = \dfrac{1}{7} \not< 0$

The number 0 is not a solution of the inequality, so the interval B is not part of the solution set. The solution set is $(-\infty, -7)$, or $\{x|x < -7\}$.

20. $(-4, \infty)$, or $\{x|x > -4\}$

21. $\dfrac{x+1}{x-3} \geq 0$

Solve the related equation.

$$\frac{x+1}{x-3} = 0$$
$$x+1 = 0$$
$$x = -1$$

Find the values that make the denominator 0.

$$x-3 = 0$$
$$x = 3$$

Use the numbers -1 and 3 to divide the number line into intervals as shown:

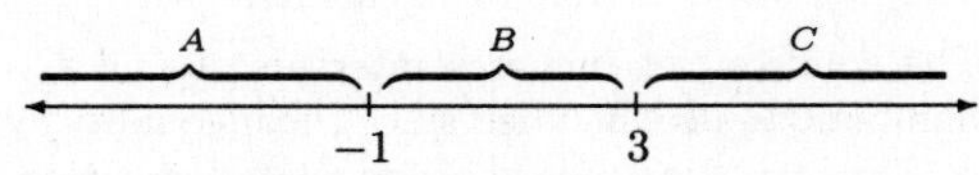

Try test numbers in each interval.

A: Test -2, $\dfrac{-2+1}{-2-3} = \dfrac{-1}{-5} = \dfrac{1}{5} > 0$

The number -2 is a solution of the inequality, so the interval A is part of the solution set.

B: Test 0, $\dfrac{0+1}{0-3} = \dfrac{1}{-3} = -\dfrac{1}{3} \not> 0$

The number 0 is not a solution of the inequality, so the interval B is not part of the solution set.

C: Test 4, $\dfrac{4+1}{4-3} = \dfrac{5}{1} = 5 > 0$

The number 4 is a solution of the inequality, so the interval C is part of the solution set.

The solution set includes intervals A and C. The number -1 is also included since the inequality symbol is $\geq$ and -1 is the solution of the related equation. The number 3 is not included since $\dfrac{x+1}{x-3}$ is undefined for $x = 3$. The solution set is $(-\infty, -1] \cup (3, \infty)$, or $\{x|x \leq -1 \text{ or } x > 3\}$.

22. $(-5, 2]$, or $\{x|-5 < x \leq 2\}$

23. $\dfrac{3x+2}{x-3} \leq 0$

Solve the related equation.

$$\frac{3x+2}{x-3} = 0$$
$$3x+2 = 0$$
$$3x = -2$$
$$x = -\frac{2}{3}$$

Find the values that make the denominator 0.

$$x-3 = 0$$
$$x = 3$$

Use the numbers $-\dfrac{2}{3}$ and 3 to divide the number line into intervals as shown:

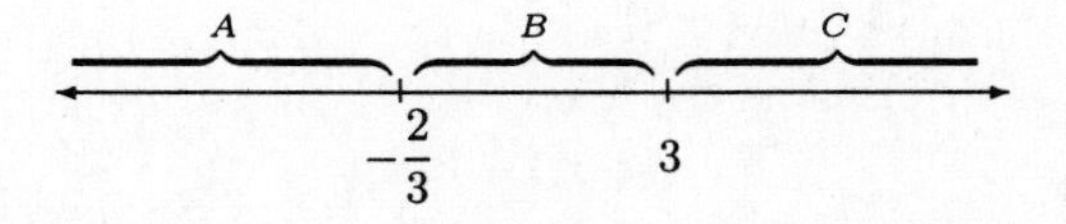

Try test numbers in each interval.

A: Test -1, $\dfrac{3(-1)+2}{-1-3} = \dfrac{-1}{-4} = \dfrac{1}{4} \not\leq 0$

The number -1 is not a solution of the inequality, so the interval A is not part of the solution set.

B: Test 0, $\dfrac{3 \cdot 0+2}{0-3} = \dfrac{2}{-3} = -\dfrac{2}{3} \leq 0$

The number 0 is a solution of the inequality, so the interval B is part of the solution set.

C: Test 4, $\dfrac{3 \cdot 4+2}{4-3} = \dfrac{14}{1} = 14 \not\leq 0$

The number 4 is not a solution of the inequality, so the interval C is not part of the solution set. The solution set includes the interval B. The number $-\dfrac{2}{3}$ is also included since the inequality symbol is $\leq$ and $-\dfrac{2}{3}$ is the solution of the related equation. The number 3 is not included since $\dfrac{3x+2}{x-3}$ is undefined for $x = 3$. The solution set is $\left[-\dfrac{2}{3}, 3\right)$, or $\left\{x\middle|-\dfrac{2}{3} \leq x < 3\right\}$.

24. $\left(-\infty, -\dfrac{3}{4}\right) \cup \left[\dfrac{5}{2}, \infty\right)$, or $\left\{x\middle|x < -\dfrac{3}{4} \text{ or } x \geq \dfrac{5}{2}\right\}$

25. $\dfrac{x+1}{2x-3} > 1$

Solve the related equation.

$$\frac{x+1}{2x-3} = 1$$
$$x+1 = 2x-3$$
$$4 = x$$

Find the values that make the denominator 0.

$$2x-3 = 0$$
$$2x = 3$$
$$x = \frac{3}{2}$$

A B C

$\dfrac{3}{2}$ 4

Try test numbers in each interval.

A: Test 0, $\dfrac{0+1}{2 \cdot 0-3} = \dfrac{1}{-3} = -\dfrac{1}{3} \not> 1$

The number 0 is not a solution of the inequality, so the interval A is not part of the solution set.

B: Test 2, $\dfrac{2+1}{2 \cdot 2-3} = \dfrac{3}{1} = 3 > 1$

The number 2 is a solution of the inequality, so the interval B is part of the solution set.

C: Test 5, $\dfrac{5+1}{2 \cdot 5-3} = \dfrac{6}{7} \not> 1$

The number 5 is not a solution of the inequality, so the interval C is not part of the solution set. The solution set is $\left(\dfrac{3}{2}, 4\right)$, or $\left\{x\middle|\dfrac{3}{2} < x < 4\right\}$.

26. $(-\infty, 2)$, or $\{x|x < 2\}$

27. $\dfrac{(x-2)(x+1)}{x-5} \leq 0$

Solve the related equation.

$$\frac{(x-2)(x+1)}{x-5} = 0$$

$$(x-2)(x+1) = 0$$

$$x = 2 \text{ or } x = -1$$

Find the values that make the denominator 0.

$$x - 5 = 0$$

$$x = 5$$

Use the numbers 2, -1, and 5 to divide the number line into intervals as shown:

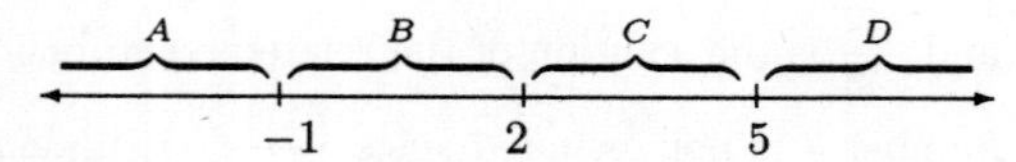

Try test numbers in each interval.

A: Test -2, $\dfrac{(-2-2)(-2+1)}{-2-5} = \dfrac{-4(-1)}{-7} =$

$-\dfrac{4}{7} \leq 0$

Interval A is part of the solution set.

B: Test 0, $\dfrac{(0-2)(0+1)}{0-5} = \dfrac{-2 \cdot 1}{-5} = \dfrac{2}{5} \not\leq 0$

Interval B is not part of the solution set.

C: Test 3, $\dfrac{(3-2)(3+1)}{3-5} = \dfrac{1 \cdot 4}{-2} = -2 \leq 0$

Interval C is part of the solution set.

D: Test 6, $\dfrac{(6-2)(6+1)}{6-5} = \dfrac{4 \cdot 7}{1} = 28 \not\leq 0$

Interval D is not part of the solution set.

The solution set includes intervals A and C. The numbers -1 and 2 are also included since the inequality symbol is $\leq$ and -1 and 2 are the solutions of the related equation. The number 5 is not included since $\dfrac{(x-2)(x+1)}{x-5}$ is undefined for $x = 5$.

The solution set is $(-\infty, -1] \cup [2, 5)$, or $\{x|x \leq -1 \text{ or } 2 \leq x < 5\}$.

28. $[-4, -3) \cup [1, \infty)$, or $\{x| -4 \leq x < -3 \text{ or } x \geq 1\}$

29. $\dfrac{x}{x+3} \geq 0$

Solve the related equation.

$$\frac{x}{x+3} = 0$$

$$x = 0$$

Find the values that make the denominator 0.

$$x + 3 = 0$$

$$x = -3$$

Use the numbers 0 and -3 to divide the number line into intervals as shown.

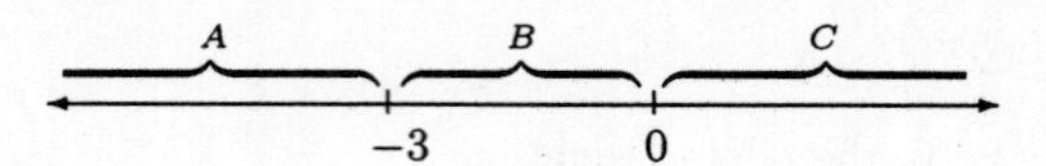

Try test numbers in each interval.

A: Test -4, $\dfrac{-4}{-4+3} = \dfrac{-4}{-1} = 4 \geq 0$

Interval A is part of the solution set.

B: Test -1, $\dfrac{-1}{-1+3} = \dfrac{-1}{2} = -\dfrac{1}{2} \not\geq 0$

Interval B is not part of the solution set.

C: Test 1, $\dfrac{1}{1+3} = \dfrac{1}{4} \geq 0$

The interval C is part of the solution set.

The solution set includes intervals A and C. The number 0 is also included since the inequality symbol is $\geq$ and 0 is the solution of the related equation. The number -3 is not included since $\dfrac{x}{x+3}$ is undefined for $x = -3$. The solution set is $(-\infty, -3) \cup [0, \infty)$, or $\{x|x < -3 \text{ or } x \geq 0\}$.

30. $(0, 2]$, or $\{x|0 < x \leq 2\}$

31. $\dfrac{x-5}{x} < 1$

Solve the related equation.

$$\frac{x-5}{x} = 1$$

$$x - 5 = x$$

$$-5 = 0$$

The related equation has no solution.

Find the values that make the denominator 0.

$$x = 0$$

Use the number 0 to divide the number line into two intervals as shown.

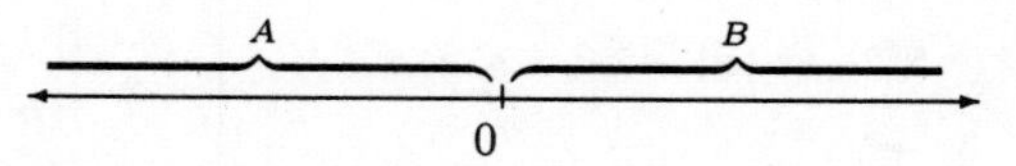

Try test numbers in each interval.

A: Test -1, $\dfrac{-1-5}{-1} = \dfrac{-6}{-1} = 6 \not< 1$

Interval A is not part of the solution set.

B: Test 1, $\dfrac{1-5}{1} = \dfrac{-4}{1} = -4 < 1$

Interval B is part of the solution set.

The solution set is $(0, \infty)$ or $\{x|x > 0\}$.

32. $(1, 2)$, or $\{x|1 < x < 2\}$

33. $\dfrac{x-1}{(x-3)(x+4)} \leq 0$

Solve the related equation.

$$\frac{x-1}{(x-3)(x+4)} = 0$$

$$x - 1 = 0$$

$$x = 1$$

Find the values that make the denominator 0.

$(x-3)(x+4)=0$

$x=3$ or $x=-4$

Use the numbers 1, 3, and -4 to divide the number line into intervals as shown:

Try test numbers in each interval.

A: Test -5, $\dfrac{-5-1}{(-5-3)(-5+4)}=\dfrac{-6}{-8(-1)}=$
$-\dfrac{3}{4}<0$

Interval A is part of the solution set.

B: Test 0, $\dfrac{0-1}{(0-3)(0+4)}=\dfrac{-1}{-3\cdot 4}=\dfrac{1}{12}\not<0$

Interval B is not part of the solution set.

C: Test 2, $\dfrac{2-1}{(2-3)(2+4)}=\dfrac{1}{-1\cdot 6}=-\dfrac{1}{6}<0$

Interval C is part of the solution set.

D: Test 4, $\dfrac{4-1}{(4-3)(4+4)}=\dfrac{3}{1\cdot 8}=\dfrac{3}{8}\not<0$

Interval D is not part of the solution set.

The solution set includes intervals A and C. The number 1 is also included since the inequality symbol is $\leq$ and 1 is the solution of the related equation. The numbers -4 and 3 are not included since $\dfrac{x-1}{(x-3)(x+4)}$ is undefined for $x=-4$ and for $x=3$.

The solution set is $(-\infty,-4)\cup[1,3)$, or $\{x|x<-4 \text{ or } 1\leq x<3\}$.

34. $(-7,-2]\cup(2,\infty)$, or $\{x|-7<x\leq -2 \text{ or } x>2\}$

35. $4<\dfrac{1}{x}$

Solve the related equation.

$4=\dfrac{1}{x}$

$x=\dfrac{1}{4}$

Find the values that make the denominator 0.

$x=0$

Use the numbers $\dfrac{1}{4}$ and 0 to divide the number line into intervals as shown.

Try test numbers in each interval.

A: Test -1, $\dfrac{1}{-1}=-1\not>4$

Interval A is not part of the solution set.

B: Test $\dfrac{1}{8}$, $\dfrac{1}{\frac{1}{8}}=8>4$

Interval B is part of the solution set.

C: Test 1, $\dfrac{1}{1}=1\not>4$

Interval C is not part of the solution set.

The solution set is $\left(0,\dfrac{1}{4}\right)$, or $\left\{x\middle|0<x<\dfrac{1}{4}\right\}$.

36. $(-\infty,0)\cup\left[\dfrac{1}{5},\infty\right)$, or $\left\{x\middle|x<0 \text{ or } x\geq\dfrac{1}{5}\right\}$

37.
$$\begin{aligned}\sqrt[5]{a^2b}\sqrt[3]{ab^2}&=(a^2b)^{1/5}(ab^2)^{1/3} \quad \text{Converting to exponential notation}\\&=a^{2/5}b^{1/5}a^{1/3}b^{2/3}\\&=a^{2/5+1/3}b^{1/5+2/3}\\&=a^{11/15}b^{13/15}\\&=(a^{11}b^{13})^{1/15}\\&=\sqrt[15]{a^{11}b^{13}} \quad \text{Converting back to radical notation}\end{aligned}$$

38. 6

39. ◈

40. ◈

41. $x^2+2x>4$

$x^2+2x-4>0$

Using the quadratic formula, we find that the solutions of the related equation are $x=-1\pm\sqrt{5}$. These numbers divide the real-number line into three intervals as shown:

We try test numbers in each interval.

A: Test -4, $f(-4)=(-4)^2+2(-4)-4=4$

B: Test 0, $f(0)=0^2+2\cdot 0-4=-4$

C: Test 2, $f(2)=2^2+2\cdot 2-4=4$

The function value will be positive for all numbers in intervals A and C. The solution set is $(-\infty,-1-\sqrt{5})\cup(-1+\sqrt{5},\infty)$, or $\{x|x<-1-\sqrt{5} \text{ or } x>-1+\sqrt{5}\}$.

42. $(-\infty,\infty)$, or the set of all real numbers

43. $x^4+3x^2\leq 0$

$x^2(x^2+3)\leq 0$

$x^2=0$ for $x=0$, $x^2>0$ for $x\neq 0$, $x^2+3>0$ for all x

The solution set is $\{0\}$.

44. $\left\{x \middle| x \le \frac{1}{4} \text{ or } x \ge \frac{5}{2}\right\}$, or $\left(-\infty, \frac{1}{4}\right] \cup \left[\frac{5}{2}, \infty\right)$

45. a) $-3x^2 + 630x - 6000 > 0$

$$x^2 - 210x + 2000 < 0 \quad \text{Multiplying by } -\frac{1}{3}$$

$$(x - 200)(x - 10) < 0$$

The solutions of $f(x) = (x - 200)(x - 10) = 0$ are 200 and 10. They divide the number line as shown:

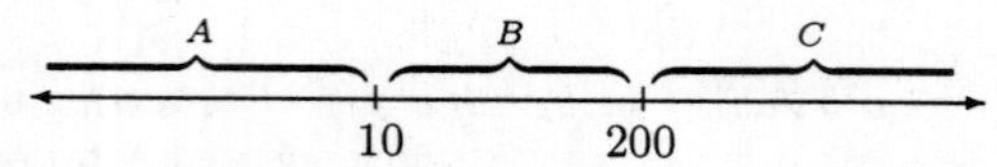

A: Test 0, $f(0) = 0^2 - 210 \cdot 0 + 2000 = 2000$

B: Test 20, $f(20) = 20^2 - 210 \cdot 20 + 2000 = -1800$

C: Test 300, $f(300) = 300^2 - 210 \cdot 300 + 2000 = 29{,}000$

The company makes a profit for values of x such that $10 < x < 200$, or for values of x in the interval $(10, 200)$.

b) See part (a). Keep in mind that x must be nonnegative since negative numbers have no meaning in this application.

The company loses money for values of x such that $0 \le x < 10$ or $x > 200$, or for values of x in the interval $[0, 10) \cup (200, \infty)$.

46. (a) $\{t|0 \text{ sec} < t < 2 \text{ sec}\}$; (b) $\{t|t > 10 \text{ sec}\}$

47. We find values of n such that $N \ge 66$ *and* $N \le 300$.

For $N \ge 66$:

$$\frac{n(n-1)}{2} \ge 66$$

$$n(n-1) \ge 132$$

$$n^2 - n - 132 \ge 0$$

$$(n - 12)(n + 11) \ge 0$$

The solutions of $f(n) = (n - 12)(n + 11) = 0$ are 12 and -11. They divide the number line as shown:

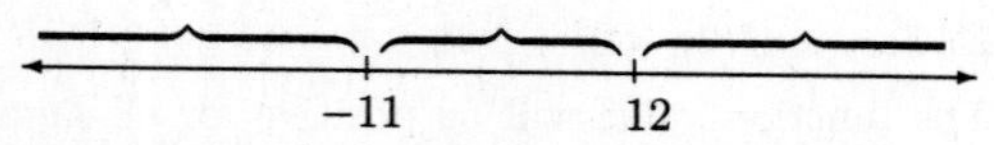

However, only positive values of n have meaning in this exercise so we need only consider the intervals shown below:

A B
0 12

A: Test 1, $f(1) = 1^2 - 1 - 132 = -132$

B: Test 20, $f(20) = 20^2 - 20 - 132 = 248$

Thus, $N \ge 66$ for $\{n|n \ge 12\}$.

For $N \le 300$:

$$\frac{n(n-1)}{2} \le 300$$

$$n(n-1) \le 600$$

$$n^2 - n - 600 \le 0$$

$$(n - 25)(n + 24) \le 0$$

The solutions of $f(n) = (n - 25)(n + 24) = 0$ are 25 and -24. They divide the number line as shown:

−24 25

However, only positive values of n have meaning in this exercise so we need only consider the intervals shown below:

A B
0 25

A: Test 1, $f(1) = 1^2 - 1 - 600 = -600$

B: Test 30, $f(30) = 30^2 - 30 - 600 = 270$

Thus, $N \le 300$ (and $n > 0$) for $\{n|0 < n \le 25\}$.

Then $66 \le N \le 300$ for $\{n|12 \le n \le 25\}$, or for all values of n in the interval $[12, 25]$.

48. $\{n|9 \le n \le 23\}$, or $[9, 23]$

49. From the graph we determine the following:

The solutions of $f(x) = 0$ are -2, 1, and 3.

The solution of $f(x) < 0$ is $(-\infty, -2) \cup (1, 3)$, or $\{x|x < -2 \text{ or } 1 < x < 3\}$.

The solution of $f(x) > 0$ is $(-2, 1) \cup (3, \infty)$, or $\{x| -2 < x < 1 \text{ or } x > 3\}$.

50. $f(x) = 0$ for $x = -2$ *or* $x = 1$;

$f(x) < 0$ for $(-\infty, -2)$, or $\{x|x < -2\}$;

$f(x) > 0$ for $(-2, 1) \cup (1, \infty)$, or $\{x| -2 < x < 1 \text{ or } x > 1\}$

51. From the graph we determine the following:

$f(x)$ has no zeros.

The solutions $f(x) < 0$ are $(-\infty, 0)$, or $\{x|x < 0\}$.

The solutions of $f(x) > 0$ are $(0, \infty)$, or $\{x|x > 0\}$.

52. $f(x) = 0$ for $x = 0$ *or* $x = 1$;

$f(x) < 0$ for $(0, 1)$, or $\{x|0 < x < 1\}$;

$f(x) > 0$ for $(1, \infty)$, or $\{x|x > 1\}$

53. From the graph we determine the following:

The solutions of $f(x) = 0$ are -2, 1, 2, and 3.

The solutions of $f(x) < 0$ are $(-2, 1) \cup (2, 3)$, or $\{x| -2 < x < 1 \text{ or } 2 < x < 3\}$.

The solutions of $f(x) > 0$ are $(-\infty, -2) \cup (1, 2) \cup (3, \infty)$, or $\{x|x < -2 \text{ or } 1 < x < 2 \text{ or } x > 3\}$.

54. $f(x) = 0$ for $x = -2,\ x = 0,\ or\ x = 1$;

$f(x) < 0$ for
$(-\infty, -3) \cup (-2, 0) \cup (1, 2)$, or
$\{x|x < -3\ \ or\ \ -2 < x < 0\ \ or\ \ 1 < x < 2\}$;

$f(x) > 0$ for
$(-3, -2) \cup (0, 1) \cup (2, \infty)$, or
$\{x| -3 < x < -2\ \ or\ \ 0 < x < 1\ \ or\ \ x > 2\}$

55. 

Chapter 9

Exponential and Logarithmic Functions

Exercise Set 9.1

1. Graph: $y = 2^x$

We compute some function values, thinking of y as $f(x)$, and keep the results in a table.

$f(0) = 2^0 = 1$

$f(1) = 2^1 = 2$

$f(2) = 2^2 = 4$

$f(-1) = 2^{-1} = \frac{1}{2^1} = \frac{1}{2}$

$f(-2) = 2^{-2} = \frac{1}{2^2} = \frac{1}{4}$

x	y, or $f(x)$
0	1
1	2
2	4
-1	$\frac{1}{2}$
-2	$\frac{1}{4}$

Next we plot these points and connect them with a smooth curve.

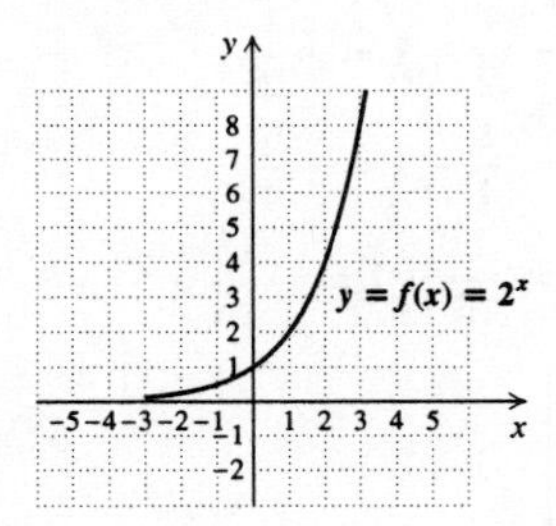

2.

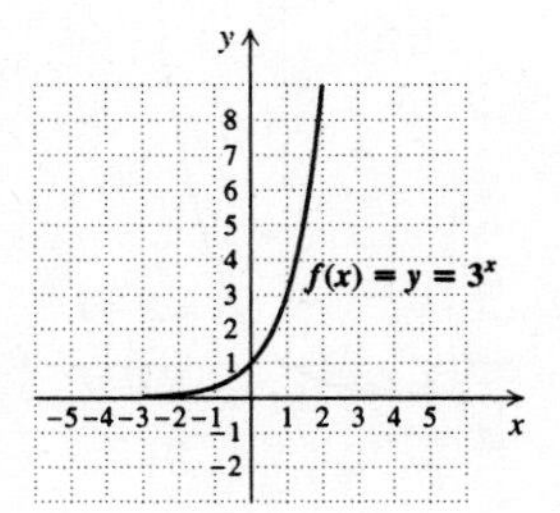

3. Graph: $y = 5^x$

We compute some function values, thinking of y as $f(x)$, and keep the results in a table.

$f(0) = 5^0 = 1$

$f(1) = 5^1 = 5$

$f(2) = 5^2 = 25$

$f(-1) = 5^{-1} = \frac{1}{5^1} = \frac{1}{5}$

$f(-2) = 5^{-2} = \frac{1}{5^2} = \frac{1}{25}$

x	y, or $f(x)$
0	1
1	5
2	25
-1	$\frac{1}{5}$
-2	$\frac{1}{25}$

Next we plot these points and connect them with a smooth curve.

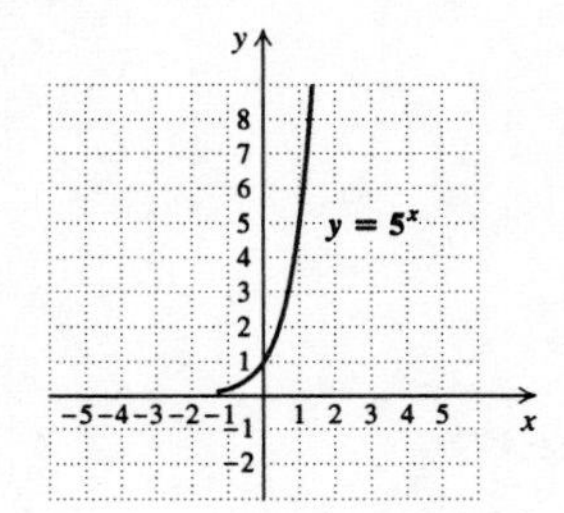

4.

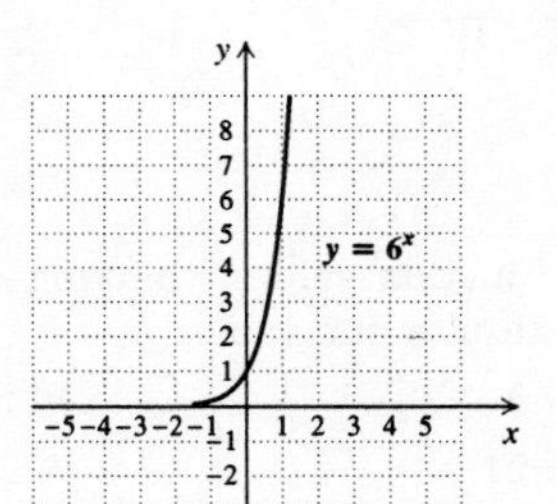

5. Graph: $y = 2^{x-1}$

We compute some function values, thinking of y as $f(x)$, and keep the results in a table.

$f(0) = 2^{0-1} = 2^{-1} = \frac{1}{2}$

$f(-1) = 2^{-1-1} = 2^{-2} = \frac{1}{2^2} = \frac{1}{4}$

$f(-2) = 2^{-2-1} = 2^{-3} = \frac{1}{2^3} = \frac{1}{8}$

$f(1) = 2^{1-1} = 2^0 = 1$

$f(2) = 2^{2-1} = 2^1 = 2$

$f(3) = 2^{3-1} = 2^2 = 4$

$f(4) = 2^{4-1} = 2^3 = 8$

x	y, or $f(x)$
0	$\frac{1}{2}$
-1	$\frac{1}{4}$
-2	$\frac{1}{8}$
1	1
2	2
3	4
4	8

Next we plot these points and connect them with a smooth curve.

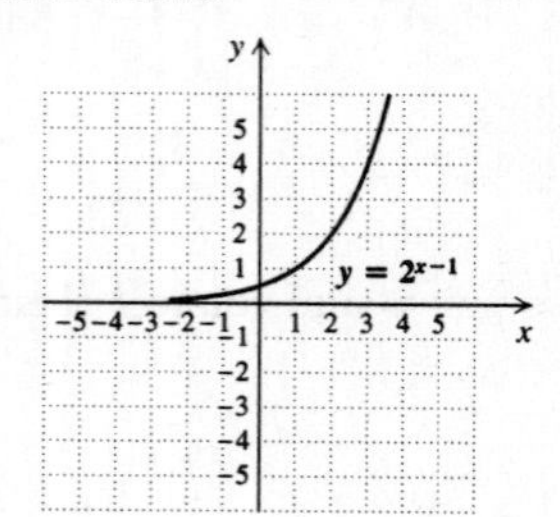

6. 

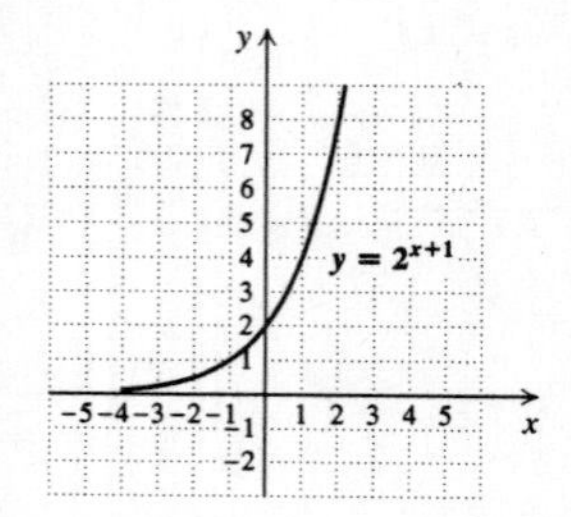

7. Graph: $y = 3^{x+2}$

We compute some function values, thinking of y as $f(x)$, and keep the results in a table.

$f(0) = 3^{0+2} = 3^2 = 9$

$f(1) = 3^{1+2} = 3^3 = 27$

$f(-1) = 3^{-1+2} = 3^1 = 3$

$f(-2) = 3^{-2+2} = 3^0 = 1$

$f(-3) = 3^{-3+2} = 3^{-1} = \frac{1}{3^1} = \frac{1}{3}$

$f(-4) = 3^{-4+2} = 3^{-2} = \frac{1}{3^2} = \frac{1}{9}$

$f(-5) = 3^{-5+2} = 3^{-3} = \frac{1}{3^3} = \frac{1}{27}$

x	y, or $f(x)$
0	9
1	27
-1	3
-2	1
-3	$\frac{1}{3}$
-4	$\frac{1}{9}$
-5	$\frac{1}{27}$

Next we plot these points and connect them with a smooth curve.

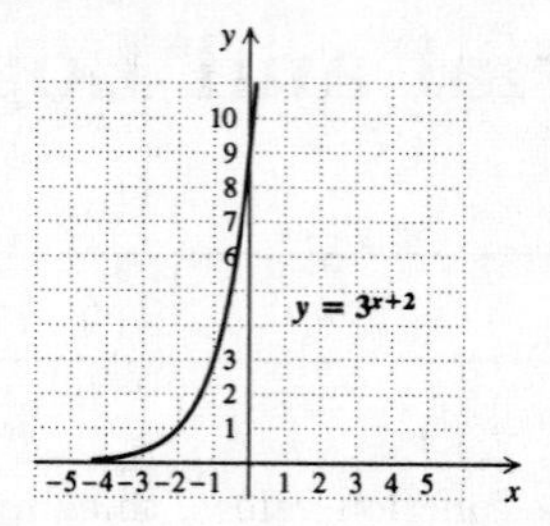

8.

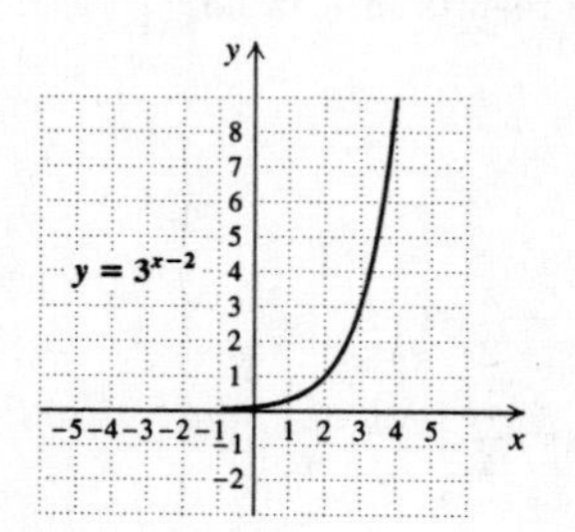

9. Graph: $y = 2^x - 1$

We construct a table of values, thinking of y as $f(x)$. Then we plot the points and connect them with a smooth curve.

$f(0) = 2^0 - 1 = 1 - 1 = 0$

$f(1) = 2^1 - 1 = 2 - 1 = 1$

$f(2) = 2^2 - 1 = 4 - 1 = 3$

$f(3) = 2^3 - 1 = 8 - 1 = 7$

$f(-1) = 2^{-1} - 1 = \frac{1}{2} - 1 = -\frac{1}{2}$

$f(-2) = 2^{-2} - 1 = \frac{1}{4} - 1 = -\frac{3}{4}$

$f(-4) = 2^{-4} - 1 = \frac{1}{16} - 1 = -\frac{15}{16}$

x	y, or $f(x)$
0	0
1	1
2	3
3	7
-1	$-\frac{1}{2}$
-2	$-\frac{3}{4}$
-4	$-\frac{15}{16}$

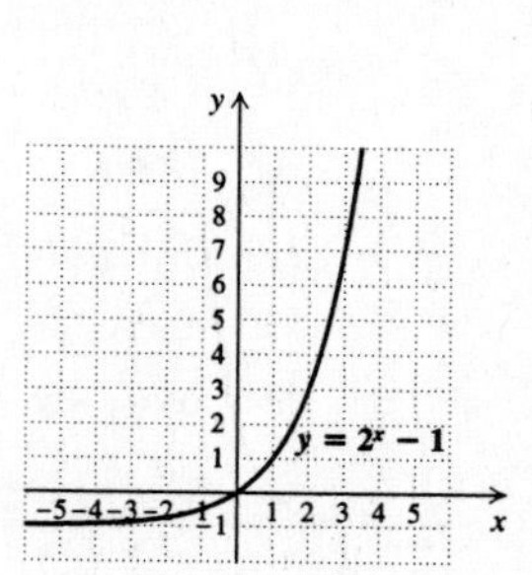

10.

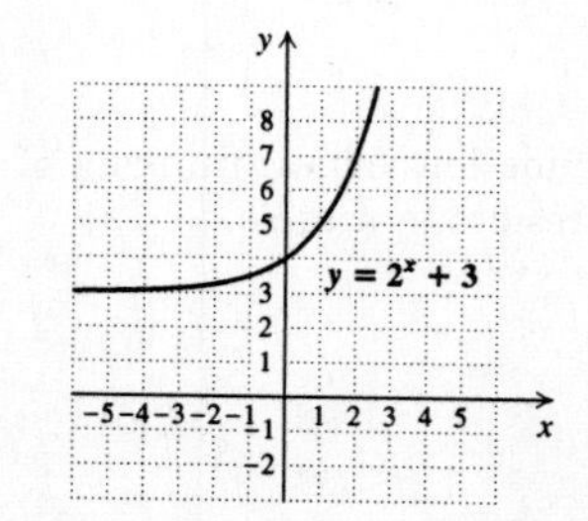

11. Graph: $y = 5^{x+3}$

We construct a table of values, thinking of y as $f(x)$. Then we plot the points and connect them with a smooth curve.

$f(0) = 5^{0+3} = 5^3 = 125$

$f(-1) = 5^{-1+3} = 5^2 = 25$

$f(-2) = 5^{-2+3} = 5^1 = 5$

$f(-3) = 5^{-3+3} = 5^0 = 1$

$f(-4) = 5^{-4+3} = 5^{-1} = \dfrac{1}{5}$

$f(-5) = 5^{-5+3} = 5^{-2} = \dfrac{1}{25}$

x	y, or $f(x)$
0	125
−1	25
−2	5
−3	1
−4	$\frac{1}{5}$
−5	$\frac{1}{25}$

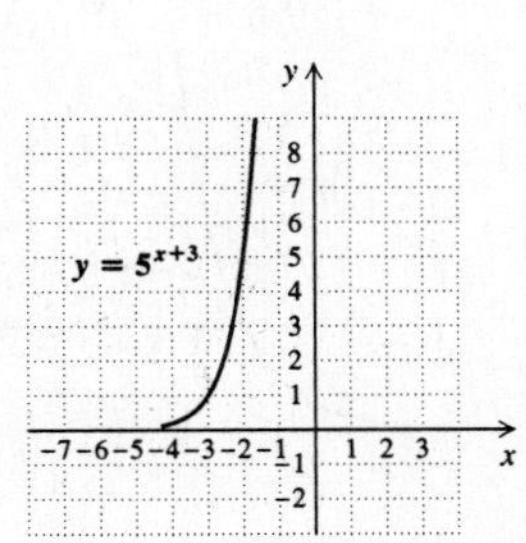

12.

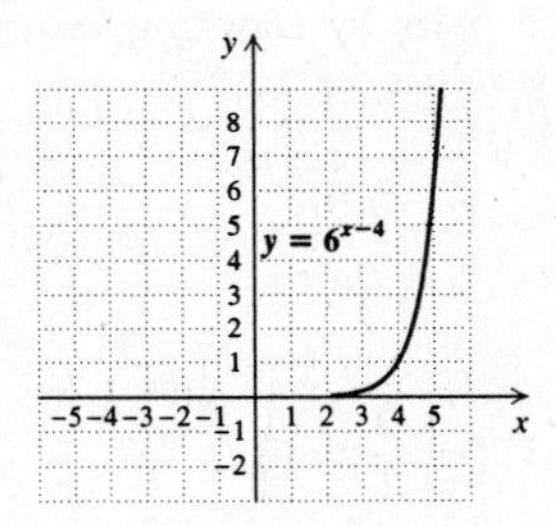

13. Graph: $y = \left(\dfrac{1}{2}\right)^x$

We construct a table of values, thinking of y as $f(x)$. Then we plot the points and connect them with a smooth curve.

$f(0) = \left(\dfrac{1}{2}\right)^0 = 1$

$f(1) = \left(\dfrac{1}{2}\right)^1 = \dfrac{1}{2}$

$f(2) = \left(\dfrac{1}{2}\right)^2 = \dfrac{1}{4}$

$f(3) = \left(\dfrac{1}{2}\right)^3 = \dfrac{1}{8}$

$f(-1) = \left(\dfrac{1}{2}\right)^{-1} = \dfrac{1}{\left(\dfrac{1}{2}\right)^1} = \dfrac{1}{\dfrac{1}{2}} = 2$

$f(-2) = \left(\dfrac{1}{2}\right)^{-2} = \dfrac{1}{\left(\dfrac{1}{2}\right)^2} = \dfrac{1}{\dfrac{1}{4}} = 4$

$f(-3) = \left(\dfrac{1}{2}\right)^{-3} = \dfrac{1}{\left(\dfrac{1}{2}\right)^3} = \dfrac{1}{\dfrac{1}{8}} = 8$

x	y, or $f(x)$
0	1
1	$\frac{1}{2}$
2	$\frac{1}{4}$
3	$\frac{1}{8}$
−1	2
−2	4
−3	8

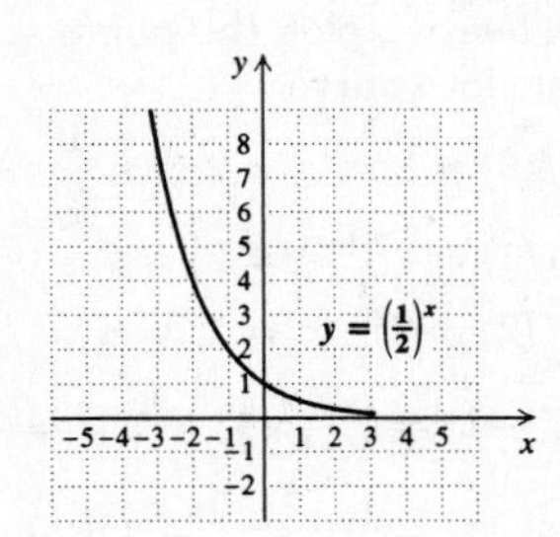

14.

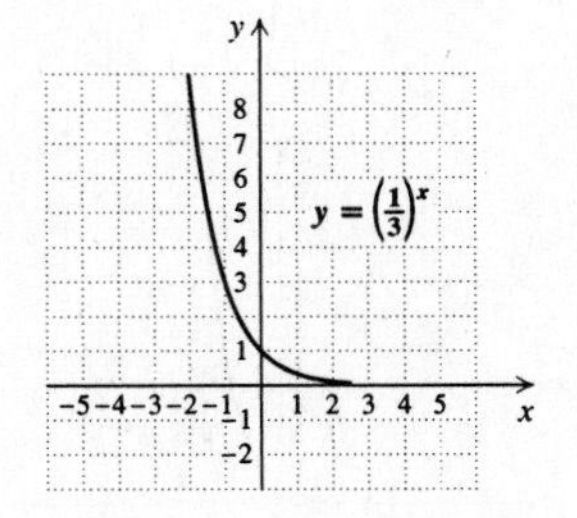

15. Graph: $y = \left(\dfrac{1}{5}\right)^x$

We construct a table of values, thinking of y as $f(x)$. Then we plot the points and connect them with a smooth curve.

$f(0) = \left(\dfrac{1}{5}\right)^0 = 1$

$f(1) = \left(\dfrac{1}{5}\right)^1 = \dfrac{1}{5}$

$f(2) = \left(\dfrac{1}{5}\right)^2 = \dfrac{1}{25}$

$f(-1) = \left(\dfrac{1}{5}\right)^{-1} = \dfrac{1}{\dfrac{1}{5}} = 5$

$f(-2) = \left(\dfrac{1}{5}\right)^{-2} = \dfrac{1}{\dfrac{1}{25}} = 25$

x	y, or $f(x)$
0	1
1	$\frac{1}{5}$
2	$\frac{1}{25}$
−1	5
−2	25

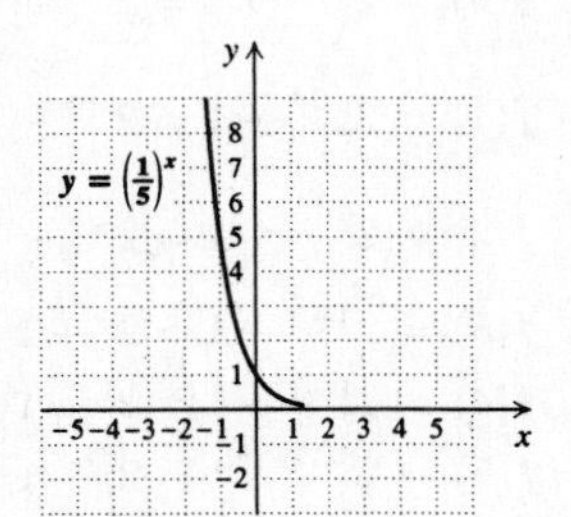

16.

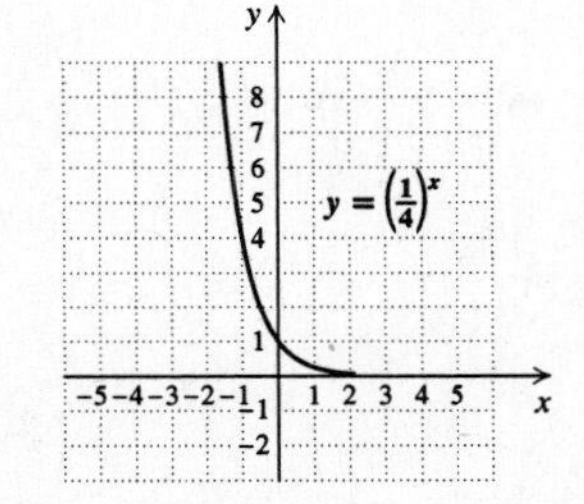

17. Graph: $y = 2^{2x-1}$

We construct a table of values, thinking of y as $f(x)$. Then we plot the points and connect them with a smooth curve.

$f(0) = 2^{2\cdot 0-1} = 2^{-1} = \frac{1}{2}$

$f(1) = 2^{2\cdot 1-1} = 2^1 = 2$

$f(2) = 2^{2\cdot 2-1} = 2^3 = 8$

$f(-1) = 2^{2(-1)-1} = 2^{-3} = \frac{1}{8}$

$f(-2) = 2^{2(-2)-1} = 2^{-5} = \frac{1}{32}$

x	y, or $f(x)$
0	$\frac{1}{2}$
1	2
2	8
-1	$\frac{1}{8}$
-2	$\frac{1}{32}$

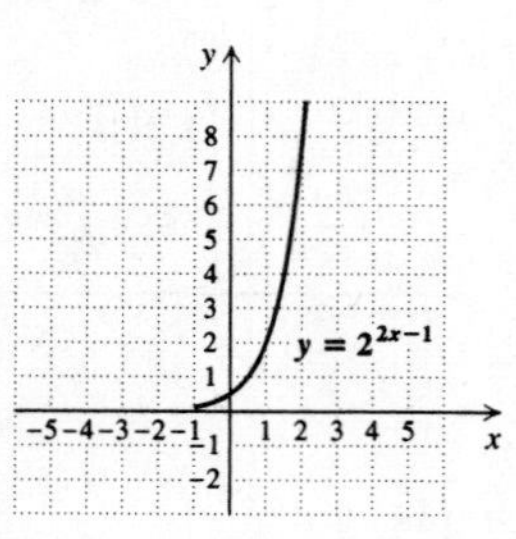

18.

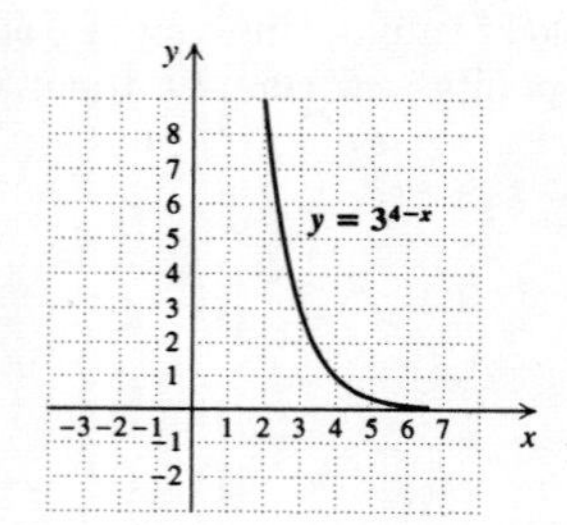

19. Graph: $y = 2^{x-3} - 1$

We construct a table of values, thinking of y as $f(x)$. Then we plot the points and connect them with a smooth curve.

$f(0) = 2^{0-3} - 1 = 2^{-3} - 1 = \frac{1}{8} - 1 = -\frac{7}{8}$

$f(1) = 2^{1-3} - 1 = 2^{-2} - 1 = \frac{1}{4} - 1 = -\frac{3}{4}$

$f(2) = 2^{2-3} - 1 = 2^{-1} - 1 = \frac{1}{2} - 1 = -\frac{1}{2}$

$f(3) = 2^{3-3} - 1 = 2^0 - 1 = 1 - 1 = 0$

$f(4) = 2^{4-3} - 1 = 2^1 - 1 = 2 - 1 = 1$

$f(5) = 2^{5-3} - 1 = 2^2 - 1 = 4 - 1 = 3$

$f(6) = 2^{6-3} - 1 = 2^3 - 1 = 8 - 1 = 7$

x	y, or $f(x)$
0	$-\frac{7}{8}$
1	$-\frac{3}{4}$
2	$-\frac{1}{2}$
3	0
4	1
5	3
6	7

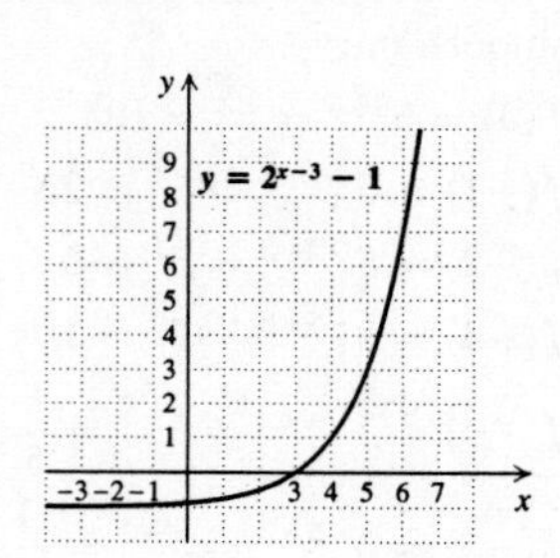

20.

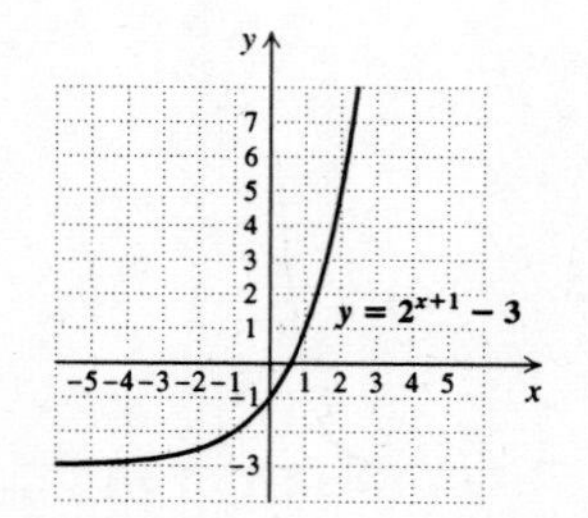

21. Graph: $x = 3^y$

We can find ordered pairs by choosing values for y and then computing values for x.

For $y = 0$, $x = 3^0 = 1$.

For $y = 1$, $x = 3^1 = 3$.

For $y = 2$, $x = 3^2 = 9$.

For $y = 3$, $x = 3^3 = 27$.

For $y = -1$, $x = 3^{-1} = \frac{1}{3^1} = \frac{1}{3}$.

For $y = -2$, $x = 3^{-2} = \frac{1}{3^2} = \frac{1}{9}$.

For $y = -3$, $x = 3^{-3} = \frac{1}{3^3} = \frac{1}{27}$.

x	y
1	0
3	1
9	2
27	3
$\frac{1}{3}$	-1
$\frac{1}{9}$	-2
$\frac{1}{27}$	-3

(1) Choose values for y.

(2) Compute values for x.

We plot the points and connect them with a smooth curve.

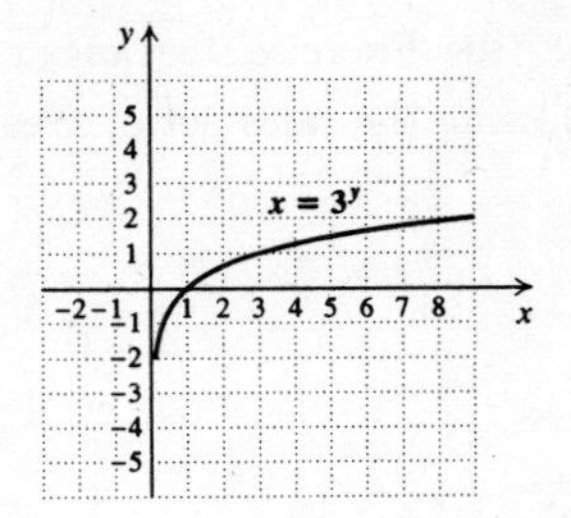

22.

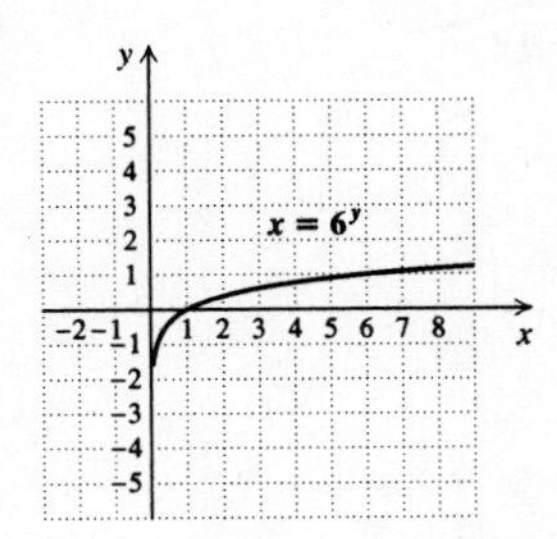

23. Graph: $x = \left(\frac{1}{2}\right)^y$

We can find ordered pairs by choosing values for y and then computing values for x. Then we plot these points and connect them with a smooth curve.

For $y = 0$, $x = \left(\frac{1}{2}\right)^0 = 1$.

For $y = 1$, $x = \left(\frac{1}{2}\right)^1 = \frac{1}{2}$.

For $y = 2$, $x = \left(\frac{1}{2}\right)^2 = \frac{1}{4}$.

For $y = 3$, $x = \left(\frac{1}{2}\right)^3 = \frac{1}{8}$.

For $y = -1$, $x = \left(\frac{1}{2}\right)^{-1} = \frac{1}{\frac{1}{2}} = 2$.

For $y = -2$, $x = \left(\frac{1}{2}\right)^{-2} = \frac{1}{\frac{1}{4}} = 4$.

For $y = -3$, $x = \left(\frac{1}{2}\right)^{-3} = \frac{1}{\frac{1}{8}} = 8$.

x	y
1	0
$\frac{1}{2}$	1
$\frac{1}{4}$	2
$\frac{1}{8}$	3
2	−1
4	−2
8	−3

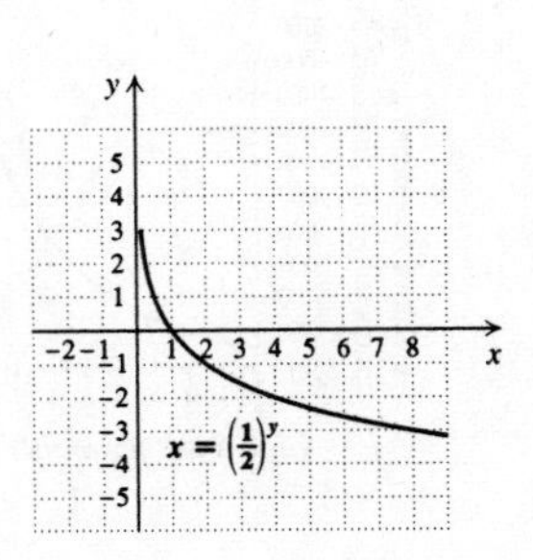

24.

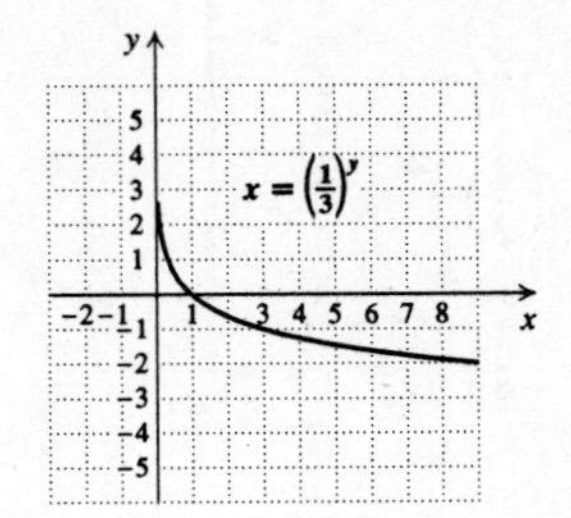

25. Graph: $x = 5^y$

We can find ordered pairs by choosing values for y and then computing values for x. Then we plot these points and connect them with a smooth curve.

For $y = 0$, $x = 5^0 = 1$.

For $y = 1$, $x = 5^1 = 5$.

For $y = 2$, $x = 5^2 = 25$.

For $y = -1$, $x = 5^{-1} = \frac{1}{5}$.

For $y = -2$, $x = 5^{-2} = \frac{1}{25}$.

x	y
1	0
5	1
25	2
$\frac{1}{5}$	−1
$\frac{1}{25}$	−2

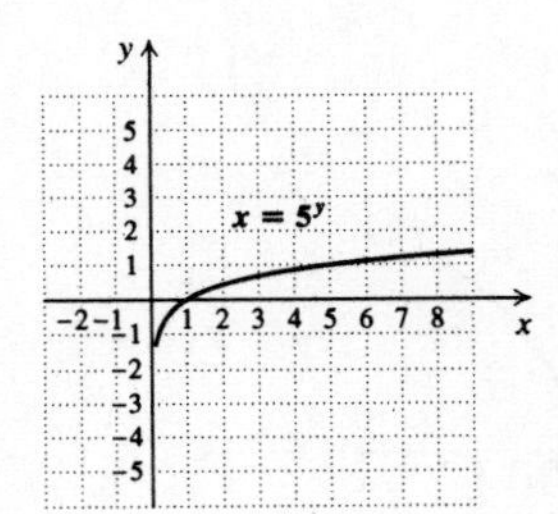

26.

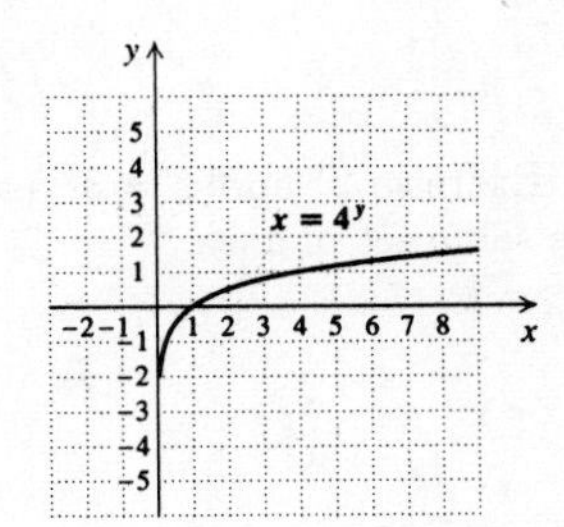

27. Graph: $x = \left(\frac{3}{2}\right)^y$

We can find ordered pairs by choosing values for y and then computing values for x. Then we plot these points and connect them with a smooth curve.

For $y = 0$, $x = \left(\frac{3}{2}\right)^0 = 1$.

For $y = 1$, $x = \left(\frac{3}{2}\right)^1 = \frac{3}{2}$.

For $y = 2$, $x = \left(\frac{3}{2}\right)^2 = \frac{9}{4}$.

For $y = 3$, $x = \left(\frac{3}{2}\right)^3 = \frac{27}{8}$.

For $y = -1$, $x = \left(\frac{3}{2}\right)^{-1} = \frac{1}{\frac{3}{2}} = \frac{2}{3}$.

For $y = -2$, $x = \left(\frac{3}{2}\right)^{-2} = \frac{1}{\frac{9}{4}} = \frac{4}{9}$.

For $y = -3$, $x = \left(\frac{3}{2}\right)^{-3} = \frac{1}{\frac{27}{8}} = \frac{8}{27}$.

x	y
1	0
$\frac{3}{2}$	1
$\frac{9}{4}$	2
$\frac{27}{8}$	3
$\frac{2}{3}$	−1
$\frac{4}{9}$	−2
$\frac{8}{27}$	−3

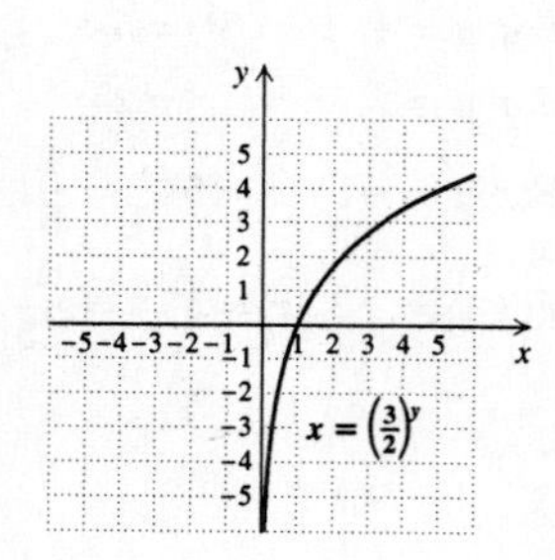

28.

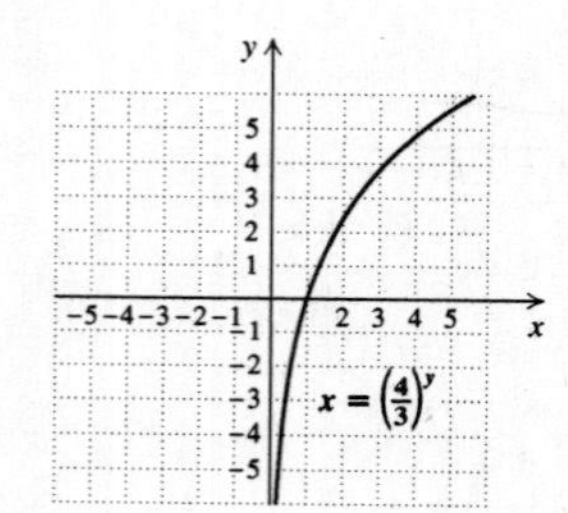

29. Graph $y = 3^x$ (see Exercise 2) and $x = 3^y$ (see Exercise 21) using the same set of axes.

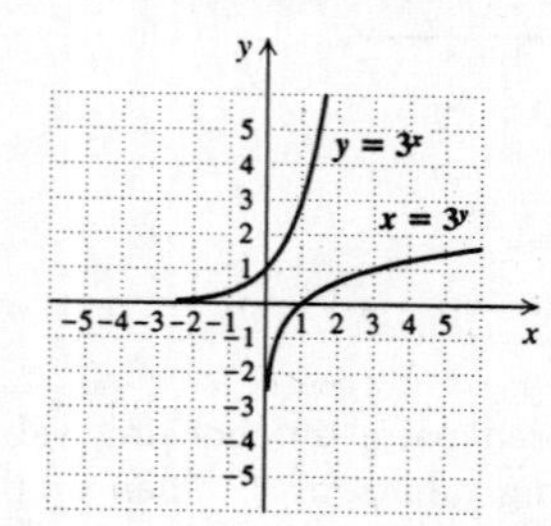

30.

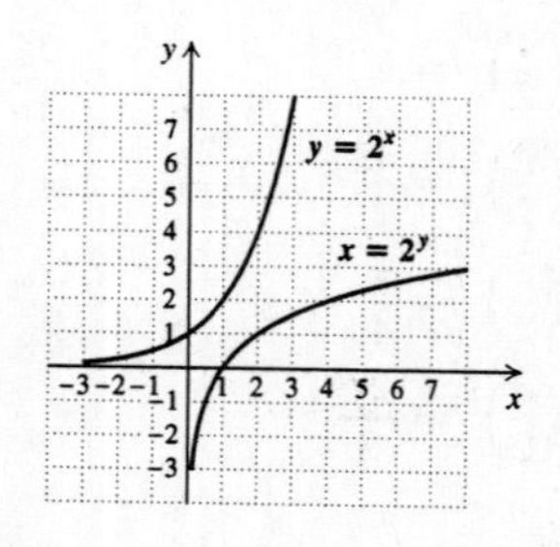

31. Graph $y = \left(\frac{1}{2}\right)^x$ (see Exercise 13) and $x = \left(\frac{1}{2}\right)^y$ (see Exercise 23) using the same set of axes.

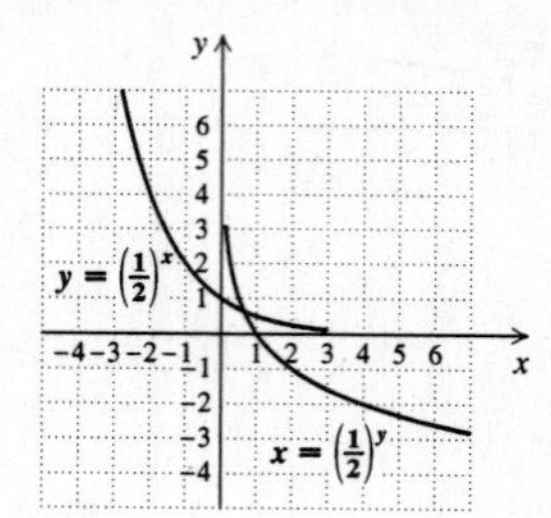

32.

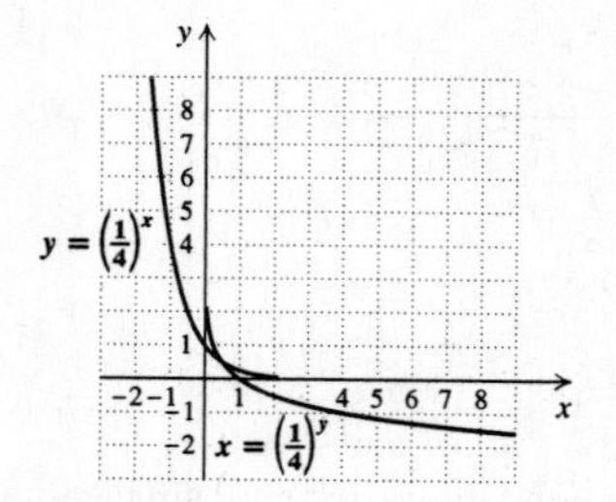

33. Keep in mind that t represents the number of years after 1989 and that N is in thousands.

a) For 1993, $t = 1993 - 1989$, or 4:

$$N(4) = 100(1.4)^4 = 384.16$$

Thus, 384.16 thousand, or 384,160 Americans had been infected as of 1993.

b) For 1998, $t = 1998 - 1989$, or 9:

$$N(9) = 100(1.4)^9 \approx 2,066.105$$

Then about 2066.105 thousand, or about 2,066,105 Americans will have been infected as of 1998.

c) We use the function values computed in parts (a) and (b), and others if we wish, to draw the graph. Note that the axes are scaled differently because of the large numbers.

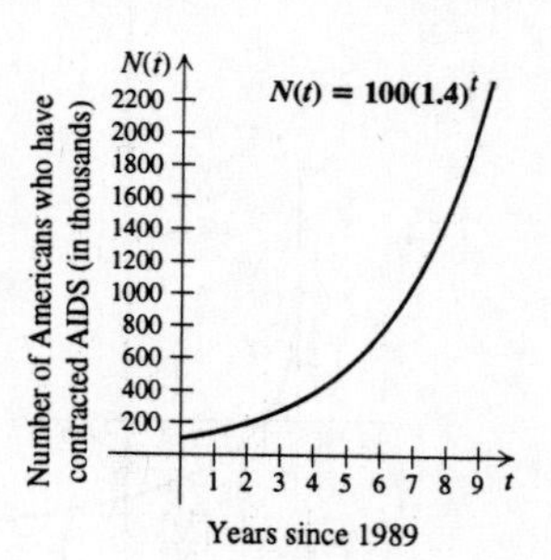

34. (a) 4243; 6000; 8485; 12,000; 24,000;

(b)

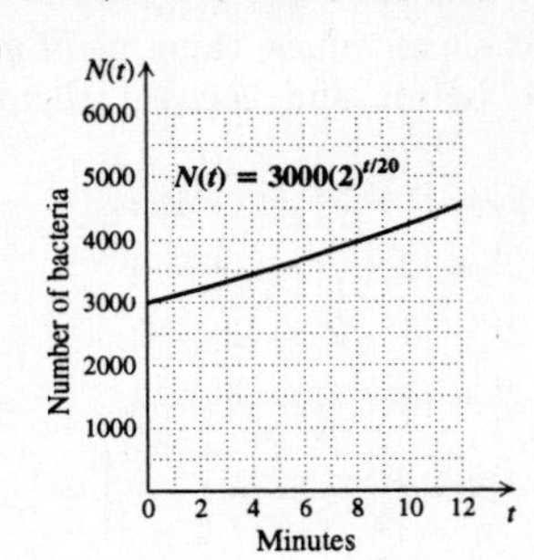

35. a) Substitute for t.

$$N(0) = 250,000\left(\frac{2}{3}\right)^0 = 250,000 \cdot 1 = 250,000;$$

$$N(1) = 250,000\left(\frac{2}{3}\right)^1 = 250,000 \cdot \frac{2}{3} = 166,667;$$

$$N(4) = 250,000\left(\frac{2}{3}\right)^4 = 250,000 \cdot \frac{16}{81} \approx 49,383;$$

$$N(10) = 250,000\left(\frac{2}{3}\right)^{10} = 250,000 \cdot \frac{1024}{59,049} \approx 4335$$

b) We use the function values computed in part (a) to draw the graph of the function. Note that the axes are scaled differently because of the large function values.

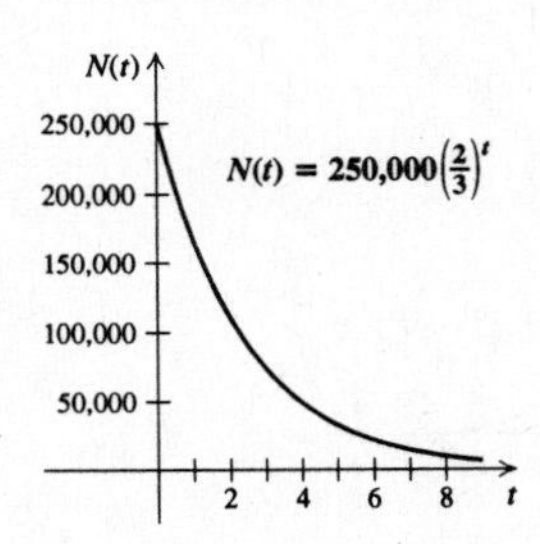

36. (a) \$5200; \$4160; \$3328; \$1703.94; \$558.35

(b)

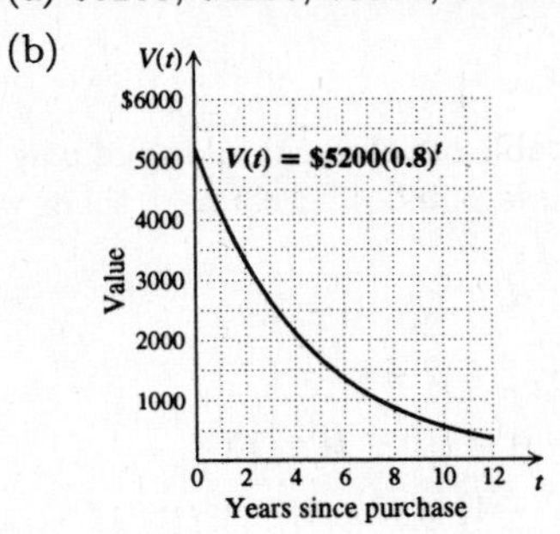

37. a) Substitute for t.

$$A(5) = 10 \cdot 34^5 = 454,354,240 \text{ cm}^2$$

$$A(7) = 10 \cdot 34^7 = 525,233,501,400 \text{ cm}^2$$

b) Use the function values computed in part (a) and others as needed to draw the graph. Note that the axes are scaled differently because of the large numbers.

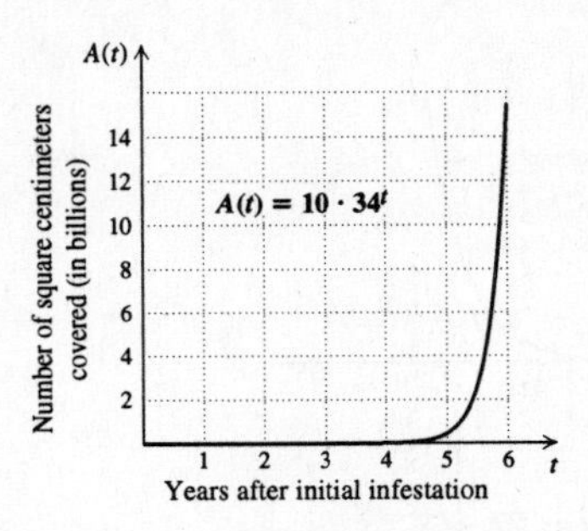

38. (a) 0.3 million, or 300,000; 3.5 million; 39.7 million; 457.0 million; 5258.6 million;

(b)

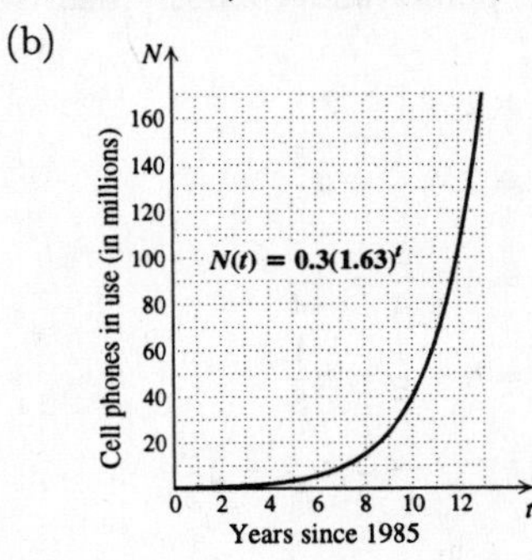

39. $x^{-5} \cdot x^3 = x^{-5+3} = x^{-2}$, or $\frac{1}{x^2}$

40. x^{-12}, or $\frac{1}{x^{12}}$

41. $\frac{x^{-3}}{x^4} = x^{-3-4} = x^{-7}$, or $\frac{1}{x^7}$

42. 1

43. ◈

44. ◈

45. ◈

46. ◈

47. Since the bases are the same, the one with the larger exponent is the larger number. Thus $\pi^{2.4}$ is larger.

48. $8^{\sqrt{3}}$

49. Graph: $f(x) = 3.8^x$

Use a calculator with a power key to construct a table of values. (We will round values of $f(x)$ to the nearest hundredth.) Then plot these points and connect them with a smooth curve.

x	y
0	1
1	3.8
2	14.44
3	54.872
−1	0.26
−2	0.7

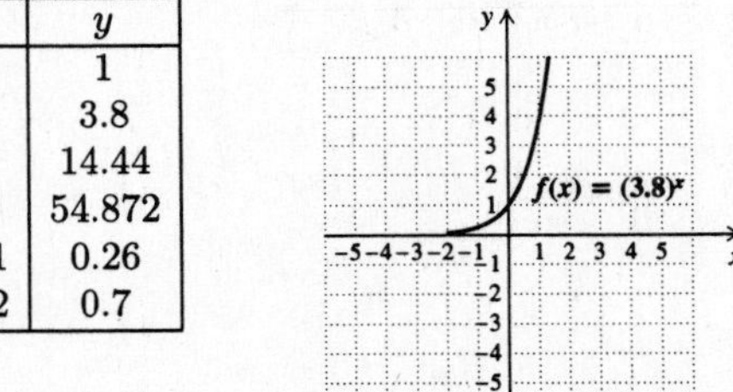

50.

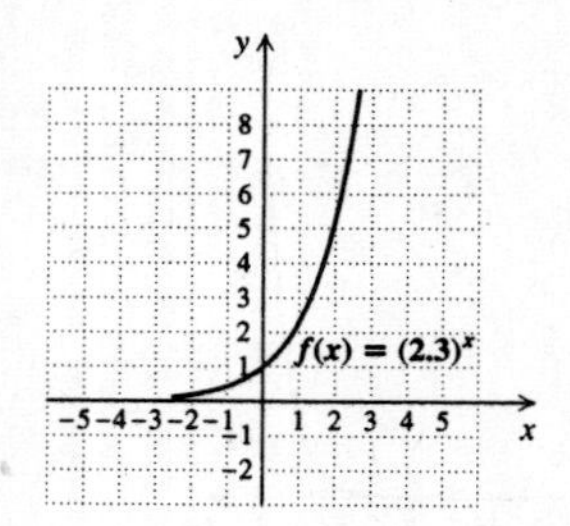

51. Graph: $y = 2^x + 2^{-x}$

Construct a table of values, thinking of y as $f(x)$. Then plot these points and connect them with a curve.

$f(0) = 2^0 + 2^{-0} = 1 + 1 = 2$

$f(1) = 2^1 + 2^{-1} = 2 + \frac{1}{2} = 2\frac{1}{2}$

$f(2) = 2^2 + 2^{-2} = 4 + \frac{1}{4} = 4\frac{1}{4}$

$f(3) = 2^3 + 2^{-3} = 8 + \frac{1}{8} = 8\frac{1}{8}$

$f(-1) = 2^{-1} + 2^{-(-1)} = \frac{1}{2} + 2 = 2\frac{1}{2}$

$f(-2) = 2^{-2} + 2^{-(-2)} = \frac{1}{4} + 4 = 4\frac{1}{4}$

$f(-3) = 2^{-3} + 2^{-(-3)} = \frac{1}{8} + 8 = 8\frac{1}{8}$

x	y, or $f(x)$
0	2
1	$2\frac{1}{2}$
2	$4\frac{1}{4}$
3	$8\frac{1}{8}$
−1	$2\frac{1}{2}$
−2	$4\frac{1}{4}$
−3	$8\frac{1}{8}$

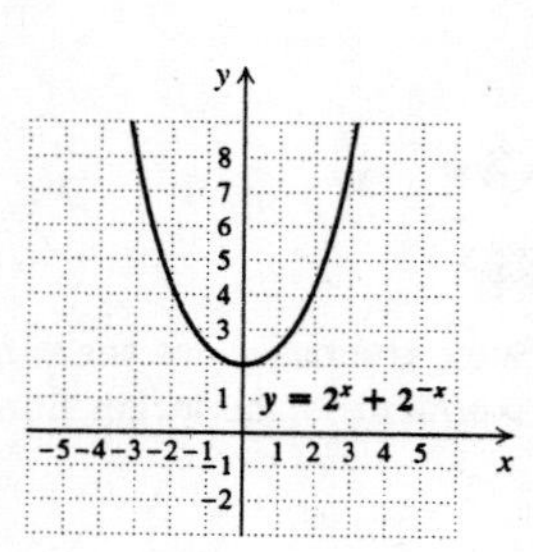

52.

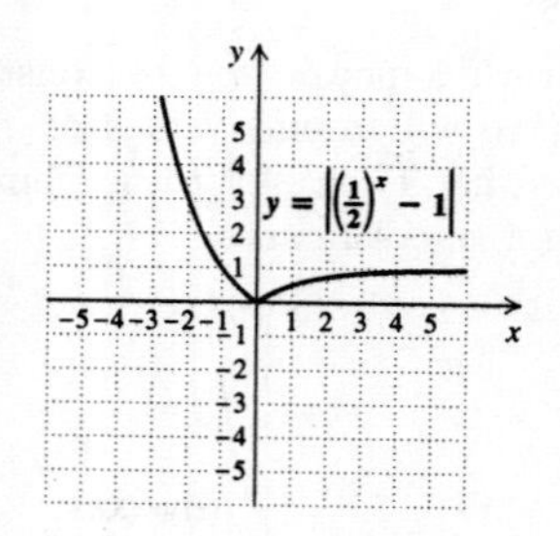

53. Graph: $y = |2^x - 2|$

We construct a table of values, thinking of y as $f(x)$. Then plot these points and connect them with a curve.

$f(0) = |2^0 - 2| = |1 - 2| = |-1| = 1$

$f(1) = |2^1 - 2| = |2 - 2| = |0| = 0$

$f(2) = |2^2 - 2| = |4 - 2| = |2| = 2$

$f(3) = |2^3 - 2| = |8 - 2| = |6| = 6$

$f(-1) = |2^{-1} - 2| = \left|\frac{1}{2} - 2\right| = \left|-\frac{3}{2}\right| = \frac{3}{2}$

$f(-3) = |2^{-3} - 2| = \left|\frac{1}{8} - 2\right| = \left|-\frac{15}{8}\right| = \frac{15}{8}$

$f(-5) = |2^{-5} - 2| = \left|\frac{1}{32} - 2\right| = \left|-\frac{63}{32}\right| = \frac{63}{32}$

x	y, or $f(x)$
0	1
1	0
2	2
3	6
−1	$\frac{3}{2}$
−3	$\frac{15}{8}$
−5	$\frac{63}{32}$

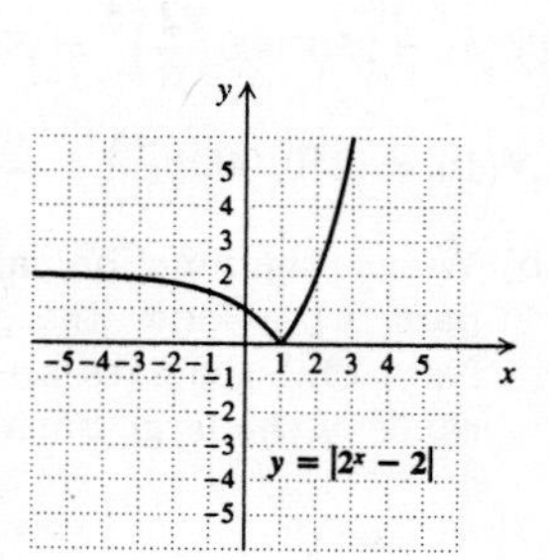

54.

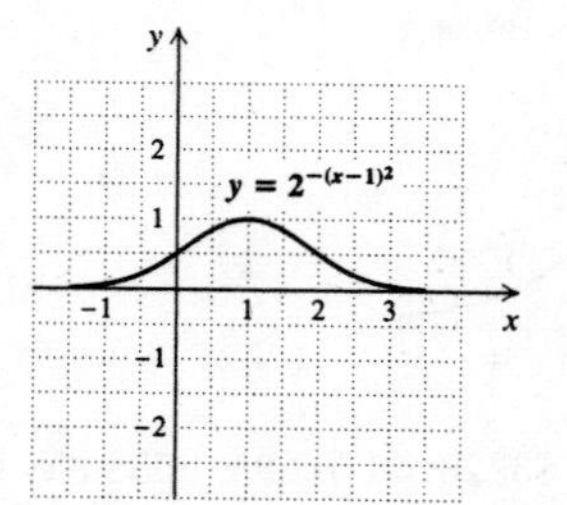

55. Graph: $y = |2x^2 - 1|$

We construct a table of values, thinking of y as $f(x)$. Then we plot these points and connect them with a curve.

$f(0) = |2^{0^2} - 1| = |1 - 1| = 0$

$f(1) = |2^{1^2} - 1| = |2 - 1| = 1$

$f(2) = |2^{2^2} - 1| = |16 - 1| = 15$

$f(-1) = |2^{(-1)^2} - 1| = |2 - 1| = 1$

$f(-2) = |2^{(-2)^2} - 1| = |16 - 1| = 15$

x	y, or $f(x)$
0	0
1	1
2	15
−1	1
−2	15

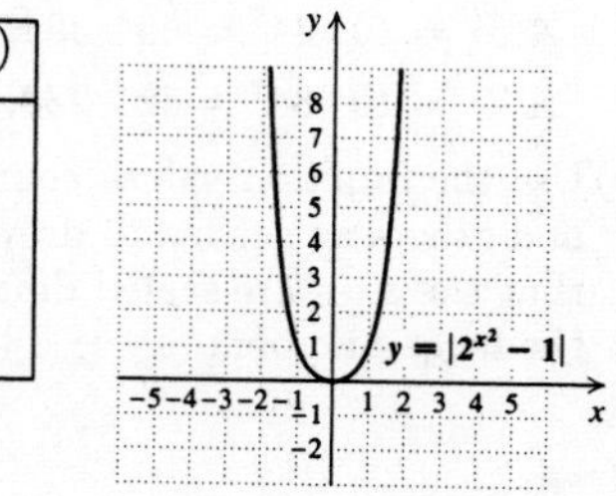

56.

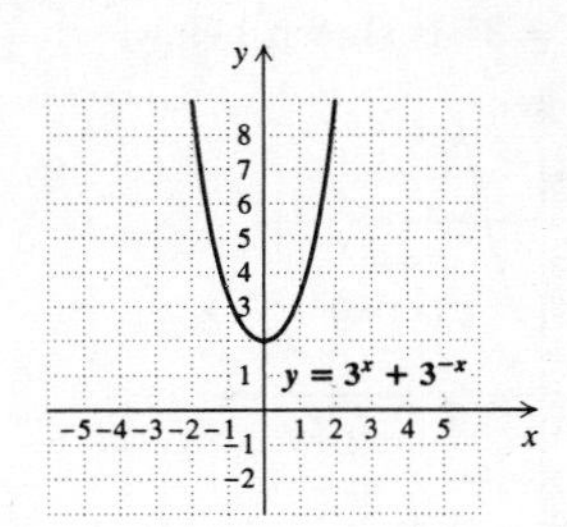

57. $y = 3^{-(x-1)}$ $\quad$ $x = 3^{-(y-1)}$

x	y
0	3
1	1
2	$\frac{1}{3}$
3	$\frac{1}{9}$
-1	9

x	y
3	0
1	1
$\frac{1}{3}$	2
$\frac{1}{9}$	3
9	-1

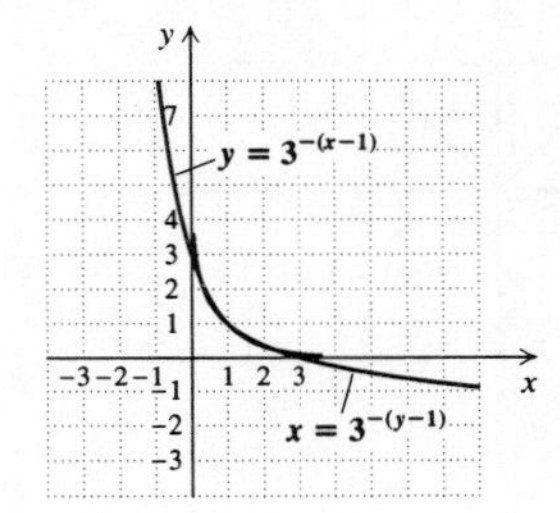

58.

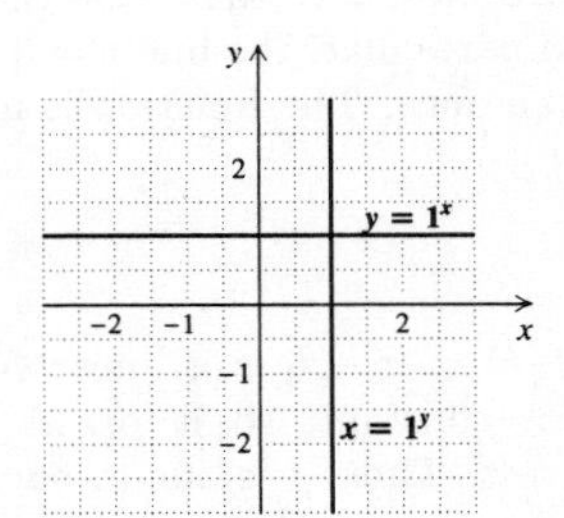

59. Enter the data points $(0, 1.9)$, $(1, 3.8)$, and $(2, 10.2)$ and then use the exponential regression feature of the grapher to find an exponential function that models the data: $y = 1.809071941(2.316985337)^t$, where t is the number of years after 1993 and y is in millions. To find the number of printers that will be sold in 1998, find the value of y when $t = 1998 - 1993$, or 5.

$y = 1.809071941(2.316985337)^5 \approx 120.8$ million

60. $N(t) = 9047.179795(1.393338213)^t$, where t is the number of years after 1985; 17,564

61. Enter the function on the grapher and then use the graph or a table of values set in ASK mode.

$S(10) \approx 19$ words per minute

$S(40) \approx 66$ words per minute

$S(80) \approx 110$ words per minute

Exercise Set 9.2

1. $f \circ g(x) = f(g(x)) = f(2x+3) =$
$3(2x+3)^2 - 1 =$
$3(4x^2 + 12x + 9) - 1 =$
$12x^2 + 36x + 27 - 1 =$
$12x^2 + 36x + 26$

$g \circ f(x) = g(f(x)) = g(3x^2 - 1) =$
$2(3x^2 - 1) + 3 = 6x^2 - 2 + 3 =$
$6x^2 + 1$

2. $f \circ g(x) = 8x^2 - 17$; $g \circ f(x) = 32x^2 + 48x + 13$

3. $f \circ g(x) = f(g(x)) = f\left(\frac{2}{x}\right) =$
$4\left(\frac{2}{x}\right)^2 - 1 =$
$4\left(\frac{4}{x^2}\right) - 1 = \frac{16}{x^2} - 1$

$g \circ f(x) = g(f(x)) = g(4x^2 - 1) = \frac{2}{4x^2 - 1}$

4. $f \circ g(x) = \frac{3}{2x^2 + 3}$; $g \circ f(x) = \frac{18}{x^2} + 3$

5. $f \circ g(x) = f(g(x)) = f(x^2 + 1) = (x^2 + 1)^2 - 3 =$
$x^4 + 2x^2 + 1 - 3 = x^4 + 2x^2 - 2$

$g \circ f(x) = g(f(x)) = g(x^2 - 3) = (x^2 - 3)^2 + 1 =$
$x^4 - 6x^2 + 9 + 1 = x^4 - 6x^2 + 10$

6. $f \circ g(x) = \frac{1}{x^2 + 4x + 4}$; $g \circ f(x) = \frac{1}{x^2} + 2$

7. $h(x) = (7 - 5x)^2$
This is $7 - 5x$ raised to the second power, so the two most obvious functions are $f(x) = x^2$ and $g(x) = 7 - 5x$.

8. $f(x) = 4x^2 + 9$, $g(x) = 3x - 1$

9. $h(x) = (3x^2 - 7)^5$
This is $3x^2 - 7$ to the fifth power, so the two most obvious functions are $f(x) = x^5$ and $g(x) = 3x^2 - 7$.

10. $f(x) = \sqrt{x}$, $g(x) = 5x + 2$

11. $h(x) = \frac{2}{x - 3}$
This is 2 divided by $x - 3$, so two functions that can be used are $f(x) = \frac{2}{x}$ and $g(x) = x - 3$.

12. $f(x) = x + 4$, $g(x) = \frac{3}{x}$

13. $h(x) = \frac{1}{\sqrt{7x+2}}$

This is the reciprocal of the square root of $7x+2$. Two functions that can be used are $f(x) = \frac{1}{\sqrt{x}}$ and $g(x) = 7x+2$.

14. $f(x) = \sqrt{x} - 3,\ g(x) = x - 7$

15. $h(x) = \frac{x^3+1}{x^3-1}$

Two functions that can be used are $f(x) = \frac{x+1}{x-1}$ and $g(x) = x^3$.

16. $f(x) = x^4,\ g(x) = \sqrt{x} + 5$

17. The graph of $f(x) = x - 5$ is shown below.

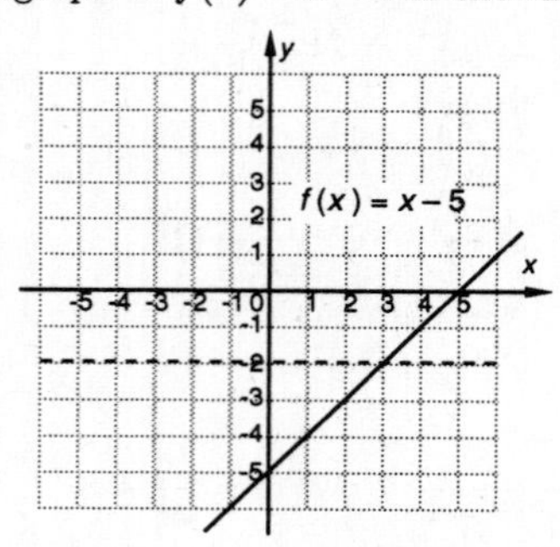

Since there is no horizontal line that crosses the graph more than once, the function is one-to-one.

18. Yes

19. The graph of $f(x) = x^2 + 1$ is shown below.

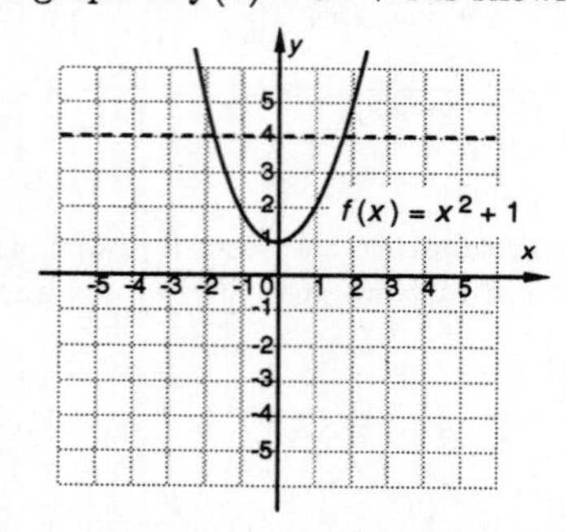

There are many horizontal lines that cross the graph more than once. In particular, the line $y = 4$ crosses the graph more than once. The function is not one-to-one.

20. No

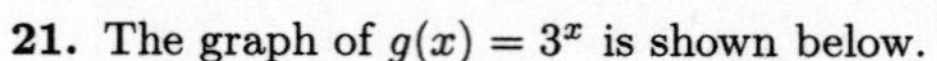

21. The graph of $g(x) = 3^x$ is shown below.

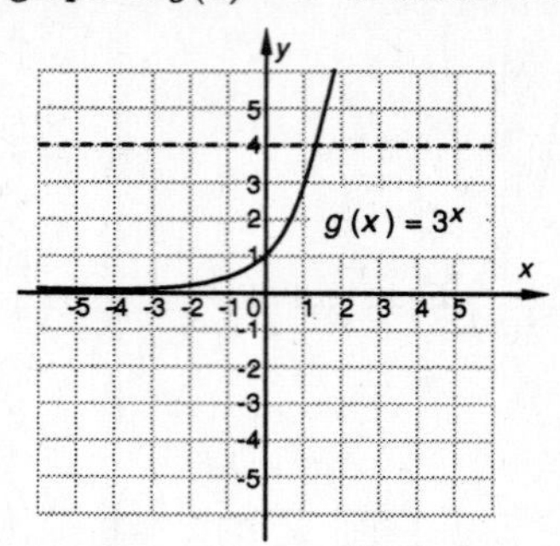

Since no horizontal line crosses the graph more than once, the function is one-to-one.

22. Yes

23. The graph of $g(x) = |x|$ is shown below.

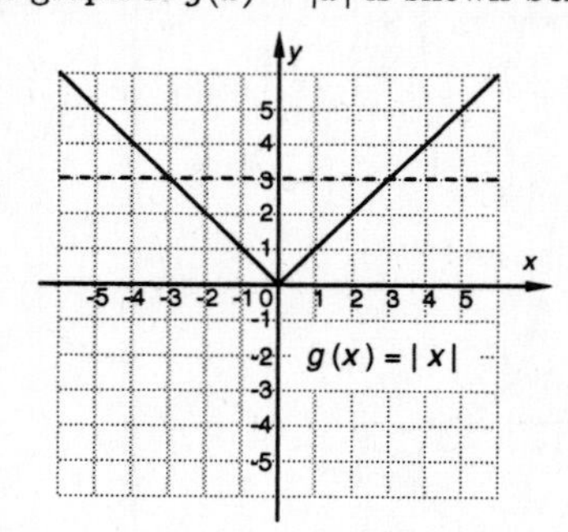

There are many horizontal lines that cross the graph more than once. In particular, the line $y = 3$ crosses the graph more than once. The function is not one-to-one.

24. No

25. a) The function $f(x) = x + 6$ is a linear function that is not constant, so it passes the horizontal-line test. Thus, f is one-to-one.

b) Replace $f(x)$ by y: $y = x + 6$

Interchange x and y: $x = y + 6$

Solve for y: $x - 6 = y$

Replace y by $f^{-1}(x)$: $f^{-1}(x) = x - 6$

26. (a) Yes; (b) $f^{-1}(x) = x - 7$

27. a) The function $f(x) = 3 - x$ is a linear function that is not constant, so it passes the horizontal-line test. Thus, f is one-to-one.

b) Replace $f(x)$ by y: $y = 3 - x$

Interchange x and y: $x = 3 - y$

Solve for y: $y = 3 - x$

Replace y by $f^{-1}(x)$: $f^{-1}(x) = 3 - x$

28. (a) Yes; (b) $f^{-1}(x) = 9 - x$

29. a) The function $g(x) = x - 5$ is a linear function that is not constant, so it passes the horizontal-line test. Thus, g is one-to-one.

b) Replace $g(x)$ by y: $y = x - 5$

Interchange x and y: $x = y - 5$

Solve for y: $x + 5 = y$

Replace y by $g^{-1}(x)$: $g^{-1}(x) = x + 5$

30. (a) Yes; (b) $g^{-1}(x) = x + 8$

31. a) The function $f(x) = 4x$ is a linear function that is not constant, so it passes the horizontal-line test. Thus, f is one-to-one.

b) Replace $f(x)$ by y: $y = 4x$

Interchange x and y: $x = 4y$

Solve for y: $\dfrac{x}{4} = y$

Replace y by $f^{-1}(x)$: $f^{-1}(x) = \dfrac{x}{4}$

32. (a) Yes; (b) $f^{-1}(x) = \dfrac{x}{7}$

33. a) The function $g(x) = 4x + 3$ is a linear function that is not constant, so it passes the horizontal-line test. Thus, g is one-to-one.

b) Replace $g(x)$ by y: $y = 4x + 3$

Interchange variables: $x = 4y + 3$

Solve for y: $x - 3 = 4y$

$$\frac{x-3}{4} = y$$

Replace y by $g^{-1}(x)$: $g^{-1}(x) = \dfrac{x-3}{4}$

34. (a) Yes; (b) $g^{-1}(x) = \dfrac{x-7}{4}$

35. a) The graph of $h(x) = 5$ is shown below. The horizontal line $y = 5$ crosses the graph more than once, so the function is not one-to-one.

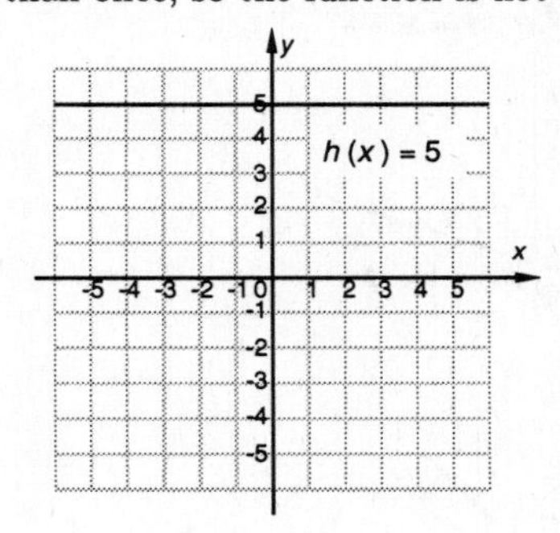

36. (a) No

37. a) The graph of $f(x) = \dfrac{1}{x}$ is shown below. It passes the horizontal-line test, so the function is one-to-one.

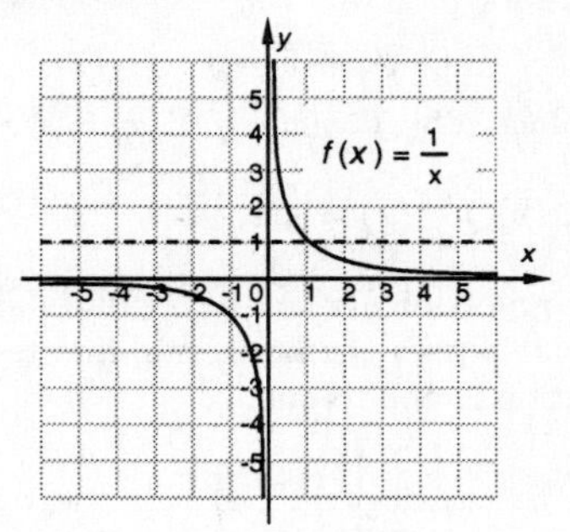

b) Replace $f(x)$ by y: $y = \dfrac{1}{x}$

Interchange x and y: $x = \dfrac{1}{y}$

Solve for y: $xy = 1$

$$y = \frac{1}{x}$$

Replace y by $f^{-1}(x)$: $f^{-1}(x) = \dfrac{1}{x}$

38. (a) Yes; (b) $f^{-1}(x) = \dfrac{3}{x}$

39. a) The function $f(x) = \dfrac{2x+1}{3} = \dfrac{2}{3}x + \dfrac{1}{3}$ is a linear function that is not constant, so it passes the horizontal-line test. Thus, f is one-to-one.

b) Replace $f(x)$ by y: $y = \dfrac{2x+1}{3}$

Interchange x and y: $x = \dfrac{2y+1}{3}$

Solve for y: $3x = 2y + 1$

$$3x - 1 = 2y$$

$$\frac{3x-1}{2} = y$$

Replace y by $f^{-1}(x)$: $f^{-1}(x) = \dfrac{3x-1}{2}$

40. (a) Yes; (b) $f^{-1}(x) = \dfrac{5x-2}{3}$

41. a) The graph of $f(x) = x^3 - 5$ is shown below. It passes the horizontal-line test, so the function is one-to-one.

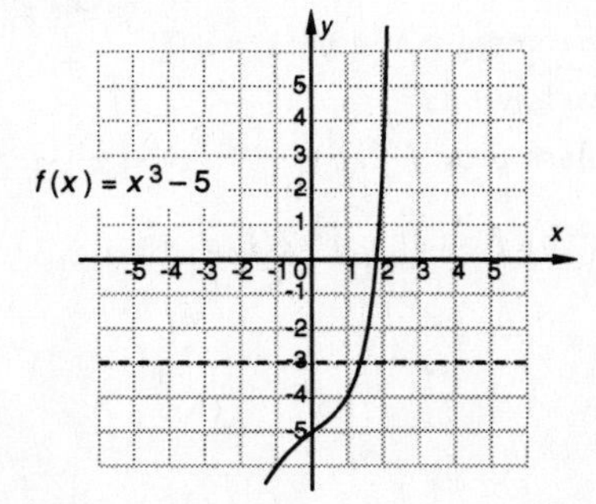

b) Replace $f(x)$ by y: $y = x^3 - 5$
Interchange x and y: $x = y^3 - 5$
Solve for y: $x + 5 = y^3$
$\sqrt[3]{x+5} = y$
Replace y by $f^{-1}(x)$: $f^{-1}(x) = \sqrt[3]{x+5}$

42. (a) Yes; (b) $f^{-1}(x) = \sqrt[3]{x-2}$

43. a) The graph of $g(x) = (x-2)^3$ is shown below. It passes the horizontal-line test, so the function is one-to-one.

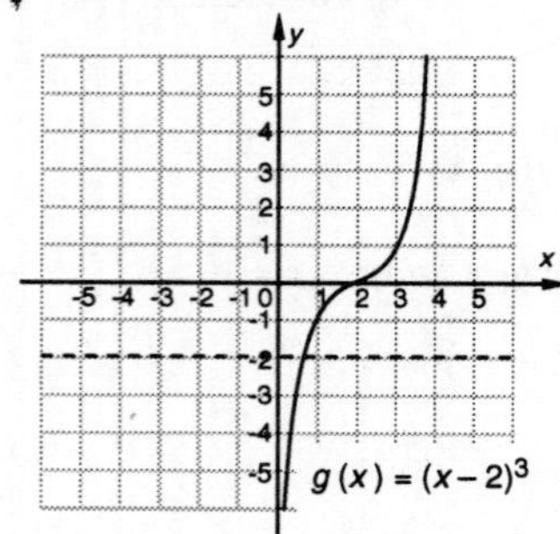

b) Replace $g(x)$ by y: $y = (x-2)^3$
Interchange x and y: $x = (y-2)^3$
Solve for y: $\sqrt[3]{x} = y - 2$
$\sqrt[3]{x} + 2 = y$
Replace y by $g^{-1}(x)$: $g^{-1}(x) = \sqrt[3]{x} + 2$

44. (a) Yes; (b) $g^{-1}(x) = \sqrt[3]{x} - 7$

45. a) The graph of $f(x) = \sqrt{x}$ is shown below. It passes the horizontal-line test, so the function is one-to-one.

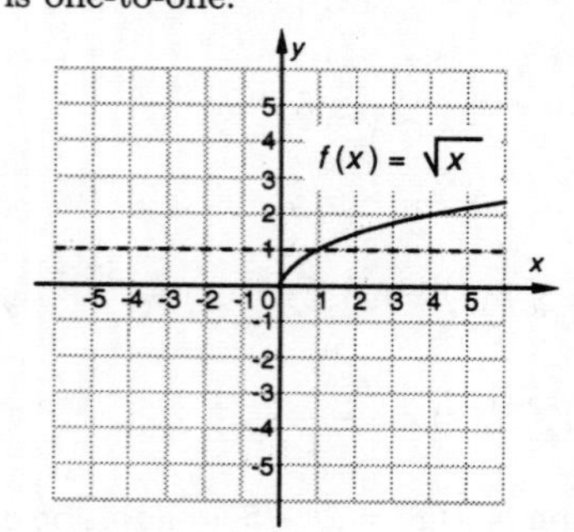

b) Replace $f(x)$ by y: $y = \sqrt{x}$ (Note that $f(x) \geq 0$.)
Interchange x and y: $x = \sqrt{y}$
Solve for y: $x^2 = y$
Replace y by $f^{-1}(x)$: $f^{-1}(x) = x^2$, $x \geq 0$

46. (a) Yes; (b) $f^{-1}(x) = x^2 + 1$, $x \geq 0$

47. a) The graph of $f(x) = 2x^2 + 1$, $x \geq 0$, is shown below. It passes the horizontal-line test, so the function is one-to-one.

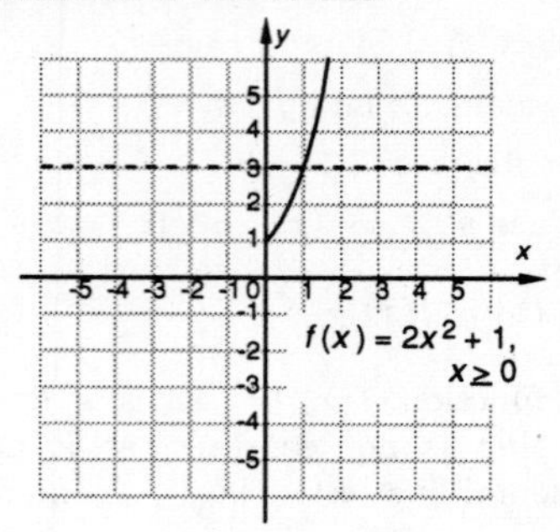

b) Replace $f(x)$ by y: $y = 2x^2 + 1$
Interchange x and y: $x = 2y^2 + 1$
Solve for y: $x - 1 = 2y^2$
$\frac{x-1}{2} = y^2$
$\sqrt{\frac{x-1}{2}} = y$
(We take the principal square root since $y \geq 0$.)
Replace y by $f^{-1}(x)$: $f^{-1}(x) = \sqrt{\frac{x-1}{2}}$

48. (a) Yes; (b) $f^{-1}(x) = \sqrt{\frac{x+2}{3}}$

49. First graph $f(x) = \frac{1}{3}x - 2$. Then graph the inverse function by reflecting the graph of $f(x) = \frac{1}{3}x - 2$ across the line $y = x$. The graph of the inverse function can also be found by first finding a formula for the inverse, substituting to find function values, and then plotting points.

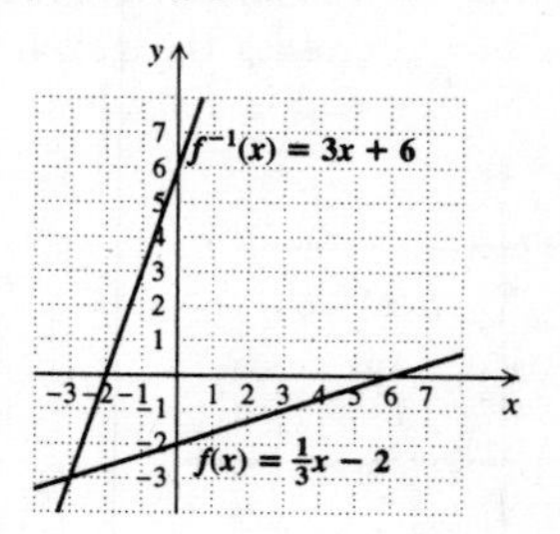

50.

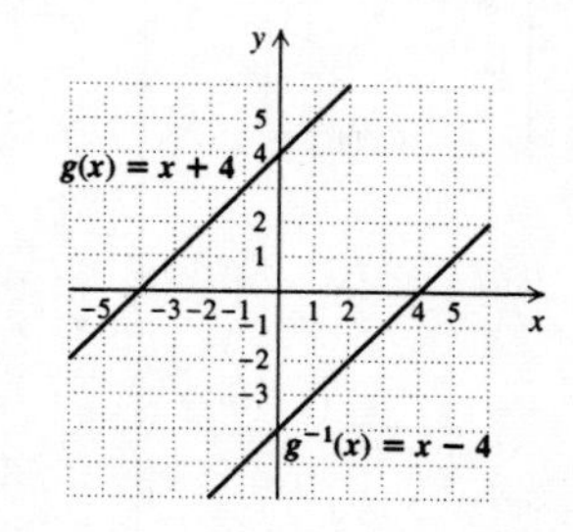

51. Follow the procedure described in Exercise 49 to graph the function and its inverse.

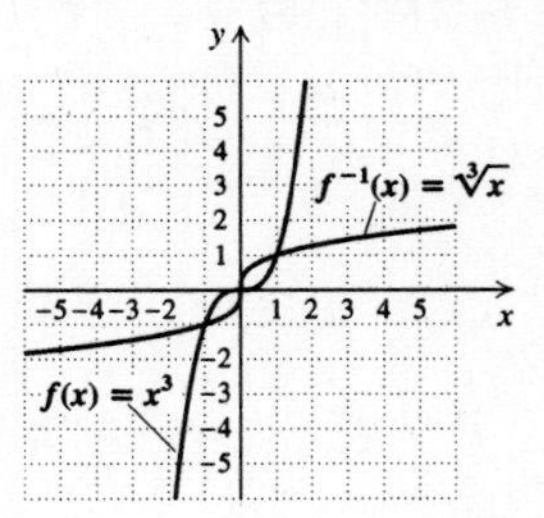

52.

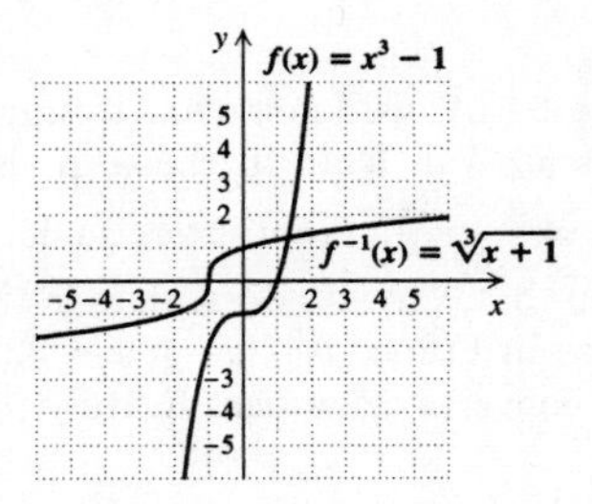

53. Use the procedure described in Exercise 49 to graph the function and its inverse.

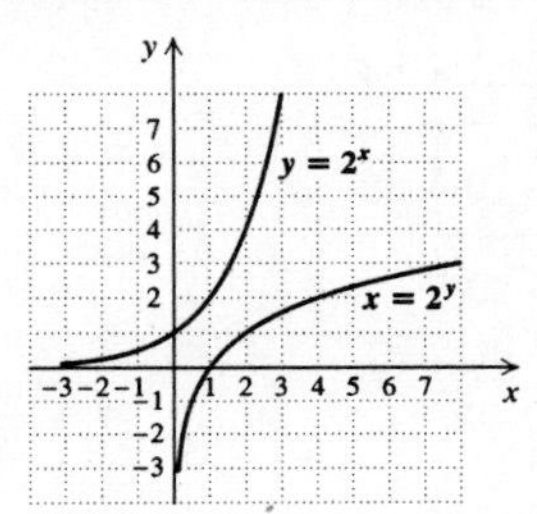

54.

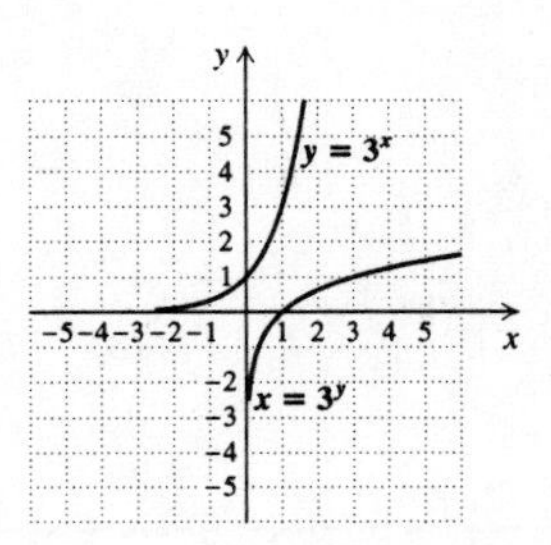

55. Use the procedure described in Exercise 49 to graph the function and its inverse.

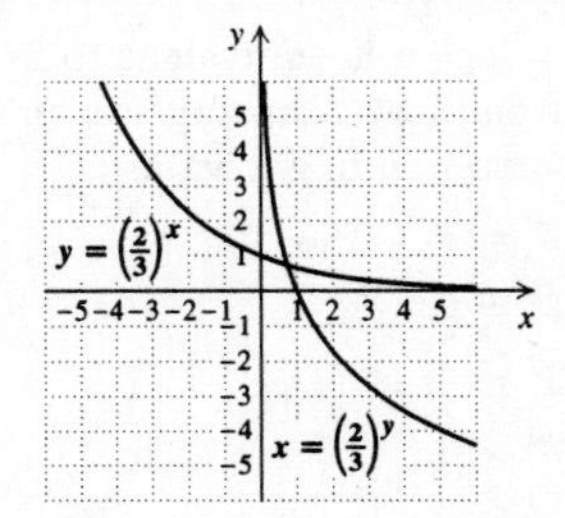

56.

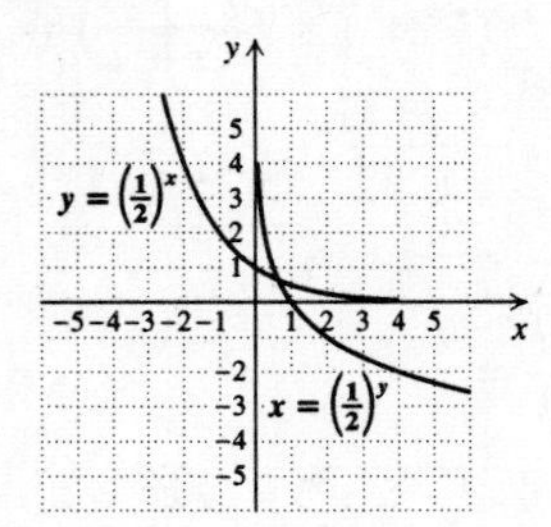

57. Use the procedure described in Exercise 49 to graph the function and its inverse.

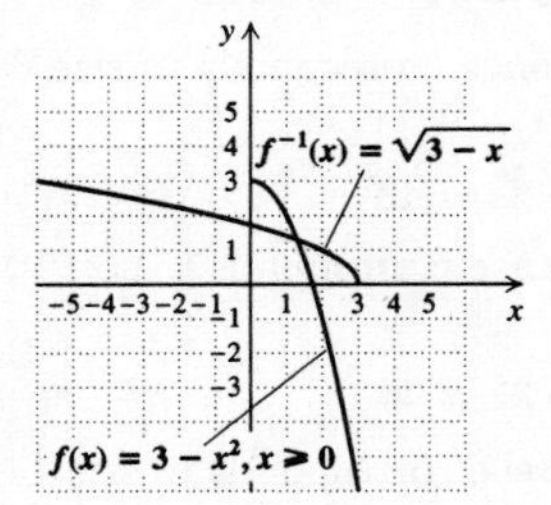

58.

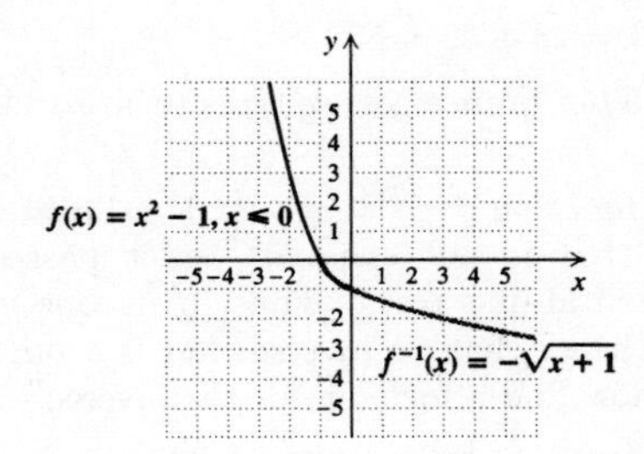

59. We check to see that $f^{-1} \circ f(x) = x$ and $f \circ f^{-1}(x) = x$.

a) $f^{-1} \circ f(x) = f^{-1}(f(x)) = f^{-1}\left(\frac{4}{5}x\right) =$
$\frac{5}{4} \cdot \frac{4}{5}x = x$

b) $f \circ f^{-1}(x) = f(f^{-1}(x)) = f\left(\frac{5}{4}x\right) =$
$\frac{4}{5} \cdot \frac{5}{4}x = x$

60. a) $f^{-1} \circ f(x) = 3\left(\frac{x+7}{3}\right) - 7 = x + 7 - 7 = x$

b) $f \circ f^{-1}(x) = \frac{(3x-7)+7}{3} = \frac{3x}{3} = x$

61. We check to see that $f^{-1} \circ f(x) = x$ and $f \circ f^{-1}(x) = x$.

a) $f^{-1} \circ f(x) = f^{-1}(f(x)) = f^{-1}\left(\frac{1-x}{x}\right) =$

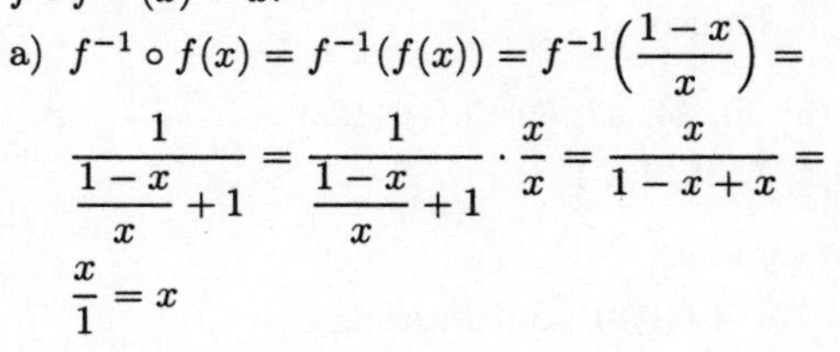

$$\frac{1}{\frac{1-x}{x} + 1} = \frac{1}{\frac{1-x}{x} + 1} \cdot \frac{x}{x} = \frac{x}{1 - x + x} =$$

$\frac{x}{1} = x$

b) $f \circ f^{-1}(x) = f(f^{-1}(x)) = f\left(\frac{1}{x+1}\right) =$

$$\frac{1-\frac{1}{x+1}}{\frac{1}{x+1}} = \frac{1-\frac{1}{x+1}}{\frac{1}{x+1}} \cdot \frac{x+1}{x+1} =$$

$$\frac{x+1-1}{1} = \frac{x}{1} = x$$

62. a) $f^{-1} \circ f(x) = \sqrt[3]{x^3 - 5 + 5} = \sqrt[3]{x^3} = x$

b) $f \circ f^{-1}(x) = (\sqrt[3]{x+5})^3 - 5 = x + 5 - 5 = x$

63. a) $f(8) = 8 + 32 = 40$

Size 40 in France corresponds to size 8 in the U.S.

$f(10) = 10 + 32 = 42$

Size 42 in France corresponds to size 10 in the U.S.

$f(14) = 14 + 32 = 46$

Size 46 in France corresponds to size 14 in the U.S.

$f(18) = 18 + 32 = 50$

Size 50 in France corresponds to size 18 in the U.S.

b) The function $f(x) = x + 32$ is a linear function that is not constant, so it passes the horizontal-line test. Thus, f is one-to-one and, hence, has an inverse that is a function. We now find a formula for the inverse.

Replace $f(x)$ by y: $y = x + 32$

Interchange x and y: $x = y + 32$

Solve for y: $x - 32 = y$

Replace y by $f^{-1}(x)$: $f^{-1}(x) = x - 32$

c) $f^{-1}(40) = 40 - 32 = 8$

Size 8 in the U.S. corresponds to size 40 in France.

$f^{-1}(42) = 42 - 32 = 10$

Size 10 in the U.S. corresponds to size 42 in France.

$f^{-1}(46) = 46 - 32 = 14$

Size 14 in the U.S. corresponds to size 46 in France.

$f^{-1}(50) = 50 - 32 = 18$

Size 18 in the U.S. corresponds to size 50 in France.

64. (a) 40, 44, 52, 60; (b) $f^{-1}(x) = \frac{x-24}{2}$, or $\frac{x}{2} - 12$; (c) 8, 10, 14, 18

65.

$y = kx$	
$7.2 = k(0.8)$	Substituting
$9 = k$	Variation constant
$y = 9x$	Equation of variation

66. $y = \frac{21.35}{x}$

67. $(a^3b^2)^5(a^2b^7) = (a^{3\cdot5}b^{2\cdot5})(a^2b^7) =$
$a^{15}b^{10}a^2b^7 = a^{15+2}b^{10+7} = a^{17}b^{17}$

68. $x^{11}y^6z^8$

69. ◈

70. ◈

71. ◈

72. ◈

73. From Exercise 64(b), we know that a function that converts dress sizes in Italy to those in the United States is $g(x) = \frac{x-24}{2}$. From Exercise 63, we know that a function that converts dress sizes in the United States to those in France is $f(x) = x + 32$. Then a function that converts dress sizes in Italy to those in France is

$$h(x) = f \circ g(x)$$
$$h(x) = f\left(\frac{x-24}{2}\right)$$
$$h(x) = \frac{x-24}{2} + 32$$
$$h(x) = \frac{x}{2} - 12 + 32$$
$$h(x) = \frac{x}{2} + 20.$$

74. ◈

75. No

76. Yes

77. Yes

78. No

79. (1) C; (2) A; (3) B; (4) D

80. ◈ 

Exercise Set 9.3

1. Graph: $y = \log_2 x$

The equation $y = \log_2 x$ is equivalent to $2^y = x$. We can find ordered pairs by choosing values for y and computing the corresponding x-values.

For $y = 0$, $x = 2^0 = 1$.

For $y = 1$, $x = 2^1 = 2$.

For $y = 2$, $x = 2^2 = 4$.

For $y = 3$, $x = 2^3 = 8$.

For $y = -1$, $x = 2^{-1} = \frac{1}{2}$.

For $y = -2$, $x = 2^{-2} = \frac{1}{4}$.

x, or 2^y	y
1	0
2	1
4	2
8	3
$\frac{1}{2}$	-1
$\frac{1}{4}$	-2

(1) Select y.

(2) Compute x.

We plot the set of ordered pairs and connect the points with a smooth curve.

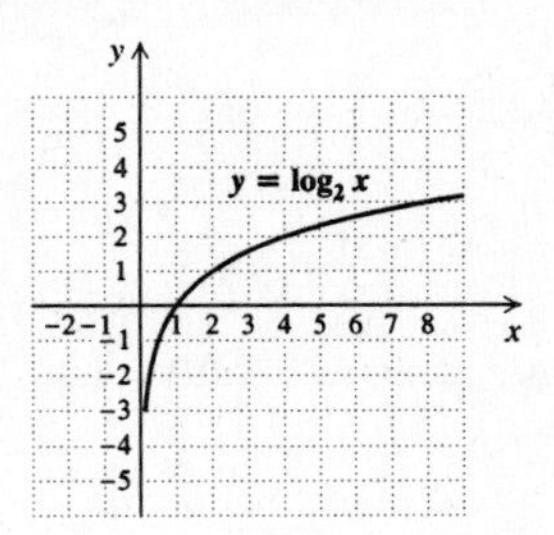

2.

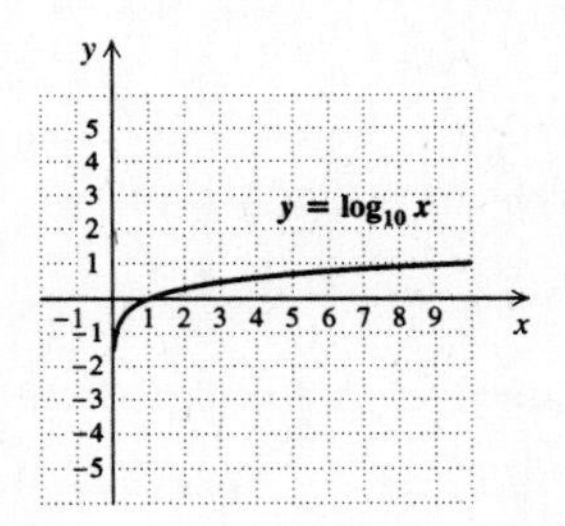

3. Graph: $y = \log_7 x$

The equation $y = \log_7 x$ is equivalent to $7^y = x$. We can find ordered pairs by choosing values for y and computing the corresponding x-values.

For $y = 0$, $x = 7^0 = 1$.

For $y = 1$, $x = 7^1 = 7$.

For $y = 2$, $x = 7^2 = 49$.

For $y = -1$, $x = 7^{-1} = \frac{1}{7}$.

For $y = -2$, $x = 7^{-2} = \frac{1}{49}$.

x, or 7^y	y
1	0
7	1
49	2
$\frac{1}{7}$	-1
$\frac{1}{49}$	-2

We plot the set of ordered pairs and connect the points with a smooth curve.

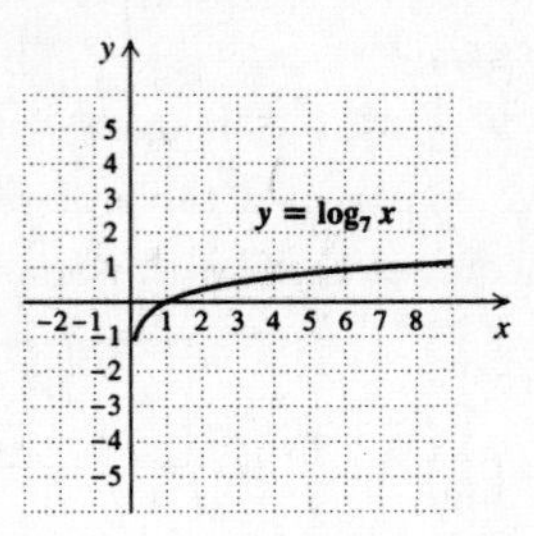

4.

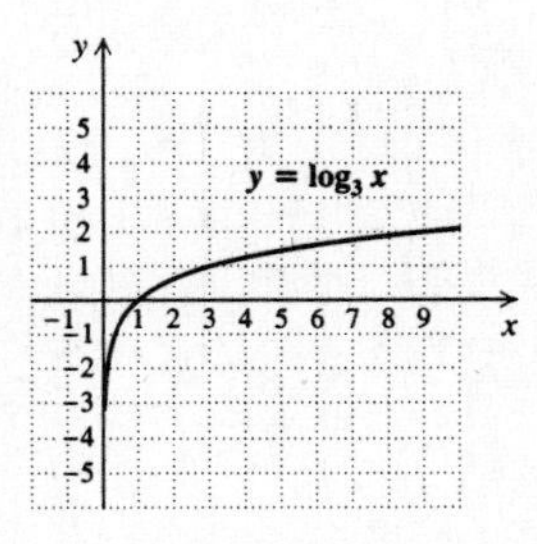

5. Graph: $f(x) = \log_4 x$

Think of $f(x)$ as y. Then $y = \log_4 x$ is equivalent to $4^y = x$. We find ordered pairs by choosing values for y and computing the corresponding x-values. Then we plot the points and connect them with a smooth curve.

For $y = 0$, $x = 4^0 = 1$.

For $y = 1$, $x = 4^1 = 4$.

For $y = 2$, $x = 4^2 = 16$.

For $y = -1$, $x = 4^{-1} = \frac{1}{4}$.

For $y = -2$, $x = 4^{-2} = \frac{1}{16}$.

x, or 4^y	y
1	0
4	1
16	2
$\frac{1}{4}$	-1
$\frac{1}{16}$	-2

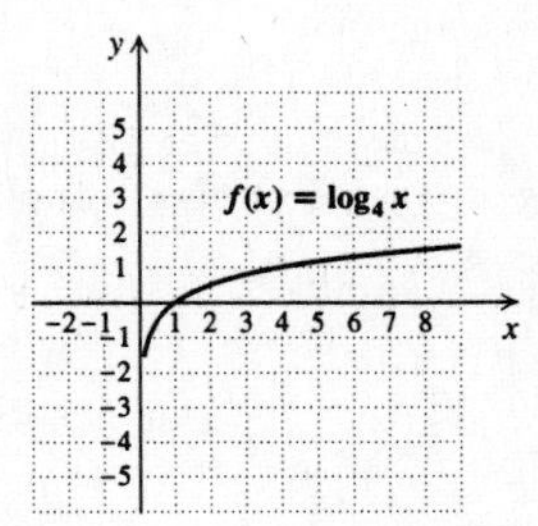

6.

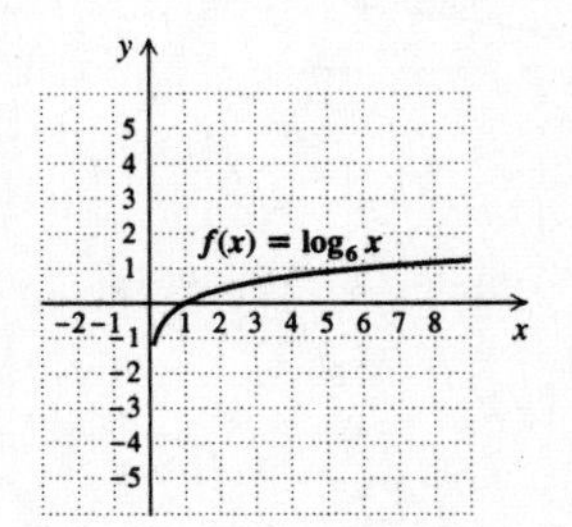

7. Graph: $f(x) = \log_{1/2} x$

Think of $f(x)$ as y. Then $y = \log_{1/2} x$ is equivalent to $\left(\frac{1}{2}\right)^y = x$. We construct a table of values, plot these points and connect them with a smooth curve.

For $y = 0$, $x = \left(\frac{1}{2}\right)^0 = 1$.

For $y = 1$, $x = \left(\frac{1}{2}\right)^1 = \frac{1}{2}$.

For $y = 2$, $x = \left(\frac{1}{2}\right)^2 = \frac{1}{4}$.

For $y = -1$, $x = \left(\frac{1}{2}\right)^{-1} = 2$.

For $y = -2$, $x = \left(\frac{1}{2}\right)^{-2} = 4$.

For $y = -3$, $x = \left(\frac{1}{2}\right)^{-3} = 8$.

x, or $\left(\frac{1}{2}\right)^y$	y
1	0
$\frac{1}{2}$	1
$\frac{1}{4}$	2
2	−1
4	−2
8	−3

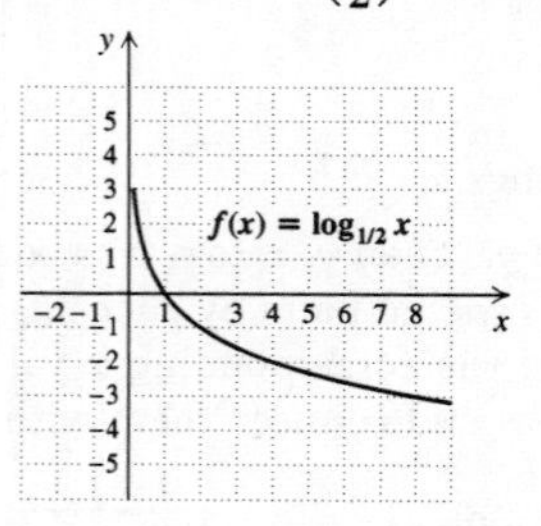

8.

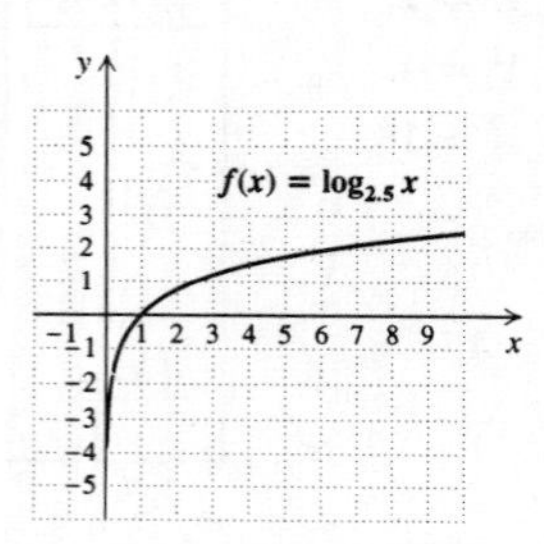

9. Graph $f(x) = 3^x$ (see Exercise Set 12.1, Exercise 2) and $f^{-1}(x) = \log_3 x$ (see Exercise 4 above) on the same set of axes.

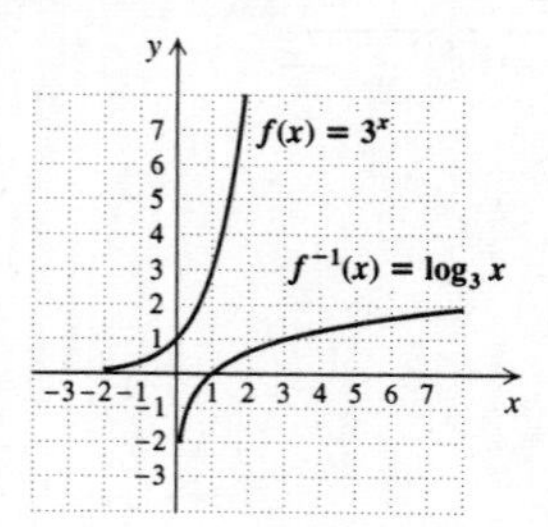

10.

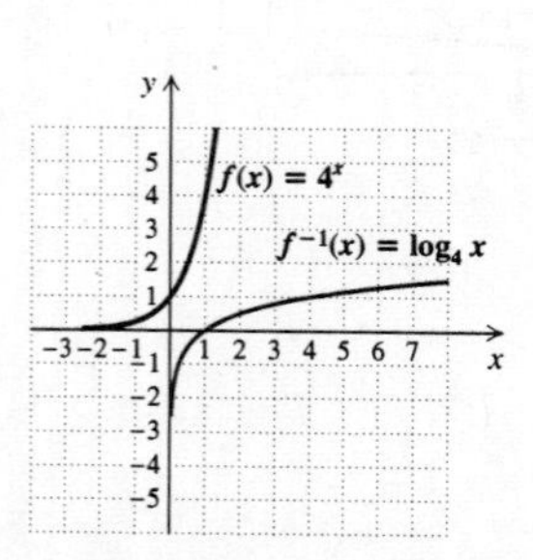

11. The exponent is the logarithm.

$10^4 = 10,000 \Rightarrow 4 = \log_{10} 10,000$

The base remains the same.

12. $2 = \log_{10} 100$

13. The exponent is the logarithm.

$5^{-3} = \frac{1}{125} \Rightarrow -3 = \log_5 \frac{1}{125}$

The base remains the same.

14. $-5 = \log_4 \frac{1}{1024}$

15. $8^{1/3} = 2$ is equivalent to $\frac{1}{3} = \log_8 2$.

16. $\frac{3}{4} = \log_{16} 8$

17. $10^{0.3010} = 2$ is equivalent to $0.3010 = \log_{10} 2$.

18. $0.4771 = \log_{10} 3$

19. $m^n = r$ is equivalent to $n = \log_m r$.

20. $k = \log_p 3$

21. $Q^t = x$ is equivalent to $t = \log_Q x$.

22. $m = \log_p V$

23. $e^2 = 7.3891$ is equivalent to $2 = \log_e 7.3891$.

24. $3 = \log_e 20.0855$

25. $e^{-2} = 0.1353$ is equivalent to $-2 = \log_e 0.1353$.

26. $-4 = \log_e 0.0183$

27. The base remains the same.

$t = \log_3 8 \Rightarrow 3^t = 8$

The logarithm is the exponent.

28. $7^h = 10$

29. The logarithm is the exponent.

$\log_5 25 = 2 \Rightarrow 5^2 = 25$

The base remains the same.

30. $6^1 = 6$

31. $\log_{10} 0.1 = -1$ is equivalent to $10^{-1} = 0.1$.

32. $10^{-2} = 0.01$

33. $\log_{10} 7 = 0.845$ is equivalent to $10^{0.845} = 7$.

34. $10^{0.4771} = 3$

35. $\log_c m = 17$ is equivalent to $c^{17} = m$.

36. $b^{23} = n$

37. $\log_t Q = k$ is equivalent to $t^k = Q$.

38. $m^a = P$

39. $\log_e 0.25 = -1.3863$ is equivalent to $e^{-1.3863} = 0.25$.

40. $e^{-0.0111} = 0.989$

41. $\log_r T = -x$ is equivalent to $r^{-x} = T$.

42. $c^{-w} = M$

43. $\log_3 x = 4$

$3^4 = x$ Converting to an exponential equation

$81 = x$ Computing 3^4

44. 16

45. $\log_x 125 = 3$

$x^3 = 125$ Converting to an exponential equation

$x = 5$ Taking cube roots

46. 4

47. $\log_2 16 = x$

$2^x = 16$ Converting to an exponential equation

$2^x = 2^4$

$x = 4$ The exponents must be the same.

48. 2

49. $\log_3 27 = x$

$3^x = 27$ Converting to an exponential equation

$3^x = 3^3$

$x = 3$ The exponents must be the same.

50. 2

51. $\log_x 8 = 1$

$x^1 = 8$ Converting to an exponential equation

$x = 8$ Simplifying x^1

52. 7

53. $\log_6 x = 0$

$6^0 = x$ Converting to an exponential equation

$1 = x$ Computing 6^0

54. 9

55. $\log_2 x = -1$

$2^{-1} = x$ Converting to an exponential equation

$\frac{1}{2} = x$ Simplifying

56. $\frac{1}{9}$

57. $\log_8 x = \frac{2}{3}$

$$8^{2/3} = x$$
$$(2^3)^{2/3} = x$$
$$2^2 = x$$
$$4 = x$$

58. 4

59. Let $\log_{10} 10,000 = x$. Then

$$10^x = 10,000$$
$$10^x = 10^4$$
$$x = 4.$$

Thus, $\log_{10} 10,000 = 4$.

60. 5

61. Let $\log_{10} 1 = x$. Then

$$10^x = 1$$
$$10^x = 10^0 \quad (10^0 = 1)$$
$$x = 0.$$

Thus, $\log_{10} 1 = 0$.

62. 1

63. Let $\log_5 625 = x$. Then

$$5^x = 625$$
$$5^x = 5^4$$
$$x = 4.$$

Thus, $\log_5 625 = 4$.

64. 0

65. Let $\log_5 \frac{1}{25} = x$. Then

$$5^x = \frac{1}{25}$$
$$5^x = 5^{-2}$$
$$x = -2.$$

Thus, $\log_5 \frac{1}{25} = -2$.

66. 3

67. Let $\log_3 3 = x$. Then

$3^x = 3$

$3^x = 3^1$

$x = 1$.

Thus, $\log_3 3 = 1$.

68. -4

69. Let $\log_7 1 = x$. Then

$7^x = 1$

$7^x = 7^0 \qquad (7^0 = 1)$

$x = 0$.

Thus, $\log_7 1 = 0$.

70. 1

71. Let $\log_6 15 = x$. Then

$6^x = 15$

$6^{\log_6 15} = 15. \qquad (x = \log_6 15)$

Thus, $6^{\log_6 15} = 15$.

72. 23

73. Let $\log_{27} 9 = x$. Then

$27^x = 9$

$(3^3)^x = 3^2$

$3^{3x} = 3^2$

$3x = 2$

$x = \frac{2}{3}$.

Thus, $\log_{27} 9 = \frac{2}{3}$.

74. $\frac{1}{3}$

75. Let $\log_b b^7 = x$. Then

$b^x = b^7$

$x = 7$.

Thus, $\log_b b^7 = 7$.

76. 8

77. $\dfrac{\frac{3}{x} - \frac{2}{xy}}{\frac{2}{x^2} + \frac{1}{xy}}$

The LCD of all the denominators is x^2y. We multiply numerator and denominator by the LCD.

$$\frac{\frac{3}{x} - \frac{2}{xy}}{\frac{2}{x^2} + \frac{1}{xy}} \cdot \frac{x^2y}{x^2y} = \frac{\left(\frac{3}{x} - \frac{2}{xy}\right)x^2y}{\left(\frac{2}{x^2} + \frac{1}{xy}\right)x^2y}$$

$$= \frac{\frac{3}{x} \cdot x^2y - \frac{2}{xy} \cdot x^2y}{\frac{2}{x^2} \cdot x^2y + \frac{1}{xy} \cdot x^2y}$$

$$= \frac{3xy - 2x}{2y + x}, \text{ or } \frac{x(3y-2)}{2y+x}$$

78. $\dfrac{x+2}{x+1}$

79. $8^{-4} = \dfrac{1}{8^4}$, or $\dfrac{1}{4096}$

80. $\sqrt[5]{x}$

81. $t^{-1/3} = \dfrac{1}{t^{1/3}} = \dfrac{1}{\sqrt[3]{t}}$

82. 5

83.

84.

85.

86.

87. Graph: $y = \left(\frac{3}{2}\right)^x$

x	y, or $\left(\frac{3}{2}\right)^x$
0	1
1	$\frac{3}{2}$
2	$\frac{9}{4}$
3	$\frac{27}{8}$
-1	$\frac{2}{3}$
-2	$\frac{4}{9}$

Graph: $y = \log_{3/2} x$, or $x = \left(\frac{3}{2}\right)^y$

x, or $\left(\frac{3}{2}\right)^y$	y
1	0
$\frac{3}{2}$	1
$\frac{9}{4}$	2
$\frac{27}{8}$	3
$\frac{2}{3}$	-1
$\frac{4}{9}$	-2

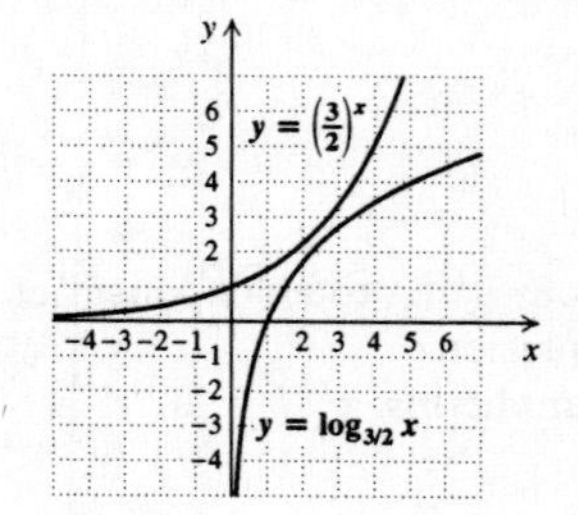

88.

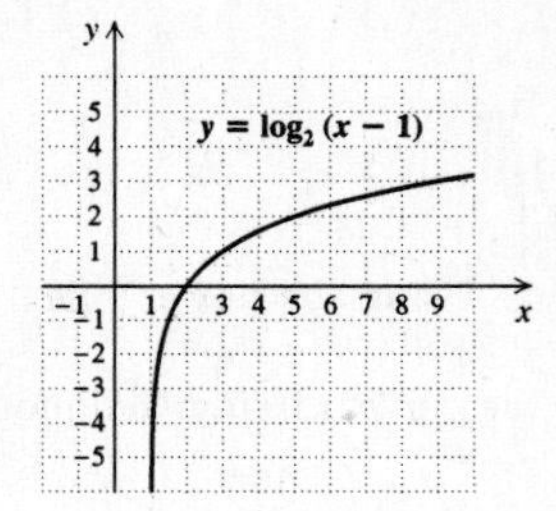

89. Graph: $y = \log_3 |x+1|$

x	y
0	0
2	1
8	2
−2	0
−4	1
−9	2

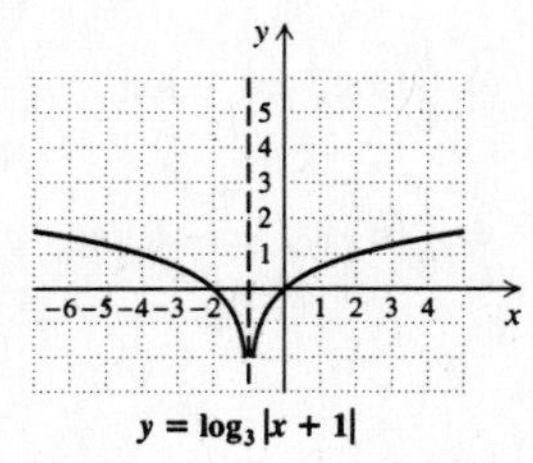
$y = \log_3 |x + 1|$

90. $\frac{1}{9}$, 9

91.
$$\log_{125} x = \frac{2}{3}$$
$$125^{2/3} = x$$
$$(5^3)^{2/3} = x$$
$$5^2 = x$$
$$25 = x$$

92. 6

93.
$$\log_8(2x+1) = -1$$
$$8^{-1} = 2x+1$$
$$\frac{1}{8} = 2x+1$$
$$1 = 16x+8 \quad \text{Multiplying by 8}$$
$$-7 = 16x$$
$$-\frac{7}{16} = x$$

94. −25, 4

95. Let $\log_{1/4} \frac{1}{64} = x$. Then
$$\left(\frac{1}{4}\right)^x = \frac{1}{64}$$
$$\left(\frac{1}{4}\right)^x = \left(\frac{1}{4}\right)^3$$
$$x = 3.$$

Thus, $\log_{1/4} \frac{1}{64} = 3$.

96. $= -2$

97.
$$\log_{81} 3 \cdot \log_3 81$$
$$= \frac{1}{4} \cdot 4 \qquad \left(\log_{81} 3 = \frac{1}{4},\ \log_3 81 = 4\right)$$
$$= 1$$

98. 0

99.
$$\log_2(\log_2(\log_4 256))$$
$$= \log_2(\log_2 4) \qquad (\log_4 256 = 4)$$
$$= \log_2 2 \qquad (\log_2 4 = 2)$$
$$= 1$$

100. Let $b = 0$, $x = 1$, and $y = 2$. Then $0^1 = 0^2$, but $1 \neq 2$. Let $b = 1$, $x = 1$, and $y = 2$. Then $1^1 = 1^2$, but $1 \neq 2$.

Exercise Set 9.4

1. $\log_3 (81 \cdot 27) = \log_3 81 + \log_3 27$ Using the product rule

2. $\log_2 16 + \log_2 32$

3. $\log_4 (64 \cdot 16) = \log_4 64 + \log_4 16$ Using the product rule

4. $\log_5 25 + \log_5 125$

5.
$$\log_c xyz$$
$$= \log_c x + \log_c y + \log_c z \quad \text{Using the product rule}$$

6. $\log_t 3 + \log_t a + \log_t b$

7. $\log_a 5 + \log_a 14 = \log_a (5 \cdot 14)$ Using the product rule

The result can also be expressed as $\log_a 70$.

8. $\log_b (65 \cdot 2)$, or $\log_b 130$

9. $\log_c t + \log_c y = \log_c (t \cdot y)$ Using the product rule

10. $\log_t HM$

11. $\log_a t^7 = 7 \log_a t$ Using the power rule

12. $5 \log_b t$

13. $\log_c y^6 = 6 \log_c y$ Using the power rule

14. $7 \log_{10} y$

15. $\log_b C^{-3} = -3 \log_b C$ Using the power rule

16. $-5 \log_c M$

17. $\log_2 \frac{64}{16} = \log_2 64 - \log_2 16$ Using the quotient rule

18. $\log_3 27 - \log_3 9$

19. $\log_b \frac{m}{n} = \log_b m - \log_b n$ Using the quotient rule

20. $\log_a y - \log_a x$

21. $\log_a 15 - \log_a 7 = \log_a \frac{15}{7}$ Using the quotient rule

22. $\log_b \frac{42}{7}$, or $\log_b 6$

23. $\log_a x^2y^3z$

$= \log_a x^2 + \log_a y^3 + \log_a z$ Using the product rule

$= 2\log_a x + 3\log_a y + \log_a z$ Using the power rule

24. $\log_a x + 4\log_a y + 3\log_a z$

25. $\log_b \frac{xy^2}{z^3}$

$= \log_b xy^2 - \log_b z^3$ Using the quotient rule

$= \log_b x + \log_b y^2 - \log_b z^3$ Using the product rule

$= \log_b x + 2\log_b y - 3\log_b z$ Using the power rule

26. $2\log_b x + 5\log_b y - 4\log_b w - 7\log_b z$

27. $\log_a \frac{x^2}{y^3z}$

$= \log_a x^2 - \log_a y^3z$ Using the quotient rule

$= \log_a x^2 - (\log_a y^3 + \log_a z)$ Using the product rule

$= \log_a x^2 - \log_a y^3 - \log_a z$ Removing parentheses

$= 2\log_a x - 3\log_a y - \log_a z$ Using the power rule

28. $4\log_a x - \log_a y - 2\log_a z$

29. $\log_b \frac{xy^2}{wz^3}$

$= \log_b xy^2 - \log_b wz^3$ Using the quotient rule

$= \log_b x + \log_b y^2 - (\log_b w + \log_b z^3)$
Using the product rule

$= \log_b x + \log_b y^2 - \log_b w - \log_b z^3$
Removing parentheses

$= \log_b x + 2\log_b y - \log_b w - 3\log_b z$
Using the power rule

30. $2\log_b w + \log_b x - 3\log_b y - \log_b z$

31. $\log_a \sqrt{\frac{x^6}{y^5z^8}}$

$= \log_a \left(\frac{x^6}{y^5z^8}\right)^{1/2}$

$= \frac{1}{2}\log_a \frac{x^6}{y^5z^8}$ Using the power rule

$= \frac{1}{2}(\log_a x^6 - \log_a y^5z^8)$ Using the quotient rule

$= \frac{1}{2}\left[\log_a x^6 - (\log_a y^5 + \log_a z^8)\right]$
Using the product rule

$= \frac{1}{2}(\log_a x^6 - \log_a y^5 - \log_a z^8)$
Removing parentheses

$= \frac{1}{2}(6\log_a x - 5\log_a y - 8\log_a z)$
Using the power rule

32. $\frac{1}{3}(4\log_c x - 3\log_c y - 2\log_c z)$

33. $\log_a \sqrt[3]{\frac{x^6y^3}{a^2z^7}}$

$= \log_a \left(\frac{x^6y^3}{a^2z^7}\right)^{1/3}$

$= \frac{1}{3}\log_a \frac{x^6y^3}{a^2z^7}$ Using the power rule

$= \frac{1}{3}(\log_a x^6y^3 - \log_a a^2z^7)$ Using the quotient rule

$= \frac{1}{3}[\log_a x^6 + \log_a y^3 - (\log_a a^2 + \log_a z^7)]$
Using the product rule

$= \frac{1}{3}(\log_a x^6 + \log_a y^3 - \log_a a^2 - \log_a z^7)$
Removing parentheses

$= \frac{1}{3}(\log_a x^6 + \log_a y^3 - 2 - \log_a z^7)$
2 is the number to which we raise a to get a^2.

$= \frac{1}{3}(6\log_a x + 3\log_a y - 2 - 7\log_a z)$
Using the power rule

34. $\frac{1}{4}(8\log_a x + 12\log_a y - 3 - 5\log_a z)$

35. $4\log_a x + 3\log_a y$

$= \log_a x^4 + \log_a y^3$ Using the power rule

$= \log_a x^4y^3$ Using the product rule

36. $\log_b m^2n^{1/2}$, or $\log_b m^2\sqrt{n}$

37. $\log_a x^2 - 2\log_a \sqrt{x}$

$= \log_a x^2 - \log_a (\sqrt{x})^2$ Using the power rule

$= \log_a x^2 - \log_a x$ $(\sqrt{x})^2 = x$

$= \log_a \frac{x^2}{x}$ Using the quotient rule

$= \log_a x$ Simplifying

38. $\log_a \dfrac{\sqrt{a}}{x}$

39. $\dfrac{1}{2}\log_a x + 3\log_a y - 2\log_a x$

$= \log_a x^{1/2} + \log_a y^3 - \log_a x^2$ Using the power rule

$= \log_a x^{1/2}y^3 - \log_a x^2$ Using the product rule

$= \log_a \dfrac{x^{1/2}y^3}{x^2}$ Using the quotient rule

The result can also be expressed as $\log_a \dfrac{\sqrt{x}y^3}{x^2}$ or as $\log_a \dfrac{y^3}{x^{3/2}}$.

40. $\log_a \dfrac{2x^4}{y^3}$

41. $\log_a(x^2-4) - \log_a(x-2)$

$= \log_a \dfrac{x^2-4}{x-2}$ Using the quotient rule

$= \log_a \dfrac{(x+2)(x-2)}{x-2}$

$= \log_a \dfrac{(x+2)\cancel{(x-2)}}{\cancel{x-2}}$ Simplifying

$= \log_a(x+2)$

42. $\log_a \dfrac{2}{x-5}$

43. $\log_b 15 = \log_b (3\cdot 5)$

$= \log_b 3 + \log_b 5$ Using the product rule

$= 1.099 + 1.609$

$= 2.708$

44. 0.51

45. $\log_b \dfrac{3}{5} = \log_b 3 - \log_b 5$ Using the quotient rule

$= 1.099 - 1.609$

$= -0.51$

46. -1.099

47. $\log_b \dfrac{1}{5} = \log_b 1 - \log_b 5$ Using the quotient rule

$= 0 - 1.609$ $(\log_b 1 = 0)$

$= -1.609$

48. $\dfrac{1}{2}$

49. $\log_b \sqrt{b^3} = \log_b b^{3/2} = \dfrac{3}{2}$ 3/2 is the number to which we raise b to get $b^{3/2}$.

50. 2.099

51. $\log_b 6$

Since 6 cannot be expressed using the numbers 1, 3, and 5, we cannot find $\log_b 6$ using the given information.

52. 3.807

53. $\log_b 75$

$= \log_b(3\cdot 5^2)$

$= \log_b 3 + \log_b 5^2$ Using the product rule

$= \log_b 3 + 2\log_b 5$ Using the power rule

$= 1.099 + 2(1.609)$

$= 4.317$

54. Cannot be found using the given information.

55. $\log_t t^9 = 9$ 9 is the number to which we raise t to get t^9.

56. 4

57. $\log_e e^m = m$

58. -2

59. $i^{29} = i^{28}\cdot i = (i^4)^7\cdot i = 1^7\cdot i = 1\cdot i = i$, or $0+i$

60. $3+4i$

61. $5i(2-i) = 10i - 5i^2 = 10i - 5(-1) = 10i + 5$, or $5+10i$

62. -1

63. $(5-3i)^2 = 25 - 30i + 9i^2 = 25 - 30i + 9(-1) = 25 - 30i - 9 = 16 - 30i$

64. $-7-28i$

65. ◈

66. ◈

67. ◈

68. $\log_a (x^6 - x^4y^2 + x^2y^4 - y^6)$

69. $\log_a (x+y) + \log_a (x^2 - xy + y^2) = \log_a (x+y)(x^2-xy+y^2) = \log_a (x^3+y^3)$

70. $\dfrac{1}{2}\log_a (1-s) + \dfrac{1}{2}\log_a (1+s)$

71. $\log_a \dfrac{c-d}{\sqrt{c^2-d^2}}$

$= \log_a (c-d) - \dfrac{1}{2}\log_a (c+d)(c-d)$

$= \log_a (c-d) - \dfrac{1}{2}\log_a (c+d) - \dfrac{1}{2}\log_a (c-d)$

$= \dfrac{1}{2}\log_a (c-d) - \dfrac{1}{2}\log_a (c+d)$

72. $\dfrac{10}{3}$

73. $\log_a \left(\dfrac{1}{x}\right) = \log_a x^{-1} = -1 \cdot \log_a x = -1 \cdot 2 = -2$

74. -2

75. False. Let $a = 10$, $P = 100$, $Q = 10$, and $x = 2$.
Then $\log_{10}\left(\dfrac{100}{10}\right)^2 = \log_{10} 10^2 = 2\log_{10} 10 = 2 \cdot 1 = 2$, but $2\log_{10} 100 - \log_{10} 10 = 2 \cdot 2 - 1 = 3$.

76. True

77. $y_1 = \log x^2$, $y_2 = \log x \cdot \log x$

Exercise Set 9.5

1. 0.6021

2. 0.6990

3. 1.7300

4. 1.8686

5. 2.6405

6. 2.4698

7. 4.1271

8. 4.9689

9. -0.2782

10. -0.3072

11. 199.5262

12. 2511.8864

13. 1.4894

14. 1.7660

15. 0.0011

16. 79,104.2833

17. 1.6094

18. 0.6931

19. 4.1271

20. 3.4012

21. 8.3814

22. 6.8037

23. -5.0832

24. -7.2225

25. 15.0293

26. 21.3276

27. 0.0305

28. 0.0714

29. 109.9472

30. 3.4212

31. We will use common logarithms for the conversion. Let $a = 10$, $b = 6$, and $M = 100$ and substitute in the change-of-base formula.

$$\log_b M = \frac{\log_a M}{\log_a b}$$
$$\log_6 100 = \frac{\log_{10} 100}{\log_{10} 6}$$
$$\approx \frac{2}{0.7782}$$
$$\approx 2.5702$$

32. 4.1918

33. We will use common logarithms for the conversion. Let $a = 10$, $b = 2$, and $M = 100$ and substitute in the change-of-base formula.

$$\log_2 100 = \frac{\log_{10} 100}{\log_{10} 2}$$
$$\approx \frac{2}{0.3010}$$
$$\approx 6.6439$$

34. 2.3666

35. We will use natural logarithms for the conversion. Let $a = e$, $b = 7$, and $M = 65$ and substitute in the change-of-base formula.

$$\log_7 65 = \frac{\ln 65}{\ln 7}$$
$$\approx \frac{4.1744}{1.9459}$$
$$\approx 2.1452$$

36. 2.3223

37. We will use natural logarithms for the conversion. Let $a = e$, $b = 0.5$, and $M = 5$ and substitute in the change-of-base formula.

$$\log_{0.5} 5 = \frac{\ln 5}{\ln 0.5}$$
$$\approx \frac{1.6094}{-0.6931}$$
$$\approx -2.3219$$

38. -0.4771

39. We will use common logarithms for the conversion. Let $a = 10$, $b = 2$, and $M = 0.2$ and substitute in the change-of-base formula.

$$\log_2 0.2 = \frac{\log_{10} 0.2}{\log_{10} 2} \approx \frac{-0.6990}{0.3010} \approx -2.3219$$

40. -3.6439

41. We will use natural logarithms for the conversion. Let $a = e$, $b = \pi$, and $M = 58$ and substitute in the change-of-base formula.

$$\log_\pi 58 = \frac{\ln 58}{\ln \pi} \approx \frac{4.0604}{1.1447} \approx 3.5471$$

42. 4.6284

43. Graph: $f(x) = e^x$

We find some function values with a calculator. We use these values to plot points and draw the graph.

x	e^x
0	1
1	2.7
2	7.4
3	20.1
-1	0.4
-2	0.1

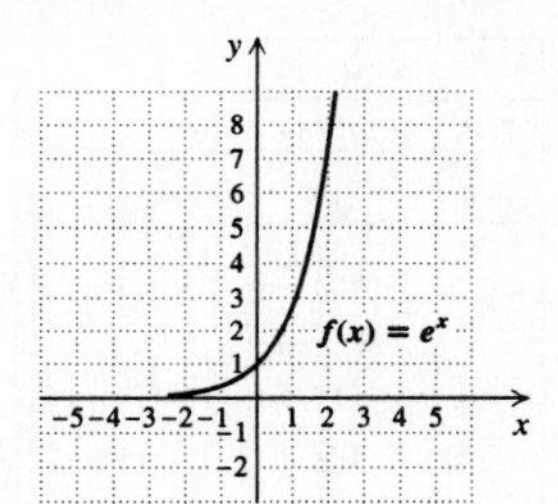

44.

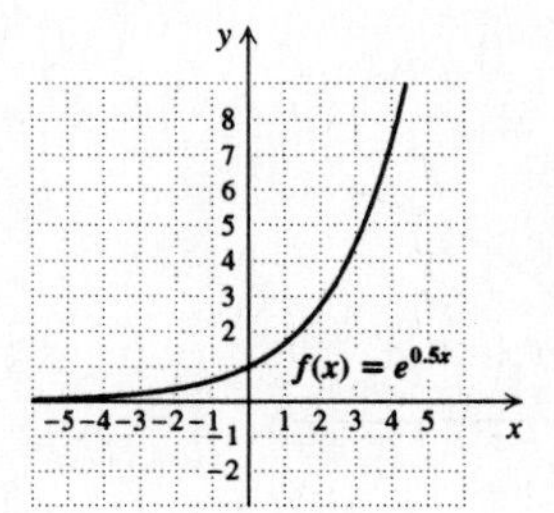

45. Graph: $f(x) = e^{-0.5x}$

We find some function values, plot points, and draw the graph.

x	$e^{-0.5x}$
0	1
1	0.61
2	0.37
-1	1.65
-2	2.72
-3	4.48
-4	7.39

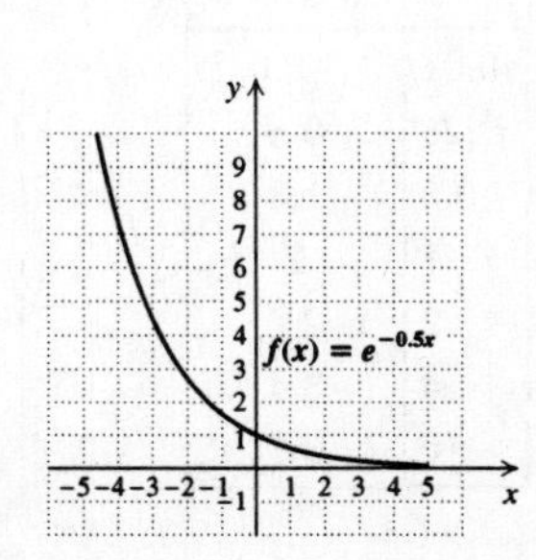

46.

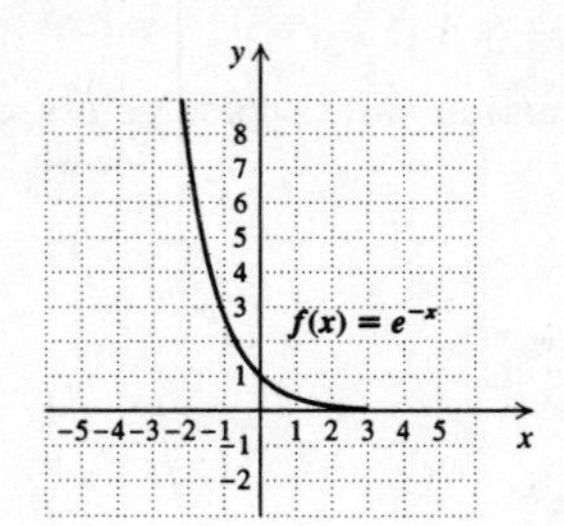

47. Graph: $f(x) = e^{x-1}$

We find some function values, plot points, and draw the graph.

x	e^{x-1}
0	0.4
1	1
2	2.7
3	7.4
4	20.1
-1	0.1
-2	0.05

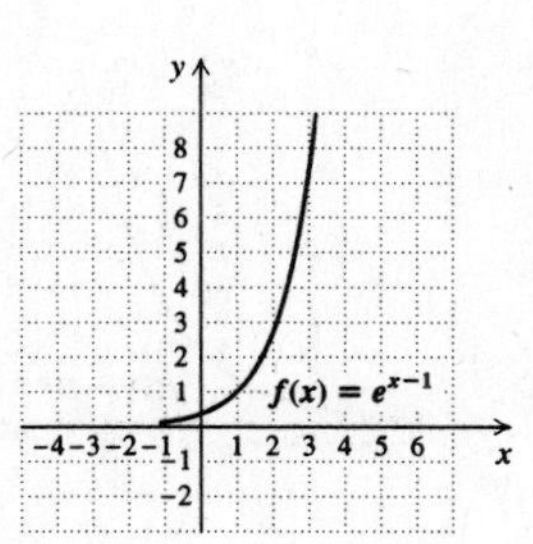

48.

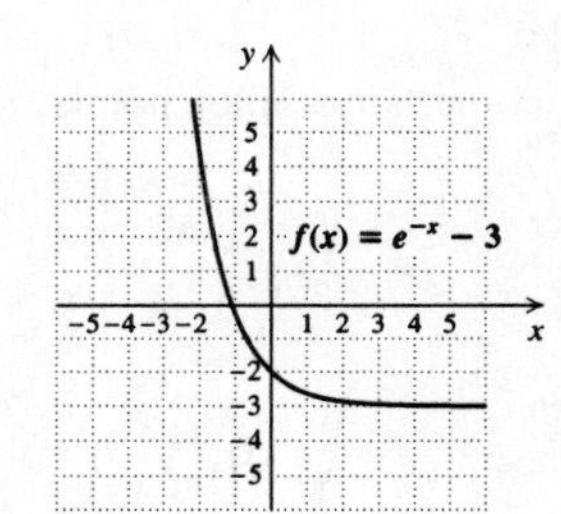

49. Graph: $f(x) = e^{x+5}$

We find some function values, plot points, and draw the graph.

x	e^{x+5}
0	148.4
-1	54.6
-3	7.4
-5	1
-7	0.1

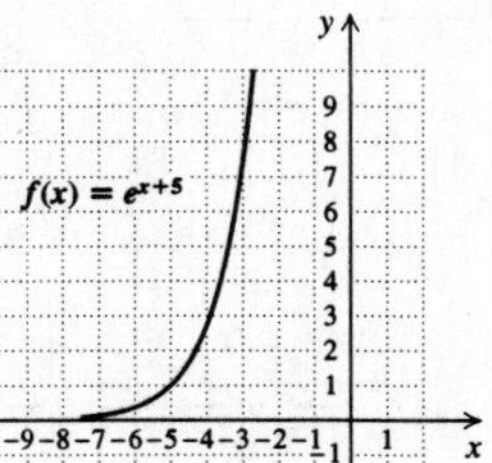

50.

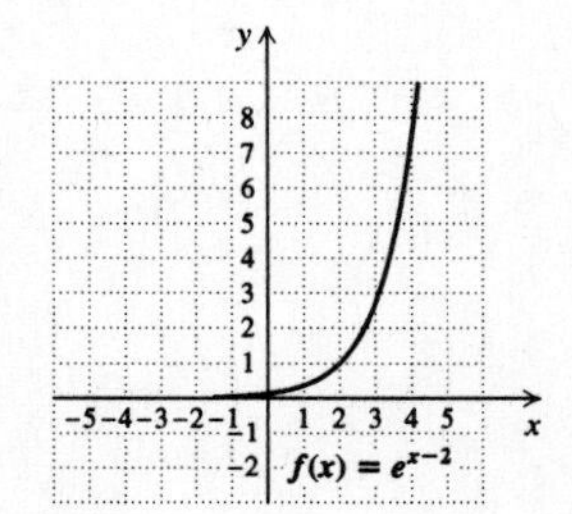

51. Graph: $f(x) = e^x + 3$

We find some function values, plot points, and draw the graph.

x	$e^x + 3$
0	4
1	5.72
2	10.39
3	23.09
−1	3.37
−2	3.14
−4	3.02

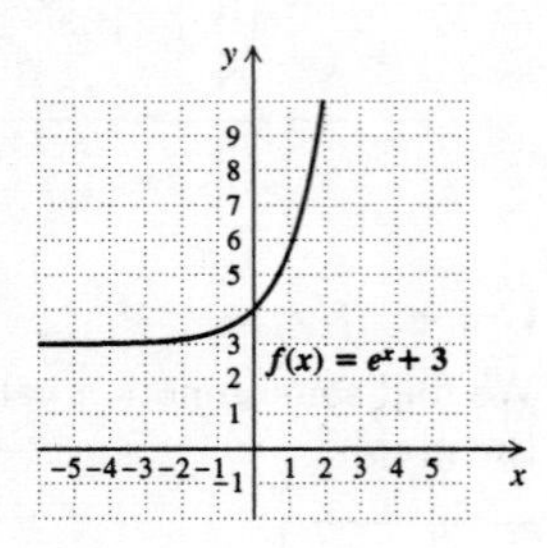

52.

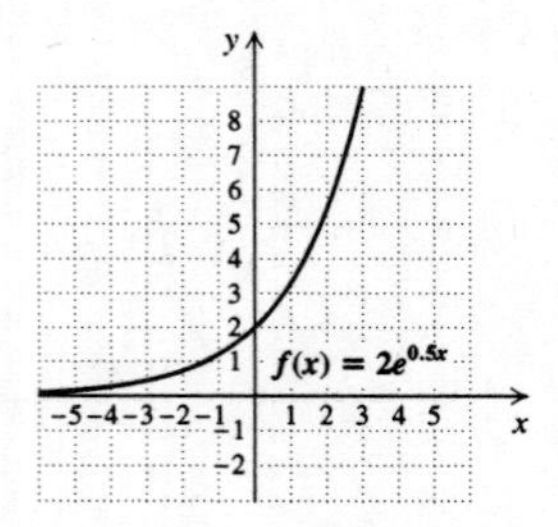

53. Graph: $f(x) = 2e^{-0.5x}$

We find some function values, plot points, and draw the graph.

x	$2e^{-0.5x}$
0	2
1	1.21
2	0.74
3	0.45
−1	3.30
−2	5.44
−3	8.96

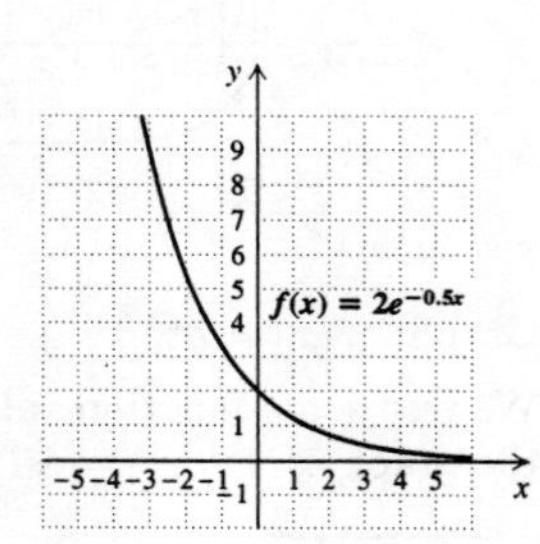

54.

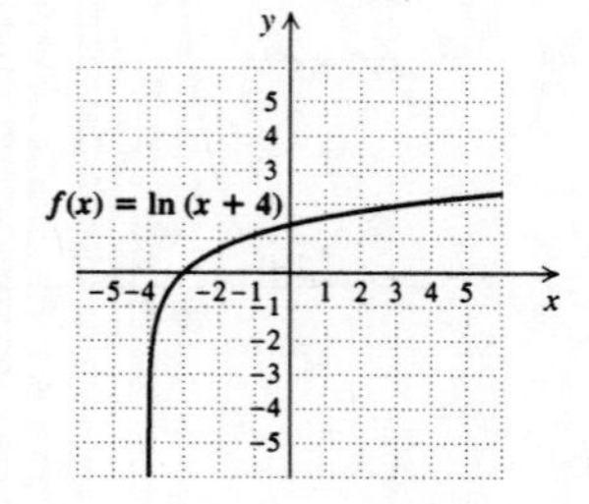

55. Graph: $f(x) = \ln(x + 1)$

We find some function values, plot points, and draw the graph.

x	$\ln(x+1)$
0	0
1	0.69
2	1.10
4	1.61
6	1.95
−1	Undefined

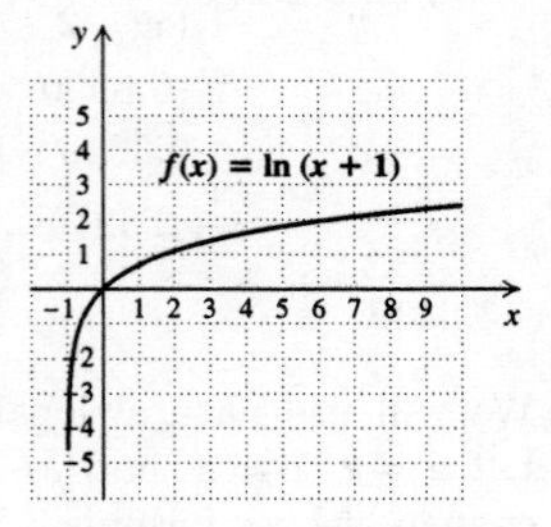

56.

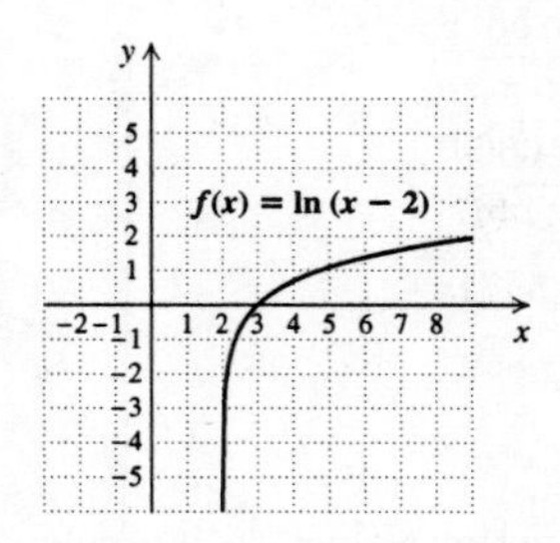

57. Graph: $f(x) = 2 \ln x$

x	$2 \ln x$
0.5	−1.4
1	0
2	1.4
3	2.2
4	2.8
5	3.2
6	3.6

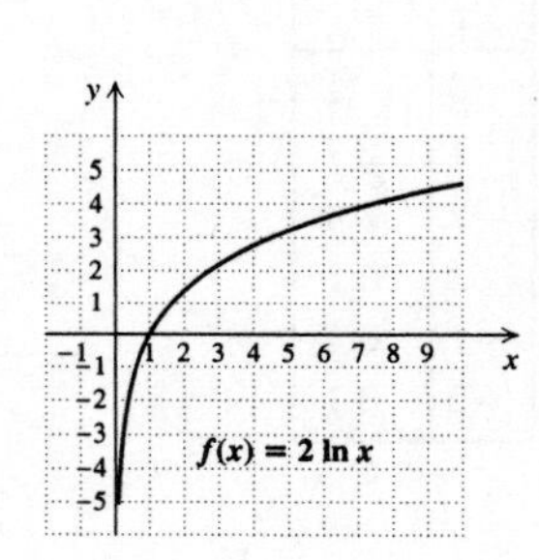

58.

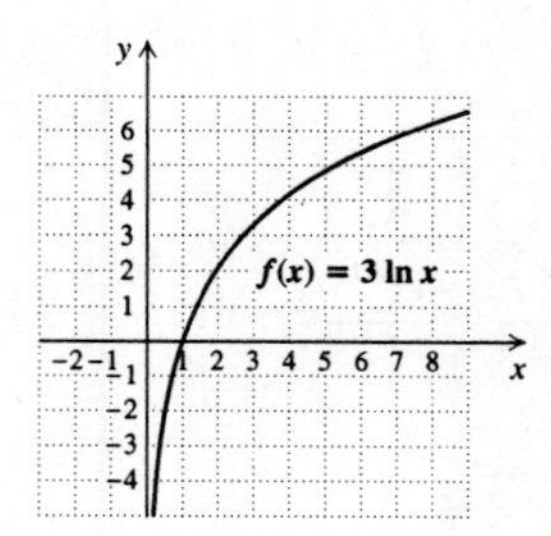

59. Graph: $f(x) = \ln x + 2$

x	$\ln x + 2$
0.01	−2.6
0.25	0.6
0.5	1.3
1	2
2	2.7
4	3.4
6	3.8

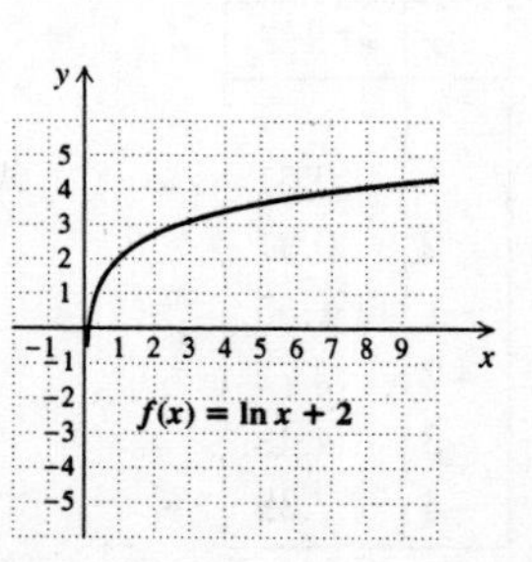

60.

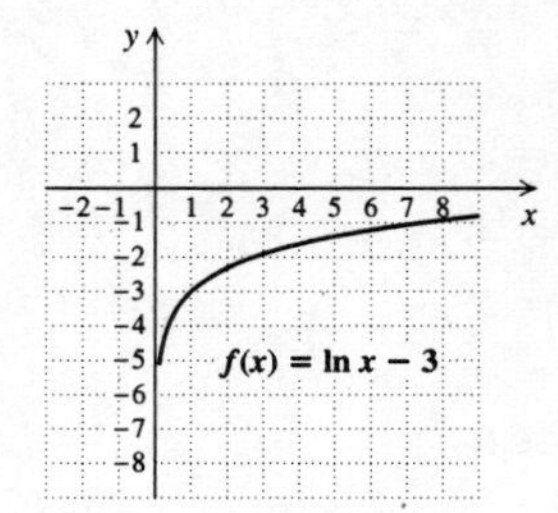

61.
$$\begin{aligned} 4x^2 - 25 &= 0 \\ (2x+5)(2x-5) &= 0 \\ 2x+5 = 0 \quad &\text{or} \quad 2x-5=0 \\ 2x = -5 \quad &\text{or} \quad 2x = 5 \\ x = -\frac{5}{2} \quad &\text{or} \quad x = \frac{5}{2} \end{aligned}$$
The solutions are $-\frac{5}{2}$ and $\frac{5}{2}$.

62. $0, \frac{7}{5}$

63.
$$\begin{aligned} 17x - 15 &= 0 \\ 17x &= 15 \\ x &= \frac{15}{17} \end{aligned}$$
The solution is $\frac{15}{17}$.

64. $\frac{9}{13}$

65.
$x^{1/2} - 6x^{1/4} + 8 = 0$

Let $u = x^{1/4}$.

$u^2 - 6u + 8 = 0$ Substituting

$(u-4)(u-2) = 0$

$u = 4$ or $u = 2$

$x^{1/4} = 4$ or $x^{1/4} = 2$

$x = 256$ or $x = 16$ Raising both sides to the fourth power

Both numbers check. The solutions are 256 and 16.

66. $\frac{1}{4}, 9$

67.

68.

69.

70.

71. Use the change-of-base formula with $a = 10$ and $b = e$. We obtain
$$\ln M = \frac{\log M}{\log e}.$$

72. $\log M = \frac{\ln M}{\ln 10}$

73.
$$\begin{aligned} \log(275x^2) &= 38 \\ 10^{38} &= 275x^2 \\ \frac{10^{38}}{275} &= x^2 \\ \pm\sqrt{\frac{10^{38}}{275}} &= x \\ \pm 6.0302 \times 10^{17} &\approx x \end{aligned}$$

74. 1086.5129

75.
$$\begin{aligned} \frac{3.01}{\ln x} &= \frac{28}{4.31} \\ 4.31(3.01) &= 28 \ln x \quad \text{Multiplying by } 4.31 \ln x \\ \frac{4.31(3.01)}{28} &= \ln x \\ 0.463325 &= \ln x \\ 1.5893 &\approx x \end{aligned}$$

76. 4.9855

77. (a) $\{x|x > 0\}$, or $(0, \infty)$;

(b) $[-3, 10, -100, 1000]$, Xscl = 1, Yscl = 100;

(c)

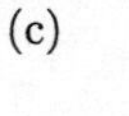

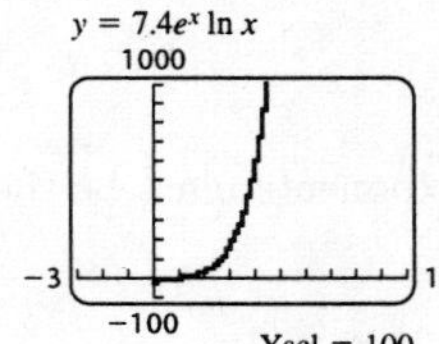

78. (a) $\{x|x > 0\}$, or $(0, \infty)$;

(b) $[-1, 5, -10, 5]$;

(c)

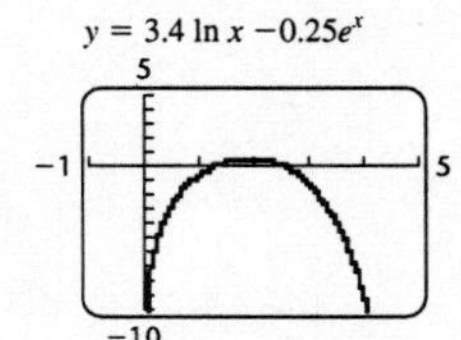

79. (a) $\{x|x > 2.1\}$, or $(2.1, \infty)$;

(b) $[-1, 10, -20, 20]$, Xscl = 1, Yscl = 5;

(c)

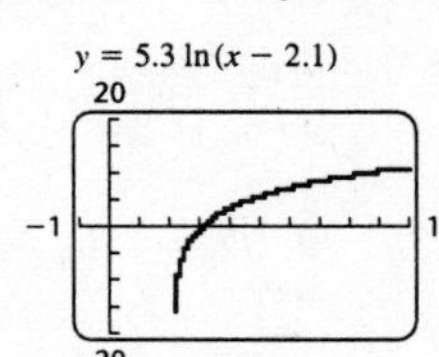

80. (a) $\{x|x > 0\}$, or $(0, \infty)$;

(b) $[-1, 5, -1, 10]$;

(c) 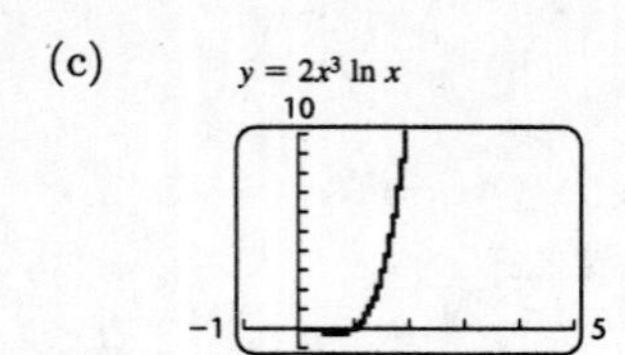

81.

82.

83.

Exercise Set 9.6

1. $3^x = 81$

$3^x = 3^4$

$x = 4$ The exponents must be the same.

The solution is 4.

2. 3

3. $4^x = 256$

$4^x = 4^4$

$x = 4$ The exponents must be the same.

The solution is 4.

4. 3

5.

$$2^{x+3} = 32$$
$$2^{x+3} = 2^5$$
$$x + 3 = 5$$
$$x = 2$$

The solution is 2.

6. 1

7.

$$5^{3x} = 625$$
$$5^{3x} = 5^4$$
$$3x = 4$$
$$x = \frac{4}{3}$$

The solution is $\frac{4}{3}$.

8. 2

9.

$$4^{2x-1} = 64$$
$$4^{2x-1} = 4^3$$
$$2x - 1 = 3$$
$$2x = 4$$
$$x = 2$$

The solution is 2.

10. $\frac{5}{2}$

11.

$$3^{2x^2} \cdot 3^{5x} = 27$$
$$3^{2x^2+5x} = 3^3$$
$$2x^2 + 5x = 3$$
$$2x^2 + 5x - 3 = 0$$
$$(2x - 1)(x + 3) = 0$$
$$x = \frac{1}{2} \text{ or } x = -3$$

The solutions are $\frac{1}{2}$ and -3.

12. -3, -1

13.

$$2^x = 13$$
$$\log 2^x = \log 13$$
$$x \log 2 = \log 13$$
$$x = \frac{\log 13}{\log 2}$$
$$x \approx 3.700$$

The solution is $\log 13/\log 2$, or approximately 3.700.

14. $\frac{\log 19}{\log 2} \approx 4.248$

15.

$$4^x = 7$$
$$\log 4^x = \log 7$$
$$x \log 4 = \log 7$$
$$x = \frac{\log 7}{\log 4}$$
$$x \approx 1.404$$

The solution is $\log 7/\log 4$, or approximately 1.404.

16. $\frac{1}{\log 8} \approx 1.107$

17.

$e^t = 100$

$\ln e^t = \ln 100$ Taking ln on both sides

$t = \ln 100$ Finding the logarithm of the base to a power

$t \approx 4.605$ Using a calculator

18. $\ln 1000 \approx 6.908$

19.

$e^{-0.07t} = 0.08$

$\ln e^{-0.07t} = \ln 0.08$ Taking ln on both sides

$-0.07t = \ln 0.08$ Finding the logarithm of the base to a power

$$t = \frac{\ln 0.08}{-0.07}$$
$$t \approx 36.082$$

20. $\frac{\ln 5}{0.03} \approx 53.648$

21.
$$\begin{aligned} 2^x &= 3^{x-1} \\ \log 2^x &= \log 3^{x-1} \\ x \log 2 &= (x-1) \log 3 \\ x \log 2 &= x \log 3 - \log 3 \\ \log 3 &= x \log 3 - x \log 2 \\ \log 3 &= x(\log 3 - \log 2) \\ \frac{\log 3}{\log 3 - \log 2} &= x \\ 2.710 &\approx x \end{aligned}$$

22. $\dfrac{\log 3}{\log 5 - \log 3} \approx 2.151$

23.
$$\begin{aligned} 4^{x+1} &= 5^x \\ \log 4^{x+1} &= \log 5^x \\ (x+1)\log 4 &= x \log 5 \\ x \log 4 + \log 4 &= x \log 5 \\ \log 4 &= x \log 5 - x \log 4 \\ \log 4 &= x(\log 5 - \log 4) \\ \frac{\log 4}{\log 5 - \log 4} &= x \\ 6.213 &\approx x \end{aligned}$$

24. $\dfrac{3 \log 2}{\log 7 - \log 2} \approx 1.660$

25.
$$\begin{aligned} 20 - (1.7)^x &= 0 \\ 20 &= (1.7)^x \\ \log 20 &= \log (1.7)^x \\ \log 20 &= x \log 1.7 \\ \frac{\log 20}{\log 1.7} &= x \\ 5.646 &\approx x \end{aligned}$$

26. $\dfrac{\log 125}{\log 4.5} \approx 3.210$

27. $\log_5 x = 4$

$x = 5^4$ Writing an equivalent exponential equation

$x = 625$

28. 27

29. $\log_4 x = \dfrac{1}{2}$

$x = 4^{1/2}$ Writing an equivalent exponential equation

$x = 2$

30. $\dfrac{1}{8}$

31. $\log x = 3$ The base is 10.

$x = 10^3$

$x = 1000$

32. 10

33. $2 \log x = -6$

$\log x = -3$ The base is 10.

$x = 10^{-3}$

$x = \dfrac{1}{1000}$, or 0.001

34. 0.01

35. $\ln x = 1$

$x = e \approx 2.718$

36. $e^2 \approx 7.389$

37. $5 \ln x = -15$

$\ln x = -3$

$x = e^{-3} \approx 0.050$

38. $e^{-1} \approx 0.368$

39.
$$\begin{aligned} \log_2(8 - 6x) &= 5 \\ 8 - 6x &= 2^5 \\ 8 - 6x &= 32 \\ -6x &= 24 \\ x &= -4 \end{aligned}$$

The answer checks. The solution is -4.

40. 66

41. $\log(x+9) + \log x = 1$ The base is 10.

$\log_{10} [(x+9)(x)] = 1$ Using the product rule

$$\begin{aligned} x(x+9) &= 10^1 \\ x^2 + 9x &= 10 \\ x^2 + 9x - 10 &= 0 \\ (x-1)(x+10) &= 0 \end{aligned}$$

$x = 1$ or $x = -10$

Check: For 1:

$\log(x+9) + \log x = 1$	
$\log(1+9) + \log 1$	? 1
$\log 10 + \log 1$	
$1 + 0$	
1	1 TRUE

For -10:

$\log(x+9) + \log x = 1$	
$\log(-10+9) + \log(-10)$	? 1 FALSE

The number -10 does not check, because negative numbers do not have logarithms. The solution is 1.

42. 10

43. $\log x - \log(x+7) = 1$ The base is 10.

$$\log_{10} \frac{x}{x+7} = 1 \quad \text{Using the quotient rule}$$
$$\frac{x}{x+7} = 10^1$$
$$x = 10(x+7)$$
$$x = 10x + 70$$
$$-9x = 70$$
$$x = -\frac{70}{9}$$

The number $-\frac{70}{9}$ does not check. The equation has no solution.

44. $\frac{1}{3}$

45. $\log_4(x+3) - \log_4(x-5) = 2$

$$\log_4 \frac{x+3}{x-5} = 2 \quad \text{Using the quotient rule}$$
$$\frac{x+3}{x-5} = 4^2$$
$$\frac{x+3}{x-5} = 16$$
$$x+3 = 16(x-5)$$
$$x+3 = 16x - 80$$
$$83 = 15x$$
$$\frac{83}{15} = x$$

The number $\frac{83}{15}$ checks. It is the solution.

46. 5

47. $\log_7(x+2) + \log_7(x+1) = \log_7 6$

$$\log_7[(x+2)(x+1)] = \log_7 6 \quad \text{Using the product rule}$$
$$\log_7(x^2+3x+2) = \log_7 6$$
$$x^2 + 3x + 2 = 6 \quad \text{Using the property of logarithmic equality}$$
$$x^2 + 3x - 4 = 0$$
$$(x+4)(x-1) = 0$$
$$x = -4 \quad or \quad x = 1$$

The number 1 checks, but -4 does not. The solution is 1.

48. 2

49. $\log_5 (x+4) + \log_5 (x-4) = 2$

$$\log_5 [(x+4)(x-4)] = 2$$
$$(x+4)(x-4) = 5^2$$
$$x^2 - 16 = 25$$
$$x^2 = 41$$
$$x = \pm\sqrt{41}$$

The number $\sqrt{41}$ checks, but $-\sqrt{41}$ does not. The solution is $\sqrt{41}$.

50. 4

51. $\log_{12}(x-4) - \log_{12}(x+5) = \log_{12} 3$

$$\log_{12} \frac{x-4}{x+5} = \log_{12} 3$$
$$\frac{x-4}{x+5} = 3 \quad \text{Using the property of logarithmic equality}$$
$$x - 4 = 3(x+5)$$
$$x - 4 = 3x + 15$$
$$-19 = 2x$$
$$-\frac{19}{2} = x$$

The number $-\frac{19}{2}$ does not check. The equation has no solution.

52. $\frac{17}{4}$

53. $\log_2(x-2) + \log_2 x = 3$

$$\log_2[(x-2)(x)] = 3$$
$$x(x-2) = 2^3$$
$$x^2 - 2x = 8$$
$$x^2 - 2x - 8 = 0$$
$$(x-4)(x+2) = 0$$
$$x = 4 \quad or \quad x = -2$$

The number 4 checks, but -2 does not. The solution is 4.

54. $\frac{2}{5}$

55. $(125x^7y^{-2}z^6)^{-2/3} =$

$(5^3)^{-2/3}(x^7)^{-2/3}(y^{-2})^{-2/3}(z^6)^{-2/3} =$

$5^{-2}x^{-14/3}y^{4/3}z^{-4} = \frac{1}{25}x^{-14/3}y^{4/3}z^{-4}$, or

$\frac{y^{4/3}}{25x^{14/3}z^4}$

56. $-i$

57. $(3+5i)^2 = 9+30i+25i^2 = 9+30i-25 = -16+30i$

58. $c = \sqrt{\frac{E}{m}}$

59.

$$x^4 + 400 = 104x^2$$
$$x^4 - 104x^2 + 400 = 0$$

Let $u = x^2$.

$$u^2 - 104u + 400 = 0$$
$$(u-100)(u-4) = 0$$
$$u = 100 \quad \text{or} \quad u = 4$$
$$x^2 = 100 \quad \text{or} \quad x^2 = 4 \quad \text{Replacing } u \text{ with } x^2$$
$$x = \pm 10 \quad \text{or} \quad x = \pm 2$$

The solutions are ± 10 and ± 2.

60. $\dfrac{3 \pm \sqrt{29}}{2}$

61. ◈

62. ◈

63. ◈

64. ◈

65.
$$
\begin{aligned}
27^x &= 81^{2x-3} \\
(3^3)^x &= (3^4)^{2x-3} \\
3^{3x} &= 3^{8x-12} \\
3x &= 8x - 12 \\
12 &= 5x \\
\frac{12}{5} &= x
\end{aligned}
$$
The solution is $\dfrac{12}{5}$.

66. -4

67.
$$
\begin{aligned}
\log_x (\log_3 27) &= 3 \\
\log_3 27 &= x^3 \\
3 &= x^3 \qquad (\log_3 27 = 3) \\
\sqrt[3]{3} &= x
\end{aligned}
$$
The solution is $\sqrt[3]{3}$.

68. 2

69.
$$
\begin{aligned}
x \cdot \log \frac{1}{8} &= \log 8 \\
x \cdot \log 8^{-1} &= \log 8 \\
x(-\log 8) &= \log 8 \quad \text{Using the power rule} \\
x &= -1
\end{aligned}
$$
The solution is -1.

70. $\pm\sqrt{34}$

71.
$$
\begin{aligned}
2^{x^2+4x} &= \frac{1}{8} \\
2^{x^2+4x} &= \frac{1}{2^3} \\
2^{x^2+4x} &= 2^{-3} \\
x^2 + 4x &= -3 \\
x^2 + 4x + 3 &= 0 \\
(x+3)(x+1) &= 0
\end{aligned}
$$
$x = -3$ or $x = -1$

The solutions are -3 and -1.

72. $10^{100,000}$

73.
$$
\begin{aligned}
\log_5 |x| &= 4 \\
|x| &= 5^4 \\
|x| &= 625
\end{aligned}
$$
$x = 625$ or $x = -625$

The solutions are 625 and -625.

74. 1, 100

75.
$$
\begin{aligned}
\log \sqrt{2x} &= \sqrt{\log 2x} \\
\log (2x)^{1/2} &= \sqrt{\log 2x} \\
\frac{1}{2} \log 2x &= \sqrt{\log 2x}
\end{aligned}
$$
$\frac{1}{4} (\log 2x)^2 = \log 2x$ Squaring both sides

$\frac{1}{4}(\log 2x)^2 - \log 2x = 0$

Let $u = \log 2x$.
$$
\begin{aligned}
\frac{1}{4}u^2 - u &= 0 \\
u\left(\frac{1}{4}u - 1\right) &= 0
\end{aligned}
$$
$u = 0$ *or* $\frac{1}{4}u - 1 = 0$

$u = 0$ *or* $\frac{1}{4}u = 1$

$u = 0$ *or* $u = 4$

$\log 2x = 0$ *or* $\log 2x = 4$ Replacing u with $\log 2x$

$2x = 10^0$ *or* $2x = 10^4$

$2x = 1$ *or* $2x = 10,000$

$x = \frac{1}{2}$ *or* $x = 5000$

Both numbers check. The solutions are $\frac{1}{2}$ and 5000.

76. $\dfrac{1}{100,000}$, 100,000

77. $3^{2x} - 8 \cdot 3^x + 15 = 0$

Let $u = 3^x$ and substitute.
$$
\begin{aligned}
u^2 - 8u + 15 &= 0 \\
(u-5)(u-3) &= 0
\end{aligned}
$$
$u = 5$ *or* $u = 3$

$3^x = 5$ *or* $3^x = 3$ Replacing u with 3^x

$\log 3^x = \log 5$ *or* $3^x = 3^1$

$x \log 3 = \log 5$ *or* $x = 1$

$x = \dfrac{\log 5}{\log 3}$ *or* $x = 1$

$x \approx 1.465$ *or* $x = 1$

Both numbers check. The solutions are $\log 5/\log 3$ and 1, or 1 and approximately 1.465.

78. $-\frac{1}{3}$

79.
$$\begin{aligned} 3^{2x}-3^{2x-1}&=18\\ 3^{2x}(1-3^{-1})&=18 \quad \text{Factoring}\\ 3^{2x}\left(1-\frac{1}{3}\right)&=18\\ 3^{2x}\left(\frac{2}{3}\right)&=18\\ 3^{2x}&=27 \quad \text{Multiplying by } \frac{3}{2}\\ 3^{2x}&=3^3\\ 2x&=3\\ x&=\frac{3}{2}\end{aligned}$$

80. 38

81. $\log_5 125 = 3$ and $\log_{125} 5 = \frac{1}{3}$, so $x = (log_{125}5)^{log_5 125}$ is equivalent to $x = \left(\frac{1}{3}\right)^3 = \frac{1}{27}$. Then $\log_3 x = \log_3 \frac{1}{27} = -3$.

82. 1

83. [graph]

Exercise Set 9.7

1. a) Replace $N(t)$ with 50 and solve for t.

$$\begin{aligned} N(t)&=0.3(1.63)^t\\ 50&=0.3(1.63)^t\\ 166.667&\approx(1.63)^t \quad \text{Dividing by 0.3 on both sides}\\ \ln 166.667&\approx\ln(1.63)^t \quad \text{Taking the natural logarithm on both sides}\\ 5.116&\approx t\ln 1.63\\ \frac{5.116}{\ln 1.63}&\approx t\\ 10.47&\approx t\end{aligned}$$

Rounding up to 11 years, we see that by the end of 1985 + 11, or 1996, 50 million cellular phones would be in use.

b) To find the doubling time we replace $N(t)$ with $2(0.3)$, or 0.6 and solve for t.

$$\begin{aligned} 0.6&=0.3(1.63)^t\\ 2&=(1.63)^t\\ \ln 2&=\ln(1.63)^t\\ \ln 2&=t\ln 1.63\\ \frac{\ln 2}{\ln 1.63}&=t\\ 1.4&\approx t\end{aligned}$$

The doubling time is about 1.4 years.

2. (a) 7.7 yr; (b) 2.8 yr

3. a) Replace $A(t)$ with 40,000 and solve for t.

$$\begin{aligned} A(t)&=29,000(1.08)^t\\ 40,000&=29,000(1.08)^t\\ 1.379&\approx(1.08)^t\\ \log 1.379&\approx\log(1.08)^t\\ \log 1.379&\approx t\log 1.08\\ \frac{\log 1.379}{\log 1.08}&\approx t\\ 4.2&\approx t\end{aligned}$$

The amount due will reach \$40,000 after about 4.2 years.

b) Replace $A(t)$ with 2(29,000), or 58,000, and solve for t.

$$\begin{aligned} 58,000&=29,000(1.08)^t\\ 2&=(1.08)^t\\ \log 2&=\log(1.08)^t\\ \log 2&=t\log 1.08\\ \frac{\log 2}{\log 1.08}&=t\\ 9.0&\approx t\end{aligned}$$

The doubling time is about 9.0 years.

4. (a) 3.6 days; (b) 0.6 days

5. a) We replace $N(t)$ with 60,000 and solve for t:

$$\begin{aligned} 60,000&=250,000\left(\frac{2}{3}\right)^t\\ \frac{60,000}{250,000}&=\left(\frac{2}{3}\right)^t\\ 0.24&=\log\left(\frac{2}{3}\right)^t\\ \log\ 0.24&=\log\left(\frac{2}{3}\right)^t\\ \log\ 0.24&=t\ \log\frac{2}{3}\\ t&=\frac{\log\ 0.24}{\log\frac{2}{3}}\approx\frac{-0.61979}{-0.17609}\approx 3.5\end{aligned}$$

After about 3.5 years 60,000 cans will still be in use.

b) We replace $N(t)$ with 1000 and solve for t.

$$1000 = 250,000\left(\frac{2}{3}\right)^t$$

$$\frac{1000}{250,000} = \left(\frac{2}{3}\right)^t$$

$$0.004 = \log\left(\frac{2}{3}\right)^t$$

$$\log 0.004 = \log\left(\frac{2}{3}\right)^t$$

$$\log 0.004 = t \log\frac{2}{3}$$

$$t = \frac{\log 0.004}{\log\frac{2}{3}} \approx \frac{-2.39794}{-0.17609} \approx 13.6$$

After about 13.6 years 1000 cans will still be in use.

6. (a) 6.6 yr; (b) 3.1 yr

7. $\text{pH} = -\log[H^+]$

$= -\log[1.3 \times 10^{-5}]$

$\approx -(-4.886057)$ Using a calculator

≈ 4.9

The pH of fresh-brewed coffee is about 4.9.

8. 6.8

9. $\text{pH} = -\log[H^+]$

$7.0 = -\log[H^+]$

$-7.0 = \log[H^+]$

$10^{-7.0} = [H^+]$ Converting to an exponential equation

The hydrogen ion concentration is 10^{-7} moles per liter.

10. 1.58×10^{-8} moles per liter

11. $L = 10 \cdot \log\frac{I}{I_0}$

$= 10 \cdot \log\frac{3.2 \times 10^{-6}}{10^{-12}}$

$= 10 \cdot \log(3.2 \times 10^6)$

$\approx 10(6.5)$

≈ 65

The intensity of sound in normal conversation is about 65 decibels.

12. 95 dB

13. $L = 10 \cdot \log\frac{I}{I_0}$

$105 = 10 \cdot \log\frac{I}{10^{-12}}$

$10.5 = \log\frac{I}{10^{-12}}$

$10.5 = \log I - \log 10^{-12}$ Using the quotient rule

$10.5 = \log I - (-12)$ $(\log 10^a = a)$

$10.5 = \log I + 12$

$-1.5 = \log I$

$10^{-1.5} = I$ Converting to an exponential equation

$3.2 \times 10^{-2} \approx I$

The intensity of the sound is $10^{-1.5}$ W/m^2, or about 3.2×10^{-2} W/m^2.

14. $10^{-9.2}$ W/m^2, or 6.3×10^{-10} W/m^2

15. a) Substitute 0.06 for k:

$P(t) = P_0\, e^{0.06t}$

b) To find the balance after one year, replace P_0 with 5000 and t with 1. We find $P(1)$:

$P(1) = 5000\, e^{0.06(1)} = 5000\, e^{0.06} \approx$

$5000(1.061836547) \approx \$5309.18$

To find the balance after 2 years, replace P_0 with 5000 and t with 2. We find $P(2)$:

$P(2) = 5000\, e^{0.06(2)} = 5000\, e^{0.12} \approx$

$5000(1.127496852) \approx \$5637.48$

c) To find the doubling time, replace P_0 with 5000 and $P(t)$ with 10,000 and solve for t.

$10,000 = 5000\, e^{0.06t}$

$2 = e^{0.06t}$

$\ln 2 = \ln e^{0.06t}$ Taking the natural logarithm on both sides

$\ln 2 = 0.06t$ Finding the logarithm of the base to a power

$\frac{\ln 2}{0.06} = t$

$11.6 \approx t$

The investment will double in about 11.6 years.

16. (a) $P(t) = P_0e^{0.05t}$; (b) \$1051.27; \$1105.17; (c) 13.9 years

17. a) $P(t) = 264e^{0.01t}$, where $P(t)$ is in millions and t is the number of years after 1995.

b) In 2001, $t = 2001 - 1995 = 6$. Replace t with 6 and compute $P(6)$.

$P(6) = 264e^{0.01(6)}$

$= 264e^{0.06}$

≈ 280

The U.S. population in 2001 will be about 280 million.

c) Replace $P(t)$ with 300 and solve for t.

$$\begin{aligned} 300 &= 264e^{0.01t} \\ 1.136 &\approx e^{0.01t} \\ \ln 1.136 &\approx \ln e^{0.01t} \\ \ln 1.136 &\approx 0.01t \\ \frac{\ln 1.136}{0.01} &\approx t \\ 12.8 &\approx t \end{aligned}$$

Rounding up to 13, we see that the U.S. population will reach 300 million in 1995 + 13, or 2008.

18. (a) $P(t) = 5.7e^{0.015t}$, where $P(t)$ is in billions and t is the numbers of years after 1995; (b) 6.2 billion; (c) 2018

19. a) Replace $N(t)$ with 60,000 and solve for t.

$$\begin{aligned} 60,000 &= 3000(2)^{t/20} \\ 20 &= (2)^{t/20} \\ \log 20 &= \log(2)^{t/20} \\ \log\ 20 &= \frac{t}{20} \log\ 2 \\ 20 \log 20 &= t \log 2 \\ \frac{20 \log^{20}}{\log 2} &= t \\ 86.4 &\approx t \end{aligned}$$

There will be 60,000 bacteria after about 86.4 minutes.

b) Replace $N(t)$ with 100,000,000 and solve for t.

$$\begin{aligned} 100,000,000 &= 3000(2)^{t/20} \\ 33,333.333 &= (2)^{t/20} \\ \log 33,333.333 &= \log(2)^{t/20} \\ \log 33,333.333 &= \frac{t}{20} \log\ 2 \\ 20 \log 33,333.333 &= t \log 2 \\ \frac{20 \log 33,333.333}{\log 2} &= t \\ 300.5 &\approx t \end{aligned}$$

About 300.5 minutes would have to pass in order for a possible infection to occur.

c) Replace $P(t)$ with 6000 and solve for t.

$$\begin{aligned} 6000 &= 3000(2)^{t/20} \\ 2 &= (2)^{t/20} \\ 1 &= \frac{t}{20} \quad \text{The exponents must be the same.} \\ t &= 20 \end{aligned}$$

The doubling time is 20 minutes.

20. 19.8 yr

21. a) Replace a with 1 and compute $N(1)$.

$$\begin{aligned} N(a) &= 2000 + 500 \log a \\ N(1) &= 2000 + 500 \log 1 \\ N(1) &= 2000 + 500 \cdot 0 \\ N(1) &= 2000 \end{aligned}$$

2000 units were sold after \$1000 was spent.

b) Find $N(8)$.

$$\begin{aligned} N(8) &= 2000 + 500 \log 8 \\ N(8) &\approx 2451.5 \end{aligned}$$

About 2452 units were sold after \$8000 was spent.

c) Using the values we computed in parts (a) and (b) and any others we wish to calculate, we sketch the graph:

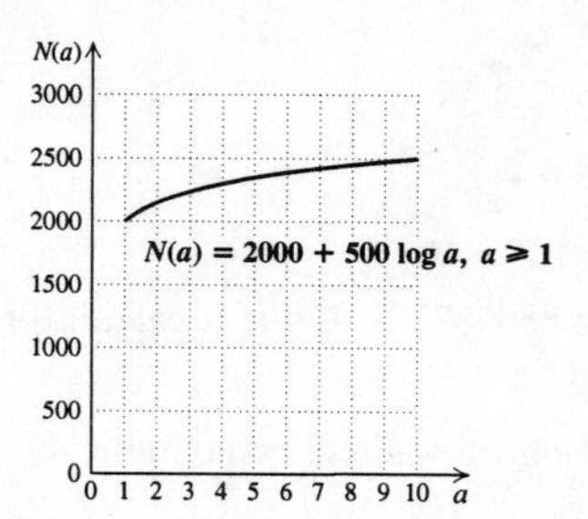

d) Replace $N(a)$ with 5000 and solve for a.

$$\begin{aligned} 5000 &= 2000 + 500 \log a \\ 3000 &= 500 \log a \\ 6 &= \log a \\ a &= 10^6 = 1,000,000 \end{aligned}$$

\$1,000,000 thousand, or \$1,000,000,000 would have to be spent.

22. (a) 68%; (b) 54%, 40%;

(c)

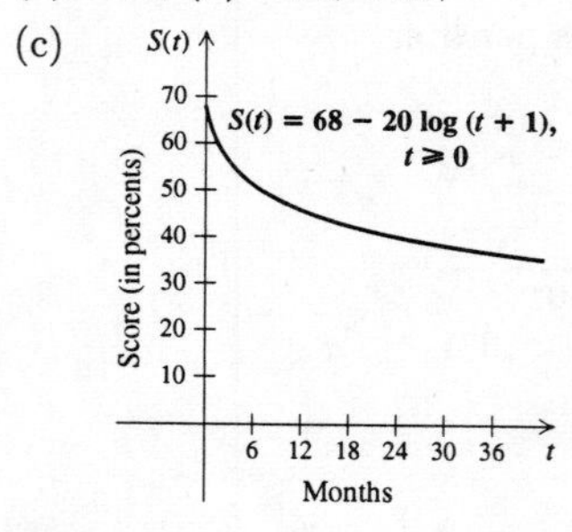

d) 6.9 months

23. a) We use the growth equation $N(t) = N_0 e^{kt}$, where t is the number of years since 1995. In 1995, at $t = 0$, 17 people were infected. We substitute 17 for N_0:

$$N(t) = 17e^{kt}.$$

To find the exponential growth rate k, observe that 1 year later 29 people were infected.

$$N(1) = 17e^{k\cdot 1} \quad \text{Substituting 1 for } t$$
$$29 = 17e^k \quad \text{Substituting 29 for } N(1)$$
$$1.706 \approx e^k$$
$$\ln 1.706 \approx \ln e^k$$
$$\ln 1.706 \approx k$$
$$0.534 \approx k$$

The exponential function is $N(t) = 17e^{0.534t}$, where t is the number of years since 1995.

b) In 2001, $t = 2001 - 1995$, or 6. Find $N(6)$.

$$N(6) = 17e^{0.534(6)}$$
$$= 17e^{3.204}$$
$$\approx 418.7$$

Approximately 419 people will be infected in 2001.

24. (a) $N(t) = e^{0.363t}$, where t is the number of years after 1967; (b) 329,391

25. We start with the exponential growth equation

$D(t) = D_0\, e^{kt}$, where t is the number of years after 1995.

Substitute 2 D_0 for $D(t)$ and 0.1 for k and solve for t.

$$2\, D_0 = D_0\, e^{0.1t}$$
$$2 = e^{0.1t}$$
$$\ln 2 = 0.1t$$
$$\ln 2 = \ln\ e^{0.1t}$$
$$\frac{\ln 2}{0.1} = t$$
$$6.9 \approx t$$

The demand will be double that of 1995 in 1995 + 7, or 2002.

26. 2013

27. a) We use the decay equation $S(t) = S_0 e^{-kt}$, where t is the number of years since 1983 and $S(t)$ is in millions. In 1983, at $t = 0$, 205 million records were sold. We substitute 205 for S_0.

$$S(t) = 205e^{-kt}.$$

To find the exponential decay rate k, observe that 10 years later, in 1993, 1.2 million records were sold.

$$1.2 = 205e^{-k\cdot 10} \quad \text{Substituting}$$
$$0.00585 \approx e^{10k}$$
$$\ln 0.00585 \approx \ln e^{-10k}$$
$$-5.14 \approx -10k$$
$$0.514 \approx k$$

Then $k \approx 0.514$ and the exponential function is $S(t) = 205e^{-0.514t}$, where $S(t)$ in millions and t is the number of years since 1983.

b) In 2001, $t = 2001 - 1983$, or 18.

$$S(18) = 205e^{-0.514(18)}$$
$$= 205e^{-9.252}$$
$$\approx 0.019664$$

In 2001, about 0.019664 million, or 19,664 records will be sold.

c) 1 is equivalent to 0.000001 million.

$$0.000001 = 205e^{-0.514t}$$
$$4.89 \times 10^{-9} = e^{-0.514t}$$
$$\ln(4.89 \times 10^{-9}) = \ln e^{-0.514t}$$
$$-19.1 \approx -0.514t$$
$$37.2 \approx t$$

Only one record will be sold in 1983+38, or 2021.

28. (a) $C(t) = 80e^{-0.016t}$, where t is the number of years after 1985; (b) 62 lb; (c) 2072

29. We will use the function derived in Example 7:

$$P(t) = P_0 e^{-0.00012t}$$

If the scrolls had lost 22.3% of their carbon-14 from an initial amount P_0, then $77.7\%(P_0)$ is the amount present. To find the age t of the scrolls, we substitute $77.7\%(P_0)$, or $0.777P_0$, for $P(t)$ in the function above and solve for t.

$$0.777P_0 = P_0 e^{-0.00012t}$$
$$0.777 = e^{-0.00012t}$$
$$\ln 0.777 = \ln\ e^{-0.00012t}$$
$$-0.2523 \approx -0.00012t$$
$$t \approx \frac{-0.2523}{-0.00012} \approx 2103$$

The scrolls are about 2103 years old.

30. 1654 yr

31. The function $P(t) = P_0\, e^{-kt}$, $k > 0$, can be used to model decay. For iodine-131, $k = 9.6\%$, or 0.096. To find the half-life we substitute 0.096 for k and $\frac{1}{2}\, P_0$ for $P(t)$, and solve for t.

$$\frac{1}{2}\, P_0 = P_0\, e^{-0.096t}, \text{ or } \frac{1}{2} = e^{-0.096t}$$
$$\ln \frac{1}{2} = \ln\ e^{-0.096t} = -0.096t$$
$$t = \frac{\ln\ 0.5}{-0.096} \approx \frac{-0.6931}{-0.096} \approx 7.2 \text{ days}$$

32. 11 yr

33. The function $P(t) = P_0 e^{-kt}$, $k > 0$, can be used to model decay. We substitute $\frac{1}{2}P_0$ for $P(t)$ and 1 for t and solve for the decay rate k.

$$\frac{1}{2}P_0 = P_0e^{-k\cdot 1}$$
$$\frac{1}{2} = e^{-k}$$
$$\ln\frac{1}{2} = \ln e^{-k}$$
$$-0.693 \approx -k$$
$$0.693 \approx k$$

The decay rate is 0.693, or 69.3% per year.

34. 3.15% per year

35. a) We start with the exponential growth equation

$V(t) = V_0\, e^{kt}$, where t is the number of years after 1991.

Substituting 451,000 for V_0, we have

$V(t) = 451,000\, e^{kt}$.

To find the exponential growth rate k, observe that the card sold for $640,500 in 1996, or 5 years after 1991. We substitute and solve for k.

$$V(5) = 451,000\, e^{k\cdot 5}$$
$$640,500 = 451,000\, e^{5k}$$
$$1.42 = e^{5k}$$
$$\ln 1.42 = \ln e^{5k}$$
$$\ln 1.42 = 5k$$
$$\frac{\ln 1.42}{5} = k$$
$$0.07 \approx k$$

Thus the exponential growth function is $V(t) = 451,000\, e^{0.07t}$, where t is the number of years after 1991.

b) In 2002, $t = 2002 - 1991$, or 11.

$$V(11) = 451,000\, e^{0.07(11)} \approx 974,055$$

The card's value in 2002 will be about $974,055.

c) Substitute $902,000 for $V(t)$ and solve for t.

$$902,000 = 451,000\, e^{0.07t}$$
$$2 = e^{0.07t}$$
$$\ln\ 2 = \ln\ e^{0.07t}$$
$$\ln\ 2 = 0.07t$$
$$\frac{\ln\ 2}{0.07} = t$$
$$9.9 \approx t$$

The doubling time is about 9.9 years.

d) Substitute $1,000,000 for $V(t)$ and solve for t.

$$1,000,000 = 451,000\, e^{0.07t}$$
$$2.217 \approx e^{0.07t}$$
$$\ln 2.217 \approx \ln\ e^{0.07t}$$
$$\ln 2.217 \approx 0.07t$$
$$\frac{\ln 2.217}{0.07} \approx t$$
$$11.4 \approx t$$

The value of the card will be $1,000,000 in 1991 + 12, or 2003.

36. (a) $k \approx 0.16$, $V(t) = 84e^{0.16t}$, where $V(t)$ is in thousands of dollars and t is the number of years after 1947; (b) $474,880 thousand, or $447,880,000; (c) 4.3 yr; (d) 58.7 yr

37. $$\frac{\frac{x-5}{x+3}}{\frac{x}{x-3}+\frac{2}{x+3}}$$

$$= \frac{\frac{x-5}{x+3}}{\frac{x}{x-3}\cdot\frac{x+3}{x+3}+\frac{2}{x+3}\cdot\frac{x-3}{x-3}}$$ Finding the LCD and adding in the denominator

$$= \frac{\frac{x-5}{x+3}}{\frac{x^2+3x+2x-6}{(x-3)(x+3)}}$$

$$= \frac{\frac{x-5}{x+3}}{\frac{x^2+5x-6}{(x-3)(x+3)}}$$

$$= \frac{x-5}{x+3}\cdot\frac{(x-3)(x+3)}{x^2+5x-6}$$ Multiplying by the reciprocal of the denominator

$$= \frac{(x-5)(x-3)(x+3)}{(x+3)\ (x+6)(x-1)}$$ Factoring x^2+5x-6 and multiplying

$$= \frac{(x-5)(x-3)\cancel{(x+3)}}{\cancel{(x+3)}(x+6)(x-1)}$$

$$= \frac{(x-5)(x-3)}{(x+6)(x-1)},\ \text{or}\ \frac{(x-5)(x-3)}{x^2+5x-6}$$

38. $\dfrac{3ab^2+5a^2b}{2b^2-4a^2}$

39. $\dfrac{6a^3b^{-7}}{8a^{-5}b^{-10}} = \dfrac{3a^{3-(-5)}b^{-7-(-10)}}{4} = \dfrac{3a^8b^3}{4}$

40. $\dfrac{ab^2+a^2}{b^2-a^2b}$

41. ◈

42. 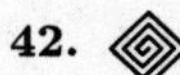

43.

44.

45. Set $S(x) = D(x)$, and solve for x.

$$e^x = 162,755\, e^{-x}$$

$$e^{2x} = 162,755 \quad \text{Multiplying by } e^x \text{ on both sides}$$

$$\ln\, e^{2x} = \ln\, 162,755$$

$$2x = \ln\, 162,755$$

$$x = \frac{\ln\, 162,755}{2}$$

$$x \approx 6$$

To find the second coordinate of the equilibrium point, find $S(6)$ or $D(6)$. We will find $S(6)$.

$$S(6) = e^6 \approx 403$$

The equilibrium point is $(6, \$403)$.

46. $P(t) = 0.13e^{0.479t}$, where t is the number of years after 1985 and $P(t)$ is a percent.

47.

Chapter 10

Conic Sections

Exercise Set 10.1

1. $y = -x^2$

a) This is equivalent to $y = -(x-0)^2 + 0$. The vertex is $(0,0)$.

b) We choose some x-values on both sides of the vertex and compute the corresponding values of y. The graph opens down, because the coefficient of x^2, -1, is negative.

x	y
0	0
1	-1
2	-4
-1	-1
-2	-4

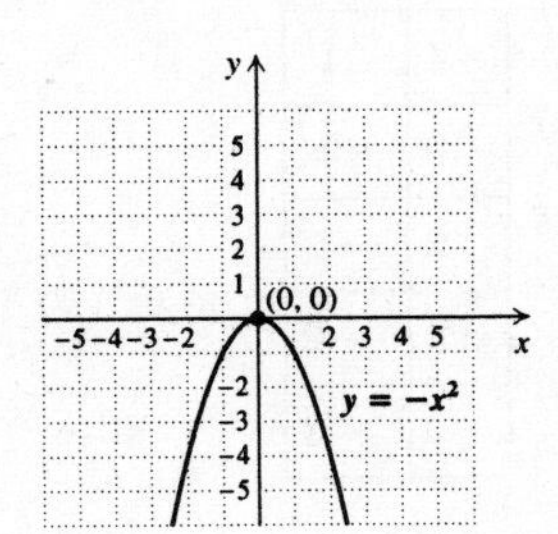

2.

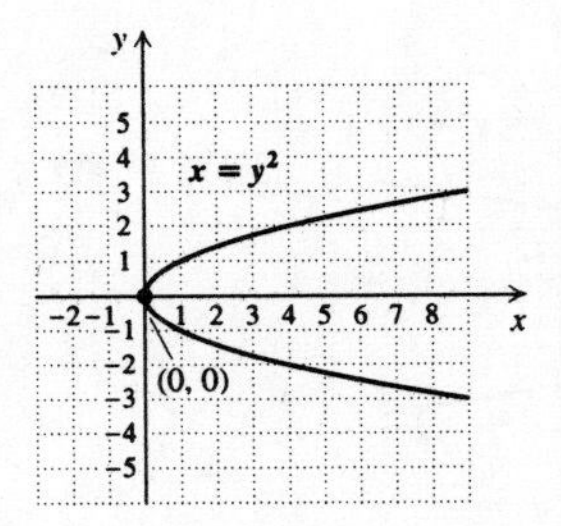

3. $x = y^2 - 4y + 1$

a) We find the vertex by completing the square.

$x = (y^2 - 4y + 4) + 1 - 4$

$x = (y-2)^2 - 3$

The vertex is $(-3, 2)$.

b) To find ordered pairs, we choose values for y and compute the corresponding values of x. The graph opens to the right, because the coefficient of y^2, 1, is positive.

x	y
6	-1
1	0
-2	1
-3	2
-2	3

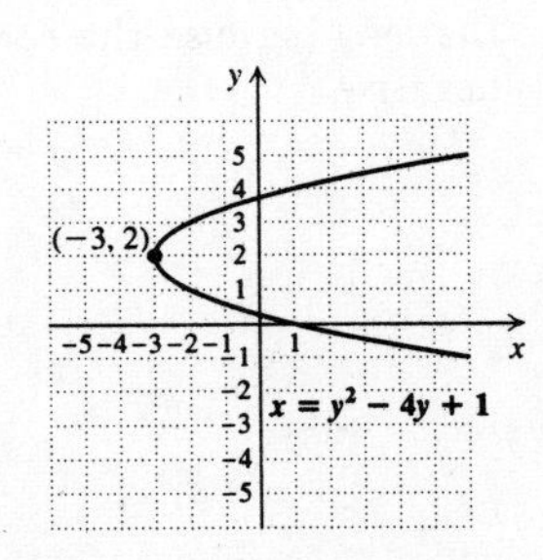

4.

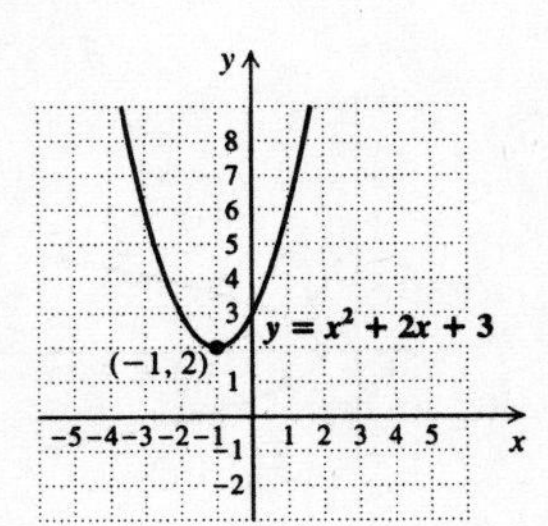

5. $y = -x^2 + 4x - 5$

a) We can find the vertex by computing the first coordinate, $x = -b/2a$, and then substituting to find the second coordinate:

$$x = -\frac{b}{2a} = -\frac{4}{2(-1)} = 2$$

$$y = -x^2 + 4x - 5 = -(2)^2 + 4(2) - 5 = -1$$

The vertex is $(2, -1)$.

b) We choose some x-values and compute the corresponding values for y. The graph opens downward because the coefficient of x^2, -1, is negative.

x	y
2	-1
3	-2
4	-5
1	-2
0	-5

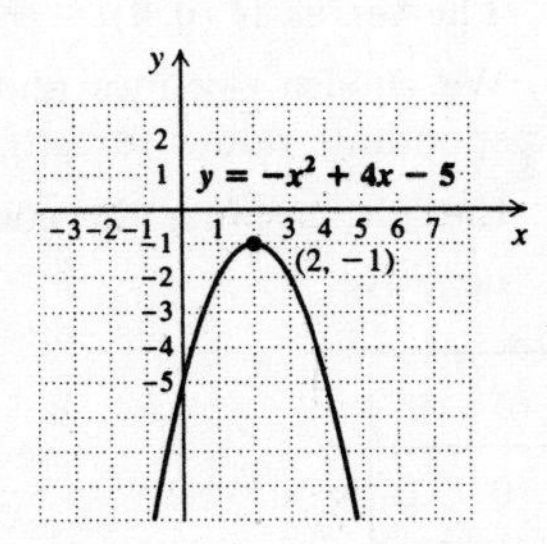

6.

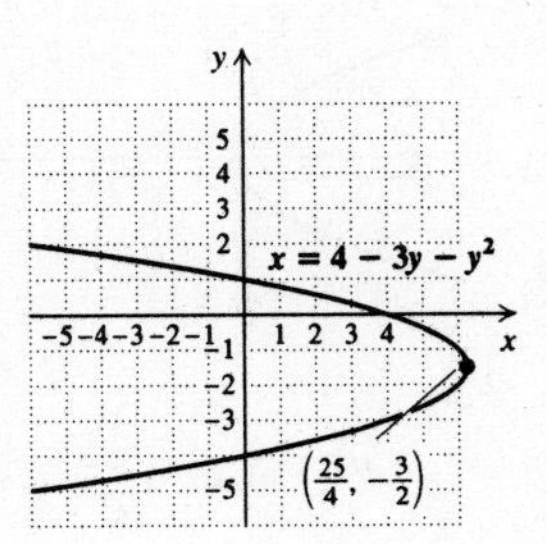

7. $x = y^2 + 1$

a) $x = (y-0)^2 + 1$

The vertex is $(1, 0)$.

b) To find the ordered pairs, we choose y-values and compute the corresponding values for x. The graph opens to the right, because the coefficient of y^2, 1, is positive.

x	y
1	0
2	1
5	2
2	-1
5	-2

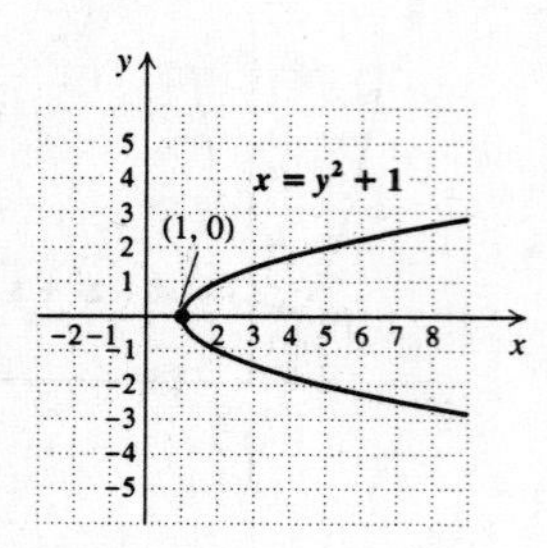

8.

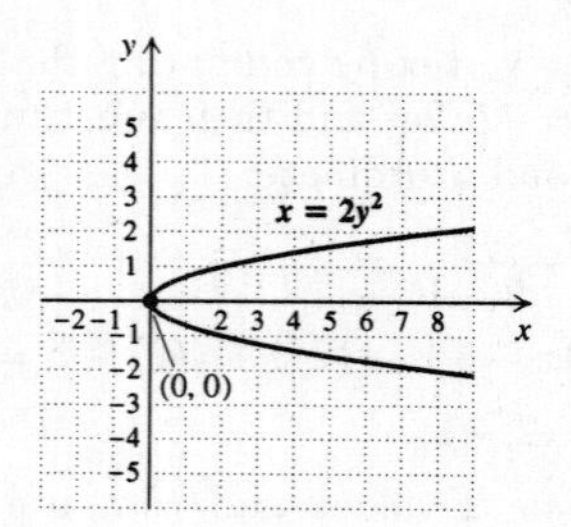

9. $x = -\dfrac{1}{2}y^2$

a) $x = -\dfrac{1}{2}(y-0)^2 + 0$

The vertex is $(0,0)$.

b) We choose y-values and compute the corresponding values for x. The graph opens to the left, because the coefficient of y^2, $-\dfrac{1}{2}$, is negative.

x	y
0	0
-2	2
-8	4
-2	-2
-8	-4

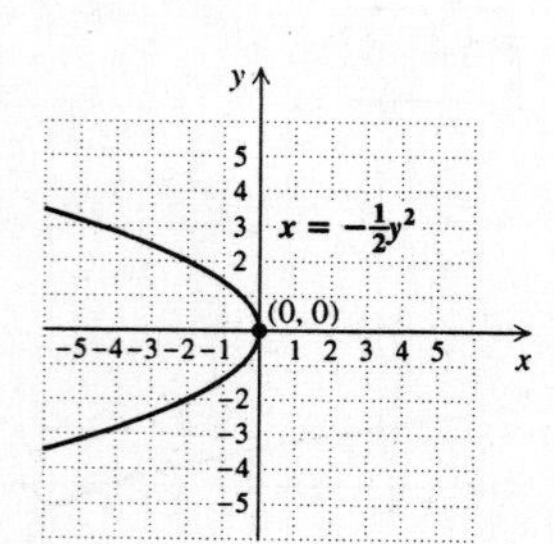

10.

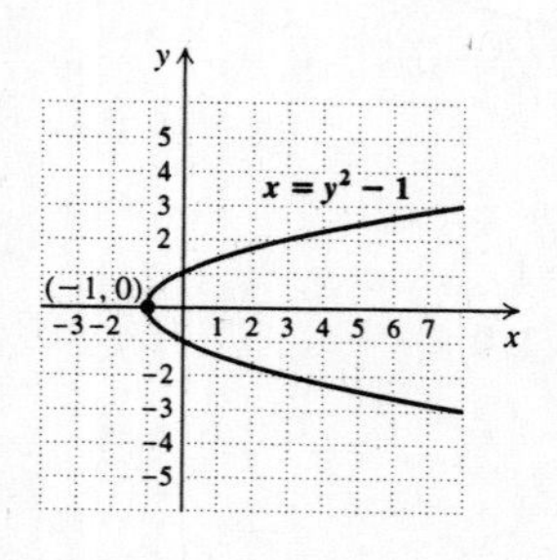

11. $x = -y^2 - 3y$

a) We find the vertex by computing the second coordinate, $y = -b/2a$, and then substituting to find the first coordinate:

$$y = -\frac{b}{2a} = -\frac{-3}{2(-1)} = -\frac{3}{2}$$

$$x = -y^2 - 3y = -\left(-\frac{3}{2}\right)^2 - 3\left(-\frac{3}{2}\right) = \frac{9}{4}$$

The vertex is $\left(\dfrac{9}{4}, -\dfrac{3}{2}\right)$.

b) We choose y-values and compute the corresponding values for x. The graph opens to the left, because the coefficient of y^2, -1, is negative.

x	y
-4	1
0	0
2	-1
2	-2
0	-3

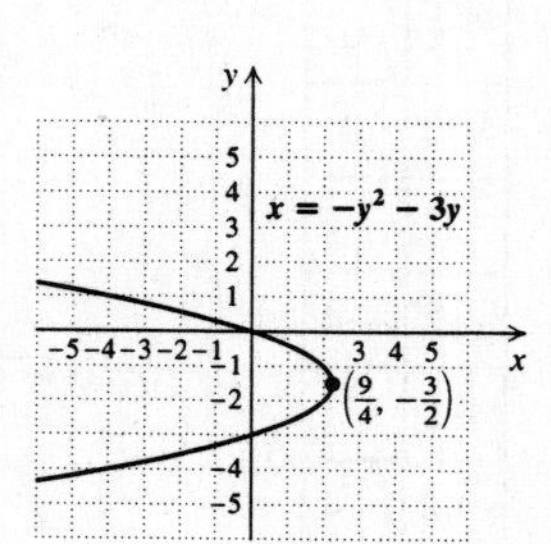

12.

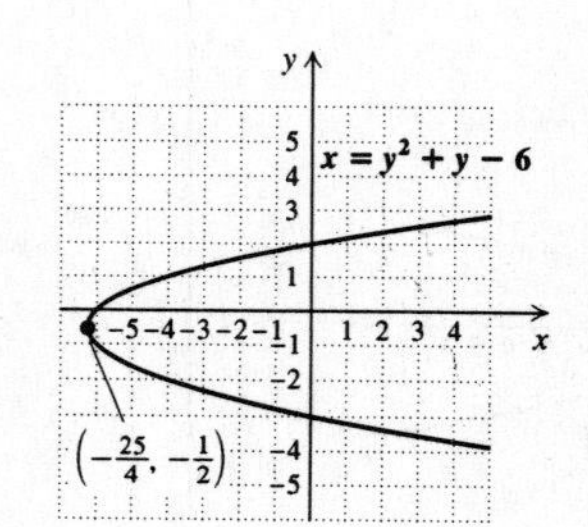

13. $x = 8 - y - y^2$

a) We find the vertex by completing the square.

$$x = -(y^2 + y) + 8$$

$$x = -\left(y^2 + y + \frac{1}{4}\right) + 8 + \frac{1}{4}$$

$$x = -\left(y + \frac{1}{2}\right)^2 + \frac{33}{4}$$

The vertex is $\left(\dfrac{33}{4}, -\dfrac{1}{2}\right)$.

b) We choose y-values and compute the corresponding values for x. The graph opens to the left, because the coefficient of y^2, -1, is negative.

x	y
$\frac{33}{4}$	$-\frac{1}{2}$
8	0
6	1
2	2
8	-1
6	-2
2	-3

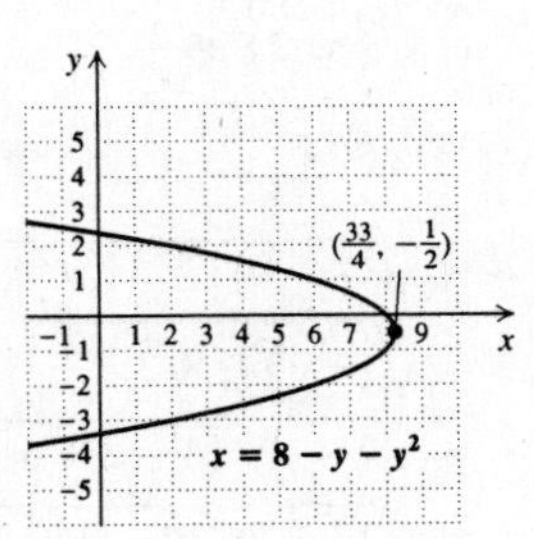

14.

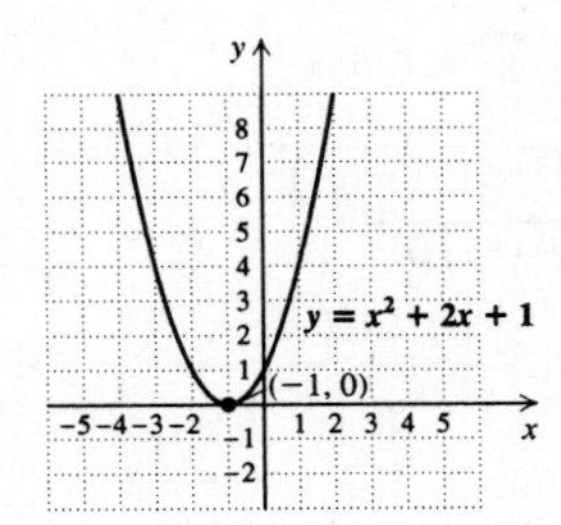

15. $y = x^2 - 2x + 1$

a) $y = (x - 1)^2 + 0$

The vertex is $(1, 0)$.

b) We choose x-values and compute the corresponding values for y. The graph opens upward, because the coefficient of x^2, 1, is positive.

x	y
1	0
0	1
-1	4
2	1
3	4

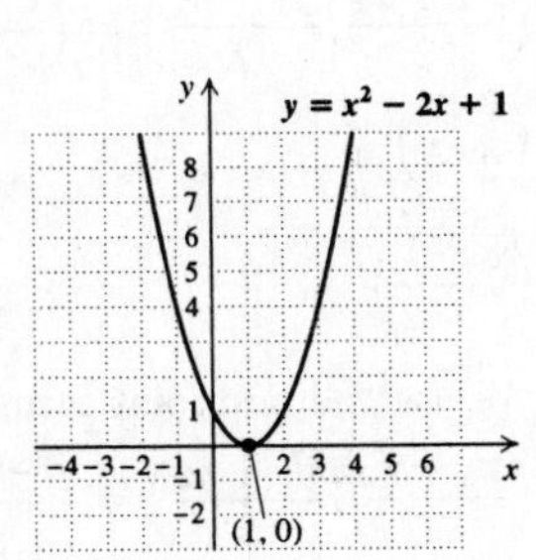

16.

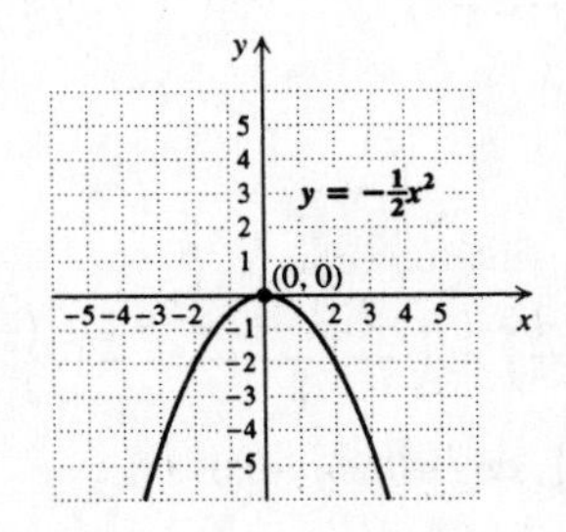

17. $x = -y^2 + 2y - 1$

a) We find the vertex by computing the second coordinate, $y = -b/2a$, and then substituting to find the first coordinate.

$$y = -\frac{b}{2a} = -\frac{2}{2(-1)} = 1$$
$$x = -y^2 + 2y - 1 = -(1)^2 + 2(1) - 1 = 0$$

The vertex is $(0, 1)$.

b) We choose y-values and compute the corresponding values for x. The graph opens to the left, because the coefficient of y^2, -1, is negative.

x	y
-4	3
-1	2
-1	0
-4	-1
-4	3

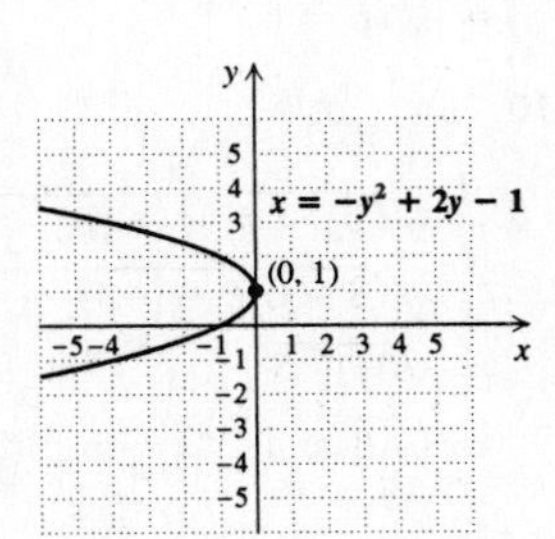

18.

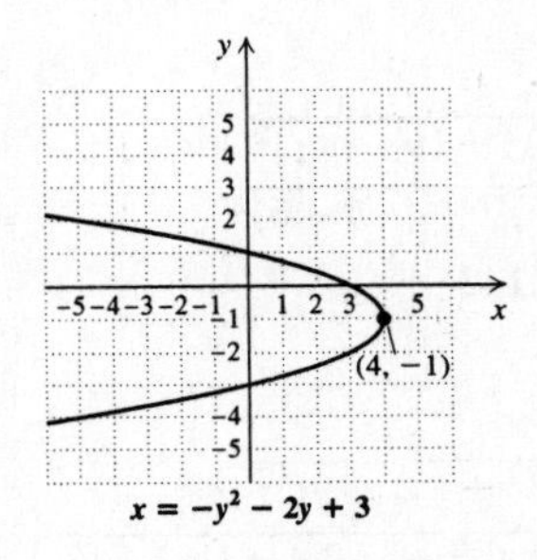

19. $x = -2y^2 - 4y + 1$

a) We find the vertex by completing the square.

$$x = -2(y^2 + 2y) + 1$$
$$x = -2(y^2 + 2y + 1) + 1 + 2$$
$$x = -2(y + 1)^2 + 3$$

The vertex is $(-3, -1)$.

b) We choose y-values and compute the corresponding values for x. The graph opens to the left, because the coefficient of y^2, -2, is negative.

x	y
3	-1
1	-2
-5	-3
1	0
-5	1

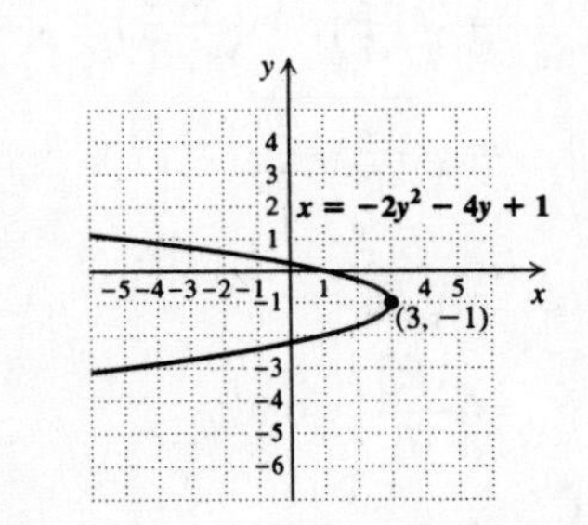

20.

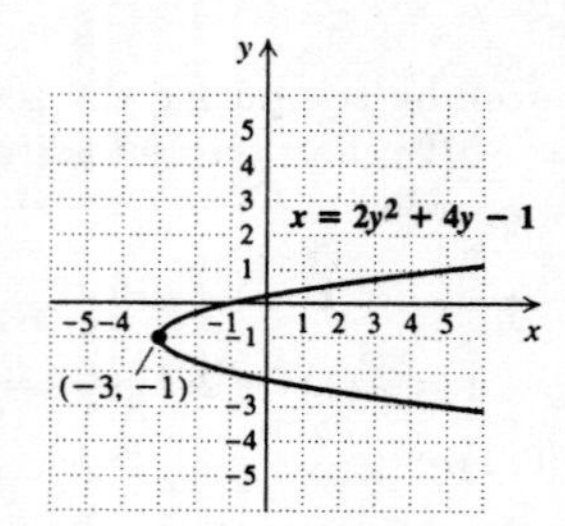

21. $d = \sqrt{(x_2 - x_1)^2 + (y_2 + y_1)^2}$ Distance formula

$= \sqrt{(6-2)^2 + (10-7)^2}$ Substituting

$= \sqrt{4^2 + 3^2}$

$= \sqrt{25} = 5$

22. 10

23. $d = \sqrt{(x_2 - x_1)^2 + (y_2 - y_1)^2}$ Distance formula

$= \sqrt{(3-0)^2 + [-4-(-7)]^2}$ Substituting

$= \sqrt{3^2 + 3^2}$

$= \sqrt{18} \approx 4.243$ Simplifying and approximating

24. 10

25. $d = \sqrt{(x_2 - x_1)^2 + (y_2 - y_1)^2}$

$= \sqrt{[5-(-5)]^2 + (-5-5)^2}$

$= \sqrt{200} \approx 14.142$

26. $\sqrt{464} \approx 21.541$

27. $d = \sqrt{(x_2 - x_1)^2 + (y_2 - y_1)^2}$

$= \sqrt{(-9.2 - 8.6)^2 + [-3.4 - (-3.4)]^2}$

$= \sqrt{(-17.8)^2 + 0^2}$

$= \sqrt{316.84} = 17.8$

(Since these points are on a horizontal line, we could have found the distance between them by finding $|x_2 - x_1| = |-9.2 - 8.6| = |-17.8| = 17.8$.)

28. $\sqrt{98.93} \approx 9.946$

29. $d = \sqrt{(x_2 - x_1)^2 + (y_2 - y_1)^2}$

$d = \sqrt{\left(\frac{5}{7} - \frac{1}{7}\right)^2 + \left(\frac{1}{14} - \frac{11}{14}\right)^2}$

$= \sqrt{\left(\frac{4}{7}\right)^2 + \left(-\frac{5}{7}\right)^2}$

$= \sqrt{\frac{16}{49} + \frac{25}{49}}$

$= \sqrt{\frac{41}{49}}$

$= \frac{\sqrt{41}}{7} \approx 0.915$

30. $\sqrt{13} \approx 3.606$

31. $d = \sqrt{(x_2 - x_1)^2 + (y_2 - y_1)^2}$

$d = \sqrt{[0 - (-\sqrt{6})]^2 + (0 - \sqrt{2})^2}$

$= \sqrt{6 + 2}$

$= \sqrt{8} \approx 2.828$

32. $\sqrt{8} \approx 2.828$

33. $d = \sqrt{(x_2 - x_1)^2 + (y_2 - y_1)^2}$

$= \sqrt{(-\sqrt{7} - \sqrt{2})^2 + [\sqrt{5} - (-\sqrt{3})]^2}$

$= \sqrt{7 + 2\sqrt{14} + 2 + 5 + 2\sqrt{15} + 3}$

$= \sqrt{17 + 2\sqrt{14} + 2\sqrt{15}} \approx 5.677$

34. $\sqrt{22 + 2\sqrt{40} + 2\sqrt{18}} \approx 6.568$

35. $d = \sqrt{(x_2 - x_1)^2 + (y_2 - y_1)^2}$

$d = \sqrt{(s-0)^2 + (t-0)^2}$

$= \sqrt{s^2 + t^2}$

36. $\sqrt{p^2 + q^2}$

37. We use the midpoint formula:

$\left(\frac{x_1 + x_2}{2}, \frac{y_1 + y_2}{2}\right) = \left(\frac{2+6}{2}, \frac{9+(-7)}{2}\right)$, or

$\left(\frac{8}{2}, \frac{2}{2}\right)$, or $(4, 1)$

38. $\left(\frac{13}{2}, -1\right)$

39. We use the midpoint formula:

$\left(\frac{x_1 + x_2}{2}, \frac{y_1 + y_2}{2}\right) = \left(\frac{8+(-1)}{2}, \frac{5+2}{2}\right)$, or

$\left(\frac{7}{2}, \frac{7}{2}\right)$

40. $\left(0, -\frac{1}{2}\right)$

41. We use the midpoint formula:

$\left(\frac{x_1 + x_2}{2}, \frac{y_1 + y_2}{2}\right) = \left(\frac{-8+6}{2}, \frac{-5+(-1)}{2}\right)$, or

$\left(\frac{-2}{2}, \frac{-6}{2}\right)$, or $(-1, -3)$

42. $\left(\frac{5}{2}, 1\right)$

43. We use the midpoint formula:

$\left(\frac{x_1 + x_2}{2}, \frac{y_1 + y_2}{2}\right) = \left(\frac{-3.4+2.9}{2}, \frac{8.1+(-8.7)}{2}\right)$,

or $\left(\frac{-0.5}{2}, \frac{-0.6}{2}\right)$, or $(-0.25, -0.3)$

44. $(4.65, 0)$

45. We use the midpoint formula:

$$\left(\frac{x_1+x_2}{2}, \frac{y_1+y_2}{2}\right) = \left(\frac{\frac{1}{6}+\left(-\frac{1}{3}\right)}{2}, \frac{-\frac{3}{4}+\frac{5}{6}}{2}\right),$$

or $\left(\frac{-\frac{1}{6}}{2}, \frac{\frac{1}{12}}{2}\right)$, or $\left(-\frac{1}{12}, \frac{1}{24}\right)$

46. $\left(-\frac{27}{80}, \frac{1}{24}\right)$

47. We use the midpoint formula:

$$\left(\frac{x_1+x_2}{2}, \frac{y_1+y_2}{2}\right) = \left(\frac{\sqrt{2}+\sqrt{3}}{2}, \frac{-1+4}{2}\right), \text{ or}$$

$\left(\frac{\sqrt{2}+\sqrt{3}}{2}, \frac{3}{2}\right)$

48. $\left(\frac{5}{2}, \frac{7\sqrt{3}}{2}\right)$

49.
$$
\begin{aligned}
(x-h)^2+(y-k)^2 &= r^2 && \text{Standard form}\\
(x-0)^2+(y-0)^2 &= 6^2 && \text{Substituting}\\
x^2+y^2 &= 36 && \text{Simplifying}
\end{aligned}
$$

50. $x^2+y^2=25$

51.
$$
\begin{aligned}
(x-h)^2+(y-k)^2 &= r^2 && \text{Standard form}\\
(x-7)^2+(y-3)^2 &= (\sqrt{5})^2 && \text{Substituting}\\
(x-7)^2+(y-3)^2 &= 5
\end{aligned}
$$

52. $(x-5)^2+(y-6)^2=2$

53.
$$
\begin{aligned}
(x-h)^2+(y-k)^2 &= r^2 && \text{Standard form}\\
[x-(-4)]^2+(y-3)^2 &= (4\sqrt{3})^2 && \text{Substituting}\\
(x+4)^2+(y-3)^2 &= 48\\
&[(4\sqrt{3})^2 = 16\cdot 3 = 48]
\end{aligned}
$$

54. $(x+2)^2+(y-7)^2=20$

55.
$$
\begin{aligned}
(x-h)^2+(y-k)^2 &= r^2\\
[x-(-7)]^2+[y-(-2)]^2 &= (5\sqrt{2})^2\\
(x+7)^2+(y+2)^2 &= 50
\end{aligned}
$$

56. $(x+5)^2+(y+8)^2=18$

57. Since the center is $(0,0)$, we have

$(x-0)^2+(y-0)^2=r^2$ or $x^2+y^2=r^2$

The circle passes through $(-3,4)$. We find r^2 by substituting -3 for x and 4 for y.

$$
\begin{aligned}
(-3)^2+4^2 &= r^2\\
9+16 &= r^2\\
25 &= r^2
\end{aligned}
$$

Then $x^2+y^2=25$ is an equation of the circle.

58. $(x-3)^2+(y+2)^2=64$

59. Since the center is $(-4,1)$, we have

$[x-(-4)]^2+(y-1)^2=r^2$, or

$(x+4)^2+(y-1)^2=r^2$.

The circle passes through $(-2,5)$. We find r^2 by substituting -2 for x and 5 for y.

$$
\begin{aligned}
(-2+4)^2+(5-1)^2 &= r^2\\
4+16 &= r^2\\
20 &= r^2
\end{aligned}
$$

Then $(x+4)^2+(y-1)^2=20$ is an equation of the circle.

60. $(x+1)^2+(y+3)^2=34$

61. We write standard form.

$(x-0)^2+(y-0)^2=7^2$

The center is $(0,0)$, and the radius is 7.

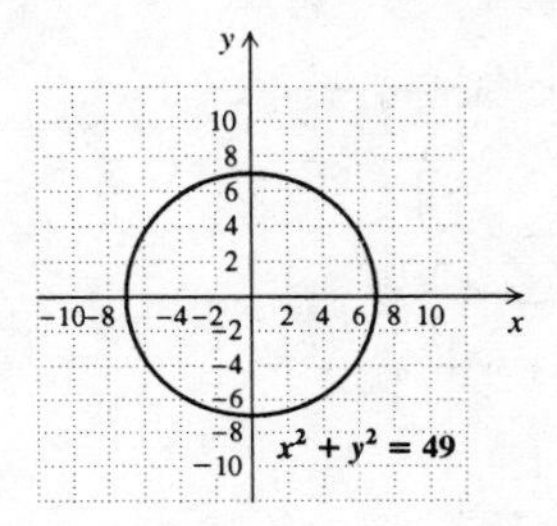

62. Center: $(0,0)$; radius: 6

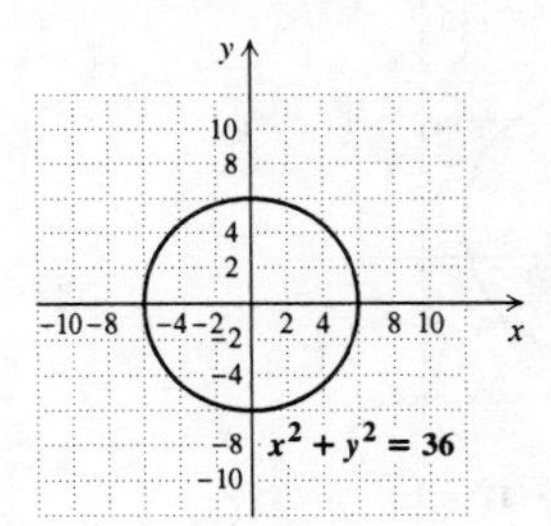

63.
$$
\begin{aligned}
(x+1)^2+(y+3)^2 &= 4\\
[x-(-1)]^2+[y-(-3)]^2 &= 2^2 && \text{Standard form}
\end{aligned}
$$

The center is $(-1,-3)$, and the radius is 2.

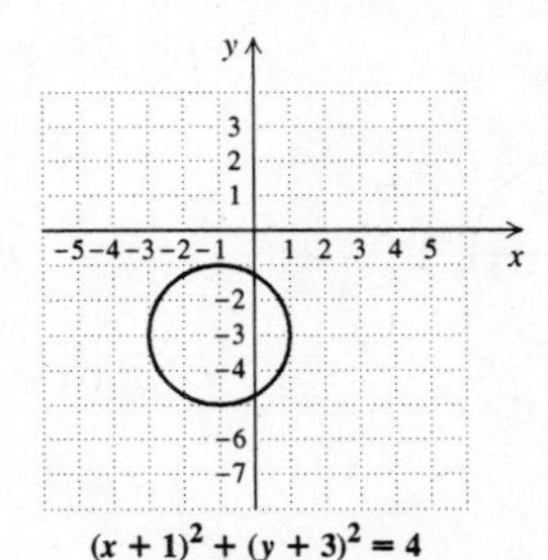

64. Center: $(2, -3)$

Radius: 1

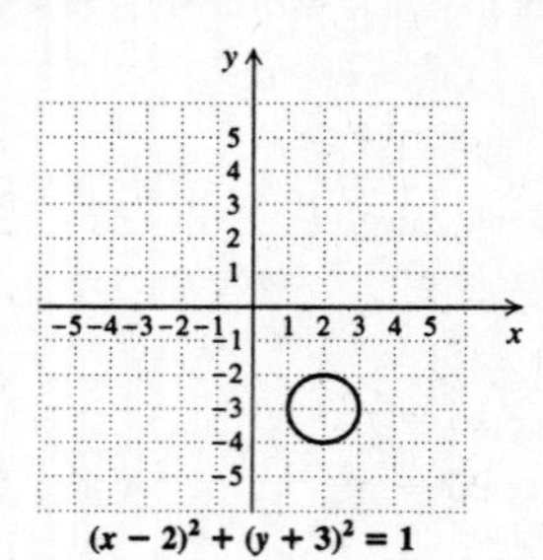

65.
$$(x-4)^2+(y+3)^2 = 10$$
$$(x-4)^2+[y-(-3)]^2 = (\sqrt{10})^2$$

The center is $(4, -3)$, and the radius is $\sqrt{10}$.

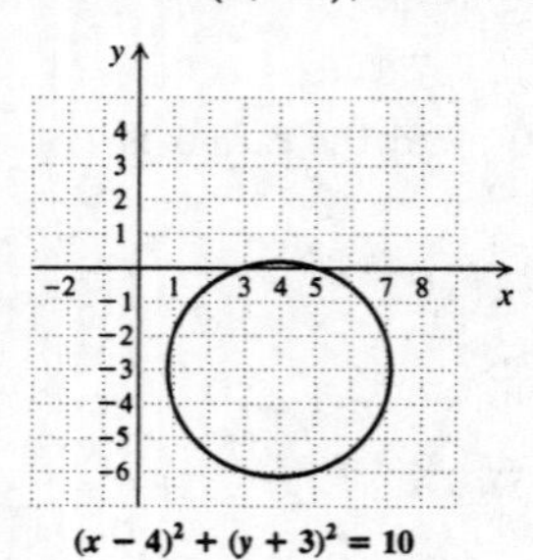

66. Center: $(-5, 1)$

Radius: $\sqrt{15}$

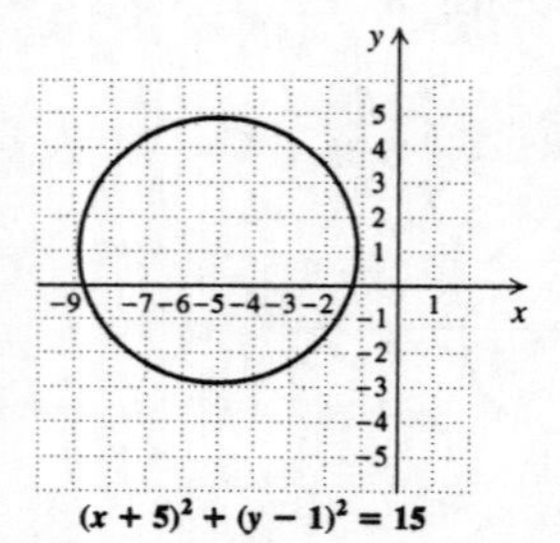

67.
$$x^2+y^2 = 7$$
$$(x-0)^2+(y-0)^2 = (\sqrt{7})^2 \quad \text{Standard form}$$

The center is $(0, 0)$, and the radius is $\sqrt{7}$.

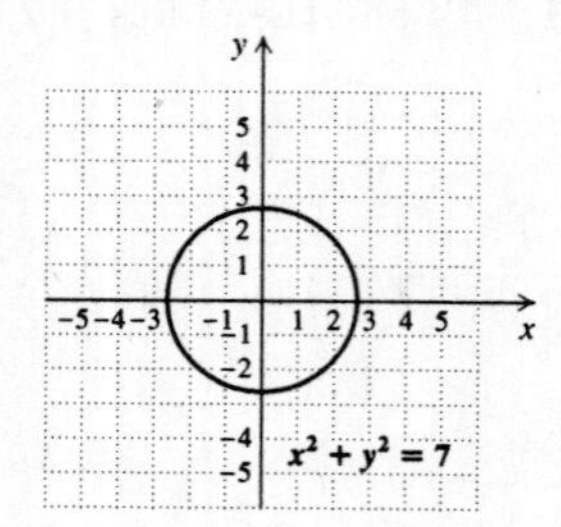

68. Center: $(0, 0)$

Radius: $\sqrt{8}$, or $2\sqrt{2}$

69.
$$(x-5)^2+y^2 = \frac{1}{4}$$
$$(x-5)^2+(y-0)^2 = \left(\frac{1}{2}\right)^2 \quad \text{Standard form}$$

The center is $(5, 0)$, and the radius is $\frac{1}{2}$.

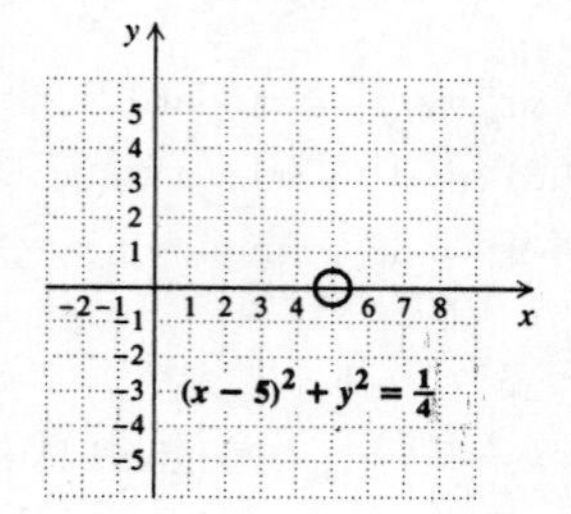

70. Center: $(0, 1)$

Radius: $\frac{1}{5}$

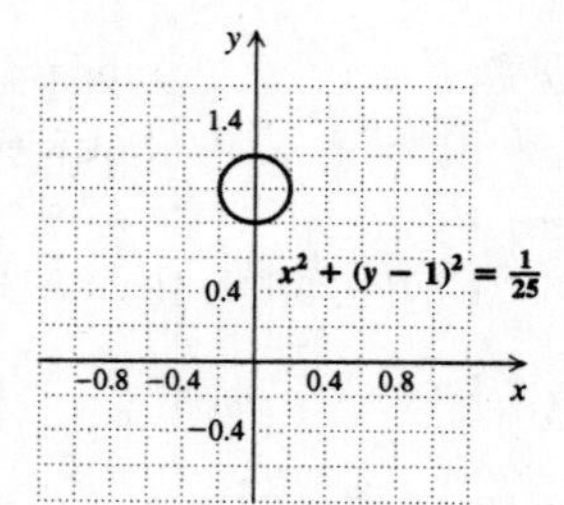

71.
$$x^2+y^2+8x-6y-15 = 0$$
$$x^2+8x+y^2-6y = 15$$
$$(x^2+8x+16)+(y^2-6y+9) = 15+16+9$$

Completing the square twice

$$(x+4)^2+(y-3)^2 = 40$$
$$[x-(-4)]^2+(y-3)^2 = (\sqrt{40})^2$$

Standard form

The center is $(-4, 3)$, and the radius is $\sqrt{40}$, or $2\sqrt{10}$.

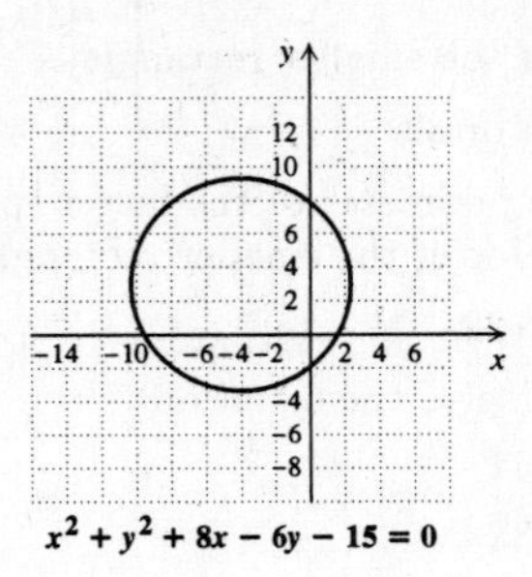

$x^2 + y^2 + 8x - 6y - 15 = 0$

72. Center: $(-3, 2)$

Radius: $\sqrt{28}$, or $2\sqrt{7}$

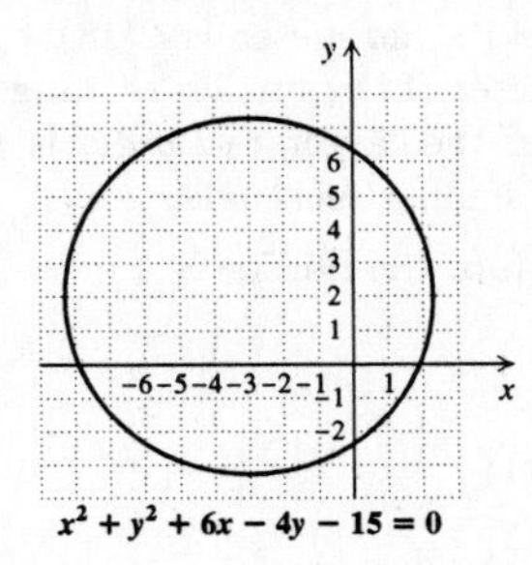

$x^2 + y^2 + 6x - 4y - 15 = 0$

73.

$$\begin{aligned}
x^2 + y^2 - 8x + 2y + 13 &= 0 \\
x^2 - 8x + y^2 + 2y &= -13 \\
(x^2 - 8x + 16) + (y^2 + 2y + 1) &= -13 + 16 + 1 \\
&\quad \text{Completing the square twice} \\
(x-4)^2 + (y+1)^2 &= 4 \\
(x-4)^2 + [y-(-1)]^2 &= 2^2 \\
&\quad \text{Standard form}
\end{aligned}$$

The center is $(4, -1)$, and the radius is 2.

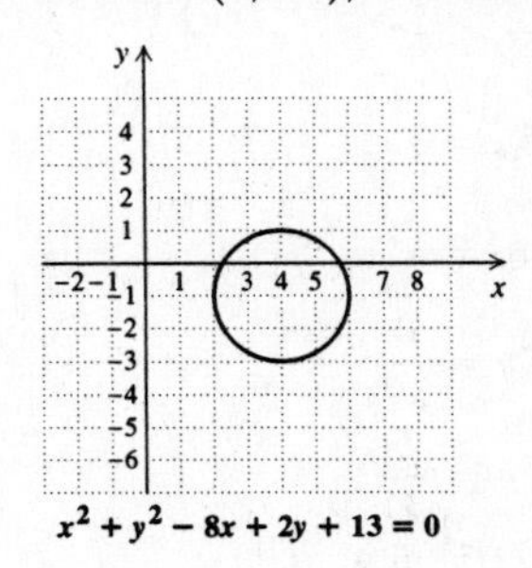

$x^2 + y^2 - 8x + 2y + 13 = 0$

74. Center: $(-3, -2)$

Radius: 1

$x^2 + y^2 + 6x + 4y + 12 = 0$

75.

$$\begin{aligned}
x^2 + y^2 + 10y - 75 &= 0 \\
x^2 + y^2 + 10y &= 75 \\
x^2 + (y^2 + 10y + 25) &= 75 + 25 \\
(x-0)^2 + (y+5)^2 &= 100 \\
(x-0)^2 + [y-(-5)]^2 &= 10^2
\end{aligned}$$

The center is $(0, -5)$, and the radius is 10.

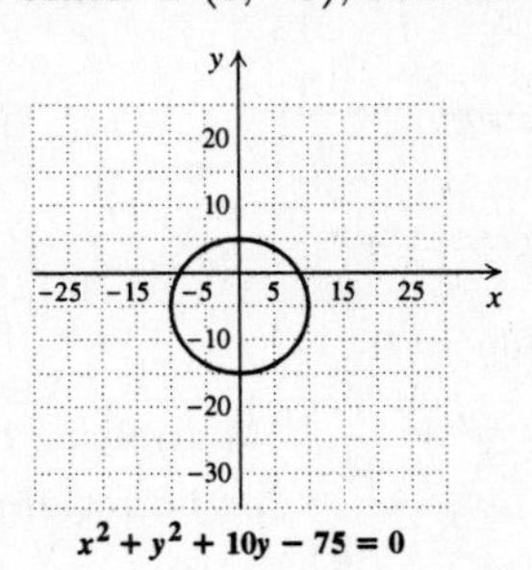

$x^2 + y^2 + 10y - 75 = 0$

76. Center: $(4, 0)$

Radius: 10

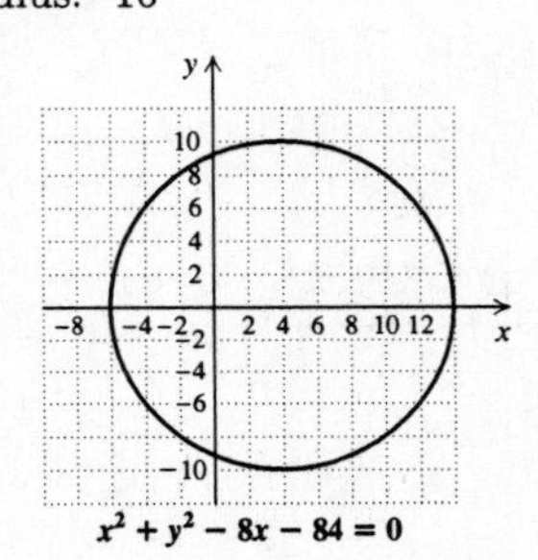

$x^2 + y^2 - 8x - 84 = 0$

77.

$$\begin{aligned}
x^2 + y^2 + 7x - 3y - 10 &= 0 \\
x^2 + 7x + y^2 - 3y &= 10 \\
\left(x^2 + 7x + \frac{49}{4}\right) + \left(y^2 - 3y + \frac{9}{4}\right) &= 10 + \frac{49}{4} + \frac{9}{4} \\
\left(x + \frac{7}{2}\right)^2 + \left(y - \frac{3}{2}\right)^2 &= \frac{98}{4} \\
\left[x - \left(-\frac{7}{2}\right)\right]^2 + \left(y - \frac{3}{2}\right)^2 &= \left(\sqrt{\frac{98}{4}}\right)^2
\end{aligned}$$

The center is $\left(-\frac{7}{2}, \frac{3}{2}\right)$, and the radius is $\sqrt{\frac{98}{4}}$, or $\frac{\sqrt{98}}{2}$, or $\frac{7\sqrt{2}}{2}$.

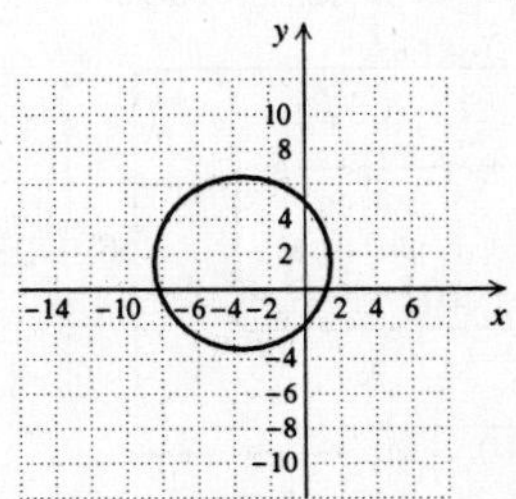

$x^2 + y^2 + 7x - 3y - 10 = 0$

78. Center: $\left(\frac{21}{2}, \frac{33}{2}\right)$; radius: $\frac{\sqrt{1462}}{2}$

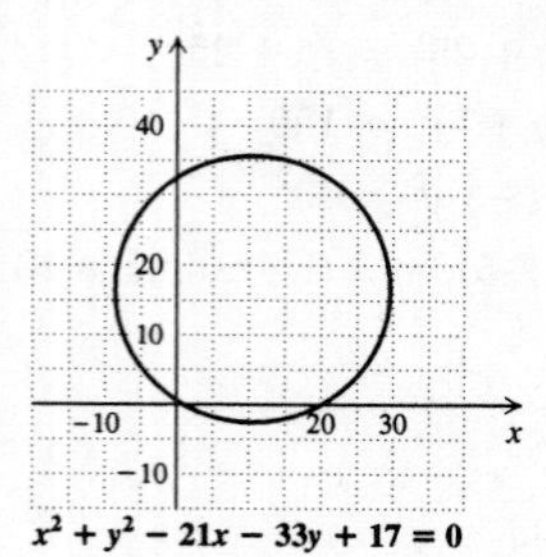

79.

$$36x^2 + 36y^2 = 1$$

$$x^2 + y^2 = \frac{1}{36} \quad \text{Multiplying by } \frac{1}{36} \text{ on both sides}$$

$$(x-0)^2 + (y-0)^2 = \left(\frac{1}{6}\right)^2$$

The center is $(0, 0)$, and the radius is $\frac{1}{6}$.

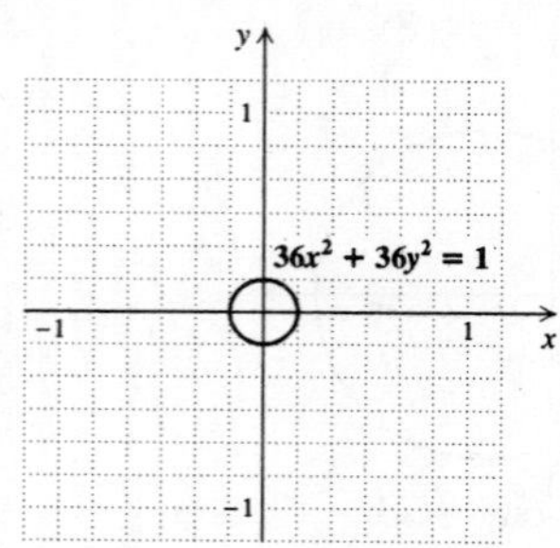

80. Center: $(0, 0)$; radius: $\frac{1}{2}$

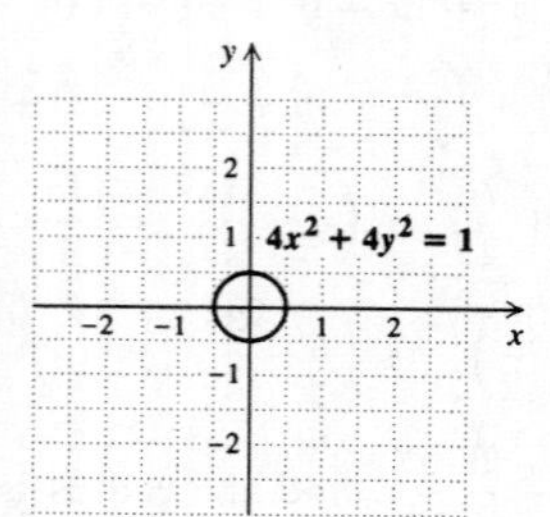

81. ***Familiarize.*** We make a drawing and label it. Let x represent the width of the border.

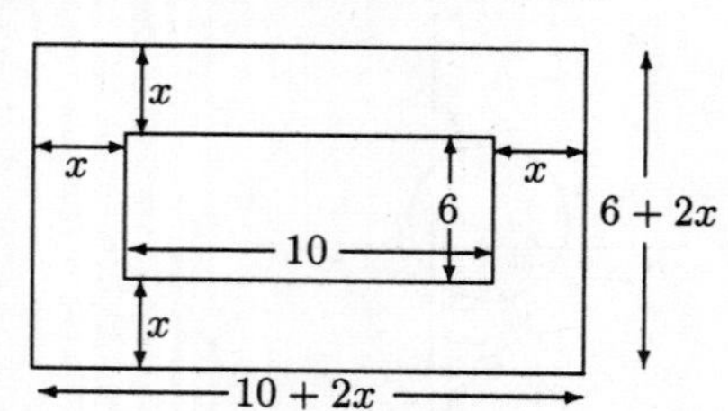

The perimeter of the larger rectangle is

$$2(10 + 2x) + 2(6 + 2x), \text{ or } 8x + 32.$$

The perimeter of the smaller rectangle is

$$2(10) + 2(6), \text{ or } 32.$$

Translate. The perimeter of the larger rectangle is twice the perimeter of the smaller rectangle.

$$8x + 32 = 2 \cdot 32$$

Carry out. We solve the equation.

$$8x + 32 = 64$$
$$8x = 32$$
$$x = 4$$

Check. If the width of the border is 4 in., then the length and width of the larger rectangle are 18 in. and 14 in. Thus its perimeter is $2(18) + 2(14)$, or 64 in. The perimeter of the smaller rectangle is 32 in. The perimeter of the larger rectangle is twice the perimeter of the smaller rectangle.

State. The width of the border is 4 in.

82. 2640 mi

83. $3x - 8y = 5$, (1)

$2x + 6y = 5$ (2)

Multiply Equation (1) by 3, multiply Equation (2) by 4, and add.

$$\begin{aligned} 9x - 24y &= 15 \\ 8x + 24y &= 20 \\ \hline 17x &= 35 \\ x &= \frac{35}{17} \end{aligned}$$

Now substitute $\frac{35}{17}$ for x in one of the original equations and solve for y. We use Equation (2).

$$2x + 6y = 5$$
$$2\left(\frac{35}{17}\right) + 6y = 5$$
$$\frac{70}{17} + 6y = 5$$
$$6y = \frac{15}{17}$$
$$y = \frac{5}{34}$$

The solution is $\left(\frac{35}{17}, \frac{5}{34}\right)$.

84. $\left(0, -\frac{9}{5}\right)$

85.

86.

87.

88.

89. We make a drawing of the circle with center $(3, -5)$ and tangent to the y-axis.

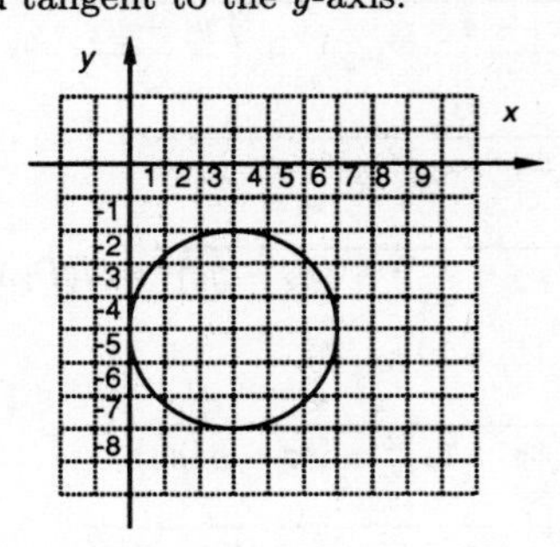

We see that the circle touches the y-axis at $(0, -5)$. Hence the radius is the distance between $(0, -5)$ and $(3, -5)$, or $\sqrt{(3-0)^2 + [-5-(-5)]^2}$, or 3. Now we write the equation of the circle.

$$(x-h)^2 + (y-k)^2 = r^2$$
$$(x-3)^2 + [y-(-5)]^2 = 3^2$$
$$(x-3)^2 + (y+5)^2 = 9$$

90. $(x+7)^2 + (y+4)^2 = 16$

91. First we use the midpoint formula to find the center:

$$\left(\frac{7+(-1)}{2}, \frac{3+(-3)}{2}\right), \text{ or } \left(\frac{6}{2}, \frac{0}{2}\right), \text{ or } (3, 0)$$

The length of the radius is the distance between the center $(3, 0)$ and either endpoint of a diameter. We will use endpoint $(7, 3)$ in the distance formula:

$r = \sqrt{(7-3)^2 + (3-0)^2} = \sqrt{25} = 5$

Now we write the equation of the circle:

$$(x-h)^2 + (y-k)^2 = r^2$$
$$(x-3)^2 + (y-0)^2 = 5^2$$
$$(x-3)^2 + y^2 = 25$$

92. $(x+3)^2 + (y-5)^2 = 16$

93. Let $(0, y)$ be the point on the y-axis that is equidistant from $(2, 10)$ and $(6, 2)$. Then the distance between $(2, 10)$ and $(0, y)$ is the same as the distance between $(6, 2)$ and $(0, y)$.

$$\sqrt{(0-2)^2 + (y-10)^2} = \sqrt{(0-6)^2 + (y-2)^2}$$
$$(-2)^2 + (y-10)^2 = (-6)^2 + (y-2)^2$$

Squaring both sides

$$4 + y^2 - 20y + 100 = 36 + y^2 - 4y + 4$$
$$64 = 16y$$
$$4 = y$$

This number checks. The point is $(0, 4)$.

94. $(-5, 0)$

95. a) Use the fact that the center of the circle $(0, k)$ is equidistant from the points $(-575, 0)$ and $(0, 19.5)$.

$$\sqrt{(-575-0)^2+(0-k)^2} = \sqrt{(0-0)^2+(19.5-k)^2}$$
$$\sqrt{330{,}625 + k^2} = \sqrt{380.25 - 39k + k^2}$$
$$330{,}625 + k^2 = 380.25 - 39k + k^2$$

Squaring both sides

$$330{,}244.75 = -39k$$
$$-8467.8 \approx k$$

Then the center of the circle is about $(0, -8467.8)$.

b) To find the radius we find the distance from the center, $(0, -8467.8)$ to any one of the points $(-575, 0)$, $(0, 19.5)$, or $(575, 0)$. We use $(0, 19.5)$.

$r = \sqrt{(0-0)^2 + [19.5 - (-8467.8)]^2} \approx$ 8487.3 mm

96. 8186.6 mm

97. a) Use the fact that the center of the circle, $(0, k)$ is equidistant from the points $(0, 2.1)$ and $(80, 0)$.

$$\sqrt{(80-0)^2+(0-k)^2} = \sqrt{(0-0)^2+(2.1-k)^2}$$
$$\sqrt{6400 + k^2} = \sqrt{4.41 - 4.2k + k^2}$$
$$6400 + k^2 = 4.41 - 4.2k + k^2$$

Squaring both sides

$$6395.59 = -4.2k$$
$$-1522.8 \approx k$$

Then the center of the circle is about $(0, -1522.8)$.

b) To find the radius we find the distance from the center, $(0, -1522.8)$, to either of the points $(0, 2.1)$ or $(80, 0)$. We use $(0, 2.1)$.

$r = \sqrt{(0-0)^2 + [2.1 - (-1522.8)]^2} \approx$ 1524.9 cm

98. (a) $(0, -3)$; (b) 5 ft

99. Position a coordinate system as shown below so that the center of the top of the bowl is at the origin. Let r represent the radius of the circle.

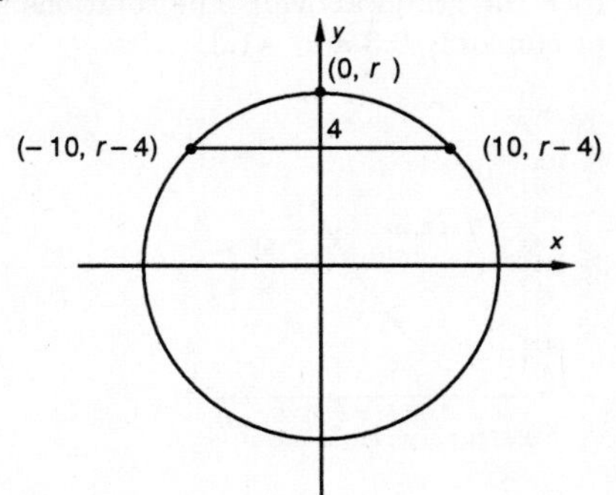

The equation of the circle is $x^2 + y^2 = r^2$, and one point on the circle is $(10, r-4)$. We substitute 10 for x and $r-4$ for y and solve for r.

$$10^2 + (r-4)^2 = r^2$$
$$100 + r^2 - 8r + 16 = r^2$$
$$r^2 - 8r + 116 = r^2$$
$$116 = 8r$$
$$14.5 = r$$

Then the original diameter of the bowl is $2r = 2(14.5)$, or 29 cm.

100. $x^2 + (y - 30.6)^2 = 590.49$

101. First we graph $x = y^2 - y - 6$, $x = 2$, and $x = -3$ on the same set of axes.

$$x = y^2 - y - 6$$
$$x = \left(y^2 - y + \frac{1}{4}\right) - 6 - \frac{1}{4}$$
$$x = \left(y - \frac{1}{2}\right)^2 - \frac{25}{4}$$

The vertex is $\left(-\frac{25}{4}, \frac{1}{2}\right)$.

x	y
$-\frac{25}{4}$	$\frac{1}{2}$
-6	1
-4	2
0	3
-6	0
-4	-1
0	-2

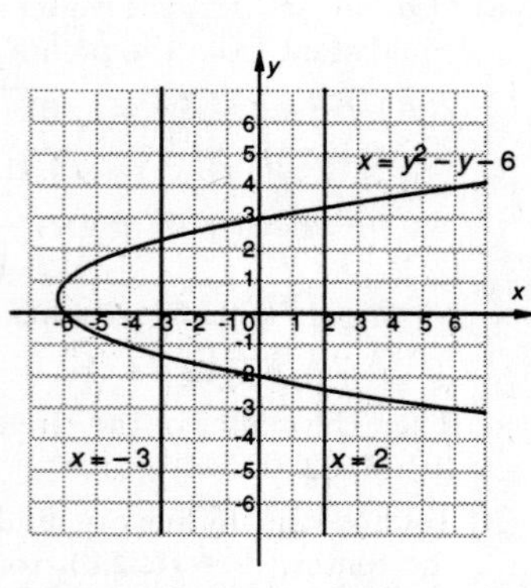

a) Graph $x = 2$ on the same set of axes as $x = y^2 - y - 6$ and approximate the y-coordinates of the points of intersection. (See the graph above.) The solutions are approximately 3.4 and -2.4.

b) Graph $x = -3$ on the same set of axes as $x = y^2 - y - 6$ and approximate the y-coordinates of the points of intersection. (See the graph above.) The solutions are approximately 2.3 and -1.3.

102.

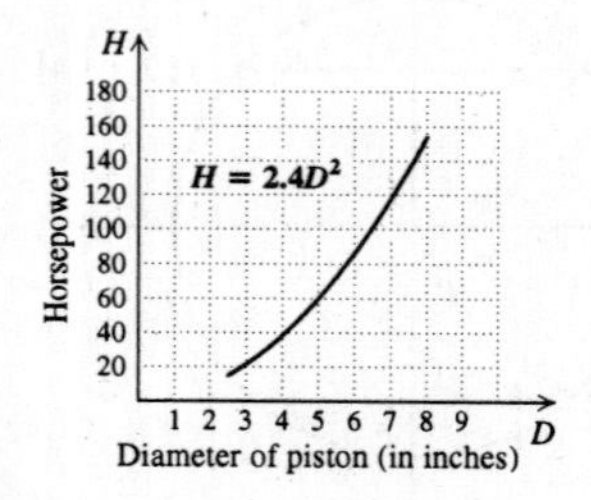

103. Let $P_1 = (x_1, y_1)$, $P_2 = (x_2, y_2)$, and $M = \left(\frac{x_1 + x_2}{2}, \frac{y_1 + y_2}{2}\right)$. Let $d(AB)$ denote the distance from point A to point B.

i) $d(P_1M)$

$$= \sqrt{\left(\frac{x_1 + x_2}{2} - x_1\right)^2 + \left(\frac{y_1 + y_2}{2} - y_1\right)^2}$$
$$= \frac{1}{2}\sqrt{(x_2 - x_1)^2 + (y_2 - y_1)^2};$$

$d(P_2M)$

$$= \sqrt{\left(\frac{x_1 + x_2}{2} - x_2\right)^2 + \left(\frac{y_1 + y_2}{2} - y_2\right)^2}$$
$$= \frac{1}{2}\sqrt{(x_1 - x_2)^2 + (y_1 - y_2)^2}$$
$$= \frac{1}{2}\sqrt{(x_2 - x_1)^2 + (y_2 - y_1)^2} = d(P_1M).$$

ii) $d(P_1M) + d(P_2M)$

$$= \frac{1}{2}\sqrt{(x_2 - x_1)^2 + (y_2 - y_1)^2} + \frac{1}{2}\sqrt{(x_2 - x_1)^2 + (y_2 - y_1)^2}$$
$$= \sqrt{(x_2 - x_1)^2 + (y_2 - y_1)^2}$$
$$= d(P_1P_2).$$

104. (a) $y = -1 \pm \sqrt{-x^2 + 6x + 7}$;

b)

105.

106.

Exercise Set 10.2

1. $\frac{x^2}{4} + \frac{y^2}{1} = 1$

$$\frac{x^2}{2^2} + \frac{y^2}{1^2} = 1$$

The x-intercepts are $(2, 0)$ and $(-2, 0)$, and the y-intercepts are $(0, 1)$ and $(0, -1)$. We plot these points and connect them with an oval-shaped curve.

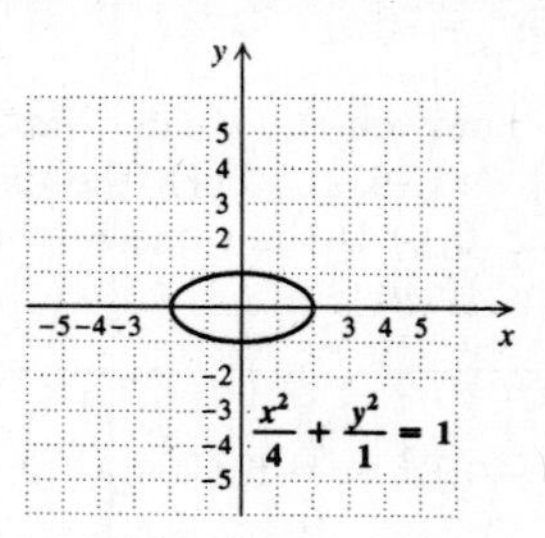

2.

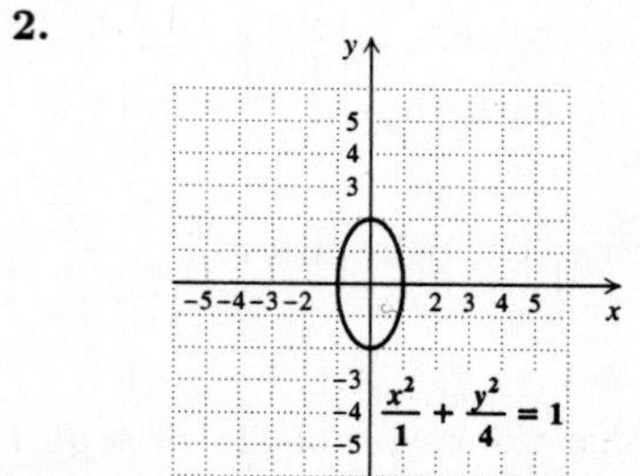

3. $\dfrac{x^2}{9}+\dfrac{y^2}{25}=1$

$\dfrac{x^2}{3^2}+\dfrac{y^2}{5^2}=1$

The x-intercepts are $(3,0)$ and $(-3,0)$, and the y-intercepts are $(0,5)$ and $(0,-5)$. We plot these points and connect them with an oval-shaped curve.

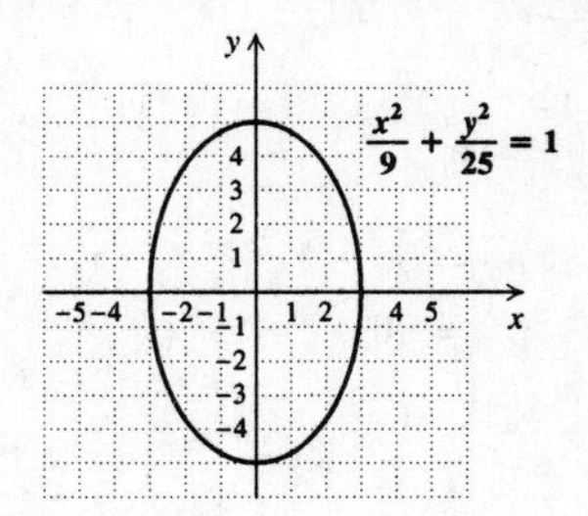

4.

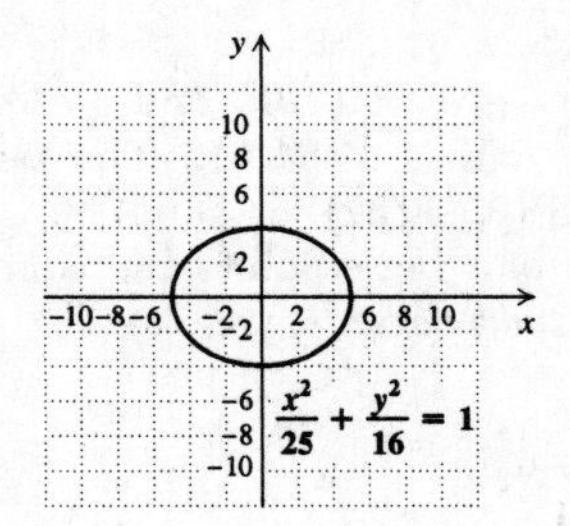

5. $4x^2+9y^2=36$

$\dfrac{1}{36}(4x^2+9y^2)=\dfrac{1}{36}(36)$ Multiplying by $\dfrac{1}{36}$

$\dfrac{x^2}{9}+\dfrac{y^2}{4}=1$

$\dfrac{x^2}{3^2}+\dfrac{y^2}{2^2}=1$

The x-intercepts are $(-3,0)$ and $(3,0)$, and the y-intercepts are $(0,-2)$ and $(0,2)$. We plot these points and connect them with an oval-shaped curve.

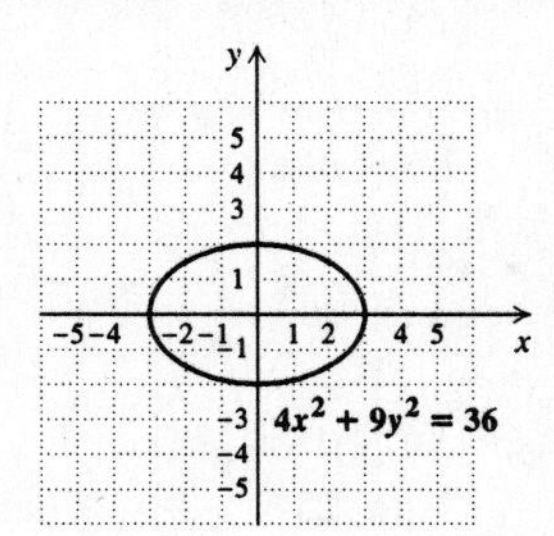

6.

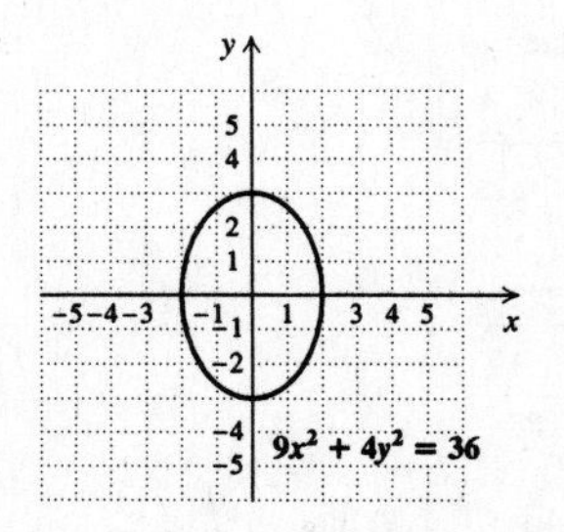

7. $16x^2+9y^2=144$

$\dfrac{x^2}{9}+\dfrac{y^2}{16}=1$ Multiplying by $\dfrac{1}{144}$

$\dfrac{x^2}{3^2}+\dfrac{y^2}{4^2}=1$

The x-intercepts are $(3,0)$ and $(-3,0)$, and the y-intercepts are $(0,4)$ and $(0,-4)$. We plot these points and connect them with an oval-shaped curve.

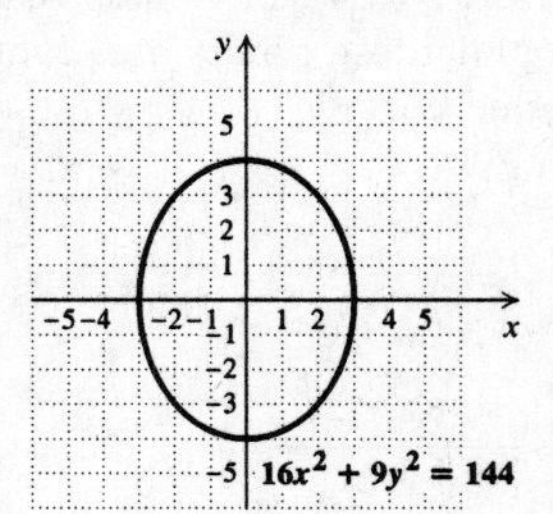

8.

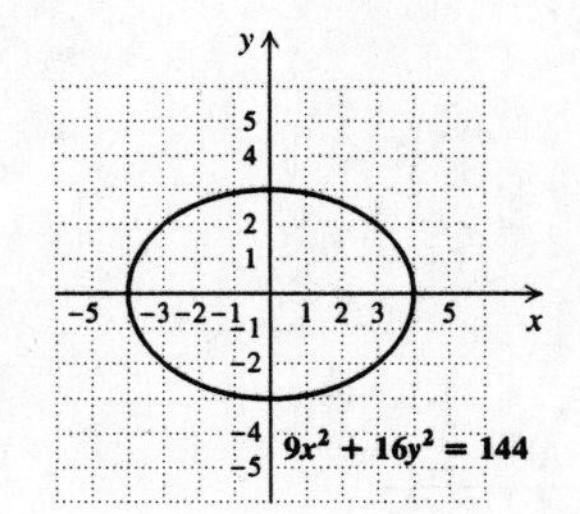

9. $2x^2+3y^2=6$

$\dfrac{x^2}{3}+\dfrac{y^2}{2}=1$ Multiplying by $\dfrac{1}{6}$

$\dfrac{x^2}{(\sqrt{3})^2}+\dfrac{y^2}{(\sqrt{2})^2}=1$

The x-intercepts are $(\sqrt{3},0)$ and $(-\sqrt{3},0)$, and the y-intercepts are $(0,\sqrt{2})$ and $(0,-\sqrt{2})$. We plot these points and connect them with an oval-shaped curve.

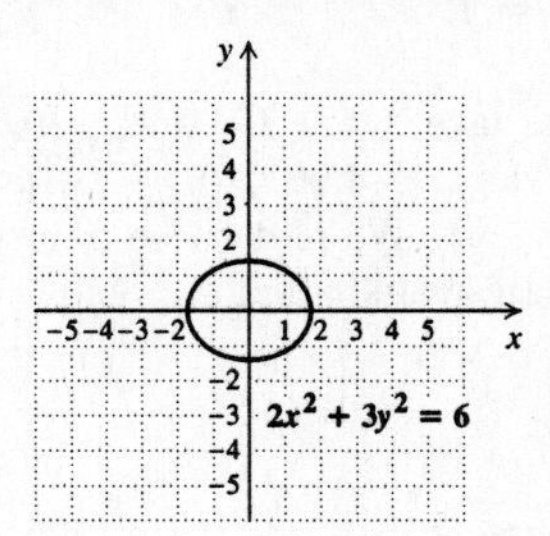

10.

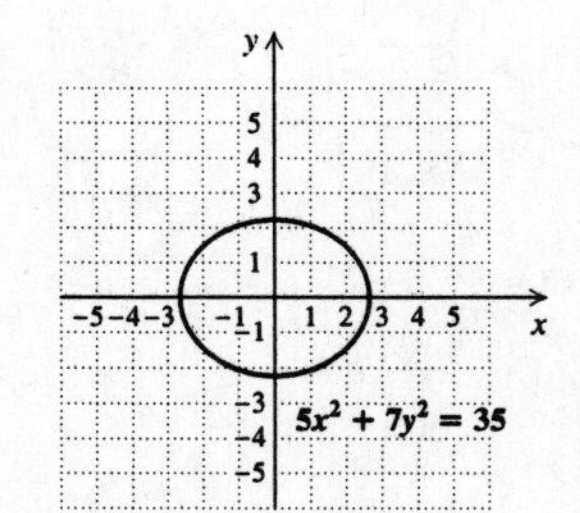

11.
$$12x^2 + 5y^2 = 120$$
$$\frac{x^2}{10} + \frac{y^2}{24} = 1 \quad \text{Multiplying by } \frac{1}{120}$$
$$\frac{x^2}{(\sqrt{10})^2} + \frac{y^2}{(\sqrt{24})^2} = 1$$

The x-intercepts are $(\sqrt{10}, 0)$ and $(-\sqrt{10}, 0)$, or about $(3.162, 0)$ and $(-3.162, 0)$. The y-intercepts are $(0, \sqrt{24})$ and $(0, -\sqrt{24})$, or about $(0, 4.899)$ and $(0, -4.899)$. We plot these points and connect them with an oval-shaped curve.

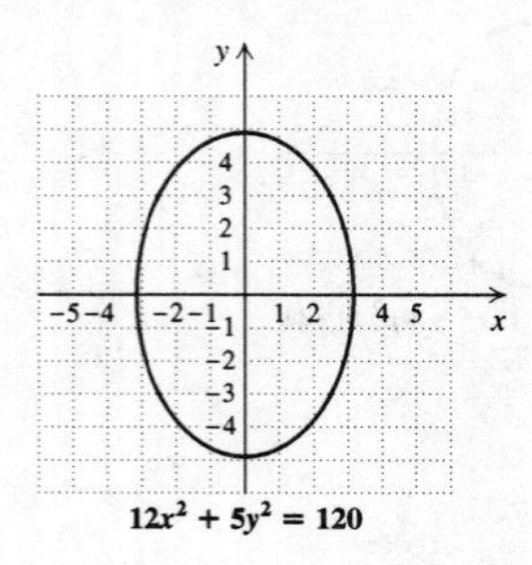

12.

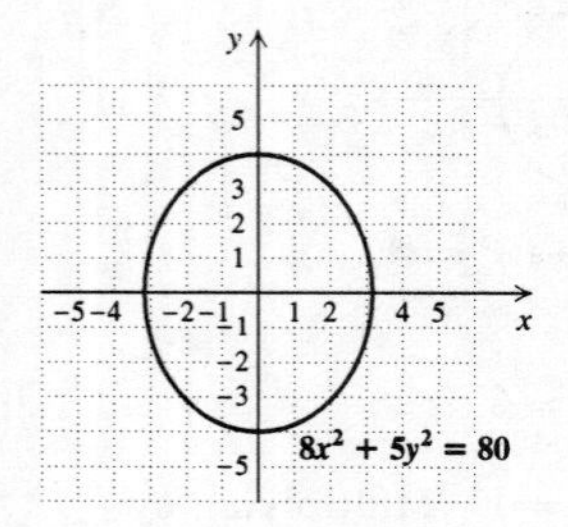

13.
$$3x^2 + 7y^2 - 63 = 0$$
$$3x^2 + 7y^2 = 63$$
$$\frac{x^2}{21} + \frac{y^2}{9} = 1 \quad \text{Multiplying by } \frac{1}{63}$$
$$\frac{x^2}{(\sqrt{21})^2} + \frac{y^2}{3^2} = 1$$

The x-intercepts are $(\sqrt{21}, 0)$ and $(-\sqrt{21}, 0)$, or about $(4.583, 0)$ and $(-4.583, 0)$. The y-intercepts are $(0, 3)$ and $(0, -3)$. We plot these points and connect them with an oval-shaped curve.

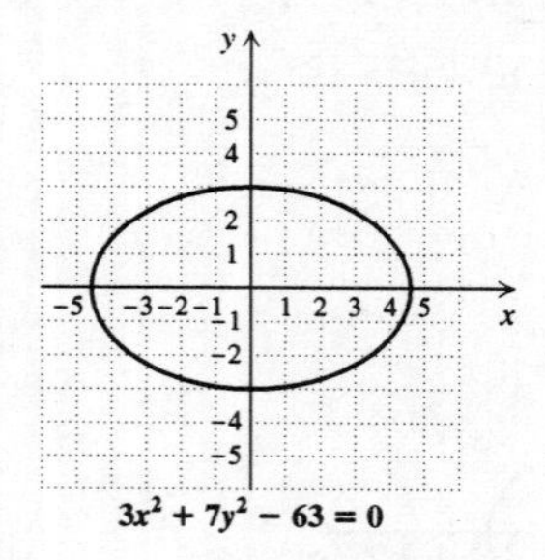

14.

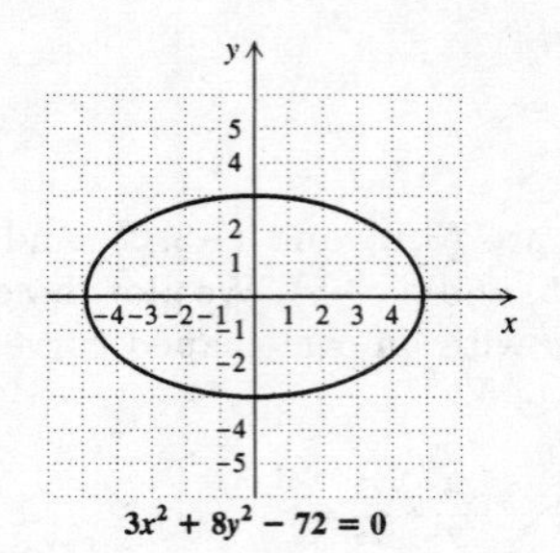

15.
$$8x^2 = 96 - 3y^2$$
$$8x^2 + 3y^2 = 96$$
$$\frac{x^2}{12} + \frac{y^2}{32} = 1$$
$$\frac{x^2}{(\sqrt{12})^2} + \frac{y^2}{(\sqrt{32})^2} = 1$$

The x-intercepts are $(\sqrt{12}, 0)$ and $(-\sqrt{12}, 0)$, or about $(3.464, 0)$ and $(-3.464, 0)$. The y-intercepts are $(0, \sqrt{32})$ and $(0, -\sqrt{32})$, or about $(0, 5.657)$ and $(0, -5.657)$. We plot these points and connect them with an oval-shaped curve.

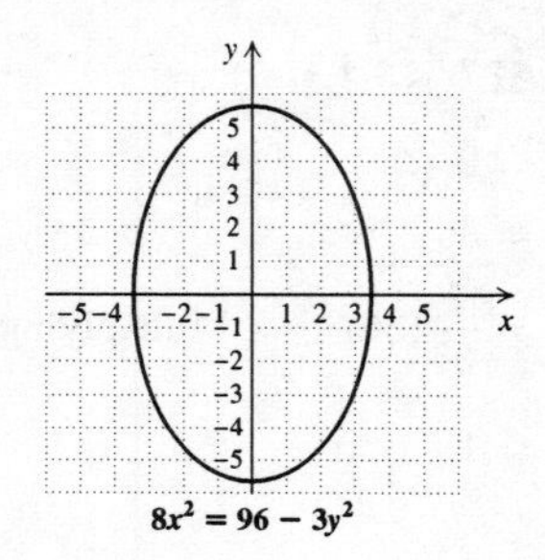

16.

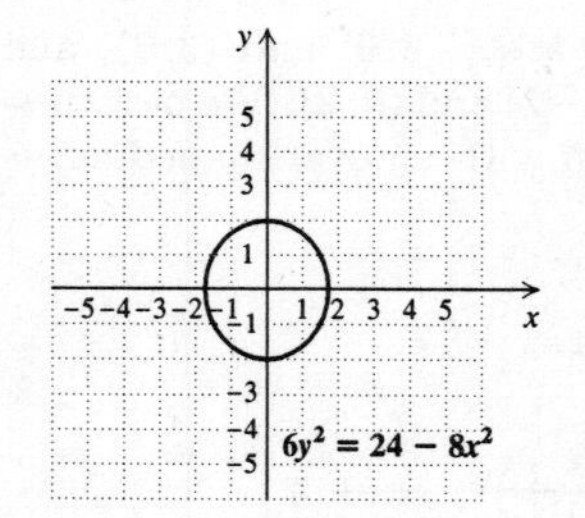

17. $16x^2 + 25y^2 = 1$

Note that $16 = \frac{1}{\frac{1}{16}}$ and $25 = \frac{1}{\frac{1}{25}}$. Thus, we can rewrite the equation:
$$\frac{x^2}{\frac{1}{16}} + \frac{y^2}{\frac{1}{25}} = 1$$
$$\frac{x^2}{\left(\frac{1}{4}\right)^2} + \frac{y^2}{\left(\frac{1}{5}\right)^2} = 1$$

The x-intercepts are $\left(\frac{1}{4}, 0\right)$ and $\left(-\frac{1}{4}, 0\right)$, and the

y-intercepts are $\left(0, \dfrac{1}{5}\right)$ and $\left(0, -\dfrac{1}{5}\right)$. We plot these points and connect them with an oval-shaped curve.

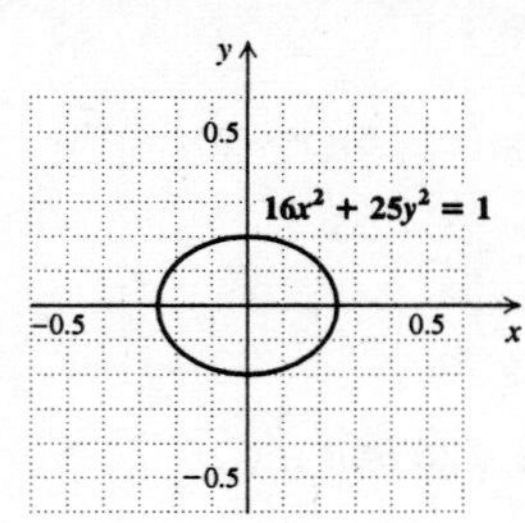

18.

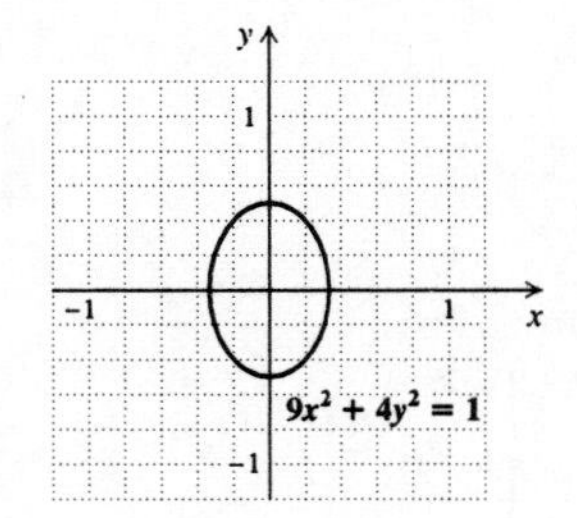

19. $$\frac{(x-2)^2}{9} + \frac{(y-1)^2}{25} = 1$$
$$\frac{(x-2)^2}{3^2} + \frac{(y-1)^2}{5^2} = 1$$

The center of the ellipse is $(2, 1)$. Note that $a = 3$ and $b = 5$. We locate the center and then plot the points $(2+3, 1)$ $(2-3, 1)$, $(2, 1+5)$, and $(2, 1-5)$, or $(5, 1)$, $(-1, 1)$, $(2, 6)$, and $(2, -4)$. Connect these points with an oval-shaped curve.

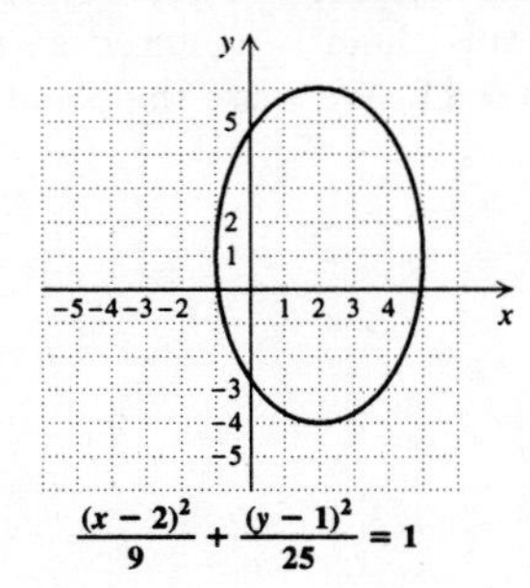

20.

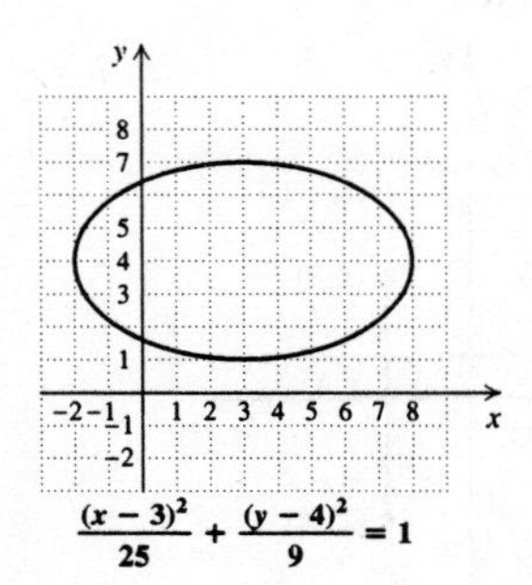

21. $$\frac{(x+4)^2}{16} + \frac{(y-3)^2}{49} = 1$$
$$\frac{(x-(-4))^2}{4^2} + \frac{(y-3)^2}{7^2} = 1$$

The center of the ellipse is $(-4, 3)$. Note that $a = 4$ and $b = 7$. We locate the center and then plot the points $(-4+4, 3)$, $(-4-4, 3)$, $(-4, 3+7)$, and $(-4, 3-7)$, or $(0, 3)$, $(-8, 3)$, $(-4, 10)$, and $(-4, -4)$. Connect these points with an oval-shaped curve.

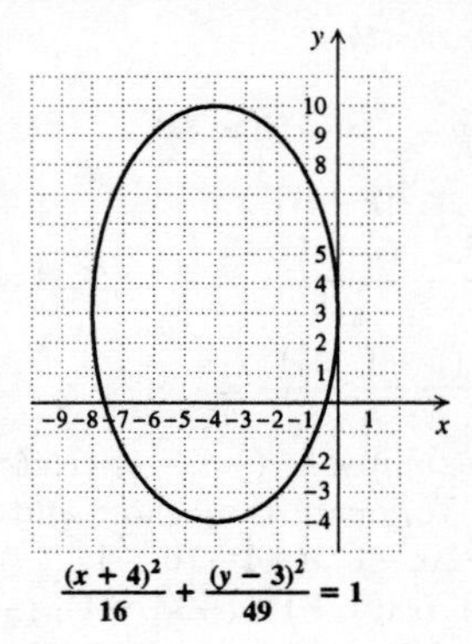

22.

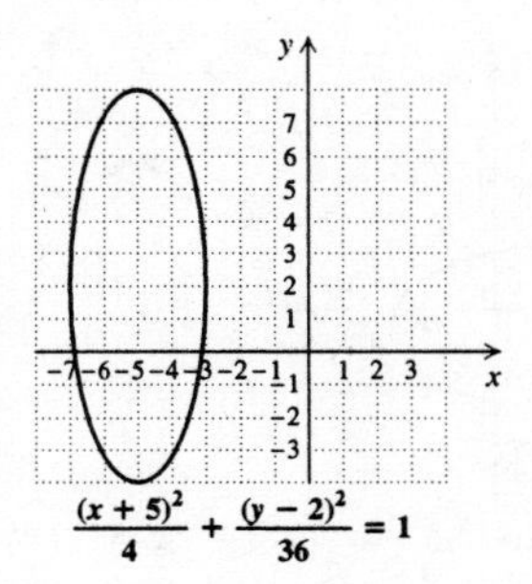

23. $$12(x-1)^2 + 3(y+4)^2 = 48$$
$$\frac{(x-1)^2}{4} + \frac{(y+4)^2}{16} = 1$$
$$\frac{(x-1)^2}{2^2} + \frac{(y-(-4))^2}{4^2} = 1$$

The center of the ellipse is $(1, -4)$. Note that $a = 2$ and $b = 4$. We locate the center and then plot the points $(1+2, -4)$, $(1-2, -4)$, $(1, -4+4)$, and $(1, -4-4)$, or $(3, -4)$, $(-1, -4)$, $(1, 0)$, and $(1, -8)$. Connect these points with an oval-shaped curve.

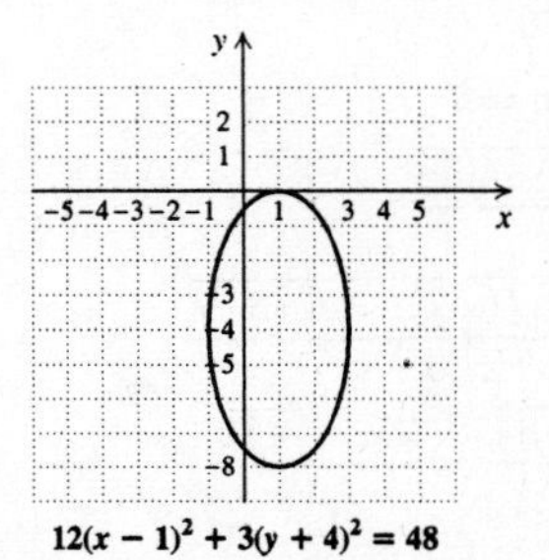

24.

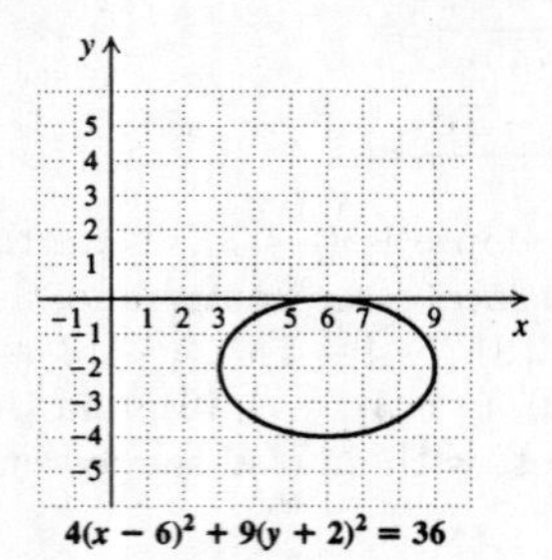

$4(x-6)^2+9(y+2)^2=36$

25.
$$\begin{aligned}(x+4)^2+4(y+1)^2-10&=90\\(x+4)^2+4(y+1)^2&=100\\\frac{(x+4)^2}{100}+\frac{(y+1)^2}{25}&=1\\\frac{(x-(-4))^2}{10^2}+\frac{(y-(-1))^2}{5^2}&=1\end{aligned}$$

The center of the ellipse is $(-4,-1)$. Note that $a=10$ and $b=5$. We locate the center and then plot the points $(-4+10,-1)$, $(-4-10,-1)$, $(-4,-1+5)$, and $(-4,-1-5)$, or $(6,-1)$, $(-14,-1)$, $(-4,4)$, and $(-4,-6)$. Connect these points with an oval-shaped curve.

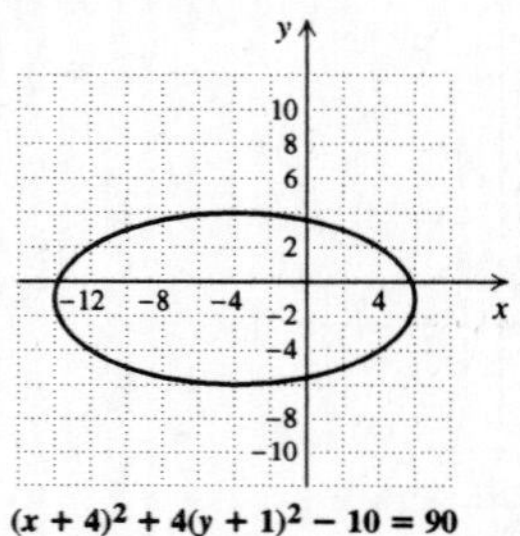

$(x+4)^2+4(y+1)^2-10=90$

26.

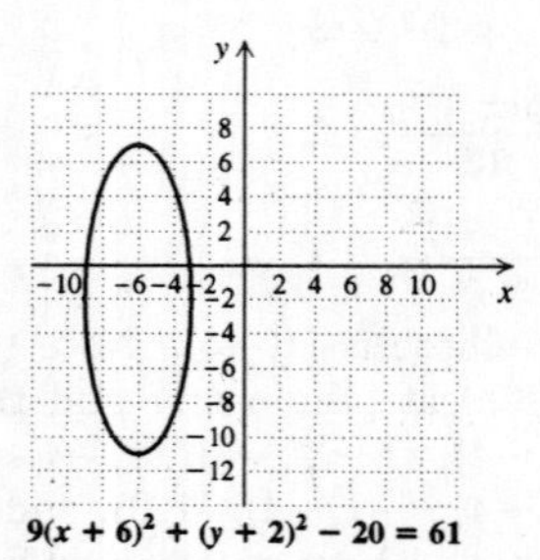

$9(x+6)^2+(y+2)^2-20=61$

27. $3x^2-2x+7=0$

$a=3$, $b=-2$, $c=7$

$$x=\frac{-b\pm\sqrt{b^2-4ac}}{2a}$$
$$x=\frac{-(-2)\pm\sqrt{(-2)^2-4\cdot3\cdot7}}{2\cdot3}$$
$$x=\frac{2\pm\sqrt{4-84}}{6}=\frac{2\pm\sqrt{-80}}{6}$$
$$x=\frac{2\pm4i\sqrt{5}}{6}=\frac{1\pm2i\sqrt{5}}{3}$$

The solutions are $\frac{1+2i\sqrt{5}}{3}$ and $\frac{1-2i\sqrt{5}}{3}$.

28. $\pm\frac{2}{3}i$

29.
$$\begin{aligned}\frac{1}{x-5}&=x+5\\(x-5)\cdot\frac{1}{x-5}&=(x-5)(x+5)\\1&=x^2-25\\26&=x^2\\\pm\sqrt{26}&=x\end{aligned}$$

The solutions are $\sqrt{26}$ and $-\sqrt{26}$.

30. $\frac{3}{2}$

31. ◈

32. ◈

33. Plot the given points.

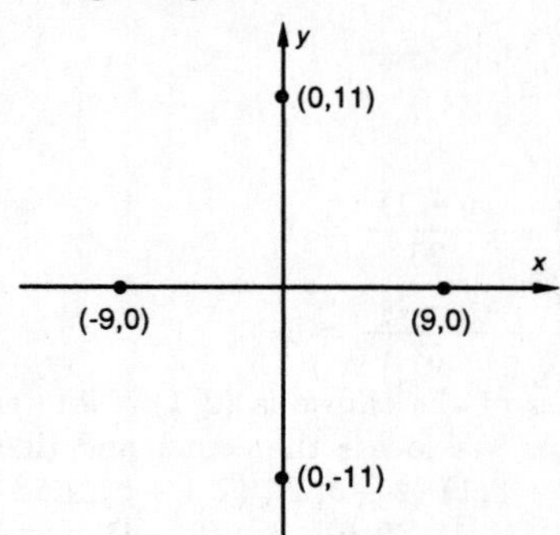

From the location of these points, we see that the ellipse that contains them is centered at the origin with $a=9$ and $b=11$. We write the equation of the ellipse:

$$\frac{x^2}{9^2}+\frac{y^2}{11^2}=1$$
$$\frac{x^2}{81}+\frac{y^2}{121}=1$$

34. $\frac{x^2}{49}+\frac{y^2}{25}=1$

35. Plot the given points.

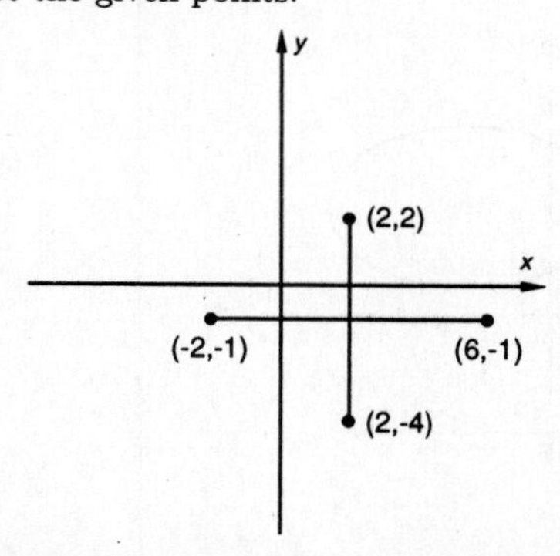

The midpoint of the segment from $(-2,-1)$ to $(6,-1)$ is $\left(\frac{-2+6}{2},\frac{-1-1}{2}\right)$, or $(2,-1)$. The midpoint of the segment from $(2, -4)$ to $(2, 2)$ is $\left(\frac{2+2}{2}, \frac{-4+2}{2}\right)$, or $(2, -1)$. Thus, we can conclude that $(2, -1)$ is the center of the ellipse. The distance from $(-2, -1)$ to $(2, -1)$ is $\sqrt{[2-(-2)]^2 + [-1-(-1)]^2} = \sqrt{16} = 4$, so $a = 4$. The distance from $(2, 2)$ to $(2, -1)$ is $\sqrt{(2-2)^2 + (-1-2)^2} = \sqrt{9} = 3$, so $b = 3$. We write the equation of the ellipse.

$$\frac{(x-2)^2}{4^2} + \frac{(y-(-1))^2}{3^2} = 1$$

$$\frac{(x-2)^2}{16} + \frac{(y+1)^2}{9} = 1$$

36. $\dfrac{(x+1)^2}{25} + \dfrac{(y-3)^2}{16} = 1$

37. We make a drawing.

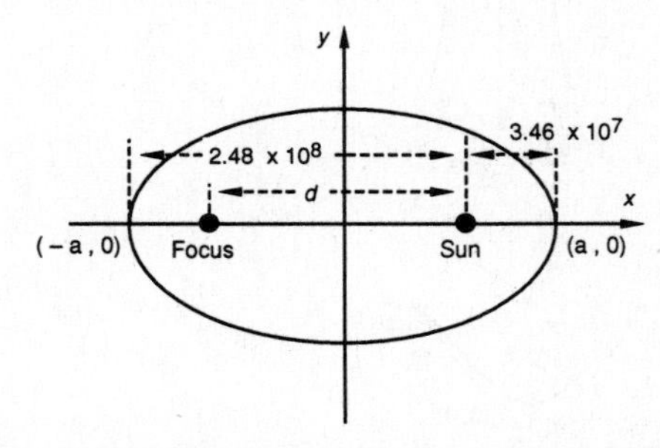

The distance between vertex $(a, 0)$ and the sun is the same as the distance between vertex $(-a, 0)$ and the other focus. Then

$$d = 2.48 \times 10^8 - 3.46 \times 10^7 =$$
$$2.48 \times 10^8 - 0.346 \times 10^8 = 2.134 \times 10^8 \text{ mi.}$$

38. a) Let $F_1 = (-c, 0)$ and $F_2 = (c, 0)$. Then the sum of the distances from the foci to P is $2a$. By the distance formula,

$$\sqrt{(x+c)^2 + y^2} + \sqrt{(x-c)^2 + y^2} = 2a, \text{ or}$$
$$\sqrt{(x+c)^2 + y^2} = 2a - \sqrt{(x-c)^2 + y^2}.$$

Squaring, we get

$$(x+c)^2 + y^2 = 4a^2 - 4a\sqrt{(x-c)^2 + y^2} + (x-c)^2 + y^2,$$

or $x^2 + 2cx + c^2 + y^2$

$$= 4a^2 - 4a\sqrt{(x-c)^2 + y^2} + x^2 - 2cx + c^2 + y^2.$$

Thus

$$-4a^2 + 4cx = -4a\sqrt{(x-c)^2 + y^2}$$
$$a^2 - cx = a\sqrt{(x-c)^2 + y^2}.$$

Squaring again, we get

$$a^4 - 2a^2cx + c^2x^2 = a^2(x^2 - 2cx + c^2 + y^2)$$
$$a^4 - 2a^2cx + c^2x^2 = a^2x^2 - 2a^2cx + a^2c^2 + a^2y^2,$$

or

$$x^2(a^2 - c^2) + a^2y^2 = a^2(a^2 - c^2)$$
$$\frac{x^2}{a^2} + \frac{y^2}{a^2 - c^2} = 1.$$

b) When P is at $(0, b)$, it follows that $b^2 = a^2 - c^2$. Substituting, we have

$$\frac{x^2}{a^2} + \frac{y^2}{b^2} = 1.$$

39. Position the ellipse on a coordinate system as shown below.

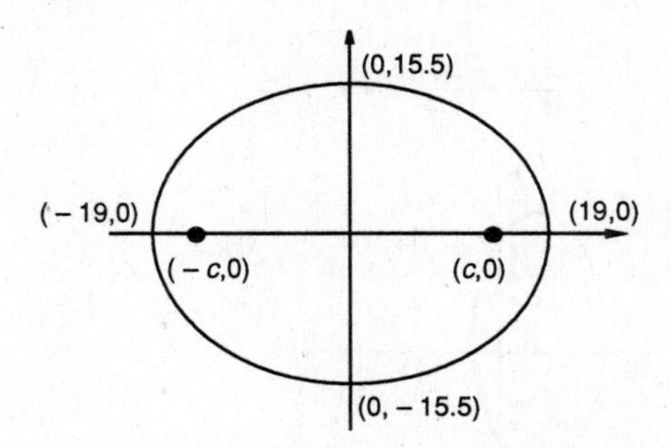

In order to best use the room's acoustics, the President and the advisor should be seated at the foci of the ellipse, or at $(-c, 0)$ and $(c, 0)$. We use the equation relating the coordinates of the foci and the intercepts to find c:

$$b^2 = a^2 - c^2$$
$$(15.5)^2 = (19)^2 - c^2$$
$$240.25 = 361 - c^2$$
$$c^2 = 120.75$$
$$c \approx 11$$

We make a sketch.

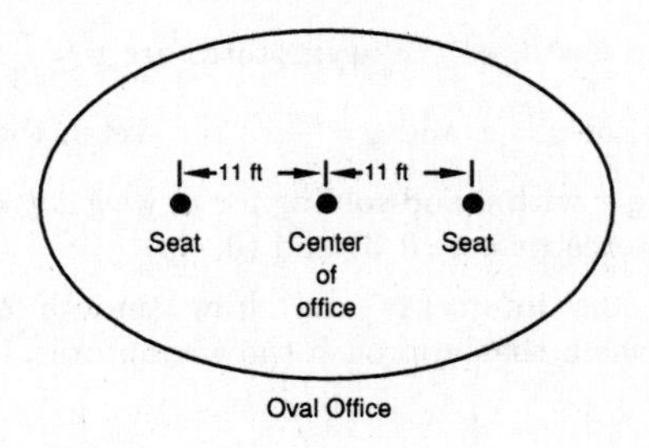

40. 5.66 ft

41.

$$x^2 - 4x + 4y^2 + 8y - 8 = 0$$
$$x^2 - 4x + 4y^2 + 8y = 8$$
$$x^2 - 4x + 4(y^2 + 2y) = 8$$
$$(x^2 - 4x + 4 - 4) + 4(y^2 + 2y + 1 - 1) = 8$$
$$(x^2 - 4x + 4) + 4(y^2 + 2y + 1) = 8 + 4 + 4 \cdot 1$$
$$(x-2)^2 + 4(y+1)^2 = 16$$
$$\frac{(x-2)^2}{16} + \frac{(y+1)^2}{4} = 1$$
$$\frac{(x-2)^2}{4^2} + \frac{(y-(-1))^2}{2^2} = 1$$

The center of the ellipse is $(2, -1)$. Note that $a = 4$ and $b = 2$. We locate the center and then plot the points $(2+4, -1)$, $(2-4, -1)$, $(2, -1+2)$, $(2, -1-2)$, or $(6, -1)$, $(-2, -1)$, $(2, 1)$, and $(2, -3)$. Connect these points with an oval-shaped curve.

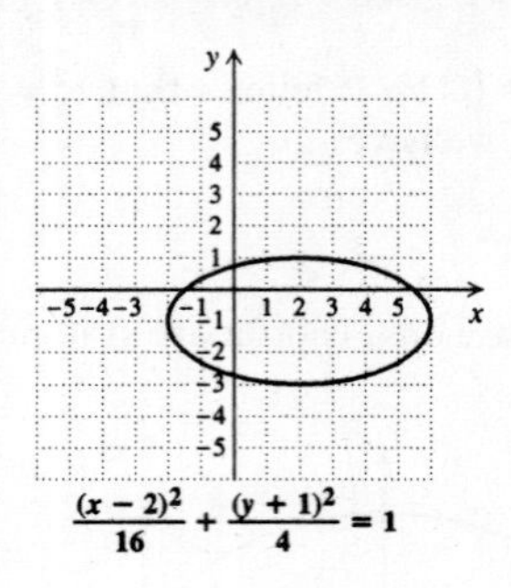

42.

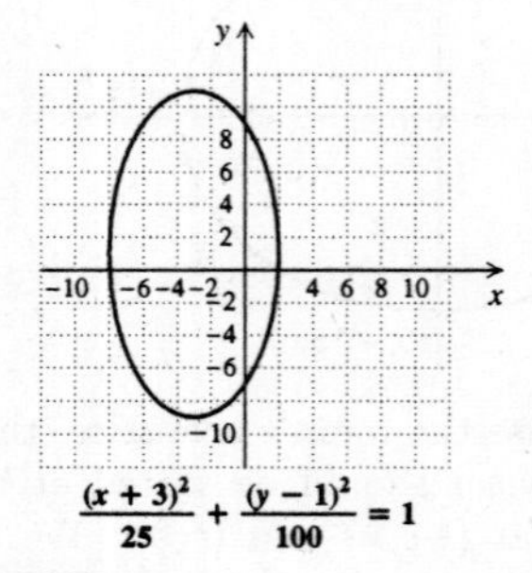

43.

Exercise Set 10.3

1. $\dfrac{y^2}{9} - \dfrac{x^2}{9} = 1$

$\dfrac{y^2}{3^2} - \dfrac{x^2}{3^2} = 1$

$a = 3$ and $b = 3$, so the asymptotes are $y = \dfrac{3}{3}x$ and $y = -\dfrac{3}{3}x$, or $y = x$ and $y = -x$. We sketch them.

Replacing x with 0 and solving for y, we get $y = \pm 3$, so the intercepts are $(0, 3)$ and $(0, -3)$.

We plot the intercepts and draw smooth curves through them that approach the asymptotes.

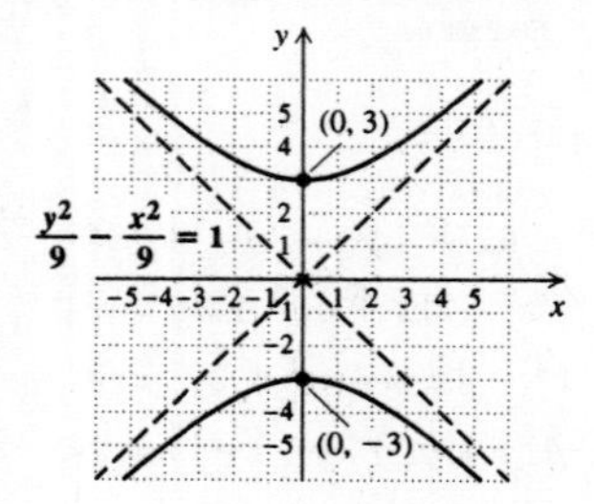

2.

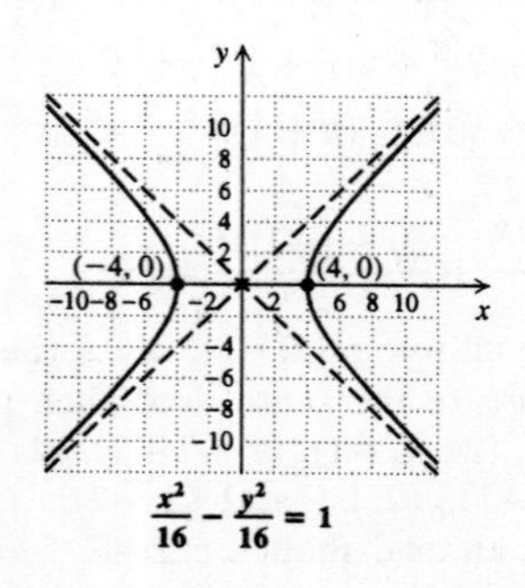

3. $\dfrac{x^2}{4} - \dfrac{y^2}{25} = 1$

$\dfrac{x^2}{2^2} - \dfrac{y^2}{5^2} = 1$

$a = 2$ and $b = 5$, so the asymptotes are $y = \dfrac{5}{2}x$ and $y = -\dfrac{5}{2}x$. We sketch them.

Replacing y with 0 and solving for x, we get $x = \pm 2$, so the intercepts are $(2, 0)$ and $(-2, 0)$.

We plot the intercepts and draw smooth curves through them that approach the asymptotes.

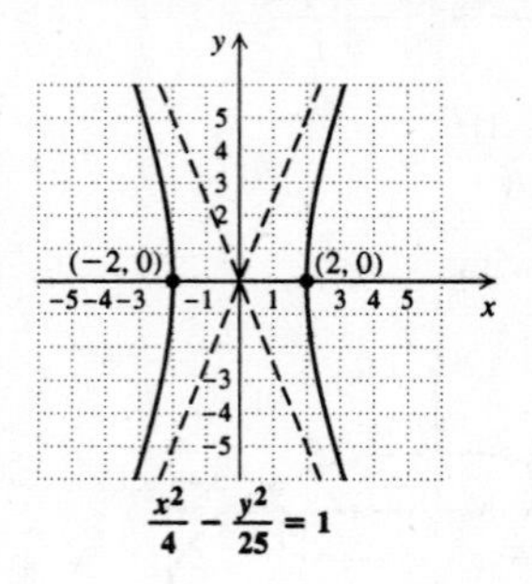

4.

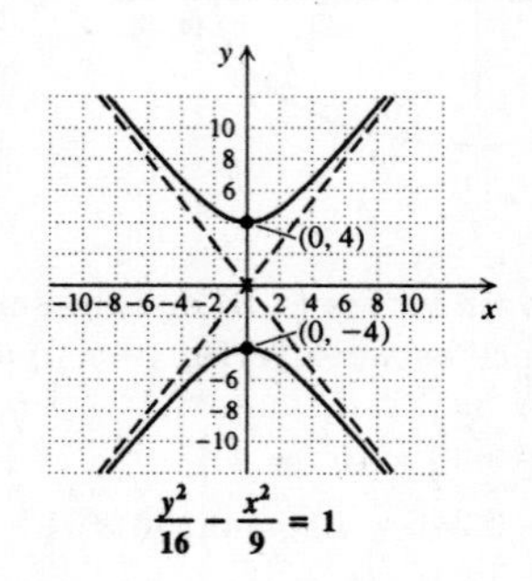

5. $\dfrac{y^2}{36} - \dfrac{x^2}{9} = 1$

$\dfrac{y^2}{6^2} - \dfrac{x^2}{3^2} = 1$

$a = 3$ and $b = 6$, so the asymptotes are $y = \dfrac{6}{3}x$ and $y = -\dfrac{6}{3}x$, or $y = 2x$ and $y = -2x$. We sketch them.

Replacing x with 0 and solving for y, we get $y = \pm 6$, so the intercepts are $(0, 6)$ and $(0, -6)$.

We plot the intercepts and draw smooth curves through them that approach the asymptotes.

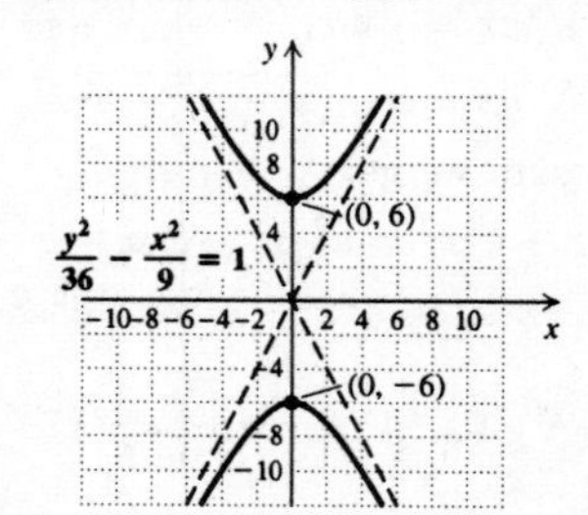

6.

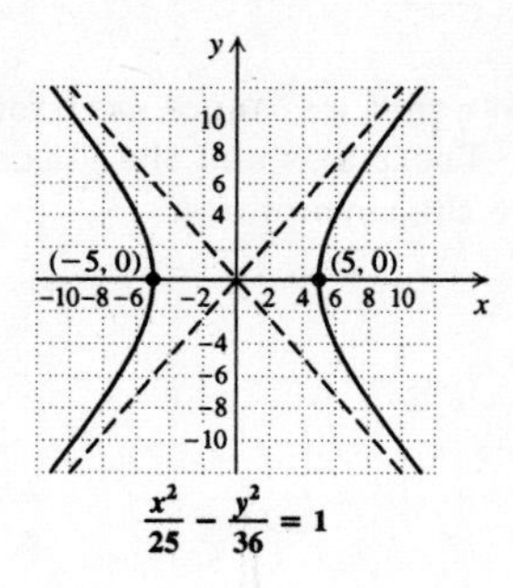

7. $y^2 - x^2 = 25$

$\frac{y^2}{25} - \frac{x^2}{25} = 1$

$\frac{y^2}{5^2} - \frac{x^2}{5^2} = 1$

$a = 5$ and $b = 5$, so the asymptotes are $y = \frac{5}{5}x$ and $y = -\frac{5}{5}x$, or $y = x$ and $y = -x$. We sketch them.

Replacing x with 0 and solving for y, we get $y = \pm 5$, so the intercepts are $(0, 5)$ and $(0, -5)$.

We plot the intercepts and draw smooth curves through them that approach the asymptotes.

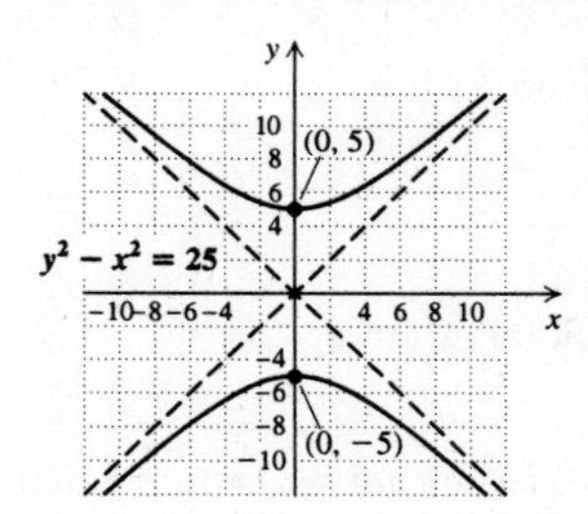

8.

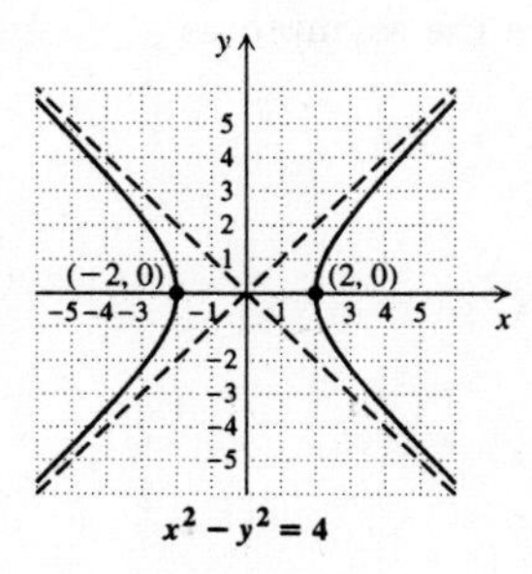

9. $25x^2 - 16y^2 = 400$

$\frac{x^2}{16} - \frac{y^2}{25} = 1$ Multiplying by $\frac{1}{400}$

$\frac{x^2}{4^2} - \frac{y^2}{5^2} = 1$

$a = 4$ and $b = 5$, so the asymptotes are $y = \frac{5}{4}x$ and $y = -\frac{5}{4}x$. We sketch them.

Replacing y with 0 and solving for x, we get $x = \pm 4$, so the intercepts are $(4, 0)$ and $(-4, 0)$.

We plot the intercepts and draw smooth curves through them that approach the asymptotes.

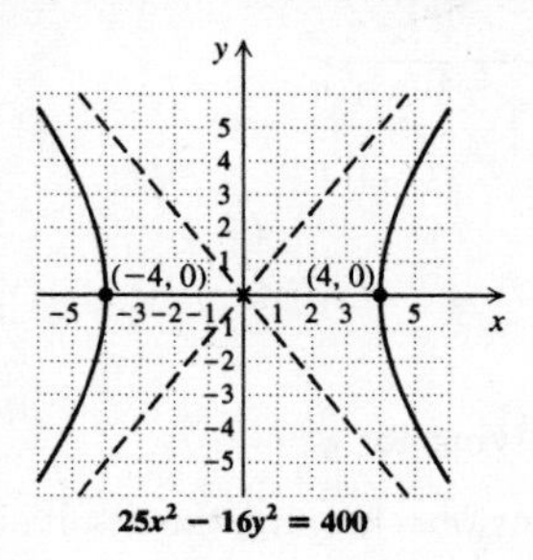

10.

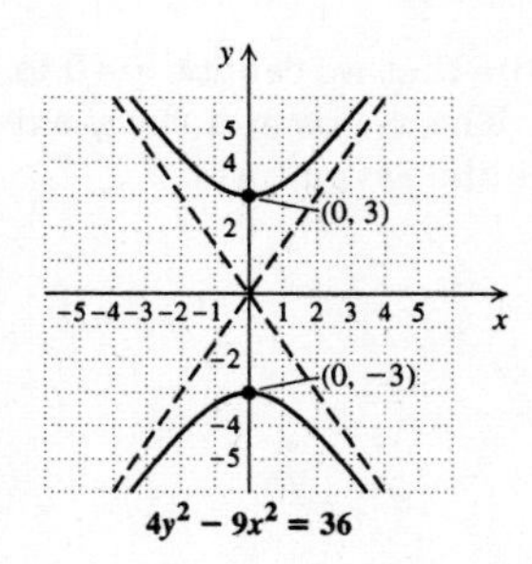

11. $xy = -4$

$y = -\frac{4}{x}$ Solving for y

We find some solutions, keeping the results in a table.

x	y
$\frac{1}{2}$	-8
1	-4
2	-2
4	-1
8	$-\frac{1}{2}$
$-\frac{1}{2}$	8
-1	4
-2	2
-8	$\frac{1}{2}$

Note that we cannot use 0 for x. The x-axis and the y-axis are the asymptotes.

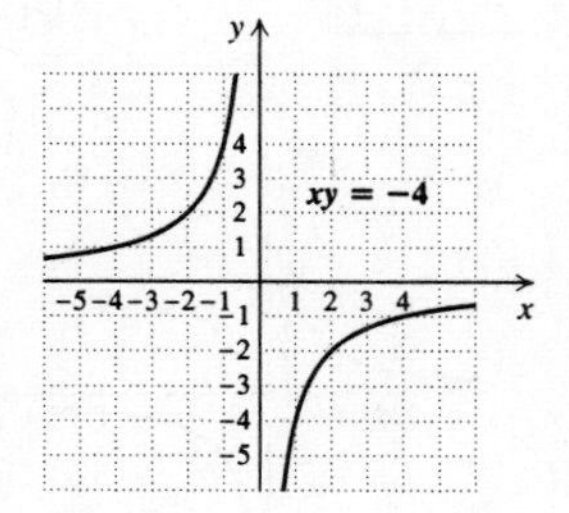

12.

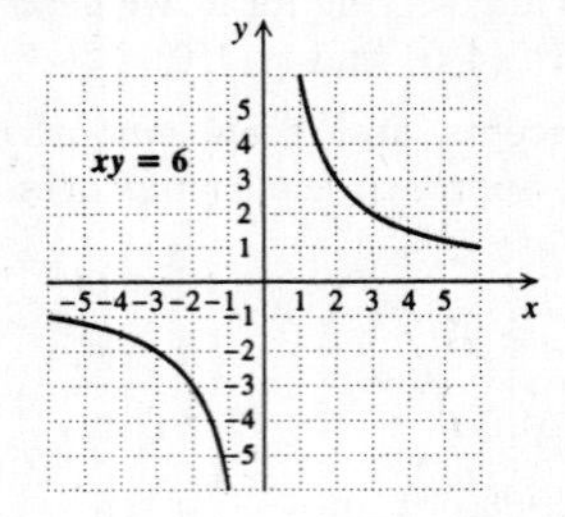

13. $xy = 3$

$y = \frac{3}{x}$ Solving for y

We find some solutions, keeping the results in a table.

x	y
$\frac{1}{3}$	9
$\frac{1}{2}$	6
1	3
3	1
6	$\frac{1}{2}$
9	$\frac{1}{3}$
$-\frac{1}{3}$	-9
$-\frac{1}{2}$	-6
-1	-3
-3	-1
-6	$-\frac{1}{2}$
-9	$-\frac{1}{3}$

Note that we cannot use 0 for x. The x-axis and the y-axis are the asymptotes.

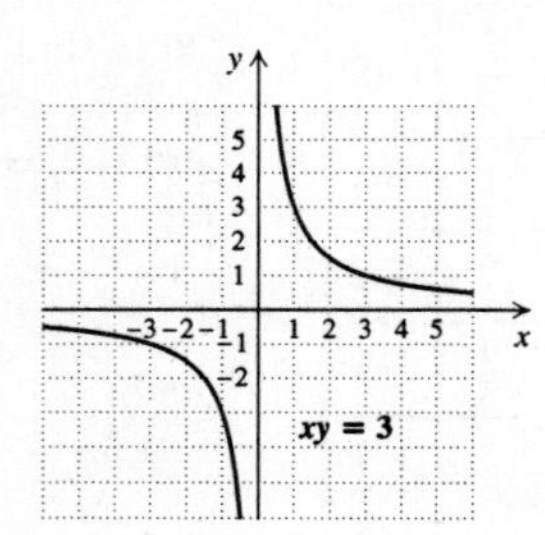

14.

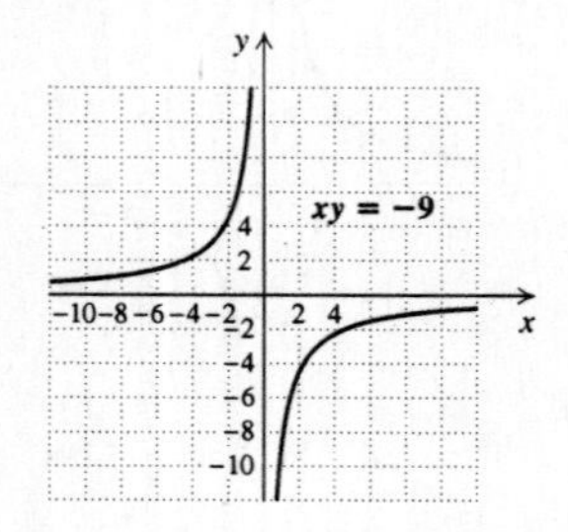

15. $xy = -2$

$y = -\frac{2}{x}$ Solving for y

x	y
$\frac{1}{2}$	-4
1	-2
2	-1
4	$-\frac{1}{2}$
$-\frac{1}{2}$	4
-1	2
-2	1
-4	$\frac{1}{2}$

Note that we cannot use 0 for x. The x-axis and the y-axis are the asymptotes.

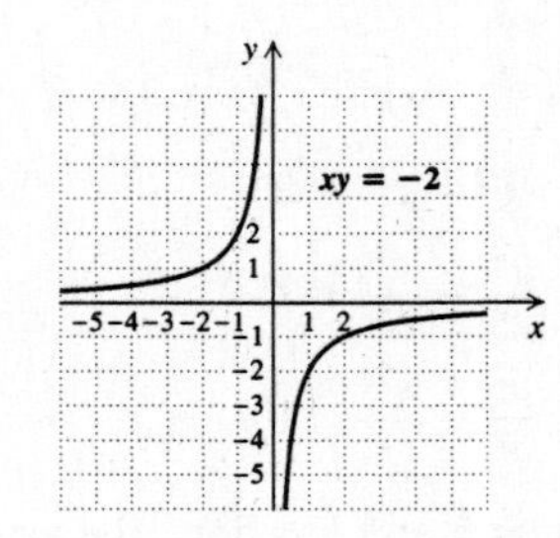

16.

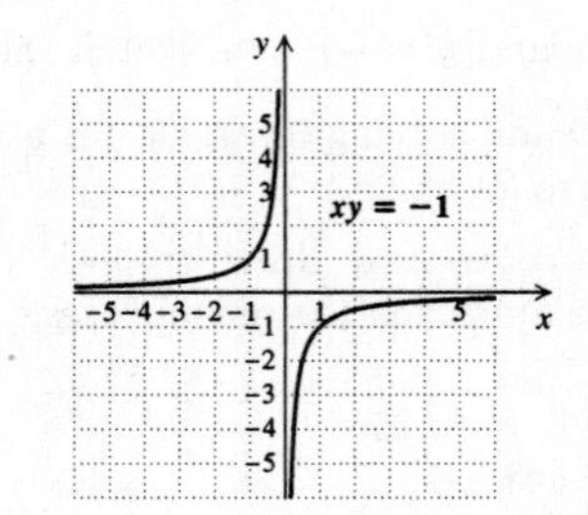

17. $xy = 1$

$y = \frac{1}{x}$ Solving for y

x	y
$\frac{1}{4}$	4
$\frac{1}{2}$	2
1	1
2	$\frac{1}{2}$
4	$\frac{1}{4}$
$-\frac{1}{4}$	-4
$-\frac{1}{2}$	-2
-1	-1
-2	$-\frac{1}{2}$
-4	$-\frac{1}{4}$

Note that we cannot use 0 for x. The x-axis and the y-axis are the asymptotes.

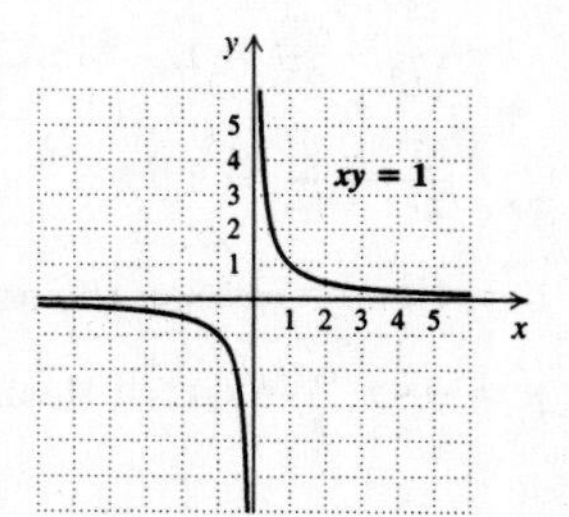

18.

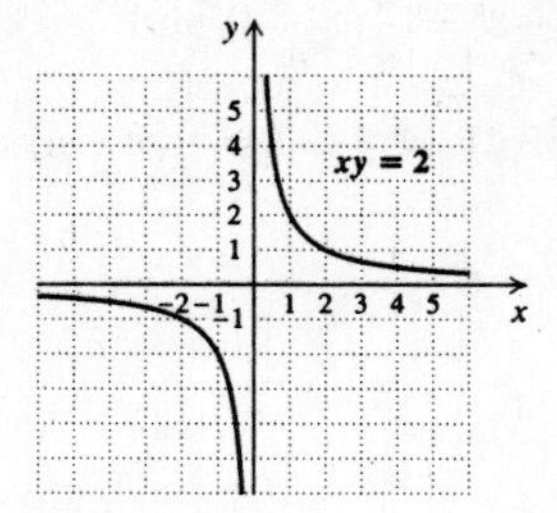

19. $x^2 + y^2 - 10x + 8y - 40 = 0$

Completing the square twice, we obtain an equivalent equation:

$$(x^2 - 10x) + (y^2 + 8y) = 40$$
$$(x^2 - 10x + 25) + (y^2 + 8y + 16) = 40 + 25 + 16$$
$$(x-5)^2 + (y+4)^2 = 81$$

The graph is a circle.

20. Parabola

21. $9x^2 + 4y^2 - 36 = 0$

$$9x^2 + 4y^2 = 36$$
$$\frac{x^2}{4} + \frac{y^2}{9} = 1$$

The graph is an ellipse.

22. Parabola

23. $4x^2 - 9y^2 - 100 = 0$

$$4x^2 - 9y^2 = 100$$
$$\frac{x^2}{25} - \frac{y^2}{100/9} = 1$$

The graph is a hyperbola.

24. Circle

25.

$$x^2 + y^2 = 2x + 4y + 4$$
$$x^2 - 2x + y^2 - 4y = 4$$
$$(x^2 - 2x + 1) + (y^2 - 4y + 4) = 4 + 1 + 4$$
$$(x-1)^2 + (y-2)^2 = 9$$

The graph is a circle.

26. Circle

27.

$$4x^2 = 64 - y^2$$
$$4x^2 + y^2 = 64$$
$$\frac{x^2}{16} + \frac{y^2}{64} = 1$$

The graph is an ellipse.

28. Hyperbola

29. $x - \frac{3}{y} = 0$

$$x = \frac{3}{y}$$
$$xy = 3$$

The graph is a hyperbola.

30. Parabola

31. $y + 6x = x^2 + 6$

$$y = x^2 - 6x + 6$$

The graph is a parabola.

32. Hyperbola

33.

$$9y^2 = 36 + 4x^2$$
$$9y^2 - 4x^2 = 36$$
$$\frac{y^2}{4} - \frac{x^2}{9} = 1$$

The graph is a hyperbola.

34. Circle

35.

$$3x^2 + y^2 - x = 2x^2 - 9x + 10y + 40$$
$$x^2 + y^2 + 8x - 10y = 40$$

Both variables are squared, so the graph is not a parabola. The plus sign between x^2 and y^2 indicates that we have either a circle or an ellipse. Since the coefficients of x^2 and y^2 are the same, the graph is a circle.

36. Ellipse

37. $16x^2 + 5y^2 - 12x^2 + 8y^2 - 3x + 4y = 568$

$$4x^2 + 13y^2 - 3x + 4y = 568$$

Both variables are squared, so the graph is not a parabola. The plus sign between x^2 and y^2 indicates that we have either a circle or an ellipse. Since the coefficients of x^2 and y^2 are different, the graph is an ellipse.

38. Ellipse

39. $\sqrt[3]{125t^{15}} = \sqrt[3]{5^3 \cdot (t^5)^3} = 5t^5$

40. $\pm i\sqrt{5}$

41.

$$\frac{4\sqrt{2} - 5\sqrt{3}}{6\sqrt{3} - 8\sqrt{2}} = \frac{4\sqrt{2} - 5\sqrt{3}}{6\sqrt{3} - 8\sqrt{2}} \cdot \frac{6\sqrt{3} + 8\sqrt{2}}{6\sqrt{3} + 8\sqrt{2}}$$
$$= \frac{24\sqrt{6} + 32 \cdot 2 - 30 \cdot 3 - 40\sqrt{6}}{36 \cdot 3 - 64 \cdot 2}$$
$$= \frac{-26 - 16\sqrt{6}}{-20}$$
$$= \frac{-2(13 + 8\sqrt{6})}{-2 \cdot 10}$$
$$= \frac{13 + 8\sqrt{6}}{10}$$

42. Smaller plane: 400 mph, larger plane: 720 mph

43. ◈

44. ◈

45. Since the intercepts are $(0,6)$ and $(0,-6)$, we know that the hyperbola is of the form $\dfrac{y^2}{b^2}-\dfrac{x^2}{a^2}=1$ and that $b=6$. The equations of the asymptotes tell us that $b/a=3$, so

$$\frac{6}{a}=3$$
$$a=2.$$

The equation is $\dfrac{y^2}{6^2}-\dfrac{x^2}{2^2}=1$, or $\dfrac{y^2}{36}-\dfrac{x^2}{4}=1$.

46. $\dfrac{x^2}{64}-\dfrac{y^2}{1024}=1$

47.
$$\frac{(x-5)^2}{36}-\frac{(y-2)^2}{25}=1$$
$$\frac{(x-5)^2}{6^2}-\frac{(y-2)^2}{5^2}=1$$

$h=5$, $k=2$, $a=6$, $b=5$

Center: $(5,2)$

Vertices: $(5-6,2)$ and $(5+6,2)$, or $(-1,2)$ and $(11,2)$

Asymptotes: $y-2=\dfrac{5}{6}(x-5)$ and $y-2=-\dfrac{5}{6}(x-5)$

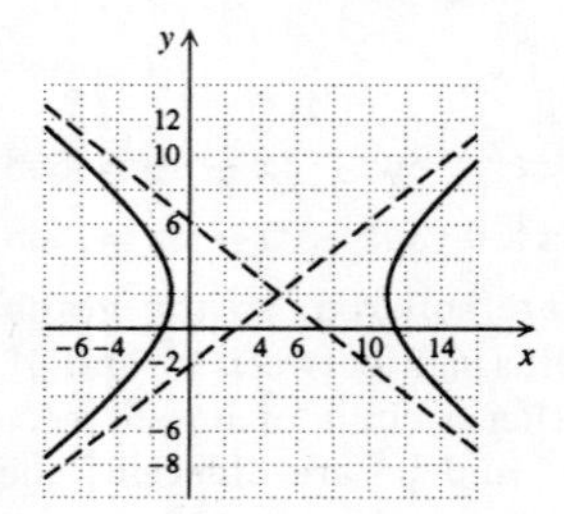

48. Center: $(2,1)$

Vertices: $(2-3,1)$ and $(2+3,1)$, or $(-1,1)$ and $(5,1)$

Asymptotes: $y-1=\dfrac{2}{3}(x-2)$ and $y-1=-\dfrac{2}{3}(x-2)$

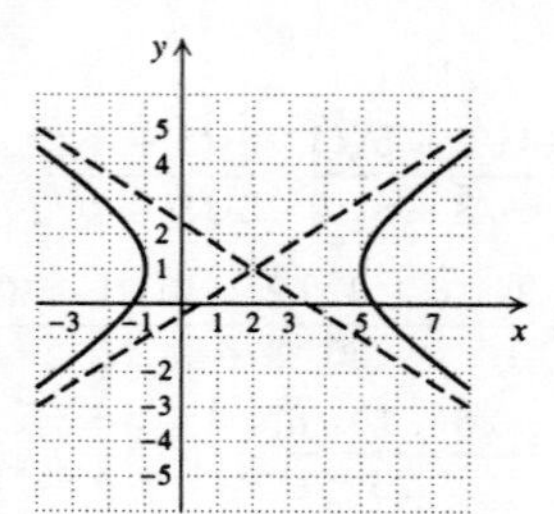

49.
$$8(y+3)^2-2(x-4)^2=32$$
$$\frac{(y+3)^2}{4}-\frac{(x-4)^2}{16}=1$$
$$\frac{(y-(-3))^2}{2^2}-\frac{(x-4)^2}{4^2}=1$$

$h=4$, $k=-3$, $a=4$, $b=2$

Center: $(4,-3)$

Vertices: $(4,-3+2)$ and $(4,-3-2)$, or $(4,-1)$ and $(4,-5)$

Asymptotes: $y-(-3)=\dfrac{2}{4}(x-4)$ and $y-(-3)=-\dfrac{2}{4}(x-4)$, or $y+3=\dfrac{1}{2}(x-4)$ and $y+3=-\dfrac{1}{2}(x-4)$

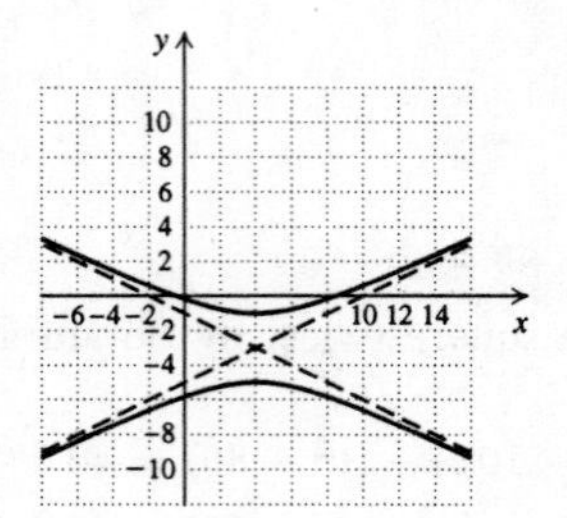

50. $\dfrac{(x-4)^2}{4}-\dfrac{(y+5)^2}{25}=1$

Center: $(4,-5)$

Vertices: $(4-2,-5)$ and $(4+2,-5)$, or $(2,-5)$ and $(6,-5)$

Asymptotes: $y+5=\dfrac{5}{2}(x-4)$ and $y+5=-\dfrac{5}{2}(x-4)$

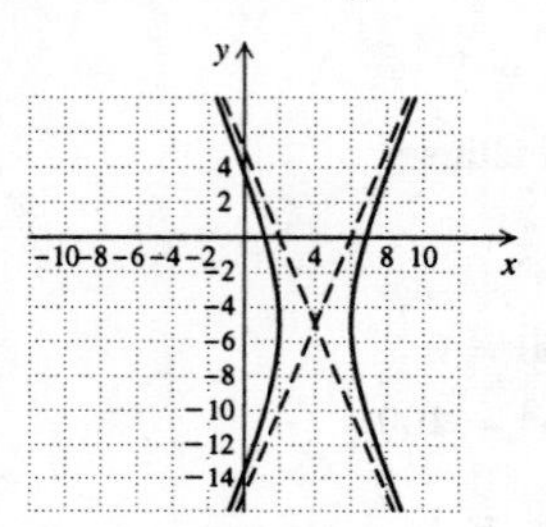

51.
$$4x^2-y^2+24x+4y+28=0$$
$$4(x^2+6x)-(y^2-4y)=-28$$
$$4(x^2+6x+9-9)-(y^2-4y+4-4)=-28$$
$$4(x^2+6x+9)-(y^2-4y+4)=-28+4\cdot 9-4$$
$$4(x+3)^2-(y-2)^2=4$$
$$\frac{(x+3)^2}{1}-\frac{(y-2)^2}{4}=1$$
$$\frac{(x-(-3))^2}{1^2}-\frac{(y-2)^2}{2^2}=1$$

$h=-3$, $k=2$, $a=1$, $b=2$

Center: $(-3,2)$

Vertices: $(-3-1,2)$, and $(-3+1,2)$, or $(-4,2)$ and $(-2,2)$

Asymptotes: $y-2=\dfrac{2}{1}(x-(-3))$ and $y-2=-\dfrac{2}{1}(x-(-3))$, or $y-2=2(x+3)$ and $y-2=-2(x+3)$

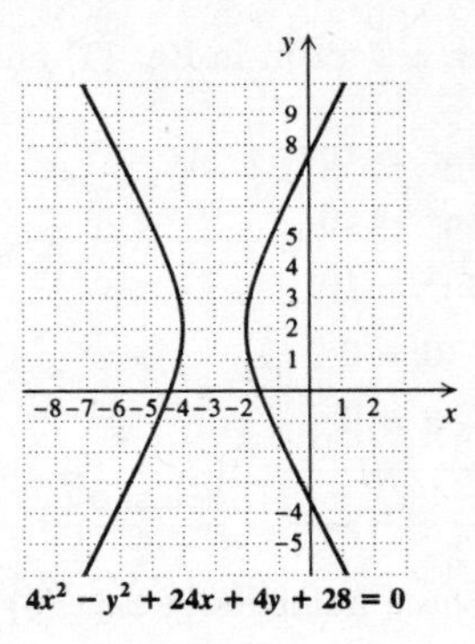

52. $\dfrac{(y-1)^2}{25} - \dfrac{(x+2)^2}{4} = 1$

Center: $(-2, 1)$

Vertices: $(-2, 1+5)$ and $(-2, 1-5)$, or $(-2, 6)$ and $(-2, -4)$

Asymptotes: $y - 1 = \dfrac{5}{2}(x+2)$ and $y - 1 = -\dfrac{5}{2}(x+2)$

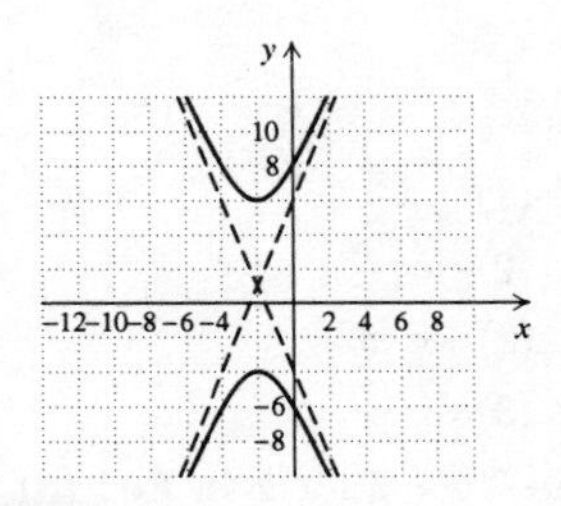

53.

Exercise Set 10.4

1. $x^2 + y^2 = 100,$ (1)

$y - x = 2$ (2)

First solve Eq. (2) for y.

$y = x + 2$ (3)

Then substitute $x + 2$ for y in Eq. (1) and solve for x.

$$\begin{aligned} x^2 + y^2 &= 100 \\ x^2 + (x+2)^2 &= 100 \\ x^2 + x^2 + 4x + 4 &= 100 \\ 2x^2 + 4x - 96 &= 0 \\ x^2 + 2x - 48 &= 0 \quad \text{Multiplying by } \tfrac{1}{2} \\ (x+8)(x-6) &= 0 \quad \text{Factoring} \end{aligned}$$

$x + 8 = 0$ *or* $x - 6 = 0$ Principle of zero products

$x = -8$ *or* $x = 6$

Now substitute these numbers into Eq. (3) and solve for y.

$y = -8 + 2 = -6$

$y = 6 + 2 = 8$

The pairs $(-8, -6)$ and $(6, 8)$ check, so they are the solutions.

2. $(-4, -3), (3, 4)$

3. $9x^2 + 4y^2 = 36,$ (1)

$3x + 2y = 6$ (2)

First solve Eq. (2) for x.

$3x = 6 - 2y$

$x = 2 - \dfrac{2}{3}y$ (3)

Then substitute $2 - \dfrac{2}{3}y$ for x in Eq. (1) and solve for y.

$$\begin{aligned} 9x^2 + 4y^2 &= 36 \\ 9\left(2 - \tfrac{2}{3}y\right)^2 + 4y^2 &= 36 \\ 9\left(4 - \tfrac{8}{3}y + \tfrac{4}{9}y^2\right) + 4y^2 &= 36 \\ 36 - 24y + 4y^2 + 4y^2 &= 36 \\ 8y^2 - 24y &= 0 \\ y^2 - 3y &= 0 \\ y(y-3) &= 0 \end{aligned}$$

$y = 0$ *or* $y = 3$

Now substitute these numbers in Eq. (3) and solve for x.

$x = 2 - \dfrac{2}{3}(0) = 2$

$x = 2 - \dfrac{2}{3}(3) = 0$

The pairs $(2, 0)$ and $(0, 3)$ check, so they are the solutions.

4. $(0, 2), (3, 0)$

5. $y = x^2,$ (1)

$3x = y + 2$ (2)

First solve Eq. (2) for y.

$y = 3x - 2$ (3)

Then substitute $3x - 2$ for y in Eq. (1) and solve for x.

$$\begin{aligned} y &= x^2 \\ 3x - 2 &= x^2 \\ 0 &= x^2 - 3x + 2 \\ 0 &= (x-2)(x-1) \end{aligned}$$

$x = 2$ *or* $x = 1$

Now substitute these numbers in Eq. (3) and solve for y.

$y = 3 \cdot 2 - 2 = 4$

$y = 3 \cdot 1 - 2 = 1$

The pairs $(2, 4)$ and $(1, 1)$ check, so they are the solutions.

6. $(-2, 1)$

7. $2y^2 + xy + x^2 = 7, \quad (1)$

$x - 2y = 5 \quad (2)$

First solve Eq. (2) for x.

$x = 2y + 5 \quad (3)$

Then substitute $2y + 5$ for x in Eq. (1) and solve for y.

$$\begin{aligned} 2y^2 + xy + x^2 &= 7 \\ 2y^2 + (2y+5)y + (2y+5)^2 &= 7 \\ 2y^2 + 2y^2 + 5y + 4y^2 + 20y + 25 &= 7 \\ 8y^2 + 25y + 18 &= 0 \\ (8y+9)(y+2) &= 0 \end{aligned}$$

$y = -\frac{9}{8}$ *or* $y = -2$

Now substitute these numbers in Eq. (3) and solve for x.

$x = 2\left(-\frac{9}{8}\right) + 5 = \frac{11}{4}$

$x = 2(-2) + 5 = 1$

The pairs $\left(\frac{11}{4}, -\frac{9}{8}\right)$ and $(1, -2)$ check, so they are the solutions.

8. $\left(\frac{5+\sqrt{70}}{3}, \frac{-1+\sqrt{70}}{3}\right), \left(\frac{5-\sqrt{70}}{3}, \frac{-1-\sqrt{70}}{3}\right)$

9. $y^2 - x^2 = 16, \quad (1)$

$2x - y = 1 \quad (2)$

First solve Eq. (2) for y.

$2x - 1 = y \quad (3)$

Then substitute $2x - 1$ for y in Eq. (1) and solve for x.

$$\begin{aligned} y^2 - x^2 &= 16 \\ (2x-1)^2 - x^2 &= 16 \\ 4x^2 - 4x + 1 - x^2 &= 16 \\ 3x^2 - 4x - 15 &= 0 \\ (3x+5)(x-3) &= 0 \end{aligned}$$

$x = -\frac{5}{3}$ *or* $x = 3$

Now substitute these numbers in Eq. (3) and solve for y.

$y = 2\left(-\frac{5}{3}\right) - 1 = -\frac{13}{3}$

$y = 2(3) - 1 = 5$

The pairs $\left(-\frac{5}{3}, -\frac{13}{3}\right)$ and $(3, 5)$ check, so they are the solutions.

10. $\left(4, \frac{3}{2}\right), (3, 2)$

11. $m^2 + 3n^2 = 10, \quad (1)$

$m - n = 2 \quad (2)$

First solve Eq. (2) for m.

$m = n + 2 \quad (3)$

Then substitute $n + 2$ for m in Eq. (1) and solve for n.

$$\begin{aligned} m^2 + 3n^2 &= 10 \\ (n+2)^2 + 3n^2 &= 10 \\ n^2 + 4n + 4 + 3n^2 &= 10 \\ 4n^2 + 4n - 6 &= 0 \\ 2n^2 + 2n - 3 &= 0 \end{aligned}$$

$$n = \frac{-2 \pm \sqrt{2^2 - 4(2)(-3)}}{2 \cdot 2} = \frac{-1 \pm \sqrt{7}}{2}$$

Now substitute these numbers in Eq. (3) and solve for m.

$m = \frac{-1+\sqrt{7}}{2} + 2 = \frac{3+\sqrt{7}}{2}$

$m = \frac{-1-\sqrt{7}}{2} + 2 = \frac{3-\sqrt{7}}{2}$

The pairs $\left(\frac{3+\sqrt{7}}{2}, \frac{-1+\sqrt{7}}{2}\right)$ and $\left(\frac{3-\sqrt{7}}{2}, \frac{-1-\sqrt{7}}{2}\right)$ check, so they are the solutions.

12. $\left(\frac{7}{3}, \frac{1}{3}\right), (1, -1)$

13. $2y^2 + xy = 5, \quad (1)$

$4y + x = 7 \quad (2)$

First solve Eq. (2) for x.

$x = -4y + 7 \quad (3)$

Then substitute $-4y + 7$ for x in Eq. (3) and solve for y.

$$\begin{aligned} 2y^2 + xy &= 5 \\ 2y^2 + (-4y+7)y &= 5 \\ 2y^2 - 4y^2 + 7y &= 5 \\ 0 &= 2y^2 - 7y + 5 \\ 0 &= (2y-5)(y-1) \end{aligned}$$

$y = \frac{5}{2}$ *or* $y = 1$

Now substitute these numbers in Eq. (3) and solve for x.

$x = -4\left(\frac{5}{2}\right) + 7 = -3$

$x = -4(1) + 7 = 3$

The pairs $\left(-3, \frac{5}{2}\right)$ and $(3, 1)$ check, so they are the solutions.

14. $\left(\frac{11}{4}, -\frac{5}{4}\right), (1, 4)$

15. $p + q = -6, \quad (1)$

$pq = -7 \quad (2)$

First solve Eq. (1) for p.

$p = -q - 6 \quad (3)$

Then substitute $-q-6$ for p in Eq. (2) and solve for q.

$$pq = -7$$
$$(-q-6)q = -7$$
$$-q^2 - 6q = -7$$
$$0 = q^2 + 6q - 7$$
$$0 = (q+7)(q-1)$$

$q = -7$ *or* $q = 1$

Now substitute these numbers in Eq. (3) and solve for p.

$p = -(-7) - 6 = 1$

$p = -1 - 6 = -7$

The pairs $(1, -7)$ and $(-7, 1)$ check, so they are the solutions.

16. $\left(\frac{7-\sqrt{33}}{2}, \frac{7+\sqrt{33}}{2}\right), \left(\frac{7+\sqrt{33}}{2}, \frac{7-\sqrt{33}}{2}\right)$

17. $4x^2 + 9y^2 = 36,\quad (1)$

$x + 3y = 3 \quad (2)$

First solve Eq. (1) for x.

$x = -3y + 3 \quad (3)$

Then substitute $-3y+3$ for x in Eq. (1) and solve for y.

$$4x^2 + 9y^2 = 36$$
$$4(-3y+3)^2 + 9y^2 = 36$$
$$4(9y^2 - 18y + 9) + 9y^2 = 36$$
$$36y^2 - 72y + 36 + 9y^2 = 36$$
$$45y^2 - 72y = 0$$
$$5y^2 - 8y = 0$$
$$y(5y-8) = 0$$

$y = 0$ *or* $y = \frac{8}{5}$

Now substitute these numbers in Eq. (3) and solve for x.

$x = -3 \cdot 0 + 3 = 3$

$x = -3\left(\frac{8}{5}\right) + 3 = -\frac{9}{5}$

The pairs $(3, 0)$ and $\left(-\frac{9}{5}, \frac{8}{5}\right)$ check, so they are the solutions.

18. $(3, -5)$, $(-1, 3)$

19. $xy = 4,\quad (1)$

$x + y = 5 \quad (2)$

First solve Eq. (2) for x.

$x = -y + 5 \quad (3)$

Substitute $-y+5$ for x in Eq. (1) and solve for y.

$$xy = 4$$
$$(-y+5)y = 4$$
$$-y^2 + 5y = 4$$
$$0 = y^2 - 5y + 4$$
$$0 = (y-4)(y-1)$$

$y = 4$ *or* $y = 1$

Then substitute these numbers in Eq. (3) and solve for x.

$x = -4 + 5 = 1$

$x = -1 + 5 = 4$

The pairs $(1, 4)$ and $(4, 1)$ check, so they are the solutions.

20. $(-5, -8)$, $(8, 5)$

21. $y = x^2,\quad (1)$

$x = y^2 \quad (2)$

Eq. (1) is already solved for y. Substitute x^2 for y in Eq. (2) and solve for x.

$x = y^2$

$x = (x^2)^2$

$x = x^4$

$0 = x^4 - x$

$0 = x(x^3 - 1)$

$0 = x(x-1)(x^2 + x + 1)$

$x = 0$ *or* $x = 1$ *or* $x = \frac{-1 \pm \sqrt{1^2 - 4 \cdot 1 \cdot 1}}{2}$

$x = 0$ *or* $x = 1$ *or* $x = -\frac{1}{2} \pm \frac{\sqrt{3}}{2}i$

Substitute these numbers in Eq. (1) and solve for y.

$y = 0^2 = 0$

$y = 1^2 = 1$

$y = \left(-\frac{1}{2} + \frac{\sqrt{3}}{2}i\right)^2 = -\frac{1}{2} - \frac{\sqrt{3}}{2}i$

$y = \left(-\frac{1}{2} - \frac{\sqrt{3}}{2}i\right)^2 = -\frac{1}{2} + \frac{\sqrt{3}}{2}i$

The pairs $(0, 0)$, $(1, 1)$, $\left(-\frac{1}{2} + \frac{\sqrt{3}}{2}i, -\frac{1}{2} - \frac{\sqrt{3}}{2}i\right)$, and $\left(-\frac{1}{2} - \frac{\sqrt{3}}{2}i, -\frac{1}{2} + \frac{\sqrt{3}}{2}i\right)$ check, so they are the solutions.

22. $(-5, 0)$, $(4, 3)$, $(4, -3)$

23. $x^2 + y^2 = 9,\quad (1)$

$x^2 - y^2 = 9 \quad (2)$

Here we use the elimination method.

$$\begin{aligned} x^2 + y^2 &= 9 & (1) \\ x^2 - y^2 &= 9 & (2) \\ \hline 2x^2 &= 18 & \text{Adding} \\ x^2 &= 9 \\ x &= \pm 3 \end{aligned}$$

If $x = 3$, $x^2 = 9$, and if $x = -3$, $x^2 = 9$, so substituting 3 or -3 in Eq. (1) gives us

$$\begin{aligned} x^2 + y^2 &= 9 \\ 9 + y^2 &= 9 \\ y^2 &= 0 \\ y &= 0. \end{aligned}$$

The pairs $(3, 0)$ and $(-3, 0)$ check. They are the solutions.

24. $(0, 2)$, $(0, -2)$

25. $x^2 + y^2 = 25$, (1)

$xy = 12$ (2)

First we solve Eq. (2) for y.

$$\begin{aligned} xy &= 12 \\ y &= \frac{12}{x} \end{aligned}$$

Then we substitute $\frac{12}{x}$ for y in Eq. (1) and solve for x.

$$\begin{aligned} x^2 + y^2 &= 25 \\ x^2 + \left(\frac{12}{x}\right)^2 &= 25 \\ x^2 + \frac{144}{x^2} &= 25 \\ x^4 + 144 &= 25x^2 \quad \text{Multiplying by } x^2 \\ x^4 - 25x^2 + 144 &= 0 \\ u^2 - 25u + 144 &= 0 \quad \text{Letting } u = x^2 \\ (u - 9)(u - 16) &= 0 \\ u = 9 \ \text{ or } \ u &= 16 \end{aligned}$$

We now substitute x^2 for u and solve for x.

$$x^2 = 9 \quad \text{or} \quad x^2 = 16$$
$$x = \pm 3 \quad \text{or} \quad x = \pm 4$$

Since $y = 12/x$, if $x = 3$, $y = 4$; if $x = -3$, $y = -4$; if $x = 4$, $y = 3$; and if $x = -4$, $y = -3$. The pairs $(3, 4)$, $(-3, -4)$, $(4, 3)$, and $(-4, -3)$ check. They are the solutions.

26. $(-5, 3)$, $(-5, -3)$, $(4, 0)$

27. $x^2 + y^2 = 4$, (1)

$16x^2 + 9y^2 = 144$ (2)

$$\begin{aligned} -9x^2 - 9y^2 &= -36 \quad \text{Multiplying (1) by } -9 \\ 16x^2 + 9y^2 &= 144 \\ \hline 7x^2 \qquad &= 108 \quad \text{Adding} \\ x^2 &= \frac{108}{7} \\ x &= \pm\sqrt{\frac{108}{7}} = \pm 6\sqrt{\frac{3}{7}} \\ x &= \pm\frac{6\sqrt{21}}{7} \quad \text{Rationalizing the denominator} \end{aligned}$$

Substituting $\frac{6\sqrt{21}}{7}$ or $-\frac{6\sqrt{21}}{7}$ for x in Eq. (1) gives us

$$\begin{aligned} \frac{36 \cdot 21}{49} + y^2 &= 4 \\ y^2 &= 4 - \frac{108}{7} \\ y^2 &= -\frac{80}{7} \\ y &= \pm\sqrt{-\frac{80}{7}} = \pm 4i\sqrt{\frac{5}{7}} \\ y &= \pm\frac{4i\sqrt{35}}{7}. \quad \text{Rationalizing the denominator} \end{aligned}$$

The pairs $\left(\frac{6\sqrt{21}}{7}, \frac{4i\sqrt{35}}{7}\right)$, $\left(\frac{6\sqrt{21}}{7}, -\frac{4i\sqrt{35}}{7}\right)$, $\left(-\frac{6\sqrt{21}}{7}, \frac{4i\sqrt{35}}{7}\right)$, and $\left(-\frac{6\sqrt{21}}{7}, -\frac{4i\sqrt{35}}{7}\right)$ check. They are the solutions.

28. $\left(\frac{16}{3}, \frac{5\sqrt{7}}{3}i\right)$, $\left(\frac{16}{3}, -\frac{5\sqrt{7}}{3}i\right)$, $\left(-\frac{16}{3}, \frac{5\sqrt{7}}{3}i\right)$, $\left(-\frac{16}{3}, -\frac{5\sqrt{7}}{3}i\right)$

29. $x^2 + y^2 = 16$, $\qquad x^2 + y^2 = 16$, (1)

or

$y^2 - 2x^2 = 10 \qquad -2x^2 + y^2 = 10$ (2)

Here we use the elimination method.

$$\begin{aligned} 2x^2 + 2y^2 &= 32 \quad \text{Multiplying (1) by 2} \\ -2x^2 + y^2 &= 10 \\ \hline 3y^2 &= 42 \quad \text{Adding} \\ y^2 &= 14 \\ y &= \pm\sqrt{14} \end{aligned}$$

Substituting $\sqrt{14}$ or $-\sqrt{14}$ for y in Eq. (1) gives us

$$\begin{aligned} x^2 + 14 &= 16 \\ x^2 &= 2 \\ x &= \pm\sqrt{2} \end{aligned}$$

The pairs $(-\sqrt{2}, -\sqrt{14})$, $(-\sqrt{2}, \sqrt{14})$, $(\sqrt{2}, -\sqrt{14})$, and $(\sqrt{2}, \sqrt{14})$ check. They are the solutions.

30. $(-3, -\sqrt{5})$, $(-3, \sqrt{5})$, $(3, -\sqrt{5})$, $(3, \sqrt{5})$

31. $x^2 + y^2 = 5$, (1)

$xy = 2$ (2)

First we solve Eq. (2) for y.

$$\begin{aligned} xy &= 2 \\ y &= \frac{2}{x} \end{aligned}$$

Then we substitute $\frac{2}{x}$ for y in Eq. (1) and solve for x.

$$\begin{aligned} x^2 + y^2 &= 5 \\ x^2 + \left(\frac{2}{x}\right)^2 &= 5 \\ x^2 + \frac{4}{x^2} &= 5 \\ x^4 + 4 &= 5x^2 \quad \text{Multiplying by } x^2 \\ x^4 - 5x^2 + 4 &= 0 \end{aligned}$$

$u^2 - 5u + 4 = 0$ Letting $u = x^2$

$(u-4)(u-1) = 0$

$u = 4$ *or* $u = 1$

We now substitute x^2 for u and solve for x.

$x^2 = 4$ *or* $x^2 = 1$

$x = \pm 2$ *or* $x = \pm 1$

Since $y = 2/x$, if $x = 2$, $y = 1$; if $x = -2$, $y = -1$; if $x = 1$, $y = 2$; and if $x = -1$, $y = -2$. The pairs $(2,1)$, $(-2,-1)$, $(1,2)$, and $(-1,-2)$ check. They are the solutions.

32. $(4,2)$, $(-4,-2)$, $(2,4)$, $(-2,-4)$

33. $x^2 + y^2 = 13$, (1)

$xy = 6$ (2)

First we solve Eq. (2) for y.

$$xy = 6$$

$$y = \frac{6}{x}$$

Then we substitute $\frac{6}{x}$ for y in Eq. (1) and solve for x.

$$x^2 + y^2 = 13$$

$$x^2 + \left(\frac{6}{x}\right)^2 = 13$$

$$x^2 + \frac{36}{x^2} = 13$$

$x^4 + 36 = 13x^2$ Multiplying by x^2

$x^4 - 13x^2 + 36 = 0$

$u^2 - 13u + 36 = 0$ Letting $u = x^2$

$(u-9)(u-4) = 0$

$u = 9$ *or* $u = 4$

We now substitute x^2 for u and solve for x.

$x^2 = 9$ *or* $x^2 = 4$

$x = \pm 3$ *or* $x = \pm 2$

Since $y = 6/x$, if $x = 3$, $y = 2$; if $x = -3$, $y = -2$; if $x = 2$, $y = 3$; and if $x = -2$, $y = -3$. The pairs $(3,2)$, $(-3,-2)$, $(2,3)$, and $(-2,-3)$ check. They are the solutions.

34. $(4,1)$, $(-4,-1)$, $(2,2)$, $(-2,-2)$

35. $3xy + x^2 = 34$, (1)

$2xy - 3x^2 = 8$ (2)

$6xy + 2x^2 = 68$ Multiplying (1) by 2

$-6xy + 9x^2 = -24$ Multiplying (2) by -3

$11x^2 = 44$ Adding

$x^2 = 4$

$x = \pm 2$

Substitute for x in Eq. (1) and solve for y.

When $x = 2$: $3 \cdot 2 \cdot y + 2^2 = 34$

$6y + 4 = 34$

$6y = 30$

$y = 5$

When $x = -2$: $3(-2)(y) + (-2)^2 = 34$

$-6y + 4 = 34$

$-6y = 30$

$y = -5$

The pairs $(2,5)$ and $(-2,-5)$ check. They are the solutions.

36. $(2,1)$, $(-2,-1)$

37. $xy - y^2 = 2$, (1)

$2xy - 3y^2 = 0$ (2)

$-2xy + 2y^2 = -4$ Multiplying (1) by -2

$2xy - 3y^2 = 0$

$-y^2 = -4$ Adding

$y^2 = 4$

$y = \pm 2$

We substitute for y in Eq. (1) and solve for x.

When $y = 2$: $x \cdot 2 - 2^2 = 2$

$2x - 4 = 2$

$2x = 6$

$x = 3$

When $y = -2$: $x(-2) - (-2)^2 = 2$

$-2x - 4 = 2$

$-2x = 6$

$x = -3$

The pairs $(3,2)$ and $(-3,-2)$ check. They are the solutions.

38. $\left(2, -\frac{4}{5}\right)$, $\left(-2, -\frac{4}{5}\right)$, $(5,2)$, $(-5,2)$

39. $x^2 - y = 5$, (1)

$x^2 + y^2 = 25$ (2)

We solve Eq. (1) for y.

$x^2 - 5 = y$ (3)

Substitute $x^2 - 5$ for y in Eq. (2) and solve for x.

$x^2 + (x^2 - 5)^2 = 25$

$x^2 + x^4 - 10x^2 + 25 = 25$

$x^4 - 9x^2 = 0$

$u^2 - 9u = 0$ Letting $u = x^2$

$u(u-9) = 0$

$u = 0$ *or* $u = 9$

$x^2 = 0$ *or* $x^2 = 9$

$x = 0$ *or* $x = \pm 3$

Substitute in Eq. (3) and solve for y.

When $x = 0$: $y = 0^2 - 5 = -5$

When $x = 3$ or -3: $y = 9 - 5 = 4$

The pairs $(0,-5)$, $(3,4)$, and $(-3,4)$ check. They are the solutions.

(This exercise could also be solved using the elimination method.)

40. $(-\sqrt{2}, \sqrt{2})$, $(\sqrt{2}, -\sqrt{2})$

41. ***Familiarize***. We first make a drawing. We let l and w represent the length and width, respectively.

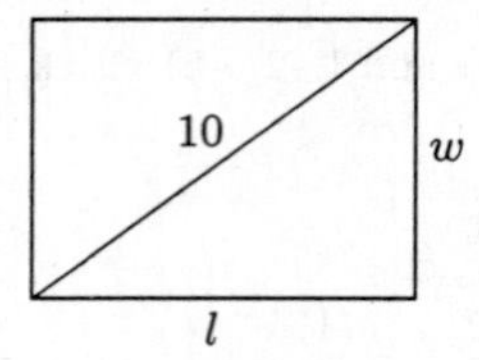

Translate. The perimeter is 28 cm.

$2l + 2w = 28$, or $l + w = 14$

Using the Pythagorean theorem we have another equation.

$l^2 + w^2 = 10^2$, or $l^2 + w^2 = 100$

Carry out. We solve the system:

$l + w = 14$, (1)

$l^2 + w^2 = 100$ (2)

First solve Eq. (1) for w.

$w = 14 - l$ (3)

Then substitute $14 - l$ for w in Eq. (2) and solve for l.

$$\begin{aligned} l^2 + w^2 &= 100 \\ l^2 + (14 - l)^2 &= 100 \\ l^2 + 196 - 28l + l^2 &= 100 \\ 2l^2 - 28l + 96 &= 0 \\ l^2 - 14l + 48 &= 0 \\ (l - 8)(l - 6) &= 0 \end{aligned}$$

$l = 8$ *or* $l = 6$

If $l = 8$, then $w = 14 - 8$, or 6. If $l = 6$, then $w = 14 - 6$, or 8. Since the length is usually considered to be longer than the width, we have the solution $l = 8$ and $w = 6$, or $(8, 6)$.

Check. If $l = 8$ and $w = 6$, then the perimeter is $2\cdot 8 + 2\cdot 6$, or 28. The length of a diagonal is $\sqrt{8^2 + 6^2}$, or $\sqrt{100}$, or 10. The numbers check.

State. The length is 8 cm, and the width is 6 cm.

42. 2 in. by 1 in.

43. ***Familiarize***. We first make a drawing. Let l = the length and w = the width of the rectangle.

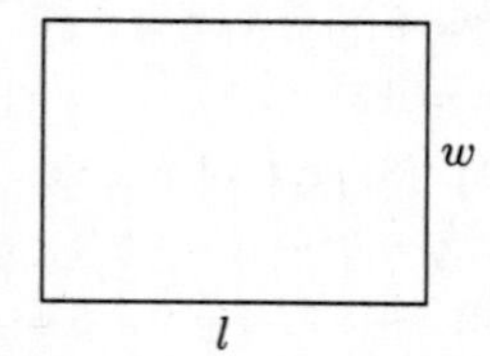

Translate.

Area: $lw = 20$

Perimeter: $2l + 2w = 18$, or $l + w = 9$

Carry out. We solve the system:

Solve the second equation for l: $l = 9 - w$

Substitute $9 - w$ for l in the first equation and solve for w.

$$\begin{aligned} (9 - w)w &= 20 \\ 9w - w^2 &= 20 \\ 0 &= w^2 - 9w + 20 \\ 0 &= (w - 5)(w - 4) \end{aligned}$$

$w = 5$ *or* $w = 4$

If $w = 5$, then $l = 9 - w$, or 4. If $w = 4$, then $l = 9 - 4$, or 5. Since length is usually considered to be longer than width, we have the solution $l = 5$ and $w = 4$, or $(5, 4)$.

Check. If $l = 5$ and $w = 4$, the area is $5 \cdot 4$, or 20. The perimeter is $2 \cdot 5 + 2 \cdot 4$, or 18. The numbers check.

State. The length is 5 in. and the width is 4 in.

44. 2 yd by 1 yd

45. ***Familiarize***. We make a drawing of the field. Let l = the length and w = the width.

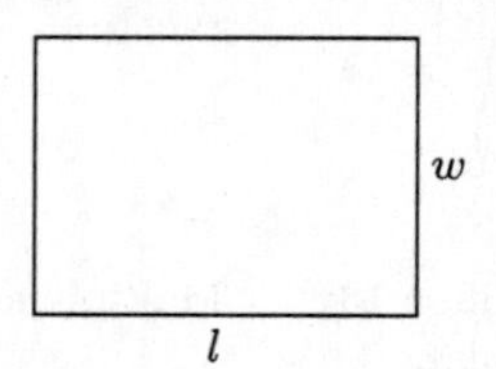

Since it takes 210 yd of fencing to enclose the field, we know that the perimeter is 210 yd.

Translate.

Perimeter: $2l + 2w = 210$, or $l + w = 105$

Area: $lw = 2250$

Carry out. We solve the system:

Solve the first equation for l: $l = 105 - w$

Substitute $105 - w$ for l in the second equation and solve for w.

$$\begin{aligned} (105 - w)w &= 2250 \\ 105w - w^2 &= 2250 \\ 0 &= w^2 - 105w + 2250 \\ 0 &= (w - 30)(w - 75) \end{aligned}$$

$w = 30$ *or* $w = 75$

If $w = 30$, then $l = 105 - 30$, or 75. If $w = 75$, then $l = 105 - 75$, or 30. Since length is usually considered to be longer than width, we have the solution $l = 75$ and $w = 30$, or $(75, 30)$.

Check. If $l = 75$ and $w = 30$, the perimeter is $2 \cdot 75 + 2 \cdot 30$, or 210. The area is $75(30)$, or 2250. The numbers check.

State. The length is 75 yd and the width is 30 yd.

46. 20 ft by 15 ft

47. ***Familiarize.*** We make a drawing and label it. Let x and y represent the lengths of the legs of the triangle.

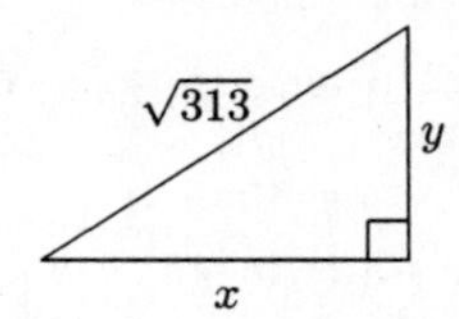

Translate. The product of the lengths of the legs is 156, so we have:

$$xy = 156$$

We use the Pythagorean theorem to get a second equation:

$$x^2 + y^2 = (\sqrt{313})^2, \text{ or } x^2 + y^2 = 313$$

Carry out. We solve the system of equations.

$$xy = 156, \quad (1)$$
$$x^2 + y^2 = 313 \quad (2)$$

First solve Equation (1) for y.

$$xy = 156$$
$$y = \frac{156}{x}$$

Then we substitute $\frac{156}{x}$ for y in Eq. (2) and solve for x.

$$x^2 + y^2 = 313 \quad (2)$$
$$x^2 + \left(\frac{156}{x}\right)^2 = 313$$
$$x^2 + \frac{24,336}{x^2} = 313$$
$$x^4 + 24,336 = 313x^2$$
$$x^4 - 313x^2 + 24,336 = 0$$
$$u^2 - 313u + 24,336 = 0 \quad \text{Letting } u = x^2$$
$$(u - 169)(u - 144) = 0$$
$$u - 169 = 0 \quad or \quad u - 144 = 0$$
$$u = 169 \quad or \quad u = 144$$

We now substitute x^2 for u and solve for x.

$$x = \pm 13 \quad \text{or} \quad x = \pm 12$$

Since $y = 156/x$, if $x = 13$, $y = 12$; if $x = -13$, $y = -12$; if $x = 12$, $y = 13$; and if $x = -12$, $y = -13$. The possible solutions are $(13, 12)$, $(-13, -12)$, $(12, 13)$, and $(-12, -13)$.

Check. Since measurements cannot be negative, we consider only $(13, 12)$ and $(12, 13)$. Since both possible solutions give the same pair of legs, we only need to check $(13, 12)$. If $x = 13$ and $y = 12$, their product is 156. Also, $\sqrt{13^2 + 12^2} = \sqrt{313}$. The numbers check.

State. The lengths of the legs are 13 and 12.

48. 6 and 10, -6 and -10

49. ***Familiarize.*** We let x = the length of a side of one peanut bed and y = the length of a side of the other peanut bed. Make a drawing.

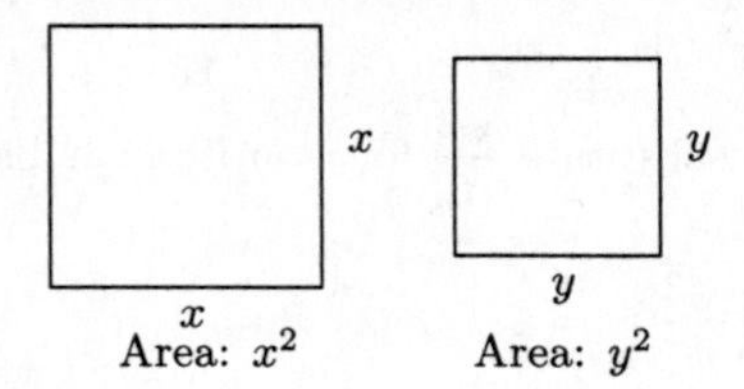

Area: x^2 Area: y^2

Translate.

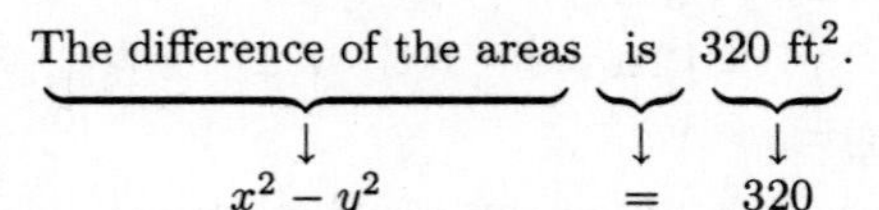

The sum of the areas is 832 ft^2.

$$x^2 + y^2 = 832$$

The difference of the areas is 320 ft^2.

$$x^2 - y^2 = 320$$

Carry out. We solve the system of equations.

$$x^2 + y^2 = 832$$
$$x^2 - y^2 = 320$$
$$2x^2 = 1152 \quad \text{Adding}$$
$$x^2 = 576$$
$$x = \pm 24$$

Since measurements cannot be negative, we consider only $x = 24$. Substitute 24 for x in the first equation and solve for y.

$$24^2 + y^2 = 832$$
$$576 + y^2 = 832$$
$$y^2 = 256$$
$$y = \pm 16$$

Again, we consider only the positive value, 16. The possible solution is $(24, 16)$.

Check. The areas of the peanut beds are 24^2, or 576, and 16^2, or 256. The sum of the areas is $576 + 256$, or 832. The difference of the areas is $576 - 256$, or 320. The values check.

State. The lengths of the beds are 24 ft and 16 ft.

50. \$3750, 6%

51. ***Familiarize.*** We first make a drawing. Let l = the length and w = the width.

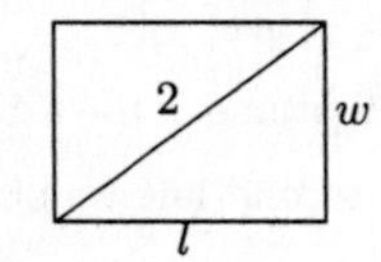

Translate.

Area: $lw = \sqrt{3}$ (1)

From the Pythagorean theorem: $l^2 + w^2 = 2^2$ (2)

Carry out. We solve the system of equations.

We first solve Eq. (1) for w.

$$lw = \sqrt{3}$$
$$w = \frac{\sqrt{3}}{l}$$

Then we substitute $\frac{\sqrt{3}}{l}$ for w in Eq. 2 and solve for l.

$$l^2 + \left(\frac{\sqrt{3}}{l}\right)^2 = 4$$
$$l^2 + \frac{3}{l^2} = 4$$
$$l^4 + 3 = 4l^2$$
$$l^4 - 4l^2 + 3 = 0$$
$$u^2 - 4u + 3 = 0 \quad \text{Letting } u = l^2$$
$$(u-3)(u-1) = 0$$
$$u = 3 \text{ or } u = 1$$

We now substitute l^2 for u and solve for l.

$$l^2 = 3 \quad \text{or} \quad l^2 = 1$$
$$l = \pm\sqrt{3} \quad \text{or} \quad l = \pm 1$$

Measurements cannot be negative, so we only need to consider $l = \sqrt{3}$ and $l = 1$. Since $w = \sqrt{3}/l$, if $l = \sqrt{3}$, $w = 1$ and if $l = 1$, $w = \sqrt{3}$. Length is usually considered to be longer than width, so we have the solution $l = \sqrt{3}$ and $w = 1$, or $(\sqrt{3}, 1)$.

Check. If $l = \sqrt{3}$ and $w = 1$, the area is $\sqrt{3}\cdot 1 = \sqrt{3}$. Also $(\sqrt{3})^2 + 1^2 = 3 + 1 = 4 = 2^2$. The numbers check.

State. The length is $\sqrt{3}$ m, and the width is 1 m.

52. $\sqrt{2}$ m by 1 m

53. $\sqrt{48} = \sqrt{16\cdot 3} = \sqrt{16}\sqrt{3} = 4\sqrt{3}$

54. $2a^6d^2\sqrt[4]{2d}$

55. ***Familiarize***. Let r = the speed of the boat in still water and t = the time of the trip upstream. Organize the information in a table.

	Speed	Time	Distance
Upstream	$r-2$	t	4
Downstream	$r+2$	$3-t$	4

Recall that $rt = d$, or $t = d/r$.

Translate. From the first line of the table we obtain $t = \frac{4}{r-2}$. From the second line we obtain $3 - t = \frac{4}{r+2}$.

Carry out. Substitute $\frac{4}{r-2}$ for t in the second equation and solve for r.

$$3 - \frac{4}{r-2} = \frac{4}{r+2},$$

LCD is $(r-2)(r+2)$

$$(r-2)(r+2)\left(3 - \frac{4}{r-2}\right) = (r-2)(r+2)\cdot\frac{4}{r+2}$$
$$3(r-2)(r+2) - 4(r+2) = 4(r-2)$$
$$3(r^2-4) - 4r - 8 = 4r - 8$$
$$3r^2 - 12 - 4r - 8 = 4r - 8$$
$$3r^2 - 8r - 12 = 0$$
$$r = \frac{-(-8) \pm \sqrt{(-8)^2 - 4(3)(-12)}}{2\cdot 3}$$
$$r = \frac{8 \pm \sqrt{208}}{6} = \frac{8 \pm 4\sqrt{13}}{6}$$
$$r = \frac{4 \pm 2\sqrt{13}}{3}$$

Since negative speed has no meaning in this problem, we consider only the positive square root.

$$r = \frac{4 + 2\sqrt{13}}{3} \approx 3.7$$

Check. The value checks. The check is left to the student.

State. The speed of the boat in still water is approximately 3.7 mph.

56. $\dfrac{x - 2\sqrt{xh} + h}{x - h}$

57. ◈

58. ◈

59. Let x = the length of the longer piece and y = the length of the shorter piece. Then the lengths of the sides of the squares are $\frac{x}{4}$ and $\frac{y}{4}$. Solve the system:

$$x + y = 100,$$
$$\left(\frac{x}{4}\right)^2 = \left(\frac{y}{4}\right)^2 + 144$$

The solution is $(61.52, 38.48)$. One piece should be 61.52 cm long, and then the other will be 38.48 cm long.

60. $(x+2)^2 + (y-1)^2 = 4$

61. $\dfrac{x^2}{a^2} + \dfrac{y^2}{b^2} = 1$ Standard form

Substitute the coordinates of the given points:

$$\frac{2^2}{a^2} + \frac{(-3)^2}{b^2} = 1,$$
$$\frac{1^2}{a^2} + \frac{(\sqrt{13})^2}{b^2} = 1, \text{ or}$$
$$\frac{4}{a^2} + \frac{9}{b^2} = 1, \quad (1)$$
$$\frac{1}{a^2} + \frac{13}{b^2} = 1 \quad (2)$$

Solve Eq. (2) for $\frac{1}{a^2}$:

$$\frac{1}{a^2} = 1 - \frac{13}{b^2}$$

$$\frac{1}{a^2} = \frac{b^2 - 13}{b^2} \qquad (3)$$

Substitute $\frac{b^2-13}{b^2}$ for $\frac{1}{a^2}$ in Eq. (1) and solve for b^2.

$$4\left(\frac{b^2-13}{b^2}\right) + \frac{9}{b^2} = 1$$

$$\frac{4b^2-52}{b^2} + \frac{9}{b^2} = 1$$

$$4b^2 - 52 + 9 = b^2$$

$$3b^2 = 43$$

$$b^2 = \frac{43}{3}$$

Substitute $\frac{43}{3}$ for b^2 in Eq. (3) and solve for a^2.

$$\frac{1}{a^2} = \frac{\frac{43}{3} - 13}{\frac{43}{3}} = \frac{\frac{43}{3} - 13}{\frac{43}{3}} \cdot \frac{3}{3}$$

$$\frac{1}{a^2} = \frac{43 - 3\cdot 13}{43} = \frac{43-39}{43}$$

$$\frac{1}{a^2} = \frac{4}{43}$$

$$a^2 = \frac{43}{4}$$

The equation of the ellipse is

$$\frac{x^2}{\frac{43}{4}} + \frac{y^2}{\frac{43}{3}} = 1, \text{ or}$$

$$\frac{4x^2}{43} + \frac{3y^2}{43} = 1, \text{ or}$$

$$4x^2 + 3y^2 = 43.$$

62. 10 in. by 7 in. by 5 in.

63.

$$R = C$$

$$100x + x^2 = 80x + 1500$$

$$x^2 + 20x - 1500 = 0$$

$$(x-30)(x+50) = 0$$

$$x = 30 \text{ or } x = -50$$

Since the number of units cannot be negative, the solution of the problem is 30. Thus, 30 units must be sold in order to break even.

64. $(-2,3)$, $(2,-3)$, $(-3,2)$, $(3,-2)$

65. $a + b = \frac{5}{6}$, (1)

$$\frac{a}{b} + \frac{b}{a} = \frac{13}{6} \qquad (2)$$

$$b = \frac{5}{6} - a = \frac{5-6a}{6} \qquad \text{Solving Eq. (1) for } b$$

$$\frac{a}{\frac{5-6a}{6}} + \frac{\frac{5-6a}{6}}{a} = \frac{13}{6} \qquad \text{Substituting for } b \text{ in Eq. (2)}$$

$$\frac{6a}{5-6a} + \frac{5-6a}{6a} = \frac{13}{6}$$

$$36a^2 + 25 - 60a + 36a^2 = 65a - 78a^2$$

$$150a^2 - 125a + 25 = 0$$

$$6a^2 - 5a + 1 = 0$$

$$(3a-1)(2a-1) = 0$$

$$a = \frac{1}{3} \text{ or } a = \frac{1}{2}$$

Substitute for a and solve for b.

When $a = \frac{1}{3}$, $b = \frac{5 - 6\left(\frac{1}{3}\right)}{6} = \frac{1}{2}$.

When $a = \frac{1}{2}$, $b = \frac{5 - 6\left(\frac{1}{2}\right)}{6} = \frac{1}{3}$.

The pairs $\left(\frac{1}{3}, \frac{1}{2}\right)$ and $\left(\frac{1}{2}, \frac{1}{3}\right)$ check. They are the solutions.

66.

Chapter 11

Sequences, Series, and the Binomial Theorem

Exercise Set 11.1

1. $a_n = 5n - 3$

$a_1 = 5 \cdot 1 - 3 = 2,$

$a_2 = 5 \cdot 2 - 3 = 7,$

$a_3 = 5 \cdot 3 - 3 = 12,$

$a_4 = 5 \cdot 4 - 3 = 17;$

$a_{10} = 5 \cdot 10 - 3 = 47;$

$a_{15} = 5 \cdot 15 - 3 = 72$

2. 7, 9, 11, 13; 25; 35

3. $a_n = \dfrac{n}{n+2}$

$a_1 = \dfrac{1}{1+2} = \dfrac{1}{3},$

$a_2 = \dfrac{2}{2+2} = \dfrac{2}{4} = \dfrac{1}{2},$

$a_3 = \dfrac{3}{3+2} = \dfrac{3}{5},$

$a_4 = \dfrac{4}{4+2} = \dfrac{4}{6} = \dfrac{2}{3};$

$a_{10} = \dfrac{10}{10+2} = \dfrac{10}{12} = \dfrac{5}{6};$

$a_{15} = \dfrac{15}{15+2} = \dfrac{15}{17}$

4. 2, 5, 10, 17; 101; 226

5. $a_n = n^2 - 2n$

$a_1 = 1^2 - 2 \cdot 1 = -1,$

$a_2 = 2^2 - 2 \cdot 2 = 0,$

$a_3 = 3^2 - 2 \cdot 3 = 3,$

$a_4 = 4^2 - 2 \cdot 4 = 8;$

$a_{10} = 10^2 - 2 \cdot 10 = 80;$

$a_{15} = 15^2 - 2 \cdot 15 = 195$

6. $0, \dfrac{3}{5}, \dfrac{4}{5}, \dfrac{15}{17}; \dfrac{99}{101}; \dfrac{112}{113}$

7. $a_n = n + \dfrac{1}{n}$

$a_1 = 1 + \dfrac{1}{1} = 2,$

$a_2 = 2 + \dfrac{1}{2} = 2\dfrac{1}{2},$

$a_3 = 3 + \dfrac{1}{3} = 3\dfrac{1}{3},$

$a_4 = 4 + \dfrac{1}{4} = 4\dfrac{1}{4};$

$a_{10} = 10 + \dfrac{1}{10} = 10\dfrac{1}{10};$

$a_{15} = 15 + \dfrac{1}{15} = 15\dfrac{1}{15}$

8. $1, -\dfrac{1}{2}, \dfrac{1}{4}, -\dfrac{1}{8}; -\dfrac{1}{512}; \dfrac{1}{16,384}$

9. $a_n = (-1)^n n^2$

$a_1 = (-1)^1 1^2 = -1,$

$a_2 = (-1)^2 2^2 = 4,$

$a_3 = (-1)^3 3^2 = -9,$

$a_4 = (-1)^4 4^2 = 16;$

$a_{10} = (-1)^{10} 10^2 = 100;$

$a_{15} = (-1)^{15} 15^2 = -225$

10. −4, 5, −6, 7; 13; −18

11. $a_n = (-1)^{n+1}(3n - 5)$

$a_1 = (-1)^{1+1}(3 \cdot 1 - 5) = -2,$

$a_2 = (-1)^{2+1}(3 \cdot 2 - 5) = -1,$

$a_3 = (-1)^{3+1}(3 \cdot 3 - 5) = 4,$

$a_4 = (-1)^{4+1}(3 \cdot 4 - 5) = -7;$

$a_{10} = (-1)^{10+1}(3 \cdot 10 - 5) = -25;$

$a_{15} = (-1)^{15+1}(3 \cdot 15 - 5) = 40$

12. 0, 7, −26, 63; 999; −3374

13. $a_n = 3n - 5$

$a_7 = 3 \cdot 7 - 5 = 21 - 5 = 16$

14. 42

15. $a_n = (3n + 4)(2n - 5)$

$a_9 = (3 \cdot 9 + 4)(2 \cdot 9 - 5) = 31 \cdot 13 = 403$

16. 400

17. $a_n = (-1)^{n-1}(3.4n - 17.3)$

$a_{12} = (-1)^{12-1}[3.4(12) - 17.3] = -23.5$

18. −37, 916, 508.16

19. $a_n = 3n^2(9n - 100)$

$a_{11} = 3 \cdot 11^2(9 \cdot 11 - 100) = 3 \cdot 121(-1) = -363$

20. 9680

21. $a_n = \left(1 + \dfrac{1}{n}\right)^2$

$a_{20} = \left(1 + \dfrac{1}{20}\right)^2 = \left(\dfrac{21}{20}\right)^2 = \dfrac{441}{400}$

22. $\dfrac{2744}{3375}$

23. $a_n = \log 10^n$

$a_{43} = \log 10^{43} = 43$

24. 67

25. 1, 3, 5, 7, 9, . . .

These are odd integers, so the general term could be $2n - 1$.

26. 3^n

27. $-2, 6, -18, 54, \ldots$

We can see a pattern if we write the sequence as

$-1\cdot 2\cdot 1,\ 1\cdot 2\cdot 3,\ -1\cdot 2\cdot 9,\ 1\cdot 2\cdot 27, \ldots$

The general term could be $(-1)^n 2(3)^{n-1}$.

28. $5n - 7$

29. $\dfrac{1}{2}, \dfrac{2}{3}, \dfrac{3}{4}, \dfrac{4}{5}, \dfrac{5}{6}, \ldots$

These are fractions in which the denominator is 1 greater than the numerator. Also, each numerator is 1 greater than the preceding numerator. The general term could be $\dfrac{n}{n+1}$.

30. $\sqrt{2n-1}$

31. $\sqrt{3}, 3, 3\sqrt{3}, 9, 9\sqrt{3}, \ldots$

These are powers of $\sqrt{3}$. The general term could be $(\sqrt{3})^n$, or $3^{n/2}$.

32. $n(n+1)$

33. $-1, -4, -7, -10, -13, \ldots$

Each term is 3 less than the preceding term. The general term may be $-1 - 3(n-1)$. After removing parentheses and simplifying, we can express the general term as $-3n + 2$, or $-(3n - 2)$.

34. $\log 10^{n-1}$, or $n - 1$

35. $1, -2, 3, -4, 5, -6, \ldots$

$S_7 = 1 - 2 + 3 - 4 + 5 - 6 + 7 = 4$

36. -8

37. 2, 4, 6, 8, . . .

$S_5 = 2 + 4 + 6 + 8 + 10 = 30$

38. $\dfrac{5269}{3600}$

39.
$$\sum_{k=1}^{5} \frac{1}{2k} = \frac{1}{2\cdot 1} + \frac{1}{2\cdot 2} + \frac{1}{2\cdot 3} + \frac{1}{2\cdot 4} + \frac{1}{2\cdot 5}$$
$$= \frac{1}{2} + \frac{1}{4} + \frac{1}{6} + \frac{1}{8} + \frac{1}{10}$$
$$= \frac{60}{120} + \frac{30}{120} + \frac{20}{120} + \frac{15}{120} + \frac{12}{120}$$
$$= \frac{137}{120}$$

40. $1 + \dfrac{1}{3} + \dfrac{1}{5} + \dfrac{1}{7} + \dfrac{1}{9} + \dfrac{1}{11} = \dfrac{6508}{3465}$

41.
$$\sum_{k=0}^{4} 3^k = 3^0 + 3^1 + 3^2 + 3^3 + 3^4$$
$$= 1 + 3 + 9 + 27 + 81$$
$$= 121$$

42. $\sqrt{9} + \sqrt{11} + \sqrt{13} + \sqrt{15} \approx 13.7952$

43.
$$\sum_{k=1}^{8} \frac{k}{k+1} = \frac{1}{1+1} + \frac{2}{2+1} + \frac{3}{3+1} + \frac{4}{4+1} +$$
$$\frac{5}{5+1} + \frac{6}{6+1} + \frac{7}{7+1} + \frac{8}{8+1}$$
$$= \frac{1}{2} + \frac{2}{3} + \frac{3}{4} + \frac{4}{5} + \frac{5}{6} + \frac{6}{7} + \frac{7}{8} + \frac{8}{9}$$
$$= \frac{15{,}551}{2520}$$

44. $-\dfrac{1}{4} + 0 + \dfrac{1}{6} + \dfrac{2}{7} = \dfrac{17}{84}$

45.
$$\sum_{k=1}^{5} (-1)^k$$
$$= (-1)^1 + (-1)^2 + (-1)^3 + (-1)^4 + (-1)^5$$
$$= -1 + 1 - 1 + 1 - 1$$
$$= -1$$

46. $1 - 1 + 1 - 1 + 1 = 1$

47.
$$\sum_{k=1}^{8} (-1)^{k+1} 2^k = (-1)^2 2^1 + (-1)^3 2^2 + (-1)^4 2^3 +$$
$$(-1)^5 2^4 + (-1)^6 2^5 + (-1)^7 2^6 +$$
$$(-1)^8 2^7 + (-1)^9 2^8$$
$$= 2 - 4 + 8 - 16 + 32 - 64 +$$
$$128 - 256$$
$$= -170$$

48. $-4^2 + 4^3 - 4^4 + 4^5 - 4^6 + 4^7 - 4^8 = -52{,}432$

49.
$$\sum_{k=0}^{5} (k^2 - 2k + 3)$$
$$= (0^2 - 2\cdot 0 + 3) + (1^2 - 2\cdot 1 + 3) +$$
$$(2^2 - 2\cdot 2 + 3) + (3^2 - 2\cdot 3 + 3) +$$
$$(4^2 - 2\cdot 4 + 3) + (5^2 - 2\cdot 5 + 3)$$
$$= 3 + 2 + 3 + 6 + 11 + 18$$
$$= 43$$

50. $4 + 2 + 2 + 4 + 8 + 14 = 34$

51.
$$\sum_{k=3}^{5} \frac{(-1)^k}{k(k+1)} = \frac{(-1)^3}{3(3+1)} + \frac{(-1)^4}{4(4+1)} + \frac{(-1)^5}{5(5+1)}$$
$$= \frac{-1}{3\cdot 4} + \frac{1}{4\cdot 5} + \frac{-1}{5\cdot 6}$$
$$= -\frac{1}{12} + \frac{1}{20} - \frac{1}{30}$$
$$= -\frac{4}{60} = -\frac{1}{15}$$

52. $\frac{3}{8}+\frac{4}{16}+\frac{5}{32}+\frac{6}{64}+\frac{7}{128}=\frac{119}{128}$

53. $\frac{2}{3}+\frac{3}{4}+\frac{4}{5}+\frac{5}{6}+\frac{6}{7}$
This is a sum of fractions in which the denominator is one greater than the numerator. Also, each numerator is 1 greater than the preceding numerator. Sigma notation is

$$\sum_{k=1}^{5}\frac{k+1}{k+2}.$$

54. $\sum_{k=1}^{5} 3k$

55. $1+4+9+16+25+36$

This is the sum of the squares of the first six natural numbers. Sigma notation is

$$\sum_{k=1}^{6} k^2.$$

56. $\sum_{k=1}^{5}\frac{1}{k^2}$

57. $4-9+16-25+\ldots+(-1)^n n^2$

This is a sum of terms of the form $(-1)^k k^2$, beginning with $k=2$ and continuing through $k=n$. Sigma notation is

$$\sum_{k=2}^{n}(-1)^k k^2.$$

58. $\sum_{k=3}^{n}(-1)^{k+1}k^2$

59. $5+10+15+20+25+\ldots$

This is a sum of multiples of 5, and it is an infinite series. Sigma notation is

$$\sum_{k=1}^{\infty} 5k.$$

60. $\sum_{k=1}^{\infty} 7k$

61. $\frac{1}{1\cdot 2}+\frac{1}{2\cdot 3}+\frac{1}{3\cdot 4}+\frac{1}{4\cdot 5}+\ldots$
This is a sum of fractions in which the numerator is 1 and the denominator is a product of two consecutive integers. The larger integer in each product is the smaller integer in the succeeding product. It is an infinite series. Sigma notation is

$$\sum_{k=1}^{\infty}\frac{1}{k(k+1)}.$$

62. $\sum_{k=1}^{\infty}\frac{1}{k(k+1)^2}$

63. Note that $\log_6 29$ is the power to which 6 is raised to get 29. Then

$$6^{\log_6 29}=29.$$

64. 43

65. $\log_3 3=1$

1 is the power to which you raise 3 to get 3.

66. 0

67. $\log_3 3^7=7$

7 is the power to which you raise 3 to get 3^7.

68. 1

69. ◈

70. ◈

71. ◈

72. 1, 3, 13, 63, 313, 1563

73. $a_1=0,\ a_{n+1}=a_n^2+3$

$a_1=0$

$a_2=0^2+3=3$

$a_3=3^2+3=12$

$a_4=12^2+3=147$

$a_5=147^2+3=21,612$

$a_6=21,612^2+3=467,078,547$

74. 1, 2, 4, 8, 16, 32, 64, 128, 256, 512, 1024, 2048, 4096, 8192, 16,384, 32,768, 65,536

75. Find each term by multiplying the preceding term by 0.75:

\$5200, \$3900, \$2925, \$2193.75, \$1645.31, \$1233.98, \$925.49, \$694.12, \$520.59, \$390.44

76. \$8.20, \$8.60, \$9.00, \$9.40, \$9.80, \$10.20, \$10.60, \$11.00, \$11.40, \$11.80

77. $a_n=\frac{1}{2^n}\log 1000^n$

$a_1=\frac{1}{2^1}\log 1000^1=\frac{1}{2}\log 10^3=\frac{1}{2}\cdot 3=\frac{3}{2}$

$a_2=\frac{1}{2^2}\log 1000^2=\frac{1}{4}\log (10^3)^2=\frac{1}{4}\log 10^6=$
$\frac{1}{4}\cdot 6=\frac{3}{2}$

$a_3=\frac{1}{2^3}\log 1000^3=\frac{1}{8}\log (10^3)^3=\frac{1}{8}\log 10^9=$
$\frac{1}{8}\cdot 9=\frac{9}{8}$

$a_4=\frac{1}{2^4}\log 1000^4=\frac{1}{16}\log (10^3)^4=$
$\frac{1}{16}\log 10^{12}=\frac{1}{16}\cdot 12=\frac{3}{4}$

$a_5 = \frac{1}{2^5}\log 1000^5 = \frac{1}{32}\log (10^3)^5 =$
$\frac{1}{32}\log 10^{15} = \frac{1}{32}\cdot 15 = \frac{15}{32}$

$S_5 = \frac{3}{2}+\frac{3}{2}+\frac{9}{8}+\frac{3}{4}+\frac{15}{32} = \frac{171}{32}$

78. $i, -1, -i, 1, i; i$

79. $\sum_{k=1}^{x} i^k = -1$

Note that $i+i^2+i^3 = i-1-i = -1$. Also $i^4+i^5+i^6+i^7 = 1+i-1-i = 0$, $i^8+i^9+i^{10}+i^{11} = 1+i-1-i = 0$, and so on.

Thus, the sum is -1 when $x = 3, 7, 11, \cdots$, or for $\{x|x = 4n-1$, where n is a natural number$\}$.

80. 11th term

81. We get the recursion formula $a_2 = 1$, $a_n = a_{n-1} + n - 1$. Enter $U_n = U_{n-1} + n - 1$ and use the table, set in ASK mode, to find that $U_{50} = 1225$. Thus, 1225 handshakes will occur if a group of 50 people shake hands with one another.

Exercise Set 11.2

1. 3, 8, 13, 18, . . .

$a_1 = 3$

$d = 5 \quad (8-3=5,\ 13-8=5,\ 18-13=5)$

2. $a_1 = 1.06$, $d = 0.06$

3. 6, 2, −2, −6, . . .

$a_1 = 6$

$d = -4 \quad (2-6=-4, -2-2=-4, -6-(-2)=-4)$

4. $a_1 = -9$, $d = 3$

5. $\frac{3}{2}, \frac{9}{4}, 3, \frac{15}{4}, \ldots$

$a_1 = \frac{3}{2}$

$d = \frac{3}{4} \quad \left(\frac{9}{4}-\frac{3}{2}=\frac{3}{4},\ 3-\frac{9}{4}=\frac{3}{4}\right)$

6. $a_1 = \frac{3}{5}$, $d = -\frac{1}{2}$

7. \$2.12, \$2.24, \$2.36, \$2.48, . . .

$a_1 = \$2.12$

$d = \$0.12$ (\$2.24 − \$2.12 = \$0.12, \$2.36 − \$2.24 = \$0.12, \$2.48 − \$2.36 = \$0.12)

8. $a_1 = \$214$, $d = -\$3$

9. 3, 7, 11, . . .

$a_1 = 3,\ d = 4,$ and $n = 12$

$a_n = a_1 + (n-1)d$

$a_{12} = 3 + (12-1)4 = 3 + 11\cdot 4 = 3 + 44 = 47$

10. 0.57

11. 7, 4, 1, . . .

$a_1 = 7,\ d = -3,$ and $n = 17$

$a_n = a_1 + (n-1)d$

$a_{17} = 7 + (17-1)(-3) = 7 + 16(-3) = 7 - 48 = -41$

12. $-\frac{17}{3}$

13. \$1200, \$964.32, \$728.64, . . .

$a_1 = \$1200,\ d = \$964.32 - \$1200 = -\$235.68,$ and $n = 13$

$a_n = a_1 + (n-1)d$

$a_{13} = \$1200 + (13-1)(-\$235.68) = \$1200 + 12(-\$235.68) = \$1200 - \$2828.16 = -\$1628.16$

14. \$7941.62

15. $a_1 = 3,\ d = 4$

$a_n = a_1 + (n-1)d$

Let $a_n = 107$, and solve for n.

$107 = 3 + (n-1)(4)$

$107 = 3 + 4n - 4$

$107 = 4n - 1$

$108 = 4n$

$27 = n$

The 27th term is 107.

16. 33rd

17. $a_1 = 7,\ d = -3$

$a_n = a_1 + (n-1)d$

$-296 = 7 + (n-1)(-3)$

$-296 = 7 - 3n + 3$

$-306 = -3n$

$102 = n$

The 102nd term is −296.

18. 46th

19. $a_n = a_1 + (n-1)d$

$a_{17} = 2 + (17-1)5$ Substituting 17 for n, 2 for a_1, and 5 for d

$= 2 + 16\cdot 5$

$= 2 + 80$

$= 82$

20. -43

21. $a_n = a_1 + (n-1)d$

$33 = a_1 + (8-1)4$ Substituting 33 for a_8, 8 for n, and 4 for d

$33 = a_1 + 28$

$5 = a_1$

(Note that this procedure is equivalent to subtracting d from a_8 seven times to get a_1: $33-7(4) = 33-28 = 5$)

22. -54

23. $a_n = a_1 + (n-1)d$

$-76 = 5 + (n-1)(-3)$ Substituting -76 for a_n, 5 for a_1, and -3 for d

$-76 = 5 - 3n + 3$

$-76 = 8 - 3n$

$-84 = -3n$

$28 = n$

24. 39

25. We know that $a_{17} = -40$ and $a_{28} = -73$. We would have to add d eleven times to get from a_{17} to a_{28}. That is,

$-40 + 11d = -73$

$11d = -33$

$d = -3.$

Since $a_{17} = -40$, we subtract d sixteen times to get to a_1.

$$a_1 = -40 - 16(-3) = -40 + 48 = 8$$

We write the first five terms of the sequence:

$8, 5, 2, -1, -4$

26. $\frac{1}{3}, \frac{5}{6}, \frac{4}{3}, \frac{11}{6}, \frac{7}{3}$

27. $1 + 5 + 9 + 13 + \ldots$

Note that $a_1 = 1$, $d = 4$, and $n = 20$. Before using the formula for S_n, we find a_{20}:

$a_{20} = 1 + (20-1)4$ Substituting into the formula for a_n

$= 1 + 19 \cdot 4$

$= 77$

Then

$S_{20} = \frac{20}{2}(1+77)$ Using the formula for S_n

$= 10(78)$

$= 780.$

28. -210

29. The sum is $1 + 2 + 3 + \ldots + 299 + 300$. This is the sum of the arithmetic sequence for which $a_1 = 1$, $a_n = 300$, and $n = 300$. We use the formula for S_n.

$$S_n = \frac{n}{2}(a_1 + a_n)$$

$$S_{300} = \frac{300}{2}(1+300) = 150(301) = 45,150$$

30. 80,200

31. The sum is $2 + 4 + 6 + \ldots + 98 + 100$. This is the sum of the arithmetic sequence for which $a_1 = 2$, $a_n = 100$, and $n = 50$. We use the formula for S_n.

$$S_n = \frac{n}{2}(a_1 + a_n)$$

$$S_{50} = \frac{50}{2}(2+100) = 25(102) = 2550$$

32. 2500

33. The sum is $6 + 12 + 18 + \ldots + 96 + 102$. This is the sum of the arithmetic sequence for which $a_1 = 6$, $a_n = 102$, and $n = 17$. We use the formula for S_n.

$$S_n = \frac{n}{2}(a_1 + a_n)$$

$$S_{17} = \frac{17}{2}(6+102) = \frac{17}{2}(108) = 918$$

34. 34,036

35. Before using the formula for S_n, we find a_{20}:

$a_{20} = 2 + (20-1)5$ Substituting into the formula for a_n

$= 2 + 19 \cdot 5 = 97$

Then

$S_{20} = \frac{20}{2}(2+97)$ Using the formula for S_n

$= 10(99) = 990.$

36. -1264

37. ***Familiarize.*** We want to find the fifteenth term and the sum of an arithmetic sequence with $a_1 = 14$, $d = 2$, and $n = 15$. We will first use the formula for a_n to find a_{15}. This result is the number of marchers in the last row. Then we will use the formula for S_n to find S_{15}. This is the total number of marchers.

Translate. Substituting into the formula for a_n, we have

$$a_{15} = 14 + (15-1)2.$$

Carry out. We first find a_{15}.

$$a_{15} = 14 + 14 \cdot 2 = 42$$

Then use the formula for S_n to find S_{15}.

$$S_{15} = \frac{15}{2}(14+42) = \frac{15}{2}(56) = 420$$

Check. We can do the calculations again. We can also do the entire addition.

$$14 + 16 + 18 + \cdots + 42.$$

State. There are 42 marchers in the last row, and there are 420 marchers altogether.

38. 3; 210

39. ***Familiarize.*** We go from 50 poles in a row, down to six poles in the top row, so there must be 45 rows. We want the sum $50 + 49 + 48 + \ldots + 6$. Thus we want the sum of an arithmetic sequence. We will use the formula $S_n = \frac{n}{2}(a_1 + a_n)$.

Translate. We want to find the sum of the first 45 terms of an arithmetic sequence with $a_1 = 50$ and $a_{45} = 6$.

Carry out. Substituting into the formula for S_n, we have

$$S_{45} = \frac{45}{2}(50 + 6)$$
$$= \frac{45}{2} \cdot 56 = 1260$$

Check. We can do the calculation again, or we can do the entire addition:

$50 + 49 + 48 + \ldots + 6.$

State. There will be 1260 poles in the pile.

40. \$49.60

41. ***Familiarize.*** We want to find the sum of an arithmetic sequence with $a_1 = \$600$, $d = \$100$, and $n = 20$. We will use the formula for a_n to find a_{20}, and then we will use the formula for S_n to find S_{20}.

Translate. Substituting into the formula for a_n, we have

$$a_{20} = 600 + (20 - 1)(100).$$

Carry out. We first find a_{20}.

$$a_{20} = 600 + 19 \cdot 100 = 600 + 1900 = 2500$$

Then we use the formula for S_n to find S_{20}.

$$S_{20} = \frac{20}{2}(600 + 2500) = 10(3100) = 31,000$$

Check. We can do the calculation again.

State. They save \$31,000 (disregarding interest).

42. \$10,230

43. ***Familiarize.*** We want to find the sum of an arithmetic sequence with $a_1 = 20$, $d = 2$, and $n = 19$. We will use the formula for a_n to find a_{19}, and then we will use the formula for S_n to find S_{19}.

Translate. Substituting into the formula for a_n, we have

$$a_{19} = 20 + (19 - 1)(2).$$

Carry out. We find a_{19}.

$$a_{19} = 20 + 18 \cdot 2 = 56$$

Then we use the formula for S_n to find S_{19}.

$$S_{19} = \frac{19}{2}(20 + 56) = 722$$

Check. We can do the calculation again.

State. There are 722 seats.

44. \$462,500

45. The logarithm is the exponent.

$\log_a P = k \qquad a^k = P$

The base does not change.

46. $e^a = t$

47. Standard form for the equation of a circle with center (h, k) and radius r is

$$(x - h)^2 + (y - k)^2 = r^2.$$

We substitute 0 for h, 0 for k, and 9 for r:

$$(x - 0)^2 + (y - 0)^2 = 9^2$$
$$x^2 + y^2 = 81$$

48. $(x + 2)^2 + (y - 5)^2 = 18$

49. ◈

50. ◈

51. $a_1 = 1,\ d = 2,\ n = n$

$a_n = 1 + (n - 1)2 = 1 + 2n - 2 = 2n - 1$

$S_n = \frac{n}{2}[1 + (2n - 1)] = \frac{n}{2} \cdot 2n = n^2$

Thus, the formula $S_n = n^2$ can be used to find the sum of the first n consecutive odd numbers starting with 1.

52. 3, 5, 7

53.

$$a_1 = \$8760$$
$$a_2 = \$8760 + (-\$798.23) = \$7961.77$$
$$a_3 = \$8760 + 2(-\$798.23) = \$7163.54$$
$$a_4 = \$8760 + 3(-\$798.23) = \$6365.31$$
$$a_5 = \$8760 + 4(-\$798.23) = \$5567.08$$
$$a_6 = \$8760 + 5(-\$798.23) = \$4768.85$$
$$a_7 = \$8760 + 6(-\$798.23) = \$3970.62$$
$$a_8 = \$8760 + 7(-\$798.23) = \$3172.39$$
$$a_9 = \$8760 + 8(-\$798.23) = \$2374.16$$
$$a_{10} = \$8760 + 9(-\$798.23) = \$1575.93$$

54. \$51,679.65

55. See the answer section in the text.

56. (a) $a_t = \$5200 - \$512.50t$; (b) \$5200, \$4687.50, \$4175, \$3662.50, \$3150, \$1612.50, \$1100; (c) $a_0 = \$5200$, $a_t = a_{t-1} - \$512.50$

Exercise Set 11.3

1. 5, 10, 20, 40, . . .

$\frac{10}{5} = 2,\ \frac{20}{10} = 2,\ \frac{40}{20} = 2$

$r = 2$

2. 3

3. 5, −5, 5, −5, . . .

$\frac{-5}{5} = -1,\ \frac{5}{-5} = -1,\ \frac{-5}{5} = -1$

$r = -1$

4. 0.1

5. $\frac{1}{2}, -\frac{1}{4}, \frac{1}{8}, -\frac{1}{16}, \ldots$

$\frac{-\frac{1}{4}}{\frac{1}{2}} = -\frac{1}{4} \cdot \frac{2}{1} = -\frac{2}{4} = -\frac{1}{2}$

$\frac{\frac{1}{8}}{-\frac{1}{4}} = \frac{1}{8} \cdot \left(-\frac{4}{1}\right) = -\frac{4}{8} = -\frac{1}{2}$

$\frac{-\frac{1}{16}}{\frac{1}{8}} = -\frac{1}{16} \cdot \frac{8}{1} = -\frac{8}{16} = -\frac{1}{2}$

$r = -\frac{1}{2}$

6. −2

7. $75, 15, 3, \frac{3}{5}, \ldots$

$\frac{15}{75} = \frac{1}{5},\ \frac{3}{15} = \frac{1}{5},\ \frac{\frac{3}{5}}{3} = \frac{3}{5} \cdot \frac{1}{3} = \frac{1}{5}$

$r = \frac{1}{5}$

8. $-\frac{1}{3}$

9. $\frac{1}{m}, \frac{3}{m^2}, \frac{9}{m^3}, \frac{27}{m^4}, \ldots$

$\frac{\frac{3}{m^2}}{\frac{1}{m}} = \frac{3}{m^2} \cdot \frac{m}{1} = \frac{3}{m}$

$\frac{\frac{9}{m^3}}{\frac{3}{m^2}} = \frac{9}{m^3} \cdot \frac{m^2}{3} = \frac{3}{m}$

$\frac{\frac{27}{m^4}}{\frac{9}{m^3}} = \frac{27}{m^4} \cdot \frac{m^3}{9} = \frac{3}{m}$

$r = \frac{3}{m}$

10. $\frac{m}{5}$

11. 5, 10, 20, . . .

$a_1 = 5$, $n = 7$, and $r = \frac{10}{5} = 2$

We use the formula $a_n = a_1 r^{n-1}$.

$a_7 = 5 \cdot 2^{7-1} = 5 \cdot 2^6 = 5 \cdot 64 = 320$

12. 131, 072

13. $3, 3\sqrt{2}, 6, \ldots$

$a_1 = 3$, $n = 9$, and $r = \frac{3\sqrt{2}}{3} = \sqrt{2}$

$a_n = a_1 r^{n-1}$

$a_9 = 3(\sqrt{2})^{9-1} = 3(\sqrt{2})^8 = 3 \cdot 16 = 48$

14. $108\sqrt{3}$

15. $-\frac{8}{243}, \frac{8}{81}, -\frac{8}{27}, \ldots$

$a_1 = -\frac{8}{243}$, $n = 10$, and $r = \frac{\frac{8}{81}}{-\frac{8}{243}} =$

$\frac{8}{81}\left(-\frac{243}{8}\right) = -3$

$a_n = a_1 r^{n-1}$

$a_{10} = -\frac{8}{243}(-3)^{10-1} = -\frac{8}{243}(-3)^9 =$

$-\frac{8}{243}(-19,683) = 648$

16. 2, 734, 375

17. \$1000, \$1080, \$1166.40, . . .

$a_1 = \$1000$, $n = 12$, and $r = \frac{\$1080}{\$1000} = 1.08$

$a_n = a_1 r^{n-1}$

$a_{12} = \$1000(1.08)^{12-1} \approx \$1000(2.331638997) \approx$

\$2331.64

18. \$1967.15

19. 1, 3, 9, . . .

$a_1 = 1$ and $r = \frac{3}{1}$, or 3

$a_n = a_1 r^{n-1}$

$a_n = 1(3)^{n-1} = 3^{n-1}$

20. 5^{3-n}

21. 1, −1, 1, −1, . . .

$a_1 = 1$ and $r = \frac{-1}{1} = -1$

$a_n = a_1 r^{n-1}$

$a_n = 1(-1)^{n-1} = (-1)^{n-1}$

22. 2^n

23. $\frac{1}{x}, \frac{1}{x^2}, \frac{1}{x^2}, \ldots$

$$a_1 = \frac{1}{x} \text{ and } r = \frac{\frac{1}{x^2}}{\frac{1}{x}} = \frac{1}{x^2} \cdot \frac{x}{1} = \frac{1}{x}$$

$$a_n = a_1 r^{n-1}$$

$$a_n = \frac{1}{x}\left(\frac{1}{x}\right)^{n-1} = \frac{1}{x} \cdot \frac{1}{x^{n-1}} = \frac{1}{x^{1+n-1}} = \frac{1}{x^n}$$

24. $5\left(\frac{m}{2}\right)^{n-1}$

25. $7 + 14 + 28 + \ldots$

$$a_1 = 7,\ n = 7, \text{ and } r = \frac{14}{7} = 2$$

$$S_n = \frac{a_1(1 - r^n)}{1 - r}$$

$$S_7 = \frac{7(1 - 2^7)}{1 - 2} = \frac{7(1 - 128)}{-1} = \frac{7(-127)}{-1} = 889$$

26. 10.5

27. $\frac{1}{18} - \frac{1}{6} + \frac{1}{2} - \ldots$

$$a_1 = \frac{1}{18},\ n = 7, \text{ and } r = \frac{-\frac{1}{6}}{\frac{1}{18}} = -\frac{1}{6} \cdot \frac{18}{1} = -3$$

$$S_n = \frac{a_1(1 - r^n)}{1 - r}$$

$$S_7 = \frac{\frac{1}{18}\left[1 - (-3)^7\right]}{1 - (-3)} = \frac{\frac{1}{18}(1 + 2187)}{4} = \frac{\frac{1}{18}(2188)}{4} =$$

$$\frac{1}{18}(2188)\left(\frac{1}{4}\right) = \frac{547}{18}$$

28. 6.6666

29. $1 + x + x^2 + x^3 + \ldots$

$$a_1 = 1,\ n = 8, \text{ and } r = \frac{x}{1}, \text{ or } x$$

$$S_n = \frac{a_1(1 - r^n)}{1 - r}$$

$$S_8 = \frac{1(x - x^8)}{1 - x} = \frac{(1 + x^4)(1 - x^4)}{1 - x} =$$

$$\frac{(1 + x^4)(1 + x^2)(1 - x^2)}{1 - x} =$$

$$\frac{(1 + x^4)(1 + x^2)(1 + x)(1 - x)}{1 - x} =$$

$$(1 + x^4)(1 + x^2)(1 + x)$$

30. $\frac{1 - x^{20}}{1 - x^2}$

31. \$200, \$200(1.06), \$200(1.06)^2, . . .

$$a_1 = \$200,\ n = 16, \text{ and } r = \frac{\$200(1.06)}{\$200} = 1.06$$

$$S_n = \frac{a_1(1 - r^n)}{1 - r}$$

$$S_{16} = \frac{\$200[1 - (1.06)^{16}]}{1 - 1.06} \approx$$

$$\frac{\$200(1 - 2.540351685)}{-0.06} \approx \$5134.51$$

32. \$60,893.30

33. $9+3+1+\ldots$

$|r| = \left|\frac{3}{9}\right| = \left|\frac{1}{3}\right| = \frac{1}{3}$, and since $|r| < 1$, the series does have a sum.

$$S_\infty = \frac{a_1}{1 - r} = \frac{9}{1 - \frac{1}{3}} = \frac{9}{\frac{2}{3}} = 9 \cdot \frac{3}{2} = \frac{27}{2}$$

34. 16

35. $7+3+\frac{9}{7}+\ldots$

$|r| = \left|\frac{3}{7}\right| = \frac{3}{7}$, and since $|r| < 1$, the series does have a sum.

$$S_\infty = \frac{a_1}{1 - r} = \frac{7}{1 - \frac{3}{7}} = \frac{7}{\frac{4}{7}} = 7 \cdot \frac{7}{4} = \frac{49}{4}$$

36. 48

37. $3 + 15 + 75 + \ldots$

$|r| = \left|\frac{15}{3}\right| = |5| = 5$, and since $|r| \not< 1$ the series does not have a sum.

38. No

39. $4 - 6 + 9 - \frac{27}{2} + \ldots$

$|r| = \left|\frac{-6}{4}\right| = \left|-\frac{3}{2}\right| = \frac{3}{2}$, and since $|r| \not< 1$ the series does not have a sum.

40. -4

41. $0.43 + 0.0043 + 0.000043 + \ldots$

$|r| = \left|\frac{0.0043}{0.43}\right| = |0.01| = 0.01$, and since $|r| < 1$, the series does have a sum.

$$S_\infty = \frac{a_1}{1 - r} = \frac{0.43}{1 - 0.01} = \frac{0.43}{0.99} = \frac{43}{99}$$

42. $\frac{37}{99}$

43. $\$500(1.02)^{-1} + \$500(1.02)^{-2} + \$500(1.02)^{-3} + \ldots$

$|r| = \left|\frac{\$500(1.02)^{-2}}{\$500(1.02)^{-1}}\right| = |(1.02)^{-1}| = (1.02)^{-1}$, or $\frac{1}{1.02}$, and since $|r| < 1$, the series does have a sum.

$$S_\infty = \frac{a_1}{1-r} = \frac{\$500(1.02)^{-1}}{1-\left(\frac{1}{1.02}\right)} = \frac{\frac{\$500}{1.02}}{\frac{0.02}{1.02}} =$$

$$\frac{\$500}{1.02} \cdot \frac{1.02}{0.02} = \$25,000$$

44. \$12,500

45. $0.7777... = 0.7 + 0.07 + 0.007 + 0.0007 + ...$

This is an infinite geometric series with $a_1 = 0.7$.

$|r| = \left|\frac{0.07}{0.7}\right| = |0.1| = 0.1 < 1$, so the series has a sum.

$$S_\infty = \frac{a_1}{1-r} = \frac{0.7}{1-0.1} = \frac{0.7}{0.9} = \frac{7}{9}$$

Fractional notation for 0.7777... is $\frac{7}{9}$.

46. $\frac{2}{9}$

47. $8.3838... = 8.3 + 0.083 + 0.00083 + ...$

This is an infinite geometric series with $a_1 = 8.3$.

$|r| = \left|\frac{0.083}{8.3}\right| = |0.01| = 0.01 < 1$, so the series has a sum.

$$S_\infty = \frac{a_1}{1-r} = \frac{8.3}{1-0.01} = \frac{8.3}{0.99} = \frac{830}{99}$$

Fractional notation for 8.3838... is $\frac{830}{99}$.

48. $\frac{740}{99}$

49. $0.15151515... = 0.15 + 0.0015 + 0.000015 + ...$

This is an infinite geometric series with $a_1 = 0.15$.

$|r| = \left|\frac{0.0015}{0.15}\right| = |0.01| = 0.01 < 1$, so the series has a sum.

$$S_\infty = \frac{a_1}{1-r} = \frac{0.15}{1-0.01} = \frac{0.15}{0.99} = \frac{15}{99} = \frac{5}{33}$$

Fractional notation for 0.15151515... is $\frac{5}{33}$.

50. $\frac{4}{33}$

51. ***Familiarize.*** The rebound distances form a geometric sequence:

$$\frac{1}{4} \times 20, \quad \left(\frac{1}{4}\right)^2 \times 20, \quad \left(\frac{1}{4}\right)^3 \times 20, \ldots,$$

or $5, \quad \frac{1}{4} \times 5, \quad \left(\frac{1}{4}\right)^2 \times 5, \ldots$

The height of the 6th rebound is the 6th term of the sequence.

Translate. We will use the formula $a_n = a_1 r^{n-1}$, with $a_1 = 5$, $r = \frac{1}{4}$, and $n = 6$:

$$a_6 = 5\left(\frac{1}{4}\right)^{6-1}$$

Carry out. We calculate to obtain $a_6 = \frac{5}{1024}$.

Check. We can do the calculation again.

State. It rebounds $\frac{5}{1024}$ ft the 6th time.

52. $6\frac{2}{3}$ ft

53. ***Familiarize.*** In one year, the population will be $100,000 + 0.03(100,000)$, or $(1.03)100,000$. In two years, the population will be $(1.03)100,000 + 0.03(1.03)100,000$, or $(1.03)^2 100,000$. Thus the populations form a geometric sequence:

$100{,}000, \quad (1.03)100,000, \quad (1.03)^2 100,000, \ldots$

The population in 15 years will be the 16th term of the sequence.

Translate. We will use the formula $a_n = a_1 r^{n-1}$ with $a_1 = 100,000$, $r = 1.03$, and $n = 16$:

$$a_{16} = 100,000(1.03)^{16-1}$$

Carry out. We calculate to obtain $a_{16} \approx 155,797$.

Check. We can do the calculation again.

State. In 15 years the population will be about 155,797.

54. About 24 years

55. ***Familiarize.*** The amounts owed at the beginning of successive years form a geometric sequence:

$\$15{,}000, \quad (1.085)\$15,000, \quad (1.085)^2\$15,000, \quad (1.085)^3\$15,000, \ldots$

The amount to be repaid at the end of 13 years is the amount owed at the beginning of the 14th year.

Translate. We use the formula $a_n = a_1 r^{n-1}$ with $a_1 = 15,000$, $r = 1.085$, and $n = 14$:

$$a_{14} = 15,000(1.085)^{14-1}$$

Carry out. We calculate to obtain $a_{14} \approx 43,318.94$.

Check. We can do the calculation again.

State. At the end of 13 years, \$43,318.94 will be repaid.

56. 2710

57. We have a geometric sequence

$5000, 5000(0.96), 5000(0.96)^2, \ldots$

where the general term $5000(0.96)^n$ represents the number of fruit flies remaining alive after n minutes. We find the value of n for which the general term is 1800.

$$\begin{aligned} 1800 &= 5000(0.96)^n \\ 0.36 &= (0.96)^n \\ \log 0.36 &= \log(0.96)^n \\ \log 0.36 &= n \log 0.96 \\ \frac{\log 0.36}{\log 0.96} &= n \\ 25 &\approx n \end{aligned}$$

It will take about 25 minutes for only 1800 fruit flies to remain alive.

58. $213,609.57

59. ***Familiarize.*** The lengths of the falls form a geometric sequence:

$556,\ \left(\frac{3}{4}\right)556,\ \left(\frac{3}{4}\right)^2 556,\ \left(\frac{3}{4}\right)^3 556, \ldots$

The total length of the first 6 falls is the sum of the first six terms of this sequence. The heights of the rebounds also form a geometric sequence:

$\left(\frac{3}{4}\right)556,\ \left(\frac{3}{4}\right)^2 556,\ \left(\frac{3}{4}\right)^3 556, \ldots,$ or

$417,\ \left(\frac{3}{4}\right)417,\ \left(\frac{3}{4}\right)^2 417, \ldots$

When the ball hits the ground for the 6th time, it will have rebounded 5 times. Thus the total length of the rebounds is the sum of the first five terms of this sequence.

Translate. We use the formula $S_n = \frac{a_1(1-r^n)}{1-r}$ twice, once with $a_1 = 556$, $r = \frac{3}{4}$, and $n = 6$ and a second time with $a_1 = 417$, $r = \frac{3}{4}$, and $n = 5$.

D = Length of falls + length of rebounds

$$= \frac{556\left[1-\left(\frac{3}{4}\right)^6\right]}{1-\frac{3}{4}} + \frac{417\left[1-\left(\frac{3}{4}\right)^5\right]}{1-\frac{3}{4}}.$$

Carry out. We use a calculator to obtain $D \approx 3100.35$.

Check. We can do the calculations again.

State. The ball will have traveled about 3100.35 ft.

60. 3892 ft

61. ***Familiarize.*** The heights of the stack form a geometric sequence:

$0.02, 0.02(2), 0.02(2^2), \ldots$

The height of the stack after it is doubled 10 times is given by the 11th term of this sequence.

Translate. We have a geometric sequence with $a_1 = 0.02$, $r = 2$, and $n = 11$. We use the formula

$a_n = a_1 r^{n-1}$.

Carry out. We substitute and calculate.

$a_{11} = 0.02(2^{11-1})$

$a_{11} = 0.02(1024) = 20.48$

Check. We can do the calculation again.

State. The final stack will be 20.48 in. high.

62. $2,684,354.55

63. $5x - 2y = -3,$ (1)

$2x + 5y = -24$ (2)

Multiply Eq. (1) by 5 and Eq. (2) by 2 and add.

$$\begin{aligned} 25x - 10y &= -15 \\ 4x + 10y &= -48 \\ \hline 29x &= -63 \\ x &= -\frac{63}{29} \end{aligned}$$

Substitute $-\frac{63}{29}$ for x in the second equation and solve for y.

$$\begin{aligned} 2\left(-\frac{63}{29}\right) + 5y &= -24 \\ -\frac{126}{29} + 5y &= -24 \\ 5y &= -\frac{570}{29} \\ y &= -\frac{114}{29} \end{aligned}$$

The solution is $\left(-\frac{63}{29}, -\frac{114}{29}\right)$.

64. $(-1, 2, 3)$

65. ◈

66. ◈

67. ◈

68. $\frac{x^2[1-(-x)^n]}{1+x}$

69. $1 + x + x^2 + \ldots$

This is a geometric series with $a_1 = 1$ and $r = x$.

$$S_n = \frac{a_1(1-r^n)}{1-r} = \frac{1(1-x^n)}{1-x} = \frac{1-x^n}{1-x}$$

70. 512 cm^2

71. ◈

72. ◈

Exercise Set 11.4

1. $8! = 8 \cdot 7 \cdot 6 \cdot 5 \cdot 4 \cdot 3 \cdot 2 \cdot 1 = 40,320$

2. $362,880$

3. $10! = 10 \cdot 9 \cdot 8 \cdot 7 \cdot 6 \cdot 5 \cdot 4 \cdot 3 \cdot 2 \cdot 1 = 3,628,800$

4. $39,916,800$

5. $\frac{7!}{4!} = \frac{7 \cdot 6 \cdot 5 \cdot 4!}{4!} = 7 \cdot 6 \cdot 5 = 210$

6. 56

7. $\frac{10!}{7!} = \frac{10 \cdot 9 \cdot 8 \cdot 7!}{7!} = 10 \cdot 9 \cdot 8 = 720$

8. 3024

9. $\binom{8}{2} = \frac{8!}{(8-2)!2!} = \frac{8!}{6!2!} = \frac{8\cdot7\cdot6!}{6!\cdot2\cdot1} = \frac{8\cdot7}{2} =$
$4\cdot7 = 28$

10. 35

11. $\binom{10}{6} = \frac{10!}{(10-6)!6!} = \frac{10!}{4!6!} = \frac{10\cdot9\cdot8\cdot7\cdot6!}{4\cdot3\cdot2\cdot6!} =$
$\frac{10\cdot9\cdot8\cdot7}{4\cdot3\cdot2} = 10\cdot3\cdot7 = 210$

12. 126

13. $\binom{20}{18} = \frac{20!}{(20-18)!18!} = \frac{20!}{2!18!} = \frac{20\cdot19\cdot18!}{2\cdot1\cdot18!} =$
$\frac{20\cdot19}{2} = 10\cdot19 = 190$

14. 4060

15. $\binom{35}{2} = \frac{35!}{(35-2)!2!} = \frac{35!}{33!2!} = \frac{35\cdot34\cdot33!}{33!\cdot2\cdot1} =$
$\frac{35\cdot34}{2} = 35\cdot17 = 595$

16. 780

17. Expand $(m+n)^5$.

Form 1: The expansion of $(m+n)^5$ has $5+1$, or 6 terms. The sum of the exponents in each term is 5. The exponents of m start with 5 and decrease to 0. The last term has no factor of m. The first term has no factor of n. The exponents of n start in the second term with 1 and increase to 5. We get the coefficients from the 6th row of Pascal's triangle.

1
1 1
1 2 1
1 3 3 1
1 4 6 4 1
1 5 10 10 5 1

$$(m+n)^5 = 1\cdot m^5 + 5\cdot m^4n^1 + 10\cdot m^3\cdot n^2 + 10\cdot m^2\cdot n^3 + 5\cdot m\cdot n^4 + 1\cdot n^5$$
$$= m^5 + 5m^4n + 10m^3n^2 + 10m^2n^3 + 5mn^4 + n^5$$

Form 2: We have $a = m$, $b = n$, and $n = 5$.

$$(m+n)^5 = \binom{5}{0}m^5 + \binom{5}{1}m^4n + \binom{5}{2}m^3n^2 + \binom{5}{3}m^2n^3 + \binom{5}{4}mn^4 + \binom{5}{5}n^5$$
$$= \frac{5!}{5!0!}m^5 + \frac{5!}{4!1!}m^4n + \frac{5!}{3!2!}m^3n^2 + \frac{5!}{2!3!}m^2n^3 + \frac{5!}{1!4!}mn^4 + \frac{5!}{0!5!}m^5$$
$$= m^5 + 5m^4n + 10m^3n^2 + 10m^2n^3 + 5mn^4 + n^5$$

18. $a^4 - 4a^3b + 6a^2b^2 - 4ab^3 + b^4$

19. Expand $(x-y)^6$.

Form 1: The expansion of $(x-y)^6$ has $6+1$, or 7 terms. The sum of the exponents in each term is 6. The exponents of x start with 6 and decrease to 0. The last term has no factor of x. The first term has no factor of $-y$. The exponents of $-y$ start in the second term with 1 and increase to 6. We get the coefficients from the 7th row of Pascal's triangle.

1
1 1
1 2 1
1 3 3 1
1 4 6 4 1
1 5 10 10 5 1
1 6 15 20 15 6 1

$$(x-y)^6 = 1\cdot x^6 + 6\cdot x^5\cdot(-y) + 15\cdot x^4\cdot(-y)^2 + 20\cdot x^3\cdot(-y)^3 + 15\cdot x^2\cdot(-y)^4 + 6\cdot x\cdot(-y)^5 + 1\cdot(-y)^6$$
$$= x^6 - 6x^5y + 15x^4y^2 - 20x^3y^3 + 15x^2y^4 - 6xy^5 + y^6$$

Form 2: We have $a = x$, $b = -y$, and $n = 6$.

$$(x-y)^6 = \binom{6}{0}x^6 + \binom{6}{1}x^5(-y) + \binom{6}{2}x^4(-y)^2 + \binom{6}{3}x^3(-y)^3 + \binom{6}{4}x^2(-y)^4 + \binom{6}{5}x(-y)^5 + \binom{6}{6}(-y)^6$$
$$= \frac{6!}{6!0!}x^6 + \frac{6!}{5!1!}x^5(-y) + \frac{6!}{4!2!}x^4y^2 + \frac{6!}{3!3!}x^3(-y^3) + \frac{6!}{2!4!}x^2y^4 + \frac{6!}{1!5!}x(-y^5) + \frac{6!}{0!6!}y^6$$
$$= x^6 - 6x^5y + 15x^4y^2 - 20x^3y^3 + 15x^2y^4 - 6xy^5 + y^6$$

20. $p^7 + 7p^6q + 21p^5q^2 + 35p^4q^3 + 35p^3q^4 + 21p^2q^5 + 7pq^6 + q^7$

21. Expand $(x^2 - 3y)^5$.

We have $a = x^2$, $b = -3y$, and $n = 5$.

Form 1: We get the coefficients from the 6th row of Pascal's triangle. From Exercise 17 we know that the coefficients are

1 5 10 10 5 1.

$$(x^2-3y)^5 = 1\cdot(x^2)^5 + 5\cdot(x^2)^4\cdot(-3y) + 10\cdot(x^2)^3\cdot(-3y)^2 + 10\cdot(x^2)^2\cdot(-3y)^3 + 5\cdot(x^2)\cdot(-3y)^4 + 1\cdot(-3y)^5$$
$$= x^{10} - 15x^8y + 90x^6y^2 - 270x^4y^3 + 405x^2y^4 - 243y^5$$

Form 2:

$$(x^2+3y)^5 = \binom{5}{0}(x^2)^5 + \binom{5}{1}(x^2)^4(-3y) + \binom{5}{2}(x^2)^3(-3y)^2 + \binom{5}{3}(x^2)^2(-3y)^3 + \binom{5}{4}x^2(-3y)^4 + \binom{5}{5}(-3y)^5$$

$$= \frac{5!}{5!0!}x^{10} + \frac{5!}{4!1!}x^8(-3y) + \frac{5!}{3!2!}x^6(9y^2) + \frac{5!}{2!3!}x^4(-27y^3) + \frac{5!}{1!4!}x^2(81y^4) + \frac{5!}{0!5!}(-243y^5)$$

$$= x^{10} - 15x^8y + 90x^6y^2 - 270x^4y^3 + 405x^2y^4 - 243y^5$$

22. $2187c^7 - 5103c^6d + 5103c^5d^2 - 2835c^4d^3 + 945c^3d^4 - 189c^2d^5 + 21cd^6 - d^7$

23. Expand $(3c-d)^6$.

We have $a = 3c$, $b = -d$, and $n = 6$.

Form 1: We get the coefficients from the 7th row of Pascal's triangle. From Exercise 19 we know that the coefficients are

1 6 15 20 15 6 1.

$$(3c-d)^6 = 1\cdot(3c)^6 + 6\cdot(3c)^5\cdot(-d) + 15\cdot(3c)^4\cdot(-d)^2 + 20\cdot(3c)^3\cdot(-d)^3 + 15\cdot(3c)^2\cdot(-d)^4 + 6\cdot(3c)\cdot(-d)^5 + 1\cdot(-d)^6$$

$$= 3^6c^6 - 6\cdot 3^5c^5d + 15\cdot 3^4c^4d^2 - 20\cdot 3^3c^3d^3 + 15\cdot 3^2c^2d^4 - 6\cdot 3cd^5 + d^6$$

$$= 729c^6 - 6\cdot 243c^5d + 15\cdot 81c^4d^2 - 20\cdot 27c^3d^3 + 15\cdot 9c^2d^4 - 6\cdot 3cd^5 + d^6$$

$$= 729c^6 - 1458c^5d + 1215c^4d^2 - 540c^3d^3 + 135c^2d^4 - 18cd^5 + d^6$$

Form 2:

$$(3c-d)^6 = \binom{6}{0}(3c)^6 + \binom{6}{1}(3c)^5(-d) + \binom{6}{2}(3c)^4(-d)^2 + \binom{6}{3}(3c)^3(-d)^3 + \binom{6}{4}(3c)^2(-d)^4 + \binom{6}{5}(3c)(-d)^5 + \binom{6}{6}(-d)^6$$

$$= \frac{6!}{6!0!}(729c^6) + \frac{6!}{5!1!}(243c^5)(-d) + \frac{6!}{4!2!}(81c^4)(d^2) + \frac{6!}{3!3!}(27c^3)(-d^3) + \frac{6!}{2!4!}(9c^2)(d^4) + \frac{6!}{1!5!}(3c)(-d^5) + \frac{6!}{0!6!}d^6$$

$$= 729c^6 - 1458c^5d + 1215c^4d^2 - 540c^3d^3 + 135c^2d^4 - 18cd^5 + d^6$$

24. $t^{-12} + 12t^{-10} + 60t^{-8} + 160t^{-6} + 240t^{-4} + 192t^{-2} + 64$

25. Expand $(x-y)^3$.

We have $a = x$, $b = -y$, and $n = 3$.

Form 1: We get the coefficients from the 4th row of Pascal's triangle.

1

1 1

1 2 1

1 3 3 1

$$(x-y)^3 = 1\cdot x^3 + 3x^2(-y) + 3x(-y)^2 + 1\cdot(-y)^3 = x^3 - 3x^2y + 3xy^2 - y^3$$

Form 2:

$$(x-y)^3 = \binom{3}{0}x^3 + \binom{3}{1}x^2(-y) + \binom{3}{2}x(-y)^2 + \binom{3}{3}(-y)^3$$

$$= \frac{3!}{3!0!}x^3 + \frac{3!}{2!1!}x^2(-y) + \frac{3!}{1!2!}xy^2 + \frac{3!}{0!3!}(-y^3)$$

$$= x^3 - 3x^2y + 3xy^2 - y^3$$

26. $x^5 - 5x^4y + 10x^3y^2 - 10x^2y^3 + 5xy^4 - y^5$

27. Expand $\left(x + \frac{2}{y}\right)^9$.

We have $a = x$, $b = \frac{2}{y}$, and $n = 9$.

Form 1: We get the coefficients from the 10th row of Pascal's triangle.

1
1 1
1 2 1
1 3 3 1
1 4 6 4 1
1 5 10 10 5 1
1 6 15 20 15 6 1
1 7 21 35 35 21 7 1
1 8 28 56 70 56 28 8 1
1 9 36 84 126 126 84 36 9 1

$$\left(x+\frac{2}{y}\right)^9 = 1\cdot x^9 + 9x^8\left(\frac{2}{y}\right) + 36x^7\left(\frac{2}{y}\right)^2 + 84x^6\left(\frac{2}{y}\right)^3 + 126x^5\left(\frac{2}{y}\right)^4 + 126x^4\left(\frac{2}{y}\right)^5 + 84x^3\left(\frac{2}{y}\right)^6 + 36x^2\left(\frac{2}{y}\right)^7 + 9x\left(\frac{2}{y}\right)^8 + 1\cdot\left(\frac{2}{y}\right)^9$$

$$= x^9 + \frac{18x^8}{y} + \frac{144x^7}{y^2} + \frac{672x^6}{y^3} + \frac{2016x^5}{y^4} + \frac{4032x^4}{y^5} + \frac{5376x^3}{y^6} + \frac{4608x^2}{y^7} + \frac{2304x}{y^8} + \frac{512}{y^9}$$

Form 2:

$$\left(x-\frac{2}{y}\right)^9$$

$$= \binom{9}{0}x^9 + \binom{9}{1}x^8\left(\frac{2}{y}\right) + \binom{9}{2}x^7\left(\frac{2}{y}\right)^2 + \binom{9}{3}x^6\left(\frac{2}{y}\right)^3 + \binom{9}{4}x^5\left(\frac{2}{y}\right)^4 + \binom{9}{5}x^4\left(\frac{2}{y}\right)^5 + \binom{9}{6}x^3\left(\frac{2}{y}\right)^6 + \binom{9}{7}x^2\left(\frac{2}{y}\right)^7 + \binom{9}{8}x\left(\frac{2}{y}\right)^8 + \binom{9}{9}\left(\frac{2}{y}\right)^9$$

$$= \frac{9!}{9!0!}x^9 + \frac{9!}{8!1!}x^8\left(\frac{2}{y}\right) + \frac{9!}{7!2!}x^7\left(\frac{4}{y^2}\right) + \frac{9!}{6!3!}x^6\left(\frac{8}{y^3}\right) + \frac{9!}{5!4!}x^5\left(\frac{16}{y^4}\right) + \frac{9!}{4!5!}x^4\left(\frac{32}{y^5}\right) + \frac{9!}{3!6!}x^3\left(\frac{64}{y^6}\right) + \frac{9!}{2!7!}x^2\left(\frac{128}{y^7}\right) + \frac{9!}{1!8!}x\left(\frac{256}{y^8}\right) + \frac{9!}{0!9!}\left(\frac{512}{y^9}\right)$$

$$= x^9 + 9x^8\left(\frac{2}{y}\right) + 36x^7\left(\frac{4}{y^2}\right) + 84x^6\left(\frac{8}{y^3}\right) + 126x^5\left(\frac{16}{y^4}\right) + 126x^4\left(\frac{32}{y^5}\right) + 84x^3\left(\frac{64}{y^6}\right) + 36x^2\left(\frac{128}{y^7}\right) + 9x\left(\frac{256}{y^8}\right) + \frac{512}{y^9}$$

$$= x^9 + \frac{18x^8}{y} + \frac{144x^7}{y^2} + \frac{672x^6}{y^3} + \frac{2016x^5}{y^4} + \frac{4032x^4}{y^5} + \frac{5376x^3}{y^6} + \frac{4608x^2}{y^7} + \frac{2304x}{y^8} + \frac{512}{y^9}$$

28. $19,683s^9 + \frac{59,049s^8}{t} + \frac{78,732s^7}{t^2} + \frac{61,236s^6}{t^3} + \frac{30,618s^5}{t^4} + \frac{10,206s^4}{t^5} + \frac{2268s^3}{t^6} + \frac{324s^2}{t^7} + \frac{27s}{t^8} + \frac{1}{t^9}$

29. Expand $(a^2-b^3)^5$.

We have $a=a^2$, $b=-b^3$, and $n=5$.

Form 1: We get the coefficient from the 6th row of Pascal's triangle. From Exercise 17 we know that the coefficients are

1 5 10 10 5 1.

$$(a^2-b^3)^5$$
$$= 1\cdot(a^2)^5 + 5(a^2)^4(-b^3) + 10(a^2)^3(-b^3)^2 + 10(a^2)^2(-b^3)^3 + 5(a^2)(-b^3)^4 + 1\cdot(-b^3)^5$$
$$= a^{10} - 5a^8b^3 + 10a^6b^6 - 10a^4b^9 + 5a^2b^{12} - b^{15}$$

Form 2:

$$(a^2-b^3)^5$$
$$= \binom{5}{0}(a^2)^5 + \binom{5}{1}(a^2)^4(-b^3) + \binom{5}{2}(a^2)^3(-b^3)^2 + \binom{5}{3}(a^2)^2(-b^3)^3 + \binom{5}{4}(a^2)(-b^3)^4 + \binom{5}{5}(-b^3)^5$$
$$= \frac{5!}{5!0!}a^{10} + \frac{5!}{4!1!}a^8(-b^3) + \frac{5!}{3!2!}a^6(b^6) + \frac{5!}{2!3!}a^4(-b^9) + \frac{5!}{1!4!}a^2(b^{12}) + \frac{5!}{0!5!}(-b^{15})$$
$$= a^{10} - 5a^8b^3 + 10a^6b^6 - 10a^4b^9 + 5a^2b^{12} - b^{15}$$

30. $x^{15} - 10x^{12}y + 40x^9y^2 - 80x^6y^3 + 80x^3y^4 - 32y^5$

31. Expand $(\sqrt{3}-t)^4$.

We have $a=\sqrt{3}$, $b=-t$, and $n=4$.

Form 1: We get the coefficients from the 5th row of Pascal's triangle.

$$\begin{array}{ccccccccc} & & & & 1 & & & & \\ & & & 1 & & 1 & & & \\ & & 1 & & 2 & & 1 & & \\ & 1 & & 3 & & 3 & & 1 & \\ 1 & & 4 & & 6 & & 4 & & 1 \end{array}$$

$$\begin{aligned}(\sqrt{3}-t)^4 &= 1\cdot(\sqrt{3})^4 + 4(\sqrt{3})^3(-t) + \\ &\quad 6(\sqrt{3})^2(-t)^2 + 4(\sqrt{3})(-t)^3 + 1\cdot(-t)^4 \\ &= 9 - 12\sqrt{3}t + 18t^2 - 4\sqrt{3}t^3 + t^4\end{aligned}$$

Form 2:

$$\begin{aligned}(\sqrt{3}-t)^4 &= \binom{4}{0}(\sqrt{3})^4 + \binom{4}{1}(\sqrt{3})^3(-t) + \\ &\quad \binom{4}{2}(\sqrt{3})^2(-t)^2 + \binom{4}{3}(\sqrt{3})(-t)^3 + \\ &\quad \binom{4}{4}(-t)^4 \\ &= \frac{4!}{4!0!}(9) + \frac{4!}{3!1!}(3\sqrt{3})(-t) + \\ &\quad \frac{4!}{2!2!}(3)(t^2) + \frac{4!}{1!3!}(\sqrt{3})(-t^3) + \\ &\quad \frac{4!}{0!4!}(t^4) \\ &= 9 - 12\sqrt{3}t + 18t^2 - 4\sqrt{3}t^3 + t^4\end{aligned}$$

32. $125 + 150\sqrt{5}\,t + 375t^2 + 100\sqrt{5}\,t^3 + 75t^4 + 6\sqrt{5}\,t^5 + t^6$

33. Expand $(x^{-2} + x^2)^4$.

We have $a = x^{-2}$, $b = x^2$, and $n = 4$.

Form 1: We get the coefficients from the fifth row of Pascal's triangle. From Exercise 31 we know that the coefficients are

$$1 \quad 4 \quad 6 \quad 4 \quad 1.$$

$$\begin{aligned}&(x^{-2} + x^2)^4 \\ &= 1\cdot(x^{-2})^4 + 4(x^{-2})^3(x^2) + 6(x^{-2})^2(x^2)^2 + \\ &\quad 4(x^{-2})(x^2)^3 + 1\cdot(x^2)^4 \\ &= x^{-8} + 4x^{-4} + 6 + 4x^4 + x^8\end{aligned}$$

Form 2:

$$\begin{aligned}&(x^{-2} + x^2)^4 \\ &= \binom{4}{0}(x^{-2})^4 + \binom{4}{1}(x^{-2})^3(x^2) + \\ &\quad \binom{4}{2}(x^{-2})^2(x^2)^2 + \binom{4}{3}(x^{-2})(x^2)^3 + \\ &\quad \binom{4}{4}(x^2)^4 \\ &= \frac{4!}{4!0!}(x^{-8}) + \frac{4!}{3!1!}(x^{-6})(x^2) + \frac{4!}{2!2!}(x^{-4})(x^4) + \\ &\quad \frac{4!}{1!3!}(x^{-2})(x^6) + \frac{4!}{0!4!}(x^8) \\ &= x^{-8} + 4x^{-4} + 6 + 4x^4 + x^8\end{aligned}$$

34. $x^{-3} - 6x^{-2} + 15x^{-1} - 20 + 15x - 6x^2 + x^3$

35. Find the 3rd term of $(a+b)^6$.

First, we note that $3 = 2+1$, $a = a$, $b = b$, and $n = 6$. Then the 3rd term of the expansion of $(a+b)^6$ is

$$\binom{6}{2}a^{6-2}b^2, \text{ or } \frac{6!}{4!2!}a^4b^2, \text{ or } 15a^4b^2.$$

36. $21x^2y^5$

37. Find the 12th term of $(a-3)^{14}$.

First, we note that $12 = 11 + 1$, $a = a$, $b = -3$, and $n = 14$. Then the 12th term of the expansion of $(a-3)^{14}$ is

$$\begin{aligned}\binom{14}{11}a^{14-11}\cdot(-3)^{11} &= \frac{14!}{3!11!}a^3(-177,147) \\ &= 364a^3(-177,147) \\ &= -64,481,508a^3\end{aligned}$$

38. $67,584x^2$

39. Find the 5th term of $(2x^3 - \sqrt{y})^8$.

First, we note that $5 = 4 + 1$, $a = 2x^3$, $b = -\sqrt{y}$, and $n = 8$. Then the 5th term of the expansion of $(2x^3 - \sqrt{y})^8$ is

$$\begin{aligned}&\binom{8}{4}(2x^3)^{8-4}(-\sqrt{y})^4 \\ &= \frac{8!}{4!4!}(2x^3)^4(-\sqrt{y})^4 \\ &= 70(16x^{12})(y^2) \\ &= 1120x^{12}y^2\end{aligned}$$

40. $\dfrac{35c^3}{b^8}$

41. The expansion of $(2u - 3v^2)^{10}$ has 11 terms so the 6th term is the middle term. Note that $6 = 5 + 1$, $a = 2u$, $b = -3v^2$, and $n = 10$. Then the 6th term of the expansion of $(2u - 3v^2)^{10}$ is

$$\begin{aligned}&\binom{10}{5}(2u)^{10-5}(-3v^2)^5 \\ &= \frac{10!}{5!5!}(2u)^5(-3v^2)^5 \\ &= 252(32u^5)(-243v^{10}) \\ &= -1,959,552u^5v^{10}\end{aligned}$$

42. $30x\sqrt{x}$, $30x\sqrt{3}$

43.
$$\begin{aligned}\log_2 x + \log_2(x-2) &= 3 \\ \log_2 x(x-2) &= 3 \\ x(x-2) &= 2^3 \\ x^2 - 2x &= 8 \\ x^2 - 2x - 8 &= 0 \\ (x-4)(x+2) &= 0\end{aligned}$$

$x = 4$ *or* $x = -2$

Only 4 checks.

44. $\frac{5}{2}$

45.
$$\begin{aligned} e^t &= 280 \\ \ln e^t &= \ln 280 \\ t &= \ln 280 \\ t &\approx 5.6348 \end{aligned}$$

46. ± 5

47.

48. Consider a set of 5 elements, $\{A, B, C, D, E\}$. List all the subsets of size 3:

$\{A, B, C\}$, $\{A, B, D\}$, $\{A, B, E\}$, $\{A, C, D\}$, $\{A, C, E\}$, $\{A, D, E\}$, $\{B, C, D\}$, $\{B, C, E\}$, $\{B, D, E\}$, $\{C, D, E\}$.

There are exactly 10 subsets of size 3 and $\binom{5}{3} = 10$, so there are exactly $\binom{5}{3}$ ways of forming a subset of size 3 from a set of 5 elements.

49. Find the third term of $(0.313 + 0.687)^5$:

$\binom{5}{2}(0.313)^{5-2}(0.687)^2 = \frac{5!}{3!2!}(0.313)^3(0.687)^2 \approx$ 0.145

50. $\binom{8}{5}(0.15)^3(0.85)^5 \approx 0.084$

51. Find and add the 3rd through 6th terms of $(0.313 + 0.687)^5$:

$\binom{5}{2}(0.313)^3(0.687)^2 + \binom{5}{3}(0.313)^2(0.687)^3 + \binom{5}{4}(0.313)(0.687)^4 + \binom{5}{5}(0.687)^5 \approx 0.964$

52. $\binom{8}{6}(0.15)^2(0.85)^6 + \binom{8}{7}(0.15)(0.85)^7 + \binom{8}{8}(0.85)^8 \approx 0.89$

53. See the answer section in the text.

54. $\frac{55}{144}$

55. The expansion of $(x^2 - 6y^{3/2})^6$ has 7 terms, so the 4th term is the middle term.

$\binom{6}{3}(x^2)^3(-6y^{3/2})^3 = \frac{6!}{3!3!}(x^6)(-216y^{9/2}) =$ $-4320x^6y^{9/2}$

56. $\frac{\sqrt[3]{q}}{2p}$

57. The $(r+1)$st term of $\left(\sqrt[3]{x} - \frac{1}{\sqrt{x}}\right)^7$ is

$\binom{7}{r}(\sqrt[3]{x})^{7-r}\left(-\frac{1}{\sqrt{x}}\right)^r$. The term containing $\frac{1}{x^{1/6}}$ is the term in which the sum of the exponents is $-1/6$. That is,

$$\begin{aligned} \left(\frac{1}{3}\right)(7-r) + \left(-\frac{1}{2}\right)(r) &= -\frac{1}{6} \\ \frac{7}{3} - \frac{r}{3} - \frac{r}{2} &= -\frac{1}{6} \\ -\frac{5r}{6} &= -\frac{15}{6} \\ r &= 3 \end{aligned}$$

Find the $(3+1)$st, or 4th term.

$\binom{7}{3}(\sqrt[3]{x})^4\left(-\frac{1}{\sqrt{x}}\right)^3 = \frac{7!}{4!3!}(x^{4/3})(-x^{-3/2}) =$ $-35x^{-1/6}$, or $-\frac{35}{x^{1/6}}$.

58. 8